SCHAUM'S OUTLINE OF

THEORY AND PROBLEMS

OF

ADVANCED MATHEMATICS

for

Engineers and Scientists

•

MURRAY R. SPIEGEL, Ph.D.

Former Professor and Chairman,
Mathematics Department
Rensselaer Polytechnic Institute
Hartford Graduate Center

SCHAUM'S OUTLINE SERIES

McGRAW-HILL

New York   San Francisco   Washington, D.C.   Auckland   Bogotá   Caracas   Lisbon
London   Madrid   Mexico City   Milan   Montreal   New Dehli
San Juan   Singapore   Sydney   Tokyo   Toronto

ISBN 0-07-060216-6

33 34 35 CUS CUS 0 9 8 7

McGraw-Hill

A Division of The McGraw-Hill Companies

Preface

In recent years the number of topics in mathematics required of engineers and scientists has greatly increased. This is to be expected since mathematics plays a vital role as a language in the formulation and solution of problems involving science and engineering and as these problems become more complex it is natural that the mathematical methods needed for their solution should increase in number and complexity.

It is the purpose of this book to provide important advanced mathematical concepts and methods needed by engineers and scientists as well as mathematicians who are interested in the applications of their field. The book has been designed as a supplement to all current standard textbooks or as a textbook for a formal course in the mathematical methods of engineering and science.

Each chapter begins with a clear statement of pertinent definitions, principles and theorems together with illustrative and other descriptive material. This is followed by graded sets of solved and supplementary problems. The solved problems serve to illustrate and amplify the theory, bring into sharp focus those fine points without which the student continually feels himself on unsafe ground, and provide the repetition of basic principles so vital to effective learning. Numerous proofs of theorems and derivations of basic results are included among the solved problems. The large number of supplementary problems serve as a review and possible extension of the material of each chapter.

Topics covered include ordinary differential equations, Laplace transforms, vector analysis, Fourier series, Fourier integrals, gamma, beta and other special functions, Bessel functions, Legendre and other orthogonal functions, partial differential equations, complex variables and conformal mapping, matrices and calculus of variations. The first chapter which provides a review of fundamental concepts of algebra, trigonometry, analytic geometry and calculus may either be read at the beginning or referred to as needed depending on the background of the student.

Considerably more material has been included here than can be covered in most courses. This has been done to make the book more flexible, to provide a more useful book of reference and to stimulate further interest in the topics.

I wish to take this opportunity to thank Daniel Schaum, Nicola Monti and Hank Hayden for their splendid cooperation.

M. R. Spiegel

Rensselear Polytechnic Institute

CONTENTS

CONTENTS

CONTENTS

Chapter 1

Review of Fundamental Concepts

REAL NUMBERS

At the very foundations of mathematics is the concept of a *set* or *collection* of objects and, in particular, sets of *numbers* on which we base our quantitative work in science and engineering. The student is already familiar with the following important sets of numbers.

1. **Natural Numbers** $1, 2, 3, 4, \ldots$ or *positive integers* used in counting.

2. **Integers** $0, \pm 1, \pm 2, \pm 3, \ldots$. These numbers arose in order to provide meaning to *subtraction* [inverse of *addition*] of any two natural numbers. Thus $2 - 6 = -4$, $8 - 8 = 0$, etc.

3. **Rational Numbers** such as $2/3$, $-10/7$, etc. arose in order to provide meaning to *division* [inverse of *multiplication*] or *quotient* of any two integers with the exception that division by zero is not defined.

4. **Irrational Numbers** such as $\sqrt{2}$, π, etc. are numbers which cannot be expressed as the quotient of two integers.

Note that the set of natural numbers is a *subset*, i.e. a part, of the set of integers which in turn is a subset of the set of rational numbers.

The set of numbers which are either rational or irrational is called the set of *real numbers* [to distinguish them from *imaginary* or *complex numbers* on page 11] and is composed of *positive* and *negative numbers* and *zero*. The real numbers can be represented as *points* on a line as indicated in Fig. 1-1. For this reason we often use *point* and *number* interchangeably.

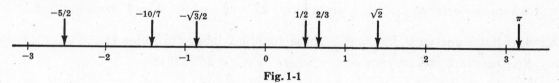

Fig. 1-1

The student is also familiar with the concept of *inequality*. Thus we say that the real number a is *greater than* or *less than* b [symbolized by $a > b$ or $a < b$] if $a - b$ is a positive or negative number respectively. For any real numbers a and b we must have $a > b$, $a = b$ or $a < b$.

RULES OF ALGEBRA

If a, b, c are any real numbers, the following rules of algebra hold.

1.	$a + b = b + a$	Commutative law for addition
2.	$a + (b + c) = (a + b) + c$	Associative law for addition
3.	$ab = ba$	Commutative law for multiplication
4.	$a(bc) = (ab)c$	Associative law for multiplication
5.	$a(b + c) = ab + ac$	Distributive law

It is from these rules [if we accept them as *axioms* or *postulates*] that we can prove the usual rules of signs, as for example $(-5)(3) = -15$, $(-2)(-3) = 6$, etc.

The student is also familiar with the usual rules of *exponents*:

$$a^m \cdot a^n = a^{m+n}, \quad a^m/a^n = a^{m-n}, \ a \neq 0, \quad (a^m)^n = a^{mn} \tag{1}$$

FUNCTIONS

Another important concept is that of *function*. The student will recall that a function f is a rule which assigns to each object x, also called *member* or *element*, of a set A an element y of a set B. To indicate this correspondence we write $y = f(x)$ where $f(x)$ is called the *value* of the function at x.

> **Example 1.** If $f(x) = x^2 - 3x + 2$, then $f(2) = 2^2 - 3(2) + 2 = 0$.

The student is also familiar with the process of "graphing functions" by obtaining number pairs (x, y) and considering these as points plotted on an xy coordinate system. In general $y = f(x)$ is represented graphically by a *curve*. Because y is usually determined from x, we sometimes call x the *independent variable* and y the *dependent variable*.

SPECIAL TYPES OF FUNCTIONS

1. **Polynomials** $f(x) = a_0 x^n + a_1 x^{n-1} + a_2 x^{n-2} + \cdots + a_n$. If $a_0 \neq 0$, n is called the *degree* of the polynomial. The polynomial equation $f(x) = 0$ has exactly n *roots* provided we count repetitions. For example $x^3 - 3x^2 + 3x - 1 = 0$ can be written $(x-1)^3 = 0$ so that the 3 roots are $1, 1, 1$. Note that here we have used the *binomial theorem*

$$(a + x)^n = a^n + \binom{n}{1} a^{n-1} x + \binom{n}{2} a^{n-2} x^2 + \cdots + x^n \tag{2}$$

where the *binomial coefficients* are given by

$$\binom{n}{k} = \frac{n!}{k!(n-k)!} \tag{3}$$

and where *factorial n*, i.e. $n!, = n(n-1)(n-2) \cdots 1$ while $0! = 1$ by definition.

2. **Exponential Functions** $f(x) = a^x$. These functions obey the rules (1).

An important special case occurs where $a = e = 2.7182818 \cdots$.

3. **Logarithmic Functions** $f(x) = \log_a x$. These functions are *inverses* of the exponential functions, i.e. if $a^x = y$ then $x = \log_a y$ where a is called the *base* of the logarithm. Interchanging x and y gives $y = \log_a x$. If $a = e$, which is often called the *natural base* of logarithms, we denote $\log_e x$ by $\ln x$, called *the natural logarithm* of x. The fundamental rules satisfied by natural logarithms [or logarithms to any base] are

$$\ln(mn) = \ln m + \ln n, \quad \ln \frac{m}{n} = \ln m - \ln n, \quad \ln m^p = p \ln m \tag{4}$$

4. **Trigonometric Functions** $\sin x$, $\cos x$, $\tan x$, $\cot x$, $\sec x$, $\csc x$.

Some fundamental relationships among these functions are as follows.

(a) $\quad \sin x = \cos\left(\frac{\pi}{2} - x\right), \quad\quad \cos x = \sin\left(\frac{\pi}{2} - x\right), \quad\quad \tan x = \frac{\sin x}{\cos x},$

$\quad\quad \cot x = \frac{\cos x}{\sin x} = \frac{1}{\tan x}, \quad\quad \sec x = \frac{1}{\cos x}, \quad\quad\quad\quad \csc x = \frac{1}{\sin x}$

(b) $\sin^2 x + \cos^2 x = 1,\qquad \sec^2 x - \tan^2 x = 1,\qquad \csc^2 x - \cot^2 x = 1$

(c) $\sin(-x) = -\sin x,\qquad \cos(-x) = \cos x,\qquad \tan(-x) = -\tan x$

(d) $\sin(x \pm y) = \sin x \cos y \pm \cos x \sin y,\qquad \cos(x \pm y) = \cos x \cos y \mp \sin x \sin y$

$$\tan(x \pm y) = \frac{\tan x \pm \tan y}{1 \mp \tan x \tan y}$$

(e) $A \cos x + B \sin x = \sqrt{A^2 + B^2}\,\sin(x + \alpha)$ where $\tan \alpha = A/B$

The trigonometric functions are *periodic*. For example $\sin x$ and $\cos x$, shown in Fig. 1-2 and 1-3 respectively, have period 2π.

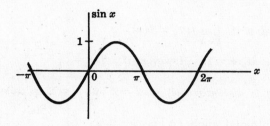

Fig. 1-2

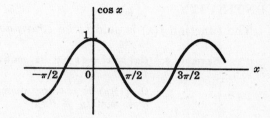

Fig. 1-3

5. **Inverse Trigonometric Functions** $\sin^{-1} x$, $\cos^{-1} x$, $\tan^{-1} x$, $\cot^{-1} x$, $\sec^{-1} x$, $\csc^{-1} x$. These are *inverses* of the trigonometric functions. For example if $\sin x = y$ then $x = \sin^{-1} y$, or on interchanging x and y, $y = \sin^{-1} x$.

6. **Hyperbolic Functions.** These are defined in terms of exponential functions as follows.

(a) $\sinh x = \dfrac{e^x - e^{-x}}{2},\qquad\qquad \cosh x = \dfrac{e^x + e^{-x}}{2},$

$\tanh x = \dfrac{\sinh x}{\cosh x} = \dfrac{e^x - e^{-x}}{e^x + e^{-x}}\qquad \coth x = \dfrac{\cosh x}{\sinh x} = \dfrac{1}{\tanh x} = \dfrac{e^x + e^{-x}}{e^x - e^{-x}},$

$\operatorname{sech} x = \dfrac{1}{\cosh x} = \dfrac{2}{e^x + e^{-x}},\qquad \operatorname{csch} x = \dfrac{1}{\sinh x} = \dfrac{2}{e^x - e^{-x}}$

Some fundamental identities analogous to those for trigonometric functions are

(b) $\cosh^2 x - \sinh^2 x = 1,\qquad \operatorname{sech}^2 x + \tanh^2 x = 1,\qquad \coth^2 x - \operatorname{csch}^2 x = 1$

(c) $\sinh(x \pm y) = \sinh x \cosh y \pm \cosh x \sinh y$

$\cosh(x \pm y) = \cosh x \cosh y \pm \sinh x \sinh y$

$$\tanh(x \pm y) = \frac{\tanh x \pm \tanh y}{1 \pm \tanh x \tanh y}$$

The inverse hyperbolic functions, given by $\sinh^{-1} x$, $\cosh^{-1} x$, etc. can be expressed in terms of logarithms [see Problem 1.9, for example].

LIMITS

The function $f(x)$ is said to have the *limit l* as x approaches a, abbreviated $\lim\limits_{x \to a} f(x) = l$, if given any number $\epsilon > 0$ we can find a number $\delta > 0$ such that $|f(x) - l| < \epsilon$ whenever $0 < |x - a| < \delta$.

Note that $|p|$, i.e. the *absolute value* of p, is equal to p if $p > 0$, $-p$ if $p < 0$ and 0 if $p = 0$.

Example 2. $\lim\limits_{x \to 1} (x^2 - 4x + 8) = 5,\quad \lim\limits_{x \to 2} \dfrac{x^2 - 4}{x - 2} = 4,\quad \lim\limits_{x \to 0} \dfrac{\sin x}{x} = 1$

If $\lim\limits_{x \to a} f_1(x) = l_1$, $\lim\limits_{x \to a} f_2(x) = l_2$ then we have the following *theorems on limits*.

(a) $\lim\limits_{x \to a} [f_1(x) \pm f_2(x)] = \lim\limits_{x \to a} f_1(x) \pm \lim\limits_{x \to a} f_2(x) = l_1 \pm l_2$

(b) $\lim\limits_{x \to a} [f_1(x) f_2(x)] = \left[\lim\limits_{x \to a} f_1(x)\right]\left[\lim\limits_{x \to a} f_2(x)\right] = l_1 l_2$

(c) $\lim\limits_{x \to a} \dfrac{f_1(x)}{f_2(x)} = \dfrac{\lim\limits_{x \to a} f_1(x)}{\lim\limits_{x \to a} f_2(x)} = \dfrac{l_1}{l_2}$ if $l_2 \neq 0$

CONTINUITY

The function $f(x)$ is said to be *continuous* at a if $\lim\limits_{x \to a} f(x) = f(a)$.

Example 3. $f(x) = x^2 - 4x + 8$ is continuous at $x = 1$.   However, if $f(x) = \begin{cases} \dfrac{x^2 - 4}{x - 2} & x \neq 2 \\ 6 & x = 2 \end{cases}$

then $f(x)$ is not continuous [or is *discontinuous*] at $x = 2$ and $x = 2$ is called a *discontinuity* of $f(x)$.

If $f(x)$ is continuous at each point of an interval such as $x_1 \leqq x \leqq x_2$ or $x_1 < x \leqq x_2$, etc., it is said to be *continuous in the interval*.

If $f_1(x)$ and $f_2(x)$ are continuous in an interval then $f_1(x) \pm f_2(x)$, $f_1(x) f_2(x)$ and $f_1(x)/f_2(x)$ where $f_2(x) \neq 0$ are also continuous in the interval.

DERIVATIVES

The *derivative* of $y = f(x)$ at a point x is defined as

$$f'(x) = \lim_{h \to 0} \frac{f(x+h) - f(x)}{h} = \lim_{\Delta x \to 0} \frac{\Delta y}{\Delta x} = \frac{dy}{dx} \tag{5}$$

where $h = \Delta x$, $\Delta y = f(x+h) - f(x) = f(x + \Delta x) - f(x)$ provided the limit exists.

The *differential* of $y = f(x)$ is defined by

$$dy = f'(x)\,dx \quad \text{where} \quad dx = \Delta x \tag{6}$$

The process of finding derivatives is called *differentiation*. By taking derivatives of $y' = dy/dx = f'(x)$ we can find second, third and higher order derivatives, denoted by $y'' = d^2y/dx^2 = f''(x)$, $y''' = d^3y/dx^3 = f'''(x)$, etc.

Geometrically the derivative of a function $f(x)$ at a point represents the *slope of the tangent line* drawn to the curve $y = f(x)$ at the point.

If a function has a derivative at a point, then it is continuous at the point. However, the converse is not necessarily true.

DIFFERENTIATION FORMULAS

In the following u, v represent functions of x while a, c, p represent constants. We assume of course that the derivatives of u and v exist, i.e. u and v are *differentiable*.

1. $\dfrac{d}{dx}(u \pm v) = \dfrac{du}{dx} \pm \dfrac{dv}{dx}$ 4. $\dfrac{d}{dx}\left(\dfrac{u}{v}\right) = \dfrac{v(du/dx) - u(dv/dx)}{v^2}$

2. $\dfrac{d}{dx}(cu) = c\dfrac{du}{dx}$ 5. $\dfrac{d}{dx}u^p = pu^{p-1}\dfrac{du}{dx}$

3. $\dfrac{d}{dx}(uv) = u\dfrac{dv}{dx} + v\dfrac{du}{dx}$ 6. $\dfrac{d}{dx}(a^u) = a^u \ln a$

7. $\dfrac{d}{dx} e^u = e^u \dfrac{du}{dx}$

8. $\dfrac{d}{dx} \ln u = \dfrac{1}{u} \dfrac{du}{dx}$

9. $\dfrac{d}{dx} \sin u = \cos u \dfrac{du}{dx}$

10. $\dfrac{d}{dx} \cos u = -\sin u \dfrac{du}{dx}$

11. $\dfrac{d}{dx} \tan u = \sec^2 u \dfrac{du}{dx}$

12. $\dfrac{d}{dx} \cot u = -\csc^2 u \dfrac{du}{dx}$

13. $\dfrac{d}{dx} \sec u = \sec u \tan u \dfrac{du}{dx}$

14. $\dfrac{d}{dx} \csc u = -\csc u \cot u \dfrac{du}{dx}$

15. $\dfrac{d}{dx} \sin^{-1} u = \dfrac{1}{\sqrt{1-u^2}} \dfrac{du}{dx}$

16. $\dfrac{d}{dx} \cos^{-1} u = \dfrac{-1}{\sqrt{1-u^2}} \dfrac{du}{dx}$

17. $\dfrac{d}{dx} \tan^{-1} u = \dfrac{1}{1+u^2} \dfrac{du}{dx}$

18. $\dfrac{d}{dx} \cot^{-1} u = \dfrac{-1}{1+u^2} \dfrac{du}{dx}$

19. $\dfrac{d}{dx} \sinh u = \cosh u \dfrac{du}{dx}$

20. $\dfrac{d}{dx} \cosh u = \sinh u \dfrac{du}{dx}$

In the special case where $u = x$, the above formulas are simplified since in such case $du/dx = 1$.

INTEGRALS

If $dy/dx = f(x)$, then we call y an *indefinite integral* or *anti-derivative* of $f(x)$ and denote it by

$$\int f(x)\,dx \tag{7}$$

Since the derivative of a constant is zero, all indefinite integrals of $f(x)$ can differ only by a constant.

The *definite integral* of $f(x)$ between $x = a$ and $x = b$ is defined as

$$\int_a^b f(x)\,dx = \lim_{h \to 0} h[f(a) + f(a+h) + f(a+2h) + \cdots + f(a+(n-1)h)] \tag{8}$$

provided this limit exists. Geometrically if $f(x) \geqq 0$, this represents the area under the curve $y = f(x)$ bounded by the x axis and the ordinates at $x = a$ and $x = b$. The integral will exist if $f(x)$ is continuous in $a \leqq x \leqq b$.

Definite and indefinite integrals are related by the following theorem.

Theorem 1-1 [**Fundamental Theorem of Calculus**].   If $f(x) = \dfrac{d}{dx} g(x)$, then

$$\int_a^b f(x)\,dx = \int_a^b \dfrac{d}{dx} g(x)\,dx = g(x) \Big|_a^b = g(b) - g(a)$$

Example 4. $\displaystyle\int_1^2 x^2\,dx = \int_1^2 \dfrac{d}{dx}\left(\dfrac{x^3}{3}\right) dx = \dfrac{x^3}{3}\Big|_1^2 = \dfrac{2^3}{3} - \dfrac{1^3}{3} = \dfrac{7}{3}$

The process of finding integrals is called *integration*.

INTEGRATION FORMULAS

In the following u, v represent functions of x while a, b, c, p represent constants. In all cases we omit the constant of integration, which nevertheless is implied.

1. $\displaystyle\int (u \pm v)\,dx = \int u\,dx \pm \int v\,dx$ 2. $\displaystyle\int cu\,dx = c \int u\,dx$

3. $\displaystyle\int u\left(\dfrac{dv}{dx}\right) dx = uv - \int v\left(\dfrac{du}{dx}\right) dx$ or $\displaystyle\int u\,dv = uv - \int v\,du$

This is called *integration by parts*.

4. $\int F[u(x)]\,dx = \int F(w)\,\dfrac{dw}{w'}$ where $w = u(x)$ and $w' = dw/dx$ expressed as a function of w.   This is called *integration by substitution* or *transformation*.

5. $\int u^p\,du = \dfrac{u^{p+1}}{p+1},\quad p \neq -1$

14. $\int \csc u\,du = \ln\,(\csc u - \cot u)$

6. $\int u^{-1}\,du = \int \dfrac{du}{u} = \ln u$

15. $\int e^{au} \sin bu\,du = \dfrac{e^{au}(a \sin bu - b \cos bu)}{a^2 + b^2}$

7. $\int a^u\,du = \dfrac{a^u}{\ln a},\quad a \neq 0, 1$

16. $\int e^{au} \cos bu\,du = \dfrac{e^{au}(a \cos bu + b \sin bu)}{a^2 + b^2}$

8. $\int e^u\,du = e^u$

17. $\int \dfrac{du}{\sqrt{a^2 - u^2}} = \sin^{-1}\dfrac{u}{a}$

9. $\int \sin u\,du = -\cos u$

18. $\int \dfrac{du}{u^2 + a^2} = \dfrac{1}{a}\tan^{-1}\dfrac{u}{a}$

10. $\int \cos u\,du = \sin u$

19. $\int \dfrac{du}{\sqrt{u^2 - a^2}} = \ln\,(u + \sqrt{u^2 - a^2})$

11. $\int \tan u\,du = -\ln \cos u$

20. $\int \dfrac{du}{\sqrt{u^2 + a^2}} = \ln\,(u + \sqrt{u^2 + a^2})$

12. $\int \cot u\,du = \ln \sin u$

21. $\int \sinh u\,du = \cosh u$

13. $\int \sec u\,du = \ln\,(\sec u + \tan u)$

22. $\int \cosh u\,du = \sinh u$

SEQUENCES AND SERIES

A *sequence*, indicated by $u_1, u_2, \ldots$ or briefly by $\langle u_n \rangle$, is a function defined on the set of natural numbers. The sequence is said to have the *limit* l or to *converge* to l, if given any $\epsilon > 0$ there exists a number $N > 0$ such that $|u_n - l| < \epsilon$ for all $n > N$, and in such case we write $\lim\limits_{n \to \infty} u_n = l$. If the sequence does not converge we say that it *diverges*.

Consider the sequence $u_1,\ u_1 + u_2,\ u_1 + u_2 + u_3, \ldots$ or $S_1, S_2, S_3, \ldots$ where $S_n = u_1 + u_2 + \cdots + u_n$. We call $\langle S_n \rangle$ the sequence of *partial sums* of the sequence $\langle u_n \rangle$. The symbol

$$u_1 + u_2 + u_3 + \cdots \quad \text{or} \quad \sum_{n=1}^{\infty} u_n \quad \text{or briefly} \quad \sum u_n \tag{9}$$

is defined as synonymous with $\langle S_n \rangle$ and is called an *infinite series*. This series will converge or diverge according as $\langle S_n \rangle$ converges or diverges. If it converges to S we call S the *sum* of the series.

The following are some important theorems concerning infinite series.

Theorem 1-2.   The series $\displaystyle\sum_{n=1}^{\infty} \dfrac{1}{n^p}$ converges if $p > 1$ and diverges if $p \leqq 1$.

Theorem 1-3.   If $\Sigma |u_n|$ converges and $|v_n| \leqq |u_n|$, then $\Sigma |v_n|$ converges.

Theorem 1-4.   If $\Sigma |u_n|$ converges, then Σu_n converges.

In such case we say that Σu_n *converges absolutely* or is *absolutely convergent*. A property of such series is that the terms can be rearranged without affecting the sum.

Theorem 1-5.   If $\Sigma |u_n|$ diverges and $v_n \geqq |u_n|$, then Σv_n diverges.

Theorem 1-6.   The series $\Sigma |u_n|$, where $|u_n| = f(n) \geqq 0$, converges or diverges according as
$$\int_1^\infty f(x)\, dx = \lim_{M \to \infty} \int_1^M f(x)\, dx \text{ exists or does not exist.}$$

This theorem is often called the *integral test*.

Theorem 1-7.   The series $\Sigma |u_n|$ diverges if $\lim_{n \to \infty} |u_n| \neq 0$.   However, if $\lim_{n \to \infty} |u_n| = 0$ the series may or may not converge [see Problem 1.31].

Theorem 1-8.   Suppose that $\lim_{n \to \infty} \left| \dfrac{u_{n+1}}{u_n} \right| = r$.   Then the series Σu_n converges (absolutely) if $r < 1$ and diverges if $r > 1$.   If $r = 1$, no conclusion can be drawn.

This theorem is often referred to as the *ratio test*.

The above ideas can be extended to the case where the u_n are functions of x denoted by $u_n(x)$.   In such case the sequences or series will converge or diverge according to the particular values of x.   The set of values of x for which a sequence or series converges is called the *region of convergence*, denoted by $\mathcal{R}$.

> **Example 5.**   The series $1 + x + x^2 + x^3 + \cdots$ has a region of convergence $\mathcal{R}$ [in this case an interval] given by $-1 < x < 1$ if we restrict ourselves to real values of x.

UNIFORM CONVERGENCE

We can say that the series $u_1(x) + u_2(x) + \cdots$ converges to the sum $S(x)$ in a region $\mathcal{R}$, if given $\epsilon > 0$ there exists a number N, which in general depends on both ϵ and x, such that $|S(x) - S_n(x)| < \epsilon$ whenever $n > N$ where $S_n(x) = u_1(x) + \cdots + u_n(x)$.   If we can find N depending only on ϵ and not on x, we say that the series converges *uniformly* to $S(x)$ in $\mathcal{R}$.   Uniformly convergent series have many important advantages as indicated in the following theorems.

Theorem 1-9.   If $u_n(x)$, $n = 1, 2, 3, \ldots$ are continuous in $a \leqq x \leqq b$ and $\Sigma u_n(x)$ is uniformly convergent to $S(x)$ in $a \leqq x \leqq b$, then $S(x)$ is continuous in $a \leqq x \leqq b$.

Theorem 1-10.   If $\Sigma u_n(x)$ converges uniformly to $S(x)$ in $a \leqq x \leqq b$ and $u_n(x)$, $n = 1, 2, 3, \ldots$ are integrable in $a \leqq x \leqq b$, then
$$\int_a^b S(x)\, dx = \int_a^b \{ u_1(x) + u_2(x) + \cdots \}\, dx = \int_a^b u_1(x)\, dx + \int_a^b u_2(x)\, dx + \cdots$$

Theorem 1-11.   If $u_n(x)$, $n = 1, 2, 3, \ldots$ are continuous and have continuous derivatives in $a \leqq x \leqq b$ and if $\Sigma u_n(x)$ converges to $S(x)$ while $\Sigma u_n'(x)$ is uniformly convergent in $a \leqq x \leqq b$, then
$$S'(x) = \frac{d}{dx} \{ u_1(x) + u_2(x) + \cdots \} = u_1'(x) + u_2'(x) + \cdots$$

An important test for uniform convergence, often called the *Weierstrass M test*, is given by the following.

Theorem 1-12.   If there is a set of positive constants M_n, $n = 1, 2, 3, \ldots$ such that $|u_n(x)| \leqq M_n$ in $\mathcal{R}$ and ΣM_n converges, then $\Sigma u_n(x)$ is uniformly convergent [and also absolutely convergent] in $\mathcal{R}$.

TAYLOR SERIES

The *Taylor series* for $f(x)$ about $x = a$ is defined as

$$f(x) = f(a) + f'(a)(x-a) + \frac{f''(a)(x-a)^2}{2!} + \cdots + \frac{f^{(n-1)}(a)(x-a)^{n-1}}{(n-1)!} + R_n \quad (10)$$

where
$$R_n = \frac{f^{(n)}(x_0)(x-a)^n}{n!}, \quad x_0 \text{ between } a \text{ and } x \quad (11)$$

is called the *remainder* and where it is supposed that $f(x)$ has derivatives of order n at least. The case where $n = 1$ is often called the *law of the mean* or *mean-value theorem* and can be written as

$$\frac{f(x) - f(a)}{x - a} = f'(x_0) \quad x_0 \text{ between } a \text{ and } x \quad (12)$$

The infinite series corresponding to (10), also called the *formal Taylor series* for $f(x)$, will converge in some interval if $\lim_{n \to \infty} R_n = 0$ in this interval. Some important Taylor series together with their intervals of convergence are as follows.

1. $e^x = 1 + x + \dfrac{x^2}{2!} + \dfrac{x^3}{3!} + \dfrac{x^4}{4!} + \cdots \qquad -\infty < x < \infty$

2. $\sin x = x - \dfrac{x^3}{3!} + \dfrac{x^5}{5!} - \dfrac{x^7}{7!} + \cdots \qquad -\infty < x < \infty$

3. $\cos x = 1 - \dfrac{x^2}{2!} + \dfrac{x^4}{4!} - \dfrac{x^6}{6!} + \cdots \qquad -\infty < x < \infty$

4. $\ln(1+x) = x - \dfrac{x^2}{2} + \dfrac{x^3}{3} - \dfrac{x^4}{4} + \cdots \qquad -1 < x \leq 1$

5. $\tan^{-1} x = x - \dfrac{x^3}{3} + \dfrac{x^5}{5} - \dfrac{x^7}{7} + \cdots \qquad -1 \leq x \leq 1$

A series of the form $\sum_{n=0}^{\infty} c_n(x-a)^n$ is often called a *power series*. Such power series are uniformly convergent in any interval which lies entirely within the interval of convergence [see Problem 1.120].

FUNCTIONS OF TWO OR MORE VARIABLES

The concept of function of one variable given on page 2 can be extended to functions of two or more variables. Thus for example $z = f(x, y)$ defines a function f which assigns to the number pair (x, y) the number z.

> **Example 6.** If $f(x, y) = x^2 + 3xy + 2y^2$, then $f(-1, 2) = (-1)^2 + 3(-1)(2) + 2(2)^2 = 3$.

The student is familiar with graphing $z = f(x, y)$ in a 3-dimensional xyz coordinate system to obtain a *surface*. We sometimes call x and y *independent* variables and z a *dependent variable*. Occasionally we write $z = z(x, y)$ rather than $z = f(x, y)$, using the symbol z in two different senses. However, no confusion should result.

The ideas of limits and continuity for functions of two or more variables pattern closely those for one variable.

PARTIAL DERIVATIVES

The *partial derivatives* of $f(x, y)$ with respect to x and y are defined by

$$\frac{\partial f}{\partial x} = \lim_{h \to 0} \frac{f(x+h, y) - f(x, y)}{h}, \quad \frac{\partial f}{\partial y} = \lim_{k \to 0} \frac{f(x, y+k) - f(x, y)}{k} \quad (13)$$

if these limits exist. We often write $h = \Delta x$, $k = \Delta y$. Note that $\partial f/\partial x$ is simply the ordinary derivative of f with respect to x *keeping y constant*, while $\partial f/\partial y$ is the ordinary derivative of f with respect to y *keeping x constant*. Thus the usual differentiation formulas on pages 4 and 5 apply.

> **Example 7.** If $f(x, y) = 3x^2 - 4xy + 2y^2$ then $\dfrac{\partial f}{\partial x} = 6x - 4y$, $\dfrac{\partial f}{\partial y} = -4x + 4y$.

Higher derivatives are defined similarly. For example, we have the second order derivatives

$$\frac{\partial}{\partial x}\left(\frac{\partial f}{\partial x}\right) = \frac{\partial^2 f}{\partial x^2}, \quad \frac{\partial}{\partial x}\left(\frac{\partial f}{\partial y}\right) = \frac{\partial^2 f}{\partial x\,\partial y}, \quad \frac{\partial}{\partial y}\left(\frac{\partial f}{\partial x}\right) = \frac{\partial^2 f}{\partial y\,\partial x}, \quad \frac{\partial}{\partial y}\left(\frac{\partial f}{\partial y}\right) = \frac{\partial^2 f}{\partial y^2} \qquad (14)$$

The derivatives in (13) are sometimes denoted by f_x and f_y. In such case $f_x(a, b)$, $f_y(a, b)$ denote these partial derivatives evaluated at (a, b). Similarly the derivatives in (14) are denoted by $f_{xx}, f_{xy}, f_{yx}, f_{yy}$ respectively. The second and third results in (14) will be the same if f has continuous partial derivatives of second order at least.

The *differential* of $f(x, y)$ is defined as

$$df = \frac{\partial f}{\partial x}\,dx + \frac{\partial f}{\partial y}\,dy \qquad (15)$$

where $h = \Delta x = dx$, $k = \Delta y = dy$.

Generalizations of these results are easily made.

TAYLOR SERIES FOR FUNCTIONS OF TWO OR MORE VARIABLES

The ideas involved in Taylor series for functions of one variable can be generalized. For example, the Taylor series for $f(x, y)$ about $x = a$, $y = b$ can be written

$$\begin{aligned} f(x, y) = \ & f(a, b) + f_x(a, b)(x - a) + f_y(a, b)(y - b) \\ & + \frac{1}{2!}\,[f_{xx}(a, b)(x - a)^2 + 2f_{xy}(a, b)(x - a)(y - b) + f_{yy}(a, b)(y - b)^2] + \cdots \end{aligned} \qquad (16)$$

LINEAR EQUATIONS AND DETERMINANTS

Consider the system of linear equations

$$\left.\begin{aligned} a_1 x + b_1 y &= c_1 \\ a_2 x + b_2 y &= c_2 \end{aligned}\right\} \qquad (17)$$

These represent two lines in the xy plane, and in general will meet in a point whose coordinates (x, y) are found by solving (17) simultaneously. We find

$$x = \frac{c_1 b_2 - b_1 c_2}{a_1 b_2 - b_1 a_2}, \qquad y = \frac{a_1 c_2 - c_1 a_2}{a_1 b_2 - b_1 a_2} \qquad (18)$$

It is convenient to write these in *determinant form* as

$$x = \frac{\begin{vmatrix} c_1 & b_1 \\ c_2 & b_2 \end{vmatrix}}{\begin{vmatrix} a_1 & b_1 \\ a_2 & b_2 \end{vmatrix}}, \qquad y = \frac{\begin{vmatrix} a_1 & c_1 \\ a_2 & c_2 \end{vmatrix}}{\begin{vmatrix} a_1 & b_1 \\ a_2 & b_2 \end{vmatrix}} \qquad (19)$$

where we define a *determinant of the second order* or *order 2* to be

$$\begin{vmatrix} a & b \\ c & d \end{vmatrix} = ad - bc \qquad (20)$$

Note that the denominator for x and y in (19) is the determinant consisting of the coefficients of x and y in (17). The numerator for x is found by replacing the first column of the denominator by the constants c_1, c_2 on the right side of (17). Similarly the numerator for y is found by replacing the second column of the denominator by c_1, c_2. This procedure is often called *Cramer's rule*. In case the denominator in (19) is zero, the two lines represented by (17) do not meet in one point but are either coincident or parallel.

The ideas are easily extended. Thus consider the equations

$$\left.\begin{array}{l} a_1x + b_1y + c_1z = d_1 \\ a_2x + b_2y + c_2z = d_2 \\ a_3x + b_3y + c_3z = d_3 \end{array}\right\} \tag{21}$$

representing 3 planes. If they intersect in a point, the coordinates (x, y, z) of this point are found from Cramer's rule to be

$$x = \frac{\begin{vmatrix} d_1 & b_1 & c_1 \\ d_2 & b_2 & c_2 \\ d_3 & b_3 & c_3 \end{vmatrix}}{\begin{vmatrix} a_1 & b_1 & c_1 \\ a_2 & b_2 & c_2 \\ a_3 & b_3 & c_3 \end{vmatrix}}, \quad y = \frac{\begin{vmatrix} a_1 & d_1 & c_1 \\ a_2 & d_2 & c_2 \\ a_3 & d_3 & c_3 \end{vmatrix}}{\begin{vmatrix} a_1 & b_1 & c_1 \\ a_2 & b_2 & c_2 \\ a_3 & b_3 & c_3 \end{vmatrix}}, \quad z = \frac{\begin{vmatrix} a_1 & b_1 & d_1 \\ a_2 & b_2 & d_2 \\ a_3 & b_3 & d_3 \end{vmatrix}}{\begin{vmatrix} a_1 & b_1 & c_1 \\ a_2 & b_2 & c_2 \\ a_3 & b_3 & c_3 \end{vmatrix}} \tag{22}$$

where we define the determinant of order 3 by

$$\begin{vmatrix} a_1 & b_1 & c_1 \\ a_2 & b_2 & c_2 \\ a_3 & b_3 & c_3 \end{vmatrix} = a_1b_2c_3 + b_1c_2a_3 + c_1a_2b_3 - (b_1a_2c_3 + a_1c_2b_3 + c_1b_2a_3) \tag{23}$$

The result (23) can be remembered by using the scheme of repeating the first two columns as follows

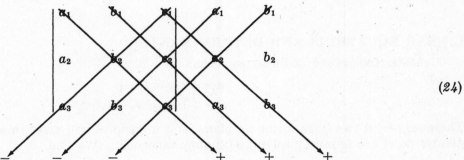

$$\tag{24}$$

and taking the sums of the products of terms as indicated by the arrows marked $+$ and $-$.

The determinant can also be evaluated in terms of second order determinants as follows

$$a_1\begin{vmatrix} b_2 & c_2 \\ b_3 & c_3 \end{vmatrix} - b_1\begin{vmatrix} a_2 & c_2 \\ a_3 & c_3 \end{vmatrix} + c_1\begin{vmatrix} a_2 & b_2 \\ a_3 & b_3 \end{vmatrix} \tag{25}$$

where it is noted that a_1, b_1, c_1 are the elements in the first row and the corresponding second order determinants are those obtained from the given third order determinant by removing the row and column in which the element appears.

The general theory of determinants, of which the above results are special cases, is considered in Chapter 15.

MAXIMA AND MINIMA

If for all x such that $|x - a| < \delta$ we have $f(x) \leqq f(a)$ [or $f(x) \geqq f(a)$], we say that $f(a)$ is a *relative maximum* [or *relative minimum*]. For $f(x)$ to have a relative maximum or minimum at $x = a$, we must have $f'(a) = 0$. Then if $f''(a) < 0$ it is a relative maximum while if $f''(a) > 0$ it is a relative minimum. Note that possible points at which $f(x)$ has a relative maxima or minima are obtained by solving $f'(x) = 0$, i.e. by finding the values of x where the *slope* of the graph of $f(x)$ is equal to zero.

Similarly $f(x, y)$ has a relative maximum or minimum at $x = a$, $y = b$ if $f_x(a, b) = 0$, $f_y(a, b) = 0$. Thus possible points at which $f(x, y)$ has relative maxima or minima are obtained by solving simultaneously the equations

$$\partial f / \partial x = 0, \quad \partial f / \partial y = 0 \qquad (26)$$

Extensions to functions of more than two variables are similar.

METHOD OF LAGRANGE MULTIPLIERS

Sometimes we wish to find the relative maxima or minima of $f(x, y) = 0$ subject to some *constraint condition* $\phi(x, y) = 0$. To do this we form the function $h(x, y) = f(x, y) + \lambda \phi(x, y)$ and set

$$\partial h / \partial x = 0, \quad \partial h / \partial y = 0 \qquad (27)$$

The constant λ is called a *Lagrange multiplier* and the method is called the *method of Lagrange multipliers*. Generalizations can be made [see Problems 1.54 and 1.150].

LEIBNITZ'S RULE FOR DIFFERENTIATING AN INTEGRAL

Let
$$I(\alpha) = \int_a^b f(x, \alpha) \, dx \qquad (28)$$

where f is supposed continuous and differentiable. Then *Leibnitz's rule* states that if a and b are differentiable functions of α,

$$\frac{dI}{d\alpha} = \int_a^b \frac{\partial f}{\partial \alpha} dx + f(b, \alpha) \frac{db}{d\alpha} - f(a, \alpha) \frac{da}{d\alpha} \qquad (29)$$

MULTIPLE INTEGRALS

A generalization of the integral for functions of one variable leads to the idea of *multiple integrals* for functions of two or more variables. Because several ideas involved in the theory may be new to some students, we postpone consideration of this topic to Chapter 6.

COMPLEX NUMBERS

Complex numbers arose in order to solve polynomial equations such as $x^2 + 1 = 0$ or $x^2 + x + 1 = 0$ which are not satisfied by real numbers. We assume that a complex number has the form $a + bi$ where a, b are real numbers and i, called the *imaginary unit*, has the property that $i^2 = -1$. We define operations with complex numbers as follows.

1. **Addition.** $(a + bi) + (c + di) = (a + c) + (b + d)i$

2. **Subtraction.** $(a + bi) - (c + di) = (a - c) + (b - d)i$

3. **Multiplication.** $(a + bi)(c + di) = ac + adi + bci + bdi^2 = (ac - bd) + (ad + bc)i$

4. **Division.** $\dfrac{a + bi}{c + di} = \dfrac{a + bi}{c + di} \cdot \dfrac{c - di}{c - di} = \dfrac{ac + bd}{c^2 + d^2} + \dfrac{bc - ad}{c^2 + d^2} i$

Note that we have used the ordinary rules of algebra except that we replace i^2 by -1 wherever it occurs. The commutative, associative and distributive laws of page 1 also apply to complex numbers. We call a and b of $a + bi$ the *real* and *imaginary parts* respectively. Two complex numbers are *equal* if and only if their real and imaginary parts are respectively equal.

A complex number $z = x + iy$ can be considered as a point P with coordinates (x, y) on a rectangular xy plane called in this case the *complex plane* or *Argand diagram* [Fig. 1-4]. If we construct the line from origin O to P and let ρ be the distance OP and ϕ the angle made by OP with the positive x axis, we have from Fig. 1-4

$$x = \rho \cos \phi, \quad y = \rho \sin \phi, \quad \rho = \sqrt{x^2 + y^2} \qquad (30)$$

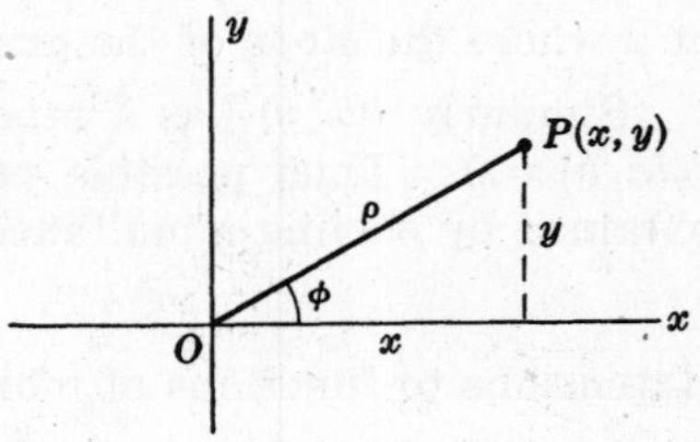

and can write the complex number in so-called *polar form* as

Fig. 1-4

$$z = x + iy = \rho(\cos \phi + i \sin \phi) = \rho \operatorname{cis} \phi \qquad (31)$$

We often call ρ the *modulus* or *absolute value* of z and denote it by $|z|$. The angle ϕ is called the *amplitude* or *argument* of z abbreviated arg z. We can also write $\rho = \sqrt{z\bar{z}}$ where $\bar{z} = x - iy$ is called the *conjugate* of $z = x + iy$.

If we write two complex numbers in polar form as

$$z_1 = \rho_1(\cos \phi_1 + i \sin \phi_1), \quad z_2 = \rho_2(\cos \phi_2 + i \sin \phi_2) \qquad (32)$$

then

$$z_1 z_2 = \rho_1 \rho_2 [\cos(\phi_1 + \phi_2) + i \sin(\phi_1 + \phi_2)] \qquad (33)$$

$$\frac{z_1}{z_2} = \frac{\rho_1}{\rho_2} [\cos(\phi_1 - \phi_2) + i \sin(\phi_1 - \phi_2)] \qquad (34)$$

Also if n is any real number, we have

$$z^n = [\rho(\cos \phi + i \sin \phi)]^n = \rho^n(\cos n\phi + i \sin n\phi) \qquad (35)$$

which is often called *De Moivre's theorem*. We can use this to determine roots of complex numbers. For example if n is a positive integer,

$$z^{1/n} = [\rho(\cos \phi + i \sin \phi)]^{1/n}$$

$$= \rho^{1/n} \left\{ \cos\left(\frac{\phi + 2k\pi}{n}\right) + i \sin\left(\frac{\phi + 2k\pi}{n}\right) \right\} \quad k = 0, 1, 2, \ldots, n-1 \qquad (36)$$

Using the series for e^x, $\sin x$, $\cos x$ on page 8, we are led to define

$$e^{i\phi} = \cos \phi + i \sin \phi, \quad e^{-i\phi} = \cos \phi - i \sin \phi \qquad (37)$$

which are called *Euler's formulas* and which enable us to rewrite equations (31)-(36) in terms of exponentials.

Many of the ideas presented in this chapter involving real numbers can be extended to complex numbers. These ideas are developed in Chapter 13.

Solved Problems

REAL NUMBERS AND LAWS OF ALGEBRA

1.1. Prove that $\sqrt{2}$ is an irrational number.

Assume the contrary, i.e. suppose that $\sqrt{2} = p/q$ where p and q are positive integers having no common integer factor other than 1 [in such case we say that p/q is a fraction in *lowest terms*]. Squaring, $2 = p^2/q^2$ or $p^2 = 2q^2$. Then p^2 is even and so p must also be even. Thus $p = 2m$ where m is a positive integer. Substituting into $p^2 = 2q^2$ we find $q^2 = 2m^2$, so that q^2 is even and thus q is even, i.e. $q = 2n$ where n is a positive integer. Since p and q are both even they have 2 as a common factor, violating our assumption that they have no common integer factor other than 1. This contradiction shows that the assumption that $\sqrt{2}$ is rational is incorrect and proves that $\sqrt{2}$ is irrational.

1.2. Which is larger $\sqrt{2}$ or $\sqrt[3]{3}$?

Assume $\sqrt{2} \geqq \sqrt[3]{3}$. Then raising each side to the 6th power we have $2^3 \geqq 3^2$ which is wrong. It follows that $\sqrt{2} < \sqrt[3]{3}$.

1.3. Assuming that the real numbers a, b, c satisfy the rules of algebra on page 1, prove that $(b+c)a = ba + ca$.

From rule 5 on page 1 we have $a(b+c) = ab + ac$. But by rule 3 it follows that $a(b+c) = (b+c)a$, $ab = ba$, $ac = ca$. Thus $(b+c)a = ba + ca$.

FUNCTIONS

1.4. If $f(x) = 2x^3 - 3x + 5$ find (a) $f(-1)$, (b) $f(0)$, (c) $f(x+h)$.

(a) $f(-1) = 2(-1)^3 - 3(-1) + 5 = 2(-1) + 3 + 5 = 6$

(b) $f(0) = 2(0)^3 - 3(0) + 5 = 0 + 0 + 5 = 5$

(c) $f(x+h) = 2(x+h)^3 - 3(x+h) + 5 = 2(x^3 + 3x^2h + 3xh^2 + h^3) - 3x - 3h + 5$
$$= 2x^3 + 6x^2h + 6xh^2 + 2h^3 - 3x - 3h + 5$$

1.5. Using the rules of exponents (*1*) on page 2, prove the rules of logarithms (*4*) on page 2.

By definition, if $e^x = m$ then $x = \ln m$. Similarly if $e^y = n$ then $y = \ln n$.

Since $e^x \cdot e^y = e^{x+y}$ we have $mn = e^{x+y}$ or $x + y = \ln(mn)$, i.e. $\ln(mn) = \ln m + \ln n$.

Since $e^x/e^y = e^{x-y}$ we have $m/n = e^{x-y}$ or $x - y = \ln(m/n)$, i.e. $\ln(m/n) = \ln m - \ln n$.

Since $(e^x)^p = e^{xp}$ we have $m^p = e^{xp}$ or $xp = \ln m^p$, i.e. $\ln m^p = p \ln m$.

1.6. Prove that (a) $\sin^2 x = \frac{1}{2}(1 - \cos 2x)$, (b) $\cos^2 x = \frac{1}{2}(1 + \cos 2x)$.

Since $\cos(x+y) = \cos x \cos y - \sin x \sin y$ we have on letting $y = x$,

$$\cos^2 x - \sin^2 x = \cos 2x \tag{1}$$

Also we have

$$\cos^2 x + \sin^2 x = 1 \tag{2}$$

From (*1*) and (*2*) the required results follow by subtraction and addition respectively.

1.7. Prove that $A \cos x + B \sin x = \sqrt{A^2 + B^2} \sin(x + \alpha)$ where $\tan \alpha = A/B$.

We have

$$A \cos x + B \sin x \; = \; \sqrt{A^2 + B^2} \left[\frac{A}{\sqrt{A^2 + B^2}} \cos x \; + \; \frac{B}{\sqrt{A^2 + B^2}} \right] \tag{1}$$

Letting

$$\sin \alpha \; = \; \frac{A}{\sqrt{A^2 + B^2}}, \quad \cos \alpha \; = \; \frac{B}{\sqrt{A^2 + B^2}} \quad \text{or} \quad \tan \alpha \; = \; \frac{A}{B}$$

[see Fig. 1-5], (1) becomes

$$A \cos x + B \sin x \; = \; \sqrt{A^2 + B^2} \left[\sin \alpha \cos x \; + \; \cos \alpha \sin x \right]$$
$$= \; \sqrt{A^2 + B^2} \sin(x + \alpha)$$

as required.

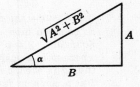

Fig. 1-5

1.8. Prove that (a) $\cosh^2 x - \sinh^2 x = 1$, (b) $\operatorname{sech}^2 x + \tanh^2 x = 1$.

(a) By definition, $\cosh x = \frac{1}{2}(e^x + e^{-x})$, $\sinh x = \frac{1}{2}(e^x - e^{-x})$. Then

$$\cosh^2 x - \sinh^2 x \; = \; \tfrac{1}{4}(e^x + e^{-x})^2 - \tfrac{1}{4}(e^x - e^{-x})^2$$
$$= \; \tfrac{1}{4}(e^{2x} + 2 + e^{-2x}) - \tfrac{1}{4}(e^{2x} - 2 + e^{-2x}) \; = \; 1$$

(b) Dividing both sides of the result in (a) by $\cosh^2 x$, we have

$$\frac{\cosh^2 x}{\cosh^2 x} - \frac{\sinh^2 x}{\cosh^2 x} \; = \; \frac{1}{\cosh^2 x} \quad \text{or} \quad 1 - \tanh^2 x \; = \; \operatorname{sech}^2 x$$

Since $\sinh x / \cosh x = \tanh x$, $1/\cosh x = \operatorname{sech} x$. From this we obtain $\operatorname{sech}^2 x + \tanh^2 x = 1$.

1.9. Prove that $\cosh^{-1} x = \pm \ln(x + \sqrt{x^2 - 1})$.

By definition, if $y = \cosh^{-1} x$ then $x = \cosh y = \frac{1}{2}(e^y + e^{-y})$. Thus $e^y + e^{-y} = 2x$ or $e^{2y} - 2xe^y + 1 = 0$. Solving this as a quadratic equation in e^y, we find

$$e^y \; = \; \frac{2x \pm \sqrt{4x^2 - 4}}{2} \; = \; x \pm \sqrt{x^2 - 1}$$

or $y = \ln(x \pm \sqrt{x^2 - 1})$. Since

$$\ln(x - \sqrt{x^2 - 1}) \; = \; \ln\left(\frac{x^2 - (x^2 - 1)}{x + \sqrt{x^2 - 1}} \right) \; = \; -\ln(x + \sqrt{x^2 - 1})$$

the required result follows. Note that we must have $x \geqq 1$ if $y = \cosh^{-1} x$ is to be real.

LIMITS AND CONTINUITY

1.10. If (a) $f(x) = x^2$, (b) $f(x) = \begin{cases} x^2 & x \neq 2 \\ 0 & x = 2 \end{cases}$, prove that $\lim\limits_{x \to 2} f(x) = 4$.

(a) We must show that given any $\epsilon > 0$, we can find $\delta > 0$ [depending on ϵ in general] such that $|x^2 - 4| < \epsilon$ when $0 < |x - 2| < \delta$.

Choose $\delta \leqq 1$ so that $0 < |x - 2| < 1$. Then

$$|x^2 - 4| \; = \; |(x-2)(x+2)| \; = \; |x-2|\,|x+2| \; = \; |x-2|\,|(x-2) + 4|$$
$$\leqq \; |x-2|\{|x-2| + 4\} \; < \; 5\delta$$

where we have used the result that $|a + b| \leqq |a| + |b|$.

Take δ as 1 or $\epsilon/5$, whichever is smaller. Then we have $|x^2 - 4| < \epsilon$ whenever $0 < |x - 2| < \delta$, and the required result is proved.

(b) There is no difference between the proof for this case and the proof in (a) since in both cases we exclude $x = 2$.

1.11. Prove that if $\lim\limits_{x \to a} f_1(x) = l_1$ and $\lim\limits_{x \to a} f_2(x) = l_2$ then $\lim\limits_{x \to a} [f_1(x) + f_2(x)] = l_1 + l_2$.

We must show that for any $\epsilon > 0$ we can find $\delta > 0$ such that

$$|[f_1(x) + f_2(x)] - (l_1 + l_2)| < \epsilon \qquad \text{when} \quad 0 < |x - a| < \delta$$

We have

$$|[f_1(x) + f_2(x)] - (l_1 + l_2)| = |[f_1(x) - l_1] + [f_2(x) - l_2]| \leq |f_1(x) - l_1| + |f_2(x) - l_2| \qquad (1)$$

Now by hypothesis, given $\epsilon > 0$ we can find $\delta_1 > 0$ and $\delta_2 > 0$ such that

$$|f_1(x) - l_1| < \epsilon/2 \qquad \text{when} \quad 0 < |x - a| < \delta_1 \qquad (2)$$

$$|f_2(x) - l_2| < \epsilon/2 \qquad \text{when} \quad 0 < |x - a| < \delta_2 \qquad (3)$$

Then from (1), (2) and (3) we have

$$|[f_1(x) + f_2(x)] - (l_1 + l_2)| < \epsilon/2 + \epsilon/2 = \epsilon \qquad \text{when} \quad 0 < |x - x_0| < \delta$$

where δ is chosen as the smaller of δ_1 and δ_2.

In a similar manner we can prove other theorems on limits such as $\lim\limits_{x \to a} [f_1(x) - f_2(x)] = l_1 - l_2$, $\lim\limits_{x \to a} [f_1(x) f_2(x)] = l_1 l_2$ and $\lim\limits_{x \to a} f_1(x)/f_2(x) = l_1/l_2$ if $l_2 \neq 0$. See Problem 1.75.

1.12. Prove that (a) $f(x) = x^2$ is continuous at $x = 2$ while (b) $f(x) = \begin{cases} x^2 & x \neq 2 \\ 0 & x = 2 \end{cases}$ is not continuous at $x = 2$.

(a) **Method 1.**

By Problem 1.10(a), $\lim\limits_{x \to 2} f(x) = f(2) = 4$ and so $f(x)$ is continuous at $x = 2$.

Method 2.

We must show that given any $\epsilon > 0$, we can find $\delta > 0$ [depending on ϵ in general] such that $|f(x) - f(2)| = |x^2 - 4| < \epsilon$ when $|x - 2| < \delta$. The proof patterns that given in Problem 1.10.

(b) Since $f(2) = 0$ we have $\lim\limits_{x \to 2} f(x) \neq f(2)$, so that $f(x)$ is not continuous [or is discontinuous] at $x = 2$. We can also give a proof involving ϵ, δ methods by showing that given any $\epsilon > 0$ we *cannot* find $\delta > 0$ such that $|f(x) - f(2)| < \epsilon$ when $|x - 2| < \delta$.

DERIVATIVES

1.13. Prove that if u and v are differentiable functions of x, then

$$(a)\ \frac{d}{dx}(u + v) = \frac{du}{dx} + \frac{dv}{dx}, \qquad (b)\ \frac{d}{dx}(uv) = u\frac{dv}{dx} + v\frac{du}{dx}$$

(a) Write $\Delta u = u(x + h) - u(x)$, $\Delta v = v(x + h) - v(x)$. Denote $u(x)$, $v(x)$ briefly by u, v respectively. Then by definition, if $h = \Delta x$ we have

$$\frac{d}{dx}(u + v) = \lim_{h \to 0} \frac{u(x + h) + v(x + h) - [u(x) + v(x)]}{h}$$

$$= \lim_{h \to 0} \frac{u(x + h) - u(x)}{h} + \lim_{h \to 0} \frac{v(x + h) - v(x)}{h}$$

$$= \lim_{\Delta x \to 0} \frac{\Delta u}{\Delta x} + \lim_{\Delta x \to 0} \frac{\Delta v}{\Delta x} = \frac{du}{dx} + \frac{dv}{dx}$$

(b)

$$\frac{d}{dx}(uv) = \lim_{h \to 0} \frac{u(x + h)\,v(x + h) - u(x)\,v(x)}{h}$$

$$= \lim_{\Delta x \to 0} \frac{(u + \Delta u)(v + \Delta v) - uv}{\Delta x}$$

$$= \lim_{\Delta x \to 0} \left(u\frac{\Delta v}{\Delta x} + v\frac{\Delta u}{\Delta x} + \Delta v\frac{\Delta u}{\Delta x} \right) = u\frac{dv}{dx} + v\frac{du}{dx}$$

since $\lim\limits_{\Delta x \to 0} \Delta v\frac{\Delta u}{\Delta x} = \left(\lim\limits_{\Delta x \to 0} \Delta v \right)\frac{du}{dx} = 0$.

1.14. Prove that if $f(x)$ has a derivative at the point a, then $f(x)$ is continuous at a.

We have $f(a+h) - f(a) = \dfrac{f(a+h) - f(a)}{h} \cdot h$ if $h \neq 0$. Then if the derivative exists,

$$\lim_{h \to 0} [f(a+h) - f(a)] = \lim_{h \to 0} \left[\frac{f(a+h) - f(a)}{h} \right] \cdot \lim_{h \to 0} h = 0$$

Thus $\lim_{h \to 0} f(a+h) = f(a)$ and the required result follows.

1.15. Prove that if p is any positive integer and u is a differentiable function of x, then $\dfrac{d}{dx} u^p = p u^{p-1} \dfrac{du}{dx}$.

By definition

$$\frac{d}{dx} u^p = \lim_{h \to 0} \frac{[u(x+h)]^p - [u(x)]^p}{h} = \lim_{\Delta x \to 0} \frac{(u + \Delta u)^p - u^p}{\Delta x}$$

$$= \lim_{\Delta x \to 0} \frac{u^p + p u^{p-1} \Delta u + \frac{1}{2} p(p-1) u^{p-2} (\Delta u)^2 + \cdots - u^p}{\Delta x}$$

$$= \lim_{\Delta x \to 0} \left[p u^{p-1} \frac{\Delta u}{\Delta x} + \frac{1}{2} p(p-1) u^{p-2} \frac{\Delta u}{\Delta x} \Delta u + \cdots \right] = p u^{p-1} \frac{du}{dx}$$

The result can also be shown to hold for all values of p.

1.16. Assuming that $\lim\limits_{b \to 0} \dfrac{\sin b}{b} = 1$, prove that (a) $\dfrac{d}{dx} \sin u = \cos u \dfrac{du}{dx}$, (b) $\dfrac{d}{dx} \cos u = -\sin u \dfrac{du}{dx}$, (c) $\dfrac{d}{dx} \tan u = \sec^2 u \dfrac{du}{dx}$.

(a)
$$\frac{d}{dx} \sin u = \lim_{h \to 0} \frac{\sin u(x+h) - \sin u(x)}{h} = \lim_{\Delta x \to 0} \frac{\sin (u + \Delta u) - \sin u}{\Delta x}$$

$$= \lim_{\Delta x \to 0} \frac{\sin u \cos \Delta u + \cos u \sin \Delta u - \sin u}{\Delta x}$$

$$= \lim_{\Delta x \to 0} \left[\cos u \frac{\sin \Delta u}{\Delta x} - \sin u \left(\frac{1 - \cos \Delta u}{\Delta x} \right) \right]$$

$$= \lim_{\Delta x \to 0} \left[\cos u \frac{\sin \Delta u}{\Delta u} \frac{\Delta u}{\Delta x} - \frac{\sin u}{2} \left(\frac{\sin^2 (\Delta u/2)}{(\Delta u/2)^2} \right) \frac{\Delta u}{\Delta x} \cdot \Delta u \right]$$

$$= \cos u \frac{du}{dx}$$

where we have used the result of Problem 1.6(a).

(b) From part (a) we have $\dfrac{d}{dx} \sin v = \cos v \dfrac{dv}{dx}$. Then letting $v = \dfrac{\pi}{2} - u$ so that $\sin v = \cos u$, we find

$$\frac{d}{dx} \cos u = \cos \left(\frac{\pi}{2} - u \right) \frac{d}{dx} \left(\frac{\pi}{2} - u \right) = - \sin u \frac{du}{dx}$$

(c) By differentiation formula 4, page 4, we have

$$\frac{d}{dx} \tan u = \frac{d}{dx} \left(\frac{\sin u}{\cos u} \right) = \frac{(\cos u) \dfrac{d}{dx} \sin u - (\sin u) \dfrac{d}{dx} \cos u}{\cos^2 u}$$

$$= \frac{(\cos u)(\cos u) \dfrac{du}{dx} - (\sin u)(-\sin u) \dfrac{du}{dx}}{\cos^2 u}$$

$$= \frac{\cos^2 u + \sin^2 u}{\cos^2 u} \frac{du}{dx} = \frac{1}{\cos^2 u} \frac{du}{dx} = \sec^2 u \frac{du}{dx}$$

1.17. Prove that $\dfrac{d}{dx}\tan^{-1}u = \dfrac{1}{1+u^2}\dfrac{du}{dx}$.

If $v = \tan^{-1}u$, then $u = \tan v$. Then by Problem 1.16(c),

$$\frac{du}{dx} = \sec^2 v\,\frac{dv}{dx} = (1+\tan^2 v)\,\frac{dv}{dx} = (1+u^2)\,\frac{dv}{dx}$$

Thus

$$\frac{dv}{dx} = \frac{1}{1+u^2}\frac{du}{dx}$$

1.18. Given that $\lim\limits_{b \to 0}(1+b)^{1/b} = e = 2.7182818\ldots$, prove that

$$(a)\quad \frac{d}{dx}\ln u = \frac{1}{u}\frac{du}{dx}, \qquad (b)\quad \frac{d}{dx}e^u = e^u\frac{du}{dx}$$

(a)
$$\frac{d}{dx}\ln u = \lim_{\Delta x \to 0}\left[\frac{\ln(u+\Delta u)-\ln u}{\Delta x}\right] = \lim_{\Delta x \to 0}\left\{\frac{u}{\Delta u}\left[\ln\left(\frac{u+\Delta u}{u}\right)\right]\frac{\Delta u}{u\,\Delta x}\right\}$$

$$= \frac{1}{u}\left[\lim_{\Delta x \to 0}\ln\left(1+\frac{\Delta u}{u}\right)^{u/\Delta u}\right]\frac{du}{dx} = \frac{1}{u}\ln e\,\frac{du}{dx} = \frac{1}{u}\frac{du}{dx}$$

We have assumed here that

$$\lim_{\Delta x \to 0}\ln\left(1+\frac{\Delta u}{u}\right)^{u/\Delta u} = \ln\left[\lim_{\Delta x \to 0}\left(1+\frac{\Delta u}{u}\right)^{u/\Delta u}\right]$$

which can be demonstrated by showing that $\ln u$ is continuous.

(b) If $v = e^u$, then $u = \ln v$. Thus by part (a),

$$\frac{du}{dx} = \frac{1}{v}\frac{dv}{dx} \qquad\text{or}\qquad \frac{dv}{dx} = v\,\frac{du}{dx}$$

i.e.
$$\frac{d}{dx}(e^u) = e^u\frac{du}{dx}$$

1.19. Find (a) $\dfrac{d}{dx}\sqrt{x^4+2x}$, (b) $\dfrac{d}{dx}\sin(\ln x)$, (c) $\dfrac{d}{dx}\ln(e^{3x}+\cos 2x)$.

(a)
$$\frac{d}{dx}\sqrt{x^4+2x} = \frac{d}{dx}(x^4+2x)^{1/2} = \tfrac{1}{2}(x^4+2x)^{-1/2}\frac{d}{dx}(x^4+2x)$$

$$= \tfrac{1}{2}(x^4+2x)^{-1/2}(4x^3+2) = (x^4+2x)^{-1/2}(2x^3+1)$$

(b)
$$\frac{d}{dx}\sin(\ln x) = \cos(\ln x)\frac{d}{dx}\ln x = \frac{\cos(\ln x)}{x}$$

(c)
$$\frac{d}{dx}\ln(e^{3x}+\cos 2x) = \frac{1}{e^{3x}+\cos 2x}\frac{d}{dx}(e^{3x}+\cos 2x)$$

$$= \frac{1}{e^{3x}+\cos 2x}\left[e^{3x}\frac{d}{dx}(3x)-\sin 2x\frac{d}{dx}(2x)\right] = \frac{3e^{3x}-2\sin 2x}{e^{3x}+\cos 2x}$$

1.20. If $x^2y - e^{2x} = \sin y$ find dy/dx.

Consider y as a function of x and differentiate both sides with respect to x. Then

$$\frac{d}{dx}(x^2y) - \frac{d}{dx}(e^{2x}) = \frac{d}{dx}\sin y$$

or
$$x^2\frac{dy}{dx} + 2xy - 2e^{2x} = \cos y\,\frac{dy}{dx}$$

i.e.
$$(x^2-\cos y)\frac{dy}{dx} = 2e^{2x} - 2xy$$

and
$$\frac{dy}{dx} = \frac{2e^{2x}-2xy}{x^2-\cos y}$$

1.21. Show that if $y = 3x^2 + \sin 2x$, then $\dfrac{d^2y}{dx^2} + 4y = 12x^2 + 6$.

We have
$$y \;=\; 3x^2 + \sin 2x, \qquad \frac{dy}{dx} \;=\; 6x + 2\cos 2x, \qquad \frac{d^2y}{dx^2} \;=\; 6 - 4\sin 2x$$

Then
$$\frac{d^2y}{dx^2} + 4y \;=\; (6 - 4\sin 2x) + 4(3x^2 + \sin 2x) \;=\; 12x^2 + 6$$

1.22. Find the differentials of (a) $y = x^2 - \ln x$, (b) $y = e^{-2x} + \cos 3x$.

(a) $dy \;=\; \dfrac{dy}{dx} dx \;=\; \dfrac{d}{dx}(x^2 - \ln x)\, dx \;=\; \left(2x - \dfrac{1}{x}\right) dx$

(b) $dy \;=\; \dfrac{dy}{dx} dx \;=\; \dfrac{d}{dx}(e^{-2x} + \cos 3x) \;=\; -(2e^{-2x} + 3\sin 3x)\, dx$

INTEGRALS

1.23. Prove that (a) $\displaystyle\int u^p\, du = \frac{u^{p+1}}{p+1}, \; p \neq -1$, (b) $\displaystyle\int \frac{du}{u} = \ln u$, (c) $\displaystyle\int (u+v)\, dx =$ $\displaystyle\int u\, dx + \int v\, dx$, (d) $\displaystyle\int \cos u\, du = \sin u$ where constants of integration are omitted.

(a) Since $\dfrac{d}{du}\left(\dfrac{u^{p+1}}{p+1}\right) = \dfrac{(p+1)u^p}{p+1} = u^p$ if $p \neq -1$, we have $\displaystyle\int u^p\, du = \dfrac{u^{p+1}}{p+1}$ if $p \neq -1$.

(b) Since $\dfrac{d}{du}\ln u = \dfrac{1}{u}$, we have $\displaystyle\int \dfrac{du}{u} = \ln u$.

(c) If $F = \displaystyle\int (u+v)\, dx$, $G = \displaystyle\int u\, dx$, $H = \displaystyle\int v\, dx$ then by definition $\dfrac{dF}{dx} = u+v$, $\dfrac{dG}{dx} = u$, $\dfrac{dH}{dx} = v$. Thus $\dfrac{dF}{dx} = \dfrac{dG}{dx} + \dfrac{dH}{dx} = \dfrac{d}{dx}(G+H)$ and so $F = G + H$ apart from a constant of integration and the required result follows.

(d) Since $\dfrac{d}{du}\sin u = \cos u$, we have $\displaystyle\int \cos u\, du = \sin u$.

1.24. Find (a) $\displaystyle\int x\sqrt{x^2+1}\, dx$, (b) $\displaystyle\int e^{3x}\cos(e^{3x})\, dx$, (c) $\displaystyle\int \frac{1-\cos x}{x - \sin x}\, dx$.

(a) Let $u = x^2 + 1$ so that $du = 2x\, dx$. Then
$$\int x\sqrt{x^2+1}\, dx \;=\; \int \sqrt{u}\, \frac{du}{2}$$
$$=\; \frac{1}{2}\int u^{1/2}\, du \;=\; \frac{1}{2}\cdot\frac{u^{3/2}}{3/2} + c \;=\; \frac{u^{3/2}}{3} + c \;=\; \frac{(x^2+1)^{3/2}}{3} + c$$

(b) Let $u = e^{3x}$ so that $du = 3e^{3x}\, dx$. Then
$$\int e^{3x}\cos(e^{3x})\, dx \;=\; \int \cos u\, \frac{du}{3} \;=\; \frac{1}{3}\int \cos u\, du$$
$$=\; \frac{1}{3}\sin u + c \;=\; \frac{1}{3}\sin(e^{3x}) + c$$

(c) Let $u = x - \sin x$ so that $du = (1 - \cos x)\, dx$. Then
$$\int \frac{1-\cos x}{x - \sin x}\, dx \;=\; \int \frac{du}{u} \;=\; \ln u + c \;=\; \ln(x - \sin x) + c$$

1.25. (a) Prove the formula for integration by parts, and (b) use the formula to find $\int x \sin 2x \, dx$.

(a) Since $d(uv) = u \, dv + v \, du$ we have

$$\int d(uv) = \int u \, dv + \int v \, dv \quad \text{or} \quad uv = \int u \, dv + \int v \, du$$

so that

$$\int u \, dv = uv - \int v \, du$$

(b) Let $u = x$, $dv = \sin 2x \, dx$. Then $du = dx$, $v = -\frac{1}{2}\cos 2x$. Thus applying integration by parts,

$$\int x \sin 2x \, dx = (x)(-\tfrac{1}{2}\cos 2x) - \int (-\tfrac{1}{2}\cos 2x) \, dx = -\tfrac{1}{2}x \cos 2x + \tfrac{1}{4}\sin 2x + c$$

1.26. Evaluate (a) $\int_1^2 \dfrac{x \, dx}{x^2 + 1}$, (b) $\int_0^{\pi/2} \cos 3x \, dx$.

(a) Let $u = x^2 + 1$ so that $du = 2x \, dx$. Then

$$\int \frac{x \, dx}{x^2 + 1} = \frac{1}{2}\int \frac{du}{u} = \tfrac{1}{2}\ln u + c = \tfrac{1}{2}\ln (x^2 + 1) + c$$

Thus

$$\int_1^2 \frac{x \, dx}{x^2 + 1} = \tfrac{1}{2}\ln (x^2 + 1) + c \Big|_1^2 = [\tfrac{1}{2}\ln 5 + c] - [\tfrac{1}{2}\ln 2 + c]$$

$$= \tfrac{1}{2}\ln 5 - \tfrac{1}{2}\ln 2 = \tfrac{1}{2}\ln \tfrac{5}{2}$$

(b) Let $u = 3x$ so that $du = 3 \, dx$. Then

$$\int \cos 3x \, dx = \frac{1}{3}\int \cos u \, du = \tfrac{1}{3}\sin u + c = \tfrac{1}{3}\sin 3x + c$$

Thus

$$\int_0^{\pi/2} \cos 3x \, dx = \tfrac{1}{3}\sin 3x + c \Big|_0^{\pi/2} = -\tfrac{1}{3}$$

In practice when evaluating definite integrals, we omit the constant of integration c since it does not enter anyway.

1.27. Find the area under the curve $y = \sin x$ between $x = 0$ and $x = \pi$ [see Fig. 1-2, page 3].

$$\text{Area} = \int_0^{\pi} \sin x \, dx = -\cos x \Big|_0^{\pi} = 2$$

1.28. (a) By using definition (8) page 5 for the definite integral, obtain an approximate value for $\int_1^2 \dfrac{dx}{x}$ and give a geometric interpretation.

(b) Show how the result in (a) can be improved.

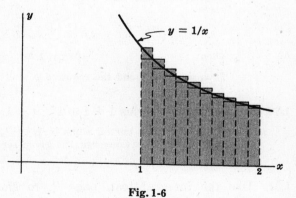

Fig. 1-6

(a) Let $a = 1$, $b = 2$ and divide the interval from a to b of length $b - a = 1$ into 10 equal parts so that $n = 10$, $h = \dfrac{b-a}{n} = \dfrac{1}{10} = .1$ [see Fig. 1-6]. We have approximately

$$\int_1^2 \frac{dx}{x} = .1[f(1) + f(1.1) + f(1.2) + \cdots + f(1.9)]$$

$$= .1\left[\frac{1}{1} + \frac{1}{1.1} + \frac{1}{1.2} + \cdots + \frac{1}{1.9}\right]$$

$$= .1[1.0000 + .9091 + .8333 + .7692 + .7143 + .6667 + .6250 + .5882 + .5556 + .5263]$$

$$= .7188$$

The correct value is $\ln a = .6932$. Note that each term of the approximating sum represents one of the 10 rectangular shaped areas shaded in Fig. 1-6.

(b) The result obtained in (a) *overestimates* the true value since the upper edge of each rectangle lies above the curve $y = 1/x$. To obtain an *underestimate* we use the lower rectangles. In this way we find approximately

$$\int_1^2 \frac{dx}{x} = .1[f(1.1) + f(1.2) + f(1.3) + \cdots + f(2.0)]$$

$$= .1\left[\frac{1}{1.1} + \frac{1}{1.2} + \frac{1}{1.3} + \cdots + \frac{1}{2.0}\right]$$

$$= .1[.9091 + .8333 + .7692 + .7143 + .6667 + .6250 + .5882 + .5556 + .5263 + .5000]$$

$$= .6688$$

As a better estimate of the integral we take the arithmetic mean of the overestimate and the underestimate, i.e. $\frac{1}{2}(.7188 + .6688) = .6938$ which compares very well with the correct value .6932. Since the arithmetic mean of the areas of any upper rectangle and corresponding lower rectangle is the area of a trapezoid, this method of averaging is often called the *trapezoidal rule*.

SEQUENCES AND SERIES

1.29. (a) Find the limit of the sequence .3, .33, .333, ... and (b) justify your conclusions.

(a) The nth term of the sequence can be written

$$u_n = \frac{3}{10} + \frac{3}{10^2} + \frac{3}{10^3} + \cdots + \frac{3}{10^n} = \frac{3}{10}\left[1 + \frac{1}{10} + \frac{1}{10^2} + \cdots + \frac{1}{10^{n-1}}\right]$$

Now if

$$S = 1 + \frac{1}{10} + \frac{1}{10^2} + \cdots + \frac{1}{10^{n-1}}$$

then

$$\frac{1}{10}S = \frac{1}{10} + \frac{1}{10^2} + \cdots + \frac{1}{10^{n-1}} + \frac{1}{10^n}$$

so that by subtracting we have

$$\frac{9}{10}S = 1 - \frac{1}{10^n} \quad \text{or} \quad S = \frac{10}{9}\left(1 - \frac{1}{10^n}\right)$$

Thus the nth term is $u_n = \frac{1}{3}\left(1 - \frac{1}{10^n}\right)$ and as $n \to \infty$, $u_n \to \frac{1}{3}$.

(b) To give a proof that 1/3 is in fact the required limit we must show that given $\epsilon > 0$ we can find N [depending on ϵ] such that $|u_n - \frac{1}{3}| < \epsilon$ for $n > N$. Now

$$\left|u_n - \frac{1}{3}\right| = \left|\frac{1}{3}\left(1 - \frac{1}{10^n}\right) - \frac{1}{3}\right| = \left|\frac{1}{3 \cdot 10^n}\right| = \frac{1}{3 \cdot 10^n} < \epsilon$$

when

$$10^n > 1/3\epsilon \quad \text{i.e.} \quad n > \log_{10}(1/3\epsilon) = N$$

Thus we have found the required N and the statement that the limit is 1/3 is proved.

1.30. Show that the series $1 - 1 + 1 - 1 + 1 - 1 + \cdots$ does not converge.

The sequence of partial sums is $1, 1-1, 1-1+1, 1-1+1-1, \ldots$ or $1, 0, 1, 0, \ldots$. Since this sequence does not converge, the given series does not converge.

1.31. Use the integral test, page 7, to show that $\sum_{n=1}^{\infty} \frac{1}{n^p}$ (a) converges if $p > 1$ and (b) diverges if $p \leq 1$.

(a) Using $f(n) = 1/n^p$ we have $f(x) = 1/x^p$ so that if $p \neq 1$,

$$\int_1^{\infty} \frac{dx}{x^p} = \lim_{M \to \infty} \int_1^M x^{-p}\,dx = \lim_{M \to \infty} \frac{x^{1-p}}{1-p}\Big|_1^M = \lim_{M \to \infty}\left[\frac{M^{1-p}}{1-p} - \frac{1}{1-p}\right]$$

Now if $p > 1$ this limit exists and the corresponding series converges.

(b) If $p < 1$ the limit does not exist and the series diverges. If $p = 1$ then

$$\int_1^\infty \frac{dx}{x} = \lim_{M \to \infty} \int_1^M \frac{dx}{x} = \lim_{M \to \infty} \ln x \Big|_1^M = \lim_{M \to \infty} \ln M$$

which does not exist and so the corresponding series for $p = 1$ diverges.

This shows that $1 + \frac{1}{2} + \frac{1}{3} + \cdots$ diverges even though the nth term approaches zero. [See Theorem 1-7, page 7.]

1.32. Investigate the convergence of $1 - \frac{1}{2^2} + \frac{1}{3^2} - \frac{1}{4^2} + \frac{1}{5^2} - \cdots$.

The series of absolute values is $1 + \frac{1}{2^2} + \frac{1}{3^2} + \frac{1}{4^2} + \frac{1}{5^2} + \cdots$ and converges by Problem 1.31 since $p = 2$. Thus the series converges absolutely and so converges by Theorem 1-4, page 6.

1.33. Prove that the series $\frac{x}{1^2} - \frac{x^2}{2^2} + \frac{x^3}{3^2} - \frac{x^4}{4^2} + \cdots$ converges for $-1 \leqq x \leqq 1$.

The nth term of the series is $u_n = (-1)^{n-1} \frac{x^n}{n^2}$. Now

$$\lim_{n \to \infty} \left| \frac{u_{n+1}}{u_n} \right| = \lim_{n \to \infty} \left| \frac{(-1)^n x^{n+1}/(n+1)^2}{(-1)^{n-1} x^n/n^2} \right| = \lim_{n \to \infty} \left| \frac{n^2}{(n+1)^2} \right| |x| = |x|$$

Then by Theorem 1-8, page 7, the series converges if $|x| < 1$, i.e. $-1 < x < 1$ and diverges for $|x| > 1$. For $|x| = 1$, i.e. $x = \pm 1$, no conclusion can be drawn. However, for $x = 1$ and -1 the series becomes

$$\frac{1}{1^2} - \frac{1}{2^2} + \frac{1}{3^2} - \frac{1}{4^2} + \cdots, \qquad \frac{-1}{1^2} - \frac{1}{2^2} - \frac{1}{3^2} - \frac{1}{4^2} - \cdots$$

which are absolutely convergent and thus convergent. It follows that the series converges for $-1 \leqq x \leqq 1$.

UNIFORM CONVERGENCE

1.34. Investigate the uniform convergence of

$$\frac{x^2}{1 + x^2} + \frac{x^2}{(1 + x^2)^2} + \frac{x^2}{(1 + x^2)^3} + \cdots$$

for $-1 \leqq x \leqq 1$.

Method 1. Let $S_n(x) = \frac{x^2}{1 + x^2} + \frac{x^2}{(1 + x^2)^2} + \cdots + \frac{x^2}{(1 + x^2)^n}$

If $x = 0$, then $S_n(0) = 0$.

If $x \neq 0$, then $S_n(x) = 1 - \frac{1}{(1 + x^2)^{n+1}}$ using the fact that the series is a geometric series [see Problem 1.110(a)].

Then

$$S(x) = \lim_{n \to \infty} S_n(x) = \begin{cases} 0 & \text{if } x = 0 \\ 1 & \text{if } x \neq 0 \end{cases}$$

Since each term of the series is continuous while the sum function $S(x)$ is discontinuous at $x = 0$, it follows by Theorem 1-9, page 7, that the series cannot be uniformly convergent in $-1 \leqq x \leqq 1$.

Method 2. We have from the results of Method 1,

$$|S_n(x) - S(x)| = \begin{cases} \dfrac{1}{(1 + x^2)^{n+1}} & \text{if } x \neq 0 \\ 0 & \text{if } x = 0 \end{cases}$$

Then $|S_n(x) - S(x)| < \epsilon$ if $(1 + x^2)^{n+1} > 1/\epsilon$ or $n > \dfrac{\ln (1/\epsilon)}{\ln (1 + x^2)} - 1 = N$. However, since we cannot find an N which holds for all x in $-1 \leqq x \leqq 1$ [consider $x = 0$ for example], it follows that the series cannot be uniformly convergent in $-1 \leqq x \leqq 1$.

1.35. Prove Theorem 1-9, page 7.

We must show that $S(x)$ is continuous in $a \leqq x \leqq b$.

Now $S(x) = S_n(x) + R_n(x)$, so that $S(x+h) = S_n(x+h) + R_n(x+h)$ and thus

$$S(x+h) - S(x) = S_n(x+h) - S_n(x) + R_n(x+h) - R_n(x) \tag{1}$$

where we choose h so that both x and $x+h$ lie in $a \leqq x \leqq b$ [if $x = b$, for example, this will require $h < 0$].

Since $S_n(x)$ is a sum of a finite number of continuous functions, it must also be continuous. Then given $\epsilon > 0$, we can find δ so that

$$|S_n(x+h) - S_n(x)| < \epsilon/3 \qquad \text{whenever} \quad |h| < \delta \tag{2}$$

Since the series, by hypothesis, is uniformly convergent, we can choose N so that

$$|R_n(x)| < \epsilon/3 \quad \text{and} \quad |R_n(x+h)| < \epsilon/3 \qquad \text{for} \quad n > N \tag{3}$$

Then from (1), (2) and (3),

$$|S(x+h) - S(x)| \leqq |S_n(x+h) - S_n(x)| + |R_n(x+h)| + |R_n(x)| < \epsilon$$

for $|h| < \delta$, and so the continuity is established.

1.36. Prove Theorem 1-10, page 7.

If a function is continuous in $a \leqq x \leqq b$, its integral exists. Then since $S(x)$, $S_n(x)$ and $R_n(x)$ are continuous,

$$\int_a^b S(x)\, dx = \int_a^b S_n(x)\, dx + \int_a^b R_n(x)\, dx$$

To prove the theorem we must show that

$$\left| \int_a^b S(x)\, dx - \int_a^b S_n(x)\, dx \right| = \left| \int_a^b R_n(x)\, dx \right|$$

can be made arbitrarily small by choosing n large enough. This, however, follows at once, since by the uniform convergence of the series we can make $|R_n(x)| < \epsilon/(b-a)$ for $n > N$ independent of x in $[a, b]$, and so

$$\left| \int_a^b R_n(x)\, dx \right| \leqq \int_a^b |R_n(x)|\, dx < \int_a^b \frac{\epsilon}{b-a}\, dx = \epsilon$$

This is equivalent to the statements

$$\int_a^b S(x)\, dx = \lim_{n \to \infty} \int_a^b S_n(x)\, dx \qquad \text{or} \qquad \lim_{n \to \infty} \int_a^b S_n(x)\, dx = \int_a^b \left\{ \lim_{n \to \infty} S_n(x) \right\} dx$$

1.37. Prove the Weierstrass M test [Theorem 1-12, page 7].

Since ΣM_n converges, we have

$$M_{n+1} + M_{n+2} + \cdots < \epsilon \quad \text{for} \quad n > N$$

Then
$$|R_n(x)| = |u_{n+1}(x) + u_{n+2}(x) + \cdots| \leqq |u_{n+1}(x)| + |u_{n+2}(x)| + \cdots$$
$$\leqq M_{n+1} + M_{n+2} + \cdots < \epsilon$$

Since N is independent of x, the series must be uniformly convergent.

1.38. Prove that $\displaystyle\sum_{n=1}^{\infty} \frac{\sin nx}{n^3}$ is uniformly convergent for $-\pi \leqq x \leqq \pi$.

We have $\left| \dfrac{\sin nx}{n^3} \right| \leqq \dfrac{1}{n^3}$. Since $\displaystyle\sum_{n=1}^{\infty} \frac{1}{n^3}$ converges, it follows by the Weierstrass M test that the given series is uniformly convergent for $-\pi \leqq x \leqq \pi$ [and in fact any finite interval].

TAYLOR SERIES

1.39. Obtain the formal Taylor series for $\sin x$ about $x = 0$.

Let $f(x) = \sin x$.   Then

$$f'(x) = \cos x, \quad f''(x) = -\sin x, \quad f'''(x) = -\cos x, \quad f^{(iv)}(x) = \sin x, \quad f^{(v)}(x) = \cos x, \quad \ldots$$

so that

$$f(0) = 0, \quad f'(0) = 1, \quad f''(0) = 0, \quad f'''(0) = -1, \quad f^{(iv)}(0) = 0, \quad f^{(v)}(0) = 1, \quad \ldots$$

Thus from

$$f(x) = f(a) + f'(a)(x-a) + \frac{f''(a)(x-a)^2}{2!} + \frac{f'''(a)(x-a)^3}{3!} + \cdots$$

with $a = 0$, we have

$$\sin x = x - \frac{x^3}{3!} + \frac{x^5}{5!} - \frac{x^7}{7!} + \cdots$$

1.40. Use series methods to obtain the approximate value of $\int_0^1 \frac{1 - e^{-x}}{x}\,dx$.

From the expansion for e^x on page 8 we have on replacing x by $-x$,

$$e^{-x} = 1 - x + \frac{x^2}{2!} - \frac{x^3}{3!} + \frac{x^4}{4!} - \cdots$$

Then

$$\frac{1 - e^{-x}}{x} = 1 - \frac{x}{2!} + \frac{x^2}{3!} - \frac{x^3}{4!} + \cdots$$

Since this series is uniformly convergent for $0 \leqq x \leqq 1$ [as can be proved by using Problem 1.120], we can integrate term by term to obtain

$$\int_0^1 \frac{1 - e^{-x}}{x}\,dx = x - \frac{x^2}{2 \cdot 2!} + \frac{x^3}{3 \cdot 3!} - \frac{x^4}{4 \cdot 4!} + \cdots \Big|_0^1$$

$$= 1 - \frac{1}{2 \cdot 2!} + \frac{1}{3 \cdot 3!} - \frac{1}{4 \cdot 4!} + \cdots$$

$$= .7966 \text{ approx.}$$

FUNCTIONS OF TWO OR MORE VARIABLES AND PARTIAL DERIVATIVES

1.41. If $f(x, y) = 3x^2 + 4xy - 2y^2$, find (a) $f(2, -3)$, (b) $f_x(2, -3)$, (c) $f_y(2, -3)$, (d) $f_{xx}(2, -3)$, (e) $f_{xy}(2, -3)$, (f) $f_{yx}(2, -3)$, (g) $f_{yy}(2, -3)$.

(a) $f(2, -3) = 3(2)^2 + 4(2)(-3) - 2(-3)^2 = -30$

(b) $f_x(x, y) = \dfrac{\partial f}{\partial x} = 6x + 4y$.   Then $f_x(2, -3) = 6(2) + 4(-3) = 0$.

(c) $f_y(x, y) = \dfrac{\partial f}{\partial y} = 4x - 4y$.   Then $f_y(2, -3) = 4(2) - 4(-3) = 20$.

(d) $f_{xx}(x, y) = \dfrac{\partial^2 f}{\partial x^2} = \dfrac{\partial}{\partial x}\left(\dfrac{\partial f}{\partial x}\right) = \dfrac{\partial}{\partial x}(6x + 4y) = 6$.   Then $f_{xx}(2, -3) = 6$.

(e) $f_{xy}(x, y) = \dfrac{\partial^2 f}{\partial x\,\partial y} = \dfrac{\partial}{\partial x}\left(\dfrac{\partial f}{\partial y}\right) = \dfrac{\partial}{\partial x}(4x - 4y) = 4$.   Then $f_{xy}(2, -3) = 4$.

(f) $f_{yx}(x, y) = \dfrac{\partial^2 f}{\partial y\,\partial x} = \dfrac{\partial}{\partial y}\left(\dfrac{\partial f}{\partial x}\right) = \dfrac{\partial}{\partial y}(6x + 4y) = 4$.   Then $f_{yx}(2, -3) = 4$.

(g) $f_{yy}(x, y) = \dfrac{\partial^2 f}{\partial y^2} = \dfrac{\partial}{\partial y}\left(\dfrac{\partial f}{\partial y}\right) = \dfrac{\partial}{\partial y}(4x - 4y) = -4$.   Then $f_{yy}(2, -3) = -4$.

Note that the results in (e) and (f) illustrate the fact that $f_{xy} = f_{yx}$ since in this case f is continuous and has continuous partial derivatives up to and including the second order.

1.42. If $f(x, y) = \sin(x^2 + 2y)$, find (a) f_x, (b) f_y, (c) f_{xx}, (d) f_{xy}, (e) f_{yx}, (f) f_{yy}.

(a) $$f_x = \frac{\partial}{\partial x} \sin(x^2 + 2y) = \cos(x^2 + 2y) \frac{\partial}{\partial x}(x^2 + 2y) = 2x \cos(x^2 + 2y)$$

(b) $$f_y = \frac{\partial}{\partial y} \sin(x^2 + 2y) = \cos(x^2 + 2y) \frac{\partial}{\partial y}(x^2 + 2y) = 2 \cos(x^2 + 2y)$$

(c) $$f_{xx} = \frac{\partial}{\partial x} f_x = \frac{\partial}{\partial x}[2x \cos(x^2 + 2y)]$$
$$= 2x \frac{\partial}{\partial x} \cos(x^2 + 2y) + \cos(x^2 + 2y) \frac{\partial}{\partial x}(2x)$$
$$= 2x\left[-\sin(x^2 + 2y) \frac{\partial(x^2 + 2y)}{\partial x}\right] + 2 \cos(x^2 + 2y)$$
$$= -4x^2 \sin(x^2 + 2y) + 2 \cos(x^2 + 2y)$$

(d) $$f_{xy} = \frac{\partial}{\partial x} f_y = \frac{\partial}{\partial x}[2 \cos(x^2 + 2y)] = -2 \sin(x^2 + 2y) \frac{\partial}{\partial x}(x^2 + 2y) = -4x \sin(x^2 + 2y)$$

(e) $$f_{yx} = \frac{\partial}{\partial y} f_x = \frac{\partial}{\partial y}[2x \cos(x^2 + 2y)] = 2x \frac{\partial}{\partial y} \cos(x^2 + 2y)$$
$$= 2x\left[-\sin(x^2 + 2y) \frac{\partial}{\partial y}(x^2 + 2y)\right] = -4x \sin(x^2 + 2y)$$

(f) $$f_{yy} = \frac{\partial}{\partial y} f_y = \frac{\partial}{\partial y}[2 \cos(x^2 + 2y)] = -2 \sin(x^2 + 2y) \frac{\partial}{\partial y}(x^2 + 2y) = -4 \sin(x^2 + 2y)$$

Note that $f_{xy} = f_{yx}$ in (d) and (e).

1.43. If $f(x, y) = 3x^2 + 4xy - 2y^2$, find df.

Method 1. $$df = \frac{\partial f}{\partial x} dx + \frac{\partial f}{\partial y} dy = (6x + 4y) dx + (4x - 4y) dy$$

Method 2. $$df = d(3x^2) + d(4xy) + d(-2y^2)$$
$$= 6x \, dx + 4(x \, dy + y \, dx) - 4y \, dy = (6x + 4y) dx + (4x - 4y) dy$$

1.44. If $z = f(y/x)$, show that $x \dfrac{\partial z}{\partial x} + y \dfrac{\partial z}{\partial y} = 0$.

Method 1.

     Let $u = y/x$ so that $z = f(u)$. Then

$$\frac{\partial z}{\partial x} = \frac{\partial z}{\partial u} \frac{\partial u}{\partial x} = f'(u) \frac{\partial}{\partial x}(y/x) = -\frac{yf'(u)}{x^2}$$

$$\frac{\partial z}{\partial y} = \frac{\partial z}{\partial u} \frac{\partial u}{\partial y} = f'(u) \frac{\partial}{\partial y}(y/x) = \frac{f'(u)}{x}$$

Then $$x \frac{\partial z}{\partial x} + y \frac{\partial z}{\partial y} = x\left[-\frac{yf'(u)}{x^2}\right] + y\left[\frac{f'(u)}{x}\right] = 0$$

Method 2. $$dz = f'(y/x) \, d(y/x) = f'(y/x)\left[\frac{x \, dy - y \, dx}{x^2}\right]$$

Then since $dz = \dfrac{\partial z}{\partial x} dx + \dfrac{\partial z}{\partial y} dy$, it follows on equating coefficients of dx and dy that

$$\frac{\partial z}{\partial x} = -\frac{yf'(y/x)}{x^2}, \qquad \frac{\partial z}{\partial y} = \frac{f'(y/x)}{x}$$

and the result follows as in Method 1.

1.45. If $dz = M(x, y)\,dx + N(x, y)\,dy$, prove that $\partial M/\partial y = \partial N/\partial x$ where M and N are assumed to have continuous partial derivatives.

Since $dz = (\partial z/\partial x)\,dx + (\partial z/\partial y)\,dy = M\,dx + N\,dy$, we must have $\partial z/\partial x = M$, $\partial z/\partial y = N$ since x and y are independent variables.

Then since $\dfrac{\partial^2 z}{\partial y\,\partial x} = \dfrac{\partial^2 z}{\partial x\,\partial y}$ [if these derivatives are continuous], we have

$$\frac{\partial^2 z}{\partial y\,\partial x} = \frac{\partial M}{\partial y}, \qquad \frac{\partial^2 z}{\partial x\,\partial y} = \frac{\partial N}{\partial x} \quad \text{or} \quad \frac{\partial M}{\partial y} = \frac{\partial N}{\partial x}$$

Similarly we can prove that if $\partial M/\partial y = \partial N/\partial x$ then $M\,dx + N\,dy$ can be written as the differential of z, i.e. dz, often called an *exact differential*.

1.46. If $z = z(u, v)$ where $u = u(x, y)$, $v = v(x, y)$, prove that (a) $\dfrac{\partial z}{\partial x} = \dfrac{\partial z}{\partial u}\dfrac{\partial u}{\partial x} + \dfrac{\partial z}{\partial v}\dfrac{\partial v}{\partial x}$, (b) $\dfrac{\partial z}{\partial y} = \dfrac{\partial z}{\partial u}\dfrac{\partial u}{\partial y} + \dfrac{\partial z}{\partial v}\dfrac{\partial v}{\partial y}$.

From $z = z(u, v)$, $u = u(x, y)$, $v = v(x, y)$ we have

$$dz = \frac{\partial z}{\partial u}\,du + \frac{\partial z}{\partial v}\,dv \quad (1) \qquad du = \frac{\partial u}{\partial x}\,dx + \frac{\partial u}{\partial y}\,dy \quad (2) \qquad dv = \frac{\partial v}{\partial x}\,dx + \frac{\partial v}{\partial y}\,dy \quad (3)$$

Then using (2) and (3) in (1) we find on combining terms,

$$dz = \left(\frac{\partial z}{\partial u}\frac{\partial u}{\partial x} + \frac{\partial z}{\partial v}\frac{\partial v}{\partial x}\right) dx + \left(\frac{\partial z}{\partial u}\frac{\partial u}{\partial y} + \frac{\partial z}{\partial v}\frac{\partial v}{\partial y}\right) dy \tag{4}$$

But considering that z is a function of x and y since u, v are functions of x and y, we must have

$$dz = (\partial z/\partial x)\,dx + (\partial z/\partial y)\,dy \tag{5}$$

Equating corresponding coefficients of dx and dy in (4) and (5) yields the required results.

TAYLOR SERIES FOR FUNCTIONS OF TWO OR MORE VARIABLES

1.47. Verify the result (16) page 9 for the function of Problem 1.41.

From Problem 1.41 we have $f(2, -3) = -30$, $f_x(2, -3) = 0$, $f_y(2, -3) = 20$, $f_{xx}(2, -3) = 6$, $f_{xy}(2, -3) = f_{yx}(2, -3) = 4$, $f_{yy}(2, -3) = -4$. Also the higher derivatives are all zero. Then from (16), page 9, we should have

$$3x^2 + 4xy - 2y^2 = -30 + 0(x - 2) + 20(y + 3) + \frac{1}{2!}\,[6(x - 2)^2 + 8(x - 2)(y + 3) - 4(y + 3)^2]$$

Now the right side can be written

$$-30 + 20y + 60 + 3(x^2 - 4x + 4) + 4(xy - 2y + 3x - 6) - 2(y^2 + 6y + 9) = 3x^2 + 4xy - 2y^2$$

and the result is verified in this special case.

LINEAR EQUATIONS AND DETERMINANTS

1.48. Verify that the solutions of (17), page 9, are given by (18).

We have $(1)\quad a_1 x + b_1 y = c_1,\qquad (2)\quad a_2 x + b_2 y = c_2$

Multiply (1) by b_2 to obtain $a_1 b_2 x + b_1 b_2 y = c_1 b_2$

Multiply (2) by b_1 to obtain $b_1 a_2 x + b_1 b_2 y = b_1 c_2$

Then by subtraction, $(a_1 b_2 - b_1 a_2)x = c_1 b_2 - b_1 c_2$ which gives the required result for x. Similarly on multiplying (1) by a_2, (2) by a_1 and subtracting we obtain the result for y.

1.49. Solve $\begin{cases} 3x + 4y = 6 \\ 2x - 5y = 8 \end{cases}$ using determinants.

By Cramer's rule we have

$$x = \frac{\begin{vmatrix} 6 & 4 \\ 8 & -5 \end{vmatrix}}{\begin{vmatrix} 3 & 4 \\ 2 & -5 \end{vmatrix}} = \frac{(6)(-5) - (4)(8)}{(3)(-5) - (4)(2)} = \frac{62}{23} \qquad y = \frac{\begin{vmatrix} 3 & 6 \\ 2 & 8 \end{vmatrix}}{\begin{vmatrix} 3 & 4 \\ 2 & -5 \end{vmatrix}} = \frac{(3)(8) - (6)(2)}{(3)(-5) - (4)(2)} = -\frac{12}{23}$$

A check is supplied by substituting these results in the given equations.

1.50. Solve $\begin{cases} 3x + y - 2z = -2 \\ x - 2y + 3z = 9 \\ 2x + 3y + z = 1 \end{cases}$ for z.

By Cramer's rule we have

$$z = \frac{\begin{vmatrix} 3 & 1 & -2 \\ 1 & -2 & 9 \\ 2 & 3 & 1 \end{vmatrix}}{\begin{vmatrix} 3 & 1 & -2 \\ 1 & -2 & 3 \\ 2 & 3 & 1 \end{vmatrix}} = \frac{-84}{-42} = 2$$

1.51. (a) For what value of k will the system of equations

$$\begin{cases} (1-k)x + y = 0 \\ kx - 2y = 0 \end{cases}$$

have solutions other than the trivial one $x = 0$, $y = 0$. (b) Find two non-trivial solutions.

(a) By Cramer's rule the solution will be

$$x = \frac{\begin{vmatrix} 0 & 1 \\ 0 & -2 \end{vmatrix}}{\begin{vmatrix} 1-k & 1 \\ k & -2 \end{vmatrix}}, \qquad y = \frac{\begin{vmatrix} 1-k & 0 \\ k & 0 \end{vmatrix}}{\begin{vmatrix} 1-k & 1 \\ k & -2 \end{vmatrix}}.$$

Now since the numerators are equal to zero, these can be non-trivial [i.e. nonzero] solutions only if the denominator is also equal to zero, i.e.

$$\begin{vmatrix} 1-k & 1 \\ k & -2 \end{vmatrix} = (1-k)(-2) - (1)(k) = k - 2 = 0 \quad \text{or} \quad k = 2$$

(b) If $k = 2$ the equations become $-x + y = 0$, $2x - 2y = 0$ and are identical, i.e. $x = y$. Then solutions are $x = 2$, $y = 2$, $x = 3$, $y = 3$ for example. Actually there are infinitely many such non-trivial solutions.

MAXIMA AND MINIMA. METHOD OF LAGRANGE MULTIPLIERS

1.52. Find the relative maxima and minima of $f(x) = x^4 - 8x^3 + 22x^2 - 24x + 20$.

The relative maxima and minima occur where

$$f'(x) = 4x^3 - 24x^2 + 44x - 24 = 0 \quad \text{or} \quad (x-1)(x-2)(x-3) = 0, \quad \text{i.e. } x = 1, 2, 3$$

Then since $f''(x) = 12x^2 - 48x + 44$, we have $f''(1) = 8 > 0$, $f''(2) = -4 < 0$, $f''(3) = 8 > 0$. Thus a relative minimum of 11 occurs at $x = 1$, a relative maximum of 12 occurs at $x = 2$ and a relative minimum of 11 occurs at $x = 3$.

1.53. Determine the dimensions of the largest rectangular parallelepiped which can be inscribed in a hemisphere of radius a [Fig. 1-7].

The volume of the parallelepiped is

$$V = (2x)(2y)(z) = 4xyz$$

and the equation of the surface of the hemisphere is $x^2 + y^2 + z^2 = a^2$ or $z = \sqrt{a^2 - x^2 - y^2}$.

The volume is a maximum where $V^2 = U = x^2y^2z^2 = x^2y^2(a^2 - x^2 - y^2)$ is a maximum. To find this we solve simultaneously the equations

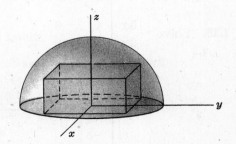

Fig. 1-7

$$\partial U/\partial x = 2a^2xy^2 - 4x^3y^2 - 2xy^4 = 0, \qquad \partial U/\partial y = 2a^2x^2y - 4x^2y^3 - 2x^4y = 0$$

Since $x \neq 0$, $y \neq 0$, these give

$$2x^2 + y^2 = a^2, \qquad x^2 + 2y^2 = a^2$$

from which $x = y = a/\sqrt{3}$ and so $z = a/\sqrt{3}$.

Then the required volume is $4a^3/3\sqrt{3}$.

1.54. (a) Prove the method of Lagrange multipliers for the case where the function $f(x, y)$ is to be made a maximum or minimum [or extremum] subject to the constraint condition $\phi(x, y) = 0$.

(b) Show how to generalize the result of (a) to the case where the function to be made an extremum is $f(x, y, z)$ and the constraint condition is $\phi(x, y, z) = 0$.

(c) Illustrate the method of Lagrange multipliers by working Problem 1.53.

(a) Assume that $\phi(x, y) = 0$ defines y as a unique function of x, i.e. $y = g(x)$, having a continuous derivative $g'(x)$. Then we must find the maximum or minimum [extremum] of

$$f(x, y) = f(x, g(x))$$

But as in elementary calculus this can be found by setting the derivative with respect to x equal to zero, i.e.

$$\frac{\partial f}{\partial x} + \frac{\partial f}{\partial y}\frac{dy}{dx} = 0 \qquad \text{or} \qquad f_x + f_y g'(x) = 0 \tag{1}$$

Also from $\phi(x, y) = 0$ we have the identity $\phi(x, g(x)) = 0$ so that

$$\frac{\partial \phi}{\partial x} + \frac{\partial \phi}{\partial y}\frac{dy}{dx} = 0 \qquad \text{or} \qquad \phi_x + \phi_y g'(x) = 0 \tag{2}$$

Eliminating $g'(x)$ between (1) and (2) we find

$$f_x - \frac{f_y}{\phi_y}\phi_x = 0 \tag{3}$$

assuming $\phi_y \neq 0$. Now if we define $\lambda = -f_y/\phi_y$ or

$$f_y + \lambda\phi_y = 0 \tag{4}$$

(3) becomes
$$f_x + \lambda f_y = 0 \tag{5}$$

But (4) and (5) are obtained by considering $h(x, y) = f(x, y) + \lambda\phi(x, y)$ and setting

$$\partial h/\partial x = 0, \qquad \partial h/\partial y = 0 \tag{6}$$

so that the method is proved. We call λ the *Lagrange multiplier*.

(b) In this case we assume that $\phi(x, y, z) = 0$ can be solved to yield $z = g(x, y)$ so that

$$f(x, y, z) = f(x, y, g(x, y))$$

But the extremum of this function of x and y can be found by setting the partial derivatives with respect to x and y equal to zero, i.e.

$$\frac{\partial f}{\partial x} + \frac{\partial f}{\partial z}\frac{\partial z}{\partial x} = 0 \qquad \text{or} \qquad f_x + f_z g_x = 0 \tag{7}$$

$$\frac{\partial f}{\partial y} + \frac{\partial f}{\partial z}\frac{\partial z}{\partial y} = 0 \qquad \text{or} \qquad f_y + f_z g_y = 0 \tag{8}$$

Also the constraint condition yields the identity $\phi(x, y, g(x, y)) = 0$ so that by differentiation with respect to x and y we have

$$\frac{\partial \phi}{\partial x} + \frac{\partial \phi}{\partial z}\frac{\partial z}{\partial x} = 0 \qquad \text{or} \qquad \phi_x + \phi_z g_x = 0 \tag{9}$$

$$\frac{\partial \phi}{\partial y} + \frac{\partial \phi}{\partial z}\frac{\partial z}{\partial y} = 0 \qquad \text{or} \qquad \phi_y + \phi_z g_y = 0 \tag{10}$$

Eliminating g_x and g_y among the equations (7), (8), (9), (10) we find, assuming $\phi_z \neq 0$,

$$f_x - \frac{f_z}{\phi_z}\phi_x = 0, \qquad f_y - \frac{f_z}{\phi_z}\phi_y = 0 \qquad (11)$$

Then calling $\lambda = -f_z/\phi_z$ or

$$f_z + \lambda\phi_z = 0 \qquad (12)$$

equations (11) yield

$$f_x + \lambda\phi_x = 0, \qquad f_y + \lambda\phi_y = 0 \qquad (13)$$

But (12) and (13) can be obtained by letting $h(x,y,z) = f(x,y,z) + \lambda\phi(x,y,z)$ and setting the derivative with respect to x,y,z equal to zero so that the required result is proved.

(c) We must find the maximum of $4xyz$ subject to the constraint condition $x^2 + y^2 + z^2 - a^2 = 0$. To do this we form the function

$$h(x,y,z) = 4xyz + \lambda(x^2 + y^2 + z^2 - a^2) \qquad (14)$$

and set the partial derivatives of h with respect to x,y,z equal to zero. Thus we obtain

$$\partial h/\partial x = 4yz + 2\lambda x = 0, \qquad \partial h/\partial y = 4xz + 2\lambda y = 0, \qquad \partial h/\partial z = 4xy + 2\lambda z = 0 \qquad (15)$$

Multiplying these by x,y,z respectively and adding, we find

$$12xyz + 2\lambda(x^2 + y^2 + z^2) = 12xyz + 2\lambda a^2 = 0$$

or $\lambda = -6xyz/a^2$. Then substituting this value of λ in equations (15) and solving simultaneously with $x^2 + y^2 + z^2 = a^2$, we find

$$x = y = z = a/\sqrt{3}$$

as in Problem 1.53.

LEIBNITZ'S RULE

1.55. If $I(\alpha) = \displaystyle\int_{\alpha}^{\alpha^2} \frac{\sin \alpha x}{x}\,dx$, find $\dfrac{dI}{d\alpha}$.

$$\frac{dI}{d\alpha} = \int_{\alpha}^{\alpha^2} \frac{\partial}{\partial\alpha}\left(\frac{\sin\alpha x}{x}\right)dx + \frac{\sin\alpha^3}{\alpha^2}\frac{d\alpha^2}{d\alpha} - \frac{\sin\alpha^2}{\alpha}\frac{d\alpha}{d\alpha}$$

$$= \int_{\alpha}^{\alpha^2} \cos\alpha x\,dx + \frac{2\sin\alpha^3}{\alpha} - \frac{\sin\alpha^2}{\alpha}$$

$$= \frac{\sin\alpha x}{\alpha}\bigg|_{\alpha}^{\alpha^2} + \frac{2\sin\alpha^3}{\alpha} - \frac{\sin\alpha^2}{\alpha} = \frac{3\sin\alpha^3 - 2\sin\alpha^2}{\alpha}$$

COMPLEX NUMBERS

1.56. Perform the indicated operations.

(a) $(4-2i) + (-6+5i) = 4 - 2i - 6 + 5i = 4 - 6 + (-2+5)i = -2 + 3i$

(b) $(-7+3i) - (2-4i) = -7 + 3i - 2 + 4i = -9 + 7i$

(c) $(3-2i)(1+3i) = 3(1+3i) - 2i(1+3i) = 3 + 9i - 2i - 6i^2 = 3 + 9i - 2i + 6 = 9 + 7i$

(d) $\dfrac{-5+5i}{4-3i} = \dfrac{-5+5i}{4-3i}\cdot\dfrac{4+3i}{4+3i} = \dfrac{(-5+5i)(4+3i)}{16-9i^2} = \dfrac{-20-15i+20i+15i^2}{16+9}$

$\qquad = \dfrac{-35+5i}{25} = \dfrac{5(-7+i)}{25} = \dfrac{-7}{5} + \dfrac{1}{5}i$

(e) $\dfrac{i+i^2+i^3+i^4+i^5}{1+i} = \dfrac{i-1+(i^2)(i)+(i^2)^2+(i^2)^2 i}{1+i} = \dfrac{i-1-i+1+i}{1+i}$

$\qquad = \dfrac{i}{1+i}\cdot\dfrac{1-i}{1-i} = \dfrac{i-i^2}{1-i^2} = \dfrac{i+1}{2} = \dfrac{1}{2} + \dfrac{1}{2}i$

(f) $|3 - 4i|\,|4 + 3i| \;=\; \sqrt{(3)^2 + (-4)^2}\,\sqrt{(4)^2 + (3)^2} \;=\; (5)(5) \;=\; 25$

(g) $\left|\dfrac{1}{1+3i} - \dfrac{1}{1-3i}\right| \;=\; \left|\dfrac{1-3i}{1-9i^2} - \dfrac{1+3i}{1-9i^2}\right| \;=\; \left|\dfrac{-6i}{10}\right| \;=\; \sqrt{(0)^2 + (-\tfrac{6}{10})^2} \;=\; \dfrac{3}{5}$

1.57. If z_1 and z_2 are two complex numbers, prove that $|z_1 z_2| = |z_1|\,|z_2|$.

Let $z_1 = x_1 + iy_1$, $z_2 = x_2 + iy_2$. Then

$$|z_1 z_2| \;=\; |(x_1 + iy_1)(x_2 + iy_2)| \;=\; |x_1 x_2 - y_1 y_2 + i(x_1 y_2 + x_2 y_1)|$$

$$=\; \sqrt{(x_1 x_2 - y_1 y_2)^2 + (x_1 y_2 + x_2 y_1)^2} \;=\; \sqrt{x_1^2 x_2^2 + y_1^2 y_2^2 + x_1^2 y_2^2 + x_2^2 y_1^2}$$

$$=\; \sqrt{(x_1^2 + y_1^2)(x_2^2 + y_2^2)} \;=\; \sqrt{x_1^2 + y_1^2}\,\sqrt{x_2^2 + y_2^2} \;=\; |x_1 + iy_1|\,|x_2 + iy_2| \;=\; |z_1|\,|z_2|.$$

1.58. Solve $x^3 - 2x - 4 = 0$.

The possible rational roots are $\pm 1, \pm 2, \pm 4$. By trial we find $x = 2$ is a root. Then the given equation can be written $(x - 2)(x^2 + 2x + 2) = 0$. The solutions to the *quadratic equation* $ax^2 + bx + c = 0$ are $x = \dfrac{-b \pm \sqrt{b^2 - 4ac}}{2a}$. For $a = 1$, $b = 2$, $c = 2$ this gives $x = \dfrac{-2 \pm \sqrt{4-8}}{2} \;=\; \dfrac{-2 \pm \sqrt{-4}}{2} \;=\; \dfrac{-2 \pm 2i}{2} \;=\; -1 \pm i$.

The set of solutions is $2, -1 + i, -1 - i$.

POLAR FORM OF COMPLEX NUMBERS

1.59. Express in polar form (a) $3 + 3i$, (b) $-1 + \sqrt{3}\,i$, (c) -1, (d) $-2 - 2\sqrt{3}\,i$.

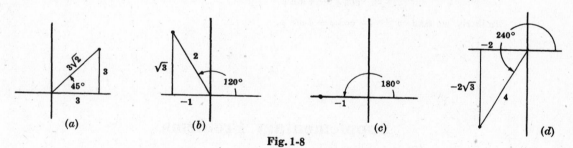

Fig. 1-8

(a) Amplitude $\phi = 45^\circ = \pi/4$ radians. Modulus $\rho = \sqrt{3^2 + 3^2} = 3\sqrt{2}$. Then
$$3 + 3i \;=\; \rho(\cos\phi + i\sin\phi) \;=\; 3\sqrt{2}(\cos \pi/4 + i\sin \pi/4) \;=\; 3\sqrt{2}\;\text{cis}\;\pi/4 \;=\; 3\sqrt{2}\,e^{\pi i/4}$$

(b) Amplitude $\phi = 120^\circ = 2\pi/3$ radians. Modulus $\rho = \sqrt{(-1)^2 + (\sqrt{3})^2} = \sqrt{4} = 2$. Then
$$-1 + \sqrt{3}\,i \;=\; 2(\cos 2\pi/3 + i\sin 2\pi/3) \;=\; 2\;\text{cis}\;2\pi/3 \;=\; 2e^{2\pi i/3}$$

(c) Amplitude $\phi = 180^\circ = \pi$ radians. Modulus $\rho = \sqrt{(-1)^2 + (0)^2} = 1$. Then
$$-1 \;=\; 1(\cos\pi + i\sin\pi) \;=\; \text{cis}\;\pi \;=\; e^{\pi i}$$

(d) Amplitude $\phi = 240^\circ = 4\pi/3$ radians. Modulus $\rho = \sqrt{(-2)^2 + (-2\sqrt{3})^2} = 4$. Then
$$-2 - 2\sqrt{3} \;=\; 4(\cos 4\pi/3 + i\sin 4\pi/3) \;=\; 4\;\text{cis}\;4\pi/3 \;=\; 4e^{4\pi i/3}$$

1.60. Evaluate (a) $(-1 + \sqrt{3}\,i)^{10}$, (b) $(-1 + i)^{1/3}$.

(c) By Problem 1.59(b) and De Moivre's theorem,
$$(-1 + \sqrt{3}\,i)^{10} \;=\; [2(\cos 2\pi/3 + i\sin 2\pi/3)]^{10} \;=\; 2^{10}(\cos 20\pi/3 + i\sin 20\pi/3)$$

$$=\; 1024[\cos(2\pi/3 + 6\pi) + i\sin(2\pi/3 + 6\pi)] \;=\; 1024(\cos 2\pi/3 + i\sin 2\pi/3)$$

$$=\; 1024(-\tfrac{1}{2} + \tfrac{1}{2}\sqrt{3}\,i) \;=\; -512 + 512\sqrt{3}\,i$$

(b) $-1 + i = \sqrt{2}\,(\cos 135° + i \sin 135°) = \sqrt{2}\,[\cos(135° + k \cdot 360°) + i \sin(135° + k \cdot 360°)]$

Then

$$(-1+i)^{1/3} = (\sqrt{2})^{1/3}\left[\cos\left(\frac{135° + k \cdot 360°}{3}\right) + i \sin\left(\frac{135° + k \cdot 360°}{3}\right)\right]$$

The results for $k = 0, 1, 2$ are

$$\sqrt[6]{2}\,(\cos 45° + i \sin 45°),$$
$$\sqrt[6]{2}\,(\cos 165° + i \sin 165°),$$
$$\sqrt[6]{2}\,(\cos 285° + i \sin 285°)$$

The results for $k = 3, 4, 5, 6, 7, \ldots$ give repetitions of these. These complex roots are represented geometrically in the complex plane by points P_1, P_2, P_3 on the circle of Fig. 1-9.

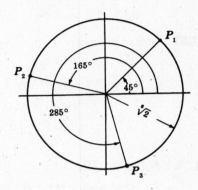

Fig. 1-9

1.61. Assuming that the series for e^x on page 8 holds for complex values of x, arrive at *Euler's formula* $e^{i\phi} = \cos\phi + i \sin\phi$.

If $x = i\phi$ we have

$$e^{i\phi} = 1 + i\phi + \frac{(i\phi)^2}{2!} + \frac{(i\phi)^3}{3!} + \frac{(i\phi)^4}{4!} + \frac{(i\phi)^5}{5!} + \cdots$$

$$= \left(1 - \frac{\phi^2}{2!} + \frac{\phi^4}{4!} - \cdots\right) + i\left(\phi - \frac{\phi^3}{3!} + \frac{\phi^5}{5!} - \cdots\right)$$

$$= \cos\phi + i \sin\phi$$

Similarly we find $e^{-i\phi} = \cos\phi - i \sin\phi$.

Supplementary Problems

REAL NUMBERS AND LAWS OF ALGEBRA

1.62. Prove that (a) $\sqrt[3]{2}$, (b) $\sqrt{3}$, (c) $\sqrt{2} + \sqrt{3}$ are irrational numbers.

1.63. Arrange in order of increasing value: $11/9$, $\sqrt[4]{2}$, $\sqrt{3}\ \sqrt[3]{5}/\sqrt[3]{2}\ \sqrt[6]{60}$.

1.64. Explain why (a) $1/0$, (b) $0/0$ cannot represent unique numbers.

1.65. Explain why we must define (a) $(-5)(3) = -15$, (b) $(-2)(-3) = 6$ if the rules of algebra on page 1 are to hold.

FUNCTIONS

1.66. If $f(x) = (3x^2 + 2x - 5)/(x - 4)$, find (a) $f(2)$, (b) $f(-1)$, (c) $f(3/2)$, (d) $f(-x)$, (e) $f(\sqrt{2})$.

1.67. An *odd function* is one for which $f(-x) = -f(x)$ while an *even function* is one for which $f(-x) = f(x)$. Classify each of the following according as they are even or odd: (a) $\cos 2x$, (b) $\sin 3x$, (c) $\tan x$, (d) e^x, (e) $e^x - e^{-x}$, (f) $e^x + e^{-x}$.

1.68. (a) Prove that $e^{a \ln b} = b^a$. (b) Find $e^{-2 \ln x}$.

1.69.   Prove that (a) $\sin 3x = 3 \sin x - 4 \sin^3 x$, (b) $\cos 3x = 4 \cos^3 x - 3 \cos x$.

1.70.   Prove that (a) $\coth^2 x - \operatorname{csch}^2 x = 1$, (b) $\cosh^2 x + \sinh^2 x = \cosh 2x$, (c) $\sinh^{-1} x = \ln(x + \sqrt{x^2 + 1})$.

1.71.   If $\cos x = 8/17$, find (a) $\sin 2x$, (b) $\cos 2x$, (c) $\sin(x/2)$.

1.72.   Find $\tanh(\ln 3)$.

LIMITS AND CONTINUITY

1.73.   Find (a) $\lim\limits_{x \to 2} (x^2 + 4x + 6)$, (b) $\lim\limits_{x \to -1} \dfrac{2x + 3}{x + 2}$, (c) $\lim\limits_{x \to 4} \dfrac{\sqrt{x} - 2}{x^2 - 16}$ and prove your conclusions.

1.74.   If $\lim\limits_{x \to a} f(x) = l$, prove that $\lim\limits_{x \to a} [f(x)]^2 = l^2$.

1.75.   If $\lim\limits_{x \to a} f_1(x) = l_1$ and $\lim\limits_{x \to x_0} f_2(x) = l_2$, prove that

(a) $\lim\limits_{x \to a} [f_1(x) - f_2(x)] = l_1 - l_2$,   (c) $\lim\limits_{x \to x_0} [1/f_2(x)] = 1/l_2, \; l_2 \neq 0$,

(b) $\lim\limits_{x \to a} [f_1(x) f_2(x)] = l_1 l_2$,   (d) $\lim\limits_{x \to a} [f_1(x)/f_2(x)] = l_1/l_2, \; l_2 \neq 0$.

1.76.   If $f(x) = |x|/x$, prove that $\lim\limits_{x \to 0} f(x)$ does not exist.

1.77.   Using the definition prove that $f(x) = 3 - x^2$ is continuous at (a) $x = 1$, (b) $x = a$.

1.78.   If $f(x)$ is continuous at a, prove that $[f(x)]^2$ is also continuous at a.

1.79.   If $f_1(x)$ and $f_2(x)$ are continuous at a prove that (a) $f_1(x) + f_2(x)$, (b) $f_1(x) - f_2(x)$, (c) $f_1(x) f_2(x)$, (d) $f_1(x)/f_2(x), \; f_2(a) \neq 0$ are also continuous at a. Revise the statement if a is replaced by an interval such as $a \leqq x \leqq b$ or $a < x < b$.

1.80.   If $f(x) = \begin{cases} x \sin(1/x) & x \neq 0 \\ 0 & x = 0 \end{cases}$ prove that $f(x)$ is continuous (a) at $x = 0$, (b) in any interval.

1.81.   Prove the statement at the end of Problem 1.12(b).

DERIVATIVES

1.82.   Using the definition, find the derivatives of (a) $f(x) = x^2 - 2x + 5$ and (b) $f(x) = (x - 1)/(x + 1)$.

1.83.   Prove the differentiation formulas (a) 2, (b) 4 on page 4.

1.84.   Assuming that $\dfrac{d}{dx} \sin u = \cos u \dfrac{du}{dx}$, $\dfrac{d}{dx} \cos u = -\sin u \dfrac{du}{dx}$, prove the formulas (a) 12, (b) 13, (c) 14 on page 5.

1.85.   Prove formulas (a) 15, (b) 16, (c) 18, (d) 19, (e) 20 on page 5.

1.86.   Find (a) $\dfrac{d}{dx} \sqrt[3]{(2x - 1)^2}$, (b) $\dfrac{d}{dx} \cos^2(e^{-3x})$, (c) $\dfrac{d}{dx} \sin(\tan^{-1} x)$, (d) $\dfrac{d}{dx} x^2 \ln^2(2x + 1)$, (e) $\dfrac{d}{dx} \left(\dfrac{\sin^{-1} x}{\cos^{-1} x} \right)$, (f) $\dfrac{d}{dx} \log_a x$.

1.87.   If $e^{xy} + y^2 = \cos x$, find dy/dx.

1.88.   If $x = a(\theta - \sin \theta)$, $y = a(1 - \cos \theta)$, find dy/dx.

1.89.   If $y = (3x + 1)/(1 - 2x)$, find $d^2 y/dx^2$ at $x = 2$.

1.90.   If $y = (c_1 \sin x + c_2 \cos x)/\sqrt{x}$, show that $xy'' + 2y' + xy = 0$ where c_1, c_2 are any constants.

1.91.   Find $d^2 y/dx^2$ for the function defined in Problem 1.88.

1.92.   Find the equation of the tangent line to the curve $xy^2 + y = 3x - 1$ at the point $(1, 1)$.

1.93.   Find the differentials of (a) $y = x^2 \ln x + 3x$, (b) $y = (2x - 1)/(x + 2)$.

1.94.   (a) Show that if $x > 0$ and Δx is numerically small compared with x, then $\sqrt{x + \Delta x}$ is approximately equal to $\sqrt{x} + (\Delta x)/2\sqrt{x}$. (b) Use this result to find the approximate values of $\sqrt{4.12}$ and $\sqrt{8.75}$. (c) By obtaining a similar result for $\sqrt[3]{x + \Delta x}$, find an approximate value for $\sqrt[3]{65}$ and $\sqrt[3]{.9}$.

1.95. Prove the law of the mean, page 8, and illustrate by an example.

INTEGRALS

1.96. Find the integrals   (a) $\int (2x^3 - 4x^2 + 3\sqrt{x})\,dx$,   (b) $\int e^{-2x}\,dx$,   (c) $\int (5\cos 2x - 3\sin 4x)\,dx$,

(d) $\int \dfrac{\ln x}{x}\,dx$,   (e) $\int \dfrac{\sin 2x}{1 - \cos 2x}\,dx$,   (f) $\int x^3 \sqrt[3]{x^4 + 4}\,dx$,   (g) $\int \dfrac{x^2\,dx}{2x^3 + 5}$,   (h) $\int \dfrac{e^x - 1}{e^x + 1}\,dx$.

1.97. Prove the integration formulas (a) 7, (b) 11, (c) 12, (d) 13, (e) 14 on page 6.

1.98. Find   (a) $\int xe^{3x}\,dx$,   (b) $\int x(\cos 3x + 4\sin 3x)\,dx$,   (c) $\int x \ln x\,dx$,   (d) $\int \tan(2x+3)\,dx$,

(e) $\int x^2 e^{-x}\,dx$.

1.99. Evaluate   (a) $\int_0^3 \sqrt{x+1}\,dx$,   (b) $\int_1^2 xe^{-x^2}\,dx$,   (c) $\int_2^4 \dfrac{dx}{2x+1}$.

1.100. Prove the integration formulas (a) 15, (b) 16, (c) 17, (d) 18, (e) 19, (f) 20, (g) 21, (h) 22 on page 6.

1.101. Find   (a) $\int \dfrac{dx}{\sqrt{4 - x^2}}$ (c) $\int \dfrac{dx}{\sqrt{4x^2 - 1}}$ (e) $\int (4\cosh 3x - 2\sinh 3x)\,dx$

(b) $\int \dfrac{dx}{x^2 + 2x + 2}$ (d) $\int \dfrac{dx}{\sqrt{x - x^2}}$

1.102. Find   (a) $\int_0^{\pi/2} \sqrt{1 - \cos x}\,dx$ (c) $\int e^x \sin x\,dx$ (e) $\int e^{\sqrt[3]{x}}\,dx$

(b) $\int \dfrac{x + 2}{\sqrt{x + 1}}\,dx$ (d) $\int_0^1 \dfrac{x\,dx}{(x+1)(x+2)}$

1.103. (a) Find the area bounded by $y = 4x - x^2$ and the x axis.

(b) Find the area bounded by the curves $y = x^2$ and $y = x$.

1.104. Using the trapezoidal rule, find the approximate value of $\int_0^1 \dfrac{dx}{x^2 + 1}$ by dividing the interval from 0 to 1 into (a) 5, (b) 10 equal parts. (c) Show that the integral has the exact value $\pi/4$ and compare with the approximate values.

1.105. Prove the *trapezoidal rule* which gives the approximate value of

$$\int_a^b f(x) = \frac{h}{2}[f(a) + 2f(a+h) + 2f(a+2h) + \cdots + 2f(a + (n-1)h) + f(a + nh)]$$

where $h = (b - a)/n$.

1.106. Assume that a function $f(x)$ is approximated by the parabolic function $c_0 + c_1 x + c_2 x^2$ in the interval $a \leqq x \leqq a + 2h$ [see Fig. 1-10]. Show that we have approximately

$$\int_a^{a+2h} f(x)\,dx = \frac{h}{3}[f(a) + 4f(a+h) + f(a+2h)]$$

The result is often called *Simpson's rule*.

1.107. (a) Show how Simpson's rule [Problem 1.106] can be used to find the approximate value of $\int_a^b f(x)\,dx$ and (b) use the result to find the approximate values of the integrals in Problems 1.28 and 1.104. Compare the accuracy of the results with those of the trapezoidal rule.

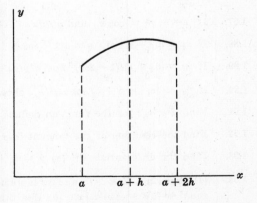

Fig. 1-10

SEQUENCES AND SERIES

1.108. Find the limit of the sequence whose nth term is $u_n = (n-1)/(2n+1)$, justifying your answer.

1.109. Show that the series $(1 - \frac{1}{2}) + (\frac{1}{2} - \frac{1}{3}) + (\frac{1}{3} - \frac{1}{4}) + \cdots$ converges and find its sum.

1.110. Prove that (a) $a + ar + ar^2 + \cdots + ar^{n-1} = \dfrac{a(1-r^n)}{1-r}$ and (b) if $|r| < 1$ then

$$a + ar + ar^2 + ar^3 + \cdots = \frac{a}{1-r}$$

1.111. Show that $.181818 \ldots = .18 + .0018 + .000018 + \cdots = 2/11$.

1.112. Investigate the convergence of the series (a) $\displaystyle\sum_{n=1}^{\infty} \frac{1}{\sqrt{n}}$, (b) $\displaystyle\sum_{n=1}^{\infty} \frac{\sqrt{n}}{\sqrt{n^4+1}}$, (c) $\displaystyle\sum_{n=1}^{\infty} ne^{-n^2}$, (d) $\displaystyle\sum_{n=1}^{\infty} \frac{\ln n}{n}$, (e) $\displaystyle\sum_{n=1}^{\infty} \frac{\sin n}{n^2}$, (f) $\displaystyle\sum_{n=1}^{\infty} \frac{(-1)^n n}{2n+1}$.

1.113. Prove that the series $x - \dfrac{x^3}{3!} + \dfrac{x^5}{5!} - \dfrac{x^7}{7!} + \cdots$ converges for $-\infty < x < \infty$.

1.114. Prove (a) Theorem 1-3, (b) Theorem 1-4, (c) Theorem 1-5, page 6.

1.115. (a) Consider the series $u_1 - u_2 + u_3 - u_4 + \cdots$ where $u_k > 0$. Prove that if $u_{n+1} \leqq u_n$ and $\lim\limits_{n \to \infty} u_n = 0$, then the series converges. (b) Thus show that $1 - \frac{1}{2} + \frac{1}{3} - \frac{1}{4} + \cdots$ is convergent but not absolutely convergent.

1.116. Find the interval of convergence of (a) $x - \dfrac{x^2}{2} + \dfrac{x^3}{3} - \dfrac{x^4}{4} + \cdots$, (b) $x - \dfrac{x^3}{3} + \dfrac{x^5}{5} - \dfrac{x^7}{7} + \cdots$.

1.117. Find the interval of convergence of (a) $\displaystyle\sum_{n=1}^{\infty} \frac{nx^n}{2^n}$, (b) $\displaystyle\sum_{n=1}^{\infty} \frac{x^n}{n^2+1}$.

UNIFORM CONVERGENCE

1.118. Prove that the series $x(1-x) + x^2(1-x) + x^3(1-x) + \cdots$ converges for $-1 < x \leqq 1$ but is not uniformly convergent in this interval. Is it uniformly convergent in $-1/2 \leqq x \leqq 1/2$? Explain.

1.119. Investigate the uniform convergence of the series (a) $\displaystyle\sum_{n=1}^{\infty} \frac{nx^n}{2^n}$ in $-1 \leqq x \leqq 1$, (b) $\displaystyle\sum_{n=1}^{\infty} \frac{1}{x^2+n^2}$ for $-\infty < x < \infty$.

1.120. Prove that a power series $c_0 + c_1(x-a) + c_2(x-a)^2 + \cdots$ converges uniformly in any interval which lies entirely within its interval of convergence. [*Hint:* Use the Weierstrass M test.]

1.121. (a) Prove that if $S_n(x)$ converges uniformly to $S(x)$ in $a \leqq x \leqq b$, then

$$\lim_{n \to \infty} \int_a^b S_n(x)\, dx = \int_a^b \lim_{n \to \infty} S_n(x)\, dx$$

(b) Show that $$\lim_{n \to \infty} \int_0^1 nxe^{-nx^2}\, dx \neq \int_0^1 \left[\lim_{n \to \infty} nxe^{-nx^2} \right] dx$$

and supply a possible explanation.

1.122. Prove Theorem 1-11, page 7.

1.123. (a) By using analogy with series, give a definition for uniform convergence and absolute convergence of the integral $\displaystyle\int_0^{\infty} f(x, \alpha)\, dx$.

(b) Prove a test for uniform [and absolute] convergence of the integral in (a) analogous to the Weierstrass M test for series.

(c) Prove theorems for integrals analogous to Theorems 1-9, 1-10 and 1-11 on page 7.

TAYLOR SERIES

1.124. Obtain the formal Taylor series (a) 1, (b) 3, (c) 4, (d) 5 given on page 8.

1.125. Expand $\sin x$ in a Taylor series about $a = \pi/6$ and use the result to find $\sin 31°$ approximately. Compare with the result from a table of sines.

1.126. Find an approximate value for (a) $\displaystyle\int_0^1 e^{-x^2}\,dx$, (b) $\displaystyle\int_0^1 \frac{1 - \cos x}{x}\,dx$.

1.127. Show that (a) $\sinh x = x + \dfrac{x^3}{3!} + \dfrac{x^5}{5!} + \dfrac{x^7}{7!} + \cdots$, (b) $\cosh x = 1 + \dfrac{x^2}{2!} + \dfrac{x^4}{4!} + \dfrac{x^6}{6!} + \cdots$ for $-\infty < x < \infty$.

PARTIAL DERIVATIVES

1.128. If $f(x, y) = x \sin(y/x)$, find (a) $f(2, \pi/2)$, (b) $f_x(2, \pi/2)$, (c) $f_y(2, \pi/2)$, (d) $f_{xx}(2, \pi/2)$, (e) $f_{xy}(2, \pi/2)$, (f) $f_{yx}(2, \pi/2)$, (g) $f_{yy}(2, \pi/2)$.

1.129. Verify that $f_{xy} = f_{yx}$ for the case where $f(x, y) = (x - y)/(x + y)$ if $x + y \neq 0$.

1.130. If $f(x, y) = x^2 \tan^{-1}(y/x)$, show that $x\dfrac{\partial f}{\partial x} + y\dfrac{\partial f}{\partial y} = 2f$. Are there any exceptional values for x and y?

1.131. If $V(x, y, z) = (x^2 + y^2 + z^2)^{-1/2}$, show that $\dfrac{\partial^2 V}{\partial x^2} + \dfrac{\partial^2 V}{\partial y^2} + \dfrac{\partial^2 V}{\partial z^2} = 0$ if x, y, z are not all zero.

1.132. If $f(x, y) = y \cos(x - 2y)$, find df if $x = \frac{1}{4}\pi$, $y = \pi$, $dx = \frac{1}{4}\pi$, $dy = \pi$.

1.133. If $f(x, y, z) = x^2 z - yz^3 + x^4$, find df.

1.134. If $z = x^2 f(y/x)$ where f is differentiable, show that $x(\partial f/\partial x) + y(\partial f/\partial y) = 2f$ and compare with Problem 1.130.

1.135. If $x = \rho \cos\phi$, $y = \rho \sin\phi$, prove that if U is a twice differentiable function of x and y,

$$\frac{\partial^2 U}{\partial x^2} + \frac{\partial^2 U}{\partial y^2} = \frac{\partial^2 U}{\partial \rho^2} + \frac{1}{\rho}\frac{\partial U}{\partial \rho} + \frac{1}{\rho^2}\frac{\partial^2 U}{\partial \phi^2}$$

TAYLOR SERIES FOR FUNCTIONS OF TWO OR MORE VARIABLES

1.136. Expand $f(x, y) = 2xy + x^2 + y^3$ in a Taylor series about $a = 1, b = 2$ and verify the expansion.

1.137. Write Taylor's series for $f(x, y, z)$ expanded about $x = a, y = b, z = c$ and illustrate by expanding $f(x, y, z) = xz + y^2$ about $x = 1, y = -1, z = 2$.

LINEAR EQUATIONS AND DETERMINANTS

1.138. Solve the systems   (a) $\begin{cases} 2x - y = -4 \\ x + 3y = 5 \end{cases}$ (b) $\begin{cases} 2x + y + z = 3 \\ x - 2y - 2z = 4 \\ 3x - y + 4z = 2 \end{cases}$.

1.139. Determine whether the system of equations

$$\begin{cases} 2x - y + z = 0 \\ 3x + 2y - 4z = 0 \\ x - y + 2z = 0 \end{cases}$$

has non-trivial solutions. If so give two such solutions. If not replace the first equation by $kx - y + z = 0$ and determine k so that the system has non-trivial solutions and give two such solutions.

1.140. Show that if two rows (or columns) of a second or third order determinant are interchanged, the sign of the determinant is changed.

1.141. Show that if the elements in two rows (or columns) of a second or third order determinant are equal or proportional, the value of the determinant is zero.

1.142. Discuss the system of equations $\begin{cases} 2x + y + z = 5 \\ x - 2y - 3z = 4 \\ 5x + 5y + 6z = 8 \end{cases}$

from a geometrical viewpoint.

MAXIMA AND MINIMA. THE METHOD OF LAGRANGE MULTIPLIERS.

1.143. Find the relative maxima and minima of $f(x) = 12 + 8x^2 - x^4$ and use the results to graph the function.

1.144. Find the relative maxima and minima of (a) $x^2 \ln x$, (b) $a \sec \theta + b \csc \theta$.

1.145. Find the area of the largest rectangle which can be inscribed in a semicircle of radius a with base on its diameter.

1.146. If $f'(a) = 0$ and $f''(a) = 0$, is it possible that $f(x)$ has a maximum or minimum at $x = a$? Explain. [*Hint*: Consider $f(x) = x^4$ or $f(x) = -x^4$.]

1.147. A box having only five rectangular sides is required to have a given volume V. Determine its dimensions so that the surface area will be a minimum.

1.148. Find the shortest and largest distance from the origin to the curve $x^2 + xy + y^2 = 16$ and give a geometric interpretation. [*Hint*: Use the method of Lagrange multipliers to find the maximum of $x^2 + y^2$.]

1.149. (a) Find the relative maximum and minimum of $x^2 + y^2 + z^2$, given that $4x^2 + 9y^2 + 16z^2 = 576$ and (b) interpret geometrically.

1.150. Generalize the method of Lagrange multipliers to the case where $f(x, y, z)$ is to be made an extremum [i.e. maximum or minimum] subject to two constraint conditions $\phi_1(x, y, z) = 0$, $\phi_2(x, y, z) = 0$. [*Hint*: Let $h(x, y, z) = f(x, y, z) + \lambda_1 \phi_1(x, y, z) + \lambda_2 \phi_2(x, y, z)$ and prove that $\partial h/\partial x = 0$, $\partial h/\partial y = 0$, $\partial h/\partial z = 0$ where λ_1, λ_2 are two Lagrange multipliers.]

LEIBNITZ'S RULE

1.151. If $I(\alpha) = \int_{\sin \alpha}^{\cos \alpha} (x^2 \sin \alpha - x^3)\, dx$, find $dI/d\alpha$ by (a) integrating first and (b) using Leibnitz's rule.

1.152. Find $dI/d\alpha$ if (a) $I(\alpha) = \int_{\alpha^2}^{\alpha^3} \tan^{-1} \frac{x}{\alpha}\, dx$, (b) $I(\alpha) = \int_{\alpha}^{\sqrt{\alpha}} e^{-\alpha x^2}\, dx$.

1.153. Prove Leibnitz's rule. [*Hint*: Consider $\dfrac{I(\alpha + \Delta\alpha) - I(\alpha)}{\Delta\alpha}$ and write the result as the sum of three integrals. Then let $\Delta\alpha \to 0$.]

COMPLEX NUMBERS. POLAR FORM

1.154. Perform each of the indicated operations: (a) $2(5 - 3i) - 3(-2 + i) + 5(i - 3)$, (b) $(3 - 2i)^3$, (c) $\dfrac{5}{3 - 4i} + \dfrac{10}{4 + 3i}$, (d) $\left(\dfrac{1 - i}{1 + i}\right)^{10}$, (e) $\left|\dfrac{2 - 4i}{5 + 7i}\right|^2$, (f) $\dfrac{(1 + i)(2 + 3i)(4 - 2i)}{(1 + 2i)^2(1 - i)}$.

1.155. If z_1 and z_2 are complex numbers, prove (a) $\left|\dfrac{z_1}{z_2}\right| = \dfrac{|z_1|}{|z_2|}$, (b) $|z_1^2| = |z_1|^2$ giving any restrictions.

1.156. Prove (a) $|z_1 + z_2| \leqq |z_1| + |z_2|$, (b) $|z_1 + z_2 + z_3| \leqq |z_1| + |z_2| + |z_3|$, (c) $|z_1 - z_2| \geqq |z_1| - |z_2|$.

1.157. Find all solutions of $2x^4 - 3x^3 - 7x^2 - 8x + 6 = 0$.

1.158. Let z_1 and z_2 be represented by points P_1 and P_2 in the Argand diagram. Construct lines OP_1 and OP_2, where O is the origin. Show that $z_1 + z_2$ can be represented by the point P_3, where OP_3 is the diagonal of a parallelogram having sides OP_1 and OP_2. This is called the *parallelogram law* of addition of complex numbers. Because of this and other properties, complex numbers can be considered as *vectors* in two dimensions. The general subject of vectors is given in Chapter 5.

1.159. Interpret geometrically the inequalities of Problem 1.156.

1.160. Express in polar form (a) $3\sqrt{3} + 3i$, (b) $-2 - 2i$, (c) $1 - \sqrt{3}\,i$, (d) 5, (e) $-5i$.

1.161. Evaluate (a) $[2(\cos 25° + i \sin 25°)][5(\cos 110° + i \sin 110°)]$, (b) $\dfrac{12 \operatorname{cis} 16°}{(3 \operatorname{cis} 44°)(2 \operatorname{cis} 62°)}$.

1.162. Determine all the indicated roots and represent them graphically:

(a) $(4\sqrt{2} + 4\sqrt{2}\,i)^{1/3}$, (b) $(-1)^{1/5}$, (c) $(\sqrt{3} - i)^{1/3}$, (d) $i^{1/4}$.

1.163. If $z_1 = \rho_1 \operatorname{cis} \phi_1$ and $z_2 = \rho_2 \operatorname{cis} \phi_2$, prove (a) $z_1 z_2 = \rho_1 \rho_2 \operatorname{cis}(\phi_1 + \phi_2)$, (b) $z_1/z_2 = (\rho_1/\rho_2) \operatorname{cis}(\phi_1 - \phi_2)$.
Interpret geometrically.

1.164. Show that De Moivre's theorem is equivalent to $(e^{i\phi})^n = e^{in\phi}$.

Answers to Supplementary Problems

1.63. $\sqrt{3} \ \sqrt[3]{5}/\sqrt[3]{2} \ \sqrt[6]{60} < \sqrt[4]{2} < 11/9$

1.66. (a) $-11/2$ (b) $4/5$ (c) $-19/10$ (d) $(5 + 2x - 3x^2)/(x + 4)$ (e) $-(8 + 9\sqrt{2})/14$

1.67. (a) even (b) odd (c) odd (d) neither (e) odd (f) even

1.68. (b) $1/x^2$ **1.71.** (a) $240/289$ (b) $-161/289$ (c) $\sqrt{17}/17$

1.72. $4/5$

1.73. (a) 18 (b) 1 (c) $1/32$

1.86. (a) $\frac{4}{3}(2x - 1)^{-1/3}$ (d) $2x \ln^2(2x + 1) + 4x^2 \ln(2x + 1)/(2x + 1)$

(b) $6e^{-3x} \sin(e^{-3x}) \cos(e^{-3x})$ (e) $(\sin^{-1} x + \cos^{-1} x)/\sqrt{1 - x^2}\,(\cos^{-1} x)^2$

(c) $\cos(\tan^{-1} x)/(1 + x^2)$ (f) $1/x \ln a$ or $\log_a e/x$

1.87. $-(ye^{xy} + \sin x)/(xe^{xy} + 2y)$ **1.88.** $\sin\theta/(1 - \cos\theta) = \cot(\theta/2)$

1.89. $-20/27$ **1.91.** $-1/a(1 - \cos\theta)^2$ or $-\csc^4(\theta/2)/4a$ **1.92.** $3y - 2x = 1$

1.93. (a) $(x + 2x \ln x + 3)\,dx$ (b) $5\,dx/(x + 2)^2$

1.96. (a) $\frac{1}{4}x^4 - 4x^3 + 2x^{3/2} + c$ (e) $\frac{1}{2} \ln(1 - \cos 2x) + c$

(b) $-\frac{1}{2}e^{-2x} + c$ (f) $\frac{3}{16}(x^4 + 4)^{4/3} + c$

(c) $\frac{5}{2}\sin 2x + \frac{9}{4} \cos 4x + c$ (g) $\frac{1}{6} \ln(2x^3 + 5) + c$

(d) $\frac{1}{2} \ln^2 x + c$ (h) $\frac{1}{2} \ln(e^{x/2} + e^{-x/2}) + c$

1.98. (a) $\frac{1}{9}e^{3x}(3x - 1) + c$ (d) $-\frac{1}{2} \ln \cos(2x + 3) + c$

(b) $\frac{1}{6}[(4 + 3x) \sin 3x + (1 - 12x) \cos 3x] + c$ (e) $e^{-x}(x^2 + 2x + 2) + c$

(c) $\dfrac{x^2}{2}(\ln x - \frac{1}{2}) + c$

1.99. (a) $14/3$ (b) $\frac{1}{2}(e - e^4)$ (c) $\frac{1}{2} \ln(\frac{9}{5})$

1.101. (a) $\sin^{-1}(x/2) + c$

(b) $\tan^{-1}(x+1) + c$

(c) $\frac{1}{2}\ln(2x + \sqrt{4x^2-1}) + c$

(d) $\sin^{-1}(2x-1) + c$

(e) $\frac{4}{3}\sinh 3x - \frac{2}{3}\cosh 3x + c$

1.102. (a) $2(\sqrt{2}-1)$

(b) $\frac{2}{3}(x+1)^{3/2} + 2(x+1)^{1/2} + c$

(c) $\frac{e^x}{2}(\sin x - \cos x) + c$

(d) $\ln(9/8)$

(e) $3e^{\sqrt[3]{x}}(x^{2/3} - 2x^{1/3} + 2) + c$

1.103. (a) 32/3   (b) 1/6   **1.107.** 1

1.110. (a) diverges   (b) converges   (c) converges   (d) diverges   (e) converges   (f) diverges

1.116. (a) $-1 < x \leqq 1$ (b) $-1 \leqq x \leqq 1$

1.117. (a) $-2 \leqq x \leqq 2$ (b) $-1 \leqq x \leqq 1$

1.128. (a) $\sqrt{2}$ (b) $\frac{\sqrt{2}}{8}(4-\pi)$ (c) $\sqrt{2}/2$ (d) $\frac{\pi\sqrt{2}}{64}(8-\pi)$ (e) $\pi\sqrt{2}/16$ (f) $\pi\sqrt{2}/16$ (g) $-\sqrt{2}/4$

1.132. $\frac{1}{8}\pi\sqrt{2}(4+7\pi)$ **1.133.** $(2xz + 4x^3)\,dx - z^3\,dy + (x^2 - 3yz^2)\,dz$

1.138. (a) $x = -1,\ y = 2$ (b) $x = 2,\ y = 0,\ z = -1$

1.141. rel. min. 12 at $x = 0$,   rel. max. 28 at $x = \pm 2$

1.144. (a) rel. min. $-1/2e$ at $x = e^{-1/2}$ (b) rel. min. $(a^{2/3} + b^{2/3})^{3/2}$ for $\theta = \tan^{-1}(b/a)^{1/3}$

1.145. a^2 **1.147** $\sqrt[3]{2V},\ \sqrt[3]{2V},\ \sqrt[3]{V/4}$

1.148. max. $4\sqrt{2}$, min. $4\sqrt{6}/3$

1.151. (a) $\frac{1}{2}\ln\left(\frac{\alpha^2+1}{\alpha^4+1}\right) + 3\alpha^2\tan^{-1}\alpha^2 - 2\alpha\tan^{-1}\alpha$,   (b) $\frac{1}{2}\alpha^{-1/2}e^{-\alpha^2} - e^{-\alpha^3} - \int_{\alpha}^{\sqrt{\alpha}} x^2 e^{-\alpha x^2}\,dx$

1.154. (a) $1 - 4i$ (b) $-9 - 46i$ (c) $\frac{11}{5} - \frac{2}{5}i$ (d) -1 (e) $\frac{10}{37}$ (f) $\frac{16}{5} - \frac{2}{5}i$

1.157. 3, $\frac{1}{2}$, $-1 \pm i$

1.160. (a) $6\operatorname{cis}\pi/6$ (b) $2\sqrt{2}\operatorname{cis}5\pi/4$ (c) $2\operatorname{cis}5\pi/3$ (d) $5\operatorname{cis}0$ (e) $5\operatorname{cis}3\pi/2$

1.161. (a) $-5\sqrt{2} + 5\sqrt{2}\,i$ (b) $-2i$

1.162. (a) $2\operatorname{cis}15°$, $2\operatorname{cis}135°$, $2\operatorname{cis}255°$

(b) $\operatorname{cis}36°$, $\operatorname{cis}108°$, $\operatorname{cis}180° = -1$, $\operatorname{cis}252°$, $\operatorname{cis}324°$

(c) $\sqrt[3]{2}\operatorname{cis}110°$, $\sqrt[3]{2}\operatorname{cis}230°$, $\sqrt[3]{2}\operatorname{cis}350°$

(d) $\operatorname{cis}22.5°$, $\operatorname{cis}112.5°$, $\operatorname{cis}202.5°$, $\operatorname{cis}292.5°$

Chapter 2

DEFINITION OF A DIFFERENTIAL EQUATION

A *differential equation* is an equation involving derivatives or differentials. The following are some examples of differential equations.

Example 1. $(y'')^2 + 3x = 2(y')^3$ where $y' = dy/dx,\ y'' = d^2y/dx^2$

Example 2. $\dfrac{dy}{dx} + \dfrac{y}{x} = y^2$

Example 3. $\dfrac{d^2Q}{dt^2} - 3\dfrac{dQ}{dt} + 2Q = 4\sin 2t$

Example 4. $\dfrac{dy}{dx} = \dfrac{x+y}{x-y}$ or $(x+y)\,dx + (y-x)\,dy = 0$

Example 5. $\dfrac{\partial^2 V}{\partial x^2} + \dfrac{\partial^2 V}{\partial y^2} = 0$

Equations such as those in Examples 1-4 involving only one independent variable are called *ordinary differential equations*. Equations such as that of Example 5 with two or more independent variables are called *partial differential equations* and are treated in Chapter 12.

ORDER OF A DIFFERENTIAL EQUATION

An equation having a derivative of nth order but no higher is called an nth *order differential equation*. In Examples 1-5 above, the orders of the differential equations are 2, 1, 2, 1, 2, respectively.

ARBITRARY CONSTANTS

An arbitrary constant, often denoted by a letter at the beginning of the alphabet such as A, B, C, c_1, c_2, etc., may assume values independently of the variables involved. For example in $y = x^2 + c_1 x + c_2$, c_1 and c_2 are arbitrary constants.

The relation $y = Ae^{-4x+B}$ which can be written $y = Ae^B e^{-4x} = Ce^{-4x}$ actually involves only one arbitrary constant. We shall always assume that the minimum number of constants is present, i.e. the arbitrary constants are *essential*.

SOLUTION OF A DIFFERENTIAL EQUATION

A *solution* of a differential equation is a relation between the variables which is free of derivatives and which satisfies the differential equation identically.

Example 6. $y = x^2 + c_1 x + c_2$ is a solution of $y'' = 2$ since by substitution we obtain the identity $2 = 2$.

38

A *general solution* of an *n*th order differential equation is one involving *n* (essential) arbitrary constants.

> **Example 7.** Since $y = x^2 + c_1 x + c_2$ has two arbitrary constants and satisfies the second order differential equation $y'' = 2$, it is a general solution of $y'' = 2$.

A *particular solution* is a solution obtained from the general solution by assigning specific values to the arbitrary constants.

> **Example 8.** $y = x^2 - 3x + 2$ is a particular solution of $y'' = 2$ and is obtained from the general solution $y = x^2 + c_1 x + c_2$ by putting $c_1 = -3$ and $c_2 = 2$.

A *singular solution* is a solution which cannot be obtained from the general solution by specifying values of the arbitrary constants.

> **Example 9.** The general solution of $y = xy' - y'^2$ is $y = cx - c^2$. However, as seen by substitution another solution is $y = x^2/4$ which cannot be obtained from the general solution for any constant c. This second solution is a singular solution. For a relationship between the general and singular solutions see Problem 2.5.

DIFFERENTIAL EQUATION OF A FAMILY OF CURVES

A general solution of an *n*th order differential equation has *n* arbitrary constants (or parameters) and represents geometrically an *n parameter family of curves*. Conversely a relation with *n* arbitrary constants [sometimes called a *primitive*] has associated with it a differential equation of order *n* [of which it is a general solution] called the *differential equation of the family*. This differential equation is obtained by differentiating the primitive *n* times and then eliminating the *n* arbitrary constants among the $n+1$ resulting equations. See Problems 2.6 and 2.7.

SPECIAL FIRST ORDER EQUATIONS AND SOLUTIONS

Any first order differential equation can be put into the form

$$\frac{dy}{dx} = f(x, y) \quad \text{or} \quad M(x, y)\, dx + N(x, y)\, dy = 0 \qquad (1)$$

and the general solution of such an equation contains one arbitrary constant. Many special devices are available for finding general solutions of various types of first order differential equations. In the following table some of these types are given.

Differential Equation	General Solution (or method by which it can be obtained)
1. Separation of variables $f_1(x)\, g_1(y)\, dx + f_2(x)\, g_2(y)\, dy = 0$	Divide by $g_1(y) f_2(x) \neq 0$ and integrate to obtain $$\int \frac{f_1(x)}{f_2(x)}\, dx + \int \frac{g_2(y)}{g_1(y)}\, dy = c$$
2. Exact equation $M(x, y)\, dx + N(x, y)\, dy = 0$ where $\dfrac{\partial M}{\partial y} = \dfrac{\partial N}{\partial x}$	The equation can be written as $$M\, dx + N\, dy = dU(x, y) = 0$$ where dU is an exact differential. Thus the solution is $U(x, y) = c$ or equivalently $$\int M\, \partial x + \int \left(N - \frac{\partial}{\partial y} \int M\, \partial x \right) dy = c$$ where ∂x indicates that the integration is to be performed with respect to x keeping y constant.

Differential Equation	General Solution (or method by which it can be obtained)
3. Integrating factor $$M(x,y)\,dx + N(x,y)\,dy = 0$$ where $\dfrac{\partial M}{\partial y} \neq \dfrac{\partial N}{\partial x}$	The equation can be written as an exact differential equation $$\mu M\,dx + \mu N\,dy = 0$$ where μ is an appropriate *integrating factor* so that $\dfrac{\partial}{\partial y}(\mu M) = \dfrac{\partial}{\partial x}(\mu N)$ and then method 2 applies. The following combinations are often useful in finding integrating factors. $$(a)\quad \frac{x\,dy - y\,dx}{x^2} = d\left(\frac{y}{x}\right)$$ $$(b)\quad \frac{x\,dy - y\,dx}{y^2} = -d\left(\frac{x}{y}\right)$$ $$(c)\quad \frac{x\,dy - y\,dx}{x^2 + y^2} = d\left(\tan^{-1}\frac{y}{x}\right)$$ $$(d)\quad \frac{x\,dy - y\,dx}{x^2 - y^2} = \frac{1}{2}\,d\left\{\ln\frac{x-y}{x+y}\right\}$$ $$(e)\quad \frac{x\,dx + y\,dy}{x^2 + y^2} = \tfrac{1}{2}d\{\ln(x^2 + y^2)\}$$ See also Problem 2.17.
4. Linear equation $$\frac{dy}{dx} + P(x)y = Q(x)$$	An integrating factor is given by $$\mu = e^{\int P(x)\,dx}$$ and the equation can then be written $$\frac{d}{dx}(\mu y) = \mu Q$$ with solution $\mu y = \displaystyle\int \mu Q\,dx + c$ or $$y e^{\int P\,dx} = \int Q e^{\int P\,dx}\,dx + c$$
5. Homogeneous equation $$\frac{dy}{dx} = F\left(\frac{y}{x}\right)$$	Let $y/x = v$ or $y = vx$, and the equation becomes $$v + x\frac{dv}{dx} = F(v) \quad\text{or}\quad x\,dv + (F(v) - v)\,dx = 0$$ which is of Type 1 and has the solution $$\ln x = \int \frac{dv}{F(v) - v} + c$$ where $v = y/x$. If $F(v) = v$, the solution is $y = cx$.
6. Bernoulli's equation $$\frac{dy}{dx} + P(x)y = Q(x)y^n, \quad n \neq 0,1$$	Letting $v = y^{1-n}$, the equation reduces to Type 4 with solution $$v e^{(1-n)\int P\,dx} = (1-n)\int Q e^{(1-n)\int P\,dx}\,dx + c$$ If $n = 0$, the equation is of Type 4. If $n = 1$, it is of Type 1.

Differential Equation	General Solution (or method by which it can be obtained)
7. Equation solvable for y $y = g(x, p)$ where $p = y'$	Differentiate both sides of the equation with respect to x to obtain $$\frac{dy}{dx} = \frac{dg}{dx} = \frac{\partial g}{\partial x} + \frac{\partial g}{\partial p}\frac{dp}{dx}$$ or $$p = \frac{\partial g}{\partial x} + \frac{\partial g}{\partial p}\frac{dp}{dx}$$ Then solve this last equation to obtain $G(x, p, c) = 0$. The required solution is obtained by eliminating p between $G(x, p, c) = 0$ and $y = g(x, p)$. An analogous method exists if the equation is solvable for x.
8. Clairaut's equation $y = px + F(p)$ where $p = y'$	The equation is of Type 7 and has solution $$y = cx + F(c)$$ The equation will also have a singular solution in general.
9. Miscellaneous equations (a) $\dfrac{dy}{dx} = F(\alpha x + \beta y)$ (b) $\dfrac{dy}{dx} = F\left(\dfrac{\alpha_1 x + \beta_1 y + \gamma_1}{\alpha_2 x + \beta_2 y + \gamma_2}\right)$	(a) Letting $\alpha x + \beta y = v$, the equation reduces to Type 1. (b) Let $x = X + h$, $y = Y + k$ and choose constants h and k so that the equation reduces to Type 5. This is possible if and only if $\alpha_1/\alpha_2 \neq \beta_1/\beta_2$. If $\alpha_1/\alpha_2 = \beta_1/\beta_2$, the equation reduces to Type 9(a).

EQUATIONS OF HIGHER ORDER

If a differential equation is of order $m > 1$ and has one of the variables x or y missing explicitly from the equation, then it can be reduced to a differential equation of order $m - 1$ by letting

$$y' = p, \quad y'' = \frac{dp}{dx} \qquad \text{or} \qquad y'' = \frac{dp}{dy}\frac{dy}{dx} = p\frac{dp}{dy} \qquad (2)$$

See Problems 2.25 and 2.26.

EXISTENCE AND UNIQUENESS OF SOLUTIONS

It is often important to be able to predict directly from a differential equation and associated conditions whether a solution exists and is unique. For the case of a first order differential equation

$$y' = f(x, y) \qquad (3)$$

an answer is supplied in the following

Existence and Uniqueness Theorem. If $f(x, y)$ is continuous and has a continuous partial derivative with respect to y at each point of the region R defined by $|x - x_0| < \delta$, $|y - y_0| < \delta$, then there exists in R one and only one solution to (3) which passes through the point (x_0, y_0).

An immediate generalization of this theorem is possible for the nth order differential equation $y^{(n)} = f(x, y, y', \ldots, y^{(n-1)})$.

APPLICATIONS OF DIFFERENTIAL EQUATIONS

Many problems of science and engineering, when formulated mathematically, lead to *boundary-value problems*, i.e. differential equations and associated conditions. Solutions to these can be of great value to the scientist and engineer.

In a mathematical formulation of a physical problem, a *mathematical model* is chosen which often approximates the actual situation. For example, in treating rotation of the earth around the sun, we may consider that the sun and earth are points.

If a mathematical model and corresponding mathematical formulation leads to fairly good agreement with that predicted by observation or experiment, then the model is good. Otherwise a new model may have to be chosen.

SOME SPECIAL APPLICATIONS

The following table lists some special applications of an elementary nature which arise in science and engineering.

I. MECHANICS   The basic law of mechanics or dynamics is that of Newton, i.e.

$$F = \frac{d}{dt}(mv) \tag{4}$$

where m is the *mass* of the moving object, v is its *velocity*, t is the *time* and F is the net *force* acting on the object. The quantity mv is often called the *momentum*.

If m is constant, then this equation becomes

$$F = m\frac{dv}{dt} = ma \tag{5}$$

where a is the *acceleration*.

On or near the earth's surface, mass m is connected with weight W by $m = W/g$ or $W = mg$ where g is the acceleration due to gravity.

Various systems of units are available.

(a) C.G.S. [centimeter (cm), gram (g), second (sec)].   If m is in grams and a in cm/sec², then F is in *dynes*.

(b) M.K.S. [meter (m), kilogram (kg), second (sec)].   If m is in kilograms and a in m/sec², then F is in *newtons* (nt).

(c) F.P.S. [foot (ft), pound (lb), second (sec)].   If W is in pounds a and g in ft/sec², then $F = ma = Wa/g$ is in pounds.

On the earth's surface $g = 32$ ft/sec² $= 980$ cm/sec² $= 9.8$ m/sec²   approximately.

II. ELECTRIC CIRCUITS   A simple series electric circuit [Fig. 2-1] may consist of

(1) a *battery* or *generator* supplying an *electromotive force* or *e.m.f.* (*voltage* or *potential*) of E volts,

(2) a *resistor* having a *resistance* of R ohms,

(3) an *inductor* having an *inductance* of L henries,

(4) a *capacitor* or *condenser* having a *capacitance* of C farads.

The *current* I measured in *amperes* is the instantaneous time rate of change of charge Q on the condenser measured in *coulombs*, i.e. $I = dQ/dt$.

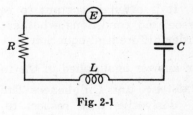

Fig. 2-1

From basic principles of electricity we have

Potential drop across resistor $= IR$

Potential drop across inductor $= L\,dI/dt$

Potential drop across capacitor $= Q/C$

Kirchhoff's Laws.

(a) The algebraic sum of the currents flowing into a junction point is zero.

(b) The algebraic sum of the potential drops around a closed loop is zero.

For single loop circuits law (a) implies that the current is the same throughout the loop.

III. ORTHOGONAL TRAJECTORIES

A curve cutting each member of a one parameter family of curves at right angles is called an *orthogonal trajectory* of the family. If $dy/dx = F(x, y)$ is the differential equation of a one parameter family (not containing the parameter), then the differential equation for the family of orthogonal trajectories is

$$dy/dx = -1/F(x, y)$$

IV. DEFLECTION OF BEAMS

A horizontal beam situated on the x axis of an xy coordinate system and supported in various ways, bends under the influence of vertical loads. The *deflection curve* of the beam, often called the *elastic curve* and shown dashed in Fig. 2-2, is given by $y = f(x)$ where y is measured as positive downward. This curve may be determined from the equation

$$\frac{EIy''}{(1 + y'^2)^{3/2}} = M(x) \qquad (6)$$

where $M(x)$ is the *bending moment* at x and is equal to the algebraic sum of the moments of all forces to one side of x, the moments being taken as positive for forces in the positive y direction and negative otherwise.

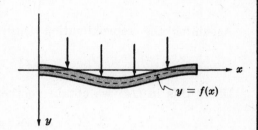

Fig. 2-2

For small deflections y' is small and the approximate equation

$$EIy'' = M(x) \qquad (7)$$

is used. The quantity EI where E is *Young's modulus* and I is the moment of inertia of a cross section of the beam about its central axis, is called the *flexural rigidity* and is generally constant.

V. MISCELLANEOUS PROBLEMS

Various problems of science and engineering pertaining to temperature, heat flow, chemistry, radioactivity, etc., can be mathematically formulated in terms of differential equations.

NUMERICAL METHODS FOR SOLVING DIFFERENTIAL EQUATIONS

Given the boundary-value problem

$$dy/dx = f(x, y) \qquad y(x_0) = y_0 \qquad (8)$$

it may not be possible to obtain an exact solution. In such case various methods are available for obtaining an approximate or numerical solution. In the following we list several such methods.

1. **Step by step or Euler method.** In this method we replace the differential equation of (8) by the approximation

$$\frac{y(x_0 + h) - y(x_0)}{h} = f(x_0, y_0) \qquad (9)$$

so that

$$y(x_0 + h) = y(x_0) + hf(x_0, y_0) \qquad (10)$$

By continuing in this manner we can then find $y(x_0 + 2h)$, $y(x_0 + 3h)$, etc. We choose h sufficiently small so as to obtain good approximations.

A modified procedure of this method can also be used. See Problem 2.45.

2. **Taylor series method.**   By successive differentiation of the differential equation in (8) we can find $y'(x_0), y''(x_0), y'''(x_0), \ldots$.   Then the solution is given by the Taylor series

$$y(x) \;=\; y(x_0) + y'(x_0)(x-x_0) + \frac{y''(x_0)(x-x_0)^2}{2!} + \cdots \tag{11}$$

assuming that the series converges.   If it does we can obtain $y(x_0+h)$ to any desired accuracy.   See Problem 2.46.

3. **Picard's method.**   By integrating the differential equation in (8) and using the boundary condition, we find

$$y(x) \;=\; y_0 + \int_{x_0}^{x} f(u, y)\, du \tag{12}$$

Assuming the approximation $y_1(x) = y_0$, we obtain from (12) a new approximation

$$y_2(x) \;=\; y_0 + \int_{x_0}^{x} f(u, y_1)\, du \tag{13}$$

Using this in (12) we obtain another approximation

$$y_3(x) \;=\; y_0 + \int_{x_0}^{x} f(u, y_2)\, du \tag{14}$$

Continuing in this manner we obtain a sequence of approximations $y_1, y_2, y_3, \ldots$.   The limit of this sequence, if it exists, is the required solution.   However, by carrying out the procedure a few times, good approximations can often be obtained.   See Problem 2.47.

4. **Runge-Kutta method.**   This method consists of computing

$$\left.\begin{aligned}
k_1 &\;=\; hf(x_0, y_0) \\
k_2 &\;=\; hf(x_0 + \tfrac{1}{2}h,\ y_0 + \tfrac{1}{2}k_1) \\
k_3 &\;=\; hf(x_0 + \tfrac{1}{2}h,\ y_0 + \tfrac{1}{2}k_2) \\
k_4 &\;=\; hf(x_0 + h,\ y_0 + k_3)
\end{aligned}\right\} \tag{15}$$

Then $$y(x_0 + h) \;=\; y_0 + \tfrac{1}{6}(k_1 + 2k_2 + 2k_3 + k_4) \tag{16}$$

See Problem 2.48.

These methods can also be adapted for higher order differential equations by writing them as several first order equations.   See Problems 2.49-2.51.

Solved Problems

CLASSIFICATION OF DIFFERENTIAL EQUATIONS

2.1.   Classify each of the following differential equations by stating the order, the dependent and independent variables, and whether the equation is ordinary or partial.

(a) $x^2 y'' + xy' + (x^2 - n^2)y = 0$ order 2, dep. var. y, indep. var. x, ordinary

(b) $\dfrac{dx}{dy} = x^2 + y^2$ order 1, dep. var. x, indep. var. y, ordinary

(c) $\dfrac{dy}{dx} = \dfrac{1}{x^2 + y^2}$ order 1, dep. var. y, indep. var. x, ordinary

Note that this equation is identical with the equation in (b).

(d) $\left(\dfrac{d^2u}{dt^2}\right)^3 + u^4 = 1$ order 2, dep. var. u, indep. var. t, ordinary

(e) $\dfrac{\partial^2 Y}{\partial t^2} = 2\dfrac{\partial^2 Y}{\partial x^2}$ order 2, dep. var. Y, indep. var. x and t, partial

(f) $(x^2 + 2y^2)\,dx + (3x^2 - 4y^2)\,dy = 0$ order 1, dep. var. y (or x) indep. var. x (or y), ordinary

SOLUTIONS OF DIFFERENTIAL EQUATIONS

2.2. Check whether each differential equation has the indicated solution. Which solutions are general solutions?

(a) $y' - x + y = 0$; $y = Ce^{-x} + x - 1$.

Substitute $y = Ce^{-x} + x - 1$, $y' = -Ce^{-x} + 1$ in the differential equation. Then
$$y' - x + y = -Ce^{-x} + 1 - x + Ce^{-x} + x - 1 = 0$$

Hence $y = Ce^{-x} + x - 1$ is a solution.

Since the number of arbitrary constants (one) equals the order of the differential equation (one), it is a general solution.

(b) $\dfrac{dy}{dx} = \dfrac{2xy}{3y^2 - x^2}$; $x^2 y - y^3 = c$.

Differentiating $x^2 y - y^3 = c$ we have $x^2 y' + 2xy - 3y^2 y^1 = 0$, i.e. $(x^2 - 3y^2)y' + 2xy = 0$

or $y' = \dfrac{dy}{dx} = \dfrac{2xy}{3y^2 - x^2}$. Then the solution is a general solution.

(c) $\dfrac{d^2 I}{dt^2} + 2\dfrac{dI}{dt} - 3I = 2\cos t - 4\sin t$; $I = c_1 e^t + c_2 e^{-3t} + \sin t$.

Substitute $I = c_1 e^t + c_2 e^{-3t} + \sin t$, $\dfrac{dI}{dt} = c_1 e^t - 3c_2 e^{-3t} + \cos t$, $\dfrac{d^2 I}{dt^2} = c_1 e^t + 9c_2 e^{-3t} - \sin t$ in the differential equation. Then

$$\dfrac{d^2 I}{dt^2} + 2\dfrac{dI}{dt} - 3I = (c_1 e^t + 9c_2 e^{-3t} - \sin t) + 2(c_1 e^t - 3c_2 e^{-3t} + \cos t) - 3(c_1 e^t + c_2 e^{-3t} + \sin t)$$
$$= 2\cos t - 4\sin t$$

and so the given relation is a general solution since the order of the differential equation and the number of arbitrary constants in the solution are both equal to 2.

(d) $x^3\left(\dfrac{d^2 v}{dx^2}\right)^2 = 2v\dfrac{dv}{dx}$; $v = cx^2$.

Substituting $v = cx^2$, $\dfrac{dv}{dx} = 2cx$, $\dfrac{d^2 v}{dx^2} = 2c$ in the differential equation, we find $x^3(2c)^2 = 2(cx^2)(2cx)$ or $4c^2 x^3 = 4c^2 x^3$. Thus $v = cx^2$ is a solution. However, it is not a general solution since the number of arbitrary constants (one) is not equal to the order of the equation (two).

2.3. Determine the particular solution of the differential equation of Problem 2.2(c) satisfying the conditions $I(0) = 2$, $I'(0) = -5$.

From Problem 2.2(c) the general solution is
$$I = I(t) = c_1 e^t + c_2 e^{-3t} + \sin t$$

At $t = 0$, $I(0) = c_1 + c_2 = 2$ i.e. $(1)\quad c_1 + c_2 = 2$

Differentiating with respect to t yields

$$I'(t) = c_1 e^t - 3c_2 e^{-3t} + \cos t$$

so that at $t = 0$

$$I'(0) = c_1 - 3c_2 + 1 = -5 \quad \text{i.e.} \quad (2) \quad c_1 - 3c_2 = -6$$

Solving (1) and (2) simultaneously, we find $c_1 = 0$, $c_2 = 2$ and the required particular solution is

$$I = 2e^{-3t} + \sin t$$

2.4. Show that the solution of the boundary-value problem

$$Q''(t) + 4Q'(t) + 20Q(t) = 16e^{-2t} \qquad t \geqq 0$$

$$Q(0) = 2, \qquad Q'(0) = 0$$

is $Q(t) = e^{-2t}(1 + \sin 4t + \cos 4t)$.

We have
$$\begin{aligned}
Q(t) &= e^{-2t}(1 + \sin 4t + \cos 4t) \\
Q'(t) &= e^{-2t}(4 \cos 4t - 4 \sin 4t) - 2e^{-2t}(1 + \sin 4t + \cos 4t) \\
&= e^{-2t}(2 \cos 4t - 6 \sin 4t - 2) \\
Q''(t) &= e^{-2t}(-8 \sin 4t - 24 \cos 4t) - 2e^{-2t}(2 \cos 4t - 6 \sin 4t - 2) \\
&= e^{-2t}(4 \sin 4t - 28 \cos 4t + 4)
\end{aligned}$$

Then
$$\begin{aligned}
Q''(t) + 4Q'(t) + 20Q(t) &= e^{-2t}(4 \sin 4t - 28 \cos 4t + 4) \\
&\quad + 4e^{-2t}(2 \cos 4t - 6 \sin 4t - 2) \\
&\quad + 20e^{-2t}(1 + \sin 4t + \cos 4t) \\
&= 16e^{-2t}
\end{aligned}$$

Furthermore $Q(0) = 2$ and $Q'(0) = 0$. Thus the given relation is a solution to the boundary-value problem.

2.5. Determine graphically a relationship between the general solution $y = cx - c^2$ and the singular solution $y = x^2/4$ of the differential equation $y = xy' - y'^2$.

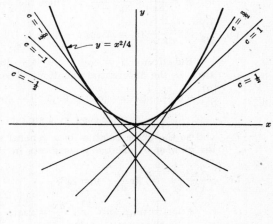

Fig. 2-3

Referring to Fig. 2-3, it is seen that $y = cx - c^2$ represents a family of straight lines tangent to the parabola $y = x^2/4$. The parabola is the *envelope* of the family of straight lines.

The envelope of a family of curves $G(x, y, c) = 0$, if it exists, can be found by solving simultaneously the equations $\partial G/\partial c = 0$ and $G = 0$. In this example $G(x, y, c) = y - cx + c^2$ and $\partial G/\partial c = -x + 2c$. Solving simultaneously $-x + 2c = 0$ and $y - cx + c^2 = 0$, we find $x = 2c$, $y = c^2$ or $y = x^2/4$.

DIFFERENTIAL EQUATION OF A FAMILY OF CURVES

2.6. (a) Graph various members of the one parameter family of curves $y = cx^3$. (b) Obtain the differential equation for the family in (a).

(a) The graphs of various members of the family are indicated in Fig. 2-4 below.

(b) From $y = cx^3$ we have $dy/dx = 3cx^2$. Then since $c = y/x^3$, the required differential equation of the family is

$$\frac{dy}{dx} = 3\left(\frac{y}{x^3}\right)x^2 \quad \text{or} \quad \frac{dy}{dx} = \frac{3y}{x}$$

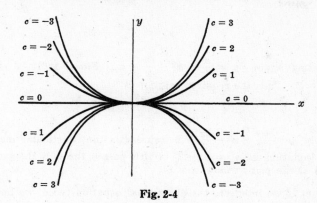

Fig. 2-4

2.7. Find the differential equation of the two parameter family of conics $ax^2 + by^2 = 1$.

Differentiating with respect to x we have

$$2ax + 2byy' = 0 \quad \text{or} \quad a = -byy'/x$$

Substituting into $ax^2 + by^2 = 1$ gives

$$(-byy'/x)x^2 + by^2 = 1 \quad \text{or} \quad -bxyy' + by^2 = 1$$

where a has been eliminated. Further differentiation gives

$$-b[xyy'' + xy'^2 + yy'] + 2byy' = 0$$

Then the required differential equation obtained on division by $b \neq 0$ is

$$xyy'' + xy'^2 - yy' = 0$$

2.8. *(a)* Find a general solution of the differential equation $dy/dx = 3x^2$.

  (b) Graph the solutions obtained in *(a)*.

  (c) Determine the equation of the particular curve in *(b)* which passes through the point $(1, 3)$.

(a) We have

$$dy = 3x^2\,dx \quad \text{or} \quad \int dy = \int 3x^2\,dx$$

so that $y = x^3 + c$ is the general solution.

(b) The curves $y = x^3 + c$ for various values of c are shown in Fig. 2-5. The set of curves for all values of c is a one parameter family of curves whose differential equation is $dy/dx = 3x^2$. Through each point of the xy plane there passes one and only one member of the family.

(c) Since the curve passes through $(1, 3)$, we have $y = 3$ when $x = 1$. Thus from $y = x^3 + c$, $3 = 1^3 + c$ or $c = 2$ and the required curve has the equation $y = x^3 + 2$.

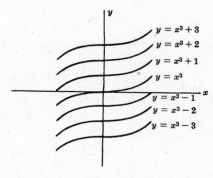

Fig. 2-5

2.9. *(a)* Solve the boundary-value problem

$$y'' = 3x - 2 \qquad y(0) = 2, \quad y'(1) = -3$$

and *(b)* give a geometric interpretation.

(a) Integrating once, we have $y' = \dfrac{3x^2}{2} - 2x + c_1$. Since $y'(1) = -3$ [i.e. $y' = -3$ when $x = 1$], we have $-3 = \frac{3}{2} - 2 + c_1$ and $c_1 = -\frac{5}{2}$. Thus

$$y' = \frac{3x^2}{2} - 2x - \frac{5}{2}$$

Integrating again, $y = \frac{x^3}{2} - x^2 - \frac{5x}{2} + c_2$. Since $y(0) = 2$ [i.e., $y = 2$ when $x = 0$], we find $c_2 = 2$. Then the required solution is

$$y = \frac{x^3}{2} - x^2 - \frac{5x}{2} + 2$$

(b) Geometrically $y = \frac{x^3}{2} - x^2 - \frac{5x}{2} + 2$ represents that particular member of the family [whose differential equation is $y'' = 3x - 2$] which passes through the point $(0, 2)$ and which has a slope of -3 at the point where $x = 1$.

Note that if we integrate the differential equation twice, we have $y = \frac{x^3}{2} - x^2 + c_1 x + c_2$ so that the solution represents a *two parameter family of curves*. The particular member of the family found above corresponds to $c_1 = -\frac{5}{2}$, $c_2 = 2$.

SEPARATION OF VARIABLES

2.10. (a) Find the general solution of $(4x + xy^2)\,dx + (y + x^2 y)\,dy = 0$.

(b) Find that particular solution for which $y(1) = 2$.

(a) The equation can be written as $x(4 + y^2)\,dx + y(1 + x^2)\,dy = 0$ or

$$\frac{x\,dx}{1 + x^2} + \frac{y\,dy}{4 + y^2} = 0$$

Integrating, $\qquad\qquad \frac{1}{2}\ln(1 + x^2) + \frac{1}{2}\ln(4 + y^2) = c_1$

i.e. $\qquad\qquad \ln[(1 + x^2)(4 + y^2)] = 2c_1 \quad$ or $\quad (1 + x^2)(4 + y^2) = e^{2c_1} = c$

Thus the required general solution is $(1 + x^2)(4 + y^2) = c$.

(b) For the particular solution where $y(1) = 2$, i.e. $y = 2$ when $x = 1$, put $x = 1, y = 2$ in $(1 + x^2)(4 + y^2) = c$ to obtain $c = 16$. Thus $(1 + x^2)(4 + y^2) = 16$.

2.11. Solve the boundary-value problem $\dfrac{dy}{dx} + 3y = 8$, $y(0) = 2$.

We have $dy/dx = 8 - 3y$ so that on separating the variables,

$$\frac{dy}{8 - 3y} = dx \quad \text{or} \quad \int \frac{dy}{8 - 3y} = \int dx$$

Thus $\qquad\qquad\qquad -\frac{1}{3}\ln(8 - 3y) = x + c_1$

Putting $x = 0$ and $y = 2$, we have $-\frac{1}{3}\ln 2 = c_1$ and so the required solution is

$$-\frac{1}{3}\ln(8 - 3y) = x - \frac{1}{3}\ln 2$$

This can also be written as $\frac{1}{3}\ln(8 - 3y) - \frac{1}{3}\ln 2 = -x$, $\ln(8 - 3y) - \ln 2 = -3x$, $\ln\left(\dfrac{8 - 3y}{2}\right) = -3x$, $\dfrac{8 - 3y}{2} = e^{-3x}$ or finally

$$y = \frac{2}{3}(4 - e^{-3x})$$

Check: If $y = \frac{2}{3}(4 - e^{-3x})$, then $y(0) = \frac{2}{3}(4 - e^0) = \frac{2}{3}(3) = 2$. Also $dy/dx = \frac{2}{3}(3e^{-3x}) = 2e^{-3x}$ so that

$$\frac{dy}{dx} + 3y = 2e^{-3x} + 3 \cdot \frac{2}{3}(4 - e^{-3x}) = 8$$

2.12. Solve the preceding problem if $y(0) = 2$ is replaced by $y(0) = 4$.

If we use the result $-\frac{1}{3} \ln(8 - 3y) = x + c_1$ of Problem 2.11, we find on formally putting $x = 0$, $y = 4$ that the logarithm of a negative number is involved. The difficulty can be avoided on noting that actually

$$\int \frac{dy}{8 - 3y} = -\frac{1}{3} \ln|8 - 3y|$$

so that the solution is $-\frac{1}{3} \ln|8 - 3y| = x - \frac{1}{3} \ln 4$. Since $8 - 3y$ is negative, the result can then be written as

$$-\frac{1}{3} \ln(3y - 8) = x - \frac{1}{3} \ln 2 \quad\text{or}\quad y = \frac{4}{3}(2 + e^{-3x})$$

2.13. Solve $dy/dx = \sec y \tan x$.

Separating the variables, we have $dy/\sec y = \tan x \, dx$

or
$$\cos y \, dy = \frac{\sin x}{\cos x} dx$$

Then integration yields
$$\int \cos y \, dy = -\int \frac{d(\cos x)}{\cos x}$$

or
$$\sin y = -\ln \cos x + c$$

i.e.
$$\ln \cos x + \sin y = c$$

EXACT DIFFERENTIAL EQUATIONS

2.14. (a) Show that $(3x^2 + y \cos x) \, dx + (\sin x - 4y^3) \, dy = 0$ is an exact differential equation and (b) find its general solution.

(a) Here $M = 3x^2 + y \cos x$, $N = \sin x - 4y^3$ so that $\partial M/\partial y = \cos x = \partial N/\partial x$ and the equation is exact.

(b) **Method 1 [Grouping of terms by inspection].** Write the equation as
$$3x^2 \, dx + (y \cos x \, dx + \sin x \, dy) - 4y^3 \, dy = 0$$

i.e.
$$d(x^3) + d(y \sin x) - d(y^4) = 0$$

or
$$d(x^3 + y \sin x - y^4) = 0$$

Then on integrating,
$$x^3 + y \sin x - y^4 = c$$

Method 2 [Direct method]. Since $M \, dx + N \, dy = dU$ an exact differential and $dU = \frac{\partial U}{\partial x} dx + \frac{\partial U}{\partial y} dy$, we must have

$$\frac{\partial U}{\partial x} = M \quad\text{and}\quad \frac{\partial U}{\partial y} = N$$

or
$$\frac{\partial U}{\partial x} = 3x^2 + y \cos x, \quad \frac{\partial U}{\partial y} = \sin x - 4y^3 \tag{1}$$

Integrating the first equation in (1) partially with respect to x [keeping y constant], we have

$$U = \int (3x^2 + y \cos x) \, \partial x = x^3 + y \sin x + F(y) \tag{2}$$

where $F(y)$ is the constant of integration which may depend on y.

Substituting (2) in the second equation of (1), we find
$$y \cos x + F'(y) = \sin x - 4y^3 \quad\text{or}\quad F'(y) = -4y^3$$

so that $F(y) = -y^4 + c_1$. Then from (2),
$$U = x^3 + y \sin x - y^4 + c_1$$

Thus
$$M \, dx + N \, dy = dU = d(x^3 + y \sin x - y^4 + c_1) = 0$$

from which we must have
$$x^3 + y \sin x - y^4 = c$$

Method 3 [Formula].   From the result 2 on page 39 we see that the solution is

$$\int (3x^2 + y \cos x) \, \partial x \; + \; \int \left[(\sin x - 4y^3) - \frac{\partial}{\partial y} \int (3x^2 + y \cos x) \, \partial x \right] dy \;\; = \;\; c$$

i.e.

$$x^3 + y \sin x \; + \; \int \left[(\sin x - 4y^3) - \frac{\partial}{\partial y} (x^3 + y \sin x) \right] dy \;\; = \;\; c$$

or

$$x^3 + y \sin x - y^4 \;\; = \;\; c$$

INTEGRATING FACTORS

2.15.   (a) Show that the equation $(3xy^2 + 2y) \, dx + (2x^2y + x) \, dy = 0$ is not exact but that (b) it becomes exact after multiplication by x.   (c) Thus solve the equation.

(a) Here $M = 3xy^2 + 2y$, $N = 2x^2y + x$ and $\dfrac{\partial M}{\partial y} = 6xy + 2$, $\dfrac{\partial N}{\partial x} = 4xy + 1$ so that the equation is not exact.

(b) Multiplying by x, the equation becomes

$$(3x^2y^2 + 2xy) \, dx \; + \; (2x^3y + x^2) \, dy \;\; = \;\; 0$$

from which $M = 3x^2y^2 + 2xy$, $N = 2x^3y + x^2$. Then $\dfrac{\partial M}{\partial y} = 6x^2y + 2x = \dfrac{\partial N}{\partial x}$ and the equation is exact.

(c) By any of the methods of Problem 2.14 we find the required solution $x^2y + x^3y^2 = c$.

2.16.   Solve (a) $(y + x^4) \, dx - x \, dy = 0$,   (b) $(x^3 + xy^2 - y) \, dx + x \, dy = 0$.

(a) Write as $y \, dx - x \, dy + x^4 \, dx = 0$. The combination $y \, dx - x \, dy$ suggests integrating factors $1/x^2$, $1/y^2$, $1/(x^2 + y^2)$ [see page 40, entry 3]. Only the first leads to favorable results, i.e.

$$\frac{y \, dx - x \, dy}{x^2} + x^2 \, dx \;\; = \;\; 0 \quad\text{or}\quad -d\!\left(\frac{y}{x}\right) + d\!\left(\frac{x^3}{3}\right) \;\; = \;\; 0$$

so that

$$-\frac{y}{x} + \frac{x^3}{3} \;\; = \;\; c$$

(b) Write as $x(x^2 + y^2) \, dx + x \, dy - y \, dx = 0$ and multiply by the integrating factor $1/(x^2 + y^2)$ to obtain

$$x \, dx + \frac{x \, dy - y \, dx}{x^2 + y^2} \;\; = \;\; 0 \quad\text{or}\quad d\!\left(\frac{x^2}{2}\right) + d\!\left(\tan^{-1}\frac{y}{x}\right) \;\; = \;\; 0$$

Then

$$\frac{x^2}{2} + \tan^{-1}\frac{y}{x} \;\; = \;\; c$$

2.17.   (a) If $M \, dx + N \, dy = 0$ has an integrating factor μ which depends only on x, show that $\mu = e^{\int f(x) \, dx}$ where $f(x) = (M_y - N_x)/N$, $M_y = \partial M/\partial y$, $N_x = \partial N/\partial x$.   (b) Write the condition that μ depend only on y.   (c) Use the result to solve Problem 2.15.

(a) By hypothesis, $\mu M \, dx + \mu N \, dy = 0$ is exact.   Then

$$\frac{\partial(\mu M)}{\partial y} \;\; = \;\; \frac{\partial(\mu N)}{\partial x}$$

Since μ depends only on x this can be written

$$\mu \frac{\partial M}{\partial y} \;\; = \;\; \mu \frac{\partial N}{\partial x} + N \frac{d\mu}{dx} \quad\text{or}\quad N \frac{d\mu}{dx} \;\; = \;\; \mu \left(\frac{\partial M}{\partial y} - \frac{\partial N}{\partial x} \right)$$

Thus

$$\frac{d\mu}{\mu} \;\; = \;\; \left(\frac{M_y - N_x}{N} \right) dx \;\; = \;\; f(x) \, dx$$

and so

$$\ln \mu \;\; = \;\; \int f(x) \, dx, \qquad \mu \;\; = \;\; e^{\int f(x) \, dx}$$

(b) By interchanging M and N, x and y we see by part (a) that there will be an integrating factor μ depending only on y if $(N_x - M_y)/M = g(y)$ and that in this case $\mu = e^{\int g(y)\,dy}$

(c) In Problem 2.15, $M = 3xy^2 + 2y$, $N = 2x^2y + x$, $M_y = 6xy + 2$, $N_x = 4xy + 1$ and $\dfrac{M_y - N_x}{N} = \dfrac{2xy + 1}{2x^2y + x} = \dfrac{1}{x}$ depends only on x. Thus by part (a), $e^{\int (1/x)\,dx} = e^{\ln x} = x$ is an integrating factor and we can proceed as in Problem 2.15(b).

LINEAR EQUATIONS

2.18. Solve the equation $\dfrac{dy}{dx} + P(x)y = Q(x)$.

Write the equation as

$$[P(x)y - Q(x)]\,dx + dy = 0$$

Then $M = P(x)y - Q(x)$, $N = 1$, $M_y = \dfrac{\partial M}{\partial y} = P(x)$, $N_x = \dfrac{\partial N}{\partial x} = 0$

Since $\dfrac{M_y - N_x}{N} = \dfrac{P(x) - 0}{1} = P(x)$ depends only on x, we see from Problem 2.17 that $e^{\int P(x)\,dx}$ is an integrating factor. Multiplying by this factor, the equation becomes

$$e^{\int P\,dx}\left(\frac{dy}{dx} + Py\right) = Qe^{\int P\,dx}$$

which can be written as

$$\frac{d}{dx}(e^{\int P\,dx}\,y) = Qe^{\int P\,dx}$$

Then on integrating we have

$$e^{\int P\,dx}\,y = \int Qe^{\int P\,dx}\,dx + c$$

or $$y = e^{-\int P\,dx}\int Qe^{\int P\,dx}\,dx + ce^{-\int P\,dx}$$

2.19. Solve $x\dfrac{dy}{dx} - 2y = x^3\cos 4x$.

Write the equation as $\dfrac{dy}{dx} - \dfrac{2}{x}y = x^2\cos 4x$, a linear equation of the form $\dfrac{dy}{dx} + Py = Q$ with $P = -2/x$, $Q = x^2\cos 4x$. As in Problem 2.18, an integrating factor is $e^{\int (-2/x)\,dx} = e^{-2\ln x} = e^{\ln x^{-2}} = x^{-2}$. Multiplying by x^{-2} we have

$$x^{-2}\frac{dy}{dx} - 2x^{-3}y = \cos 4x$$

which can be written as

$$\frac{d}{dx}(x^{-2}y) = \cos 4x$$

Then by integrating we find $x^{-2}y = \tfrac{1}{4}\sin 4x + c$ or $y = \tfrac{1}{4}x^2\sin 4x + cx^2$.

HOMOGENEOUS EQUATIONS

2.20. Solve $(2x^3 + y^3)\,dx - 3xy^2\,dy = 0$

Write as

$$\frac{dy}{dx} = \frac{2x^3 + y^3}{3xy^2}$$

The right side is seen to be a function of y/x either by writing it as

$$\frac{2x^3}{3xy^2} + \frac{y^3}{3xy^2} = \frac{2}{3(y/x)^2} + \frac{1}{3}\left(\frac{y}{x}\right)$$

or by letting $y = vx$ and showing that the right side depends only on v, i.e.

$$\frac{2x^3 + y^3}{3xy^2} \;=\; \frac{2x^3 + v^3x^3}{3x \cdot v^2x^2} \;=\; \frac{(2 + v^3)x^3}{3v^2x^3} \;=\; \frac{2 + v^3}{3v^2}$$

Thus the equation is homogeneous. From $y = vx$ we obtain

$$v + x\frac{dv}{dx} \;=\; \frac{2 + v^3}{3v^2} \quad\text{or}\quad x\frac{dv}{dx} \;=\; \frac{2 - 2v^3}{3v^2}$$

Separating the variables, we have

$$\frac{3v^2\,dv}{v^3 - 1} \;=\; -2\frac{dx}{x}$$

Integrating, $$\ln(v^3 - 1) \;=\; -2\ln x + c_1$$

or $$\ln[(v^3 - 1)x^2] \;=\; c_1$$

Thus $(v^3 - 1)x^2 = e^{c_1} = c$ and the required solution on letting $v = y/x$ is $y^3 - x^3 = cx$.

Note that in the integration we assumed $v^3 - 1 > 0$. Similar reasoning analogous to that of Problem 2.12 shows that the solution is also valid if $v^3 - 1 < 0$.

2.21. Solve $\dfrac{dy}{dx} = \dfrac{5x + 4y}{2x - y}$.

The right side is a function of y/x. Thus letting $y = vx$,

$$x\frac{dv}{dx} + v \;=\; \frac{5 + 4v}{2 - v}$$

or $$\frac{v - 2}{v^2 + 2v + 5}\,dv + \frac{dx}{x} \;=\; 0$$

Integrating, $$\int \frac{v - 2}{v^2 + 2v + 5}\,dv + \int \frac{dx}{x} \;=\; c$$

Since $$\int \frac{v - 2}{v^2 + 2v + 5}\,dv \;=\; \int \frac{(v + 1)\,dv}{v^2 + 2v + 5} + \int \frac{-3\,dv}{(v + 1)^2 + 4}$$

$$=\; \frac{1}{2}\ln(v^2 + 2v + 5) - \frac{3}{2}\tan^{-1}\left(\frac{v + 1}{2}\right)$$

the required solution is

$$\ln(y^2 + 2xy + 5x^2) - 3\tan^{-1}\left(\frac{y + x}{2x}\right) \;=\; c$$

BERNOULLI'S EQUATION

2.22. Solve $x\dfrac{dy}{dx} + y = xy^3$.

The equation written as $\dfrac{dy}{dx} + \dfrac{y}{x} = y^3$ is a *Bernoulli equation* [see page 40] with $P(x) = 1/x$, $Q(x) = 1$, $n = 3$.

Making the transformation $y^{1-n} = v$, i.e. $y^{-2} = v$, we find on differentiating with respect to x,

$$-2y^{-3}\frac{dy}{dx} \;=\; \frac{dv}{dx} \quad\text{or}\quad \frac{dy}{dx} \;=\; -\frac{1}{2}y^3\frac{dv}{dx}$$

Thus the equation becomes

$$-\frac{1}{2}y^3\frac{dv}{dx} + \frac{y}{x} \;=\; y^3 \quad\text{or}\quad \frac{dv}{dx} - \frac{2}{xy^2} \;=\; -2$$

i.e. $$\frac{dv}{dx} - 2v \;=\; -2$$

Since the general solution of this is $v = 2x + cx^2$, the solution of the required equation is $y^2 = \dfrac{1}{2x + cx^2}$.

EQUATIONS SOLVABLE FOR ONE VARIABLE

2.23. Solve $xp^2 + 2px - y = 0$ where $p = y'$.

The equation can be solved explicitly for y, i.e. $y = xp^2 + 2px$. Differentiation with respect to x yields

$$\frac{dy}{dx} = p = 2px\frac{dp}{dx} + p^2 + 2p + 2x\frac{dp}{dx}$$

or

$$p(p+1) + 2x(p+1)\frac{dp}{dx} = 0 \qquad (1)$$

Case 1, $p + 1 \neq 0$. In this case (1) becomes on division by $(p+1)$,

$$p + 2x\frac{dp}{dx} = 0$$

whose solution is $xp^2 = c$. Then

$$x = \frac{c}{p^2}, \qquad y = c + \frac{2c}{p} \qquad (2)$$

are the parametric equations of the general solution. By eliminating p from (2) we can obtain the general solution in the form

$$(y-c)^2 = 4cx \qquad (3)$$

Case 2, $p + 1 = 0$. In this case $p = -1$, and substituting into $xp^2 + 2px - y = 0$ we find $x + y = 0$ which is a solution of the differential equation, as can be checked. However, it cannot be obtained from the general solution (3) by any choice of c. Thus $x + y = 0$ is a singular solution.

CLAIRAUT'S EQUATION

2.24. Solve $y = px \pm \sqrt{p^2 + 1}$ where $p = y'$.

Use the method of Problem 2.23. Differentiating,

$$y' = p = p + x\frac{dp}{dx} \pm \frac{p}{\sqrt{p^2+1}}\frac{dp}{dx}$$

from which

$$\frac{dp}{dx}\left(x \pm \frac{p}{\sqrt{p^2+1}}\right) = 0$$

Case 1, $dp/dx = 0$. In this case $p = c$ and so the general solution is

$$y = cx \pm \sqrt{c^2+1}$$

Case 2, $dp/dx \neq 0$. In this case $x \pm p/\sqrt{p^2+1} = 0$ or $x = \mp p/\sqrt{p^2+1}$ and

$$y = \mp\frac{p^2}{\sqrt{p^2+1}} \pm \sqrt{p^2+1} = \pm\frac{1}{\sqrt{p^2+1}}$$

To eliminate p, note that $x^2 + y^2 = \frac{p^2}{p^2+1} + \frac{1}{p^2+1} = 1$. The equation $x^2 + y^2 = 1$ satisfies the given differential equation but cannot be obtained from the general solution for any choice of c and so is a singular solution.

EQUATIONS OF ORDER HIGHER THAN ONE

2.25. Solve $y'' + 2y' = 4x$.

Since y is missing from the equation, let $y' = p$, $y'' = dp/dx$. Then the equation becomes

$$\frac{dp}{dx} + 2p = 4x \qquad (1)$$

a linear first order equation having integrating factor $e^{\int 2\,dx} = e^{2x}$. Multiplying by e^{2x}, (1) can be written as

$$\frac{d}{dx}(pe^{2x}) = 4xe^{2x}$$

so that $\qquad pe^{2x} \;=\; 4 \int xe^{2x}\,dx \;+\; c \;=\; 4\left[(x)\left(\dfrac{e^{2x}}{2}\right) - (1)\left(\dfrac{e^{2x}}{4}\right)\right] + c_1$

i.e. $\qquad\qquad\qquad\qquad p \;=\; dy/dx \;=\; 2x - 1 + c_1 e^{-2x}$

Integrating again, we find the required general solution

$$y \;=\; x^2 - x - \tfrac{1}{2}c_1 e^{-2x} + c_2 \;=\; x^2 - x + Ae^{-2x} + B$$

2.26. Solve $\;1 + yy'' + y'^2 = 0.$

Since x is missing from the equation, let $y' = p$, $y'' = \dfrac{dp}{dx} = \dfrac{dp}{dy}\dfrac{dy}{dx} = p\dfrac{dp}{dy}$. Then the given equation becomes

$$1 + py\frac{dp}{dy} + p^2 \;=\; 0$$

Separating variables and integrating,

$$\int \frac{p\,dp}{1+p^2} + \int \frac{dy}{y} \;=\; c_1$$

i.e. $\qquad\quad \tfrac{1}{2}\ln(1+p^2) + \ln y \;=\; c_1 \qquad$ or $\qquad \ln[(1+p^2)y^2] \;=\; c_2$

from which $(1+p^2)y^2 = a^2$ and

$$p \;=\; dy/dx \;=\; \pm\sqrt{a^2 - y^2}/y$$

Separating variables, $\qquad\qquad\qquad \pm\dfrac{y\,dy}{\sqrt{a^2 - y^2}} \;=\; dx$

Integrating, $\qquad\qquad\qquad\qquad \mp\sqrt{a^2 - y^2} \;=\; x + b$

Squaring gives $\qquad\qquad\qquad\quad (x+b)^2 + y^2 \;=\; a^2$

which is the general solution.

MECHANICS PROBLEMS

2.27. An object is thrown vertically upward from the ground with initial velocity 1960 cm/sec. Neglecting air resistance, find (a) the maximum height reached and (b) the total time taken to return to the starting point.

Let the object of mass m be located at distance x cm from the ground after time t sec [see Fig. 2-6]. Choose the upward direction as positive. By Newton's law,

$$\text{Net force} \;=\; \text{Weight}$$

$$m\frac{d^2x}{dt^2} \;=\; -mg \qquad \text{or} \qquad \frac{d^2x}{dt^2} = -g = -980 \qquad (1)$$

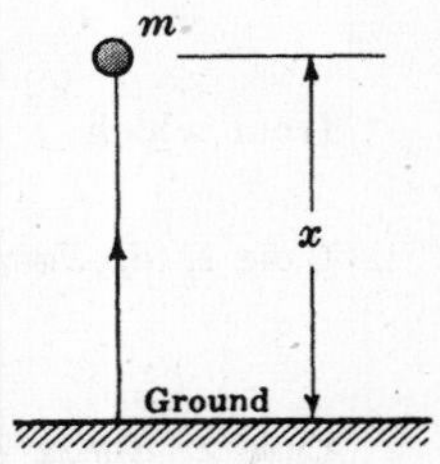

Fig. 2-6

The initial conditions are $x = 0$, $dx/dt = 1960$ at $t = 0$.

Solving (1) subject to the initial conditions, we find

$$x \;=\; 1960t - 490t^2 \qquad\qquad (2)$$

(a) The height is a maximum when $dx/dt = 1960 - 980t = 0$ or $t = 2$. Then $x = 1960(2) - 490(2)^2 = 1960$. Thus the maximum height reached is 1960 cm.

(b) $x = 0$ when $t(1960 - 490t) = 0$ i.e. $t = 0, 4$. Then the required time to return is 4 sec.

2.28. A 192 lb object falls from rest at time $t = 0$ in a medium offering a resistance in lb numerically equal to twice its instantaneous velocity in ft/sec. Find (a) the velocity and distance traveled at any time $t > 0$ and (b) the limiting velocity.

Choose the positive direction downward.

$$\text{Net force} \;=\; \text{Weight} - \text{Resistance}$$

$$\frac{192}{g}\frac{dv}{dt} \;=\; 192 \;-\; 2v \qquad \text{or} \qquad \frac{dv}{dt} = 32 - \frac{v}{3} \qquad (1)$$

(a) Solving (1) subject to the initial condition $v = 0$ at $t = 0$, we find for the velocity at any time,

$$v = 96(1 - e^{-t/3}) \qquad (2)$$

Replace v by dx/dt in (2) and use the initial condition $x = 0$ at $t = 0$ to obtain for the distance traveled,

$$x = 96(t + 3e^{-t/3} - 3)$$

(b) The limiting velocity is

$$\lim_{t \to \infty} 96(1 - e^{-t/3}) = 96\,(\text{ft/sec})$$

and can also be obtained by setting $dv/dt = 32 - v/3 = 0$.

2.29. Solve Problem 2.28 if the medium offers a resistance in lb numerically equal to $3v^2$.

$$\text{Net force} \quad = \quad \text{Weight} \quad - \quad \text{Resistance}$$

$$\frac{192}{g}\frac{dv}{dt} \quad = \quad 192 \quad - \quad 3v^2 \qquad \text{or} \qquad 2\frac{dv}{dt} = 64 - v^2 \qquad (1)$$

Then separating variables and integrating, we find

$$\int \frac{dv}{64 - v^2} = \int \frac{dt}{2} \qquad \text{or} \qquad \frac{1}{16}\ln\left(\frac{8 + v}{8 - v}\right) = \frac{t}{2} + c$$

(a) Since $v = 0$ at $t = 0$, $c = 0$. Then

$$\frac{1}{16}\ln\left(\frac{8 + v}{8 - v}\right) = \frac{t}{2} \qquad \text{or} \qquad v = \frac{dx}{dt} = 8\left(\frac{e^{4t} - e^{-4t}}{e^{4t} + e^{-4t}}\right) \qquad (2)$$

which is the velocity at any time.

Integrating the second equation in (2) subject to $x = 0$ at $t = 0$, we find for the distance traveled,

$$x = 2\ln\left(\frac{e^{4t} + e^{-4t}}{2}\right) = 2\ln\cosh 4t$$

(b) The limiting velocity is

$$\lim_{t \to \infty} 8\left(\frac{e^{4t} - e^{-4t}}{e^{4t} + e^{-4t}}\right) = \lim_{t \to \infty} 8\left(\frac{1 - e^{-8t}}{1 + e^{-8t}}\right) = 8 \text{ (ft/sec)}$$

and can also be obtained by setting $\dfrac{dv}{dt} = 32 - \dfrac{v^2}{2} = 0$.

2.30. A boat of mass m is traveling with velocity v_0. At $t = 0$ the power is shut off. Assuming water resistance proportional to v^n where n is a constant and v is the instantaneous velocity, find v as a function of distance traveled.

Let x = distance traveled after time $t > 0$.

$$\text{Net force} \quad = \quad \text{Forward thrust} \quad - \quad \text{Water resistance}$$

$$m\frac{dv}{dt} \quad = \quad 0 \quad - \quad kv^n$$

where k is the constant of proportionality. Then

$$m\frac{dv}{dt} = m\frac{dv}{dx}\cdot\frac{dx}{dt} = mv\frac{dv}{dx} = -kv^n \qquad \text{or} \qquad mv^{1-n}dv = -k\,dx$$

Case 1, $n \neq 2$. Integrating, using $v = v_0$ at $x = 0$ where v_0 is the velocity at $t = 0$,

$$v^{2-n} = v_0^{2-n} - \frac{k}{m}(2-n)x$$

Case 2, $n = 2$. Integrating, using $v = v_0$ at $x = 0$,

$$v = v_0 e^{-kx/m}$$

2.31. A uniform chain of length a is placed on a horizontal frictionless table so that a length b of the chain dangles over the side. How long will it take for the chain to slide off the table?

Suppose that at time t a length x of the chain is dangling over the side [Fig. 2-7]. Assume that the density (mass per unit length) of the chain is σ. Then

Fig. 2-7

$$\text{Net force} \quad = \quad \text{Mass being accelerated} \cdot \text{Acceleration}$$

$$(\sigma g)x \;=\; \sigma a \frac{dv}{dt} \tag{1}$$

Then since $\dfrac{dv}{dt} = \dfrac{dv}{dx} \cdot \dfrac{dx}{dt} = v\dfrac{dv}{dx}$, *(1)* becomes

$$v\frac{dv}{dx} = \frac{gx}{a} \tag{2}$$

Integrating *(2)*, using $x = b$ when $v = 0$,

$$v \;=\; \frac{dx}{dt} \;=\; \sqrt{\frac{g}{a}}\,\sqrt{x^2 - b^2} \tag{3}$$

Separating the variables in *(3)* and integrating again using $x = b$ when $t = 0$,

$$\ln\left(\frac{x + \sqrt{x^2 - b^2}}{b}\right) \;=\; \sqrt{\frac{g}{a}}\,t$$

Since the chain slides off when $x = a$, the time T taken is

$$T \;=\; \sqrt{\frac{a}{g}}\,\ln\left(\frac{a + \sqrt{a^2 - b^2}}{b}\right)$$

ELECTRIC CIRCUIT PROBLEMS

2.32. A resistor of $R = 10$ ohms, an inductor of $L = 2$ henries and a battery of E volts are connected in series with a switch S [Fig. 2-8]. At $t = 0$ the switch is closed and the current $I = 0$. Find I for $t > 0$ if (a) $E = 40$, (b) $E = 20e^{-3t}$, (c) $E = 50 \sin 5t$.

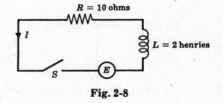

Fig. 2-8

By Kirchhoff's laws,

Potential drop across R + Potential drop across L + Potential drop across E = 0

$$10I \qquad\qquad + \qquad\qquad 2\frac{dI}{dt} \qquad\qquad + \qquad\qquad (-E) \qquad\qquad = \qquad 0$$

or

$$\frac{dI}{dt} + 5I = \frac{E}{2}$$

(a) If $E = 40$, $\dfrac{dI}{dt} + 5I = 20$. Solving this subject to $I = 0$ at $t = 0$, we have $I = 4(1 - e^{-5t})$.

(b) If $E = 20e^{-3t}$, $\dfrac{dI}{dt} + 5I = 10e^{-3t}$ or multiplying by the integrating factor e^{5t}, $\dfrac{d}{dt}(Ie^{5t}) = 10e^{2t}$ from which we obtain $I = 5(e^{-3t} - e^{-5t})$ on using $I = 0$ at $t = 0$.

(c) If $E = 50 \sin 5t$, $\dfrac{dI}{dt} + 5I = 25 \sin 5t$, $\dfrac{d}{dt}(Ie^{5t}) = 25e^{5t} \sin 5t$. Integrating,

$$Ie^{5t} \;=\; 25 \int e^{5t} \sin 5t\, dt \;=\; \frac{5e^{5t}}{2}(\sin 5t - \cos 5t) + c$$

Since $I = 0$ at $t = 0$, $c = \frac{5}{2}$ and so $I = \frac{5}{2}(\sin 5t - \cos 5t) + \frac{5}{2}e^{-5t}$.

The term $\frac{5}{2}e^{-5t}$, which approaches zero as t increases, is called the *transient* current. The remaining terms $\frac{5}{2}(\sin 5t - \cos 5t)$ comprise the *steady-state* current.

2.33. A resistor of $R = 5$ ohms and a condenser of $C = .02$ farads are connected in series with a battery of $E = 100$ volts [Fig. 2-9]. If at $t = 0$ the charge Q on the condenser is 5 coulombs, find Q and the current I for $t > 0$.

$$\text{Potential drop across } R = 5I = 5\frac{dQ}{dt}$$

$$\text{Potential drop across } C = \frac{Q}{.02} = 50Q$$

$$\text{Potential drop across } E = -E$$

Then by Kirchhoff's laws,

$$5\frac{dQ}{dt} + 50Q = E$$

If $E = 100$ volts, $\frac{dQ}{dt} + 10Q = 20$, $\frac{d}{dt}(e^{10t}Q) = 20e^{10t}$. Then integrating and solving subject to $Q = 5$ at $t = 0$, we find $Q = 2 + 3e^{-10t}$ and $I = dQ/dt = -30e^{-10t}$.

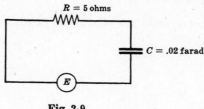

R = 5 ohms

C = .02 farad

E

Fig. 2-9

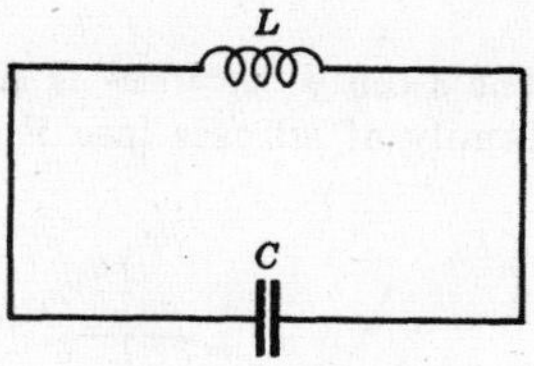

L

C

Fig. 2-10

2.34. An inductor of L henries and a condenser of C farads are connected in series [Fig. 2-10]. If $Q = Q_0$ and $I = 0$ at $t = 0$, find (a) Q and (b) I at $t > 0$.

$$\text{Potential drop across } L = L\frac{dI}{dt} = L\frac{d^2Q}{dt^2}$$

$$\text{Potential drop across } C = \frac{Q}{C}$$

Then

$$L\frac{d^2Q}{dt^2} + \frac{Q}{C} = 0 \tag{1}$$

Since $\frac{dQ}{dt} = I$, we have $\frac{d^2Q}{dt^2} = \frac{dI}{dt} = \frac{dI}{dQ}\frac{dQ}{dt} = I\frac{dI}{dQ}$ so that (1) becomes

$$LI\frac{dI}{dQ} + \frac{Q}{C} = 0 \quad \text{or} \quad LI\,dI + \frac{Q}{C}dQ = 0 \tag{2}$$

Integrating the second equation of (2) yields

$$\tfrac{1}{2}LI^2 + \frac{Q^2}{2C} = C_1 \tag{3}$$

Since $I = 0$ when $Q = Q_0$, we have $C_1 = Q_0^2/2C$. Thus (3) becomes on solving for I,

$$I = \frac{dQ}{dt} = \pm\frac{1}{\sqrt{LC}}\sqrt{Q_0^2 - Q^2} \tag{4}$$

Separating variables in (4) and integrating,

$$\int \frac{dQ}{\sqrt{Q_0^2 - Q^2}} = \pm\int\frac{dt}{\sqrt{LC}} \quad \text{or} \quad \sin^{-1}\frac{Q}{Q_0} = \pm\frac{t}{\sqrt{LC}} + C_2 \tag{5}$$

Since $Q = Q_0$ for $t = 0$, we find $C_2 = \pi/2$. Thus from (5),

$$\sin^{-1}\frac{Q}{Q_0} = \frac{\pi}{2} \pm \frac{t}{\sqrt{LC}} \quad \text{or} \quad Q = Q_0\cos\frac{t}{\sqrt{LC}}$$

and

$$I = \frac{dQ}{dt} = -\frac{Q_0}{\sqrt{LC}}\sin\frac{t}{\sqrt{LC}}$$

Note that the charge oscillates with *amplitude Q_0, period $2\pi\sqrt{LC}$* and *frequency* (1/period) $\frac{1}{2\pi\sqrt{LC}}$. The current oscillates with amplitude $Q_0/\sqrt{LC}$ and the same period and frequency as the charge.

GEOMETRY PROBLEMS

2.35. (a) Find the orthogonal trajectories of the family of curves $y = cx^2$ and (b) give a geometrical interpretation.

(a) The differential equation of the family is

$$\frac{dy}{dx} = 2cx = 2\left(\frac{y}{x^2}\right)x \quad \text{or} \quad \frac{dy}{dx} = \frac{2y}{x}$$

Since the slope of each member of the orthogonal family must be the negative reciprocal of this slope, we see that the slope of the orthogonal family is

$$\frac{dy}{dx} = -\frac{1}{2y/x} \quad \text{or} \quad \frac{dy}{dx} = -\frac{x}{2y}$$

Solving this we find that the equation of the orthogonal trajectories is

$$x^2 + 2y^2 = k$$

(b) The family $y = cx^2$ is a family of parabolas while the orthogonal family $x^2 + 2y^2 = k$ is a family of ellipses [see Fig. 2-11].

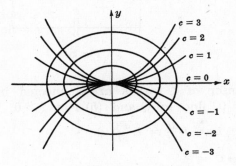

Fig. 2-11

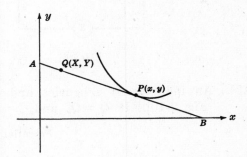

Fig. 2-12

2.36. Determine a curve such that the length of its tangent included between the x and y axes is a constant $a > 0$.

Let (x, y) be any point P on the required curve and (X, Y) any point Q on the tangent line AB [Fig. 2-12].

The equation of line AB passing through (x, y) with slope y' is

$$Y - y = y'(X - x)$$

Set $X = 0$, $Y = 0$ to obtain the y and x intercepts

$$\overline{OA} = y - xy', \quad \overline{OB} = x - y/y' = -(y - xy')/y'$$

Then the length of AB, apart from sign, is

$$\sqrt{\overline{OA}^2 + \overline{OB}^2} = (y - xy')\sqrt{1 + y'^2}/y'$$

Since this must equal $\pm a$, we have on solving for y,

$$y = xy' \pm \frac{ay'}{\sqrt{1 + y'^2}} = xp \pm \frac{ap}{\sqrt{1 + p^2}} \qquad (1)$$

where $y' = p$.

To solve (1) differentiate both sides with respect to x, so that

$$y' = p = x\frac{dp}{dx} + p \pm \frac{a}{(1 + p^2)^{3/2}}\frac{dp}{dx}$$

or

$$\frac{dp}{dx}\left[x \pm \frac{a}{(1 + p^2)^{3/2}}\right] = 0$$

Case 1, $dp/dx = 0$.   In this case $p = c$ and the general solution is

$$y = cx \pm \frac{ac}{\sqrt{1+c^2}} \qquad (2)$$

Case 2, $dp/dx \neq 0$.   In this case, using (1) we find

$$x = \mp \frac{a}{(1+p^2)^{3/2}}, \qquad y = \pm \frac{ap^3}{(1+p^2)^{3/2}}$$

Then $\qquad x^{2/3} = \dfrac{a^{2/3}}{1+p^2}, \qquad y^{2/3} = \dfrac{a^{2/3}p^2}{1+p^2}$

so that $\qquad\qquad x^{2/3} + y^{2/3} = a^{2/3} \qquad (3)$

which is a singular solution.

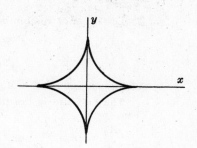

Fig. 2-13

The curve (3) which is a *hypocycloid* [Fig. 2-13] and which is the envelope of the family of lines (2), is the required curve.

FLOW PROBLEMS

2.37. A cylindrical tank has 40 gallons (gal) of a salt solution containing 2 lb dissolved salt per gallon, i.e. 2 lb salt/gal. A salt solution of concentration 3 lb salt/gal flows into the tank at 4 gal/min. How much salt is in the tank at any time if the well-stirred mixture flows out at 4 gal/min?

Let the tank contain A lb salt after t minutes.   Then

Rate of change of amount of salt $=$ Rate of entrance $-$ Rate of exit

$$\frac{dA}{dt}\frac{\text{lb}}{\text{min}} \qquad = \qquad 3\,\frac{\text{lb}}{\text{gal}} \cdot 4\,\frac{\text{gal}}{\text{min}} \quad - \quad \frac{A}{40}\frac{\text{lb}}{\text{gal}} \cdot 4\,\frac{\text{gal}}{\text{min}}$$

Solving the equation $\dfrac{dA}{dt} = 12 - \dfrac{A}{10}$ subject to $A = 40\text{ gal} \cdot 2\,\dfrac{\text{lb salt}}{\text{gal}} = 80$ lb salt at $t = 0$, we find $A = 120 - 40e^{-t/10}$.

2.38. A right circular cone [Fig. 2-14] is filled with water. In what time will the water empty through an orifice O of cross-sectional area a at the vertex? Assume velocity of exit is $v = \kappa\sqrt{2gh}$ where h is the instantaneous height ("head") of the water level above O and κ is the *discharge coefficient*.

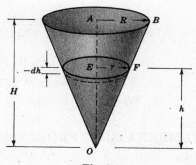

Fig. 2-14

At time t the water level is at h. At time $t + dt$, $dt > 0$, the water level is at $h + dh$ where $dh < 0$. We have

Change in volume of water $=$ Amount of water leaving

$$-\pi r^2\, dh \qquad = \qquad av\, dt \quad = \quad a\kappa\sqrt{2gh}\, dt$$

From similar triangles OAB and OEF, $r = Rh/H$.   Then

$$-\frac{\pi R^2 h^2}{H^2}\, dh = a\kappa\sqrt{2gh}\, dt$$

Solving, subject to $h = H$ at $t = 0$, we have

$$t = \frac{2\pi R^2}{5a\kappa H^2\sqrt{2g}}\,(H^{5/2} - h^{5/2})$$

Then time for emptying, when $h = 0$, is $T = \dfrac{\pi R^2}{5a\kappa}\sqrt{\dfrac{2H}{g}}$.

CHEMISTRY PROBLEMS

2.39. Radium decays at a rate proportional to the instantaneous amount present at any time. If the half life of radium is T years, determine the amount present after t years.

Let A gm of radium be present after t years. Then

$$\text{Time rate of change of } A \quad \propto \quad A$$

$$\frac{dA}{dt} \quad \propto \quad A \quad \text{or} \quad \frac{dA}{dt} = -kA$$

Solving subject to $A = A_0$ at $t = 0$, we find $A = A_0 e^{-kt}$.

The half life is the time T when the amount present is half the original amount, i.e. $A_0/2$. Then

$$\frac{A_0}{2} = A_0 e^{-kT} \quad \text{or} \quad e^{-kT} = \tfrac{1}{2}, \quad e^{-k} = (\tfrac{1}{2})^{1/T} \quad \text{and} \quad A = A_0(e^{-k})^t = A_0(\tfrac{1}{2})^{t/T}$$

Another method. $e^{-kT} = \tfrac{1}{2}$, $k = \dfrac{\ln 2}{T}$ and $A = A_0 e^{-t \ln 2/T}$.

2.40. Chemical A dissolves in solution at a rate proportional to the instantaneous amount of undissolved chemical and to the difference in concentration between the actual solution C_a and saturated solution C_s. A porous inert solid containing 10 lb of A is agitated with 100 gallons water and after an hour 4 lb of A is dissolved. If a saturated solution contains .2 lb of A per gallon, find (a) the amount of A which is undissolved after 2 hr and (b) the time to dissolve 80% of A.

Let x lb of A be undissolved after t hours. We have

$$\frac{dx}{dt} \propto x(C_s - C_a) \quad \text{or} \quad \frac{dx}{dt} = kx\left(.2 - \frac{10 - x}{100}\right)$$

Then

$$\frac{dx}{dt} = \frac{kx(x + 10)}{100} = \kappa x(x + 10)$$

Separating the variables and integrating,

$$\int \frac{dx}{x(x + 10)} = \frac{1}{10} \int \left(\frac{1}{x} - \frac{1}{x + 10}\right) dx = \kappa t + C_1$$

i.e.

$$\frac{1}{10} \ln \frac{x}{x + 10} = \kappa t + C_1$$

Using the conditions $t = 0$, $x = 10$ and $t = 1$, $x = 6$, we find

$$x = \frac{5(3/4)^t}{1 - \tfrac{1}{2}(3/4)^t}$$

(a) When $t = 2$ hr, $x = 3.91$ lb of A undissolved.

(b) When $x = 2$ lb, $(\tfrac{3}{4})^t = \tfrac{1}{3}$ and $t = 3.82$ hr.

TEMPERATURE PROBLEM

2.41. Newton's law of cooling states that the time rate of change in temperature of an object varies as the difference in temperature between object and surroundings. If an object cools from 80°C to 60°C in 20 minutes, find the temperature in 40 minutes if the surrounding temperature is 20°C.

Let $U = $ temperature of object after t minutes. Then

$$\frac{dU}{dt} \propto U - 20 \quad \text{or} \quad \frac{dU}{dt} = k(U - 20)$$

Solving, $U = 20 + ce^{kt}$. At $t = 0$, $U = 80$ so that $c = 60$ and $U = 20 + 60e^{kt}$. At $t = 20$, $U = 60$ so that $e^{20k} = 2/3$, $e^k = (2/3)^{1/20}$. Then

$$U = 20 + 60e^{kt} = 20 + 60(e^k)^t = 20 + 60(\tfrac{2}{3})^{t/20}$$

When $t = 40$, $U = 20 + 60(2/3)^2 = 46.7$°C.

BENDING OF BEAMS

2.42. A beam of length L is simply supported at both ends [Fig. 2-15]. (a) Find the deflection if the beam has constant weight W per unit length and (b) determine the maximum deflection.

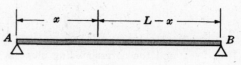

Fig. 2-15

(a) The total weight of the beam is WL, so each end supports weight $\frac{1}{2}WL$. Let x be the distance from the left end A of the beam. To find the bending moment M at x, consider forces to the left of x.

   (1) Force $\frac{1}{2}WL$ at A has moment $-(\frac{1}{2}WL)x$.

   (2) Force due to weight of beam to left of x has magnitude Wx and moment
$$Wx(x/2) = \tfrac{1}{2}Wx^2$$

Then the total bending moment at x is $\frac{1}{2}Wx^2 - \frac{1}{2}WLx$. Thus
$$EIy'' = \tfrac{1}{2}Wx^2 - \tfrac{1}{2}WLx$$

Solving this subject to $y' = 0$ at $x = L/2$ [from symmetry] and $y = 0$ at $x = 0$, we find
$$y = \frac{W}{24EI}(x^4 - 2Lx^3 + L^3x)$$

This could also be obtained using conditions $y = 0$ at $x = 0$ and $x = L$.

(b) The maximum deflection occurs at $x = L/2$ and is $5WL^4/384EI$.

Note that if forces to the right of x had been considered, the bending moment would be
$$-\tfrac{1}{2}WL(L-x) + W(L-x)\left(\frac{L-x}{2}\right) = \tfrac{1}{2}Wx^2 - \tfrac{1}{2}WLx \quad \text{as above.}$$

2.43. A cantilever beam [Fig. 2-16] has one end horizontally imbedded in concrete and a force W acting on the other end. Find (a) the deflection and (b) maximum deflection of the beam assuming its weight to be negligible.

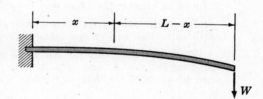

(a) Considering the portion of the beam to the right of x, the bending moment at x is $W(L-x)$. Then
$$EIy'' = W(L-x)$$

Fig. 2-16

Solving this subject to $y' = 0$ at $x = 0$ and $y = 0$ at $x = 0$, we find
$$y = \frac{W}{6EI}(3Lx^2 - x^3)$$

(b) The maximum deflection occurring at $x = L$ is $WL^3/3EI$.

NUMERICAL METHODS

2.44. If $dy/dx = 2x + y$, $y(0) = 1$, (a) find the approximate value of $y(.5)$ using the step by step or Euler method with $h = .1$ and (b) compare with the exact solution.

(a) To obtain the entries in the first line of the table below, use the fact that $x_0 = 0$, $y_0 = 1.0000$ so that $y' = 2x_0 + y_0 = 1.0000$.

x	y	$y' = 2x + y$
.0	1.0000	1.0000
.1	1.1000	1.3000
.2	1.2300	1.6300
.3	1.3930	1.9930
.4	1.5923	2.3923
.5	1.8315	

To obtain the entries in the second line use $h = .1$ and equation (10), page 43, to obtain

$$y(x_0 + h) = y(x_0) + hf(x_0, y_0)$$
$$= \text{value of } y \text{ in first line} + (.1)(\text{slope in first line})$$
$$= 1.0000 + (.1)(1.0000) = 1.1000$$

and the corresponding slope is

$$y' = 2(.1) + 1.1000 = 1.3000$$

Similarly we have

$$\text{value of } y \text{ in third line} = \text{value of } y \text{ in second line} + (.1)(\text{slope in second line})$$
$$= 1.1000 + .1(1.3000)$$
$$= 1.2300$$

and   slope in third line $= 2(.2) + 1.2300 = 1.6300$

The remainder of the entries are obtained by continuing in this manner and we find finally $y(.5) = 1.8315$.

(b) The equation is linear and has integrating factor e^{-x}. Solving, we find $y = 3e^x - 2x - 2$. Then when $x = .5$, $y = 3e^{.5} - 3 = 3(1.6487) - 3 = 1.9461$. Better accuracy can be obtained by using smaller values of h or proceeding as in Problem 2.45 which is a modification of the method.

2.45. Show how to improve the accuracy of the method of Problem 2.44.

The method which we shall use is essentially the same as that of Problem 2.44 except that we obtain improved values of y and y', which we denote by $y_1, y_1'; y_2, y_2'$, etc., as indicated in the table below. Success is achieved when there is very little difference between the improved values of y [see columns headed y_3 and y_4 respectively in the table].

The first line in the table below is the same as the first line in the table of Problem 2.44 and is found in exactly the same way.

Similarly the entries corresponding to y_1 and y_1' in line two of the table below are the same as those in line two of the table of Problem 2.44.

We now refer only to the table given below. As seen from this table, the slopes corresponding to $x = .0$ and $x = .1$ are 1.0000 and 1.3000 respectively. Then the average [mean] slope is $\frac{1}{2}(1.0000 + 1.3000) = 1.1500$. Using this modified slope, we find for the corresponding value of y

$$y_2 = 1.0000 + .1(1.1500) = 1.1150 \tag{1}$$

The slope corresponding to this value of y is given by

$$y_2' = 2(.1) + 1.1150 = 1.3150 \tag{2}$$

using the given differential equation.

From the improved slope (2) we now obtain an improved average slope

$$\tfrac{1}{2}(1.0000 + 1.3150) = 1.1575 \tag{3}$$

and from this an improved value of y given by

$$y_3 = 1.0000 + (.1)(1.1575) = 1.1158 \tag{4}$$

which in turn gives an improved slope equal to

$$y_3' = 2(.1) + 1.1158 = 1.3158 \tag{5}$$

The improved average slope using (5) is then

$$\tfrac{1}{2}(1.0000 + 1.3158) = 1.1579$$

and so $y_4 = 1.0000 + (.1)(1.1579) = 1.1158 \tag{6}$

Since this agrees with (4), the process ends.

We now use $y_4 = 1.1158$ and $y_3' = 1.3158$ to obtain the entries in the third line. Thus

$$y_1 \text{ (line 3)} = 1.1158 + (.1)(1.3158) = 1.2474$$

and $y_1' \text{ (line 3)} = 2(.2) + 1.2474 = 1.6474$

Continuing in this manner we finally obtain $y(.5) = 1.9483$, the last entry of the last line in the table. This agrees closely with the true value 1.9461.

x	y_1	y_1'	y_2	y_2'	y_3	y_3'	y_4
.0	1.0000	1.0000					
.1	1.1000	1.3000	1.1150	1.3150	1.1158	1.3158	1.1158
.2	1.2474	1.6474	1.2640	1.6640	1.2648	1.6648	1.2648
.3	1.4313	2.0313	1.4496	2.0496	1.4505	2.0505	1.4506
.4	1.6557	2.4557	1.6759	2.4759	1.6769	2.4769	1.6770
.5	1.9247	2.9247	1.9471	2.9471	1.9482	2.9482	1.9483

2.46. Solve Problem 2.44 by the Taylor series method.

By successively differentiating the given differential equation, we have

$$y' = 2x + y, \quad y'' = 2 + y', \quad y''' = y'', \quad y^{(iv)} = y''', \quad y^{(v)} = y^{(iv)}, \quad \ldots$$

Thus
$$y(0) = 1, \quad y'(0) = 1, \quad y'' = 3, \quad y''' = 3, \quad y^{(iv)} = 3, \quad y^{(v)} = 3, \quad \ldots$$

and so
$$y(x) = y(0) + y'(0)x + \frac{y''(0)}{2!}x^2 + \frac{y'''(0)}{3!}x^3 + \frac{y^{(iv)}(0)}{4!}x^4 + \frac{y^{(v)}(0)}{5!}x^5 + \cdots$$

$$= 1 + x + \frac{3x^2}{2} + \frac{3x^3}{3!} + \frac{3x^4}{4!} + \frac{3x^5}{5!} + \frac{3x^6}{6!} + \cdots$$

$$= 1 + x + \frac{3x^2}{2} + \frac{x^3}{2} + \frac{x^4}{8} + \frac{x^5}{40} + \cdots$$

Then when $x = .5$,

$$y(.5) = 1 + .5 + \frac{3(.5)^2}{2} + \frac{(.5)^3}{2} + \frac{(.5)^4}{8} + \frac{(.5)^5}{40} + \cdots$$

$$= 1 + .5 + .375 + .0625 + .0078 + .0008 + \cdots = 1.9461$$

2.47. Solve Problem 2.44 by Picard's method.

We have on integrating the differential equation using the boundary condition,

$$y(x) = 1 + \int_0^x (2u + y)\, du$$

Letting $y_1 = 1$ as a first approximation we find the second approximation

$$y_2(x) = 1 + \int_0^x (2u + 1)\, du = 1 + x + x^2$$

Then a third approximation is

$$y_3(x) = 1 + \int_0^x (2u + 1 + u + u^2)\, du = 1 + x + \frac{3x^2}{2} + \frac{x^3}{3}$$

Proceeding in this manner we find the successive approximations

$$y_4(x) = 1 + \int_0^x \left(2u + 1 + u + \frac{3u^2}{2} + \frac{u^3}{3}\right) du$$

$$= 1 + x + \frac{3x^2}{2} + \frac{x^3}{2} + \frac{x^4}{12}$$

$$y_5(x) = 1 + \int_0^x \left(2u + 1 + u + \frac{3u^2}{2} + \frac{u^3}{2} + \frac{u^4}{12}\right) du$$

$$= 1 + x + \frac{3x^2}{2} + \frac{x^3}{2} + \frac{x^4}{8} + \frac{x^5}{60}$$

$$y_6(x) = 1 + \int_0^x \left(2u + 1 + u + \frac{3u^2}{2} + \frac{u^3}{2} + \frac{u^4}{8} + \frac{u^5}{60} \right) du$$

$$= 1 + x + \frac{3x^2}{2} + \frac{x^3}{2} + \frac{x^4}{8} + \frac{x^5}{40} + \frac{x^6}{360}$$

Putting $x = .5$ in this last approximation, we find

$$y_6(.5) = 1 + .5 + .375 + .0625 + .0078 + .0008 = 1.9461$$

2.48. Solve Problem 2.44 by the Runge-Kutta method.

Using $h = .5$, $x_0 = 0$, $y_0 = 1$, $f(x,y) = 2x + y$ in the equations on page 44,

$$k_1 = .5[2(0) + 1] = .5$$
$$k_2 = .5[2(.25) + 1.25] = .875$$
$$k_3 = .5[2(.25) + 1.4375] = .96875$$
$$k_4 = .5[2(.5) + 1.9675] = 1.48375$$

Then $y(.5) = 1 + \frac{1}{6}(.5 + 1.750 + 1.9375 + 1.48375) = 1.9452$

Better approximations can be obtained by using two or more applications of the method with smaller values of h.

2.49. (a) Show how to solve numerically the system of equations

$$dy/dx = f(x, y, v), \quad dv/dx = g(x, y, v); \quad y(x_0) = y_0, \quad v(x_0) = v_0$$

by the Runge-Kutta method.

(b) Use (a) to show how to solve numerically

$$d^2y/dx^2 = g(x, y, dy/dx); \quad y(x_0) = y_0, \quad y'(x_0) = y_0$$

(a) Using the results on page 44, we write

$$k_1 = hf(x_0, y_0, v_0) \qquad\qquad l_1 = hg(x_0, y_0, v_0)$$
$$k_2 = hf(x_0 + \tfrac{1}{2}h, y_0 + \tfrac{1}{2}k_1, v_0 + \tfrac{1}{2}l_1) \qquad l_2 = hg(x_0 + \tfrac{1}{2}h, y_0 + \tfrac{1}{2}k, v_0 + \tfrac{1}{2}l_1)$$
$$k_3 = hf(x_0 + \tfrac{1}{2}h, y_0 + \tfrac{1}{2}k_2, v_0 + \tfrac{1}{2}l_2) \qquad l_3 = hg(x_0 + \tfrac{1}{2}h, y_0 + \tfrac{1}{2}k_2, v_0 + \tfrac{1}{2}l_2)$$
$$k_4 = hf(x_0 + h, y_0 + k_3, v_0 + l_3) \qquad\quad l_4 = hg(x_0 + h, y_0 + k_3, v_0 + l_3)$$

so that

$$y(x_0 + h) = y_0 + \tfrac{1}{6}(k_1 + 2k_2 + 2k_3 + k_4), \qquad v(x_0 + h) = v_0 + \tfrac{1}{6}(l_1 + 2l_2 + 2l_3 + l_4)$$

(b) The equation $d^2y/dx^2 = g(x, y, dy/dx)$ is equivalent to the system

$$dy/dx = v, \quad dv/dx = g(x, y, v)$$

and comparing with the equations of (a) we have $f(x, y, v) = v$. Then the expression for $y(x_0 + h)$ can be written in terms of l_1, l_2, l_3 as

$$y(x_0 + h) = y_0 + \tfrac{1}{6}[h_0 v_0 + 2h(v_0 + \tfrac{1}{2}l_1) + 2h(v_0 + \tfrac{1}{2}l_2) + h(v_0 + l_3)]$$
$$= y_0 + h_0 v_0 + \tfrac{1}{6}h(l_1 + l_2 + l_3)$$

2.50. If $\dfrac{d^2y}{dx^2} - 3\dfrac{dy}{dx} + 2y = x$; $y(0) = 1$, $y'(0)$ use Problem 2.49 to find an approximate value of $y(.5)$.

The differential equation is equivalent to the system

$$dy/dx = v, \quad dv/dx = x - 2y + 3v$$

and with the notation of Problem 2.49(b) we have $g(x, y, v) = x - 2y + 3v$, $x_0 = 0$, $y_0 = 1$, $v_0 = 0$. Thus if $h = .5$ we have from Problem 2.49(a),

$$l_1 = .5(-2) = -1, \quad l_2 = .5(-3.25) = -1.625, \quad l_3 = .5(-3.9375) = -1.96875$$

Then $\qquad\qquad\qquad\qquad y(.5) \;=\; 1 + \dfrac{.5}{6}(-1 - 1.625 - 1.96875) \;=\; .6172$

The equation can be solved exactly and the solution is $y = e^x - \frac{3}{4}e^{2x} + \frac{1}{2}x + \frac{3}{4}$ so that
$y(.5) = e^{.5} - \frac{3}{4}e + 1 = 1.6487 - \frac{3}{4}(2.7183) + 1 = .6100$.

2.51. Work Problem 2.50 by the Taylor series method.

We have $y' = v,\;\; v' = x - 2y + 3v$ so that by successive differentiation,

$$y'' = v', \quad v'' = 1 - 2y' + 3v', \quad y''' = v'', \quad v''' = -2y'' + 3v'', \quad \ldots$$

and we find corresponding to $x = 0$,

$$y' = 0, \;\; v' = -2, \;\; y'' = -2, \;\; v'' = -5, \;\; y''' = -5, \;\; v''' = -11, \;\; y^{(iv)} = -11, \;\; v^{(iv)} = -23$$

$$y^{(v)} = -23, \;\; v^{(v)} = -47, \;\; y^{(vi)} = -47, \;\; v^{(vi)} = -95, \;\; y^{(vii)} = -95, \;\; \ldots$$

Then $\qquad\quad y(x) \;=\; y(0) + y'(0)x + \dfrac{y''(0)x^2}{2!} + \dfrac{y'''(0)x^3}{3!} + \cdots$

$$= 1 - x^2 - \frac{5x^3}{6} - \frac{11x^4}{24} - \frac{23x^5}{120} - \frac{47x^6}{720} - \frac{95x^7}{5040} - \cdots$$

Thus we have approximately

$$y(.5) \;=\; 1 - .25 - .104167 - .028646 - .005990 - .001020 - .000147$$

$$= \;\; .61003$$

Supplementary Problems

CLASSIFICATION OF DIFFERENTIAL EQUATIONS

2.52. Classify each of the following differential equations by stating the order, the dependent and independent variables and whether the equation is ordinary or partial.

(a) $\quad \dfrac{d^3U}{dt^3} - 3\dfrac{d^2U}{dt^2} + 3\dfrac{dU}{dt} - U \;=\; e^t$ $\qquad$ (d) $\quad \dfrac{d^3s}{dt^3} \;=\; \sqrt[3]{s - 4t}$

(b) $\quad \dfrac{\partial^2 T}{\partial x^2} + \dfrac{\partial^2 T}{\partial y^2} + \dfrac{\partial^2 T}{\partial z^2} \;=\; 0$ $\qquad$ (e) $\quad y'' \;=\; [1 + y'^2]^{3/2}$

(c) $\quad xyy' \;=\; (y'')^3$ $\qquad\qquad\qquad\qquad$ (f) $\quad d(uv) \;=\; v^2\,dv$

SOLUTIONS OF DIFFERENTIAL EQUATIONS

2.53. Check whether each differential equation has the indicated solution. Determine which solutions are general solutions.

(a) $\quad y'' - 2y' + y = x; \quad y = (c_1 + c_2 x)e^x + x + 2$

(b) $\quad tI'(t) + I(t) = t^2; \quad t^3 - 3tI = c$

(c) $\quad y = xy' + \sqrt[3]{y'}; \quad 27x^4 + 256y = 0$

(d) $\quad \dfrac{d^4V}{dt^4} + 8\dfrac{d^2V}{dt^2} + 16V = 0; \quad V = c_1 \sin 2t + c_2 \cos 2t + c_3 t \sin 2t + c_4 t \cos 2t$

2.54. (a) Show that $y = e^{-x}(c_1 \cos x + c_2 \sin x)$ is a general solution of $y'' + 2y' + 2y = 0$. (b) Determine the particular solution such that $y(0) = -2$, $y'(0) = 5$.

2.55. (a) Show that $y = cx + 3c^2 + c$ and $(x+1)^2 + 12y = 0$ are solutions of the differential equation $y = xy' + 3y'^2 + y'$. (b) What is the name given to each of these solutions? (c) Explain the relationship between these solutions and illustrate graphically.

DIFFERENTIAL EQUATION OF A FAMILY OF CURVES

2.56. Find differential equations for the following families of curves: (a) $x^2 + cy^2 = 1$, (b) $y^2 = ax + b$.

2.57. Find the differential equation for (a) the family of straight lines which intersect at the point $(2, 1)$ and (b) the family of circles tangent to the x axis and having unit radius.

MISCELLANEOUS TYPES OF DIFFERENTIAL EQUATIONS

2.58. (a) Find the general solution of $dy/dx = 8x^3 - 4x + 1$. (b) Determine the particular solution such that $y(1) = 3$.

2.59. Solve each of the following boundary-value problems.

(a) $y'' = 28\sqrt[3]{x}$; $y(1) = 0$, $y'(0) = -2$

(b) $x^3 y''' = 1 + x^4$; $y(1) = y'(1) = y''(1) = 0$

(c) $d^2s/dt^2 + 12t = 16 \sin t$; $s = 2$, $ds/dt = -4$ at $t = 0$

2.60. Solve (a) $\dfrac{dy}{dx} = -2xy$, (b) $xy' + 3y = 0$; $y(1) = 2$, (c) $\dfrac{dI}{dt} + 2I = 10$; $I(0) = 0$, (d) $\dfrac{dy}{dx} = \dfrac{x\sqrt{1-y^2}}{y\sqrt{1-x^2}}$,

(e) $(1 - x^2)y' = 4y$; $y(0) = 1$.

2.61. Solve

(a) $(x + 2y)\,dx + (2x - 5y)\,dy = 0$ (c) $(ye^x - e^{-y})\,dx + (xe^{-y} + e^x)\,dy = 0$

(b) $\dfrac{dy}{dx} = \dfrac{3 - 4xy^2}{4x^2y + 6y^2}$; $y(1) = -1$ (d) $\dfrac{dy}{dx} = \dfrac{4 - 2x \cos y - 2y^3 \sec^2 2x}{3y^2 \tan 2x - x^2 \sin y}$

2.62. Each of the following differential equations has an integrating factor depending on only one variable. Find the integrating factor and solve the equation.

(a) $(4y - x^2)\,dx + x\,dy = 0$

(b) $(2xy^2 - y)\,dx + (2x - x^2y)\,dy = 0$

(c) $2\,dx + (2x - 3y - 3)\,dy = 0$; $y(2) = 0$

(d) $(2y \sin x + 3y^4 \sin x \cos x)\,dx - (4y^3 \cos^2 x + \cos x)\,dy = 0$

2.63. Solve each of the following differential equations given that each has an integrating factor of the form $x^p y^q$:

(a) $(3y - 2xy^3)\,dx + (4x - 3x^2y^2)\,dy = 0$

(b) $(2xy^3 + 2y)\,dx + (x^2y^2 + 2x)\,dy = 0$

2.64. Solve (a) $(2x^2 + 2y^2 - y)\,dx + (x^2y + y^3 + x)\,dy = 0$

(b) $(x^2 + x - y^2)\,dx - y\,dy = 0$

2.65. Solve (a) $\dfrac{dy}{dx} + \dfrac{y}{x} = 4x^2$; $y(1) = 2$, (b) $xy' - 4y = x$, (c) $\dfrac{dy}{dx} = 3(y + 2x) + 1$; $y(0) = 0$,

(d) $\dfrac{dy}{dx} + 2y \cot x = \csc x$.

2.66. Solve (a) $\dfrac{dy}{dx} = \dfrac{2y}{x} - \dfrac{y^2}{x^2}$, (b) $x^2 \dfrac{dy}{dx} = x^2 + 3xy + y^2$, (c) $\dfrac{dy}{dx} = \dfrac{2x - 3y}{4y + 3x}$, (d) $(x - y)y' + 3y - 5x = 0$.

2.67. Solve (a) $x\dfrac{dy}{dx} + y = x^3y^2$, (b) $2x^2y' = xy + y^3$.

2.68. Solve (a) $\dfrac{dy}{dx} = \dfrac{x + 2y - 4}{y - 2x + 3}$, (b) $\dfrac{dy}{dx} = \dfrac{x - y + 1}{x - y - 1}$.

2.69. Solve $dy/dx = x^2 + 2xy + y^2 + 2x + 2y$, $y(0) = 0$.

2.70. Solve (a) $y'^2 + (y - 1)y' - y = 0$, (b) $(xy' + y)^2 = x^{-3}$.

2.71. Solve each of the following, determining any singular solutions. In each case $p = y'$.

(a) $y = px + 2p^2$,   (b) $xp^2 = 2y(p+2)$,   (c) $(xp - y)^2 = p^2 - 1$,   (d) $x^2 y = x^3 p - yp^2$.

2.72. Solve   (a) $xy'' - 3y' = x^2$,   (b) $yy'' + 2y'^2 = 0$,   (c) $y'' + 4y = 0$,   (d) $(1 + y'^2)^3 = (y'')^2$.

EXISTENCE AND UNIQUENESS OF SOLUTIONS

2.73. Use the theorem on page 41 to discuss the existence and uniqueness of solutions for each of the following: (a) $y' = 2xy$; $y(0) = 1$,   (b) $y = xy' - y'^2$; $y(2) = 1$,   (c) $y' = (y + x)/(y - x)$; $y(1) = 1$.

MISCELLANEOUS APPLICATIONS

2.74. An object moves along the x axis, acted upon by a constant force. If its initial velocity in the positive direction is 40 meters/sec while 5 seconds later it is 20 meters/sec, find (a) the velocity at any time, (b) the position at any time assuming the object starts from the origin $x = 0$.

2.75. A 64 lb object falls from rest. The limiting velocity is 4 ft/sec. Find the velocity after t seconds assuming a force of resistance proportional to (a) v,   (b) v^2,   (c) $\sqrt{v}$.

2.76. A particle is located at $x = a$ $(a > 0)$ at $t = 0$. It moves toward $x = 0$ in such a way that its velocity is always proportional to x^n where n is a constant. Show that the particle will reach $x = 0$ if and only if $n < 1$.

2.77. An electric circuit contains an 8 ohm resistor in series with an inductor of .5 henries and a battery of E volts. At $t = 0$ the current is zero. Find the current at any time $t > 0$ and the maximum current if (a) $E = 64$,   (b) $E = 8te^{-16t}$,   (c) $E = 32e^{-8t}$.

2.78. (a) Solve for the current in the circuit of Problem 2.77 if $E = 64 \sin 8t$. (b) What is the transient current and steady-state current?

2.79. An electric circuit contains a 20 ohm resistor in series with a capacitor of .05 farads and a battery of E volts. At $t = 0$ there is no charge on the capacitor. Find the charge and current at any time $t > 0$ if (a) $E = 60$,   (b) $E = 100te^{-2t}$.

2.80. Find the charge and current in the circuit of Problem 2.79 if $E = 100 \cos 2t$.

2.81. Find the orthogonal trajectories of the family of curves

$$\text{(a)} \quad xy = c, \quad \text{(b)} \quad x^2 + y^2 = cx, \quad \text{(c)} \quad y^2 = cx^2 - 2y$$

2.82. Find the equation of that curve passing through $(0, 1)$ which is orthogonal to each member of the family $x^2 + y^2 = ce^x$.

2.83. A curve passing through $(1, 2)$ has the property that the length of the perpendicular drawn from the origin to the normal at any point of the curve is always equal numerically to the ordinate of the point. Find its equation.

2.84. The tangent to any point of a certain curve forms with the coordinate axes a triangle having constant area A. Find the equation of the curve.

2.85. A tank contains 100 gallons of water. A salt solution containing 2 lb of salt per gallon flows in at the rate of 3 gallons per minute and the well-stirred mixture flows out at the same rate. (a) How much salt is in the tank at any time? (b) When will the tank have 100 lb of salt?

2.86. In Problem 2.85, how much salt is in the tank at any time if the mixture flows out at (a) 2 gal/min, (b) 4 gal/min.

2.87. A right circular cylinder of radius 8 ft and height 16 ft whose axis is vertical is filled with water. How long will it take for all the water to escape through a 2 in^2 orifice at the bottom of the tank assuming the velocity of escape v in terms of the instantaneous height h is given by $v = .6\sqrt{2gh}$.

2.88. Solve Problem 2.87 if the cylinder has its axis horizontal.

2.89. The rate at which bacteria multiply is proportional to the instantaneous number present. If the original number doubles in 2 hours, in how many hours will it triple?

2.90. After 2 days, 10 grams of a radioactive chemical is present. Three days later 5 grams is present. How much of the chemical was present initially assuming the rate of disintegration is proportional to the instantaneous amount which is present?

2.91. Find the half life of a radioactive substance if three quarters of it is present after 8 hours.

2.92. Chemical A is transformed into chemical B at a rate proportional to the instantaneous amount of A which is untransformed. If 20% of chemical A is transformed in 2 hours, (a) what percentage of A is transformed in 6 hours and (b) when will 80% of A be transformed?

2.93. It takes 15 minutes for an object to warm up from 10°C to 20°C in a room whose temperature is 30°C. Assuming Newton's law of cooling, how long would it take to warm up from 20°C to 25°C?

2.94. At 1:00 P.M. the temperature of a tank of water is 200°F. At 1:30 P.M. its temperature is 160°F. Assuming the surrounding temperature is maintained at 80°F, (a) what is the temperature at 2:00 P.M. and (b) at what time will the temperature be 100°F? Assume Newton's law of cooling.

2.95. A beam of length L ft and negligible weight is simply supported at the ends and has a concentrated load W lb at the center. Find (a) the deflection and (b) the maximum deflection.

2.96. A cantilever beam of length L ft has a weight of w lb/ft. Find (a) the deflection and (b) the maximum deflection.

NUMERICAL METHODS

2.97. Use the step-by-step or Euler method to solve numerically each of the following.

(a) $dy/dx = y$, $y(0) = 1$; find $y(.2)$ using $h = .05$.

(b) $dy/dx = x + y$, $y(0) = 0$; find $y(.5)$ using $h = .1$.

(c) $dy/dx = 3x - y$, $y(1) = 0$; find $y(.5)$ using $h = .1$.

2.98. Work Problem 2.97 using the modified step-by-step or Euler method.

2.99. Work Problem 2.97 using (a) the Taylor series method, (b) Picard's method, (c) the Runge-Kutta method.

2.100. (a) Given $dy/dx = (x + y)^2$, $y(0) = 1$, find an approximate value for $y(.2)$ by using an appropriate numerical method. (b) Compare with the exact solution obtained by using the transformation $x + y = v^2$.

2.101. If $d^2y/dx^2 + y = x$, $y(0) = 1$, $y'(0) = 0$, find $y(.5)$ using the (a) Euler method, (b) Taylor series method, (c) Picard method, (d) Runge-Kutta method.

2.102. If $\dfrac{d^2y}{dx^2} - 2\dfrac{dy}{dx} + y = 2x$, $y(1) = 0$, $y'(1) = 0$ find $y(.4)$ using an appropriate numerical method.

Answers to Supplementary Problems

2.52. (a) order 3, dep. var. U, ind. var. t, ordinary. (b) order 2, dep. var. T, ind. var. x, y, z, partial.
(c) order 2, dep. var. y, ind. var. x, ordinary. (d) order 3, dep. var. s, ind. var. t, ordinary. (e) order 2, dep. var. y, ind. var. x, ordinary. (f) order 1, dep. var. u (or v), ind. var. v (or u), ordinary.

2.53. All solutions except for (c) are general solutions.

2.54. (b) $y = e^{-x}(3 \sin x - 2 \cos x)$

2.55. The first is a general solution and the second a singular solution.

2.56. (a) $(1 - x^2)y' + xy = 0$ (b) $yy'' + y'^2 = 0$

2.57. (a) $y' = (y - 1)/(x - 2)$ (b) $(y - 1)^2(1 + y'^2) = 1$

2.58. (a) $y = 2x^4 - 2x^2 + x + c$ (b) $y = 2x^4 - 2x^2 + x + 2$

2.59. (a) $y = 9x^{7/3} - 2x - 7$ (c) $s = 2 + 12t - 2t^3 - 16 \sin t$
(b) $y = \frac{1}{2} \ln x + \frac{x^4}{24} - \frac{2x}{3} + \frac{5}{8}$

2.60. (a) $y = ce^{-x^2}$ (c) $I = 5(1 - e^{-2t})$ (e) $y = (1 + x)^2/(1 - x)^2$
(b) $y = 2/x^3$ (d) $\sqrt{1 - x^2} - \sqrt{1 - y^2} = c$

2.61. (a) $x^2 + 4xy - 5y^2 = c$ (c) $ye^x - xe^{-y} = c$
(b) $2x^2y^2 - 3x + 2y^3 = -3$ (d) $x^2 \cos y + y^3 \tan 2x - 4x = c$

2.62. (a) x^3; $x^4y - \frac{1}{6}x^6 = c$ (c) e^y; $e^y(2x - 3y) = 4$
(b) $1/y^3$; $x^2y - x = cy^2$ (d) $\cos x$; $y \cos^2 x + y^4 \cos^3 x = c$

2.63. (a) $2x^3y^4 - x^4y^6 = c$; int. factor x^2y^3 (b) $3xy^2 + 2 = cx^3y^3$; int. factor $x^{-4}y^{-4}$

2.64. (a) $2 \tan^{-1}(y/x) + 4x + y^2 = c$ (b) $x^2 - y^2 = ce^{-2x}$

2.65. (a) $xy = x^4 + 1$ (c) $y = e^{3x} - 2x - 1$
(b) $x + 3y = cx^4$ (d) $y \sin^2 x = c - \cos x$

2.66. (a) $x^2 - xy = cy$ (c) $x^2 - 3xy - 2y^2 = c$
(b) $x = ce^{-x/(x+y)}$ (d) $\ln(5x^2 - 4xy + y^2) + 2 \tan^{-1}\left(\dfrac{y - 2x}{x}\right) = c$

2.67. (a) $xy(c - x^2) = 2$ (b) $xe^{x/y^2} = c$

2.68. (a) $(x - 2)^2 + 4(x - 2)(y - 1) - (y - 1)^2 = c$
(b) $x^2 - 2xy + y^2 + 2x + 2y = c$

2.69. $(x + y + 1)(1 - x) = 1$

2.70. (a) $(y - x + c)(y - ce^x) = 0$ (b) $x(xy - c)^2 = 4$

2.71. (a) $y = cx + 2c^2$; sing. solution $y = -x^2/8$.
(b) $(x - c)^2 = cy$; sing. solutions $y = 0$, $4x + y = 0$.
(c) $(cx - y)^2 = c^2 - 1$; sing. solution $x^2 - y^2 = 1$.
(d) $y^2 = cx^2 - c^2$; sing. solutions $y = \pm x^2/2$.

2.72.   (a) $y = -\frac{1}{3}x^3 + c_1 x^4 + c_2$ (c) $y = c_1 \sin 2x + c_2 \cos 2x$ or $y = a \sin (2x + b)$

         (b) $y^3 = c_1 x + c_2$ (d) $(x - a)^2 + (y - b)^2 = 1$

2.74.   (a) $v = 40 - 40t$ (b) $x = 40t - 2t^2$

2.75.   (a) $v = 4(1 - e^{-8t})$ (b) $v = 4\left(\dfrac{e^{8t} - e^{-8t}}{e^{8t} + e^{-8t}}\right) = 4 \tanh 8t$

         (c) $\sqrt{v} + 2 \ln\left(\dfrac{2 - \sqrt{v}}{2}\right) = -8t$

2.77.   (a) $I = 8(1 - e^{-16t})$,   max. current $= 8$ amp

         (b) $I = 8t^2 e^{-16t}$,   max. current $= e^{-2}/8 = .01692$ amp

         (c) $I = 8(e^{-8t} - e^{-16t})$,   max. current $= 2$ amp

2.78.   (a) $I = 6.4 \sin 8t - 3.2 \cos 8t + 3.2 e^{-16t}$

         (b) Transient current $= 3.2 e^{-16t}$,   steady-state current $= 6.4 \sin 8t - 3.2 \cos 8t$

2.79.   (a) $Q = 3(1 - e^{-t})$, $I = 3e^{-t}$

         (b) $Q = 5e^{-t} - 5e^{-2t} - 5te^{-2t}$, $I = 10te^{-2t} + 5e^{-2t} - 5e^{-t}$

2.80. $Q = 2 \sin 2t + \cos 2t - e^{-t}$, $I = 4 \cos 2t - 2 \sin 2t + e^{-t}$

2.81.   (a) $x^2 - y^2 = k$ (b) $x^2 + y^2 = ky$ (c) $y + 1 = ke^{\frac{1}{2}(x^2 + 2y + y^2)}$

2.82. $x = y \tan\left(\dfrac{1 - y}{2}\right)$ **2.83.** $x^2 + y^2 = 5x$ **2.84.** $2xy = A$

2.85.   (a) $200(1 - e^{-.03t})$ (b) after 23.1 minutes

2.86.   (a) $\dfrac{3}{2}(100 + t) - \dfrac{1.5 \times 10^8}{(100 + t)^3}$ (b) $3(100 - t) - 300\left(1 - \dfrac{t}{100}\right)^3$, $0 \leqq t \leqq 100$

2.87.   6.72 hr   **2.88.**   5.69 hr   **2.89.**   3.17 hr

2.90. $10\sqrt[3]{4}$ or   21.54 gm   **2.91.**   19.26 hr

2.92.   (a) 48.8%   (b) after 14.43 hr   **2.93.**   30 min

2.94.   (a) $133.3°$F   (b) about 3:12 P.M.

2.95.   (a) $y = \dfrac{W}{48EI}(3L^2 x - 4x^3)$ for $0 \leqq x \leqq L/2$ and is symmetrical for $L/2 \leqq x \leqq L$ (b) $WL^3/48EI$

2.96.   (a) $y = \dfrac{w}{24EI}(x^4 - 4Lx^3 + 6L^2 x^2)$ (b) $\dfrac{wL^4}{8EI}$

2.97.   (a) Exact value $= 1.2214$ (b) Exact value $= 2.2974$ (c) Exact value $= -1.5$

2.100.   Exact value $= 1.5087$

2.101.   Exact value $= .89815$

2.102.   Exact value $= .14776$

Chapter 3

Linear Differential Equations

GENERAL LINEAR DIFFERENTIAL EQUATION OF ORDER n

The general linear differential equation of order n has the form

$$a_0(x)\frac{d^ny}{dx^n} + a_1(x)\frac{d^{n-1}y}{dx^{n-1}} + \cdots + a_{n-1}(x)\frac{dy}{dx} + a_n(x)y = R(x) \tag{1}$$

A differential equation which cannot be written in this form is called *nonlinear*.

Example 1. $x\dfrac{d^2y}{dx^2} + 3\dfrac{dy}{dx} - 2xy = \sin x$ is a second order linear equation.

Example 2. $y\dfrac{d^2y}{dx^2} - x\left(\dfrac{dy}{dx}\right)^2 + x^2y = e^{-x}$ is a second order nonlinear equation.

If $R(x)$, the right side of (1), is replaced by zero the resulting equation is called the *complementary*, *reduced* or *homogeneous* equation. If $R(x) \neq 0$, the equation is called the *complete* or *nonhomogeneous* equation.

Example 3. If $x\dfrac{d^2y}{dx^2} + 3\dfrac{dy}{dx} - 2xy = \sin x$ is the complete equation, then $x\dfrac{d^2y}{dx^2} + 3\dfrac{dy}{dx} - 2xy = 0$ is the corresponding complementary, reduced or homogeneous equation.

If $a_0(x), a_1(x), \ldots, a_n(x)$ are all constants, (1) is said to have *constant coefficients*, otherwise it is said to have *variable coefficients*.

EXISTENCE AND UNIQUENESS THEOREM

If $a_0(x), a_1(x), \ldots, a_n(x)$ and $R(x)$ are continuous in the interval $|x - x_0| < \delta$ and $a_0(x) \neq 0$, then there exists one and only one solution to (1) which satisfies the conditions

$$y(x_0) = y_0, \quad y'(x_0) = y_0', \quad \ldots, \quad y^{(n-1)}(x_0) = y_0^{(n-1)} \tag{2}$$

OPERATOR NOTATION

It is sometimes convenient to adopt the notation $Dy, D^2y, \ldots, D^ny$ to denote $\dfrac{dy}{dx}, \dfrac{d^2y}{dx^2}, \ldots, \dfrac{d^ny}{dx^n}$. The symbols $D, D^2, \ldots$ are called *differential operators* and have properties analogous to those of algebraic quantities. Using this notation, we shall agree to write (1) as

$$[a_0(x)D^n + a_1(x)D^{n-1} + \cdots + a_{n-1}(x)D + a_n(x)]y = R(x) \tag{3}$$

or briefly
$$\phi(D)y = R(x)$$

where $\phi(D) = a_0(x)D^n + a_1(x)D^{n-1} + \cdots + a_{n-1}(x)D + a_n(x)$ is called an *operator polynomial in D*.

Example 4. $x\dfrac{d^2y}{dx^2} + 3\dfrac{dy}{dx} - 2xy = \sin x$ can be written $(xD^2 + 3D - 2x)y = \sin x$.

LINEAR OPERATORS

An operator L is called a *linear operator* if for any constants A, B and functions u, v to which L can be applied, we have

$$L(Au + Bv) = AL(u) + BL(v)$$

The operators $D, D^2, \ldots$ and $\phi(D)$ are linear operators [see Problem 3.3].

FUNDAMENTAL THEOREM ON LINEAR DIFFERENTIAL EQUATIONS

In order to find the general solution of

$$\phi(D)y = R(x) \tag{4}$$

where $R(x) \neq 0$, let $Y_c(x)$ be the general solution of the complementary reduced or homogeneous equation

$$\phi(D)y = 0 \tag{5}$$

We often refer to $Y_c(x)$ as the *complementary* or *homogeneous solution*. Then we have the following important theorem, sometimes referred to as the *superposition principle or theorem*.

Theorem 3-1.   The general solution of (4) is obtained by adding the complementary solution $Y_c(x)$ to a particular solution $Y_P(x)$ of (4), i.e.

$$y = Y_c(x) + Y_P(x)$$

Example 5.   The general solution of $(D^2 - 3D + 2)y = 0$ is $y = c_1 e^x + c_2 e^{2x}$ and a particular solution of $(D^2 - 3D + 2)y = 4x^2$ is $2x^2 + 6x + 7$. Then the general solution of $(D^2 - 3D + 2)y = 4x^2$ is $y = c_1 e^x + c_2 e^{2x} + 2x^2 + 6x + 7$.

Because of this theorem it is clear that we shall have to consider separately the problems of finding general solutions of homogeneous equations and particular solutions of nonhomogeneous equations.

LINEAR DEPENDENCE AND WRONSKIANS

A set of n functions $y_1(x), y_2(x), \ldots, y_n(x)$ is said to be *linearly dependent* over an interval if there exist n constants $c_1, c_2, \ldots, c_n$, not all zero, such that

$$c_1 y_1(x) + c_2 y_2(x) + \cdots + c_n y_n(x) = 0$$

identically over the interval. Otherwise the set of functions is said to be *linearly independent*.

Example 6. $2e^{3x}, 5e^{3x}, e^{-4x}$ are linearly dependent over any interval since we can find constants c_1, c_2, c_3 not all zero such that $c_1(2e^{3x}) + c_2(5e^{3x}) + c_3(e^{-4x}) = 0$ identically; for instance, $c_1 = -5$, $c_2 = 2$, $c_3 = 0$.

Example 7. e^x and xe^x are linearly independent since $c_1 e^x + c_2 x e^x = 0$ identically if and only if $c_1 = 0$, $c_2 = 0$.

Theorem 3-2.   The set of functions $y_1(x), y_2(x), \ldots, y_n(x)$ [assumed differentiable] is linearly independent on an interval if and only if the determinant

$$W(y_1, y_2, \ldots, y_n) = \begin{vmatrix} y_1(x) & y_2(x) & \ldots & y_n(x) \\ y_1'(x) & y_2'(x) & \ldots & y_n'(x) \\ \cdots\cdots & \cdots\cdots & & \cdots\cdots \\ y_1^{(n-1)}(x) & y_2^{(n-1)}(x) & \ldots & y_n^{(n-1)}(x) \end{vmatrix}$$

called the *Wronskian* of $y_1, \ldots, y_n$ is different from zero on the interval.

This last theorem is important in connection with solutions of the homogeneous or reduced equation as seen in the following

Theorem 3-3.   **[Superposition Principle].** If $y_1(x), y_2(x), \ldots, y_n(x)$ are n linearly independent solutions of the nth order linear equation $\phi(D)y = 0$, then

$$y = c_1 y_1(x) + c_2 y_2(x) + \cdots + c_n y_n(x)$$

where $c_1, c_2, \ldots, c_n$ are n arbitrary constants, is the general solution of $\phi(D)y = 0$.

SOLUTIONS OF LINEAR EQUATIONS WITH CONSTANT COEFFICIENTS

The remarks made so far have applied to the general equation (*1*). Particular simplifications occur when the equation has constant coefficients and we now turn to this case. Two general procedures are available in this case, namely those which do not involve operator techniques and those which do. For each case methods exist for finding complementary and particular solutions and use is then made of the fundamental Theorem 3-1.

NON-OPERATOR TECHNIQUES

I. THE COMPLEMENTARY OR HOMOGENEOUS SOLUTION

Let $y = e^{mx}$, $m = $ constant, in $(a_0 D^n + a_1 D^{n-1} + \cdots + a_n)y = 0$ to obtain

$$a_0 m^n + a_1 m^{n-1} + \cdots + a_n = 0 \tag{6}$$

which is called the *auxiliary equation* or *characteristic equation*. This can be factored into

$$a_0(m - m_1)(m - m_2) \cdots (m - m_n) = 0 \tag{7}$$

which has roots $m_1, m_2, \ldots, m_n$. Three cases must be considered.

Case 1. Roots all real and distinct.

Then $e^{m_1 x}, e^{m_2 x}, \ldots, e^{m_n x}$ are n linearly independent solutions so that by Theorem 3-3 the required solution is

$$y = c_1 e^{m_1 x} + c_2 e^{m_2 x} + \cdots + c_n e^{m_n x} \tag{8}$$

Case 2. Some roots are complex.

If $a_0, a_1, \ldots, a_n$ are real, then when $a + bi$ is a root of (*6*) so also is $a - bi$ [where a, b are real]. Then a solution corresponding to the roots $a + bi$ and $a - bi$ is

$$y = e^{ax}(c_1 \cos bx + c_2 \sin bx) \tag{9}$$

where use is made of Euler's formula $e^{iu} = \cos u + c \sin u$.

Case 3. Some roots are repeated.

If m_1 is a root of multiplicity k, then a solution is given by

$$y = (c_1 + c_2 x + c_3 x^2 + \cdots + c_k x^{k-1})e^{m_1 x} \tag{10}$$

II. THE PARTICULAR SOLUTION

Two important methods for finding a particular solution of $\phi(D)y = R(x)$ are available.

1. Method of Undetermined Coefficients.

In this method we assume a *trial solution* containing unknown constants (indicated by $a, b, c, \ldots$) which are to be determined by substitution in the given equation. The trial solution to be assumed in each case depends on the special form of $R(x)$ and is shown in the following table. In each case f, g, p, q are given constants and k is a given positive integer.

$R(x)$	Assumed Trial Solution
fe^{px}	ae^{px}
$f \cos px + g \sin px$	$a \cos px + b \sin px$
$f_0 x^k + f_1 x^{k-1} + \cdots + f_k$	$a_0 x^k + a_1 x^{k-1} + \cdots + a_k$
$e^{qx}(f \cos px + g \sin qx)$	$e^{qx}(a \cos px + b \sin px)$
$e^{qx}(f_0 x^k + f_1 x^{k-1} + \cdots + f_k)$	$e^{qx}(a_0 x^k + a_1 x^{k-1} + \cdots + a_k)$
$(f_0 x^k + \cdots + f_k) \cos px$ $+ (g_0 x^k + \cdots + g_k) \sin px$	$(a_0 x^k + \cdots + a_k) \cos px$ $+ (b_0 x^k + \cdots + b_k) \sin px$
$e^{qx}(f_0 x^k + \cdots + f_k) \cos px$ $+ e^{qx}(g_0 x^k + \cdots + g_k) \sin px$	$e^{qx}(a_0 x^k + \cdots + a_k) \cos px$ $+ e^{qx}(b_0 x^k + \cdots + b_k) \sin px$
Sums of any or some of the above entries.	Sums of the corresponding trial solutions.

The above method holds in case no term in the assumed trial solution appears in the complementary solution. If any term of the assumed trial solution does appear in the complementary solution, we must multiply this trial solution by the smallest positive integral power of x which is large enough so that none of the terms which are then present appear in the complementary solution.

2. Method of Variation of Parameters.

Let the complementary solution of $\phi(D)y = R(x)$ be

$$y = c_1 y_1(x) + c_2 y_2(x) + \cdots + c_n y_n(x)$$

Replace the arbitrary constants $c_1, c_2, \ldots, c_n$ by functions $K_1(x), K_2(x), \ldots, K_n(x)$ and seek to determine these so that $y = K_1 y_1 + K_2 y_2 + \cdots + K_n y_n$ is a solution of $\phi(D)y = R(x)$. Since, to determine these n functions, we must impose n restrictions on them and since one of these is that the differential equation be satisfied, it follows that the remaining $n-1$ may be taken at will. The conditions which lead to the greatest simplicity are given by the equations

$$\left. \begin{aligned} K_1' y_1 + K_2' y_2 + \cdots + K_n' y_n &= 0 \\ K_1' y_1' + K_2' y_2' + \cdots + K_n' y_n' &= 0 \\ \cdots\cdots\cdots\cdots\cdots\cdots\cdots\cdots\cdots & \\ K_1' y_1^{(n-2)} + K_2' y_2^{(n-2)} + \cdots + K_n' y_n^{(n-2)} &= 0 \\ K_1' y_1^{(n-1)} + K_2' y_2^{(n-1)} + \cdots + K_n' y_n^{(n-1)} &= R(x)/a_0 \end{aligned} \right\} \quad (11)$$

where the last equation represents the condition that the given differential equation be satisfied.

Since the determinant of the above system of equations is the Wronskian of $y_1, y_2, \ldots, y_n$ which is supposed to be different from zero, the equations can be solved for $K_1', K_2', \ldots, K_n'$. From these $K_1, K_2, \ldots, K_n$ can be found by integration leading to the required solution. The method is applicable whenever the complementary solution can be found, including cases where $a_0, \ldots, a_n$ are not constants.

OPERATOR TECHNIQUES

When $a_0, a_1, \ldots, a_n$ are constants, the equation $\phi(D)y = R(x)$ can be written in factored form as

$$a_0(D - m_1)(D - m_2) \cdots (D - m_n)y \;=\; R(x) \tag{12}$$

where $m_1, \ldots, m_n$ are constants and where the order of the factors $(D - m_1), \ldots, (D - m_n)$ is immaterial. This is not true if $a_0, a_1, \ldots, a_n$ are not constants [see Problem 3.2]. The constants $m_1, \ldots, m_n$ are the same as the roots of the auxiliary equation (6) or (7), and so the complementary solution can be written as before. To obtain particular solutions, the following operator methods will be found useful.

1. Method of Reduction of Order.

Let $a_0(D - m_2) \cdots (D - m_n)y = Y_1$. Then (12) becomes $(D - m_1)Y_1 = R(x)$ which can be solved for Y_1. Then let $a_0(D - m_3) \cdots (D - m_n)y = Y_2$ so that $(D - m_2)Y_2 = Y_1$ which can be solved for Y_2. By continuing in this manner, y can be obtained. This method yields the general solution if all arbitrary constants are kept, while if arbitrary constants are omitted it yields a particular solution.

2. Method of Inverse Operators.

Let $\dfrac{1}{\phi(D)} R(x)$ be defined as a particular solution y_P such that $\phi(D)y_P = R(x)$. We call $1/\phi(D)$ an *inverse operator*. By reference to the entries in the following table, the labor involved in finding particular solutions of $\phi(D)y = R(x)$ is often diminished considerably. In using these we often employ the theorem that

$$\frac{1}{\phi(D)} \{c_1 R_1(x) + \cdots + c_k R_k(x)\} \;=\; c_1 \frac{1}{\phi(D)} R_1(x) + \cdots + c_k \frac{1}{\phi(D)} R_k(x)$$

which is simply a statement of the fact that $1/\phi(D)$ is a linear operator.

Table of Inverse Operator Techniques

A.	$\dfrac{1}{D - m} R(x)$	$e^{mx} \displaystyle\int e^{-mx} R(x)\, dx$
B.	$\dfrac{1}{(D - m_1)(D - m_2)\cdots(D - m_n)} R(x)$	$e^{m_1 x} \displaystyle\int e^{-m_1 x} e^{m_2 x} \int \cdots$ $\displaystyle\int e^{-m_{n-1} x} e^{m_n x} \int e^{-m_n x} R(x)\, dx^n$ This can also be evaluated by expanding the inverse operator into partial fractions and then using entry A.
C.	$\dfrac{1}{\phi(D)} e^{px}$	$\dfrac{e^{px}}{\phi(p)}$ if $\phi(p) \neq 0$ $\dfrac{x^k e^{px}}{\phi^{(k)}(p)}$ if $\phi(p) = \phi'(p) = \cdots \phi^{(k-1)}(p) = 0$ but $\phi^{(k)}(p) \neq 0$
D.	$\dfrac{1}{\phi(D^2)} \cos(px + q)$ $\dfrac{1}{\phi(D^2)} \sin(px + q)$	$\dfrac{\cos(px + q)}{\phi(-p^2)}$ if $\phi(-p^2) \neq 0$ $\dfrac{\sin(px + q)}{\phi(-p^2)}$

Table of Inverse Operator Techniques (cont.)

E. $\dfrac{1}{\phi(D)}\cos(px+q) = \mathrm{Re}\left\{\dfrac{1}{\phi(D)}e^{i(px+q)}\right\}$ $\dfrac{1}{\phi(D)}\sin(px+q) = \mathrm{Im}\left\{\dfrac{1}{\phi(D)}e^{i(px+q)}\right\}$ "Re" means "real part of" and "Im" means "imaginary part of".	$\mathrm{Re}\left\{\dfrac{e^{i(px+q)}}{\phi(ip)}\right\}$ $\mathrm{Im}\left\{\dfrac{e^{i(px+q)}}{\phi(ip)}\right\}$ if $\phi(ip)\neq 0$, otherwise use entry C.
F. $\dfrac{1}{\phi(D)}x^p = (c_0+c_1D+\cdots+c_kD^k+\cdots)x^p$ by expanding $1/\phi(D)$ in powers of D. (p = positive integer)	$(c_0+c_1D+\cdots+c_pD^p)x^p$ since $D^{p+n}x^p = 0$ for $n > 0$.
G. $\dfrac{1}{\phi(D)}e^{px}F(x)$	$e^{px}\dfrac{1}{\phi(D+p)}F(x)$ called the "operator shift theorem".
H. $\dfrac{1}{\phi(D)}xF(x)$	$x\dfrac{1}{\phi(D)}F(x) - \dfrac{\phi'(D)}{[\phi(D)]^2}F(x)$

LINEAR EQUATIONS WITH VARIABLE COEFFICIENTS

Various methods are available for solving differential equations of the form (1) where $a_0, a_1, \ldots, a_n$ are not constants. In the following we list a few important methods.

I. MISCELLANEOUS TRANSFORMATIONS OF VARIABLES

1. **Cauchy or Euler Equation.** This equation has the form

$$(b_0x^nD^n + b_1x^{n-1}D^{n-1} + \cdots + b_{n-1}xD + b_n)y = R(x)$$

where $b_0, b_1, \ldots, b_n$ are constants. It can be solved by letting $x = e^t$ and using the results

$$xD = D_t, \quad x^2D^2 = D_t(D_t-1), \quad x^3D^3 = D_t(D_t-1)(D_t-2), \ldots$$

where $D_t = d/dt$, thus reducing the equation to one with constant coefficients. The case where $R(x) = 0$ can be solved by letting $y = x^p$ and determining the constant p.

2. **Case where one solution is known.** If one solution $y = Y(x)$ of $\phi(D)y = R(x)$ is known, then the substitution $y = vY(x)$ transforms the differential equation into one of order $n-1$ in v'. If $n = 2$, the equation can then be solved exactly. See Problem 3.34.

3. **Reduction to canonical form.** The general second order linear equation

$$y'' + p(x)y' + q(x)y = r(x) \tag{13}$$

can be transformed into the canonical form

$$v'' + f(x)v = g(x) \tag{14}$$

where $\qquad f(x) = q(x) - \frac{1}{4}[p(x)]^2 - \frac{1}{2}p'(x), \qquad g(x) = r(x)e^{\frac{1}{2}\int p(x)\,dx}$ $\qquad$ (15)

by using the substitution $\qquad y = ve^{-\frac{1}{2}\int p(x)\,dx}$ $\qquad$ (16)

Thus if (14) can be solved, so can (13). See Problems 3.35 and 3.36.

II. EXACT EQUATIONS

The equation $[a_0(x)D^n + a_1(x)D^{n-1} + \cdots + a_n(x)]y = R(x)$ is called *exact* if

$$a_0(x)D^n + a_1(x)D^{n-1} + \cdots + a_n(x) = D[p_0(x)D^{n-1} + \cdots + p_{n-1}(x)]$$

For example, $[a_0(x)D^2 + a_1(x)D + a_2(x)]y = R(x)$ is exact if and only if $a_0'' + a_1' + a_2 = 0$ identically. See Problems 3.37 and 3.38.

III. VARIATION OF PARAMETERS

This method, identical to that on page 74, can be used when the complementary solution is known.

IV. OPERATOR FACTORIZATION

If $\phi(D)$ can be factored into factors each having the form $p(x)D + q(x)$, then the method of reduction of order (page 75) can be used. See Problem 3.39.

V. SERIES METHODS

The equation $a_0(x)y'' + a_1(x)y' + a_2(x)y = 0$ where a_0, a_1, a_2 are polynomials, can often be solved by assuming that

$$y = x^\beta(c_0 + c_1x + c_2x^2 + \cdots) = \sum_{k=-\infty}^{\infty} c_k x^{k+\beta} \quad \text{where } c_k = 0,\ k < 0 \qquad (17)$$

where β and c_k are constants. Substituting (17) into the differential equation leads to an equation for β, called the *indicial equation,* and equations for the constants $c_0, c_1, \ldots$ in the form of a *recursion formula.* By solving for β and the other constants, a series solution can often be obtained. The series is called a *Frobenius series* and the method is often called the *method of Frobenius.* See Problem 3.40.

SIMULTANEOUS DIFFERENTIAL EQUATIONS

A system of differential equations with two or more dependent variables and one independent variable can be solved by eliminating all but one of the dependent variables, thus obtaining a single ordinary differential equation. Solutions obtained should be checked by substitution into the original differential equations to insure that the proper number of arbitrary constants are present. See Problem 3.41.

APPLICATIONS

Problems in mechanics, electricity and other fields of science and engineering often lead to linear differential equations and can be solved by the above methods. See Problems 3.42-3.47.

Solved Problems

OPERATORS

3.1. Show that $(D^2 + 3D + 2)e^{4x} = (D+2)(D+1)e^{4x} = (D+1)(D+2)e^{4x}$.

$$(D^2 + 3D + 2)e^{4x} = D^2 e^{4x} + 3D e^{4x} + 2e^{4x} = 16e^{4x} + 12e^{4x} + 2e^{4x} = 30e^{4x}$$

$$(D+2)(D+1)e^{4x} = (D+2)(De^{4x} + e^{4x}) = (D+2)(4e^{4x} + e^{4x}) = (D+2)(5e^{4x}) = 30e^{4x}$$

$$(D+1)(D+2)e^{4x} = (D+1)(De^{4x} + 2e^{4x}) = (D+1)(4e^{4x} + 2e^{4x}) = (D+1)(6e^{4x}) = 30e^{4x}$$

The result illustrates the commutative law of multiplication for operators with constant coefficients. In general, however, the commutative law for multiplication does not hold for operators with nonconstant coefficients, as seen in Problem 3.2.

3.2. Show that the operators $xD + 1$ and $D - 2$ are not commutative with respect to multiplication.

$$(xD+1)(D-2)y = (xD+1)(y'-2y) = xD(y'-2y) + (y'-2y) = xy'' - 2xy' + y' - 2y$$

$$(D-2)(xD+1)y = (D-2)(xy'+y) = D(xy'+y) - 2(xy'+y) = xy'' - 2xy' + 2y' - 2y$$

Then $(xD+1)(D-2)y \neq (D-2)(xD+1)y$ and the required result is proved.

3.3. (a) Prove that $D, D^2, D^3, \ldots$ are linear operators. (b) Prove that $\phi(D) = a_0(x)D^n + a_1(x)D^{n-1} + \cdots + a_n(x)$ is a linear operator.

(a)
$$D(Au + Bv) = \frac{d}{dx}(Au + Bv) = A\frac{du}{dx} + B\frac{dv}{dx} = A\,Du + B\,Dv$$

and so D is a linear operator [see page 72].

In a similar way we can show that $D^2, D^3, \ldots$ are linear operators.

(b)
$$\phi(D)[Au + Bv] = (a_0 D^n + a_1 D^{n-1} + \cdots + a_n)[Au + Bv]$$
$$= a_0 D^n[Au + Bv] + \cdots + a_n[Au + Bv]$$
$$= (Aa_0 D^n u + Ba_0 D^n v) + \cdots + (Aa_n u + Ba_n v)$$
$$= A(a_0 D^n + \cdots + a_n)u + B(a_0 D^n + \cdots + a_n)v$$
$$= A\phi(D)u + B\phi(D)v$$

and so $\phi(D)$ is a linear operator.

3.4. Prove that if $y_1, y_2, \ldots, y_n$ are solutions of the equation $\phi(D)y = 0$ then $c_1 y_1 + c_2 y_2 + \cdots + c_n y_n$ where $c_1, c_2, \ldots c_n$ are arbitrary constants, is also a solution.

We have
$$\phi(D)y_1 = 0, \quad \phi(D)y_2 = 0, \quad \ldots, \quad \phi(D)y_n = 0$$

Then using Problem 3.3,

$$\phi(D)[c_1 y_1 + c_2 y_2 + \cdots + c_n y_n] = c_1 \phi(D)y_1 + c_2 \phi(D)y_2 + \cdots + c_n \phi(D)y_n = 0$$

and so $c_1 y_1 + c_2 y_2 + \cdots + c_n y_n$ is a solution.

3.5. Prove Theorem 3-1, page 72.

Let $y = Y_C(x)$ be the general solution of $\phi(D)y = 0$, i.e. one having n arbitrary constants. Let $y = Y_P(x)$ be a particular solution of $\phi(D)y = R(x)$. Then $y = Y_C(x) + Y_P(x)$ is the general solution of $\phi(D)y = R(x)$, since by Problem 3.3

$$\phi(D)[Y_C(x) + Y_P(x)] = \phi(D)[Y_C(x)] + \phi(D)[Y_P(x)] = 0 + R(x) = R(x)$$

LINEAR DEPENDENCE AND WRONSKIANS

3.6.   Show that the functions $\cos 2x$, $\sin^2 x$, $\cos^2 x$ are linearly dependent.

We must show that there are constants c_1, c_2, c_3 not all zero such that $c_1 \cos 2x + c_2 \sin^2 x + c_3 \cos^2 x = 0$ identically.   Since $\cos 2x = \cos^2 x - \sin^2 x$, we can choose $c_1 = 1$, $c_2 = 1$, $c_3 = -1$ and the required result follows.

3.7.   Prove that if the Wronskian of the set of functions $y_1, \ldots y_n$ is different from zero on an interval, then the functions are linearly independent on the interval.

Suppose the contrary, i.e. that the functions are linearly dependent on the interval.   Then there are n constants $c_1, \ldots, c_n$, not all zero, such that

$$c_1 y_1 + \cdots + c_n y_n = 0 \qquad\qquad (1)$$

identically.   By successive differentiation we then have identically

$$
\begin{aligned}
c_1 y_1' \ \ &+ \cdots + c_n y_n' = 0 \\
&\cdots\cdots\cdots\cdots\cdots\cdots\cdots\cdots \\
c_1 y_1^{(n-1)} &+ \cdots + c_n y_n^{(n-1)} = 0
\end{aligned}
\qquad\qquad (2)
$$

Now in order for the system of n equations (1) and (2) to have solutions for $c_1, \ldots, c_n$ which are not all zero, we must have

$$
W = \begin{vmatrix}
y_1 & \cdots & y_n \\
y_1' & \cdots & y_n' \\
\cdots & \cdots & \cdots \\
y_1^{(n-1)} & \cdots & y_n^{(n-1)}
\end{vmatrix} = 0
$$

Since $W \neq 0$ by hypothesis, the contradiction shows that the functions cannot be linearly dependent and so must be linearly independent.

3.8.   Prove that if the functions $y_1, \ldots, y_n$ are linearly independent on an interval, then the Wronskian is different from zero.

Suppose the contrary, i.e. the Wronskian is zero for a particular value x_0 of the interval. Consider the system of equations

$$
\left.
\begin{aligned}
c_1 y_1(x_0) \ \ &+ \cdots + c_n y_n(x_0) = 0 \\
c_1 y_1'(x_0) \ \ &+ \cdots + c_n y_n'(x_0) = 0 \\
&\cdots\cdots\cdots\cdots\cdots\cdots\cdots\cdots \\
c_1 y_1^{(n-1)}(x_0) &+ \cdots + c_n y_n^{(n-1)}(x_0) = 0
\end{aligned}
\right\}
\qquad\qquad (1)
$$

Since the Wronskian is zero we see that this system has a solution $c_1, \ldots, c_n$ not all zero.   Now let

$$y = c_1 y_1(x) + \cdots + c_n y_n(x) \qquad\qquad (2)$$

If $y_1(x), \ldots, y_n(x)$ are solutions of $\phi(D)y = 0$, it follows from Problem 4 that (2) is also a solution of $\phi(D)y = 0$ which as is seen from equations (1) satisfies the conditions $y(x_0) = 0$, $y'(x_0) = 0, \ldots,$ $y^{(n-1)}(x_0) = 0$.   But $y = 0$ satisfies $\phi(D)y = 0$ and these same conditions.   Thus by the uniqueness theorem of page 41 it follows that the only solution is $y = 0$, i.e.

$$c_1 y_1(x) + \cdots + c_n y_n(x) = 0$$

from which we see that $y_1, \ldots, y_n$ are linearly dependent.   Since $y_1, \ldots, y_n$ are linearly independent by hypothesis, the contradiction shows that the Wronskian cannot be zero for x_0 and the required result is proved.

This problem and Problem 3.7 together provide a proof of Theorem 3-2.

3.9. Let $y_1(x)$, $y_2(x)$ be solutions of $y'' + p(x)y' + q(x)y = 0$.

(a) Prove that the Wronskian is $W = y_1y_2' - y_2y_1' = ce^{-\int p\,dx}$.

(b) Under what conditions on c will y_1 and y_2 be linearly independent.

(a) Since y_1 and y_2 are solutions,

$$y_1'' + py_1' + qy_1 = 0, \qquad y_2'' + py_2' + qy_2 = 0$$

Multiplying these equations by y_2 and y_1 respectively and subtracting,

$$y_1y_2'' - y_2y_1'' + p(y_1y_2' - y_2y_1') = 0 \qquad\qquad (1)$$

Then using $W = y_1y_2' - y_2y_1'$ and noting that $y_1y_2'' - y_2y_1'' = dW/dx$, (1) becomes

$$\frac{dW}{dx} + pW = 0$$

which has the solution $W = y_1y_2' - y_2y_1' = ce^{-\int p\,dx}$ $\qquad\qquad (2)$

This result is sometimes called *Abel's identity*.

(b) Since $e^{-\int p\,dx}$ is never zero, $W = 0$ if and only if $c = 0$ and $W \neq 0$ if and only if $c \neq 0$. Thus if $c \neq 0$, y_1 and y_2 are linear independent.

3.10. Use Problem 3.9 to prove that if y_1 is a known solution of $y'' + p(x)y' + q(x)y = 0$, then a linearly independent solution is

$$y_2 = y_1 \int \frac{e^{-\int p\,dx}}{y_1^2}\,dx$$

so that the general solution is $y = c_1y_1 + c_2y_2$.

From (2) of Problem 3.9 we have on dividing by $y_1^2 \neq 0$,

$$\frac{d}{dx}\left(\frac{y_2}{y_1}\right) = \frac{ce^{-\int p\,dx}}{y_1^2}$$

Then integrating and solving for y_2 leads to the required linearly independent solution.

THE REDUCED OR HOMOGENEOUS EQUATION

3.11. (a) Find three linearly independent solutions of $(D^3 - 9D)y = 0$ and (b) write the general solution.

(a) Assume that $y = e^{mx}$ is a solution where m is a constant. Then $(D^3 - 9D)e^{mx} = (m^3 - 9m)e^{mx}$ is zero when $m^3 - 9m = 0$, i.e. $m(m-3)(m+3) = 0$ or $m = 0, 3, -3$.

Then $e^{0x} = 1, e^{3x}, e^{-3x}$ are solutions. The Wronskian is

$$\begin{vmatrix} 1 & e^{3x} & e^{-3x} \\ 0 & 3e^{3x} & -3e^{-3x} \\ 0 & 9e^{3x} & 9e^{-3x} \end{vmatrix} = \begin{vmatrix} 3e^{3x} & -3e^{-3x} \\ 9e^{3x} & 9e^{-3x} \end{vmatrix} = 54$$

so that the functions are linearly independent by Theorem 3-2.

(b) The general solution is $y = c_1e^{0x} + c_2e^{3x} + c_3e^{-3x} = c_1 + c_2e^{3x} + c_3e^{-3x}$.

3.12. Solve (a) $2y'' - 5y' + 2y = 0$. (b) $(2D^3 - D^2 - 5D - 2)y = 0$.

(a) The auxiliary equation is $2m^2 - 5m + 2 = 0$ or $(2m-1)(m-2) = 0$ so that $m = 1/2, 2$. Then the general solution is $y = c_1e^{x/2} + c_2e^{2x}$.

(b) The auxiliary equation is $2m^3 - m^2 - 5m - 2 = 0$ or $(2m+1)(m+1)(m-2) = 0$ so that $m = -1/2, -1, 2$. Then the general solution is $y = c_1e^{-x/2} + c_2e^{-x} + c_3e^{2x}$.

3.13. Solve $y'' + 9y = 0$ or $(D^2 + 9)y = 0$.

The auxiliary equation is $m^2 + 9 = 0$ and $m = \pm 3i$. Then the general solution is $y = Ae^{3ix} + Be^{-3ix} = A(\cos 3x + i \sin 3x) + B(\cos 3x - i \sin 3x)$ which can be written $y = c_1 \cos 3x + c_2 \sin 3x$. Since $\cos 3x$, $\sin 3x$ are linearly independent, this is the general solution.

3.14. Solve $(D^2 + 6D + 25)y = 0$.

The auxiliary equation is $m^2 + 6m + 25 = 0$ and so $m = \dfrac{-6 \pm \sqrt{36 - 100}}{2} = \dfrac{-6 \pm 8i}{2} = -3 \pm 4i$. Then the general solution is

$$y = Ae^{(-3+4i)x} + Be^{(-3-4i)x} = e^{-3x}(Ae^{4ix} + Be^{-4ix}) = e^{-3x}(c_1 \cos 4x + c_2 \sin 4x)$$

3.15. Solve $(D^4 - 16)y = 0$.

The auxiliary equation $m^4 - 16 = 0$ or $(m^2 + 4)(m^2 - 4) = 0$ has roots $\pm 2i, \pm 2$. Then the general solution is $y = c_1 \cos 2x + c_2 \sin 2x + c_3 e^{2x} + c_4 e^{-2x}$.

3.16. Solve $y'' - 8y' + 16y = 0$.

The equation can be written $(D^2 - 8D + 16)y = 0$ or $(D - 4)^2 y = 0$. The auxiliary equation $(m - 1)^2 = 0$ has roots $m = 4, 4$ so that e^{4x}, e^{4x} are solutions corresponding to these repeated roots. However, they are clearly not linearly independent and so we cannot say that $y = c_1 e^{4x} + c_2 e^{4x}$ is the general solution, since it can be written $(c_1 + c_2)e^{4x} = ce^{4x}$ which involves only one arbitrary constant. There are two methods which can be used to get the general solution.

Method 1.

Write the given equation as $(D - 4)(D - 4)y = 0$ and let $(D - 4)y = Y_1$ so that $(D - 4)Y_1 = 0$ and so $Y_1 = A_1 e^{4x}$. Thus $(D - 4)y = A_1 e^{4x}$. Solving, we find $y = (c_1 + c_2 x)e^{4x}$.

This method, called the method of *reduction of order*, illustrates a general procedure for repeated roots. Using it we would find, for example, that if 4 were a triple repeated root, the general solution would be $y = (c_1 + c_2 x + c_3 x^2)e^{4x}$.

Method 2.

Since one solution $y_1 = e^{4x}$ is known, we can use Problem 3.10 to find the linearly independent solution

$$y_2 = e^{4x} \int \frac{e^{\int 8 \, dx}}{(e^{4x})^2} \, dx = xe^{4x}$$

Then the general solution is $y = c_1 e^{4x} + c_2 xe^{4x} = (c_1 + c_2 x)e^{4x}$.

3.17. Solve $(D + 2)^3 (D - 3)^4 (D^2 + 2D + 5)y = 0$.

The auxiliary equation $(m + 2)^3 (m - 3)^4 (m^2 + 2m + 5) = 0$ has roots $-2, -2, -2, 3, 3, 3, 3, -1 \pm 2i$. Then the general solution is

$$y = (c_1 + c_2 x + c_3 x^2)e^{-2x} + (c_4 + c_5 x + c_6 x^2 + c_7 x^3)e^{3x} + e^{-x}(c_8 \cos 2x + c_9 \sin 2x)$$

THE COMPLETE OR NONHOMOGENEOUS EQUATION. UNDETERMINED COEFFICIENTS

3.18. Solve $(D^2 + 2D + 4)y = 8x^2 + 12e^{-x}$.

The complementary solution is $e^{-x}(c_1 \cos \sqrt{3}\, x + c_2 \sin \sqrt{3}\, x)$.

To obtain a particular solution assume corresponding to $8x^2$ and $12e^{-x}$ the trial solutions $ax^2 + bx + c$ and de^{-x} respectively, since none of these terms are present in the complementary solution. Then substituting $y = ax^2 + bx + c + de^{-x}$ in the given equation, we find

$$4ax^2 + (4a + 4b)x + (2a + 2b + 4c) + 3de^{-x} = 8x^2 + 12e^{-x}$$

Equating corresponding coefficients on both sides of the equation,

$$4a = 8, \quad 4a + 4b = 0, \quad 2a + 2b + 4c = 0, \quad 3d = 12$$

Then $a = 2$, $b = -2$, $c = 0$, $d = 4$ and the particular solution is $2x^2 - 2x + 4e^{-x}$. Thus the required general solution is

$$y = e^{-x}(c_1 \cos \sqrt{3}\, x + c_2 \sin \sqrt{3}\, x) + 2x^2 - 2x + 4e^{-x}$$

3.19. Solve Problem 3.18 if the term $10 \sin 3x$ is added to the right side.

Corresponding to the additional term $10 \sin 3x$ we assume the additional trial solution $h \cos 3x + k \sin 3x$ which does not appear in the complementary solution. Substituting this into the equation $(D^2 + 2D + 4)y = 10 \sin 3x$,

$$(6k - 5h) \cos 3x - (5k + 6h) \sin 3x = 10 \sin 3x$$

Equating coefficients,

$$6k - 5h = 0, \quad 5k + 6h = -10$$

or $h = -\dfrac{12}{61}$, $k = -\dfrac{10}{61}$. Then the required general solution is

$$y = e^{-x}(c_1 \cos \sqrt{3}\, x + c_2 \sin \sqrt{3}\, x) + 2x^2 - 2x + 4e^{-x} - \frac{12}{61} \cos 3x - \frac{10}{61} \sin 3x$$

3.20. Solve $(D^2 + 4)y = 8 \sin 2x$.

The complementary solution is $c_1 \cos 2x + c_2 \sin 2x$.

For a particular solution we would normally assume a trial solution $a \cos 2x + b \sin 2x$. However, since the terms appear in the complementary solution, we multiply by x to obtain the trial solution $x(a \cos 2x + b \sin 2x)$. Then substituting in the given equation,

$$-4a \cos 2x - 4b \sin 2x = 8 \sin 2x$$

so that $4a = 0$, $-4b = 8$ and $a = 0$, $b = -2$. Then the required general solution is

$$y = c_1 \cos 2x + c_2 \sin 2x - 2x \sin 2x$$

3.21. Solve $(D^5 - 3D^4 + 3D^3 - D^2)y = x^2 + 2x + 3e^x$.

The auxiliary equation is $m^5 - 3m^4 + 3m^3 - m^2 = 0$ or $m^2(m-1)^3 = 0$. Thus $m = 0, 0, 1, 1, 1$ and the complementary solution is

$$c_1 + c_2 x + (c_3 + c_4 x + c_5 x^2)e^x$$

Corresponding to the polynomial $x^2 + 2x$, we would normally assume a trial solution $ax^2 + bx + c$. However, some of these terms appear in the complementary solution. Multiplying by x, the trial solution would be $x(ax^2 + bx + c) = ax^3 + bx^2 + cx$, but one of these terms is still in the complementary solution. Finally multiplying by x again to obtain $ax^4 + bx^3 + cx^2$, we see that this has no term in the complementary solution and so is the needed trial solution.

Similarly, corresponding to $3e^x$ we would normally assume a trial solution de^x. But since this term as well as dxe^x and dx^2e^x are in the complementary solution, we must use as trial solution dx^3e^x.

Thus the assumed trial solution is $ax^4 + bx^3 + cx^2 + dx^3e^x$. Substituting this in the given equation, we find

$$-12ax^2 + (72a - 6b)x + (18b - 72a - 2c) + 6de^x = x^2 + 2x + 3e^x$$

from which $a = -1/12$, $b = -4/3$, $c = -9$, $d = 1/2$ and the general solution is

$$y = c_1 + c_2 x + (c_3 + c_4 x + c_5 x^2)e^x + \frac{1}{2}x^3 e^x - \frac{1}{12}x^4 - \frac{4}{3}x^3 - 9x^2$$

THE COMPLETE OR NONHOMOGENEOUS EQUATION.
VARIATION OF PARAMETERS

3.22. Solve $y'' + y = \sec x$ or $(D^2 + 1)y = \sec x$.

The complementary solution is $c_1 \cos x + c_2 \sin x$. Then we assume the general solution to be

$y = K_1 \cos x + K_2 \sin x$ where K_1, K_2 are suitable functions of x to be determined.   Differentiation yields

$$y' = -K_1 \sin x + K_2 \cos x + K_1' \cos x + K_2' \sin x \tag{1}$$

Since there are two functions K_1, K_2, we must arrive at two conditions for determining them. However, one of these conditions is that the differential equation must be satisfied.   Thus we are at liberty to impose arbitrarily the second condition.   We choose this condition to be the one which simplifies (1) most, namely

$$K_1' \cos x + K_2' \sin x = 0 \tag{2}$$

Then (1) becomes $\qquad y' = -K_1 \sin x + K_2 \cos x$

Differentiating again

$$y'' = -K_1 \cos x - K_2 \sin x - K_1' \sin x + K_2' \cos x$$

Thus $\qquad y'' + y = -K_1' \sin x + K_2' \cos x = \sec x \tag{3}$

From the two equations

$$K_1' \cos x + K_2' \sin x = 0$$

$$-K_1' \sin x + K_2' \cos x = \sec x$$

we find $K_1' = 1$, $K_2' = -\tan x$. Thus by integrating, $K_1 = x + c_1$, $K_2 = -\ln \sec x + c_2$ and the required general solution is

$$y = c_1 \sin x + c_2 \cos x + x \sin x - \cos x \ln \sec x$$

3.23. Solve $(D^3 + 4D)y = 4 \cot 2x$.

The complementary solution is $c_1 + c_2 \cos 2x + c_3 \sin 2x$.   Then we are led to the following equations for determining the functions K_1, K_2, K_3 in the general solution

$$y = K_1 + K_2 \cos 2x + K_3 \sin 2x$$

$$K_1' + K_2' \cos 2x + K_3' \sin 2x = 0$$

$$0 - 2K_2' \sin 2x + 2K_3' \cos 2x = 0$$

$$0 - 4K_2' \cos 2x - 4K_3' \sin 2x = 4 \cot 2x$$

Solving this system, we find $K_1' = \cot 2x$, $K_2' = -\cos^2 2x / \sin 2x$, $K_3' = -\cos 2x$ so that

$$K_1 = \tfrac{1}{2} \ln \sin 2x + c_1, \qquad K_2 = -\tfrac{1}{2} \ln (\csc 2x - \cot 2x) - \tfrac{1}{2} \cos 2x + c_2, \qquad K_3 = -\tfrac{1}{2} \sin 2x + c_3$$

From this we find the general solution

$$y = c_1 + c_2 \cos 2x + c_3 \sin 2x + \tfrac{1}{2} \ln \sin 2x - \tfrac{1}{2} \cos 2x \ln (\csc 2x - \cot 2x)$$

OPERATOR TECHNIQUES

3.24. Evaluate $\dfrac{1}{D-2} (e^{4x})$.

Method 1.

Let $\dfrac{1}{D-2} e^{4x} = y$. Then by definition $(D-2)y = e^{4x}$ or $\dfrac{dy}{dx} - 2y = e^{4x}$. Solving this as a first order linear equation with integrating factor e^{-2x}, we obtain $y = \tfrac{1}{2} e^{4x} + c e^{2x}$. Since we are interested only in particular solutions,

$$\frac{1}{D-2} e^{4x} = \frac{1}{2} e^{4x}$$

Method 2 [using formula A, page 75, derived as in method 1].

$$\frac{1}{D-2} e^{4x} = e^{2x} \int e^{-2x} e^{4x} \, dx = e^{2x} \int e^{2x} \, dx = \frac{1}{2} e^{4x}$$

3.25. Find $\dfrac{1}{(D+1)(D-2)}(3e^{-2x})$.

Method 1.

$$\frac{1}{(D+1)(D-2)}(3e^{-2x}) = \frac{1}{D+1}\left[\frac{1}{D-2}(3e^{-2x})\right]$$

$$= \frac{1}{D+1}\left[e^{2x}\int e^{-2x}(3e^{-2x})\,dx\right]$$

$$= \frac{1}{D+1}[-\tfrac{3}{4}e^{-2x}] = e^{-x}\int e^{x}(-\tfrac{3}{4}e^{-2x})\,dx = \tfrac{3}{4}e^{-2x}$$

Method 2. Using partial fractions,

$$\frac{1}{(D+1)(D-2)}(3e^{-2x}) = \left[\frac{-1/3}{D+1} + \frac{1/3}{D-2}\right](3e^{-2x})$$

$$= -\frac{1}{D+1}(e^{-2x}) + \frac{1}{D-2}(e^{-2x})$$

$$= -e^{-x}\int e^{x}(e^{-2x})\,dx + e^{2x}\int e^{-2x}(e^{-2x})\,dx = \tfrac{3}{4}e^{-2x}$$

3.26. (a) Prove that $\dfrac{1}{\phi(D)}e^{px} = \dfrac{e^{px}}{\phi(p)}$ if $\phi(p) \neq 0$.

(b) Find the general solution of $(D^2 - 3D + 2)y = e^{5x}$ using the result in (a).

(a) By definition $\phi(D) = a_0 D^n + a_1 D^{n-1} + \cdots + a_n$ where $a_0, a_1, \ldots, a_n$ are constants. Then

$$\phi(D)e^{px} = (a_0 D^n + a_1 D^{n-1} + \cdots + a_n)e^{px} = (a_0 p^n + a_1 p^{n-1} + \cdots + a_n)e^{px}$$

$$= \phi(p)e^{px}$$

Thus if $\phi(p) \neq 0$, then $\qquad\qquad \dfrac{1}{\phi(D)}e^{px} = \dfrac{e^{px}}{\phi(p)}$

(b) $\dfrac{1}{D^2 - 3D + 2}e^{5x} = \dfrac{1}{(5)^2 - 3(5) + 2}e^{5x} = \dfrac{e^{5x}}{12}$ by part (a). Then since the complementary solution is $c_1 e^x + c_2 e^{2x}$, the general solution is $y = c_1 e^x + c_2 e^{2x} + \dfrac{e^{5x}}{12}$.

3.27. (a) Prove that $\dfrac{1}{\phi(D^2)}\cos(px+q) = \dfrac{\cos(px+q)}{\phi(-p^2)}$ if $\phi(-p^2) \neq 0$.

(b) Find the general solution of $(D^2 + 1)^2 y = \cos 2x$ by using the result in (a).

(a) $\phi(D^2) = a_0(D^2)^n + a_1(D^2)^{n-1} + \cdots + a_n$ so that

$$\phi(D^2)\cos(px+q) = [a_0(D^2)^n + a_1(D^2)^{n-1} + \cdots + a_n]\cos(px+q)$$

$$= [a_0(-p^2)^n + a_1(-p^2)^{n-1} + \cdots + a_n]\cos(px+q)$$

$$= \phi(-p^2)\cos(px+q)$$

using $\quad D^2[\cos(px+q)] = -p^2\cos(px+q), \quad (D^2)^2\cos(px+q) = (-p^2)^2\cos(px+q), \quad$ etc.

Thus if $\phi(-p^2) \neq 0$, the required result follows.

(b) $\dfrac{1}{(D^2+1)^2}\cos 2x = \dfrac{1}{(-4+1)^2}\cos 2x = \dfrac{1}{9}\cos 2x$ by part (a). Then since the complementary solution is $c_1 \cos x + c_2 \sin x + x(c_3 \cos x + c_4 \sin x)$, the required general solution is

$$y = c_1 \cos x + c_2 \sin x + x(c_3 \cos x + c_4 \sin x) + \frac{1}{9}\cos 2x$$

3.28. Evaluate $\dfrac{1}{D^3 + D^2 + 2D - 1} \cos 2x$.

We can show by a method similar to that in Problem 3.27 that D^2 can formally be replaced by $-2^2 = -4$. Thus

$$\frac{1}{D^3 + D^2 + 2D - 1} \cos 2x = \frac{1}{D(-4) - 4 + 2D - 1} \cos 2x = \frac{-1}{2D + 5} \cos 2x$$

$$= -\frac{(2D - 5)}{4D^2 - 25} \cos 2x = -\frac{(2D - 5)}{4(-4) - 25} \cos 2x$$

$$= \frac{1}{41}(2D - 5) \cos 2x = \frac{1}{41}(-4 \sin 2x - 5 \cos 2x)$$

3.29. Prove that *(a)* $D^n[e^{px}R(x)] = e^{px}(D + p)^n R(x)$, *(b)* $\phi(D)[e^{px}R(x)] = e^{px}\phi(D + p)R(x)$, *(c)* $\dfrac{1}{\phi(D)}[e^{px}R(x)] = e^{px}\dfrac{1}{\phi(D + p)}R(x)$.

(a) Use mathematical induction. The result is true for $n = 1$, since $D[e^{px}R(x)] = e^{px}DR(x) + pe^{px}R(x) = e^{px}(D + p)R(x)$. Assume the result true for $n = k$. Then $D^k[e^{px}R(x)] = e^{px}(D + p)^k R(x)$. Differentiating both sides,

$$D^{k+1}[e^{px}R(x)] = D[e^{px}(D + p)^k R(x)]$$

$$= e^{px}D(D + p)^k R(x) + pe^{px}(D + p)^k R(x)$$

$$= e^{px}[(D + p)^{k+1}R(x)]$$

Thus if the result is true for $n = k$, it is true for $n = k + 1$; but since it is true for $n = 1$, it must be true for $n = 2, \ldots$ and thus all n.

(b)
$$\phi(D)[e^{px}R(x)] = a_0 D^n[e^{px}R(x)] + a_1 D^{n-1}[e^{px}R(x)] + \cdots + a_n$$

$$= e^{px}[a_0(D + p)^n + a_1(D + p)^{n-1} + \cdots + a_n]R(x)$$

$$= e^{px}\phi(D + p)R(x)$$

(c) Let $\dfrac{1}{\phi(D)}[e^{px}R(x)] = y$ so that $\phi(D)y = e^{px}R(x)$. Then

$$\phi(D)[e^{px}(ye^{-px})] = e^{px}\phi(D + p)[ye^{-px}] = e^{px}R(x)$$

or
$$\phi(D + p)[ye^{-px}] = R(x)$$

from which
$$y = e^{px}\frac{1}{\phi(D + p)}R(x)$$

3.30. Solve $(D^3 + D)y = e^{-2x}\cos 2x$.

Using Problem 3.29,

$$\frac{1}{D^3 + D}(e^{-2x}\cos 2x) = e^{-2x}\frac{1}{(D - 2)^3 + D - 2}\cos 2x$$

$$= e^{-2x}\frac{1}{D^3 - 6D^2 + 13D - 10}\cos 2x = e^{-2x}\frac{1}{-4D + 24 + 13D - 10}\cos 2x$$

$$= e^{-2x}\frac{1}{9D + 14}\cos 2x = e^{-2x}\frac{9D - 14}{81D^2 - 196}\cos 2x$$

$$= \frac{e^{-2x}}{81(-4) - 196}(9D - 14)\cos 2x = \frac{e^{-2x}}{260}(9 \sin 2x + 7 \cos 2x)$$

Since the complementary solution is $c_1 + c_2 \cos x + c_3 \sin x$, the required general solution is

$$y = c_1 + c_2 \cos x + c_3 \sin x + \frac{e^{-2x}}{260}(9 \sin 2x + 7 \cos 2x)$$

3.31. Evaluate $\dfrac{1}{2D^3 - D + 2}(x^3 - x)$.

By formal long division in ascending powers of D,

$$\frac{1}{2D^3 - D + 2}(x^3 - x) = \frac{1}{2 - D + 2D^3}(x^3 - x) = \left(\frac{1}{2} + \frac{D}{4} + \frac{D^2}{8} - \frac{7}{16}D^3 + \cdots\right)(x^3 - x)$$

$$= \tfrac{1}{2}(x^3 - x) + \tfrac{1}{4}(3x^2 - 1) + \tfrac{1}{8}(6x) - \tfrac{7}{16}(6)$$

$$= \tfrac{1}{2}x^3 + \tfrac{3}{4}x^2 + \tfrac{1}{4}x - \tfrac{23}{8}$$

CAUCHY OR EULER EQUATION

3.32. Let $D = d/dx$ and $D_t = d/dt$. Prove that if $x = e^t$, then

$$(a)\ \ xD = D_t, \qquad (b)\ \ x^2D^2 = D_t(D_t - 1)$$

(a) $$Dy = \frac{dy}{dx} = \frac{dy}{dt} \cdot \frac{dt}{dx} = \frac{dy}{dt}\bigg/\frac{dx}{dt} = \frac{dy}{dt}\bigg/e^t = e^{-t}\frac{dy}{dt} = e^{-t}D_ty$$

Then $xDy = e^tDy = D_ty$ or $xD = D_t$.

(b) $$D^2y = \frac{d^2y}{dx^2} = \frac{d}{dx}\left(e^{-t}\frac{dy}{dt}\right) = \frac{dy}{dt}\left(e^{-t}\frac{dy}{dt}\right)\bigg/\frac{dx}{dt} = e^{-2t}(D - D_t)y$$

Then $x^2D^2y = e^{2t}D^2y = (D_t^2 - D_t)y$ or $x^2D^2 = D_t^2 - D_t = D_t(D_t - 1)$.

3.33. Solve $(x^2D^2 + xD - 4)y = x^3$.

By the transformation $x = e^t$ the equation becomes

$$[D_t(D_t - 1) + D_t - 4]y = [e^t]^3 \quad \text{or} \quad (D_t^2 - 4)y = e^{3t}$$

Then the general solution is $\ \ y = c_1e^{2t} + c_2e^{-2t} + \tfrac{1}{5}e^{3t} = c_1x^2 + c_2x^{-2} + \tfrac{1}{5}x^3$.

Another method.

Letting $y = x^p$ in the complementary equation $(x^2D^2 + xD - 4)y = 0$, we find

$$p(p-1)x^p + px^p - 4x^p = 0 \quad \text{or} \quad (p^2 - 4)x^p = 0, \quad \text{i.e.} \ \ p = \pm 2.$$

Thus x^2 and x^{-2} are solutions and the complementary solution is $\ \ y = K_1x^2 + K_2x^{-2}$.

We can now use the method of variation of parameters to find the required general solution.

CASE WHERE ONE SOLUTION IS KNOWN

3.34. Solve $(1 - x^2)y'' - 2xy' + 2y = 0$ given that $y = x$ is a solution.

Let $y = xv$. Then $y' = xv' + v$, $y'' = xv'' + 2v'$ and the given equation becomes

$$x(1 - x^2)v'' + (2 - 4x^2)v' = 0 \quad \text{or} \quad \frac{dv'}{v'} + \frac{2 - 4x^2}{x(1 - x^2)}\,dx = 0$$

Integration yields $$\int \frac{dv'}{v'} + \int \left(\frac{2}{x} - \frac{2x}{1 - x^2}\right)dx = c_1$$

i.e. $$\ln v' + 2\ln x + \ln(1 - x^2) = c_1$$

or $$v' = \frac{c_2}{x^2(1 - x^2)} = c_2\left(\frac{1}{x^2} + \frac{1}{1 - x^2}\right)$$

Then $$v = c_2\left(\frac{1}{2}\ln\frac{1 + x}{1 - x} - \frac{1}{x}\right) + c_3$$

and so the general solution is

$$y = vx = c_2\left(\frac{x}{2}\ln\frac{1 + x}{1 - x} - 1\right) + c_3x$$

We can also use the result of Problem 3.10 to solve the equation.

REDUCTION TO CANONICAL FORM

3.35. By letting $y = uv$ and choosing u appropriately, obtain a differential equation corresponding to $y'' + p(x)y' + q(x)y = r(x)$ with the term involving the first derivative removed.

Substituting $y = uv$ in the given equation, we find

$$uv'' + (2u' + pu)v' + (u'' + pu' + qu)v = r \qquad (1)$$

Let $2u' + pu = 0$ so that $u = e^{-\int (p/2)\,dx}$. Then (1) becomes

$$v'' + (q - \tfrac{1}{4}p^2 - \tfrac{1}{2}p')v = re^{\int (p/2)\,dx} \qquad (2)$$

as required. Equation (2) is called the *canonical form* of the given equation.

3.36. Solve $4x^2y'' + 4xy' + (x^2 - 1)y = 0$.

Comparing with Problem 3.35, $p = 1/x$, $q = (x^2 - 1)/4x^2$, $r = 0$. Then the canonical form is

$$v'' + \tfrac{1}{4}v = 0 \qquad \text{or} \qquad v = c_1 \cos \tfrac{1}{2}x + c_2 \sin \tfrac{1}{2}x$$

Thus

$$y = ve^{-\int (p/2)\,dx} = \frac{c_1 \cos \tfrac{1}{2}x + c_2 \sin \tfrac{1}{2}x}{\sqrt{x}}$$

EXACT EQUATIONS

3.37. Show that $[a_0(x)D^2 + a_1(x)D + a_2(x)]y = R(x)$ is exact if and only if $a_0'' - a_1' + a_2 = 0$ identically.

By definition of an exact equation, there are functions $p_0(x)$, $p_1(x)$ such that

$$[a_0D^2 + a_1D + a_2]y = D[p_0D + p_1]y = [p_0D^2 + (p_0' + p_1)D + p_1']y$$

Then we must have $\qquad a_0 = p_0, \qquad a_1 = p_0' + p_1, \qquad a_2 = p_1'$

Eliminating p_0, p_1 from these 3 equations, we find $a_0'' - a_1' + a_2 = 0$.

Conversely if $a_0'' - a_1' + a_2 = 0$, then

$$[a_0D^2 + a_1D + a_2]y = [a_0D^2 + a_1D + a_1' - a_0'']y = D[a_0D + a_1 - a_0']y$$

and the equation is exact.

3.38. Solve $(1 - x^2)y'' - 3xy' - y = 1$.

Comparing with Problem 3.37, $a_0 = 1 - x^2$, $a_1 = -3x$, $a_2 = -1$ and $a_0'' - a_1' + a_2 = 0$ so that the equation is exact and can be written

$$D[(1 - x^2)D - x]y = 1$$

Integrating this equation and solving the resulting first order linear differential equation

$$\frac{dy}{dx} - \frac{x}{1 - x^2}y = \frac{x + c_1}{1 - x^2}$$

we find that

$$y = c_1 \frac{\sin^{-1} x}{\sqrt{1 - x^2}} + \frac{c_2}{\sqrt{1 - x^2}} - 1$$

OPERATOR FACTORIZATION

3.39. (a) Show that the operator $xD^2 + (2x + 3)D + 4 = (D + 2)(xD + 2)$.

(b) Use (a) to solve $xy'' + (2x + 3)y' + 4y = e^{2x}$.

(a) $(D + 2)(xD + 2)y = (D + 2)(xDy + 2y) = D(xDy + 2y) + 2(xDy + 2y)$

$\qquad = xD^2y + Dy + 2Dy + 2xDy + 4y = [xD^2 + (2x + 3)D + 4]y$

(b) The given equation can be written $(D+2)(xD+2)y = e^{2x}$. Let $(xD+2)y = Y$. Then $(D+2)Y = e^{2x}$ and $Y = \frac{1}{4}e^{2x} + c_1 e^{-2x}$. Now solving $(xD+2)y = \frac{1}{4}e^{2x} + c_1 e^{-2x}$, we find

$$x^2 y = \frac{1}{4}\int xe^{2x}\,dx + c_1\int xe^{-2x}\,dx + c_2 = \frac{1}{8}e^{2x}\left(x - \frac{1}{2}\right) - \frac{1}{2}c_1 e^{-2x}\left(x + \frac{1}{2}\right) + c_2$$

SERIES METHODS

3.40. Solve Problem 3.36 by using the method of Frobenius.

Let $y = \sum\limits_{k=0}^{\infty} c_k x^{k+\beta} = \sum\limits_{k=-\infty}^{\infty} c_k x^{k+\beta}$ where we define $c_k = 0$ for $k < 0$. Then

$$y' = \sum (k+\beta)c_k x^{k+\beta-1}, \qquad y'' = \sum (k+\beta)(k+\beta-1)c_k x^{k+\beta-2}$$

omitting the summation limits. Thus

$$4x^2 y'' + 4xy' + (x^2-1)y = \sum 4(k+\beta)(k+\beta-1)c_k x^{k+\beta} + \sum 4(k+\beta)c_k x^{k+\beta}$$
$$+ \sum c_k x^{k+\beta+2} - \sum c_k x^{k+\beta}$$

In order to write the series on the right in terms of coefficients of $x^{k+\beta}$, we must replace the index of summation k in the third series by $k-2$ [note that this does not affect the limits $-\infty$ and ∞ of the index of summation]. Then the series on the right can be written

$$\sum [4(k+\beta)(k+\beta-1)c_k + 4(k+\beta)c_k + c_{k-2} - c_k]x^{k+\beta} = \sum \{[4(k+\beta)^2-1]c_k + c_{k-2}\}x^{k+\beta}$$

Since this must be zero, each coefficient must be zero so that

$$\{4(k+\beta)^2 - 1\}c_k + c_{k-2} = 0 \tag{1}$$

Let $k = 0$. Then since $c_{-2} = 0$, (1) becomes $(4\beta^2-1)c_0 = 0$ which is called the *indicial equation*. Assuming $c_0 \neq 0$, this leads to $4\beta^2 - 1 = 0$, $\beta = \pm 1/2$. We have two cases $\beta = 1/2, -1/2$ and we shall consider the smaller value $-1/2$ first.

Case 1, $\beta = -1/2$. In this case (1) becomes

$$\{4(k - \tfrac{1}{2})^2 - 1\}c_k + c_{k-2} = 0 \qquad \text{or} \qquad 4k(k-1)c_k + c_{k-2} = 0 \tag{2}$$

Putting $k = 1, 2, 3, 4, \ldots$ in (2), we find

$$c_{-1} = 0, \quad c_2 = \frac{-c_0}{4\cdot 2\cdot 1}, \quad c_3 = \frac{-c_1}{4\cdot 3\cdot 2}, \quad c_4 = \frac{-c_2}{4\cdot 4\cdot 3} = \frac{c_0}{4^2\cdot 4\cdot 3\cdot 2\cdot 1}$$

$$c_5 = \frac{-c_3}{4\cdot 5\cdot 4} = \frac{c_1}{4^2\cdot 5\cdot 4\cdot 3\cdot 2}, \quad c_6 = \frac{-c_4}{4\cdot 6\cdot 5} = \frac{-c_0}{4^3\cdot 6\cdot 5\cdot 4\cdot 3\cdot 2\cdot 1}$$

and from these it is clear that c_0 and c_1 are undetermined while $c_2, c_3, c_4, \ldots$ are found in terms of c_0 and c_1 as follows:

$$c_2 = \frac{-c_0}{4\cdot 2!}, \quad c_3 = \frac{-c_1}{4\cdot 3!}, \quad c_4 = \frac{c_0}{4^2\cdot 4!}, \quad c_5 = \frac{c_1}{4^2\cdot 5!}, \quad c_6 = \frac{-c_0}{4^3\cdot 6!}, \quad \cdots$$

Then the corresponding solution is

$$y = \sum c_k x^{k+\beta} = \sum c_k x^{k-1/2}$$

$$= c_0\left(x^{-1/2} - \frac{x^{3/2}}{4\cdot 2!} + \frac{x^{7/2}}{4^2\cdot 4!} - \frac{x^{11/2}}{4^3\cdot 6!} + \cdots\right)$$

$$+ c_1\left(x^{1/2} - \frac{x^{5/2}}{4\cdot 3!} + \frac{x^{9/2}}{4^2\cdot 5!} - \cdots\right)$$

$$= \frac{c_0}{\sqrt{x}}\left(1 - \frac{(x/2)^2}{2!} + \frac{(x/2)^4}{4!} - \frac{(x/2)^6}{6!} + \cdots\right)$$

$$+ \frac{2c_1}{\sqrt{x}}\left((x/2) - \frac{(x/2)^3}{3!} + \frac{(x/2)^5}{5!} - \cdots\right)$$

$$= \frac{c_0 \cos(x/2) + 2c_1 \sin(x/2)}{\sqrt{x}}$$

which is equivalent to the solution obtained in Problem 3.36.

Note that for this equation it is not necessary to consider Case 2, $\beta = 1/2$ since we have already obtained the required solution. Note also that if we had considered $\beta = 1/2$ first, we would not have obtained the general solution. This is typical in general when the roots of the indicial equation differ by any integer except zero. In other equations both cases would have to be considered, each leading to a series solution. The general solution would then be obtained from these series on multiplying each by an arbitrary constant and adding.

SIMULTANEOUS EQUATIONS

3.41. Solve the system of equations

$$\frac{d^2x}{dt^2} + \frac{dy}{dt} + 3x = e^{-t}, \qquad \frac{d^2y}{dt^2} - 4\frac{dx}{dt} + 3y = \sin 2t$$

Write the system as

$$(D^2+3)x + Dy = e^{-t}, \qquad -4Dx + (D^2+3)y = \sin 2t \tag{1}$$

Method 1, using determinants.

Formal application of Cramer's rule for solving linear equations can be used and we find

$$x = \frac{\begin{vmatrix} e^{-t} & D \\ \sin 2t & D^2+3 \end{vmatrix}}{\begin{vmatrix} D^2+3 & D \\ -4D & D^2+3 \end{vmatrix}} = \frac{(D^2+3)(e^{-t}) - D(\sin 2t)}{D^4 + 10D^2 + 9} = \frac{4e^{-t} - 2\cos 2t}{(D^2+1)(D^2+9)}$$

$$y = \frac{\begin{vmatrix} D^2+3 & e^{-t} \\ -4D & \sin 2t \end{vmatrix}}{\begin{vmatrix} D^2+3 & D \\ -4D & D^2+3 \end{vmatrix}} = \frac{(D^2+3)(\sin 2t) - (-4D)(e^{-t})}{D^4 + 10D^2 + 9} = \frac{-\sin 2t - 4e^{-t}}{(D^2+1)(D^2+9)}$$

where we have expanded the determinants so that operators precede functions.

Note that the results are equivalent to

$$(D^2+1)(D^2+9)x = 4e^{-t} - 2\cos 2t, \qquad (D^2+1)(D^2+9)y = -\sin 2t - 4e^{-t}$$

Solving these equations we find

$$x = c_1 \cos t + c_2 \sin t + c_3 \cos 3t + c_4 \sin 3t + \tfrac{1}{5}e^{-t} + \tfrac{2}{15}\cos 2t$$

$$y = c_5 \cos t + c_6 \sin t + c_7 \cos 3t + c_8 \sin 3t - \tfrac{1}{5}e^{-t} + \tfrac{1}{15}\sin 2t$$

We can show that the total number of arbitrary constants in the solution is the same as the degree of the polynomial in D obtained from the determinant

$$\begin{vmatrix} D^2+3 & D \\ -4D & D^2+3 \end{vmatrix} = D^4 + 10D^2 + 9$$

i.e. 4. Thus there is a relationship between the constants c_1, c_2, c_3, c_4 and c_5, c_6, c_7, c_8. To determine this relationship we must substitute the values of x and y obtained above in the original equations. If we do, we find

$$c_5 = 2c_2, \qquad c_6 = -2c_1, \qquad c_7 = -2c_4, \qquad c_8 = 2c_3$$

Thus the required solution is

$$x = c_1 \cos t + c_2 \sin t + c_3 \cos 3t + c_4 \sin 3t + \tfrac{1}{5}e^{-t} + \tfrac{2}{15}\cos 2t$$

$$y = 2c_2 \cos t - 2c_1 \sin t - 2c_4 \cos 3t + 2c_3 \sin 3t - \tfrac{1}{5}e^{-t} + \tfrac{1}{15}\sin 2t$$

Method 2.

We can also eliminate one of the variables, for example x, by operating on the first of equations (1) with $4D$, the second with D^2+3 and adding. To solve the resulting equations, we can then use the method of undetermined coefficients.

APPLICATIONS

3.42. A particle P of mass 2 (gm) moves on the x axis attracted toward origin O with a force numerically equal to $8x$. If it is initially at rest at $x = 10$ (cm), find its position at any later time assuming (a) no other forces act, (b) a damping force numerically equal to 8 times the instantaneous velocity acts.

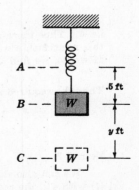

Fig. 3-1

(a) Choose the positive direction to the right [Fig. 3-1]. When $x > 0$, the net force is to the left [i.e. is negative] and so is $-8x$. When $x < 0$, the net force is to the right [i.e. is positive] and so is also $-8x$. Thus by Newton's law,

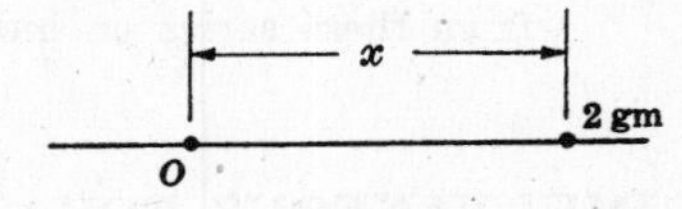

$$2\frac{d^2x}{dt^2} = -8x \quad \text{or} \quad \frac{d^2x}{dt^2} + 4x = 0 \quad \text{and} \quad x = c_1 \cos 2t + c_2 \sin 2t$$

Since $x = 10$, $dx/dt = 0$ at $t = 0$, we find $x = 10 \cos 2t$.

    The graph of the motion is shown in Fig. 3-2. The *amplitude* [maximum displacement from O] is 10 (cm). The *period* [time for a complete cycle] is π (sec). The *frequency* [number of cycles per second] is $1/\pi$ (cycles per second). The motion is often called *simple harmonic motion*.

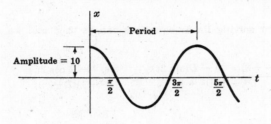

Fig. 3-2

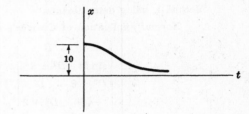

Fig. 3-3

(b) The damping force is given by $-8\,dx/dt$, regardless of where the particle is. Thus for example if $x < 0$ and $dx/dt > 0$, then the particle is to the left of O and moving to the right so the damping force must be to the left, i.e. negative. Thus by Newton's law,

$$2\frac{d^2x}{dt^2} = -8x - 8\frac{dx}{dt} \quad \text{or} \quad \frac{d^2x}{dt^2} + 4\frac{dx}{dt} + 4x = 0$$

the general solution of which is $x = e^{-2t}(c_1 + c_2 t)$. Since $x = 10$, $dx/dt = 0$ when $t = 0$, $c_1 = 10$, $c_2 = 20$ so that $x = 10e^{-2t}(1 + 2t)$. The motion is *non-oscillatory*. The particle approaches O but never reaches it [see Fig. 3-3].

3.43. A 20 lb weight suspended from the end of a vertical spring stretches it 6 inches. Assuming no external forces, find the position of the weight at any time if initially the weight is (a) pulled down 2 inches and released, (b) pulled down 3 inches and given an initial velocity of 2 ft/sec downward. Find the period and amplitude in each case.

    Let A and B [Fig. 3-4] represent the position of the end of the spring before and after the weight W is put on. B is called the *equilibrium position*. Call y the displacement of W at any position C from the equilibrium position. Assume that y is positive in the downward direction.

    By *Hooke's law*, 20 lb stretches the spring .5 ft, 40 lb stretches it 1 ft and so $40(.5 + y)$ lb stretches it $(.5 + y)$ ft. Thus when W is at C, the tension in the spring is $40(.5 + y)$ lb.

    By Newton's law,

Mass · Acceleration = Net force on W

                    = Weight downward $-$ Tension upward

or $\dfrac{20}{32}\dfrac{d^2y}{dt^2}$ =   20 $-$ $40(.5 + y)$

Fig. 3-4

which becomes
$$\frac{d^2y}{dt^2} + 64y = 0 \quad \text{or} \quad y = c_1 \cos 8t + c_2 \sin 8t$$

(a) Since $y = \frac{1}{6}$ ft, $dy/dt = 0$ at $t = 0$, we have $c_1 = \frac{1}{6}$, $c_2 = 0$ and so $y = \frac{1}{6} \cos 8t$. The amplitude is $\frac{1}{6}$ ft and the period is $2\pi/8 = \pi/4$ sec.

(b) Since $y = \frac{1}{4}$ ft, $dy/dt = 4$ ft/sec, $c_1 = 1/4$, $c_2 = 1/4$ and

$$y = \frac{1}{4} \cos 8t + \frac{1}{4} \sin 8t = \sqrt{(\tfrac{1}{4})^2 + (\tfrac{1}{4})^2} \sin\left(8t + \frac{\pi}{4}\right) = \frac{\sqrt{2}}{4} \sin\left(8t + \frac{\pi}{4}\right)$$

Then the amplitude is $\sqrt{2}/4 = .35$ ft approx. and the period is $\pi/4$ sec.

3.44. Solve Problem 3.43 taking into account an external damping force given in pounds by βv where v is the instantaneous velocity in ft/sec and (a) $\beta = 8$, (b) $\beta = 10$, (c) $\beta = 12.5$.

The equation of motion with damping force $\beta v = \beta \, dy/dt$ is

$$\frac{20}{32} \frac{d^2y}{dt^2} = -40y - \beta \frac{dy}{dt} \quad \text{or} \quad \frac{d^2y}{dt^2} + \frac{8\beta}{5} \frac{dy}{dt} + 64y = 0$$

(a) If $\beta = 8$, $\frac{d^2y}{dt^2} + 12.8 \frac{dy}{dt} + 64y = 0$. Solving subject to the conditions $y = 1/6$, $dy/dt = 0$ at $t = 0$, we find

$$y = \frac{1}{18} e^{-6.4t} (3 \cos 4.8t + 4 \sin 4.8t) = \frac{5}{18} e^{-6.4t} \sin (4.8t + 36°52')$$

The motion is *damped oscillatory* with period $2\pi/4.8 = 5\pi/12$ sec.

(b) If $\beta = 10$, $\frac{d^2y}{dt^2} + 16 \frac{dy}{dt} + 64y = 0$. Solving subject to the conditions as in (a), $y = \frac{1}{6} e^{-4t}(1 + 4t)$. The motion is called *critically damped* since any smaller value of β would produce oscillatory motion.

(c) If $\beta = 12.5$, $\frac{d^2y}{dt^2} + 20 \frac{dy}{dt} + 64y = 0$ and we find $y = \frac{1}{6} e^{-4t} - \frac{1}{24} e^{-16t}$. The motion is called *overdamped*.

3.45. (a) Work Problem 3.43 (a) if an external force given by $F(t) = 40 \cos 8t$ is applied for $t > 0$ and (b) give a physical interpretation of what happens as t increases.

(a) The equation of motion in this case becomes
$$\frac{20}{32} \frac{d^2y}{dt^2} = 20 - 40(.5 + y) + 40 \cos 8t$$

or
$$\frac{d^2y}{dt^2} + 64y = 64 \cos 8t$$

The solution of this, subject to $y = 1/6$, $dy/dt = 0$ at $t = 0$, is
$$y = \frac{1}{6} \cos 8t + 4t \sin 8t$$

(b) As t increases, the term $4t \sin 8t$ increases numerically without bound and physically the spring will ultimately break. This illustrates the phenomenon of *resonance* and shows what can happen when the frequency of the applied force is equal to the natural frequency of the system.

3.46. A rod AOB [Fig. 3-5 below] rotates in a vertical plane about a point O on it with constant angular velocity ω. A particle P of mass m is constrained to move along the rod. Assuming no frictional forces, find (a) a differential equation of motion of P, (b) the position of P at any time and (c) the condition under which P describes simple harmonic motion.

(a) Let r be the distance of P from O at time t and suppose that the rod is horizontal at $t = 0$. We have

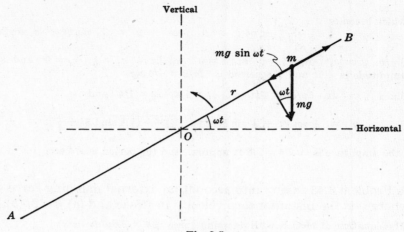

Fig. 3-5

$$\text{Net force on } P \;=\; \text{Centrifugal force} \;+\; \text{Component force due to gravity}$$

$$m\frac{d^2r}{dt^2} \;=\; m\omega^2 r \;-\; mg \sin \omega t$$

or
$$\frac{d^2r}{dt^2} - \omega^2 r \;=\; -g \sin \omega t \tag{1}$$

where
$$r = r_0, \;\; dr/dt = v_0 \quad \text{at} \quad t = 0 \tag{2}$$

i.e. r_0 and v_0 are the initial displacement and velocity of P.

(b) Solving (1) subject to (2), we find

$$r \;=\; \left(\frac{r_0}{2} + \frac{v_0}{2\omega} - \frac{g}{4\omega^2}\right)e^{\omega t} + \left(\frac{r_0}{2} - \frac{v_0}{2\omega} + \frac{g}{4\omega^2}\right)e^{-\omega t} + \frac{g \sin \omega t}{2\omega^2}$$

(c) Simple harmonic motion along the rod results if and only if $r_0 = 0$ and $v_0 = g/2\omega$.

3.47. An inductor of 2 henries, resistor of 16 ohms and capacitor of .02 farads are connected in series with a battery of e.m.f. $E = 100 \sin 3t$. At $t = 0$ the charge on the capacitor and current in the circuit are zero. Find the (a) charge and (b) current at $t > 0$.

Letting Q and I be the instantaneous charge and current at time t, we find by Kirchhoff's laws

$$2\frac{dI}{dt} + 16I + \frac{Q}{.02} \;=\; 100 \sin 3t$$

or since $I = dQ/dt$,

$$\frac{d^2Q}{dt^2} + 8\frac{dQ}{dt} + 25Q \;=\; 50 \sin 3t$$

Solving this subject to $Q = 0$, $dQ/dt = 0$ at $t = 0$, we find

(a)
$$Q \;=\; \frac{25}{52}(2 \sin 3t - 3 \cos 3t) + \frac{25}{52}e^{-4t}(3 \cos 3t + 2 \sin 3t)$$

(b)
$$I \;=\; \frac{dQ}{dt} \;=\; \frac{75}{52}(2 \cos 3t + 3 \sin 3t) - \frac{25}{52}e^{-4t}(17 \sin 3t + 6 \cos 3t)$$

The first term is the *steady-state* current and the second, which becomes negligible as time increases, is called the *transient* current.

3.48. Given the electric network of Fig. 3-6. Find the currents in the various branches if the initial currents are zero.

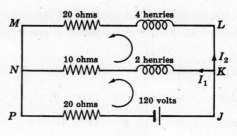

    Kirchhoff's second law states that the algebraic sum of the voltage drops around any closed loop is zero. Let us traverse loops *KLMNK* and *JKNPJ* in a counterclockwise direction as shown. In traversing these loops we consider voltage drops as positive when we go against the current. A voltage rise is considered as the negative of a voltage drop.

Fig. 3-6

    Let I be the current in *NPJKN*. This current divides at junction point K into I_1 and I_2 so that $I = I_1 + I_2$. This is equivalent to Kirchhoff's first law.

    Applying Kirchhoff's second law to loops *JKNPJ* and *KLMNK* respectively, we have

$$20I - 120 + 2\frac{dI_1}{dt} + 10I_1 = 0 \tag{1}$$

$$-10I_1 - 2\frac{dI_1}{dt} + 4\frac{dI_2}{dt} + 20I_2 = 0 \tag{2}$$

Putting $I = I_1 + I_2$ and using the operator $D = d/dt$, these become

$$(D + 15)I_1 + 10I_2 = 60 \tag{3}$$

$$-(D + 5)I_1 + (2D + 10)I_2 = 0 \tag{4}$$

Solving these subject to $I_1 = I_2 = 0$ at $t = 0$, we find

$$I_1 = 3(1 - e^{-20t}), \quad I_2 = \tfrac{3}{2}(1 - e^{-20t}), \quad I = \tfrac{9}{2}(1 - e^{-20t})$$

Supplementary Problems

OPERATORS

3.49. Write each of the following in operator form: (a) $y'' + 4y' + 5y = e^{-x}$, (b) $2y''' = x - 2y' + y$, (c) $xy' + 2y = 1$.

3.50. (a) Evaluate $(D^3 + 1)\sin 2x$ and $(D + 1)(D^2 - D + 1)\sin 2x$. (b) Are the operators $D^3 + 1$ and $(D + 1)(D^2 - D + 1)$ equivalent? Explain.

3.51. (a) Evaluate $(xD - 3)(D + 2)\{x^2 + x - 4\}$ and $(D + 2)(xD - 3)\{x^2 + x - 4\}$. (b) Are the operators $(xD - 3)(D + 2)$ and $(D + 2)(xD - 3)$ equivalent? Explain.

3.52. (a) Evaluate $(xD)(xD)(xD)\{x^2 + 2e^x\}$ which can be written $(xD)^3\{x^2 + 2e^x\}$. (b) Is this the same as $x^3D^3\{x^2 + 2e^x\}$? (c) Are the operators $(xD)^3$ and x^3D^3 equivalent? Explain.

3.53. Under what conditions will the operators $(D + a)(D + b)$ and $(D + b)(D + a)$ be equivalent? Explain.

LINEAR DEPENDENCE AND WRONSKIANS

3.54. (a) Show that the functions x^2, $3x + 2$, $x - 1$, $2x + 5$ are linearly dependent. (b) Are the functions x^2, $3x + 2$, $x - 1$ linearly dependent?

3.55. Investigate the linear dependence of e^x, xe^x, x^2e^x.

3.56. Show that $y = x$ is a solution of $xy'' + xy' - y = 0$ and find the general solution.

3.57. If y_1 is a solution of $y'' + p(x)y' + q(x)y = r(x)$, explain how to find the general solution. Illustrate by an example.

THE REDUCED OR HOMOGENEOUS SOLUTION

3.58.   (a) Find three linearly independent solutions of $(D+2)(D-1)(D-3)y = 0$ and (b) write the general solution.

3.59.   Solve   (a) $y'' + 8y' + 12y = 0$,   (b) $(D^2 - 4D - 1)y = 0$.

3.60.   Solve   (a) $(D^2 + 25)y = 0$; $y(0) = 2$, $y'(0) = -5$,   (b) $y'' - 8y' + 20y = 0$,   (c) $(D^3 + 8)y = 0$.

3.61.   Solve   (a) $y'' + 4y' + 4y = 0$ (d) $(D^8 + D^6)y = 0$

              (b) $16y'' - 8y' + y = 0$ (e) $(D^3 + 64)^2 y = 0$

              (c) $(D+6)^4(D-3)^2 y = 0$

3.62.   Solve $D^4(D+1)^2(D^2 + 4D + 5)^2(D^2 + 4)y = 0$.

THE COMPLETE OR NONHOMOGENEOUS EQUATION

3.63.   Solve   (a) $y'' - 5y' + 6y = 50 \sin 4x$,   (b) $(D^3 - 8)y = 16x + 18e^{-x} + 64 \cos 2x - 32$.

3.64.   Solve   (a) $y'' + 3y' + 2y = 4e^{-2x}$,   (b) $(D^3 + 3D^2)y = 180x^3 + 24x$.

3.65.   Solve $(D^6 - 2D^5 + D^4)y = 120x + 8e^x$.

3.66.   Solve using variation of parameters

           (a) $y'' + 2y' - 3y = xe^{-x}$ (c) $xy'' - y' = x$

           (b) $y'' + 4y = \csc 2x$ (d) $(D^3 + D)y = 4 \tan x$

OPERATOR TECHNIQUES

3.67.   Evaluate each of the following:

      (a) $\dfrac{1}{D+3}\{e^{-2x}\}$ (f) $\dfrac{1}{D^2 - 4}\{16x^3\}$

      (b) $\dfrac{1}{D^2 + D - 12}\{9e^{5x} - 4e^{-x}\}$ (g) $\dfrac{1}{D^2 + D - 2}\{x^2 e^{2x}\}$

      (c) $\dfrac{1}{(D+1)^2}\{4 \sin 2x + 3 \cos 2x\}$ (h) $\dfrac{1}{D^2 - 1}\{e^x(\sin x + \cos x)\}$

      (d) $\dfrac{D-1}{D^4 + D^2 + 1}\{8 \cos x\}$ (i) $\dfrac{1}{D^3 + D^2 + D}\{3x^4 - 2 \sin x\}$

      (e) $\dfrac{1}{(D-4)^5}\{xe^{4x}\}$ (j) $\dfrac{1}{(D-4)(D+3)(D+1)}\{e^{-2x}\cos 2x\}$

3.68.   Solve using operator techniques: (a) $(D^2 + 4D + 4)y = 18e^x - 8 \sin 2x$,   (b) $(D+1)(D-3)y = e^{3x} + x^2$,   (c) $(D+2)^2(D-2)y = e^{-2x}\cos x$.

3.69.   Prove entries   (a) A,   (b) B,   (c) E,   (d) H in the table on pages 75 and 76.

CAUCHY OR EULER EQUATION

3.70.   Solve   (a) $x^2 y'' + xy' - \lambda^2 y = 0$ (d) $x^3 y''' + 3x^2 y'' + xy' + 8y = 7x^{-1/2}$

              (b) $x^2 y'' - 2xy' + 2y = x^3$ (e) $r^2 R'' + 2rR' - n(n+1)R = 0$

              (c) $(2x^2 D^2 + 5xD + 1)y = \ln x$

MISCELLANEOUS METHODS

3.71.   (a) Show that $(xD^2 - xD + 1) = (D-1)(xD - 1)$.   (b) Thus solve $xy'' - xy' + y = x^2$.

3.72.   Work Problem 3.71(b) by noting that $y = x$ is a solution of the homogeneous equation and letting $y = vx$.

3.73.   Solve $y'' + 2xy' + x^2 y = 0$ by reducing the equation to canonical form.

3.74.   Show that $xy'' + (x+2)y' + y = 0$ is exact and find its solution.

3.75. Use the method of Frobenius to solve the equations

(a) $y'' + y = 0$,   (b) $xy'' - 2y' + xy = 0$,   (c) $y'' = xy$.

SIMULTANEOUS EQUATIONS

3.76. Solve the system $\dfrac{dx}{dt} + y = e^t$, $x - \dfrac{dy}{dt} = t$.

3.77. Solve the system $\dfrac{dx}{dt} + 2x = \dfrac{dy}{dt} + 10 \cos t$, $\dfrac{dy}{dt} + 2y = 4e^{-2t} - \dfrac{dx}{dt}$ if $x = 2$, $y = 0$ when $t = 0$.

3.78. Solve the system $(D^2 + 2)x + Dy = 2 \sin t + 3 \cos t + 5e^{-t}$, $Dx + (D^2 - 1)y = 3 \cos t - 5 \sin t - e^{-t}$ if $x = 2$, $y = -3$, $Dx = 0$, $Dy = 4$ when $t = 0$.

3.79. Solve $\dfrac{dx}{dt} = 2x + y$, $\dfrac{dy}{dt} = y + 6e^{-t}$, $\dfrac{dz}{dt} = 2 \sin t + x$ if $x = y = z = 0$ at $t = 0$.

APPLICATIONS

3.80. A particle moves on the x axis, attracted toward the origin O with a force proportional to its instantaneous distance from O. If the particle starts from rest at $x = 5$ cm and reaches $x = 2.5$ cm for the first time after 2 seconds, find (a) the position at any time t after it starts, (b) the magnitude of its velocity at $x = 0$, (c) the amplitude, period and frequency of the vibration and (d) the instantaneous acceleration.

3.81. The position of a particle moving along the x axis is determined by the equation $\dfrac{d^2x}{dt^2} + 4\dfrac{dx}{dt} + 8x = 20 \cos 2t$. If the particle starts from rest at $x = 0$, find (a) x as a function of t, (b) the amplitude, period and frequency after a long time.

3.82. A 60 lb weight hung on a vertical spring stretches it 2 inches. The weight is then pulled down 4 inches and released. (a) Find the position of the weight at any time if a damping force numerically equal to 15 times the instantaneous velocity is acting. (b) Is the motion oscillatory, overdamped or critically damped?

3.83. The weight on a vertical spring undergoes forced vibrations according to the equation

$$\frac{d^2x}{dt^2} + 4x = 8 \sin \omega t$$

where x is the displacement from the equilibrium position and $\omega > 0$ is a constant. If $x = dx/dt = 0$ when $t = 0$, find (a) x as a function of t, (b) the period of the external force for which resonance occurs.

3.84. The charge on the capacitor in the network of Fig. 3-7 is 2 coulombs. If the switch K is closed at time $t = 0$, find the charge and the current at any time $t > 0$ when (a) $E = 100$ volts, (b) $E = 100 \sin 4t$.

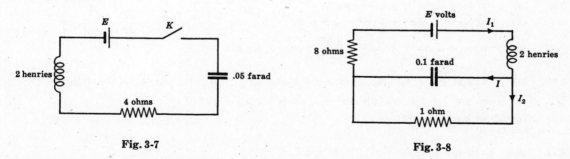

Fig. 3-7   Fig. 3-8

3.85. Find the currents I, I_1 and I_2 in the network of Fig. 3-8 and also the charge Q on the capacitor if (a) $E = 360$, (b) $E = 600e^{-5t} \sin 3t$. Assume the charge and currents are zero at $t = 0$.

Answers to Supplementary Problems

3.49.   (a) $(D^2 + 4D + 5)y = e^{-x}$,   (b) $(2D^3 + 2D - 1)y = x$,   (c) $(xD + 2)y = 1$

3.50.   (a) Both equal $\sin 2x - 8 \cos 2x$.   (b) Yes

3.51.   (a) $-2x^2 - 8x + 21$, $-2x^2 - 6x + 22$,   (b) No

3.52.   (a) $8x^2 + 2x^3 e^x + 6x^2 e^x + 2xe^x$ **3.55.**   Functions are linearly independent.

3.56. $y = c_1 x + c_2 x \int \dfrac{e^{-x}}{x^2}\,dx$ **3.58.**   (b) $y = c_1 e^{-2x} + c_2 e^x + c_3 e^{3x}$

3.59.   (a) $y = c_1 e^{-2x} + c_2 e^{-6x}$,   (b) $y = e^{2x}(c_1 e^{\sqrt{5}\,x} + c_2 e^{-\sqrt{5}\,x})$

3.60.   (a) $y = 2 \cos 5x - \sin 5x$,   (b) $y = e^{4x}(c_1 \cos 2x + c_2 \sin 2x)$

       (c) $y = c_1 e^{-2x} + e^x(c_1 \cos \sqrt{3}\,x + c_2 \sin \sqrt{3}\,x)$

3.61.   (a) $y = (c_1 + c_2 x)e^{-2x}$,   (b) $(c_1 + c_2 x)e^{x/4}$

       (c) $y = (c_1 + c_2 x + c_3 x^2 + c_4 x^3)e^{-6x} + (c_5 + c_6 x)e^{3x}$

       (d) $y = c_1 + c_2 x + c_3 x^2 + c_4 x^3 + c_5 x^4 + c_6 x^5 + c_7 \cos x + c_8 \sin x$

       (e) $y = (c_1 + c_2 x)e^{-4x} + e^{4x}(c_3 \cos 4x + c_4 \sin 4x) + xe^{4x}(c_5 \cos 4x + c_6 \sin 4x)$

3.62. $y = c_1 + c_2 x + c_3 x^2 + c_4 x^3 + (c_5 + c_6 x)e^{-x} + e^{-2x}(c_7 \cos x + c_8 \sin x)$

           $+ \, xe^{-2x}(c_9 \cos x + c_{10} \sin x) + c_{11} \cos 2x + c_{12} \sin 2x$

3.63.   (a) $y = c_1 e^{2x} + c_2 e^{3x} + 2 \cos 4x - \sin 4x$

       (b) $y = c_1 e^{2x} + e^{-x}(c_1 \cos \sqrt{3}\,x + c_2 \sin \sqrt{3}\,x) - 2x + 4 - 2e^{-x} - 4 \cos 2x - 4 \sin 2x$

3.64.   (a) $y = c_1 e^{-x} + c_2 e^{-2x} - 4xe^{-2x}$ (b) $y = c_1 e^{-3x} + c_2 + c_3 x - 8x^2 + 8x^3 - 5x^4 + 3x^5$

3.65. $y = c_1 + c_2 x + c_3 x^2 + c_4 x^3 + (c_5 + c_6 x)e^x + x^5 + 4x^2 e^x$

3.66.   (a) $y = c_1 e^{-3x} + c_2 e^x - \frac{1}{4} xe^{-x}$

       (b) $y = c_1 \cos 2x + c_2 \sin 2x - \frac{1}{2} x \cos 2x - \frac{1}{4} \sin 2x \ln \csc 2x$

       (c) $y = c_1 + c_2 x^2 + \frac{1}{2} x^2 (\ln x - \frac{1}{2})$

       (d) $y = c_1 + c_2 \cos x + c_3 \sin x - 4 \cos x \ln (\sec x + \tan x)$

3.67.   (a) e^{-2x} (d) $-8(\sin x + \cos x)$ (g) $\dfrac{e^{2x}}{32}(8x^2 - 20x + 21)$ (i) $\frac{3}{5}x^5 - 3x^4 + 36x^2 - 72x + 2 \sin x$

       (b) $\frac{1}{2}e^{5x} + \frac{1}{3}e^{-x}$ (e) $\dfrac{x^6 e^{4x}}{6!}$ (h) $\frac{1}{5}e^x(\sin x - 3 \cos x)$ (j) $\dfrac{e^{-2x}}{100}(3 \cos 2x - \sin 2x)$

       (c) $-\cos 2x$ (f) $-4x^3 - 6x$

3.68.   (a) $y = (c_1 + c_2 x)e^{-2x} + 2e^x + \cos 2x$

       (b) $y = c_1 e^{-x} + c_2 e^{3x} + \frac{1}{4} xe^{3x} - \frac{1}{3} x^2 + \frac{4}{9} x - \frac{14}{27}$

       (c) $y = (c_1 + c_2 x)e^{-2x} + c_3 e^{2x} + \dfrac{e^{-2x}}{17}(4 \cos x - \sin x)$

3.70.   (a) $y = c_1 x^\lambda + c_2 x^{-\lambda}$,   (b) $y = c_1 x + c_2 x^2 + \frac{1}{2} x^3$,   (c) $y = c_1 x + c_2 x^{-1/2} + \ln x - 3$

       (d) $y = c_1 x^{-2} + x[c_2 \cos (\sqrt{3} \ln x) + c_3 \sin (\sqrt{3} \ln x)] + \frac{8}{9} x^{-1/2}$

       (e) $R = c_1 r^n + c_2 / r^{n+1}$

3.71. $y = c_1 x + c_2 x \int \dfrac{e^x}{x^2}\,dx - x^2 - 2x \ln x + 2$

3.73. $y = e^{-x^2/2}(c_1 e^x + c_2 e^{-x})$ **3.74.** $y = (c_1 + c_2 e^{-x})/x$

3.75. (a) $y = c_1\left(1 - \dfrac{x^2}{2!} + \dfrac{x^4}{4!} - \cdots\right) + c_2\left(x - \dfrac{x^3}{3!} + \dfrac{x^5}{5!} - \cdots\right) = c_1 \cos x + c_2 \sin x$

(b) $y = c_1\left(x^{-1} + \dfrac{x}{2!} + \dfrac{x^3}{4!} + \cdots\right) + c_2\left(1 + \dfrac{x^2}{3!} + \dfrac{x^4}{5!} + \cdots\right) = \dfrac{c_1 \cosh x + c_2 \sinh x}{x}$

(c) $y = c_1\left(1 + \dfrac{x^3}{3!} + \dfrac{1 \cdot 4 x^6}{6!} + \dfrac{1 \cdot 4 \cdot 7 x^9}{9!} + \cdots\right)$

$\qquad\qquad + c_2\left(x + \dfrac{2x^4}{4!} + \dfrac{2 \cdot 5 x^7}{7!} + \dfrac{2 \cdot 5 \cdot 8 x^{10}}{10!} + \cdots\right)$

3.76. $x = c_1 \cos t - c_2 \sin t + \tfrac{1}{2}e^t + t, \quad y = c_1 \sin t + c_2 \cos t + \tfrac{1}{2}e^t - 1$

3.77. $x = 4 \cos t + 3 \sin t - 2e^{-2t} - 2e^{-t}\sin t, \quad y = \sin t - 2 \cos t + 2e^{-t}\cos t$

3.78. $x = \cos t + \sin t + e^{-t}, \quad y = 2 \sin t - \cos t - 2e^{-t}$

3.79. $x = 2e^{2t} - 3e^t + e^{-t}, \quad y = 3e^t - 3e^{-t}, \quad z = e^{2t} - e^{-t} - 3e^t + 5 - 2 \cos t$

3.80. (a) $x = 5 \cos(\pi t/6)$, (b) $5\pi/6$(cm/sec), (c) amp. $= 5$ cm, period $= 12$ sec, freq. $= \tfrac{1}{12}$ cycle per second, (d) $-25\pi^2/36$(cm/sec^2)

3.81. (a) $x = \cos 2t + 2 \sin 2t - e^{-2t}(\cos 2t + 3 \sin 2t)$

(b) amp. $= \sqrt{5}$, period $= \pi$, freq. $= 1/\pi$

3.82. (a) $x = 4e^{-4t}(4t + 1)$, (b) critically damped

3.83. (a) $x = \dfrac{8 \sin \omega t - 4\omega \sin 2t}{4 - \omega^2}$ if $\omega \neq 2$, $x = \sin 2t - 2t \cos 2t$ if $\omega = 2$

(b) $\omega = 2$ or period $= \pi$

3.84. (a) $Q = 5 - e^{-t}(3 \cos 3t + \sin 3t), \quad I = 10e^{-t}\sin 3t$

(b) $Q = 6e^{-t}(\sin 3t + \cos 3t) - 3 \sin 4t - 4 \cos 4t,$

$\quad I = 12e^{-t}(\cos 3t - 2 \sin 3t) - 12 \cos 4t + 16 \sin 4t$

3.85. (a) $Q = 4 + 5e^{-9t} - 9e^{-5t}, \quad I = 45(e^{-5t} - e^{-9t}),$

$\quad I_1 = 40 + 5e^{-9t} - 45e^{-5t}, \quad I_2 = 40 + 50e^{-9t} - 90e^{-5t}$

(b) $Q = 25e^{-5t} - 9e^{-9t} - 4e^{-5t}(4 \cos 3t + 3 \sin 3t),$

$\quad I = 81e^{-9t} - 125e^{-5t} + 4e^{-5t}(11 \cos 3t + 27 \sin 3t),$

$\quad I_1 = 125e^{-5t} - 9e^{-9t} - 4e^{-5t}(29 \cos 3t + 3 \sin 3t),$

$\quad I_2 = 250e^{-5t} - 90e^{-9t} - 40e^{-5t}(4 \cos 3t + 3 \sin 3t)$

Chapter 4

Laplace Transforms

DEFINITION OF A LAPLACE TRANSFORM

The Laplace transform of a function $f(t)$ is defined as

$$\mathcal{L}\{f(t)\} = F(s) = \int_0^\infty e^{-st} f(t)\, dt \qquad (1)$$

and is said to exist or not according as the integral in (1) exists [converges] or does not exist [diverges]. In this chapter we assume that s is real. Later [Chapter 14] we shall find it convenient to take s as complex.

Often in practice there will be a real number s_0 such that the integral (1) exists for $s > s_0$ and does not exist for $s \leqq s_0$. The set of values $s > s_0$ for which (1) exists is called the *range of convergence* or *existence* of $\mathcal{L}\{f(t)\}$. It may happen, however, that (1) does not exist for any value of s [see Problem 4.50].

The symbol $\mathcal{L}$ in (1) is called the *Laplace transform operator*. We can show that $\mathcal{L}$ is a linear operator, i.e.

$$\mathcal{L}\{c_1 f_1(t) + c_2 f_2(t)\} = c_1\mathcal{L}\{f_1(t)\} + c_2\mathcal{L}\{f_2(t)\} \qquad (2)$$

LAPLACE TRANSFORMS OF SOME ELEMENTARY FUNCTIONS

In the following table we give Laplace transforms of some special elementary functions together with the range of existence or convergence. Often, however, we shall omit this range of existence since in most instances it can easily be supplied when needed.

	$f(t)$		$\mathcal{L}\{f(t)\} = F(s)$	
1.	1		$\dfrac{1}{s}$	$s > 0$
2.	t^n	$n = 1, 2, 3, \ldots$	$\dfrac{n!}{s^{n+1}}$	$s > 0$
3.	t^p	$p > -1$	$\dfrac{\Gamma(p+1)}{s^{p+1}}$	$s > 0$
4.	e^{at}		$\dfrac{1}{s-a}$	$s > a$
5.	$\cos \omega t$		$\dfrac{s}{s^2 + \omega^2}$	$s > 0$
6.	$\sin \omega t$		$\dfrac{\omega}{s^2 + \omega^2}$	$s > 0$
7.	$\cosh at$		$\dfrac{a}{s^2 - a^2}$	$s > |a|$
8.	$\sinh at$		$\dfrac{s}{s^2 - a^2}$	$s > |a|$

In entry 3, $\Gamma(p+1)$ is the *gamma function* defined by

$$\Gamma(p+1) \;=\; \int_0^\infty x^p e^{-x}\,dx \qquad p>-1 \tag{3}$$

A study of this function is provided in Chapter 9. For our present purposes, however, we need only the following properties:

$$\Gamma(p+1) \;=\; p\Gamma(p), \quad \Gamma(\tfrac{1}{2}) = \sqrt{\pi}, \quad \Gamma(1) = 1 \tag{4}$$

The first is called a *recursion formula* for the gamma function. Note that if p is any positive integer n then $\Gamma(n+1) = n!$, thus explaining the relationship of entries 2 and 3 of the table.

SUFFICIENT CONDITIONS FOR EXISTENCE OF LAPLACE TRANSFORMS

In order to be able to state sufficient conditions on $f(t)$ under which we can guarantee the existence of $\mathcal{L}\{f(t)\}$, we introduce the concepts of *piecewise continuity* and *exponential order* as follows.

1. **Piecewise continuity.** A function $f(t)$ is said to be *piecewise continuous* in an interval if (i) the interval can be divided into a finite number of subintervals in each of which $f(t)$ is continuous and (ii) the limits of $f(t)$ as t approaches the endpoints of each subinterval are finite. Another way of stating this is to say that a piecewise continuous function is one that has only a finite number of finite discontinuities. An example of a piecewise continuous function is shown in Fig. 4-1.

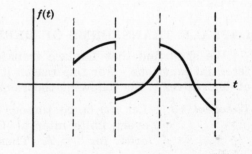

Fig. 4-1

2. **Exponential order.**   A function $f(t)$ is said to be of *exponential order* for $t > T$ if we can find constants M and α such that $|f(t)| \leqq Me^{\alpha t}$ for $t > T$.

Using these we have the following theorem,

Theorem 4-1.   If $f(t)$ is piecewise continuous in every finite interval $0 \leqq t \leqq T$ and is of exponential order for $t > T$, then $\mathcal{L}\{f(t)\}$ exists for $s > \alpha$.

It should be emphasized that these conditions are only sufficient [and not necessary], i.e. if the conditions are not satisfied $\mathcal{L}\{f(t)\}$ may still exist. For example, $\mathcal{L}\{t^{-1/2}\}$ exists even though $t^{-1/2}$ is not piecewise continuous in $0 \leqq t \leqq T$.

An interesting theorem which is related to Theorem 4-1 is the following

Theorem 4-2.   If $f(t)$ satisfies the conditions of Theorem 4-1, then

$$\lim_{s \to \infty} \mathcal{L}\{f(t)\} = \lim_{s \to \infty} F(s) = 0$$

It follows that if $\lim_{s \to \infty} F(s) \neq 0$, then $f(t)$ cannot satisfy the conditions of Theorem 4-1.

INVERSE LAPLACE TRANSFORMS

If $\mathcal{L}\{f(t)\} = F(s)$, then we call $f(t)$ the *inverse Laplace transform* of $F(s)$ and write $\mathcal{L}^{-1}\{F(s)\} = f(t)$.

Example 1. Since $\mathcal{L}\{t\} = \dfrac{1}{s^2}$, we have $\mathcal{L}^{-1}\left\{\dfrac{1}{s^2}\right\} = t$.

While it is clear that whenever a Laplace transform exists it is unique, the same is not true for an inverse Laplace transform.

> **Example 2.** If $f(t) = \begin{cases} t & t \neq 2 \\ 10 & t = 2 \end{cases}$ we can show that $\mathcal{L}\{f(t)\} = 1/s^2$. However, this function $f(t)$ differs from that of Example 1 at $t = 2$ although both have the same Laplace transform. It follows that $\mathcal{L}^{-1}(1/s^2)$ can represent two (or more) different functions.

We can show that if two functions have the same Laplace transform, then they cannot differ from each other on any interval of positive length no matter how small. This is sometimes called *Lerch's theorem*. The theorem implies that if two functions have the same Laplace transform, then they are for all practical purposes the same and so in practice we can take the inverse Laplace transform as *essentially* unique. In particular if two continuous functions have the same Laplace transform, they must be identical.

The symbol $\mathcal{L}^{-1}$ is called the *inverse Laplace transform* operator and is a linear operator, i.e.

$$\mathcal{L}^{-1}\{c_1 F_1(s) + c_2 F_2(s)\} = c_1 f_1(t) + c_2 f_2(t)$$

LAPLACE TRANSFORMS OF DERIVATIVES

We shall find that Laplace transforms provide useful means for solving linear differential equations. For this reason it will be necessary for us to find Laplace transforms of derivatives. The following theorems are fundamental.

Theorem 4-3.   Let $f(t)$ be continuous and have a piecewise continuous derivative $f'(t)$ in every finite interval $0 \leqq t \leqq T$. Suppose also that $f(t)$ is of exponential order for $t > T$. Then

$$\mathcal{L}\{f'(t)\} = s\,\mathcal{L}\{f(t)\} - f(0)$$

This can be extended as follows.

Theorem 4-4.   Let $f(t)$ be such that $f^{(n-1)}(t)$ is continuous and $f^{(n)}(t)$ piecewise continuous in every finite interval $0 \leqq t \leqq T$. Suppose also that $f(t), f'(t), \ldots, f^{(n-1)}(t)$ are of exponential order for $t > T$. Then

$$\mathcal{L}\{f^{(n)}(t)\} = s^n\,\mathcal{L}\{f(t)\} - s^{n-1}f(0) - s^{n-2}f'(0) - \cdots - f^{(n-1)}(0)$$

THE UNIT STEP FUNCTION

The *unit step function*, also called *Heaviside's unit step function*, is defined as

$$\mathcal{U}(t-a) = \begin{cases} 0 & t < a \\ 1 & t > a \end{cases}$$

and is shown graphically in Fig. 4-2.

It is possible to express various discontinuous functions in terms of the unit step function.

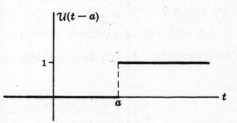

Fig. 4-2

We can show [Problem 4.17] that the Laplace transform of the unit step function is

$$\mathcal{L}\{\mathcal{U}(t-a)\} = \frac{e^{-as}}{s} \qquad s > 0$$

and similarly we have

$$\mathcal{L}^{-1}\left\{\frac{e^{-as}}{s}\right\} = \mathcal{U}(t-a)$$

SOME SPECIAL THEOREMS ON LAPLACE TRANSFORMS

Because of the relationship between Laplace transforms and inverse Laplace transforms, any theorem involving Laplace transforms will have a corresponding theorem involving inverse Laplace transforms. In the following we shall consider some of the important results involving Laplace transforms and corresponding inverse Laplace transforms. In all cases we assume that $f(t)$ satisfies the conditions of Theorem 4-1.

Theorem 4-5 [First translation theorem]. If $\mathcal{L}\{f(t)\} = F(s)$, then

$$\mathcal{L}\{e^{at}f(t)\} = F(s-a)$$

Similarly if $\mathcal{L}^{-1}\{F(s)\} = f(t)$, then

$$\mathcal{L}^{-1}\{F(s-a)\} = e^{at}f(t)$$

Theorem 4-6 [Second translation theorem]. If $\mathcal{L}\{f(t)\} = F(s)$, then

$$\mathcal{L}\{\mathcal{U}(t-a)f(t-a)\} = e^{-as}F(s)$$

Similarly if $\mathcal{L}^{-1}\{F(s)\} = f(t)$, then

$$\mathcal{L}^{-1}\{e^{-as}F(s)\} = \mathcal{U}(t-a)f(t-a)$$

Theorem 4-7. If $\mathcal{L}\{f(t)\} = F(s)$, then

$$\mathcal{L}\{f(at)\} = \frac{1}{a}F\left(\frac{s}{a}\right)$$

Similarly if $\mathcal{L}^{-1}\{F(s)\} = f(t)$, then

$$\mathcal{L}^{-1}\left\{F\left(\frac{s}{a}\right)\right\} = af(at)$$

Theorem 4-8. If $\mathcal{L}\{f(t)\} = F(s)$ then if $n = 1, 2, 3, \ldots$

$$\mathcal{L}\{t^n f(t)\} = (-1)^n \frac{d^n F}{ds^n} = (-1)^n F^{(n)}(s)$$

Similarly if $\mathcal{L}^{-1}\{F(s)\} = f(t)$, then

$$\mathcal{L}^{-1}\{F^{(n)}(s)\} = (-1)^n t^n f(t)$$

Theorem 4-9 [Periodic functions]. If $f(t)$ has period $P > 0$, i.e. if $f(t+P) = f(t)$, then

$$\mathcal{L}\{f(t)\} = \frac{\int_0^P e^{-st}f(t)\,dt}{1 - e^{-sP}}$$

Theorem 4-10 [Integration]. If $\mathcal{L}\{f(t)\} = F(s)$, then

$$\mathcal{L}\left\{\int_0^t f(u)\,du\right\} = \frac{F(s)}{s}$$

Similarly if $\mathcal{L}^{-1}\{F(s)\} = f(t)$, then

$$\mathcal{L}^{-1}\left\{\frac{F(s)}{s}\right\} = \int_0^t f(u)\,du$$

Theorem 4-11. If $\lim_{t \to 0} \frac{f(t)}{t}$ exists and $\mathcal{L}\{f(t)\} = F(s)$, then

$$\mathcal{L}\left\{\frac{f(t)}{t}\right\} = \int_s^\infty F(u)\,du$$

Theorem 4-12 [Convolution theorem]. If $\mathcal{L}\{f(t)\} = F(s)$, $\mathcal{L}\{g(t)\} = G(s)$, then

$$\mathcal{L}\left\{ \int_0^t f(u)\,g(t-u)\,du \right\} \;=\; F(s)\,G(s)$$

Similarly if $\mathcal{L}^{-1}\{F(s)\} = f(t)$, $\mathcal{L}^{-1}\{G(s)\} = g(t)$, then

$$\mathcal{L}^{-1}\{F(s)\,G(s)\} \;=\; \int_0^t f(u)\,g(t-u)\,du$$

We call the above integral the *convolution* of f and g and write

$$f*g \;=\; \int_0^t f(u)\,g(t-u)\,du$$

We have the result $f*g = g*f$, i.e. the convolution is commutative. Similarly we can prove that it is associative and distributive [see Problem 4.75].

PARTIAL FRACTIONS

Although the above theorems are often useful in finding inverse Laplace transforms, perhaps the most important single elementary method for our purposes is the method of *partial fractions*. This is because in many problems which we shall encounter it will be necessary to find the inverse of $P(s)/Q(s)$ where $P(s)$ and $Q(s)$ are polynomials and the degree of $Q(s)$ is larger than that of $P(s)$. For illustrations of the method see Problems 4.39-4.41.

SOLUTIONS OF DIFFERENTIAL EQUATIONS BY LAPLACE TRANSFORMS

The method of Laplace transforms is particularly useful for solving linear differential equations with constant coefficients and associated initial conditions. To accomplish this we take the Laplace transform of the given differential equation [or equations in the case of a system], making use of the initial conditions. This leads to an algebraic equation [or system of algebraic equations] in the Laplace transform of the required solution. By solving for this Laplace transform and then taking the inverse, the required solution is obtained. For illustrations see Problems 4.42-4.44.

APPLICATIONS TO PHYSICAL PROBLEMS

Since formulation of many physical problems leads to linear differential equations with initial conditions, the Laplace transform method is particularly suited for obtaining their solutions. For applications to various fields see Problems 4.45-4.47.

LAPLACE INVERSION FORMULAS

There exists a direct method for finding inverse Laplace transforms, called the *complex inversion formula*. This makes use of the theory of complex variables and is considered in Chapter 14.

Solved Problems

LAPLACE TRANSFORMS OF ELEMENTARY FUNCTIONS

4.1. Prove that $\mathcal{L}\{e^{at}\} = \dfrac{1}{s-a}$ if $s > a$.

$$\mathcal{L}\{e^{at}\} = \int_0^\infty e^{-st} e^{at}\, dt = \int_0^\infty e^{-(s-a)t}\, dt = \frac{e^{-(s-a)t}}{s-a}\bigg|_0^\infty$$

$$= \frac{1}{s-a} \text{ provided } s-a > 0, \text{ i.e. } s > a$$

4.2. (a) Prove that $\mathcal{L}\{t^p\} = \dfrac{\Gamma(p+1)}{s^{p+1}}$ if $s > 0$ and $p > -1$.

(b) Show that if $p = n$, a positive integer, then $\mathcal{L}\{t^n\} = \dfrac{n!}{s^{n+1}}$ where $s > 0$.

(a) $\mathcal{L}\{t^p\} = \displaystyle\int_0^\infty e^{-st} t^p\, dt$. Let $st = u$ and note that in order for the integral to converge we must have $s > 0$. Then the integral equals

$$\frac{1}{s^{p+1}}\int_0^\infty u^p e^{-u}\, du = \frac{\Gamma(p+1)}{s^{p+1}}$$

The restriction $p > -1$ occurs because the integral defining the gamma function converges if and only if $p > -1$.

(b) Integrating by parts, we have

$$\Gamma(p+1) = \int_0^\infty x^p e^{-x}\, dx = (x^p)(-e^{-x})\bigg|_0^\infty - \int_0^\infty (px^{p-1})(-e^{-x})\, dx$$

$$= p\int_0^\infty x^{p-1} e^{-x}\, dx = p\Gamma(p)$$

i.e. $$\Gamma(p+1) = p\Gamma(p)$$

If $p = n$, then

$$\Gamma(n+1) = n\Gamma(n) = n(n-1)\Gamma(n-1) = n(n-1)(n-2)\Gamma(n-2) = n(n-1)\cdots 1\Gamma(1)$$

But $\Gamma(1) = \displaystyle\int_0^\infty e^{-x}\, dx = -e^{-x}\bigg|_0^\infty = 1$. Thus $\Gamma(n+1) = n!$ and so from (a), $\mathcal{L}\{t^n\} = \dfrac{n!}{s^{n+1}}$.

4.3. Prove that $\mathcal{L}$ is a linear operator.

We must show that if c_1, c_2 are any constants and $f_1(t), f_2(t)$ any functions whose Laplace transforms exist, then

$$\mathcal{L}\{c_1 f_1(t) + c_2 f_2(t)\} = c_1\mathcal{L}\{f_1(t)\} + c_2\mathcal{L}\{f_2(t)\}$$

We have

$$\mathcal{L}\{c_1 f_1(t) + c_2 f_2(t)\} = \int_0^\infty e^{-st}\{c_1 f_1(t) + c_2 f_2(t)\}\, dt$$

$$= c_1\int_0^\infty e^{-st} f_1(t)\, dt + c_2\int_0^\infty e^{-st} f_2(t)\, dt$$

$$= c_1\mathcal{L}\{f_1(t)\} + c_2\mathcal{L}\{f_2(t)\}$$

and so $\mathcal{L}$ is a linear operator.

4.4. Prove that (a) $\mathcal{L}\{\cos \omega t\} = \dfrac{s}{s^2 + \omega^2}$, (b) $\mathcal{L}\{\sin \omega t\} = \dfrac{\omega}{s^2 + \omega^2}$.

Method 1. We have if $s > 0$,

$$\int_0^\infty e^{-st} e^{i\omega t}\, dt \;=\; \int_0^\infty e^{-(s-i\omega)t}\, dt \;=\; \frac{e^{-(s-i\omega)t}}{s - i\omega}\Big|_0^\infty \;=\; \frac{1}{s - i\omega}$$

Then taking real and imaginary parts, we have

$$\int_0^\infty e^{-st}(\cos \omega t + i \sin \omega t)\, dt \;=\; \frac{s + i\omega}{s^2 + \omega^2} \;=\; \frac{s}{s^2 + \omega^2} + i\frac{\omega}{s^2 + \omega^2}$$

or

$$\int_0^\infty e^{-st} \cos \omega t\, dt \;=\; \frac{s}{s^2 + \omega^2}, \qquad \int_0^\infty e^{-st} \sin \omega t\, dt \;=\; \frac{\omega}{s^2 + \omega^2}$$

Method 2. By direct integration,

$$\int_0^\infty e^{-st} \cos \omega t\, dt \;=\; \frac{e^{-st}(\omega \sin \omega t - s \cos \omega t)}{s^2 + \omega^2}\Big|_0^\infty \;=\; \frac{s}{s^2 + \omega^2}$$

$$\int_0^\infty e^{-st} \sin \omega t\, dt \;=\; \frac{-e^{-st}(s \sin \omega t + \omega \cos \omega t)}{s^2 + \omega^2}\Big|_0^\infty \;=\; \frac{\omega}{s^2 + \omega^2}$$

4.5. Find the Laplace transforms of each of the following:

(a) $3e^{-4t}$,   (b) $2t^2$,   (c) $4\cos 5t$,   (d) $\sin \pi t$,   (e) $-3/\sqrt{t}$.

(a) $\mathcal{L}\{3e^{-4t}\} \;=\; 3\mathcal{L}\{e^{-4t}\} \;=\; \dfrac{3}{s - (-4)} \;=\; \dfrac{3}{s + 4}, \quad s > -4$

(b) $\mathcal{L}\{2t^2\} \;=\; 2\mathcal{L}\{t^2\} \;=\; \dfrac{2\Gamma(3)}{s^3} \;=\; \dfrac{2 \cdot 2!}{s^3} \;=\; \dfrac{4}{s^3}, \quad s > 0$

(c) $\mathcal{L}\{4\cos 5t\} \;=\; 4\mathcal{L}\{\cos 5t\} \;=\; 4 \cdot \dfrac{s}{s^2 + 25} \;=\; \dfrac{4s}{s^2 + 25}, \quad s > 0$

(d) $\mathcal{L}\{\sin \pi t\} \;=\; \dfrac{\pi}{s^2 + \pi^2}, \quad s > 0$

(e) $\mathcal{L}\left\{-\dfrac{3}{\sqrt{t}}\right\} \;=\; -3\mathcal{L}\{t^{-1/2}\} \;=\; -\dfrac{3\Gamma(1/2)}{s^{1/2}} \;=\; -\dfrac{3\sqrt{\pi}}{\sqrt{s}} \;=\; -3\sqrt{\dfrac{\pi}{s}}, \quad s > 0$

4.6. Find the Laplace transforms of each of the following:

(a) $3t^4 - 2t^{3/2} + 6$,   (b) $5\sin 2t - 3\cos 2t$,   (c) $3\sqrt[3]{t} + 4e^{2t}$,   (d) $1/t^2$.

(a) $\mathcal{L}\{3t^4 - 2t^{3/2} + 6\} \;=\; 3\mathcal{L}\{t^4\} - 2\mathcal{L}\{t^{3/2}\} + 6\mathcal{L}\{1\}$

$$= \; \frac{3\Gamma(5)}{s^5} - \frac{2\Gamma(5/2)}{s^{5/2}} + \frac{6}{s}$$

$$= \; \frac{3 \cdot 4!}{s^5} - \frac{2 \cdot (3/2)(1/2)\Gamma(1/2)}{s^{5/2}} + \frac{6}{s}$$

$$= \; \frac{72}{s^5} - \frac{3\sqrt{\pi}}{2s^{5/2}} + \frac{6}{s}$$

(b) $\mathcal{L}\{5\sin 2t - 3\cos 2t\} \;=\; 5\mathcal{L}\{\sin 2t\} - 3\mathcal{L}\{\cos 2t\}$

$$= \; \frac{5 \cdot 2}{s^2 + 4} - \frac{3 \cdot s}{s^2 + 4} \;=\; \frac{10 - 3s}{s^2 + 4}$$

(c) $\mathcal{L}\{3\sqrt[3]{t} + 4e^{2t}\} \;=\; 3\mathcal{L}\{t^{1/3}\} + 4\mathcal{L}\{e^{2t}\} \;=\; \dfrac{3\Gamma(4/3)}{s^{4/3}} + \dfrac{4}{s - 2}$

$$= \; \frac{3(1/3)\Gamma(1/3)}{s^{4/3}} + \frac{4}{s - 2} \;=\; \frac{\Gamma(1/3)}{s^{4/3}} + \frac{4}{s - 2}$$

(d) $\mathcal{L}\left\{\dfrac{1}{t^2}\right\} = \displaystyle\int_0^\infty \frac{e^{-st}}{t^2}\, dt$. Since this integral does not converge, the Laplace transform does not exist.

In parts (a), (b), (c) we have omitted the range of existence which can easily be supplied.

4.7.   If $f(t) = \begin{cases} 3 & 0 < t < 2 \\ -1 & 2 < t < 4 \\ 0 & t \geqq 4 \end{cases}$ find $\mathcal{L}\{f(t)\}$.

We have

$$\mathcal{L}\{f(t)\} = \int_0^2 e^{-st} f(t)\, dt + \int_2^4 e^{-st} f(t)\, dt + \int_4^\infty e^{-st} f(t)\, dt$$

$$= \int_0^2 e^{-st}(3)\, dt + \int_2^4 e^{-st}(-1)\, dt + \int_4^\infty e^{-st}(0)\, dt$$

$$= 3\left(\frac{e^{-st}}{-s}\right)\Big|_0^2 - \left(\frac{e^{-st}}{-s}\right)\Big|_2^4 + 0$$

$$= \frac{3(1 - e^{-2s})}{s} + \frac{e^{-4s} - e^{-2s}}{s} = \frac{3 - 4e^{-2s} + e^{-4s}}{s}$$

4.8.   Find $\mathcal{L}\{\sin t \cos t\}$.

We have $\sin 2t = 2 \sin t \cos t$ so that $\sin t \cos t = \frac{1}{2} \sin 2t$.   Thus

$$\mathcal{L}\{\sin t \cos t\} = \frac{1}{2} \mathcal{L}\{\sin 2t\} = \frac{1}{2} \cdot \frac{2}{s^2 + 4} = \frac{1}{s^2 + 4}$$

EXISTENCE OF LAPLACE TRANSFORMS

4.9.   Prove (a) Theorem 4-1, and (b) Theorem 4-2, page 99.

(a) We have

$$F(s) = \mathcal{L}\{f(t)\} = \int_0^\infty e^{-st} f(t)\, dt = \int_0^T e^{-st} f(t)\, dt + \int_T^\infty e^{-st} f(t)\, dt$$

Now since $f(t)$ is piecewise continuous for $0 \leqq t \leqq T$ so also is $e^{-st} f(t)$, and thus the first integral on the right exists.

To show that the second integral on the right also exists, we use the fact that $|f(t)| \leqq Me^{\alpha t}$ so that

$$\left| \int_T^\infty e^{-st} f(t)\, dt \right| \leqq \int_T^\infty e^{-st} |f(t)|\, dt \leqq \int_T^\infty e^{-st} Me^{\alpha t}\, dt$$

$$\leqq M \int_0^\infty e^{-(s-\alpha)}\, dt = \frac{M}{s - \alpha} \tag{1}$$

for $s > \alpha$ and the required result is proved.

(b) We have as in part (a),

$$|F(s)| = |\mathcal{L}\{f(t)\}| \leqq \int_0^T e^{-st} |f(t)|\, dt + \int_T^\infty e^{-st} |f(t)|\, dt$$

Now since $f(t)$ is piecewise continuous for $0 \leqq t \leqq T$, it is bounded, i.e. $|f(t)| \leqq K$ for some constant K. Using this and the result (1), we have

$$|F(s)| \leqq \int_0^T e^{-st} K\, dt + \frac{M}{s - \alpha} \leqq \int_0^\infty e^{-st} K\, dt + \frac{M}{s - \alpha} = \frac{K + M}{s - \alpha}$$

Taking the limit as $s \to \infty$, it follows that $\lim\limits_{s \to \infty} F(s) = 0$ as required.

4.10.   Prove that (a) $\mathcal{L}\left\{\dfrac{e^{2t}}{t + 4}\right\}$ exists, (b) $\lim\limits_{s \to \infty} \mathcal{L}\left\{\dfrac{e^{2t}}{t + 4}\right\} = 0$.

(a) In every finite interval $e^{2t}/(t+4)$ is continuous [and thus certainly piecewise continuous]. Also for all $t \geqq 0$,

$$\frac{e^{2t}}{t + 4} < \frac{e^{2t}}{4}$$

so that $e^{2t}/(t+4)$ is of exponential order.   Thus by Problem 4.9(a) the Laplace transform exists.

(b) This follows at once from Problem 4.9(b) and the results of (a).

SOME ELEMENTARY INVERSE LAPLACE TRANSFORMS

4.11. Prove that $\mathcal{L}^{-1}$ is a linear operator.

We have, since $\mathcal{L}$ is a linear operator [Problem 4.3],

$$\mathcal{L}\{c_1 f_1(t) + c_2 f_2(t)\} = c_1 \mathcal{L}\{f_1(t)\} + c_2 \mathcal{L}\{f_2(t)\} = c_1 F_1(s) + c_2 F_2(s)$$

Thus by definition

$$\mathcal{L}^{-1}\{c_1 F_1(s) + c_2 F_2(s)\} = c_1 f_1(t) + c_2 f_2(t)$$
$$= c_1 \mathcal{L}^{-1}\{F_1(s)\} + c_2 \mathcal{L}^{-1}\{F_2(s)\}$$

which shows that $\mathcal{L}^{-1}$ is a linear operator.

4.12. Find (a) $\mathcal{L}^{-1}\left\{\dfrac{5}{s+2}\right\}$, (b) $\mathcal{L}^{-1}\left\{\dfrac{4s-3}{s^2+4}\right\}$, (c) $\mathcal{L}^{-1}\left(\dfrac{2s-5}{s^2}\right)$, (d) $\mathcal{L}^{-1}\left\{\dfrac{1}{s^k}\right\}$, $k > 0$.

(a) $\mathcal{L}^{-1}\left\{\dfrac{5}{s+2}\right\} = 5\mathcal{L}^{-1}\left\{\dfrac{1}{s+2}\right\} = 5e^{-2t}$

(b) $\mathcal{L}^{-1}\left\{\dfrac{4s-3}{s^2+4}\right\} = 4\mathcal{L}^{-1}\left\{\dfrac{s}{s^2+4}\right\} - \dfrac{3}{2}\mathcal{L}^{-1}\left\{\dfrac{2}{s^2+4}\right\} = 4\cos 2t - \dfrac{3}{2}\sin 2t$

(c) $\mathcal{L}^{-1}\left\{\dfrac{2s-5}{s^2}\right\} = 2\mathcal{L}^{-1}\left\{\dfrac{1}{s}\right\} - 5\mathcal{L}^{-1}\left\{\dfrac{1}{s^2}\right\} = 2 - 5t$

(d) Since $\mathcal{L}\{t^p\} = \dfrac{\Gamma(p+1)}{s^{p+1}}$,

$$\mathcal{L}\left\{\dfrac{t^p}{\Gamma(p+1)}\right\} = \dfrac{1}{s^{p+1}} \quad \text{or} \quad \mathcal{L}^{-1}\left\{\dfrac{1}{s^{p+1}}\right\} = \dfrac{t^p}{\Gamma(p+1)}$$

Then letting $p = k - 1$,

$$\mathcal{L}^{-1}\left\{\dfrac{1}{s^k}\right\} = \dfrac{t^{k-1}}{\Gamma(k)}$$

4.13. Find (a) $\mathcal{L}^{-1}\left\{\dfrac{4-5s}{s^{3/2}}\right\}$, (b) $\mathcal{L}^{-1}\left\{\dfrac{1}{s^2+2s}\right\}$.

(a) $\mathcal{L}^{-1}\left\{\dfrac{4-5s}{s^{3/2}}\right\} = 4\mathcal{L}^{-1}\left\{\dfrac{1}{s^{3/2}}\right\} - 5\mathcal{L}^{-1}\left\{\dfrac{1}{s^{1/2}}\right\}$

$\qquad\qquad\qquad\quad = 4 \cdot \dfrac{t^{1/2}}{\Gamma(3/2)} - 5 \cdot \dfrac{t^{-1/2}}{\Gamma(1/2)}$

$\qquad\qquad\qquad\quad = \dfrac{8t^{1/2}}{\sqrt{\pi}} - \dfrac{5t^{-1/2}}{\sqrt{\pi}} = \dfrac{8t^{1/2} - 5t^{-1/2}}{\sqrt{\pi}}$

(b) $\mathcal{L}^{-1}\left\{\dfrac{1}{s^2+2s}\right\} = \mathcal{L}^{-1}\left\{\dfrac{1}{s(s+2)}\right\} = \dfrac{1}{2}\mathcal{L}^{-1}\left\{\dfrac{1}{s} - \dfrac{1}{s+2}\right\}$

$\qquad\qquad\qquad\quad = \dfrac{1}{2}\mathcal{L}^{-1}\left\{\dfrac{1}{s}\right\} - \dfrac{1}{2}\mathcal{L}^{-1}\left\{\dfrac{1}{s+2}\right\}$

$\qquad\qquad\qquad\quad = \dfrac{1}{2} - \dfrac{1}{2}e^{-2t} = \dfrac{1}{2}(1 - e^{-2t})$

LAPLACE TRANSFORMS OF DERIVATIVES

4.14. Prove Theorem 4-3, page 100.

Since $f'(t)$ is piecewise continuous in $0 \leqq t \leqq T$, there exists a finite number of subintervals, say $(0, T_1), (T_1, T_2), \ldots, (T_n, T)$, in each of which $f'(t)$ is continuous and where the limits of $f'(t)$ as t approaches the endpoints of each subinterval are finite. Then

$$\int_0^T e^{-st} f'(t)\, dt = \int_0^{T_1} e^{-st} f'(t)\, dt + \int_{T_1}^{T_2} e^{-st} f'(t)\, dt + \cdots + \int_{T_n}^T e^{-st} f'(t)\, dt$$

The right side becomes on integrating by parts,

$$\left[e^{-st} f(t) \Big|_0^{T_1} + s \int_0^{T_1} e^{-st} f(t)\, dt \right] + \left[e^{-st} f(t) \Big|_{T_1}^{T_2} + s \int_{T_1}^{T_2} e^{-st} f(t)\, dt \right]$$

$$+ \cdots + \left[e^{-st} f(t) \Big|_{T_n}^{T} + s \int_{T_n}^{T} e^{-st} f(t)\, dt \right]$$

Since $f(t)$ is continuous, we can thus write

$$\int_0^T e^{-st} f'(t)\, dt \;=\; e^{-st} f(T) - f(0) + s \int_0^T e^{-st} f(t)\, dt \tag{1}$$

Now since $f(t)$ is of exponential order, we have if T is large enough,

$$|e^{-sT} f(T)| \;\leq\; |e^{-sT} M e^{\alpha T}| \;=\; M e^{-(s-\alpha)T} \tag{2}$$

Then taking the limit as $T \to \infty$ in (1), using the fact that $\lim_{T \to \infty} e^{-st} f(T) = 0$, we have as required

$$\mathcal{L}\{f'(t)\} \;=\; \int_0^\infty e^{-st} f'(t)\, dt \;=\; s \int_0^\infty e^{-st} f(t)\, dt - f(0) \;=\; s\,\mathcal{L}\{f(t)\} - f(0)$$

4.15. Prove that if $f'(t)$ is continuous and $f''(t)$ is piecewise continuous in every finite interval $0 \leq t \leq T$ and if $f(t)$ and $f'(t)$ are of exponential order for $t > T$, then

$$\mathcal{L}\{f''(t)\} \;=\; s^2 \mathcal{L}\{f(t)\} - s f(0) - f'(0)$$

Let $g(t) = f'(t)$. Then $g(t)$ satisfies the conditions of Theorem 4-3 so that

$$\mathcal{L}\{g'(t)\} \;=\; s\,\mathcal{L}\{g(t)\} - g(0)$$

Thus

$$\begin{aligned}
\mathcal{L}\{f''(t)\} &= s\,\mathcal{L}\{f'(t)\} - f'(0) \\
&= s[s\,\mathcal{L}\{f(t)\} - f(0)] - f'(0) \\
&= s^2 \mathcal{L}\{f(t)\} - s f(0) - f'(0)
\end{aligned}$$

4.16. Let $f(t) = te^{at}$. (a) Show that $f(t)$ satisfies the equation $f'(t) = af(t) + e^{at}$. (b) Use part (a) to find $\mathcal{L}\{te^{at}\}$.

(a) $f'(t) = t(ae^{at}) + e^{at} = af(t) + e^{at}$

(b) $\mathcal{L}\{f'(t)\} = \mathcal{L}\{af(t) + e^{at}\} = a\,\mathcal{L}\{f(t)\} + \mathcal{L}\{e^{at}\}$

Thus using Problem 4.14, we have since $f(0) = 0$,

$$s\,\mathcal{L}\{f(t)\} - f(0) \;=\; a\,\mathcal{L}\{f(t)\} + \frac{1}{s-a} \qquad \text{or} \qquad (s-a)\,\mathcal{L}\{f(t)\} \;=\; \frac{1}{s-a}$$

i.e.

$$\mathcal{L}\{f(t)\} \;=\; \mathcal{L}\{te^{at}\} \;=\; \frac{1}{(s-a)^2}$$

THE UNIT STEP FUNCTION

4.17. Prove that $\mathcal{L}\{\mathcal{U}(t-a)\} = \dfrac{e^{-as}}{s}$ if $s > 0$.

We have $\mathcal{U}(t-a) = \begin{cases} 0 & t < a \\ 1 & t > a \end{cases}$ so that

$$\mathcal{L}\{\mathcal{U}(t-a)\} \;=\; \int_0^a e^{-st}(0)\, dt + \int_a^\infty e^{-st}(1)\, dt \;=\; 0 + \frac{e^{-st}}{-s} \Big|_a^\infty$$

$$=\; \frac{e^{-as}}{s} \qquad \text{if } s > 0$$

4.18. (a) Express the function $f(t) = \begin{cases} 8 & t < 2 \\ 6 & t > 2 \end{cases}$ in terms of the unit step function and thus (b) obtain its Laplace transform.

(a) We have
$$f(t) = 8 + \begin{cases} 0 & t < 2 \\ -2 & t > 2 \end{cases}$$

$$= 8 - 2\begin{cases} 0 & t < 2 \\ 1 & t > 2 \end{cases}$$

$$= 8 - 2\,\mathcal{U}(t-1)$$

(b) $\mathcal{L}\{f(t)\} = \mathcal{L}\{8 - 2u(t-1)\} = \dfrac{8}{s} - \dfrac{2e^{-s}}{s} = \dfrac{8 - 2e^{-s}}{s}$

The result can also be obtained directly.

SPECIAL THEOREMS ON LAPLACE TRANSFORMS

4.19. Prove Theorem 4-5, page 101.

We have
$$\mathcal{L}\{f(t)\} = F(s) = \int_0^\infty e^{-st} f(t)\, dt$$

Then
$$\mathcal{L}\{e^{at} f(t)\} = \int_0^\infty e^{-st}[e^{at} f(t)]\, dt = \int_0^\infty e^{-(s-a)t} f(t)\, dt = F(s-a)$$

4.20. Prove Theorem 4-6, page 101.

Method 1. Since $\mathcal{U}(t-a) = \begin{cases} 0 & t < a \\ 1 & t > a \end{cases}$ we have

$$\mathcal{L}\{\mathcal{U}(t-a) f(t-a)\} = \int_0^\infty e^{-st}\, \mathcal{U}(t-a)\, f(t-a)\, dt$$

$$= \int_0^a e^{-st}(0)\, dt + \int_a^\infty e^{-st} f(t-a)\, dt$$

$$= \int_a^\infty e^{-st} f(t-a)\, dt = \int_0^\infty e^{-s(v+a)} f(v)\, dv$$

$$= e^{-as} \int_0^\infty e^{-sv} f(v)\, dv$$

$$= e^{-as} F(s)$$

Method 2. Since $F(s) = \int_0^\infty e^{-st} f(t)\, dt,$

$$e^{-as} F(s) = \int_0^\infty e^{-s(t+a)} f(t)\, dt = \int_a^\infty e^{-sv} f(v-a)\, dv$$

$$= \int_0^a e^{-sv}(0)\, dv + \int_a^\infty e^{-sv} f(v-a)\, dv$$

$$= \int_0^\infty e^{-st} \begin{cases} 0 & t < a \\ f(t-a) & t > a \end{cases} dt$$

$$= \int_0^\infty e^{-st} f(t-a) \begin{cases} 0 & t < a \\ 1 & t > a \end{cases} dt$$

$$= \mathcal{L}\{f(t-a)\, \mathcal{U}(t-a)\}$$

4.21. Prove Theorem 4-7, page 101.

$$\mathcal{L}\{f(at)\} = \int_0^\infty e^{-st} f(at)\, dt = \frac{1}{a} \int_0^\infty e^{-sv/a} f(v)\, dv = \frac{1}{a} F\left(\frac{s}{a}\right)$$

where we have used the transformation $t = v/a$.

4.22. Prove Theorem 4-8, page 101, for (a) $n = 1$, (b) any positive integer n.

(a) Since $F(s) = \int_0^\infty e^{-st} f(t)\, dt$, we have on differentiating with respect to s and using Leibnitz's rule,

$$\frac{dF}{ds} = F'(s) = \frac{d}{ds} \int_0^\infty e^{-st} f(t)\, dt = \int_0^\infty \frac{\partial}{\partial s} [e^{-st} f(t)]\, dt$$

$$= -\int_0^\infty e^{-st}\, t f(t)\, dt = -\mathcal{L}\{t f(t)\}$$

Thus $\mathcal{L}\{t f(t)\} = -F'(s)$.

(b)
$$\frac{d^n F}{ds^n} = \frac{d^n}{ds^n} \int_0^\infty e^{-st} f(t)\, dt = \int_0^\infty \frac{\partial^n}{\partial s^n} [e^{-st} f(t)]\, dt$$

$$= (-1)^n \int_0^\infty e^{-st} [t^n f(t)]\, dt = (-1)^n \mathcal{L}\{t^n f(t)\}$$

Thus $\mathcal{L}\{t^n f(t)\} = (-1)^n F^{(n)}(s)$.

4.23. Prove Theorem 4-9, page 101.

We have

$$\int_0^\infty e^{-st} f(t)\, dt = \int_0^P e^{-st} f(t)\, dt + \int_P^{2P} e^{-st} f(t)\, dt + \int_{2P}^{3P} e^{-st} f(t)\, dt + \cdots$$

$$= \int_0^P e^{-st} f(t)\, dt + \int_0^P e^{-s(v+P)} f(v+P)\, dv + \int_0^P e^{-s(v+2P)} f(v+2P)\, dv + \cdots$$

Then since $f(t)$ has period $P > 0$, $f(v+P) = f(v)$, $f(v+2P) = f(v)$, etc. Furthermore, we can replace the dummy variable v by t. Thus

$$\int_0^\infty e^{-st} f(t)\, dt = \int_0^P e^{-st} f(t)\, dt + e^{-sP} \int_0^P e^{-st} f(t)\, dt + e^{-2sP} \int_0^P e^{-st} f(t)\, dt + \cdots$$

$$= (1 + e^{-sP} + e^{-2sP} + \cdots) \int_0^P e^{-st} f(t)\, dt$$

$$= \frac{1}{1 - e^{-sP}} \int_0^P e^{-st} f(t)\, dt \qquad \text{where } s > 0$$

4.24. Prove Theorem 4-10, page 101.

Let $G(t) = \int_0^t f(u)\, du$. Then $G'(t) = f(t)$, $G(0) = 0$. Thus

$$\mathcal{L}\{G'(t)\} = s\,\mathcal{L}\{G(t)\} - G(0) \qquad \text{or} \qquad \mathcal{L}\{f(t)\} = s\,\mathcal{L}\{G(t)\}$$

and so

$$\mathcal{L}\{G(t)\} = \mathcal{L}\left\{ \int_0^t f(u)\, du \right\} = \frac{1}{s}\, \mathcal{L}\{f(t)\} = \frac{F(s)}{s}$$

From this we have

$$\mathcal{L}^{-1}\left\{ \frac{F(s)}{s} \right\} = \int_0^t f(u)\, du$$

4.25. Prove Theorem 4-11, page 101.

We shall assume that $\lim_{t \to 0} \dfrac{f(t)}{t}$ exists [as well as the conditions of Theorem 4-1, page 99], otherwise its Laplace transform may not exist. Then we have if $g(t) = f(t)/t$, or $f(t) = t g(t)$,

$$F(s) = \mathcal{L}\{f(t)\} = \mathcal{L}\{t g(t)\} = -\frac{d}{ds}\, \mathcal{L}\{g(t)\} = -\frac{d\, G(s)}{ds} \tag{1}$$

Thus
$$G(s) = -\int_c^s F(u)\, du = \int_s^c F(u)\, du \tag{2}$$

Now since $g(t)$ satisfies the conditions of Theorem 4-1, page 99, it follows that $\lim\limits_{s \to \infty} G(s) = 0$. Then from (2) we see that c must be infinite and so

$$G(s) = \mathcal{L}\left\{\frac{f(t)}{t}\right\} = \int_s^\infty F(u)\, du$$

4.26. Prove Theorem 4-12, page 102.

We have
$$F(s) = \int_0^\infty e^{-su} f(u)\, du, \qquad G(s) = \int_0^\infty e^{-sv} g(v)\, dv$$

Then
$$F(s)\, G(s) = \left[\int_0^\infty e^{-su} f(u)\, du\right]\left[\int_0^\infty e^{-sv} g(v)\, dv\right]$$

$$= \int_0^\infty \int_0^\infty e^{-s(u+v)} f(u)\, g(v)\, du\, dv$$

$$= \int_{t=0}^\infty \int_{u=0}^t e^{-st} f(u)\, g(t-u)\, du\, dt$$

$$= \int_{t=0}^\infty e^{-st}\left[\int_{u=0}^t f(u)\, g(t-u)\, du\right] dt$$

$$= \mathcal{L}\left\{\int_0^t f(u)\, g(t-u)\, du\right\}$$

where we have used the transformation $t = u + v$ from the uv plane to the ut plane.

APPLICATIONS OF THE THEOREMS TO FINDING LAPLACE TRANSFORMS

4.27. Find (a) $\mathcal{L}\{e^{3t} \sin 4t\}$, (b) $\mathcal{L}\{t^2 e^{-2t}\}$, (c) $\mathcal{L}\{e^t/\sqrt{t}\}$.

(a) Since $\mathcal{L}\{\sin 4t\} = \dfrac{4}{s^2 + 16}$, we have by Theorem 4-5,

$$\mathcal{L}\{e^{3t} \sin 4t\} = \frac{4}{(s-3)^2 + 16} = \frac{4}{s^2 - 6s + 25}$$

(b) Since $\mathcal{L}\{t^2\} = \dfrac{2!}{s^3}$, $\mathcal{L}\{t^2 e^{-2t}\} = \dfrac{2!}{(s+2)^3}$

(c) Since $\mathcal{L}\{1/\sqrt{t}\} = \mathcal{L}\{t^{-1/2}\} = \Gamma(1/2)/s^{1/2} = \sqrt{\pi/s}$, $\mathcal{L}\{e^t/\sqrt{t}\} = \sqrt{\pi/(s-1)}$

4.28. Find $\mathcal{L}\{F(t)\}$ where $f(t) = \begin{cases} \sin t & t < \pi \\ t & t > \pi \end{cases}$.

We have
$$f(t) = \sin t + \begin{cases} 0 & t < \pi \\ t - \sin t & t > \pi \end{cases}$$

$$= \sin t + (t - \sin t)\mathcal{U}(t - \pi)$$

$$= \sin t + [\pi + (t - \pi) + \sin(t - \pi)]\mathcal{U}(t - \pi)$$

Then using Theorem 4-6,

$$\mathcal{L}\{f(t)\} = \mathcal{L}\{\sin t\} + \mathcal{L}\{[\pi + (t - \pi) + \sin(t - \pi)]\mathcal{U}(t - \pi)\}$$

$$= \frac{1}{s^2 + 1} + e^{-\pi s}\left[\frac{\pi}{s} + \frac{1}{s^2} + \frac{1}{s^2 + 1}\right]$$

This can also be obtained directly without using Theorem 4-6.

4.29. Given that $\mathcal{L}\left\{\dfrac{\sin t}{t}\right\} = \tan^{-1}\left(\dfrac{1}{s}\right)$, find $\mathcal{L}\left\{\dfrac{\sin at}{t}\right\}$.

By Theorem 4-7, page 101,

$$\mathcal{L}\left\{\frac{\sin at}{at}\right\} = \frac{1}{a}\tan^{-1}\left(\frac{1}{s/a}\right) \quad \text{i.e.} \quad \mathcal{L}\left\{\frac{\sin at}{t}\right\} = \tan^{-1}\left(\frac{a}{s}\right)$$

4.30. Find　(a) $\mathcal{L}\{t \sin 2t\}$,　(b) $\mathcal{L}\{t^2 \sin 2t\}$.

By Theorem 4-8, page 101, we have since $\mathcal{L}\{\sin 2t\} = \dfrac{2}{s^2+4}$,

(a)　$\mathcal{L}\{t \sin 2t\} = -\dfrac{d}{ds}\left(\dfrac{2}{s^2+4}\right) = \dfrac{4s}{(s^2+4)^2}$

(b)　$\mathcal{L}\{t^2 \sin 2t\} = \dfrac{d^2}{ds^2}\left(\dfrac{2}{s^2+4}\right) = \dfrac{12s^2-16}{(s^2+4)^3}$

4.31. Find the Laplace transform of the function of period 2π which in the interval $0 \le t < 2\pi$ is given by $f(t) = \begin{cases} \sin t & 0 \le t < \pi \\ 0 & \pi \le t < 2\pi \end{cases}$.

The graph of this function, often called a *rectified sine wave*, is shown in Fig. 4-3.

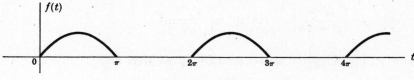

Fig. 4-3

By Theorem 4-9, page 101, the Laplace transform is given by

$$\frac{1}{1-e^{-2\pi s}}\int_0^\pi e^{-st}\sin t\, dt = \frac{1}{1-e^{-2\pi s}}\left(\frac{1+e^{-\pi s}}{s^2+1}\right) = \frac{1}{(s^2+1)(1-e^{-\pi s})}$$

4.32. Find $\mathcal{L}\left\{\dfrac{1-e^{-t}}{t}\right\}$.

Since $\displaystyle\lim_{t\to 0}\frac{1-e^{-t}}{t} = \lim_{t\to 0}\frac{e^{-t}}{1} = 1$ by L'Hospital's rule and $1-e^{-t}$ is continuous and of exponential order, the conditions of Theorem 4-11 apply. Then since $\mathcal{L}\{1-e^{-t}\} = \dfrac{1}{s} - \dfrac{1}{s+1}$, it follows that

$$\mathcal{L}\left\{\frac{1-e^{-t}}{t}\right\} = \int_s^\infty \left(\frac{1}{u} - \frac{1}{u+1}\right) du = \lim_{K\to\infty}\int_s^K \left(\frac{1}{u} - \frac{1}{u+1}\right) du$$

$$= \lim_{K\to\infty}\left[\ln u - \ln(u+1)\right]\Big|_s^K$$

$$= \lim_{K\to\infty}\left[\ln\left(1+\frac{1}{s}\right) - \ln\left(1+\frac{1}{K}\right)\right] = \ln\left(1+\frac{1}{s}\right)$$

APPLICATIONS OF THE THEOREMS TO FINDING INVERSE LAPLACE TRANSFORMS

4.33. Find　(a) $\mathcal{L}^{-1}\left\{\dfrac{2s+3}{s^2-2s+5}\right\}$,　(b) $\mathcal{L}^{-1}\left\{\dfrac{1}{\sqrt{s+3}}\right\}$,　(c) $\mathcal{L}^{-1}\left\{\dfrac{3s-4}{(2s-3)^5}\right\}$.

(a) $\mathcal{L}^{-1}\left\{\dfrac{2s+3}{s^2-2s+5}\right\} = \mathcal{L}^{-1}\left\{\dfrac{2(s-1)+5}{(s-1)^2+4}\right\} = 2\mathcal{L}^{-1}\left\{\dfrac{s-1}{(s-1)^2+4}\right\} + \dfrac{5}{2}\mathcal{L}^{-1}\left\{\dfrac{2}{(s-1)^2+4}\right\}$

Now since

$$\mathcal{L}^{-1}\left\{\frac{s}{s^2 + 4}\right\} = \cos 2t, \qquad \mathcal{L}^{-1}\left\{\frac{2}{s^2 + 4}\right\} = \sin 2t$$

we have by Theorem 4-5,

$$\mathcal{L}^{-1}\left\{\frac{s - 1}{(s - 1)^2 + 4}\right\} = e^t \cos 2t, \qquad \mathcal{L}^{-1}\left\{\frac{2}{(s - 1)^2 + 4}\right\} = e^t \sin 2t$$

Thus

$$\mathcal{L}^{-1}\left\{\frac{2s + 3}{s^2 - 2s + 5}\right\} = 2e^t \cos 2t + \tfrac{5}{2}e^t \sin 2t = \tfrac{1}{2}e^t(4 \cos 2t + 5 \sin 2t)$$

(b) From Problem 4.12(d) we have $\mathcal{L}^{-1}\left\{\dfrac{1}{s^k}\right\} = \dfrac{t^{k-1}}{\Gamma(k)}$ or if $k = \tfrac{1}{2}$, $\mathcal{L}^{-1}\left\{\dfrac{1}{\sqrt{s}}\right\} = \dfrac{t^{-1/2}}{\sqrt{\pi}}$. Then by Theorem 4-5,

$$\mathcal{L}^{-1}\left\{\frac{1}{\sqrt{s + 3}}\right\} = \frac{e^{-3t}\, t^{-1/2}}{\sqrt{\pi}}$$

(c)
$$\mathcal{L}^{-1}\left\{\frac{3s - 4}{(2s - 3)^5}\right\} = \frac{1}{2^5}\mathcal{L}^{-1}\left\{\frac{3s - 4}{(s - 3/2)^5}\right\} = \frac{1}{32}\mathcal{L}^{-1}\left\{\frac{3(s - 3/2) + 1/2}{(s - 3/2)^5}\right\}$$

$$= \frac{3}{32}\mathcal{L}^{-1}\left\{\frac{1}{(s - 3/2)^4}\right\} + \frac{1}{64}\mathcal{L}^{-1}\left\{\frac{1}{(s - 3/2)^5}\right\}$$

$$= \frac{3}{32}\cdot\frac{t^3}{3!}e^{3t/2} + \frac{1}{64}\frac{t^4}{4!}e^{3t/2} = \frac{t^3(t + 8)e^{3t/2}}{1536}$$

4.34. Find (a) $\mathcal{L}^{-1}\left\{\dfrac{se^{-2s}}{s^2 + 16}\right\}$, (b) $\mathcal{L}^{-1}\left\{\dfrac{e^{-5s}}{\sqrt{s - 2}}\right\}$.

(a) Since $\mathcal{L}^{-1}\left\{\dfrac{s}{s^2 + 16}\right\} = \cos 4t$, we have by Theorem 4-6,

$$\mathcal{L}^{-1}\left\{\frac{se^{-2s}}{s^2 + 16}\right\} = \mathcal{U}(t - 2)\cos 4(t - 2) = \begin{cases} 0 & t < 2 \\ \cos 4(t - 2) & t > 2 \end{cases}$$

(b) Since $\mathcal{L}^{-1}\left\{\dfrac{1}{\sqrt{s}}\right\} = \dfrac{t^{-1/2}}{\sqrt{\pi}}$, $\mathcal{L}^{-1}\left\{\dfrac{1}{\sqrt{s - 2}}\right\} = \dfrac{t^{-1/2}e^{2t}}{\sqrt{\pi}}$ and so by Theorem 4-6,

$$\mathcal{L}^{-1}\left\{\frac{e^{-5s}}{\sqrt{s - 2}}\right\} = \mathcal{U}(t - 5)\frac{(t - 5)^{-1/2}e^{2(t-5)}}{\sqrt{\pi}} = \begin{cases} 0 & t < 5 \\ \dfrac{(t - 5)^{-1/2}e^{2(t-5)}}{\sqrt{\pi}} & t > 5 \end{cases}$$

4.35. Find $\mathcal{L}^{-1}\left\{\ln\left(1 + \dfrac{1}{s}\right)\right\}$.

Let $F(s) = \ln\left(1 + \dfrac{1}{s}\right)$ so that $F'(s) = \dfrac{1}{s + 1} - \dfrac{1}{s}$. Then by Theorem 4-8,

$$\mathcal{L}^{-1}\{F'(s)\} = -t\,\mathcal{L}^{-1}\left\{\ln\left(1 + \frac{1}{s}\right)\right\}$$

or

$$\mathcal{L}^{-1}\left\{\ln\left(1 + \frac{1}{s}\right)\right\} = -\frac{1}{t}\mathcal{L}^{-1}\left\{\frac{1}{s + 1} - \frac{1}{s}\right\} = \frac{1 - e^{-t}}{t}$$

4.36. Find $\mathcal{L}^{-1}\left\{\dfrac{1}{s\sqrt{s + 1}}\right\}$.

Since $\mathcal{L}^{-1}\left\{\dfrac{1}{\sqrt{s + 1}}\right\} = \dfrac{t^{-1/2}e^{-t}}{\sqrt{\pi}}$, we have by Theorem 4-10,

$$\mathcal{L}^{-1}\left\{\frac{1}{s\sqrt{s + 1}}\right\} = \int_0^t \frac{u^{-1/2}e^{-u}}{\sqrt{\pi}}\,du = \frac{1}{\sqrt{\pi}}\int_0^{\sqrt{t}} e^{-v^2}\,dv$$

on letting $u = v^2$.

4.37. Find $\mathcal{L}^{-1}\left\{\dfrac{1}{(s^2+a^2)^2}\right\}$.

Let $F(s) = G(s) = \dfrac{1}{s^2+a^2}$. Then $f(t) = g(t) = \dfrac{\sin at}{a}$. Thus by the convolution theorem [Theorem 4-12],

$$\mathcal{L}^{-1}\left\{\frac{1}{(s^2+a^2)^2}\right\} = \int_0^t \frac{\sin au}{a} \cdot \frac{\sin a(t-u)}{a}\, du = \frac{1}{a^2}\int_0^t \sin au \sin a(t-u)\, du$$

$$= \frac{1}{2a^2}\int_0^t [\cos a(2u-t) - \cos at]\, du = \frac{1}{2a^3}(\sin at - at\cos at)$$

4.38. Solve for $y(t)$ the equation

$$y(t) = 1 + \int_0^t y(u)\sin(t-u)\,du$$

Taking the Laplace transform, we have by the convolution theorem,

$$Y(s) = \frac{1}{s} + \mathcal{L}\{y(t)* \sin t\} = \frac{1}{s} + \frac{Y(s)}{s^2+1}$$

Then

$$\left[1 - \frac{1}{s^2+1}\right]Y(s) = \frac{1}{s} \quad\text{or}\quad Y(s) = \frac{s^2+1}{s^3} = \frac{1}{s} + \frac{1}{s^3}$$

and so

$$y(t) = \mathcal{L}^{-1}\left\{\frac{1}{s} + \frac{1}{s^3}\right\} = 1 + \frac{t^2}{2}$$

which can be checked as a solution.

The given equation is called an *integral equation* since the unknown function occurs under the integral.

PARTIAL FRACTIONS

4.39. Find $\mathcal{L}^{-1}\left\{\dfrac{2s^2-4}{(s-2)(s+1)(s-3)}\right\}$.

$$\frac{2s^2-4}{(s-2)(s+1)(s-3)} = \frac{A}{s-2} + \frac{B}{s+1} + \frac{C}{s-3}$$

To determine constants A, B, C, multiply by $(s-2)(s+1)(s-3)$ so that

$$2s^2 - 4 = A(s+1)(s-3) + B(s-2)(s-3) + C(s-2)(s+1)$$

This must be an identity and thus must hold for all values of s. Then by letting $s = 2, -1, 3$ in succession we find $A = -4/3$, $B = -1/6$, $C = 7/2$. Thus

$$\mathcal{L}^{-1}\left\{\frac{2s^2-4}{(s-2)(s+1)(s-3)}\right\} = \mathcal{L}^{-1}\left\{\frac{-4/3}{s-2} + \frac{-1/6}{s+1} + \frac{7/2}{s-3}\right\} = -\tfrac{4}{3}e^{2t} - \tfrac{1}{6}e^{-t} + \tfrac{7}{2}e^{3t}$$

4.40. Find $\mathcal{L}^{-1}\left\{\dfrac{3s+1}{(s-1)(s^2+1)}\right\}$.

$$\frac{3s+1}{(s-1)(s^2+1)} = \frac{A}{s-1} + \frac{Bs+C}{s^2+1} \tag{1}$$

or

$$3s+1 = A(s^2+1) + (Bs+C)(s-1)$$

Letting $s = 1$, we find $A = 2$. Letting $s = 0$, we find $A - C = 1$ so that $C = 1$. Then letting s equal any other number, say -1, we find, $-2 = 2A - 2(C - B)$ and $B = -2$. Thus

$$\mathcal{L}^{-1}\left\{\frac{3s+1}{(s-1)(s^2+1)}\right\} = \mathcal{L}^{-1}\left\{\frac{2}{s-1}\right\} + \mathcal{L}^{-1}\left\{\frac{-2s+1}{s^2+1}\right\} = 2e^t - 2\cos t + \sin t$$

Another method.

Multiplying (1) by s after finding A and then letting $s \to \infty$, we find $A + B = 0$ or $B = -A = -2$. This method affords some simplification of procedure.

4.41. Find $\mathcal{L}^{-1}\left\{\dfrac{5s^2-15s+7}{(s+1)(s-2)^3}\right\}$.

We have
$$\frac{5s^2-15s+7}{(s+1)(s-2)^3} \;=\; \frac{A}{s+1}+\frac{B}{(s-2)^3}+\frac{C}{(s-2)^2}+\frac{D}{s-2} \tag{1}$$

or clearing of fractions
$$5s^2-15s+7 \;=\; A(s-2)^3 + B(s+1) + C(s+1)(s-2) + D(s+1)(s-2)^2$$

Letting $s=-1$, we find $A=-1$. Letting $s=2$, we find $B=-1$. Letting s equal two other numbers, say 0 and 1, we find $-2C+4D=0$ and $-2C+2D=-2$ from which $C=2$, $D=1$. Thus

$$\mathcal{L}^{-1}\left\{\frac{5s^2-15s+7}{(s+1)(s-2)^3}\right\} \;=\; \mathcal{L}^{-1}\left\{\frac{-1}{s+1}+\frac{-1}{(s-2)^3}+\frac{2}{(s-2)^2}+\frac{1}{s-2}\right\}$$
$$=\; -e^{-t}-\tfrac{1}{2}t^2e^{2t}+2te^{2t}+e^{2t}$$

Another method.

On multiplying (1) by s, after finding A, and letting $s \to \infty$, we find that $A+D=0$ or $D=1$, providing some simplification in the procedure.

SOLUTIONS OF DIFFERENTIAL EQUATIONS

4.42. Solve $y''(t)+y(t)=1$ given $y(0)=1$, $y'(0)=0$.

Take the Laplace transform of both sides of the given differential equation and let $Y=Y(s)=\mathcal{L}\{y(t)\}$. Then

$$\mathcal{L}\{y''(t)+y(t)\} \;=\; \mathcal{L}\{1\} \quad\text{or}\quad s^2Y-sy(0)-y'(0)+Y \;=\; 1/s$$

Since $y(0)=1$, $y'(0)=0$ this becomes

$$s^2Y-s+Y=\frac{1}{s}, \quad (s^2+1)Y=s+\frac{1}{s} \quad\text{or}\quad Y=\frac{s+1/s}{s^2+1}=\frac{1}{s}$$

Thus
$$y(t) \;=\; \mathcal{L}^{-1}\{Y\} \;=\; \mathcal{L}^{-1}\left\{\frac{1}{s}\right\} \;=\; 1$$

which is the required solution.

4.43. Solve $y''-3y'+2y=2e^{-t}$, $y(0)=2$, $y'(0)=-1$.

Taking the Laplace transform of the given differential equation,

$$[s^2Y-sy(0)-y'(0)] - 3[sY-y(0)] + 2Y \;=\; \frac{2}{s+1}$$

Then using $y(0)=2$, $y'(0)=-1$ and solving this algebraic equation for Y, we find using partial fractions,

$$Y \;=\; \frac{2s^2-5s-5}{(s+1)(s-1)(s-2)} \;=\; \frac{1/3}{s+1}+\frac{4}{s-1}+\frac{-7/3}{s-2}$$

Thus taking the inverse Laplace transform, we obtain the required solution
$$y \;=\; \tfrac{1}{3}e^{-t}+4e^t-\tfrac{7}{3}e^{2t}$$

4.44. Solve $y^{(iv)}+2y''+y=\sin t$, $y(0)=1$, $y'(0)=-2$, $y''(0)=3$, $y'''(0)=0$.

Taking the Laplace transform of the given differential equation and using the initial conditions,

$$[s^4Y-s^3(1)-s^2(-2)-s(3)-0] + 2[s^2Y-s(1)-(-2)] + Y \;=\; \frac{1}{s^2+1}$$

which can be written
$$(s^4+2s^2+1)Y \;=\; \frac{1}{s^2+1}+s^3-2s^2+5s-4$$

or
$$Y \;=\; \frac{1}{(s^2+1)^3}+\frac{s^3-2s^2+5s-4}{(s^2+1)^2} \;=\; \frac{1}{(s^2+1)^3}+\frac{(s^3+s)-2(s^2+1)+4s-2}{(s^2+1)^2}$$
$$=\; \frac{1}{(s^2+1)^3}+\frac{s}{s^2+1}-\frac{2}{s^2+1}+\frac{4s-2}{(s^2+1)^2}$$

Now by using the special theorems on Laplace transforms [see Problem 4.78],

$$\mathcal{L}^{-1}\left\{\frac{1}{(s^2+1)^3}\right\} = \tfrac{3}{8}\sin t - \tfrac{3}{8}t\cos t - \tfrac{1}{8}t^2\sin t$$

$$\mathcal{L}^{-1}\left\{\frac{4s-2}{(s^2+1)^2}\right\} = 2t\sin t - \sin t + t\cos t$$

and so we find the required solution

$$y = (1+\tfrac{5}{8}t)\cos t - (\tfrac{21}{8}-2t+\tfrac{1}{8}t^2)\sin t$$

APPLICATIONS TO PHYSICAL PROBLEMS

4.45. Solve Problem 2.32, page 56, by using Laplace transforms.

As in Problem 2.32, the differential equation is

$$\frac{dI}{dt} + 5I = \frac{E}{2}, \quad I(0) = 0 \tag{1}$$

(a) If $E = 40$ the Laplace transform of (1) is

$$[s\bar{I} - I(0)] + 5\bar{I} = \frac{20}{s}$$

where $\bar{I} = \mathcal{L}\{I\}$. Then using $I(0) = 0$ and solving for $\bar{I}$, we find

$$\bar{I} = \frac{20}{s(s+5)} = \frac{20}{5}\left(\frac{1}{s} - \frac{1}{s+5}\right) = 4\left(\frac{1}{s} - \frac{1}{s+5}\right)$$

Thus

$$I = 4(1 - e^{-5t})$$

(b) If $E = 20e^{-3t}$ then

$$[s\bar{I} - I(0)] + 5\bar{I} = \frac{10}{s+3}$$

so that

$$\bar{I} = \frac{10}{(s+3)(s+5)} = \frac{10}{2}\left(\frac{1}{s+3} - \frac{1}{s+5}\right) = 5\left(\frac{1}{s+3} - \frac{1}{s+5}\right)$$

Thus

$$I = 5(e^{-3t} - e^{-5t})$$

(c) If $E = 50\sin 5t$, then

$$[s\bar{I} - I(0)] + 5\bar{I} = \frac{125}{s^2+25}$$

so that

$$\bar{I} = \frac{125}{(s+5)(s^2+25)} = \frac{5/2}{s+5} + \frac{(-5/2)s + (25/2)}{s^2+25}$$

Then

$$I = \frac{5}{2}e^{-5t} - \frac{5}{2}\cos 5t + \frac{5}{2}\sin 5t$$

4.46. A mass m [Fig. 4-4] is suspended from the end of a vertical spring of constant κ [force required to produce unit stretch]. An external force $F(t)$ acts on the mass as well as a resistive force proportional to the instantaneous velocity. Assuming that x is the displacement of the mass at time t and that the mass starts from rest at $x = 0$, (a) set up a differential equation for the motion and (b) find x at any time t.

(a) The resistive force is given by $-\beta\dfrac{dx}{dt}$. The restoring force is given by $-\kappa x$. Then by Newton's law,

$$m\frac{d^2x}{dt^2} = -\beta\frac{dx}{dt} - \kappa x + F(t)$$

or

$$m\frac{d^2x}{dt^2} + \beta\frac{dx}{dt} + \kappa x = F(t) \tag{1}$$

where

$$x(0) = 0, \quad x'(0) = 0 \tag{2}$$

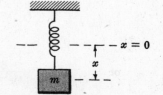

Fig. 4-4

(b) Taking the Laplace transform of (1), using $\mathcal{L}\{F(t)\} = \bar{F}(s)$, $\mathcal{L}\{x\} = X$, we obtain

$$m[s^2X - sx(0) - x'(0)] + \beta[sX - x(0)] + \kappa X = \bar{F}(s)$$

so that on using (2),

$$X = \frac{\bar{F}(s)}{ms^2 + \beta s + \kappa} = \frac{\bar{F}(s)}{m[(s + \beta/2m)^2 + R]} \tag{3}$$

where $R = \dfrac{\kappa}{m} - \dfrac{\beta^2}{4m^2}$. There are 3 cases to be considered.

Case 1, $R > 0$. In this case let $R = \omega^2$. We have

$$\mathcal{L}^{-1}\left\{\frac{1}{(s + \beta/2m)^2 + \omega^2}\right\} = e^{-\beta t/2m}\frac{\sin \omega t}{\omega}$$

Then using the convolution theorem, we find from (3)

$$x = \frac{1}{\omega m}\int_0^t F(u)\, e^{-\beta(t-u)/2m}\sin \omega(t - u)\, du$$

Case 2, $R = 0$. In this case $\mathcal{L}^{-1}\left\{\dfrac{1}{(s + \beta/2m)^2}\right\} = te^{-\beta t/2m}$ and the convolution theorem in (3) yields

$$x = \frac{1}{m}\int_0^t F(u)(t - u)e^{-\beta(t-u)/m}\, du$$

Case 3, $R < 0$. In this case let $R = -\alpha^2$. We have

$$\mathcal{L}^{-1}\left\{\frac{1}{(s + \beta/2m)^2 - \alpha^2}\right\} = e^{-\beta t/2m}\frac{\sinh \alpha t}{\alpha}$$

Then using the convolution theorem (3) yields

$$x = \frac{1}{\alpha m}\int_0^t F(u)e^{-\beta(t-u)/2m}\sinh \alpha\,(t - u)\, du$$

4.47. Solve Problem 3.48, page 93, by Laplace transforms.

From equations (1) and (2) of Problem 3.48 we have on taking Laplace transforms, using $\mathcal{L}\{I_1\} = \bar{I}_1$, $\mathcal{L}\{I_2\} = \bar{I}_2$, $I = I_1 + I_2$, $\mathcal{L}\{I\} = \bar{I}_1 + \bar{I}_2$,

$$20(\bar{I}_1 + \bar{I}_2) - \frac{120}{s} + 2[s\bar{I}_1 - I_1(0)] + 10\bar{I}_1 = 0$$

$$-10\bar{I}_1 - 2[s\bar{I}_1 - I_1(0)] + 4[s\bar{I}_2 - I_2(0)] + 20\bar{I}_2 = 0$$

Using $I_1(0) = 0$, $I_2(0) = 0$, these become

$$(30 + 2s)\bar{I}_1 + 20\bar{I}_2 = 120/s$$

$$(-10 - 2s)\bar{I}_1 + (4s + 20)\bar{I}_2 = 0$$

Solving,

$$\bar{I}_1 = \frac{\begin{vmatrix} 120/s & 20 \\ 0 & 4s + 20 \end{vmatrix}}{\begin{vmatrix} 30 + 2s & 20 \\ -10 - 2s & 4s + 20 \end{vmatrix}}, \qquad \bar{I}_2 = \frac{\begin{vmatrix} 30 + 2s & 120/s \\ -10 - 2s & 0 \end{vmatrix}}{\begin{vmatrix} 30 + 2s & 20 \\ -10 - 2s & 4s + 20 \end{vmatrix}}$$

or $\qquad \bar{I}_1 = \dfrac{60}{s(s + 20)} = 3\left(\dfrac{1}{s} - \dfrac{1}{s + 20}\right)$, $\quad \bar{I}_2 = \dfrac{30}{s(s + 20)} = \dfrac{3}{2}\left(\dfrac{1}{s} - \dfrac{1}{s + 20}\right)$

Then $\quad I_1 = 3(1 - e^{-20t})$, $\quad I_2 = \dfrac{3}{2}(1 - e^{-20t})$, $\quad I = I_1 + I_2 = \dfrac{9}{2}(1 - e^{-20t})$.

Supplementary Problems

LAPLACE TRANSFORMS OF ELEMENTARY FUNCTIONS

4.48. Find the Laplace transforms of each of the following:

(a) $4e^{2t/3}$

(b) $6t - 3$

(c) $(t+1)^2$

(d) $2 \sin 3t + 5 \cos 3t$

(e) $(e^{3t} - e^{-3t})^2$

(f) $(\sqrt{t} + 1)(2 - \sqrt{t})/\sqrt{t}$

(g) $8 \sin^2 3t$

(h) $\sin 2t \cos 2t$

(i) $(\sqrt[3]{t^2} - 1/\sqrt[3]{t})^2$

(j) $5 \sinh 2t - 5 \cosh 2t$

4.49. Find $\mathcal{L}\{f(t)\}$ in each case: (a) $f(t) = \begin{cases} -1 & 0 \leq t \leq 4 \\ 1 & t > 4 \end{cases}$ (b) $f(t) = \begin{cases} t+1 & 0 \leq t < 3 \\ 0 & t \geq 3 \end{cases}$

(c) $f(t) = \begin{cases} 0 & 0 \leq t < 2 \\ 1 & 2 \leq t < 4 \\ 0 & t \geq 4 \end{cases}$

4.50. Prove that e^{t^2} does not have a Laplace transform.

4.51. Determine whether $\sin t^2$ has a Laplace transform and justify your conclusions.

4.52. Find (a) $\mathcal{L}\{10 \sin 3t \cos 5t\}$, (b) $\mathcal{L}\{f(t)\}$ if $f(t) = \begin{cases} \sin t & 0 \leq t < \pi \\ 0 & t \geq \pi \end{cases}$.

ELEMENTARY INVERSE LAPLACE TRANSFORMS

4.53. Find (a) $\mathcal{L}^{-1}\left\{\dfrac{2}{s-3}\right\}$ (d) $\mathcal{L}^{-1}\left\{\dfrac{1}{4s^2+9}\right\}$ (g) $\mathcal{L}^{-1}\left\{\dfrac{2-s}{s^{3/2}}\right\}$ (i) $\mathcal{L}^{-1}\left\{\dfrac{(2s+1)^2}{s^5}\right\}$

(b) $\mathcal{L}^{-1}\left\{\dfrac{1}{3s+5}\right\}$ (e) $\mathcal{L}^{-1}\left\{\dfrac{2s-8}{s^2+36}\right\}$ (h) $\mathcal{L}^{-1}\left\{\dfrac{3s-16}{s^2-64}\right\}$ (j) $\mathcal{L}^{-1}\left\{\dfrac{s}{(s+3)(s+5)}\right\}$

(c) $\mathcal{L}^{-1}\left\{\dfrac{4s}{s^2+16}\right\}$ (f) $\mathcal{L}^{-1}\left\{\dfrac{s^3-s^2+s-1}{s^5}\right\}$

4.54. Find $\mathcal{L}^{-1}\left\{\dfrac{s^2+2}{(s^2+10)(s^2+20)}\right\}$.

LAPLACE TRANSFORMS OF DERIVATIVES

4.55. Verify the result $\mathcal{L}\{f'(t)\} = s\mathcal{L}\{f(s)\} - f(0)$ for each of the following: (a) $f(t) = 3e^{2t}$, (b) $f(t) = \cos 5t$, (c) $f(t) = t^2 + 2t - 4$.

4.56. Verify the result $\mathcal{L}\{f''(t)\} = s^2\mathcal{L}\{f(t)\} - sf(0) - f'(0)$ for each of the functions in Problem 4.55.

4.57. If $f(t) = t \sin at$, (a) show that $f''(t) + a^2 f(t) = 2a \cos at$ and thus (b) find $\mathcal{L}\{f(t)\}$.

4.58. Does the result $\mathcal{L}\{f'(t)\} = s\mathcal{L}\{f(t)\} - f(0)$ hold for

(a) $f(t) = \sqrt{t}$, (b) $f(t) = 1/\sqrt{t}$, (c) $f(t) = \begin{cases} t & t > 0 \\ 5 & t = 0 \end{cases}$? Explain.

4.59. Prove that $\mathcal{L}\{f'''(t)\} = s^3\mathcal{L}\{f(t)\} - s^2 f(0) - sf'(0) - f''(0)$, giving conditions under which it holds.

THE UNIT STEP FUNCTION

4.60. Find (a) $\mathcal{L}\{2\mathcal{U}(t-1) + 3\mathcal{U}(t-2)\}$, (b) $\mathcal{L}\{t\mathcal{U}(t-3)\}$ and graph each of the given functions of t.

4.61. Discuss the significance of (a) $[\mathcal{U}(t-a)]^2$, (b) $[\mathcal{U}(t-a)]^p$ where p is any positive integer.

4.62. Find (a) $\mathcal{L}^{-1}\left\{\dfrac{e^{-5s}}{s}\right\}$, (b) $\mathcal{L}^{-1}\left\{\dfrac{2e^{-s} - e^{-2s}}{s}\right\}$ and graph.

4.63. (a) Express the function $f(t) = \begin{cases} \cos t & t > \pi/2 \\ 3 & t < \pi/2 \end{cases}$ in terms of the unit step function and (b) find its Laplace transform.

MISCELLANEOUS LAPLACE TRANSFORMS

4.64. Find (a) $\mathcal{L}\{t^3 e^{-5t}\}$,   (b) $\mathcal{L}\{e^{-t}\cos 2t\}$,   (c) $\mathcal{L}\{\sqrt{t}\,e^{4t}\}$,   (d) $\mathcal{L}\{e^{-3t}/\sqrt{t}\}$.

4.65. Find (a) $\mathcal{L}\{t\sin 3t\}$,   (b) $\mathcal{L}\{t^2\cos 2t\}$.

4.66. Find $\mathcal{L}\{te^t\sin t\}$.

4.67. (a) Graph the function $f(t) = t$, $0 \le t < 4$ which is extended periodically with period 4 and (b) find the Laplace transform of this function.

4.68. Find $\mathcal{L}\{f(t)\}$ if $f(t) = e^{-t}$, $0 \le t < 2$ and $f(t+2) = f(t)$.

4.69. Verify that $\mathcal{L}\left\{\displaystyle\int_0^t \cos 2u\,du\right\} = \dfrac{1}{s}\mathcal{L}\{\cos 2t\}$.

4.70. Show that $\mathcal{L}\left\{\displaystyle\int_0^t \int_0^{t_1} f(u)\,du\right\} = \dfrac{F(s)}{s^2}$ where $F(s) = \mathcal{L}\{f(t)\}$ and generalize.

4.71. Find (a) $\mathcal{L}\left\{\dfrac{e^{-3t} - e^{-5t}}{t}\right\}$,   (b) $\mathcal{L}\left\{\dfrac{\cos 2t - \cos 3t}{t}\right\}$,   (c) $\mathcal{L}\left\{\dfrac{\sin t}{t}\right\}$.

4.72. Find $\mathcal{L}\left\{\dfrac{1 - \cos t}{t^2}\right\}$.

4.73. Show that (a) $\displaystyle\int_0^\infty \dfrac{e^{-2t} - e^{-4t}}{t}\,dt = \ln 2$, (b) $\displaystyle\int_0^\infty \dfrac{\sin t}{t}\,dt = \dfrac{\pi}{2}$.

4.74. Evaluate (a) $t*e^t$ and (b) $\mathcal{L}\{t*e^t\}$.

4.75. Prove that (a) $f*g = g*f$, (b) $f*(g+h) = f*g + f*h$, (c) $f*(g*h) = (f*g)*h$ and discuss the significance of the results.

MISCELLANEOUS INVERSE LAPLACE TRANSFORMS

4.76. Find (a) $\mathcal{L}^{-1}\left\{\dfrac{3s+9}{s^2+2s+10}\right\}$ (c) $\mathcal{L}^{-1}\left\{\dfrac{1}{(s-5)^3}\right\}$ (e) $\mathcal{L}^{-1}\left\{\dfrac{e^{-3s}}{s^2+\pi^2}\right\}$.

(b) $\mathcal{L}^{-1}\left\{\dfrac{1}{\sqrt{4s-1}}\right\}$ (d) $\mathcal{L}^{-1}\left\{\dfrac{s}{(s-5)^3}\right\}$

4.77. Find (a) $\mathcal{L}^{-1}\left\{\dfrac{1}{s^2-4}\right\}$ (c) $\mathcal{L}^{-1}\left\{\dfrac{8-10s}{(s+1)(s-2)^2}\right\}$ (e) $\mathcal{L}^{-1}\left\{\dfrac{150}{(s^2+2s+5)(s^2-4s+8)}\right\}$

(b) $\mathcal{L}^{-1}\left\{\dfrac{1}{s(s+1)(s+2)}\right\}$ (d) $\mathcal{L}^{-1}\left\{\dfrac{3s-12}{(s^2+8)(s-1)}\right\}$ (f) $\mathcal{L}^{-1}\left\{\dfrac{1}{s^4+4}\right\}$.

4.78. Find (a) $\mathcal{L}^{-1}\left\{\dfrac{1}{s^4-2s^3}\right\}$, (b) $\mathcal{L}^{-1}\left\{\dfrac{s^2}{(s^2+1)^2}\right\}$, (c) $\mathcal{L}^{-1}\left\{\dfrac{4s-2}{(s^2+1)^2}\right\}$, (d) $\mathcal{L}^{-1}\left\{\dfrac{1}{(s^2+1)^3}\right\}$.

4.79. Find $\mathcal{L}^{-1}\left\{\ln\left(\dfrac{s+6}{s+2}\right)\right\}$.

4.80. Use the series for e^u to show that $\mathcal{L}^{-1}\left\{\dfrac{e^{-1/s}}{\sqrt{s}}\right\} = \dfrac{\cos 2\sqrt{t}}{\sqrt{\pi t}}$.

4.81. Solve the integral equations (a) $y(t) + \displaystyle\int_0^t y(t-u)e^u\,du = 2t - 3$, (b) $\displaystyle\int_0^t y(u)\,y(t-u)\,du = t^2 e^{-t}$.

4.82. Use Laplace transforms to evaluate $\displaystyle\int_0^\infty \sin x^2\,dx$. [*Hint.* Consider $\displaystyle\int_0^\infty \sin tx^2\,dx$.]

SOLUTIONS OF DIFFERENTIAL EQUATIONS

4.83. Solve each of the following:

(a) $y''(t) - 3y'(t) + 2y(t) = 4$; $y(0) = 1$, $y'(0) = 0$

(b) $y''(t) + 16y(t) = 32t$; $y(0) = 3$, $y'(0) = -2$

(c) $y''(t) + 4y'(t) + 4y(t) = 6e^{-2t}$; $y(0) = -2$, $y'(0) = 8$

(d) $y'''(t) + y'(t) = t + 1$

4.84. Solve $y'''(t) + 8y(t) = 32t^3 - 16t$ if $y(0) = y'(0) = y''(0) = 0$.

4.85. Solve the simultaneous equations

$$\begin{cases} 2x(t) - y(t) - y'(t) = 4(1 - e^{-t}) \\ 2x'(t) + y(t) = 2(1 + 3e^{-2t}) \end{cases}$$

subject to the conditions $x(0) = y(0) = 0$.

4.86. Solve (a) Problem 3.76, (b) Problem 3.77, (c) Problem 3.78, (d) Problem 3.79 on page 95 by Laplace transforms.

APPLICATIONS TO PHYSICAL PROBLEMS

4.87. Solve (a) Problem 3.80, (b) Problem 3.81, (c) Problem 3.83, (d) Problem 3.84, page 95, by Laplace transforms.

4.88. Use Laplace transforms to find the charge and current at any time in a series circuit having an inductance L, capacitance C, resistance R and e.m.f. $E(t)$. Treat all cases assuming that the initial charge and current are zero.

4.89. Solve Problem 3.85, page 95, by Laplace transforms.

4.90. (a) A particle of mass m moves along the x axis in such a way that the force acting is given by $F(t) = \begin{cases} F_0/\epsilon & 0 < t < \epsilon \\ 0 & t > \epsilon \end{cases}$. Assuming that it starts from rest at the origin, find the position of the particle at any time and interpret physically. (b) Discuss the result in (a) if the limit is taken as $\epsilon \to 0$.

4.91. Let $f(t)$ be any continuous function. Suppose there exists a function $\delta(t)$ such that

$$\int_0^\infty f(t)\, \delta(t - t_0)\, dt = f(t_0)$$

Show that (a) $\mathcal{L}\{\delta(t)\} = 1$, (b) $\mathcal{L}\{\delta(t - t_0)\} = e^{-st_0}$, (c) $\mathcal{L}\{\delta'(t)\} = s$. We often call $\delta(t)$ the *Dirac delta function*).

4.92. In Problem 4.90 replace the force by $F(t) = F_0 \delta(t)$ and solve. Discuss a possible relationship between $\delta(t)$ and the function $\begin{cases} 1/\epsilon & 0 < t < \epsilon \\ 0 & t > \epsilon \end{cases}$. [*Hint.* Examine the case where $\epsilon \to 0$.]

4.93. Graph the function $F(t) = \begin{cases} 1/\epsilon^2 & 0 < t < \epsilon \\ -1/\epsilon^2 & \epsilon < t < 2\epsilon \end{cases}$. Find (a) $\mathcal{L}\{F(t)\}$ and (b) $\lim_{\epsilon \to 0} \mathcal{L}\{F(t)\}$. Discuss the relationship of this problem with $\mathcal{L}^{-1}\{s^2\}$.

Answers to Supplementary Problems

4.48. (a) $\dfrac{12}{3s - 2}$ (d) $\dfrac{5s + 6}{s^2 + 9}$ (g) $\dfrac{144}{s(s^2 + 36)}$ (i) $\dfrac{4\Gamma(1/3)}{9s^{7/3}} - \dfrac{2\Gamma(1/3)}{3s^{4/3}} + \dfrac{\Gamma(2/3)}{s^{2/3}}$

     (b) $\dfrac{6 - 3s}{s^2}$ (e) $\dfrac{72}{s(s^2 - 36)}$ (h) $\dfrac{2}{s^2 + 16}$ (j) $\dfrac{-5}{s + 2}$

     (c) $\dfrac{s^2 + 2s + 2}{s^3}$ (f) $\dfrac{1}{s} + \dfrac{2\sqrt{\pi}}{\sqrt{s}} - \dfrac{3\sqrt{\pi}}{4s^{5/2}}$

4.49. (a) $\dfrac{2e^{-4s} - 1}{s}$ (b) $\dfrac{1}{s} + \dfrac{1}{s^2} - \dfrac{4e^{-3s}}{s} - \dfrac{e^{-3s}}{s^2}$ (c) $\dfrac{e^{-2s} - e^{-4s}}{s}$

4.52. (a) $\dfrac{40}{s^2 + 64} - \dfrac{10}{s^2 + 4}$ (b) $\dfrac{1 + e^{-\pi s}}{s^2 + 1}$

4.53. (a) $2e^{3t}$ (d) $\frac{1}{6} \sin \frac{3}{2} t$ (g) $\dfrac{4t^{1/2} - t^{-1/2}}{\sqrt{\pi}}$ (i) $2t^2 + \frac{2}{3}t^3 + \dfrac{t^4}{24}$

(b) $\frac{1}{8} e^{-5t/3}$ (e) $2 \cos 6t - \frac{4}{3} \sin 6t$ (h) $3 \cosh 8t - 2 \sinh 8t$ (j) $\frac{5}{2}e^{-5t} - \frac{3}{2}e^{-3t}$

(c) $4 \cos 4t$ (f) $t - \dfrac{t^2}{2} + \dfrac{t^3}{6} - \dfrac{t^4}{24}$

4.54. $\dfrac{9}{10\sqrt{5}} \sin 2\sqrt{5}\, t - \dfrac{4}{5\sqrt{10}} \sin \sqrt{10}\, t$

4.60. (a) $\dfrac{2e^{-s} + 3e^{-2s}}{s}$ (b) $\dfrac{e^{-3s}(3s + 1)}{s^2}$

4.62. (a) $\mathcal{U}(t - 5)$ (b) $2\mathcal{U}(t - 1) - \mathcal{U}(t - 2)$

4.63. (a) $3 + (\cos t - 3)\mathcal{U}(t - \pi/2)$ (b) $\dfrac{3(1 - e^{-\pi s/2})}{s} - \dfrac{e^{-\pi s/2}}{s^2 + 1}$

4.64. (a) $\dfrac{6}{(s + 5)^4}$ (b) $\dfrac{s + 1}{s^2 + 2s + 5}$ (c) $\dfrac{\sqrt{\pi}}{2(s - 4)^{3/2}}$ (d) $\sqrt{\dfrac{\pi}{s + 3}}$

4.65. (a) $\dfrac{6s}{(s^2 + 9)^2}$ (b) $\dfrac{2s^3 - 24s}{(s^2 + 4)^3}$ **4.66.** $\dfrac{2s - 2}{(s^2 - 2s + 2)^2}$

4.67. (b) $\dfrac{1 - 4se^{-4s} - e^{-4s}}{s^2(1 - e^{-4s})}$ **4.68.** $\dfrac{1 - e^{-2(s+1)}}{(s + 1)(1 - e^{-2s})}$

4.71. (a) $\ln\left(\dfrac{s + 5}{s + 3}\right)$ (b) $\frac{1}{2} \ln\left(\dfrac{s^2 + 9}{s^2 + 4}\right)$ (c) $\tan^{-1}(1/s)$

4.72. $s \ln(s/\sqrt{s^2 + 1}) + \dfrac{\pi}{2} - \tan^{-1} s$

4.74. (a) $e^t - 1 - t$ (b) $\dfrac{1}{s^2(s - 1)}$

4.76. (a) $e^{-t}(3 \cos 3t + 2 \sin 3t)$ (c) $\frac{1}{2}t^2 e^{5t}$ (d) $te^{5t} + \frac{5}{2}t^2 e^{5t}$

(b) $t^{-1/2} e^{t/4}/2\sqrt{\pi}$

(e) $\mathcal{U}(t - 3) \sin \pi(t - 3)$ or $-\mathcal{U}(t - 3) \sin \pi t = \begin{cases} 0 & t < 3 \\ -\sin \pi t & t > 3 \end{cases}$

4.77. (a) $\frac{1}{2} \sinh 2t$ (b) $\frac{1}{2} - e^{-t} + \frac{1}{2}e^{-2t}$ (c) $2e^{-t} - 2e^{2t} - 4te^{2t}$

(d) $\cos 2\sqrt{2}\, t + \sqrt{2} \sin 2\sqrt{2}\, t - e^t$

(e) $e^{-t}(4 \cos 2t + 3 \sin 2t) + e^{2t}(3 \sin 2t - 4 \cos 2t)$

(f) $\frac{1}{4}(\sin t \cosh t - \cos t \sinh t)$

4.78. (a) $\frac{1}{8}(e^{2t} - 2t^2 - 2t - 1)$ (c) $2t \sin t - \sin t + t \cos t$

(b) $\frac{1}{2}(\sin t + t \cos t)$ (d) $\frac{1}{8}(3 \sin t - 3t \cos t - t^2 \sin t)$

4.79. $\dfrac{e^{-2t} - e^{-6t}}{t}$ **4.81.** (a) $y(t) = 5t - 3 - t^2$ (b) $y(t) = 2\sqrt{2t/\pi}\, e^{-t}$ **4.82.** $\frac{1}{2}\sqrt{\pi}/2$

4.83. (a) $y(t) = 2 - 2e^t + e^{2t}$ (c) $y(t) = (3t^2 + 4t - 2)e^{-2t}$

(b) $y(t) = 2 \sin 4t + 3 \cos 4t + 2t$ (d) $y(t) = c_1 + c_2 \cos t + c_3 \sin t + \frac{1}{2}t^2 + t$

4.84. $y(t) = 4t^3 - 2t - 3 + \frac{2}{3}e^{-2t} + \frac{1}{3}e^t(7 \cos \sqrt{3}\, t + \sqrt{3} \sin \sqrt{3}\, t)$

4.85. $x(t) = 3 - 2e^{-t} - e^{-2t},\quad y(t) = 2 - 4e^{-t} + 2e^{-2t}$

4.90. (a) $x = \dfrac{F_0 t^2}{2m\epsilon}\{1 - \mathcal{U}(t - \epsilon)\}$

Chapter 5

Vector Analysis

VECTORS AND SCALARS

There are quantities in physics characterized by both magnitude and direction, such as displacement, velocity, force and acceleration. To describe such quantities, we introduce the concept of a *vector* as a directed line segment $\overrightarrow{PQ}$ from one point P called the *initial point* to another point Q called the *terminal point*. We denote vectors by bold faced letters or letters with an arrow over them. Thus $\overrightarrow{PQ}$ is denoted by **A** or $\vec{A}$ as in Fig. 5-1. The *magnitude* or *length* of the vector is then denoted by $|\overrightarrow{PQ}|$, $\overline{PQ}$, $|\mathbf{A}|$ or $|\vec{A}|$.

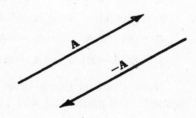

Fig. 5-1

Other quantities in physics are characterized by magnitude only, such as mass, length and temperature. Such quantities are often called *scalars* to distinguish them from vectors, but it must be emphasized that apart from units such as feet, degrees, etc., they are nothing more than real numbers. We can thus denote them by ordinary letters as usual.

VECTOR ALGEBRA

The operations of addition, subtraction and multiplication familiar in the algebra of numbers are, with suitable definition, capable of extension to an algebra of vectors. The following definitions are fundamental.

1. Two vectors **A** and **B** are *equal* if they have the same magnitude and direction regardless of their initial points. Thus **A = B** in Fig. 5-1 above.

2. A vector having direction opposite to that of vector **A** but with the same magnitude is denoted by $-\mathbf{A}$ [see Fig. 5-2].

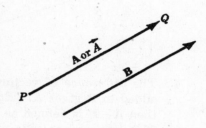

Fig. 5-2

3. The *sum* or *resultant* of vectors **A** and **B** of Fig. 5-3(a) below is a vector **C** formed by placing the initial point of **B** on the terminal point of **A** and joining the initial point of **A** to the terminal point of **B** [see Fig. 5-3(b) below]. The sum **C** is written **C = A + B**. The definition here is equivalent to the *parallelogram law* for vector addition as indicated in Fig. 5-3(c) below.

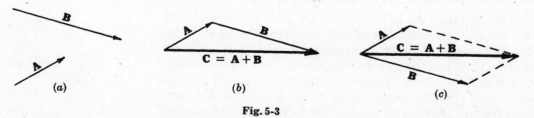

(a) (b) (c)

Fig. 5-3

Extensions to sums of more than two vectors are immediate. For example, Fig. 5-4 below shows how to obtain the sum or resultant **E** of the vectors **A, B, C** and **D**.

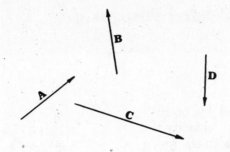

Fig. 5-4

4. The *difference* of vectors **A** and **B**, represented by **A** − **B**, is that vector **C** which added to **B** gives **A**. Equivalently, **A** − **B** may be defined as **A** + (−**B**). If **A** = **B**, then **A** − **B** is defined as the *null* or *zero vector* and is represented by the symbol **0**. This has a magnitude of zero but its direction is not defined.

5. Multiplication of a vector **A** by a scalar m produces a vector m**A** with magnitude $|m|$ times the magnitude of **A** and direction the same as or opposite to that of **A** according as m is positive or negative. If $m = 0$, m**A** = **0**, the null vector.

LAWS OF VECTOR ALGEBRA

If **A**, **B** and **C** are vectors, and m and n are scalars, then

1. **A** + **B** = **B** + **A**   Commutative Law for Addition
2. **A** + (**B** + **C**) = (**A** + **B**) + **C**   Associative Law for Addition
3. $m(n$**A**$) = (mn)$**A** $= n(m$**A**$)$ Associative Law for Multiplication
4. $(m + n)$**A** $= m$**A** $+ n$**A**   Distributive Law
5. $m($**A** + **B**$) = m$**A** $+ m$**B**   Distributive Law

Note that in these laws only multiplication of a vector by one or more scalars is defined. On pages 123 and 124 we define products of vectors.

UNIT VECTORS

Unit vectors are vectors having unit length. If **A** is any vector with length $A > 0$, then **A**$/A$ is a unit vector, denoted by **a**, having the same direction as **A**. Then **A** = A**a**.

RECTANGULAR UNIT VECTORS

The rectangular unit vectors **i, j** and **k** are unit vectors having the direction of the positive x, y and z axes of a rectangular coordinate system [see Fig. 5-5]. We use right-handed rectangular coordinate systems unless otherwise specified. Such systems derive their name from the fact that a right threaded screw rotated through 90° from Ox to Oy will advance in the positive z direction. In general,

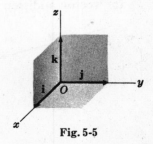

Fig. 5-5

three vectors **A**, **B** and **C** which have coincident initial points and are not coplanar are said to form a *right-handed system* or *dextral system* if a right threaded screw rotated through an angle less than 180° from **A** to **B** will advance in the direction **C** [see Fig. 5-6 below].

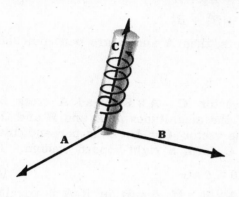

Fig. 5-6

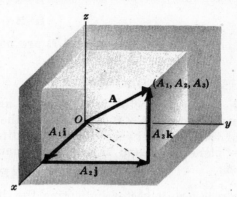

Fig. 5-7

COMPONENTS OF A VECTOR

Any vector **A** in 3 dimensions can be represented with initial point at the origin O of a rectangular coordinate system [see Fig. 5-7 above]. Let (A_1, A_2, A_3) be the rectangular coordinates of the terminal point of vector **A** with initial point at O. The vectors $A_1\mathbf{i}$, $A_2\mathbf{j}$ and $A_3\mathbf{k}$ are called the *rectangular component vectors*, or simply *component vectors*, of **A** in the x, y and z directions respectively. A_1, A_2 and A_3 are called the *rectangular components*, or simply *components*, of **A** in the x, y and z directions respectively.

The sum or resultant of $A_1\mathbf{i}$, $A_2\mathbf{j}$ and $A_3\mathbf{k}$ is the vector **A**, so that we can write

$$\mathbf{A} = A_1\mathbf{i} + A_2\mathbf{j} + A_3\mathbf{k} \tag{1}$$

The magnitude of **A** is

$$A = |\mathbf{A}| = \sqrt{A_1^2 + A_2^2 + A_3^2} \tag{2}$$

In particular, the *position vector* or *radius vector* **r** from O to the point (x, y, z) is written

$$\mathbf{r} = x\mathbf{i} + y\mathbf{j} + z\mathbf{k} \tag{3}$$

and has magnitude $r = |\mathbf{r}| = \sqrt{x^2 + y^2 + z^2}$.

DOT OR SCALAR PRODUCT

The dot or scalar product of two vectors **A** and **B**, denoted by $\mathbf{A} \cdot \mathbf{B}$ (read **A** dot **B**) is defined as the product of the magnitudes of **A** and **B** and the cosine of the angle between them. In symbols,

$$\mathbf{A} \cdot \mathbf{B} = AB \cos \theta, \quad 0 \leqq \theta \leqq \pi \tag{4}$$

Note that $\mathbf{A} \cdot \mathbf{B}$ is a scalar and not a vector.

The following laws are valid:

1. $\mathbf{A} \cdot \mathbf{B} = \mathbf{B} \cdot \mathbf{A}$ Commutative Law for Dot Products
2. $\mathbf{A} \cdot (\mathbf{B} + \mathbf{C}) = \mathbf{A} \cdot \mathbf{B} + \mathbf{A} \cdot \mathbf{C}$ Distributive Law
3. $m(\mathbf{A} \cdot \mathbf{B}) = (m\mathbf{A}) \cdot \mathbf{B} = \mathbf{A} \cdot (m\mathbf{B}) = (\mathbf{A} \cdot \mathbf{B})m,$ where m is a scalar.
4. $\mathbf{i} \cdot \mathbf{i} = \mathbf{j} \cdot \mathbf{j} = \mathbf{k} \cdot \mathbf{k} = 1, \quad \mathbf{i} \cdot \mathbf{j} = \mathbf{j} \cdot \mathbf{k} = \mathbf{k} \cdot \mathbf{i} = 0$

5. If $\mathbf{A} = A_1\mathbf{i} + A_2\mathbf{j} + A_3\mathbf{k}$ and $\mathbf{B} = B_1\mathbf{i} + B_2\mathbf{j} + B_3\mathbf{k}$, then

$$\mathbf{A} \cdot \mathbf{B} = A_1B_1 + A_2B_2 + A_3B_3$$

$$\mathbf{A} \cdot \mathbf{A} = A^2 = A_1^2 + A_2^2 + A_3^2$$

$$\mathbf{B} \cdot \mathbf{B} = B^2 = B_1^2 + B_2^2 + B_3^2$$

6. If $\mathbf{A} \cdot \mathbf{B} = 0$ and $\mathbf{A}$ and $\mathbf{B}$ are not null vectors, then $\mathbf{A}$ and $\mathbf{B}$ are perpendicular.

CROSS OR VECTOR PRODUCT

The cross or vector product of $\mathbf{A}$ and $\mathbf{B}$ is a vector $\mathbf{C} = \mathbf{A} \times \mathbf{B}$ (read $\mathbf{A}$ cross $\mathbf{B}$). The magnitude of $\mathbf{A} \times \mathbf{B}$ is defined as the product of the magnitudes of $\mathbf{A}$ and $\mathbf{B}$ and the sine of the angle between them. The direction of the vector $\mathbf{C} = \mathbf{A} \times \mathbf{B}$ is perpendicular to the plane of $\mathbf{A}$ and $\mathbf{B}$ and such that $\mathbf{A}, \mathbf{B}$ and $\mathbf{C}$ form a right-handed system. In symbols,

$$\mathbf{A} \times \mathbf{B} = AB \sin \theta \, \mathbf{u}, \qquad 0 \leqq \theta \leqq \pi \tag{5}$$

where $\mathbf{u}$ is a unit vector indicating the direction of $\mathbf{A} \times \mathbf{B}$. If $\mathbf{A} = \mathbf{B}$ or if $\mathbf{A}$ is parallel to $\mathbf{B}$, then $\sin \theta = 0$ and we define $\mathbf{A} \times \mathbf{B} = 0$.

The following laws are valid:

1. $\mathbf{A} \times \mathbf{B} = -\mathbf{B} \times \mathbf{A}$ (Commutative Law for Cross Products Fails)

2. $\mathbf{A} \times (\mathbf{B} + \mathbf{C}) = \mathbf{A} \times \mathbf{B} + \mathbf{A} \times \mathbf{C}$ Distributive Law

3. $m(\mathbf{A} \times \mathbf{B}) = (m\mathbf{A}) \times \mathbf{B} = \mathbf{A} \times (m\mathbf{B}) = (\mathbf{A} \times \mathbf{B})m$,   where m is a scalar.

4. $\mathbf{i} \times \mathbf{i} = \mathbf{j} \times \mathbf{j} = \mathbf{k} \times \mathbf{k} = 0$, $\mathbf{i} \times \mathbf{j} = \mathbf{k}$, $\mathbf{j} \times \mathbf{k} = \mathbf{i}$, $\mathbf{k} \times \mathbf{i} = \mathbf{j}$

5. If $\mathbf{A} = A_1\mathbf{i} + A_2\mathbf{j} + A_3\mathbf{k}$ and $\mathbf{B} = B_1\mathbf{i} + B_2\mathbf{j} + B_3\mathbf{k}$,   then

$$\mathbf{A} \times \mathbf{B} = \begin{vmatrix} \mathbf{i} & \mathbf{j} & \mathbf{k} \\ A_1 & A_2 & A_3 \\ B_1 & B_2 & B_3 \end{vmatrix}$$

6. $|\mathbf{A} \times \mathbf{B}|$ = the area of a parallelogram with sides $\mathbf{A}$ and $\mathbf{B}$.

7. If $\mathbf{A} \times \mathbf{B} = 0$ and $\mathbf{A}$ and $\mathbf{B}$ are not null vectors, then $\mathbf{A}$ and $\mathbf{B}$ are parallel.

TRIPLE PRODUCTS

Dot and cross multiplication of three vectors $\mathbf{A}, \mathbf{B}$ and $\mathbf{C}$ may produce meaningful products of the form $(\mathbf{A} \cdot \mathbf{B})\mathbf{C}, \mathbf{A} \cdot (\mathbf{B} \times \mathbf{C})$ and $\mathbf{A} \times (\mathbf{B} \times \mathbf{C})$. The following laws are valid:

1. $(\mathbf{A} \cdot \mathbf{B})\mathbf{C} \neq \mathbf{A}(\mathbf{B} \cdot \mathbf{C})$ in general

2. $\mathbf{A} \cdot (\mathbf{B} \times \mathbf{C}) = \mathbf{B} \cdot (\mathbf{C} \times \mathbf{A}) = \mathbf{C} \cdot (\mathbf{A} \times \mathbf{B})$ = volume of a parallelepiped having $\mathbf{A}, \mathbf{B}$, and $\mathbf{C}$ as edges, or the negative of this volume according as $\mathbf{A}, \mathbf{B}$ and $\mathbf{C}$ do or do not form a right-handed system. If $\mathbf{A} = A_1\mathbf{i} + A_2\mathbf{j} + A_3\mathbf{k}, \mathbf{B} = B_1\mathbf{i} + B_2\mathbf{j} + B_3\mathbf{k}$ and $\mathbf{C} = C_1\mathbf{i} + C_2\mathbf{j} + C_3\mathbf{k}$, then

$$\mathbf{A} \cdot (\mathbf{B} \times \mathbf{C}) = \begin{vmatrix} A_1 & A_2 & A_3 \\ B_1 & B_2 & B_3 \\ C_1 & C_2 & C_3 \end{vmatrix} \tag{6}$$

3. $\mathbf{A} \times (\mathbf{B} \times \mathbf{C}) \neq (\mathbf{A} \times \mathbf{B}) \times \mathbf{C}$ (Associative Law for Cross Products Fails)

4. $\mathbf{A} \times (\mathbf{B} \times \mathbf{C}) = (\mathbf{A} \cdot \mathbf{C})\mathbf{B} - (\mathbf{A} \cdot \mathbf{B})\mathbf{C}$
    $(\mathbf{A} \times \mathbf{B}) \times \mathbf{C} = (\mathbf{A} \cdot \mathbf{C})\mathbf{B} - (\mathbf{B} \cdot \mathbf{C})\mathbf{A}$

The product $\mathbf{A} \cdot (\mathbf{B} \times \mathbf{C})$ is sometimes called the *scalar triple product* or *box product* and may be denoted by $[\mathbf{ABC}]$. The product $\mathbf{A} \times (\mathbf{B} \times \mathbf{C})$ is called the *vector triple product*.

In $\mathbf{A} \cdot (\mathbf{B} \times \mathbf{C})$ parentheses are sometimes omitted and we write $\mathbf{A} \cdot \mathbf{B} \times \mathbf{C}$. However, parentheses must be used in $\mathbf{A} \times (\mathbf{B} \times \mathbf{C})$ (see Problem 5.25). Note that $\mathbf{A} \cdot (\mathbf{B} \times \mathbf{C}) = (\mathbf{A} \times \mathbf{B}) \cdot \mathbf{C}$. This is often expressed by stating that in a scalar triple product the dot and the cross can be interchanged without affecting the result [see Problem 5.22].

VECTOR FUNCTIONS

If corresponding to each value of a scalar u we associate a vector $\mathbf{A}$, then $\mathbf{A}$ is called a *function* of u denoted by $\mathbf{A}(u)$. In three dimensions we can write $\mathbf{A}(u) = A_1(u)\mathbf{i} + A_2(u)\mathbf{j} + A_3(u)\mathbf{k}$.

The function concept is easily extended. Thus if to each point (x, y, z) there corresponds a vector $\mathbf{A}$, then $\mathbf{A}$ is a function of (x, y, z), indicated by $\mathbf{A}(x, y, z) = A_1(x, y, z)\mathbf{i} + A_2(x, y, z)\mathbf{j} + A_3(x, y, z)\mathbf{k}$.

We sometimes say that a vector function $\mathbf{A}(x, y, z)$ defines a *vector field* since it associates a vector with each point of a region. Similarly $\phi(x, y, z)$ defines a *scalar field* since it associates a scalar with each point of a region.

LIMITS, CONTINUITY AND DERIVATIVES OF VECTOR FUNCTIONS

Limits, continuity and derivatives of vector functions follow rules similar to those for scalar functions already considered. The following statements show the analogy which exists.

1. The vector function $\mathbf{A}(u)$ is said to be *continuous* at u_0 if given any positive number ϵ, we can find some positive number δ such that $|\mathbf{A}(u) - \mathbf{A}(u_0)| < \epsilon$ whenever $|u - u_0| < \delta$. This is equivalent to the statement $\lim_{u \to u_0} \mathbf{A}(u) = \mathbf{A}(u_0)$.

2. The derivative of $\mathbf{A}(u)$ is defined as

$$\frac{d\mathbf{A}}{du} = \lim_{\Delta u \to 0} \frac{\mathbf{A}(u + \Delta u) - \mathbf{A}(u)}{\Delta u} \tag{7}$$

provided this limit exists. In case $\mathbf{A}(u) = A_1(u)\mathbf{i} + A_2(u)\mathbf{j} + A_3(u)\mathbf{k}$; then

$$\frac{d\mathbf{A}}{du} = \frac{dA_1}{du}\mathbf{i} + \frac{dA_2}{du}\mathbf{j} + \frac{dA_3}{du}\mathbf{k}$$

Higher derivatives such as $d^2\mathbf{A}/du^2$, etc., can be similarly defined.

3. If $\mathbf{A}(x, y, z) = A_1(x, y, z)\mathbf{i} + A_2(x, y, z)\mathbf{j} + A_3(x, y, z)\mathbf{k}$, then

$$d\mathbf{A} = \frac{\partial \mathbf{A}}{\partial x}dx + \frac{\partial \mathbf{A}}{\partial y}dy + \frac{\partial \mathbf{A}}{\partial z}dz \tag{8}$$

is the *differential* of $\mathbf{A}$.

4. Derivatives of products obey rules similar to those for scalar functions. However, when cross products are involved the order may be important. Some examples are:

$$(a) \quad \frac{d}{du}(\phi\mathbf{A}) = \phi\frac{d\mathbf{A}}{du} + \frac{d\phi}{du}\mathbf{A},$$

$$(b) \quad \frac{\partial}{\partial y}(\mathbf{A} \cdot \mathbf{B}) = \mathbf{A} \cdot \frac{\partial \mathbf{B}}{\partial y} + \frac{\partial \mathbf{A}}{\partial y} \cdot \mathbf{B},$$

$$(c) \quad \frac{\partial}{\partial z}(\mathbf{A} \times \mathbf{B}) = \mathbf{A} \times \frac{\partial \mathbf{B}}{\partial z} + \frac{\partial \mathbf{A}}{\partial z} \times \mathbf{B}$$

GEOMETRIC INTERPRETATION OF A VECTOR DERIVATIVE

If **r** is the vector joining the origin O of a coordinate system and the point (x, y, z), then specification of the vector function $\mathbf{r}(u)$ defines x, y and z as functions of u. As u changes, the terminal point of **r** describes a *space curve* (see Fig. 5-8) having parametric equations $x = x(u)$, $y = y(u)$, $z = z(u)$. If the parameter u is the arc length s measured from some fixed point on the curve, then

$$\frac{d\mathbf{r}}{ds} = \mathbf{T} \qquad (9)$$

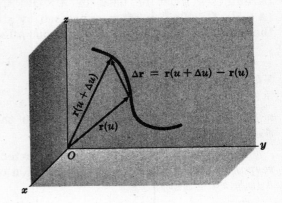

is a unit vector in the direction of the tangent to the curve and is called the *unit tangent vector*. If u is the time t, then

Fig. 5-8

$$\frac{d\mathbf{r}}{dt} = \mathbf{v} \qquad (10)$$

is the *velocity* with which the terminal point of **r** describes the curve. We have

$$\mathbf{v} = \frac{d\mathbf{r}}{dt} = \frac{d\mathbf{r}}{ds}\frac{ds}{dt} = \frac{ds}{dt}\mathbf{T} = v\mathbf{T} \qquad (11)$$

from which we see that the magnitude of **v**, often called the *speed*, is $v = ds/dt$. Similarly,

$$\frac{d^2\mathbf{r}}{dt^2} = \mathbf{a} \qquad (12)$$

is the *acceleration* with which the terminal point of **r** describes the curve. These concepts have important applications in *mechanics*.

GRADIENT, DIVERGENCE AND CURL

Consider the vector operator ∇ (*del*) defined by

$$\nabla \equiv \mathbf{i}\frac{\partial}{\partial x} + \mathbf{j}\frac{\partial}{\partial y} + \mathbf{k}\frac{\partial}{\partial z} \qquad (13)$$

Then if $\phi(x, y, z)$ and $\mathbf{A}(x, y, z)$ have continuous first partial derivatives in a region (a condition which is in many cases stronger than necessary), we can define the following.

1. Gradient. The *gradient* of ϕ is defined by

$$\text{grad } \phi = \nabla\phi = \left(\mathbf{i}\frac{\partial}{\partial x} + \mathbf{j}\frac{\partial}{\partial y} + \mathbf{k}\frac{\partial}{\partial z}\right)\phi = \mathbf{i}\frac{\partial\phi}{\partial x} + \mathbf{j}\frac{\partial\phi}{\partial y} + \mathbf{k}\frac{\partial\phi}{\partial z} \qquad (14)$$

$$= \frac{\partial\phi}{\partial x}\mathbf{i} + \frac{\partial\phi}{\partial y}\mathbf{j} + \frac{\partial\phi}{\partial z}\mathbf{k}$$

An interesting interpretation is that if $\phi(x, y, z) = c$ is the equation of a surface, then $\nabla\phi$ is a normal to this surface (see Problem 5.36).

2. Divergence. The *divergence* of **A** is defined by

$$\text{div } \mathbf{A} = \nabla \cdot \mathbf{A} = \left(\mathbf{i}\frac{\partial}{\partial x} + \mathbf{j}\frac{\partial}{\partial y} + \mathbf{k}\frac{\partial}{\partial z}\right) \cdot (A_1\mathbf{i} + A_2\mathbf{j} + A_3\mathbf{k}) \qquad (15)$$

$$= \frac{\partial A_1}{\partial x} + \frac{\partial A_2}{\partial y} + \frac{\partial A_3}{\partial z}$$

3. Curl. The *curl* of **A** is defined by

$$\text{curl } \mathbf{A} = \nabla \times \mathbf{A} = \left(\mathbf{i}\frac{\partial}{\partial x} + \mathbf{j}\frac{\partial}{\partial y} + \mathbf{k}\frac{\partial}{\partial z} \right) \times (A_1\mathbf{i} + A_2\mathbf{j} + A_3\mathbf{k}) \tag{16}$$

$$= \begin{vmatrix} \mathbf{i} & \mathbf{j} & \mathbf{k} \\ \dfrac{\partial}{\partial x} & \dfrac{\partial}{\partial y} & \dfrac{\partial}{\partial z} \\ A_1 & A_2 & A_3 \end{vmatrix}$$

$$= \mathbf{i}\begin{vmatrix} \dfrac{\partial}{\partial y} & \dfrac{\partial}{\partial z} \\ A_2 & A_3 \end{vmatrix} - \mathbf{j}\begin{vmatrix} \dfrac{\partial}{\partial x} & \dfrac{\partial}{\partial z} \\ A_1 & A_2 \end{vmatrix} + \mathbf{k}\begin{vmatrix} \dfrac{\partial}{\partial x} & \dfrac{\partial}{\partial y} \\ A_1 & A_2 \end{vmatrix}$$

$$= \left(\frac{\partial A_3}{\partial y} - \frac{\partial A_2}{\partial z} \right)\mathbf{i} + \left(\frac{\partial A_1}{\partial z} - \frac{\partial A_3}{\partial x} \right)\mathbf{j} + \left(\frac{\partial A_2}{\partial x} - \frac{\partial A_1}{\partial y} \right)\mathbf{k}$$

Note that in the expansion of the determinant, the operators $\partial/\partial x$, $\partial/\partial y$, $\partial/\partial z$ must precede A_1, A_2, A_3.

FORMULAS INVOLVING ∇

If the partial derivatives of **A**, **B**, U and V are assumed to exist, then

1. $\nabla(U + V) = \nabla U + \nabla V$ or $\text{grad }(U + V) = \text{grad } u + \text{grad } V$

2. $\nabla \cdot (\mathbf{A} + \mathbf{B}) = \nabla \cdot \mathbf{A} + \nabla \cdot \mathbf{B}$ or $\text{div }(\mathbf{A} + \mathbf{B}) = \text{div } \mathbf{A} + \text{div } \mathbf{B}$

3. $\nabla \times (\mathbf{A} + \mathbf{B}) = \nabla \times \mathbf{A} + \nabla \times \mathbf{B}$ or $\text{curl }(\mathbf{A} + \mathbf{B}) = \text{curl } \mathbf{A} + \text{curl } \mathbf{B}$

4. $\nabla \cdot (U\mathbf{A}) = (\nabla U) \cdot \mathbf{A} + U(\nabla \cdot \mathbf{A})$

5. $\nabla \times (U\mathbf{A}) = (\nabla U) \times \mathbf{A} + U(\nabla \times \mathbf{A})$

6. $\nabla \cdot (\mathbf{A} \times \mathbf{B}) = \mathbf{B} \cdot (\nabla \times \mathbf{A}) - \mathbf{A} \cdot (\nabla \times \mathbf{B})$

7. $\nabla \times (\mathbf{A} \times \mathbf{B}) = (\mathbf{B} \cdot \nabla)\mathbf{A} - \mathbf{B}(\nabla \cdot \mathbf{A}) - (\mathbf{A} \cdot \nabla)\mathbf{B} + \mathbf{A}(\nabla \cdot \mathbf{B})$

8. $\nabla(\mathbf{A} \cdot \mathbf{B}) = (\mathbf{B} \cdot \nabla)\mathbf{A} + (\mathbf{A} \cdot \nabla)\mathbf{B} + \mathbf{B} \times (\nabla \times \mathbf{A}) + \mathbf{A} \times (\nabla \times \mathbf{B})$

9. $\nabla \cdot (\nabla U) \equiv \nabla^2 U \equiv \dfrac{\partial^2 U}{\partial x^2} + \dfrac{\partial^2 U}{\partial y^2} + \dfrac{\partial^2 U}{\partial z^2}$ is called the *Laplacian* of U

    and $\nabla^2 \equiv \dfrac{\partial^2}{\partial x^2} + \dfrac{\partial^2}{\partial y^2} + \dfrac{\partial^2}{\partial z^2}$ is called the *Laplacian operator*.

10. $\nabla \times (\nabla U) = 0$. The curl of the gradient of U is zero.

11. $\nabla \cdot (\nabla \times \mathbf{A}) = 0$. The divergence of the curl of **A** is zero.

12. $\nabla \times (\nabla \times \mathbf{A}) = \nabla(\nabla \cdot \mathbf{A}) - \nabla^2 \mathbf{A}$

ORTHOGONAL CURVILINEAR COORDINATES. JACOBIANS

The *transformation equations*

$$x = f(u_1, u_2, u_3), \quad y = g(u_1, u_2, u_3), \quad z = h(u_1, u_2, u_3) \tag{17}$$

[where we assume that f, g, h are continuous, have continuous partial derivatives and have a single-valued inverse] establish a one to one correspondence between points in an xyz and $u_1u_2u_3$ rectangular coordinate system.   In vector notation the transformation (17) can be written

$$\mathbf{r} = x\mathbf{i} + y\mathbf{j} + z\mathbf{k} = f(u_1, u_2, u_3)\mathbf{i} + g(u_1, u_2, u_3)\mathbf{j} + h(u_1, u_2, u_3)\mathbf{k} \tag{18}$$

A point P in Fig. 5-9 can then be defined not only by *rectangular coordinates* (x, y, z) but by coordinates (u_1, u_2, u_3) as well. We call (u_1, u_2, u_3) the *curvilinear coordinates* of the point.

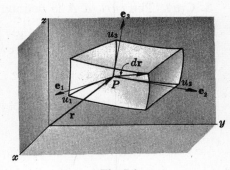

Fig. 5-9

If u_2 and u_3 are constant, then as u_1 varies, **r** describes a curve which we call the u_1 *coordinate curve*. Similarly we define the u_2 and u_3 coordinate curves through P.

From (*18*), we have

$$d\mathbf{r} = \frac{\partial \mathbf{r}}{\partial u_1} du_1 + \frac{\partial \mathbf{r}}{\partial u_2} du_2 + \frac{\partial \mathbf{r}}{\partial u_3} du_3 \qquad (19)$$

The vector $\partial \mathbf{r}/\partial u_1$ is tangent to the u_1 coordinate curve at P. If $\mathbf{e}_1$ is a unit vector at P in this direction, we can write $\partial \mathbf{r}/\partial u_1 = h_1 \mathbf{e}_1$ where $h_1 = |\partial \mathbf{r}/\partial u_1|$. Similarly we can write $\partial \mathbf{r}/\partial u_2 = h_2 \mathbf{e}_2$ and $\partial \mathbf{r}/\partial u_3 = h_3 \mathbf{e}_3$, where $h_2 = |\partial \mathbf{r}/\partial u_2|$ and $h_3 = |\partial \mathbf{r}/\partial u_3|$ respectively. Then (*19*) can be written

$$d\mathbf{r} = h_1 du_1 \mathbf{e}_1 + h_2 du_2 \mathbf{e}_2 + h_3 du_3 \mathbf{e}_3 \qquad (20)$$

The quantities h_1, h_2, h_3 are sometimes called *scale factors*.

If $\mathbf{e}_1, \mathbf{e}_2, \mathbf{e}_3$ are mutually perpendicular at any point P, the curvilinear coordinates are called *orthogonal*. In such case the element of arc length ds is given by

$$ds^2 = d\mathbf{r} \cdot d\mathbf{r} = h_1^2 du_1^2 + h_2^2 du_2^2 + h_3^2 du_3^2 \qquad (21)$$

and corresponds to the square of the length of the diagonal in the above parallelepiped.

Also, in the case of orthogonal coordinates the volume of the parallelepiped is given by

$$dV = |(h_1 du_1 \mathbf{e}_1) \cdot (h_2 du_2 \mathbf{e}_2) \times (h_3 du_3 \mathbf{e}_3)| = h_1 h_2 h_3 du_1 du_2 du_3 \qquad (22)$$

which can be written as

$$dV = \left| \frac{\partial \mathbf{r}}{\partial u_1} \cdot \frac{\partial \mathbf{r}}{\partial u_2} \times \frac{\partial \mathbf{r}}{\partial u_3} \right| du_1 du_2 du_3 = \left| \frac{\partial(x, y, z)}{\partial(u_1, u_2, u_3)} \right| du_1 du_2 du_3 \qquad (23)$$

where

$$\frac{\partial(x, y, z)}{\partial(u_1, u_2, u_3)} = \begin{vmatrix} \dfrac{\partial x}{\partial u_1} & \dfrac{\partial x}{\partial u_2} & \dfrac{\partial x}{\partial u_3} \\[2mm] \dfrac{\partial y}{\partial u_1} & \dfrac{\partial y}{\partial u_2} & \dfrac{\partial y}{\partial u_3} \\[2mm] \dfrac{\partial z}{\partial u_1} & \dfrac{\partial z}{\partial u_2} & \dfrac{\partial z}{\partial u_3} \end{vmatrix} \qquad (24)$$

is called the *Jacobian* of the transformation.

It is clear that when the Jacobian is identically zero there is no parallelepiped. In such case there is a functional relationship between x, y and z, i.e. there is a function ϕ such that $\phi(x, y, z) = 0$ identically.

GRADIENT, DIVERGENCE, CURL AND LAPLACIAN IN ORTHOGONAL CURVILINEAR COORDINATES

If Φ is a scalar function and $\mathbf{A} = A_1 \mathbf{e}_1 + A_2 \mathbf{e}_2 + A_3 \mathbf{e}_3$ a vector function of orthogonal curvilinear coordinates u_1, u_2, u_3, we have the following results.

1. $\nabla \Phi = \text{grad } \Phi = \dfrac{1}{h_1} \dfrac{\partial \Phi}{\partial u_1} \mathbf{e}_1 + \dfrac{1}{h_2} \dfrac{\partial \Phi}{\partial u_2} \mathbf{e}_2 + \dfrac{1}{h_3} \dfrac{\partial \Phi}{\partial u_3} \mathbf{e}_3$

2. $\nabla \cdot \mathbf{A} = \text{div } \mathbf{A} = \dfrac{1}{h_1 h_2 h_3} \left[\dfrac{\partial}{\partial u_1}(h_2 h_3 A_1) + \dfrac{\partial}{\partial u_2}(h_3 h_1 A_2) + \dfrac{\partial}{\partial u_3}(h_1 h_2 A_3) \right]$

3. $\quad \nabla \times \mathbf{A} \;=\; \text{curl } \mathbf{A} \;=\; \dfrac{1}{h_1 h_2 h_3} \begin{vmatrix} h_1\mathbf{e}_1 & h_2\mathbf{e}_2 & h_3\mathbf{e}_3 \\[4pt] \dfrac{\partial}{\partial u_1} & \dfrac{\partial}{\partial u_2} & \dfrac{\partial}{\partial u_3} \\[6pt] h_1 A_1 & h_2 A_2 & h_3 A_3 \end{vmatrix}$

4. $\quad \nabla^2 \Phi \;=\; \text{Laplacian of } \Phi \;=\; \dfrac{1}{h_1 h_2 h_3}\left[\dfrac{\partial}{\partial u_1}\left(\dfrac{h_2 h_3}{h_1} \dfrac{\partial \Phi}{\partial u_1}\right) + \dfrac{\partial}{\partial u_2}\left(\dfrac{h_3 h_1}{h_2} \dfrac{\partial \Phi}{\partial u_2}\right)\right.$

$$\left. + \dfrac{\partial}{\partial u_3}\left(\dfrac{h_1 h_2}{h_3} \dfrac{\partial \Phi}{\partial u_3}\right)\right]$$

These reduce to the usual expressions in rectangular coordinates if we replace (u_1, u_2, u_3) by (x, y, z), in which case $\mathbf{e}_1, \mathbf{e}_2$ and $\mathbf{e}_3$ are replaced by $\mathbf{i}, \mathbf{j}$ and $\mathbf{k}$ and $h_1 = h_2 = h_3 = 1$.

SPECIAL CURVILINEAR COORDINATES

1. Cylindrical Coordinates (ρ, ϕ, z). See Fig. 5-10.

Transformation equations:

$$x = \rho \cos\phi, \quad y = \rho \sin\phi, \quad z = z$$

where $\rho \geqq 0, \;\; 0 \leqq \phi < 2\pi, \;\; -\infty < z < \infty$.

Scale factors: $\quad h_1 = 1, \quad h_2 = \rho, \quad h_3 = 1$

Element of arc length: $\quad ds^2 = d\rho^2 + \rho^2\, d\phi^2 + dz^2$

Jacobian: $\quad \dfrac{\partial(x, y, z)}{\partial(\rho, \phi, z)} = \rho$

Element of volume: $\quad dV = \rho\, d\rho\, d\phi\, dz$

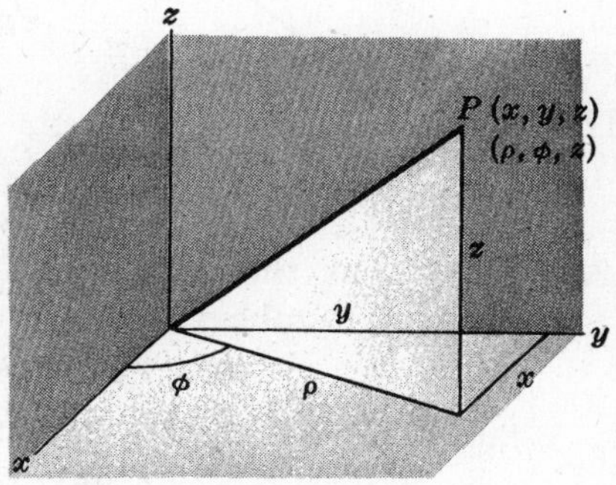

Laplacian:

Fig. 5-10

$$\nabla^2 U \;=\; \frac{1}{\rho}\frac{\partial}{\partial \rho}\left(\rho \frac{\partial U}{\partial \rho}\right) + \frac{1}{\rho^2}\frac{\partial^2 U}{\partial \phi^2} + \frac{\partial^2 U}{\partial z^2} \;=\; \frac{\partial^2 U}{\partial \rho^2} + \frac{1}{\rho}\frac{\partial U}{\partial \rho} + \frac{1}{\rho^2}\frac{\partial^2 U}{\partial \phi^2} + \frac{\partial^2 U}{\partial z^2}$$

Note that corresponding results can be obtained for polar coordinates in the plane by omitting z dependence. In such case for example, $ds^2 = d\rho^2 + \rho^2\, d\phi^2$, while the element of volume is replaced by the element of area, $dA = \rho\, d\rho\, d\phi$.

2. Spherical Coordinates (r, θ, ϕ). See Fig. 5-11.

Transformation equations:

$x = r\sin\theta \cos\phi, \;\; y = r\sin\theta \sin\phi, \;\; z = r\cos\theta$

where $r \geqq 0, \; 0 \leqq \theta \leqq \pi, \; 0 \leqq \phi < 2\pi$.

Scale factors: $\quad h_1 = 1, \; h_2 = r, \; h_3 = r\sin\theta$

Element of arc length:

$$ds^2 = dr^2 + r^2\, d\theta^2 + r^2 \sin^2\theta\, d\phi^2$$

Jacobian: $\quad \dfrac{\partial(x, y, z)}{\partial(r, \theta, \phi)} = r^2 \sin\theta$

Element of volume: $\quad dV = r^2 \sin\theta\, dr\, d\theta\, d\phi$

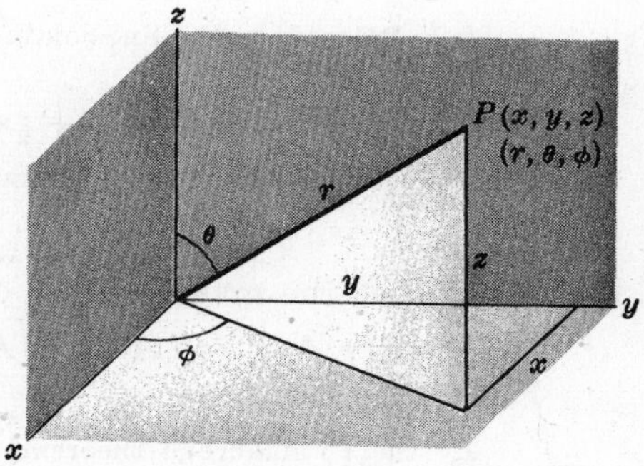

Fig. 5-11

Laplacian: $\quad \nabla^2 U \;=\; \dfrac{1}{r^2}\dfrac{\partial}{\partial r}\left(r^2 \dfrac{\partial U}{\partial r}\right) + \dfrac{1}{r^2 \sin\theta}\dfrac{\partial}{\partial \theta}\left(\sin\theta \dfrac{\partial U}{\partial \theta}\right) + \dfrac{1}{r^2 \sin^2\theta}\dfrac{\partial^2 U}{\partial \phi^2}.$

Other types of coordinate systems are possible.

Solved Problems

VECTOR ALGEBRA

5.1. Show that addition of vectors is commutative, i.e. $\mathbf{A} + \mathbf{B} = \mathbf{B} + \mathbf{A}$. See Fig. (*a*) below.

$$\mathbf{OP} + \mathbf{PQ} = \mathbf{OQ} \quad \text{or} \quad \mathbf{A} + \mathbf{B} = \mathbf{C}$$

and $\qquad\qquad \mathbf{OR} + \mathbf{RQ} = \mathbf{OQ} \quad \text{or} \quad \mathbf{B} + \mathbf{A} = \mathbf{C}$

Then $\mathbf{A} + \mathbf{B} = \mathbf{B} + \mathbf{A}$.

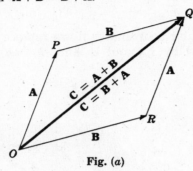

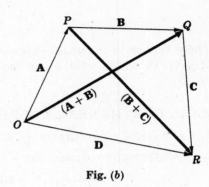

Fig. (*a*) 　　　　　　　　　　　　　　　　　　Fig. (*b*)

5.2. Show that the addition of vectors is associative, i.e. $\mathbf{A} + (\mathbf{B} + \mathbf{C}) = (\mathbf{A} + \mathbf{B}) + \mathbf{C}$.
See Fig. (*b*) above.

$$\mathbf{OP} + \mathbf{PQ} = \mathbf{OQ} = (\mathbf{A} + \mathbf{B}) \quad \text{and} \quad \mathbf{PQ} + \mathbf{QR} = \mathbf{PR} = (\mathbf{B} + \mathbf{C})$$

Since $\qquad\qquad \mathbf{OP} + \mathbf{PR} = \mathbf{OR} = \mathbf{D}, \quad \text{i.e.} \quad \mathbf{A} + (\mathbf{B} + \mathbf{C}) = \mathbf{D}$

$$\mathbf{OQ} + \mathbf{QR} = \mathbf{OR} = \mathbf{D}, \quad \text{i.e.} \quad (\mathbf{A} + \mathbf{B}) + \mathbf{C} = \mathbf{D}$$

we have $\mathbf{A} + (\mathbf{B} + \mathbf{C}) = (\mathbf{A} + \mathbf{B}) + \mathbf{C}$.

Extensions of the results of Problems 5.1 and 5.2 show that the order of addition of any number of vectors is immaterial.

5.3. Prove that the line joining the midpoint of two sides of a triangle is parallel to the third side and has half its length.

From Fig. 5-12, $\mathbf{AC} + \mathbf{CB} = \mathbf{AB}$ or $\mathbf{b} + \mathbf{a} = \mathbf{c}$.

Let $\mathbf{DE} = \mathbf{d}$ be the line joining the midpoints of sides AC and CB. Then

$$\mathbf{d} = \mathbf{DC} + \mathbf{CE} = \tfrac{1}{2}\mathbf{b} + \tfrac{1}{2}\mathbf{a} = \tfrac{1}{2}(\mathbf{b} + \mathbf{a}) = \tfrac{1}{2}\mathbf{c}$$

Thus $\mathbf{d}$ is parallel to $\mathbf{c}$ and has half its length.

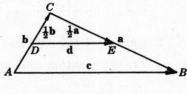

Fig. 5-12

5.4. Prove that the magnitude A of the vector $\mathbf{A} = A_1\mathbf{i} + A_2\mathbf{j} + A_3\mathbf{k}$ is $A = \sqrt{A_1^2 + A_2^2 + A_3^2}$. See Fig. 5-13.

By the Pythagorean theorem,

$$(\overline{OP})^2 = (\overline{OQ})^2 + (\overline{QP})^2$$

where $\overline{OP}$ denotes the magnitude of vector $\mathbf{OP}$, etc. Similarly, $(\overline{OQ})^2 = (\overline{OR})^2 + (\overline{RQ})^2$.

Then $(\overline{OP})^2 = (\overline{OR})^2 + (\overline{RQ})^2 + (\overline{QP})^2$ or

$$A^2 = A_1^2 + A_2^2 + A_3^2, \quad \text{i.e.} \quad A = \sqrt{A_1^2 + A_2^2 + A_3^2}$$

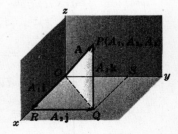

Fig. 5-13

5.5. Determine the vector having initial point $P(x_1, y_1, z_1)$ and terminal point $Q(x_2, y_2, z_2)$ and find its magnitude. See Fig. 5-14.

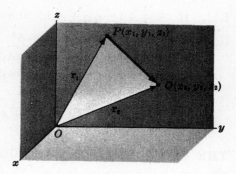

Fig. 5-14

The position vector of P is $\mathbf{r}_1 = x_1\mathbf{i} + y_1\mathbf{j} + z_1\mathbf{k}$.

The position vector of Q is $\mathbf{r}_2 = x_2\mathbf{i} + y_2\mathbf{j} + z_2\mathbf{k}$.

$\mathbf{r}_1 + \mathbf{PQ} = \mathbf{r}_2$ or

$$\mathbf{PQ} = \mathbf{r}_2 - \mathbf{r}_1 = (x_2\mathbf{i} + y_2\mathbf{j} + z_2\mathbf{k}) - (x_1\mathbf{i} + y_1\mathbf{j} + z_1\mathbf{k})$$
$$= (x_2 - x_1)\mathbf{i} + (y_2 - y_1)\mathbf{j} + (z_2 - z_1)\mathbf{k}$$

Magnitude of $\mathbf{PQ} = \overline{PQ}$
$$= \sqrt{(x_2 - x_1)^2 + (y_2 - y_1)^2 + (z_2 - z_1)^2}$$

Note that this is the distance between points P and Q.

THE DOT OR SCALAR PRODUCT

5.6. Prove that the projection of $\mathbf{A}$ on $\mathbf{B}$ is equal to $\mathbf{A} \cdot \mathbf{b}$, where $\mathbf{b}$ is a unit vector in the direction of $\mathbf{B}$.

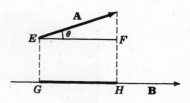

Fig. 5-15

Through the initial and terminal points of $\mathbf{A}$ pass planes perpendicular to $\mathbf{B}$ at G and H respectively as in the adjacent Fig. 5-15; then

Projection of $\mathbf{A}$ on $\mathbf{B}$ = $\overline{GH}$ = $\overline{EF}$ = $A \cos\theta$ = $\mathbf{A} \cdot \mathbf{b}$

5.7. Prove $\mathbf{A} \cdot (\mathbf{B} + \mathbf{C}) = \mathbf{A} \cdot \mathbf{B} + \mathbf{A} \cdot \mathbf{C}$.

Let $\mathbf{a}$ be a unit vector in the direction of $\mathbf{A}$; then [see Fig. 5-16]

Projection of $(\mathbf{B} + \mathbf{C})$ on $\mathbf{A}$ = projection of $\mathbf{B}$ on $\mathbf{A}$ + projection of $\mathbf{C}$ on $\mathbf{A}$

$(\mathbf{B} + \mathbf{C}) \cdot \mathbf{a}$ = $\mathbf{B} \cdot \mathbf{a} + \mathbf{C} \cdot \mathbf{a}$

Multiplying by A,

$(\mathbf{B} + \mathbf{C}) \cdot A\mathbf{a}$ = $\mathbf{B} \cdot A\mathbf{a} + \mathbf{C} \cdot A\mathbf{a}$

and $(\mathbf{B} + \mathbf{C}) \cdot \mathbf{A}$ = $\mathbf{B} \cdot \mathbf{A} + \mathbf{C} \cdot \mathbf{A}$

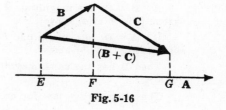

Fig. 5-16

Then by the commutative law for dot products,

$\mathbf{A} \cdot (\mathbf{B} + \mathbf{C})$ = $\mathbf{A} \cdot \mathbf{B} + \mathbf{A} \cdot \mathbf{C}$

and the distributive law is valid.

5.8. Prove that $(\mathbf{A} + \mathbf{B}) \cdot (\mathbf{C} + \mathbf{D}) = \mathbf{A} \cdot \mathbf{C} + \mathbf{A} \cdot \mathbf{D} + \mathbf{B} \cdot \mathbf{C} + \mathbf{B} \cdot \mathbf{D}$.

By Problem 5.7, $(\mathbf{A} + \mathbf{B}) \cdot (\mathbf{C} + \mathbf{D}) = \mathbf{A} \cdot (\mathbf{C} + \mathbf{D}) + \mathbf{B} \cdot (\mathbf{C} + \mathbf{D}) = \mathbf{A} \cdot \mathbf{C} + \mathbf{A} \cdot \mathbf{D} + \mathbf{B} \cdot \mathbf{C} + \mathbf{B} \cdot \mathbf{D}$.
The ordinary laws of algebra are valid for dot products.

5.9. Evaluate each of the following.

(a) $\mathbf{i} \cdot \mathbf{i}$ = $|\mathbf{i}|\, |\mathbf{i}| \cos 0°$ = $(1)(1)(1)$ = 1

(b) $\mathbf{i} \cdot \mathbf{k}$ = $|\mathbf{i}|\, |\mathbf{k}| \cos 90°$ = $(1)(1)(0)$ = 0

(c) $\mathbf{k} \cdot \mathbf{j}$ = $|\mathbf{k}|\, |\mathbf{j}| \cos 90°$ = $(1)(1)(0)$ = 0

(d) $\mathbf{j} \cdot (2\mathbf{i} - 3\mathbf{j} + \mathbf{k})$ = $2\mathbf{j} \cdot \mathbf{i} - 3\mathbf{j} \cdot \mathbf{j} + \mathbf{j} \cdot \mathbf{k}$ = $0 - 3 + 0$ = -3

(e) $(2\mathbf{i} - \mathbf{j}) \cdot (3\mathbf{i} + \mathbf{k})$ = $2\mathbf{i} \cdot (3\mathbf{i} + \mathbf{k}) - \mathbf{j} \cdot (3\mathbf{i} + \mathbf{k})$
$$= 6\mathbf{i} \cdot \mathbf{i} + 2\mathbf{i} \cdot \mathbf{k} - 3\mathbf{j} \cdot \mathbf{i} - \mathbf{j} \cdot \mathbf{k} = 6 + 0 - 0 - 0 = 6$$

5.10. If $\mathbf{A} = A_1\mathbf{i} + A_2\mathbf{j} + A_3\mathbf{k}$ and $\mathbf{B} = B_1\mathbf{i} + B_2\mathbf{j} + B_3\mathbf{k}$, prove $\mathbf{A} \cdot \mathbf{B} = A_1B_1 + A_2B_2 + A_3B_3$.

$\mathbf{A} \cdot \mathbf{B}$ = $(A_1\mathbf{i} + A_2\mathbf{j} + A_3\mathbf{k}) \cdot (B_1\mathbf{i} + B_2\mathbf{j} + B_3\mathbf{k})$
$$= A_1\mathbf{i} \cdot (B_1\mathbf{i} + B_2\mathbf{j} + B_3\mathbf{k}) + A_2\mathbf{j} \cdot (B_1\mathbf{i} + B_2\mathbf{j} + B_3\mathbf{k}) + A_3\mathbf{k} \cdot (B_1\mathbf{i} + B_2\mathbf{j} + B_3\mathbf{k})$$
$$= A_1B_1\mathbf{i} \cdot \mathbf{i} + A_1B_2\mathbf{i} \cdot \mathbf{j} + A_1B_3\mathbf{i} \cdot \mathbf{k} + A_2B_1\mathbf{j} \cdot \mathbf{i} + A_2B_2\mathbf{j} \cdot \mathbf{j} + A_2B_3\mathbf{j} \cdot \mathbf{k}$$
$$+ A_3B_1\mathbf{k} \cdot \mathbf{i} + A_3B_2\mathbf{k} \cdot \mathbf{j} + A_3B_3\mathbf{k} \cdot \mathbf{k}$$
$$= A_1B_1 + A_2B_2 + A_3B_3$$

since $\mathbf{i} \cdot \mathbf{i} = \mathbf{j} \cdot \mathbf{j} = \mathbf{k} \cdot \mathbf{k} = 1$ and all other dot products are zero.

5.11. If $\mathbf{A} = A_1\mathbf{i} + A_2\mathbf{j} + A_3\mathbf{k}$, show that $A = \sqrt{\mathbf{A} \cdot \mathbf{A}} = \sqrt{A_1^2 + A_2^2 + A_3^2}$.

$\mathbf{A} \cdot \mathbf{A} = (A)(A) \cos 0° = A^2$.   Then $A = \sqrt{\mathbf{A} \cdot \mathbf{A}}$.

Also, $\mathbf{A} \cdot \mathbf{A} = (A_1\mathbf{i} + A_2\mathbf{j} + A_3\mathbf{k}) \cdot (A_1\mathbf{i} + A_2\mathbf{j} + A_3\mathbf{k})$

$= (A_1)(A_1) + (A_2)(A_2) + (A_3)(A_3) = A_1^2 + A_2^2 + A_3^2$

by Problem 5.10, taking $\mathbf{B} = \mathbf{A}$.

Then $A = \sqrt{\mathbf{A} \cdot \mathbf{A}} = \sqrt{A_1^2 + A_2^2 + A_3^2}$ is the magnitude of $\mathbf{A}$.   Sometimes $\mathbf{A} \cdot \mathbf{A}$ is written $\mathbf{A}^2$.

THE CROSS OR VECTOR PRODUCT

5.12. Prove $\mathbf{A} \times \mathbf{B} = -\mathbf{B} \times \mathbf{A}$.

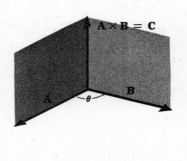   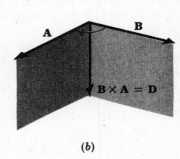

(a)   (b)

Fig. 5-17

$\mathbf{A} \times \mathbf{B} = \mathbf{C}$ has magnitude $AB \sin \theta$ and direction such that $\mathbf{A}, \mathbf{B}$ and $\mathbf{C}$ form a right-handed system [Fig. 5-17(a) above].

$\mathbf{B} \times \mathbf{A} = \mathbf{D}$ has magnitude $BA \sin \theta$ and direction such that $\mathbf{B}, \mathbf{A}$ and $\mathbf{D}$ form a right-handed system [Fig. 5-17(b) above].

Then $\mathbf{D}$ has the same magnitude as $\mathbf{C}$ but is opposite in direction, i.e. $\mathbf{C} = -\mathbf{D}$ or $\mathbf{A} \times \mathbf{B} = -\mathbf{B} \times \mathbf{A}$.

The commutative law for cross products is not valid.

5.13. Prove that $\mathbf{A} \times (\mathbf{B} + \mathbf{C}) = \mathbf{A} \times \mathbf{B} + \mathbf{A} \times \mathbf{C}$ for the case where $\mathbf{A}$ is perpendicular to $\mathbf{B}$ and also to $\mathbf{C}$.

Since $\mathbf{A}$ is perpendicular to $\mathbf{B}$, $\mathbf{A} \times \mathbf{B}$ is a vector perpendicular to the plane of $\mathbf{A}$ and $\mathbf{B}$ and having magnitude $AB \sin 90° = AB$ or magnitude of $A\mathbf{B}$. This is equivalent to multiplying vector $\mathbf{B}$ by A and rotating the resultant vector through 90° to the position shown in Fig. 5-18.

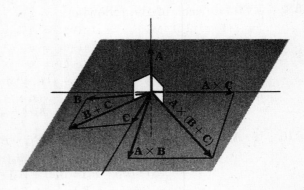

Similarly, $\mathbf{A} \times \mathbf{C}$ is the vector obtained by multiplying $\mathbf{C}$ by A and rotating the resultant vector through 90° to the position shown.

In like manner, $\mathbf{A} \times (\mathbf{B} + \mathbf{C})$ is the vector obtained by multiplying $\mathbf{B} + \mathbf{C}$ by A and rotating the resultant vector through 90° to the position shown.

Since $\mathbf{A} \times (\mathbf{B} + \mathbf{C})$ is the diagonal of the parallelogram with $\mathbf{A} \times \mathbf{B}$ and $\mathbf{A} \times \mathbf{C}$ as sides, we have $\mathbf{A} \times (\mathbf{B} + \mathbf{C}) = \mathbf{A} \times \mathbf{B} + \mathbf{A} \times \mathbf{C}$.

Fig. 5-18

5.14. Prove that $\mathbf{A} \times (\mathbf{B} + \mathbf{C}) = \mathbf{A} \times \mathbf{B} + \mathbf{A} \times \mathbf{C}$ in the general case where $\mathbf{A}, \mathbf{B}$ and $\mathbf{C}$ are non-coplanar. See Fig. 5-19.

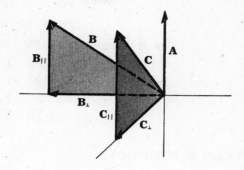

Resolve $\mathbf{B}$ into two component vectors, one perpendicular to $\mathbf{A}$ and the other parallel to $\mathbf{A}$, and denote them by $\mathbf{B}_\perp$ and $\mathbf{B}_{||}$ respectively. Then $\mathbf{B} = \mathbf{B}_\perp + \mathbf{B}_{||}$.

If θ is the angle between $\mathbf{A}$ and $\mathbf{B}$, then $B_\perp = B \sin\theta$. Thus the magnitude of $\mathbf{A} \times \mathbf{B}_\perp$ is $AB \sin\theta$, the same as the magnitude of $\mathbf{A} \times \mathbf{B}$. Also, the direction $\mathbf{A} \times \mathbf{B}$ is the same as the direction of $\mathbf{A} \times \mathbf{B}$. Hence $\mathbf{A} \times \mathbf{B}_\perp = \mathbf{A} \times \mathbf{B}$.

Similarly if $\mathbf{C}$ is resolved into two component vectors $\mathbf{C}_{||}$ and $\mathbf{C}_\perp$, parallel and perpendicular respectively to $\mathbf{A}$, then $\mathbf{A} \times \mathbf{C}_\perp = \mathbf{A} \times \mathbf{C}$.

Fig. 5-19

Also, since $\mathbf{B} + \mathbf{C} = \mathbf{B}_\perp + \mathbf{B}_{||} + \mathbf{C}_\perp + \mathbf{C}_{||} = (\mathbf{B}_\perp + \mathbf{C}_\perp) + (\mathbf{B}_{||} + \mathbf{C}_{||})$ it follows that

$$\mathbf{A} \times (\mathbf{B}_\perp + \mathbf{C}_\perp) = \mathbf{A} \times (\mathbf{B} + \mathbf{C})$$

Now $\mathbf{B}_\perp$ and $\mathbf{C}_\perp$ are vectors perpendicular to $\mathbf{A}$ and so by Problem 5.13,

$$\mathbf{A} \times (\mathbf{B}_\perp + \mathbf{C}_\perp) = \mathbf{A} \times \mathbf{B}_\perp + \mathbf{A} \times \mathbf{C}_\perp$$

Then

$$\mathbf{A} \times (\mathbf{B} + \mathbf{C}) = \mathbf{A} \times \mathbf{B} + \mathbf{A} \times \mathbf{C}$$

and the distributive law holds. Multiplying by -1, using Problem 5.12, this becomes $(\mathbf{B} + \mathbf{C}) \times \mathbf{A} = \mathbf{B} \times \mathbf{A} + \mathbf{C} \times \mathbf{A}$. Note that the order of factors in cross products is important. The usual laws of algebra apply only if proper order is maintained.

5.15. If $\mathbf{A} = A_1\mathbf{i} + A_2\mathbf{j} + A_3\mathbf{k}$ and $\mathbf{B} = B_1\mathbf{i} + B_2\mathbf{j} + B_3\mathbf{k}$, prove that $\mathbf{A} \times \mathbf{B} = \begin{vmatrix} \mathbf{i} & \mathbf{j} & \mathbf{k} \\ A_1 & A_2 & A_3 \\ B_1 & B_2 & B_3 \end{vmatrix}$.

$$\begin{aligned}
\mathbf{A} \times \mathbf{B} &= (A_1\mathbf{i} + A_2\mathbf{j} + A_3\mathbf{k}) \times (B_1\mathbf{i} + B_2\mathbf{j} + B_3\mathbf{k}) \\
&= A_1\mathbf{i} \times (B_1\mathbf{i} + B_2\mathbf{j} + B_3\mathbf{k}) + A_2\mathbf{j} \times (B_1\mathbf{i} + B_2\mathbf{j} + B_3\mathbf{k}) + A_3\mathbf{k} \times (B_1\mathbf{i} + B_2\mathbf{j} + B_3\mathbf{k}) \\
&= A_1B_1\mathbf{i} \times \mathbf{i} + A_1B_2\mathbf{i} \times \mathbf{j} + A_1B_3\mathbf{i} \times \mathbf{k} + A_2B_1\mathbf{j} \times \mathbf{i} + A_2B_2\mathbf{j} \times \mathbf{j} + A_2B_3\mathbf{j} \times \mathbf{k} \\
&\quad + A_3B_1\mathbf{k} \times \mathbf{i} + A_3B_2\mathbf{k} \times \mathbf{j} + A_3B_3\mathbf{k} \times \mathbf{k} \\
&= (A_2B_3 - A_3B_2)\mathbf{i} + (A_3B_1 - A_1B_3)\mathbf{j} + (A_1B_2 - A_2B_1)\mathbf{k} = \begin{vmatrix} \mathbf{i} & \mathbf{j} & \mathbf{k} \\ A_1 & A_2 & A_3 \\ B_1 & B_2 & B_3 \end{vmatrix}
\end{aligned}$$

5.16. If $\mathbf{A} = 3\mathbf{i} - \mathbf{j} + 2\mathbf{k}$ and $\mathbf{B} = 2\mathbf{i} + 3\mathbf{j} - \mathbf{k}$, find $\mathbf{A} \times \mathbf{B}$.

$$\begin{aligned}
\mathbf{A} \times \mathbf{B} &= \begin{vmatrix} \mathbf{i} & \mathbf{j} & \mathbf{k} \\ 3 & -1 & 2 \\ 2 & 3 & -1 \end{vmatrix} = \mathbf{i}\begin{vmatrix} -1 & 2 \\ 3 & -1 \end{vmatrix} - \mathbf{j}\begin{vmatrix} 3 & 2 \\ 2 & -1 \end{vmatrix} + \mathbf{k}\begin{vmatrix} 3 & -1 \\ 2 & 3 \end{vmatrix} \\
&= -5\mathbf{i} + 7\mathbf{j} + 11\mathbf{k}
\end{aligned}$$

5.17. Prove that the area of a parallelogram with sides $\mathbf{A}$ and $\mathbf{B}$ is $|\mathbf{A} \times \mathbf{B}|$. See Fig. 5-20.

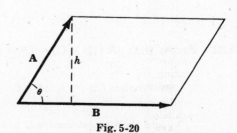

$$\begin{aligned}
\text{Area of parallelogram} &= h|\mathbf{B}| \\
&= |\mathbf{A}| \sin\theta |\mathbf{B}| \\
&= |\mathbf{A} \times \mathbf{B}|
\end{aligned}$$

Note that the area of the triangle with sides $\mathbf{A}$ and $\mathbf{B} = \frac{1}{2}|\mathbf{A} \times \mathbf{B}|$.

Fig. 5-20

5.18. Find the area of the triangle with vertices at $P(2, 3, 5)$, $Q(4, 2, -1)$, $R(3, 6, 4)$.

$$\mathbf{PQ} = (4-2)\mathbf{i} + (2-3)\mathbf{j} + (-1-5)\mathbf{k} = 2\mathbf{i} - \mathbf{j} - 6\mathbf{k}$$

$$\mathbf{PR} = (3-2)\mathbf{i} + (6-3)\mathbf{j} + (4-5)\mathbf{k} = \mathbf{i} + 3\mathbf{j} - \mathbf{k}$$

Area of triangle $= \tfrac{1}{2}|\mathbf{PQ} \times \mathbf{PR}| = \tfrac{1}{2}|(2\mathbf{i} - \mathbf{j} - 6\mathbf{k}) \times (\mathbf{i} + 3\mathbf{j} - \mathbf{k})|$

$$= \tfrac{1}{2}\left| \begin{vmatrix} \mathbf{i} & \mathbf{j} & \mathbf{k} \\ 2 & -1 & -6 \\ 1 & 3 & -1 \end{vmatrix} \right| = \tfrac{1}{2}|19\mathbf{i} - 4\mathbf{j} + 7\mathbf{k}|$$

$$= \tfrac{1}{2}\sqrt{(19)^2 + (-4)^2 + (7)^2} = \tfrac{1}{2}\sqrt{426}$$

TRIPLE PRODUCTS

5.19. Show that $\mathbf{A} \cdot (\mathbf{B} \times \mathbf{C})$ is in absolute value equal to the volume of a parallelepiped with sides $\mathbf{A}, \mathbf{B}$ and $\mathbf{C}$. See Fig. 5-21.

Let $\mathbf{n}$ be a unit normal to parallelogram I, having the direction of $\mathbf{B} \times \mathbf{C}$, and let h be the height of the terminal point of $\mathbf{A}$ above the parallelogram I.

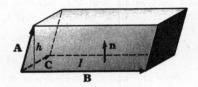

Fig. 5-21

Volume of parallelepiped $=$ (height h)(area of parallelogram I)

$$= (\mathbf{A} \cdot \mathbf{n})(|\mathbf{B} \times \mathbf{C}|)$$

$$= \mathbf{A} \cdot \{|\mathbf{B} \times \mathbf{C}|\,\mathbf{n}\} = \mathbf{A} \cdot (\mathbf{B} \times \mathbf{C})$$

If $\mathbf{A}, \mathbf{B}$ and $\mathbf{C}$ do not form a right-handed system, $\mathbf{A} \cdot \mathbf{n} < 0$ and the volume $= |\mathbf{A} \cdot (\mathbf{B} \times \mathbf{C})|$.

5.20. If $\mathbf{A} = A_1\mathbf{i} + A_2\mathbf{j} + A_3\mathbf{k}$, $\mathbf{B} = B_1\mathbf{i} + B_2\mathbf{j} + B_3\mathbf{k}$, $\mathbf{C} = C_1\mathbf{i} + C_2\mathbf{j} + C_3\mathbf{k}$ show that

$$\mathbf{A} \cdot (\mathbf{B} \times \mathbf{C}) = \begin{vmatrix} A_1 & A_2 & A_3 \\ B_1 & B_2 & B_3 \\ C_1 & C_2 & C_3 \end{vmatrix}$$

$$\mathbf{A} \cdot (\mathbf{B} \times \mathbf{C}) = \mathbf{A} \cdot \begin{vmatrix} \mathbf{i} & \mathbf{j} & \mathbf{k} \\ B_1 & B_2 & B_3 \\ C_1 & C_2 & C_3 \end{vmatrix}$$

$$= (A_1\mathbf{i} + A_2\mathbf{j} + A_3\mathbf{k}) \cdot [(B_2C_3 - B_3C_2)\mathbf{i} + (B_3C_1 - B_1C_3)\mathbf{j} + (B_1C_2 - B_2C_1)\mathbf{k}]$$

$$= A_1(B_2C_3 - B_3C_2) + A_2(B_3C_1 - B_1C_3) + A_3(B_1C_2 - B_2C_1) = \begin{vmatrix} A_1 & A_2 & A_3 \\ B_1 & B_2 & B_3 \\ C_1 & C_2 & C_3 \end{vmatrix}$$

5.21. Find the volume of a parallelepiped with sides $\mathbf{A} = 3\mathbf{i} - \mathbf{j}$, $\mathbf{B} = \mathbf{j} + 2\mathbf{k}$, $\mathbf{C} = \mathbf{i} + 5\mathbf{j} + 4\mathbf{k}$.

By Problems 5.19 and 5.20, volume of parallelepiped $= |\mathbf{A} \cdot (\mathbf{B} \times \mathbf{C})| = \left| \begin{vmatrix} 3 & -1 & 0 \\ 0 & 1 & 2 \\ 1 & 5 & 4 \end{vmatrix} \right|$

$$= |-20| = 20$$

5.22. Prove that $\mathbf{A} \cdot (\mathbf{B} \times \mathbf{C}) = (\mathbf{A} \times \mathbf{B}) \cdot \mathbf{C}$, i.e. the dot and cross can be interchanged.

By Problem 5.20: $\mathbf{A} \cdot (\mathbf{B} \times \mathbf{C}) = \begin{vmatrix} A_1 & A_2 & A_3 \\ B_1 & B_2 & B_3 \\ C_1 & C_2 & C_3 \end{vmatrix}$, $(\mathbf{A} \times \mathbf{B}) \cdot \mathbf{C} = \mathbf{C} \cdot (\mathbf{A} \times \mathbf{B}) = \begin{vmatrix} C_1 & C_2 & C_3 \\ A_1 & A_2 & A_3 \\ B_1 & B_2 & B_3 \end{vmatrix}$

Since the two determinants are equal, the required result follows.

5.23. Let $r_1 = x_1 i + y_1 j + z_1 k$, $r_2 = x_2 i + y_2 j + z_2 k$ and $r_3 = x_3 i + y_3 j + z_3 k$ be the position vectors of points $P_1(x_1, y_1, z_1)$, $P_2(x_2, y_2, z_2)$ and $P_3(x_3, y_3, z_3)$. Find an equation for the plane passing through P_1, P_2 and P_3. See Fig. 5-22.

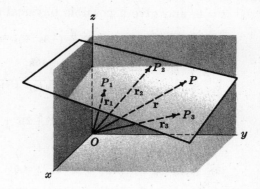

Fig. 5-22

We assume that P_1, P_2 and P_3 do not lie in the same straight line; hence they determine a plane.

Let $r = xi + yj + zk$ denote the position vector of any point $P(x, y, z)$ in the plane. Consider vectors $P_1P_2 = r_2 - r_1$, $P_1P_3 = r_3 - r_1$ and $P_1P = r - r_1$ which all lie in the plane. Then

$$P_1P \cdot P_1P_2 \times P_1P_3 = 0$$

or $$(r - r_1) \cdot (r_2 - r_1) \times (r_3 - r_1) = 0$$

In terms of rectangular coordinates this becomes

$$[(x - x_1)i + (y - y_1)j + (z - z_1)k] \cdot [(x_2 - x_1)i + (y_2 - y_1)j + (z_2 - z_1)k]$$
$$\times [(x_3 - x_1)i + (y_3 - y_1)j + (z_3 - z_1)k] = 0$$

or, using Problem 5.20,
$$\begin{vmatrix} x - x_1 & y - y_1 & z - z_1 \\ x_2 - x_1 & y_2 - y_1 & z_2 - z_1 \\ x_3 - x_1 & y_3 - y_1 & z_3 - z_1 \end{vmatrix} = 0$$

5.24. Find an equation for the plane passing through the points $P_1(3, 1, -2)$, $P_2(-1, 2, 4)$, $P_3(2, -1, 1)$.

The position vectors of P_1, P_2, P_3 and any point $P(x, y, z)$ on the plane are respectively

$$r_1 = 3i + j - 2k, \quad r_2 = -i + 2j + 4k, \quad r_3 = 2i - j + k, \quad r = xi + yj + zk$$

Then $PP_1 = r - r_1$, $P_2P_1 = r_2 - r_1$, $P_3P_1 = r_3 - r_1$, all lie in the required plane and so the required equation is $(r - r_1) \cdot (r_2 - r_1) \times (r_3 - r_1) = 0$, i.e.,

$$\{(x - 3)i + (y - 1)j + (z + 2)k\} \cdot \{-4i + j + 6k\} \times \{-i - 2j + 3k\} = 0$$
$$\{(x - 3)i + (y - 1)j + (z + 2)k\} \cdot \{15i + 6j + 9k\} = 0$$
$$15(x - 3) + 6(y - 1) + 9(z + 2) = 0 \quad \text{or} \quad 5x + 2y + 3z = 11$$

Another method. By Problem 5.23, the required equation is

$$\begin{vmatrix} x - 3 & y - 1 & z + 2 \\ -1 - 3 & 2 - 1 & 4 + 2 \\ 2 - 3 & -1 - 1 & 1 + 2 \end{vmatrix} = 0 \quad \text{or} \quad 5x + 2y + 3z = 11$$

5.25. If $A = i + j$, $B = 2i - 3j + k$, $C = 4j - 3k$, find (a) $(A \times B) \times C$, (b) $A \times (B \times C)$.

(a) $A \times B = \begin{vmatrix} i & j & k \\ 1 & 1 & 0 \\ 2 & -3 & 1 \end{vmatrix} = i - j - 5k$. Then $(A \times B) \times C = \begin{vmatrix} i & j & k \\ 1 & -1 & -5 \\ 0 & 4 & -3 \end{vmatrix} = 23i + 3j + 4k$.

(b) $B \times C = \begin{vmatrix} i & j & k \\ 2 & -3 & 1 \\ 0 & 4 & -3 \end{vmatrix} = 5i + 6j + 8k$. Then $A \times (B \times C) = \begin{vmatrix} i & j & k \\ 1 & 1 & 0 \\ 5 & 6 & 8 \end{vmatrix} = 8i - 8j + k$.

It follows that, in general, $(A \times B) \times C \neq A \times (B \times C)$.

DERIVATIVES

5.26. If $\mathbf{r} = (t^3 + 2t)\mathbf{i} - 3e^{-2t}\mathbf{j} + 2\sin 5t\,\mathbf{k}$, find $(a)\ \dfrac{d\mathbf{r}}{dt}$, $(b)\ \left|\dfrac{d\mathbf{r}}{dt}\right|$, $(c)\ \dfrac{d^2\mathbf{r}}{dt^2}$, $(d)\ \left|\dfrac{d^2\mathbf{r}}{dt^2}\right|$ at $t = 0$ and give a possible physical significance.

(a) $\dfrac{d\mathbf{r}}{dt} = \dfrac{d}{dt}(t^3 + 2t)\mathbf{i} + \dfrac{d}{dt}(-3e^{-2t})\mathbf{j} + \dfrac{d}{dt}(2\sin 5t)\mathbf{k} = (3t^2 + 2)\mathbf{i} + 6e^{-2t}\mathbf{j} + 10\cos 5t\,\mathbf{k}$

     At $t = 0$, $d\mathbf{r}/dt = 2\mathbf{i} + 6\mathbf{j} + 10\mathbf{k}$.

(b) From (a), $|d\mathbf{r}/dt| = \sqrt{(2)^2 + (6)^2 + (10)^2} = \sqrt{140} = 2\sqrt{35}$ at $t = 0$.

(c) $\dfrac{d^2\mathbf{r}}{dt^2} = \dfrac{d}{dt}\left(\dfrac{d\mathbf{r}}{dt}\right) = \dfrac{d}{dt}\{(3t^2 + 2)\mathbf{i} + 6e^{-2t}\mathbf{j} + 10\cos 5t\,\mathbf{k}\} = 6t\mathbf{i} - 12e^{-2t}\mathbf{j} - 50\sin 5t\,\mathbf{k}$

     At $t = 0$, $d^2\mathbf{r}/dt^2 = -12\mathbf{j}$.

(d) From (c), $|d^2\mathbf{r}/dt^2| = 12$ at $t = 0$.

     If t represents time, these represent respectively the velocity, magnitude of the velocity, acceleration and magnitude of the acceleration at $t = 0$ of a particle moving along the space curve $x = t^3 + 2t$, $y = -3e^{-2t}$, $z = 2\sin 5t$.

5.27. Prove that $\dfrac{d}{du}(\mathbf{A} \cdot \mathbf{B}) = \mathbf{A} \cdot \dfrac{d\mathbf{B}}{du} + \dfrac{d\mathbf{A}}{du} \cdot \mathbf{B}$, where $\mathbf{A}$ and $\mathbf{B}$ are differentiable functions of u.

Method 1. $\dfrac{d}{du}(\mathbf{A} \cdot \mathbf{B}) = \lim\limits_{\Delta u \to 0} \dfrac{(\mathbf{A} + \Delta\mathbf{A}) \cdot (\mathbf{B} + \Delta\mathbf{B}) - \mathbf{A} \cdot \mathbf{B}}{\Delta u}$

$= \lim\limits_{\Delta u \to 0} \dfrac{\mathbf{A} \cdot \Delta\mathbf{B} + \Delta\mathbf{A} \cdot \mathbf{B} + \Delta\mathbf{A} \cdot \Delta\mathbf{B}}{\Delta u}$

$= \lim\limits_{\Delta u \to 0} \left(\mathbf{A} \cdot \dfrac{\Delta\mathbf{B}}{\Delta u} + \dfrac{\Delta\mathbf{A}}{\Delta u} \cdot \mathbf{B} + \dfrac{\Delta\mathbf{A}}{\Delta u} \cdot \Delta\mathbf{B}\right) = \mathbf{A} \cdot \dfrac{d\mathbf{B}}{du} + \dfrac{d\mathbf{A}}{du} \cdot \mathbf{B}$

Method 2. Let $\mathbf{A} = A_1\mathbf{i} + A_2\mathbf{j} + A_3\mathbf{k}$, $\mathbf{B} = B_1\mathbf{i} + B_2\mathbf{j} + B_3\mathbf{k}$. Then

$\dfrac{d}{du}(\mathbf{A} \cdot \mathbf{B}) = \dfrac{d}{du}(A_1 B_1 + A_2 B_2 + A_3 B_3)$

$= \left(A_1 \dfrac{dB_1}{du} + A_2 \dfrac{dB_2}{du} + A_3 \dfrac{dB_3}{du}\right) + \left(\dfrac{dA_1}{du} B_1 + \dfrac{dA_2}{du} B_2 + \dfrac{dA_3}{du} B_3\right)$

$= \mathbf{A} \cdot \dfrac{d\mathbf{B}}{du} + \dfrac{d\mathbf{A}}{du} \cdot \mathbf{B}$

5.28. If $\phi(x, y, z) = x^2 yz$ and $\mathbf{A} = 3x^2 y\mathbf{i} + yz^2\mathbf{j} - xz\mathbf{k}$, find $\dfrac{\partial^2}{\partial y\, \partial z}(\phi\mathbf{A})$ at the point $(1, -2, 1)$.

$\phi\mathbf{A} = (x^2 yz)(3x^2 y\mathbf{i} + yz^2\mathbf{j} - xz\mathbf{k}) = 3x^4 y^2 z\mathbf{i} + x^2 y^2 z^3\mathbf{j} - x^3 yz^2\mathbf{k}$

$\dfrac{\partial}{\partial z}(\phi\mathbf{A}) = \dfrac{\partial}{\partial z}(3x^4 y^2 z\mathbf{i} + x^2 y^2 z^3\mathbf{j} - x^3 yz^2\mathbf{k}) = 3x^4 y^2\mathbf{i} + 3x^2 y^2 z^2\mathbf{j} - 2x^3 yz\mathbf{k}$

$\dfrac{\partial^2}{\partial y\, \partial z}(\phi\mathbf{A}) = \dfrac{\partial}{\partial y}(3x^4 y^2\mathbf{i} + 3x^2 y^2 z^2\mathbf{j} - 2x^3 yz\mathbf{k}) = 6x^4 y\mathbf{i} + 6x^2 yz^2\mathbf{j} - 2x^3 z\mathbf{k}$

If $x = 1$, $y = -2$, $z = -1$, this becomes $-12\mathbf{i} - 12\mathbf{j} + 2\mathbf{k}$.

5.29. If $\mathbf{A} = x^2 \sin y\,\mathbf{i} + z^2 \cos y\,\mathbf{j} - xy^2\mathbf{k}$, find $d\mathbf{A}$.

Method 1.

$\dfrac{\partial\mathbf{A}}{\partial x} = 2x \sin y\,\mathbf{i} - y^2\mathbf{k}$, $\dfrac{\partial\mathbf{A}}{\partial y} = x^2 \cos y\,\mathbf{i} - z^2 \sin y\,\mathbf{j} - 2xy\mathbf{k}$, $\dfrac{\partial\mathbf{A}}{\partial z} = 2z \cos y\,\mathbf{j}$

$$dA = \frac{\partial A}{\partial x} dx + \frac{\partial A}{\partial y} dy + \frac{\partial A}{\partial z} dz$$

$$= (2x \sin y\, \mathbf{i} - y^2 \mathbf{k})\, dx + (x^2 \cos y\, \mathbf{i} - z^2 \sin y\, \mathbf{j} - 2xy\mathbf{k})\, dy + (2z \cos y\, \mathbf{j})\, dz$$

$$= (2x \sin y\, dx + x^2 \cos y\, dy)\mathbf{i} + (2z \cos y\, dz - z^2 \sin y\, dy)\mathbf{j} - (y^2\, dx + 2xy\, dy)\mathbf{k}$$

Method 2.

$$dA = d(x^2 \sin y)\mathbf{i} + d(z^2 \cos y)\mathbf{j} - d(xy^2)\mathbf{k}$$

$$= (2x \sin y\, dx + x^2 \cos y\, dy)\mathbf{i} + (2z \cos y\, dz - z^2 \sin y\, dy)\mathbf{j} - (y^2\, dx + 2xy\, dy)\mathbf{k}$$

5.30. A particle moves along a space curve $\mathbf{r} = \mathbf{r}(t)$, where t is the time measured from some initial time. If $v = |d\mathbf{r}/dt| = ds/dt$ is the magnitude of the velocity of the particle (s is the arc length along the space curve measured from the initial position), prove that the acceleration $\mathbf{a}$ of the particle is given by

$$\mathbf{a} = \frac{dv}{dt}\mathbf{T} + \frac{v^2}{\rho}\mathbf{N}$$

where $\mathbf{T}$ and $\mathbf{N}$ are unit tangent and normal vectors to the space curve and

$$\rho = \left|\frac{d^2\mathbf{r}}{ds^2}\right|^{-1} = \left\{\left(\frac{d^2x}{ds^2}\right)^2 + \left(\frac{d^2y}{ds^2}\right)^2 + \left(\frac{d^2z}{ds^2}\right)^2\right\}^{-1/2}$$

The velocity of the particle is given by $\mathbf{v} = v\mathbf{T}$. Then the acceleration is given by

$$\mathbf{a} = \frac{d\mathbf{v}}{dt} = \frac{d}{dt}(v\mathbf{T}) = \frac{dv}{dt}\mathbf{T} + v\frac{d\mathbf{T}}{dt} = \frac{dv}{dt}\mathbf{T} + v\frac{d\mathbf{T}}{ds}\frac{ds}{dt} = \frac{dv}{dt}\mathbf{T} + v^2\frac{d\mathbf{T}}{ds} \qquad (1)$$

Since $\mathbf{T}$ has unit magnitude, we have $\mathbf{T} \cdot \mathbf{T} = 1$. Then differentiating with respect to s,

$$\mathbf{T} \cdot \frac{d\mathbf{T}}{ds} + \frac{d\mathbf{T}}{ds} \cdot \mathbf{T} = 0, \qquad 2\mathbf{T} \cdot \frac{d\mathbf{T}}{ds} = 0 \quad \text{or} \quad \mathbf{T} \cdot \frac{d\mathbf{T}}{ds} = 0$$

from which it follows that $d\mathbf{T}/ds$ is perpendicular to $\mathbf{T}$. Denoting by $\mathbf{N}$ the unit vector in the direction of $d\mathbf{T}/ds$, and called the *principal normal* to the space curve, we have

$$\frac{d\mathbf{T}}{ds} = \kappa\mathbf{N} \qquad (2)$$

where κ is the magnitude of $d\mathbf{T}/ds$. Now since $\mathbf{T} = d\mathbf{r}/ds$ [see equation (9), page 126], we have $d\mathbf{T}/ds = d^2\mathbf{r}/ds^2$. Hence

$$\kappa = \left|\frac{d^2\mathbf{r}}{ds^2}\right| = \left\{\left(\frac{d^2x}{ds^2}\right)^2 + \left(\frac{d^2y}{ds^2}\right)^2 + \left(\frac{d^2z}{ds^2}\right)^2\right\}^{1/2}$$

Defining $\rho = 1/\kappa$, (2) becomes $d\mathbf{T}/ds = \mathbf{N}/\rho$. Thus from (1) we have, as required,

$$\mathbf{a} = \frac{dv}{dt}\mathbf{T} + \frac{v^2}{\rho}\mathbf{N}$$

The components dv/dt and v^2/ρ in the direction of $\mathbf{T}$ and $\mathbf{N}$ are called the *tangential* and *normal* components of the acceleration, the latter being sometimes called the *centripetal acceleration*. The quantities ρ and κ are respectively the *radius of curvature* and *curvature* of the space curve.

GRADIENT, DIVERGENCE AND CURL

5.31. If $\phi = x^2yz^3$ and $\mathbf{A} = xz\mathbf{i} - y^2\mathbf{j} + 2x^2y\mathbf{k}$, find (a) $\nabla\phi$, (b) $\nabla \cdot \mathbf{A}$, (c) $\nabla \times \mathbf{A}$, (d) div $(\phi\mathbf{A})$, (e) curl $(\phi\mathbf{A})$.

(a)
$$\nabla\phi = \left(\frac{\partial}{\partial x}\mathbf{i} + \frac{\partial}{\partial y}\mathbf{j} + \frac{\partial}{\partial z}\mathbf{k}\right)\phi = \frac{\partial\phi}{\partial x}\mathbf{i} + \frac{\partial\phi}{\partial y}\mathbf{j} + \frac{\partial\phi}{\partial z}\mathbf{k} = \frac{\partial}{\partial x}(x^2yz^3)\mathbf{i} + \frac{\partial}{\partial y}(x^2yz^3)\mathbf{j} + \frac{\partial}{\partial z}(x^2yz^3)\mathbf{k}$$

$$= 2xyz^3\mathbf{i} + x^2z^3\mathbf{j} + 3x^2yz^2\mathbf{k}$$

(b)
$$\nabla \cdot \mathbf{A} = \left(\frac{\partial}{\partial x}\mathbf{i} + \frac{\partial}{\partial y}\mathbf{j} + \frac{\partial}{\partial z}\mathbf{k}\right) \cdot (xz\mathbf{i} - y^2\mathbf{j} + 2x^2y\mathbf{k})$$

$$= \frac{\partial}{\partial x}(xz) + \frac{\partial}{\partial y}(-y^2) + \frac{\partial}{\partial z}(2x^2y) = z - 2y$$

(c) $\nabla \times \mathbf{A} = \left(\dfrac{\partial}{\partial x}\mathbf{i} + \dfrac{\partial}{\partial y}\mathbf{j} + \dfrac{\partial}{\partial z}\mathbf{k} \right) \times (xz\mathbf{i} - y^2\mathbf{j} + 2x^2y\mathbf{k})$

$= \begin{vmatrix} \mathbf{i} & \mathbf{j} & \mathbf{k} \\ \partial/\partial x & \partial/\partial y & \partial/\partial z \\ xz & -y^2 & 2x^2y \end{vmatrix}$

$= \left(\dfrac{\partial}{\partial y}(2x^2y) - \dfrac{\partial}{\partial z}(-y^2) \right)\mathbf{i} + \left(\dfrac{\partial}{\partial z}(xz) - \dfrac{\partial}{\partial x}(2x^2y) \right)\mathbf{j} + \left(\dfrac{\partial}{\partial x}(-y^2) - \dfrac{\partial}{\partial y}(xz) \right)\mathbf{k}$

$= 2x^2\mathbf{i} + (x - 4xy)\mathbf{j}$

(d) $\mathrm{div}\,(\phi\mathbf{A}) = \nabla \cdot (\phi\mathbf{A}) = \nabla \cdot (x^3yz^4\mathbf{i} - x^2y^3z^3\mathbf{j} + 2x^4y^2z^3\mathbf{k})$

$= \dfrac{\partial}{\partial x}(x^3yz^4) + \dfrac{\partial}{\partial y}(-x^2y^3z^3) + \dfrac{\partial}{\partial z}(2x^4y^2z^3) = 3x^2yz^4 - 3x^2y^2z^3 + 6x^4y^2z^2$

(e) $\mathrm{curl}\,(\phi\mathbf{A}) = \nabla \times (\phi\mathbf{A}) = \nabla \times (x^3yz^4\mathbf{i} - x^2y^3z^3\mathbf{j} + 2x^4y^2z^3\mathbf{k})$

$= \begin{vmatrix} \mathbf{i} & \mathbf{j} & \mathbf{k} \\ \partial/\partial x & \partial/\partial y & \partial/\partial z \\ x^3yz^4 & -x^2y^3z^3 & 2x^4y^2z^3 \end{vmatrix}$

$= (4x^4yz^3 + 3x^2y^3z^2)\mathbf{i} + (4x^3yz^3 - 8x^3y^2z^3)\mathbf{j} - (2xy^3z^3 + x^3z^4)\mathbf{k}$

5.32. Prove $\nabla \cdot (\phi\mathbf{A}) = (\nabla\phi) \cdot \mathbf{A} + \phi(\nabla \cdot \mathbf{A})$.

$\nabla \cdot (\phi\mathbf{A}) = \nabla \cdot (\phi A_1\mathbf{i} + \phi A_2\mathbf{j} + \phi A_3\mathbf{k})$

$= \dfrac{\partial}{\partial x}(\phi A_1) + \dfrac{\partial}{\partial y}(\phi A_2) + \dfrac{\partial}{\partial z}(\phi A_3)$

$= \dfrac{\partial\phi}{\partial x}A_1 + \dfrac{\partial\phi}{\partial y}A_2 + \dfrac{\partial\phi}{\partial z}A_3 + \phi\left(\dfrac{\partial A_1}{\partial x} + \dfrac{\partial A_2}{\partial y} + \dfrac{\partial A_3}{\partial z} \right)$

$= \left(\dfrac{\partial\phi}{\partial x}\mathbf{i} + \dfrac{\partial\phi}{\partial y}\mathbf{j} + \dfrac{\partial\phi}{\partial z}\mathbf{k} \right) \cdot (A_1\mathbf{i} + A_2\mathbf{j} + A_3\mathbf{k})$

$\quad + \phi\left(\dfrac{\partial}{\partial x}\mathbf{i} + \dfrac{\partial}{\partial y}\mathbf{j} + \dfrac{\partial}{\partial z}\mathbf{k} \right) \cdot (A_1\mathbf{i} + A_2\mathbf{j} + A_3\mathbf{k})$

$= (\nabla\phi) \cdot \mathbf{A} + \phi(\nabla \cdot \mathbf{A})$

5.33. Prove that $\nabla\phi$ is a vector perpendicular to the surface $\phi(x, y, z) = c$, where c is a constant.

Let $\mathbf{r} = x\mathbf{i} + y\mathbf{j} + z\mathbf{k}$ be the position vector to any point $P(x, y, z)$ on the surface.

Then $d\mathbf{r} = dx\,\mathbf{i} + dy\,\mathbf{j} + dz\,\mathbf{k}$ lies in the plane tangent to the surface at P. But

$d\phi = \dfrac{\partial\phi}{\partial x}dx + \dfrac{\partial\phi}{\partial y}dy + \dfrac{\partial\phi}{\partial z}dz = 0 \quad \text{or} \quad \left(\dfrac{\partial\phi}{\partial x}\mathbf{i} + \dfrac{\partial\phi}{\partial y}\mathbf{j} + \dfrac{\partial\phi}{\partial z}\mathbf{k} \right) \cdot (dx\,\mathbf{i} + dy\,\mathbf{j} + dz\,\mathbf{k}) = 0$

i.e. $\nabla\phi \cdot d\mathbf{r} = 0$ so that $\nabla\phi$ is perpendicular to $d\mathbf{r}$ and therefore to the surface.

5.34. Find a unit normal to the surface $2x^2 + 4yz - 5z^2 = -10$ at the point $P(3, -1, 2)$.

By Problem 5.33, a vector normal to the surface is

$\nabla(2x^2 + 4yz - 5z^2) = 4x\mathbf{i} + 4z\mathbf{j} + (4y - 10z)\mathbf{k} = 12\mathbf{i} + 8\mathbf{j} - 24\mathbf{k} \quad \text{at } (3, -1, 2)$

Then a unit normal to the surface at P is $\dfrac{12\mathbf{i} + 8\mathbf{j} - 24\mathbf{k}}{\sqrt{(12)^2 + (8)^2 + (-24)^2}} = \dfrac{3\mathbf{i} + 2\mathbf{j} - 6\mathbf{k}}{7}$.

Another unit normal to the surface at P is $-\dfrac{3\mathbf{i} + 2\mathbf{j} - 6\mathbf{k}}{7}$.

5.35. If $\phi = 2x^2y - xz^3$, find (a) $\nabla\phi$ and (b) $\nabla^2\phi$.

(a) $\nabla\phi = \dfrac{\partial\phi}{\partial x}\mathbf{i} + \dfrac{\partial\phi}{\partial y}\mathbf{j} + \dfrac{\partial\phi}{\partial z}\mathbf{k} = (4xy - z^3)\mathbf{i} + 2x^2\mathbf{j} - 3xz^2\mathbf{k}$

(b) $\nabla^2\phi = \text{Laplacian of } \phi = \nabla \cdot \nabla\phi = \dfrac{\partial}{\partial x}(4xy - z^3) + \dfrac{\partial}{\partial y}(2x^2) + \dfrac{\partial}{\partial z}(-3xz^2) = 4y - 6xz$

Another method.

$$\nabla^2\phi = \frac{\partial^2\phi}{\partial x^2} + \frac{\partial^2\phi}{\partial y^2} + \frac{\partial^2\phi}{\partial z^2} = \frac{\partial^2}{\partial x^2}(2x^2y - xz^3) + \frac{\partial^2}{\partial y^2}(2x^2y - xz^3) + \frac{\partial^2}{\partial z^2}(2x^2y - xz^3)$$

$$= 4y - 6xz$$

5.36. Prove div curl $\mathbf{A} = 0$.

$$\text{div curl}\,\mathbf{A} = \nabla\cdot(\nabla\times\mathbf{A}) = \nabla\cdot\begin{vmatrix} \mathbf{i} & \mathbf{j} & \mathbf{k} \\ \partial/\partial x & \partial/\partial y & \partial/\partial z \\ A_1 & A_2 & A_3 \end{vmatrix}$$

$$= \nabla\cdot\left[\left(\frac{\partial A_3}{\partial y} - \frac{\partial A_2}{\partial z}\right)\mathbf{i} + \left(\frac{\partial A_1}{\partial z} - \frac{\partial A_3}{\partial x}\right)\mathbf{j} + \left(\frac{\partial A_2}{\partial x} - \frac{\partial A_1}{\partial y}\right)\mathbf{k}\right]$$

$$= \frac{\partial}{\partial x}\left(\frac{\partial A_3}{\partial y} - \frac{\partial A_2}{\partial z}\right) + \frac{\partial}{\partial y}\left(\frac{\partial A_1}{\partial z} - \frac{\partial A_3}{\partial x}\right) + \frac{\partial}{\partial z}\left(\frac{\partial A_2}{\partial x} - \frac{\partial A_1}{\partial y}\right)$$

$$= \frac{\partial^2 A_3}{\partial x\,\partial y} - \frac{\partial^2 A_2}{\partial x\,\partial z} + \frac{\partial^2 A_1}{\partial y\,\partial z} - \frac{\partial^2 A_3}{\partial y\,\partial x} + \frac{\partial^2 A_2}{\partial z\,\partial x} - \frac{\partial^2 A_1}{\partial z\,\partial y} = 0$$

assuming that $\mathbf{A}$ has continuous second partial derivatives so that the order of differentiation is immaterial.

5.37. Find equations for (a) the tangent plane and (b) the normal line to the surface $F(x, y, z) = 0$ at the point $P(x_0, y_0, z_0)$. See Fig. 5-23.

(a) A vector normal to the surface at P is $\mathbf{N}_0 = \nabla F|_P$. Then if $\mathbf{r}_0$ and $\mathbf{r}$ are the vectors drawn respectively from O to $P(x_0, y_0, z_0)$ and $Q(x, y, z)$ on the plane, the equation of the plane is

$$(\mathbf{r} - \mathbf{r}_0)\cdot\mathbf{N}_0 = (\mathbf{r} - \mathbf{r}_0)\cdot\nabla F|_P = 0$$

since $\mathbf{r} - \mathbf{r}_0$ is perpendicular to $\mathbf{N}_0$. In rectangular form this is

$$F_x|_P\,(x - x_0) + F_y|_P\,(y - y_0) + F_z|_P\,(z - z_0) = 0$$

(b) If $\mathbf{r}$ is the vector drawn from O in Fig. 5-23 to any point (x, y, z) on the normal line, then $\mathbf{r} - \mathbf{r}_0$ is collinear with $\mathbf{N}_0$ and so

$$(\mathbf{r} - \mathbf{r}_0)\times\mathbf{N}_0 = (\mathbf{r} - \mathbf{r}_0)\times\nabla F|_P = \mathbf{0}$$

which in rectangular form is $\quad\dfrac{x - x_0}{F_x|_P} = \dfrac{y - y_0}{F_y|_P} = \dfrac{z - z_0}{F_z|_P}$

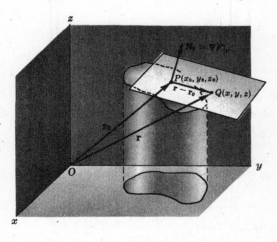

Fig. 5-23

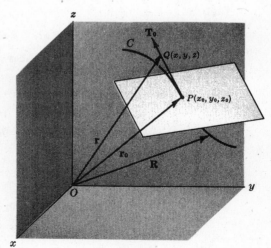

Fig. 5-24

5.38. Find equations for (a) the tangent line, (b) the normal plane to a space curve $x = f(u)$, $y = g(u)$, $z = h(u)$ at the point where $u = u_0$. See Fig. 5-24.

(a) If $\mathbf{R} = f(u)\mathbf{i} + g(u)\mathbf{j} + h(u)\mathbf{k}$, then a vector tangent to the curve C at point P is given by $\mathbf{T}_0 = \dfrac{d\mathbf{R}}{du}\Big|_P$. Then if $\mathbf{r}_0$ and $\mathbf{r}$ are the vectors drawn respectively from O to P and Q on the tangent line, we have

$$(\mathbf{r} - \mathbf{r}_0) \times \mathbf{T}_0 \;=\; (\mathbf{r} - \mathbf{r}_0) \times \dfrac{d\mathbf{R}}{du}\Big|_P \;=\; \mathbf{0}$$

since $\mathbf{r} - \mathbf{r}_0$ is collinear with $\mathbf{T}_0$. In rectangular form this becomes

$$\frac{x - x_0}{f'(u_0)} = \frac{y - y_0}{g'(u_0)} = \frac{z - z_0}{h'(u_0)}$$

(b) If $\mathbf{r}$ is the vector from O to any point (x, y, z) on the normal plane, it follows that $\mathbf{r} - \mathbf{r}_0$ is perpendicular to $\mathbf{T}_0$. Then the required equation is

$$(\mathbf{r} - \mathbf{r}_0) \cdot \mathbf{T}_0 \;=\; (\mathbf{r} - \mathbf{r}_0) \cdot \dfrac{d\mathbf{R}}{du}\Big|_P \;=\; 0$$

or in rectangular form

$$f'(u_0)(x - x_0) + g'(u_0)(y - y_0) + h'(u_0)(z - z_0) \;=\; 0$$

5.39. If $F(x, y, z)$ is defined at any point on a space curve C and s is the arclength to (x, y, z) from some given point on C, show that

$$\frac{dF}{ds} \;=\; \nabla F \cdot \frac{d\mathbf{r}}{ds} \;=\; \nabla F \cdot \mathbf{T}$$

where $\mathbf{T} = d\mathbf{r}/ds$ is a unit tangent vector to C at (x, y, z).

We have
$$\begin{aligned}
\frac{dF}{ds} &= \frac{\partial F}{\partial x}\frac{dx}{ds} + \frac{\partial F}{\partial y}\frac{dy}{ds} + \frac{\partial F}{\partial z}\frac{dz}{ds} \\
&= \left(\frac{\partial F}{\partial x}\mathbf{i} + \frac{\partial F}{\partial y}\mathbf{j} + \frac{\partial F}{\partial z}\mathbf{k}\right) \cdot \left(\frac{dx}{ds}\mathbf{i} + \frac{dy}{ds}\mathbf{j} + \frac{dz}{ds}\mathbf{k}\right) \\
&= \nabla F \cdot \frac{d\mathbf{r}}{ds} \;=\; \nabla F \cdot \mathbf{T}
\end{aligned}$$

The fact that $\mathbf{T} = d\mathbf{r}/ds$ is a unit vector follows from $dr^2 = dx^2 + dy^2 + dz^2$. We often call dF/ds the *directional derivative* of F at (x, y, z) along C.

CURVILINEAR COORDINATES AND JACOBIANS

5.40. Find ds^2 in (a) cylindrical and (b) spherical coordinates and determine the scale factors.

(a) **Method 1.** $\qquad\qquad x = \rho\cos\phi, \qquad y = \rho\sin\phi, \qquad z = z$

$$dx = -\rho\sin\phi\,d\phi + \cos\phi\,d\rho, \qquad dy = \rho\cos\phi\,d\phi + \sin\phi\,d\rho, \qquad dz = dz$$

Then
$$\begin{aligned}
ds^2 &= dx^2 + dy^2 + dz^2 = (-\rho\sin\phi\,d\phi + \cos\phi\,d\rho)^2 + (\rho\cos\phi\,d\phi + \sin\phi\,d\rho)^2 + (dz)^2 \\
&= (d\rho)^2 + \rho^2(d\phi)^2 + (dz)^2 = h_1^2(d\rho)^2 + h_2^2(d\phi)^2 + h_3^2(dz)^2
\end{aligned}$$

and $h_1 = h_\rho = 1$, $h_2 = h_\phi = \rho$, $h_3 = h_z = 1$ are the scale factors.

Method 2. The position vector is $\mathbf{r} = \rho\cos\phi\,\mathbf{i} + \rho\sin\phi\,\mathbf{j} + z\mathbf{k}$. Then

$$\begin{aligned}
d\mathbf{r} &= \frac{\partial \mathbf{r}}{\partial \rho}d\rho + \frac{\partial \mathbf{r}}{\partial \phi}d\phi + \frac{\partial \mathbf{r}}{\partial z}dz \\
&= (\cos\phi\,\mathbf{i} + \sin\phi\,\mathbf{j})\,d\rho + (-\rho\sin\phi\,\mathbf{i} + \rho\cos\phi\,\mathbf{j})\,d\phi + \mathbf{k}\,dz \\
&= (\cos\phi\,d\rho - \rho\sin\phi\,d\phi)\mathbf{i} + (\sin\phi\,d\rho + \rho\cos\phi\,d\phi)\mathbf{j} + \mathbf{k}\,dz
\end{aligned}$$

Thus
$$\begin{aligned}
ds^2 = d\mathbf{r} \cdot d\mathbf{r} &= (\cos\phi\,d\rho - \rho\sin\phi\,d\phi)^2 + (\sin\phi\,d\rho + \rho\cos\phi\,d\phi)^2 + (dz)^2 \\
&= (d\rho)^2 + \rho^2(d\phi)^2 + (dz)^2
\end{aligned}$$

(b) $\qquad\qquad\qquad x = r\sin\theta\cos\phi, \qquad y = r\sin\theta\sin\phi, \qquad z = r\cos\theta$

Then
$$dx = -r \sin \theta \sin \phi \, d\phi + r \cos \theta \cos \phi \, d\theta + \sin \theta \cos \phi \, dr$$
$$dy = r \sin \theta \cos \phi \, d\phi + r \cos \theta \sin \phi \, d\theta + \sin \theta \sin \phi \, dr$$
$$dz = -r \sin \theta \, d\theta + \cos \theta \, dr$$

and
$$(ds)^2 = (dx)^2 + (dy)^2 + (dz)^2 = (dr)^2 + r^2(d\theta)^2 + r^2 \sin^2 \theta \, (d\phi)^2$$

The scale factors are $h_1 = h_r = 1$, $h_2 = h_\theta = r$, $h_3 = h_\phi = r \sin \theta$.

5.41. Find the volume element dV in (a) cylindrical and (b) spherical coordinates and sketch.

The volume element in orthogonal curvilinear coordinates u_1, u_2, u_3 is

$$dV = h_1 h_2 h_3 \, du_1 \, du_2 \, du_3 = \left| \frac{\partial(x, y, z)}{\partial(u_1, u_2, u_3)} \right| du_1 \, du_2 \, du_3$$

(a) In cylindrical coordinates, $u_1 = \rho$, $u_2 = \phi$, $u_3 = z$, $h_1 = 1$, $h_2 = \rho$, $h_3 = 1$ [see Problem 40(a)]. Then
$$dV = (1)(\rho)(1) \, d\rho \, d\phi \, dz = \rho \, d\rho \, d\phi \, dz$$

This can also be observed directly from Fig. 5-25(a) below.

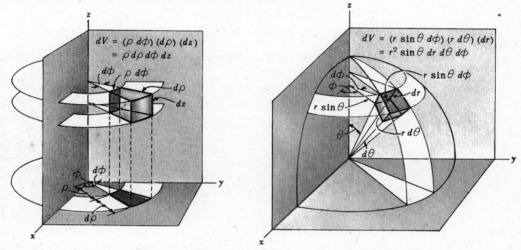

(a) Volume element in cylindrical coordinates.   (b) Volume element in spherical coordinates.

Fig. 5-25

(b) In spherical coordinates, $u_1 = r$, $u_2 = \theta$, $u_3 = \phi$, $h_1 = 1$, $h_2 = r$, $h_3 = r \sin \theta$ [see Problem 5.40(b)]. Then
$$dV = (1)(r)(r \sin \theta) \, dr \, d\theta \, d\phi = r^2 \sin \theta \, dr \, d\theta \, d\phi$$

This can also be observed directly from Fig. 5-25(b) above.

5.42. Express in cylindrical coordinates: (a) grad Φ, (b) div **A**, (c) $\nabla^2 \Phi$.

Let $u_1 = \rho$, $u_2 = \phi$, $u_3 = z$, $h_1 = 1$, $h_2 = \rho$, $h_3 = 1$ [see Problem 5.40(a)] in the results 1, 2, page 128, and 4, page 129. Then

(a) $$\text{grad } \Phi = \nabla \Phi = \frac{1}{1} \frac{\partial \Phi}{\partial \rho} \mathbf{e}_1 + \frac{1}{\rho} \frac{\partial \Phi}{\partial \phi} \mathbf{e}_2 + \frac{1}{1} \frac{\partial \Phi}{\partial z} \mathbf{e}_3 = \frac{\partial \Phi}{\partial \rho} \mathbf{e}_1 + \frac{1}{\rho} \frac{\partial \Phi}{\partial \phi} \mathbf{e}_2 + \frac{\partial \Phi}{\partial z} \mathbf{e}_3$$

where $\mathbf{e}_1, \mathbf{e}_2, \mathbf{e}_3$ are the unit vectors in the directions of increasing ρ, ϕ, z respectively.

(b) $$\text{div } \mathbf{A} = \nabla \cdot \mathbf{A} = \frac{1}{(1)(\rho)(1)} \left[\frac{\partial}{\partial \rho} ((\rho)(1)A_1) + \frac{\partial}{\partial \phi} ((1)(1)A_2) + \frac{\partial}{\partial z} ((1)(\rho)A_3) \right]$$
$$= \frac{1}{\rho} \left[\frac{\partial}{\partial \rho} (\rho A_1) + \frac{\partial A_2}{\partial \phi} + \frac{\partial A_3}{\partial z} \right]$$

where $\mathbf{A} = A_1 \mathbf{e}_1 + A_2 \mathbf{e}_2 + A_3 \mathbf{e}_3$.

(c) $\nabla^2 \Phi = \dfrac{1}{(1)(\rho)(1)} \left[\dfrac{\partial}{\partial \rho} \left(\dfrac{(\rho)(1)}{(1)} \dfrac{\partial \Phi}{\partial \rho} \right) + \dfrac{\partial}{\partial \phi} \left(\dfrac{(1)(1)}{(\rho)} \dfrac{\partial \Phi}{\partial \phi} \right) + \dfrac{\partial}{\partial z} \left(\dfrac{(1)(\rho)}{(1)} \dfrac{\partial \Phi}{\partial z} \right) \right]$

$\qquad = \dfrac{1}{\rho} \dfrac{\partial}{\partial \rho} \left(\rho \dfrac{\partial \Phi}{\partial \rho} \right) + \dfrac{1}{\rho^2} \dfrac{\partial^2 \Phi}{\partial \phi^2} + \dfrac{\partial^2 \Phi}{\partial z^2}$

5.43. If $F(x, y, u, v) = 0$ and $G(x, y, u, v) = 0$, find (a) $\partial u / \partial x$, (b) $\partial u / \partial y$, (c) $\partial v / \partial x$, (d) $\partial v / \partial y$.

The two equations in general define the dependent variables u and v as (implicit) functions of the independent variables x and y. Using the subscript notation, we have

$\qquad$ (1)$\quad dF = F_x \, dx + F_y \, dy + F_u \, du + F_v \, dv = 0$

$\qquad$ (2)$\quad dG = G_x \, dx + G_y \, dy + G_u \, du + G_v \, dv = 0$

Also, since u and v are functions of x and y,

$\qquad$ (3)$\quad du = u_x \, dx + u_y \, dy \qquad$ (4)$\quad dv = v_x \, dx + v_y \, dy$

Substituting (3) and (4) in (1) and (2) yields

$\qquad$ (5)$\quad dF = (F_x + F_u u_x + F_v v_x) \, dx + (F_y + F_u u_y + F_v v_y) \, dy = 0$

$\qquad$ (6)$\quad dG = (G_x + G_u u_x + G_v v_x) \, dx + (G_y + G_u u_y + G_v v_y) \, dy = 0$

Since x and y are independent, the coefficients of dx and dy in (5) and (6) are zero. Hence we obtain

$\qquad$ (7)$\quad \begin{cases} F_u u_x + F_v v_x = -F_x \\ G_u u_x + G_v v_x = -G_x \end{cases} \qquad\qquad$ (8)$\quad \begin{cases} F_u u_y + F_v v_y = -F_y \\ G_u u_y + G_v v_y = -G_y \end{cases}$

Solving (7) and (8) gives

(a) $u_x = \dfrac{\partial u}{\partial x} = \dfrac{\begin{vmatrix} -F_x & F_v \\ -G_x & G_v \end{vmatrix}}{\begin{vmatrix} F_u & F_v \\ G_u & G_v \end{vmatrix}} = -\dfrac{\dfrac{\partial(F, G)}{\partial(x, v)}}{\dfrac{\partial(F, G)}{\partial(u, v)}}$
$\qquad$ (b) $v_x = \dfrac{\partial v}{\partial x} = \dfrac{\begin{vmatrix} F_u & -F_x \\ G_u & -G_x \end{vmatrix}}{\begin{vmatrix} F_u & F_v \\ G_u & G_v \end{vmatrix}} = -\dfrac{\dfrac{\partial(F, G)}{\partial(u, x)}}{\dfrac{\partial(F, G)}{\partial(u, v)}}$

(c) $u_y = \dfrac{\partial u}{\partial y} = \dfrac{\begin{vmatrix} -F_y & F_v \\ -G_y & G_v \end{vmatrix}}{\begin{vmatrix} F_u & F_v \\ G_u & G_v \end{vmatrix}} = -\dfrac{\dfrac{\partial(F, G)}{\partial(y, v)}}{\dfrac{\partial(F, G)}{\partial(u, v)}}$
$\qquad$ (d) $v_y = \dfrac{\partial v}{\partial y} = \dfrac{\begin{vmatrix} F_u & -F_y \\ G_u & -G_y \end{vmatrix}}{\begin{vmatrix} F_u & F_v \\ G_u & G_v \end{vmatrix}} = -\dfrac{\dfrac{\partial(F, G)}{\partial(u, y)}}{\dfrac{\partial(F, G)}{\partial(u, v)}}$

The functional determinant $\begin{vmatrix} F_u & F_v \\ G_u & G_v \end{vmatrix}$, denoted by $\dfrac{\partial(F, G)}{\partial(u, v)}$ or $J\left(\dfrac{F, G}{u, v} \right)$, is the *Jacobian* of F and G with respect to u and v and is supposed $\neq 0$.

Note that it is possible to devise mnemonic rules for writing at once the required partial derivatives in terms of Jacobians.

Supplementary Problems

VECTOR ALGEBRA

5.44. Given any two vectors **A** and **B**, illustrate geometrically the equality $4\mathbf{A} + 3(\mathbf{B} - \mathbf{A}) = \mathbf{A} + 3\mathbf{B}$.

5.45. A man travels 25 miles northeast, 15 miles due east and 10 miles due south. By using an appropriate scale determine graphically (a) how far and (b) in what direction he is from his starting position. Is it possible to determine the answer analytically?

5.46. If **A** and **B** are any two non-zero vectors which do not have the same direction, prove that $m\mathbf{A} + n\mathbf{B}$ is a vector lying in the plane determined by **A** and **B**.

5.47. If **A, B** and **C** are non-coplanar vectors (vectors which do not all lie in the same plane) and $x_1\mathbf{A} + y_1\mathbf{B} + z_1\mathbf{C} = x_2\mathbf{A} + y_2\mathbf{B} + z_2\mathbf{C}$, prove that necessarily $x_1 = x_2$, $y_1 = y_2$, $z_1 = z_2$.

5.48. Let $ABCD$ be any quadrilateral and points P, Q, R and S the midpoints of successive sides. Prove (a) that $PQRS$ is a parallelogram and (b) that the perimeter of $PQRS$ is equal to the sum of the lengths of the diagonals of $ABCD$.

5.49. Prove that the medians of a triangle intersect at a point which is a trisection point of each median.

5.50. Find a unit vector in the direction of the resultant of vectors $\mathbf{A} = 2\mathbf{i} - \mathbf{j} + \mathbf{k}$, $\mathbf{B} = \mathbf{i} + \mathbf{j} + 2\mathbf{k}$, $\mathbf{C} = 3\mathbf{i} - 2\mathbf{j} + 4\mathbf{k}$.

THE DOT OR SCALAR PRODUCT

5.51. Evaluate $|(\mathbf{A} + \mathbf{B}) \cdot (\mathbf{A} - \mathbf{B})|$ if $\mathbf{A} = 2\mathbf{i} - 3\mathbf{j} + 5\mathbf{k}$ and $\mathbf{B} = 3\mathbf{i} + \mathbf{j} - 2\mathbf{k}$.

5.52. Prove the law of cosines for a triangle. [*Hint.* Take the sides as $\mathbf{A}, \mathbf{B}, \mathbf{C}$ where $\mathbf{C} = \mathbf{A} - \mathbf{B}$. Then use $\mathbf{C} \cdot \mathbf{C} = (\mathbf{A} - \mathbf{B}) \cdot (\mathbf{A} - \mathbf{B})$.]

5.53. Find a so that $2\mathbf{i} - 3\mathbf{j} + 5\mathbf{k}$ and $3\mathbf{i} + a\mathbf{j} - 2\mathbf{k}$ are perpendicular.

5.54. If $\mathbf{A} = 2\mathbf{i} + \mathbf{j} + \mathbf{k}$, $\mathbf{B} = \mathbf{i} - 2\mathbf{j} + 2\mathbf{k}$ and $\mathbf{C} = 3\mathbf{i} - 4\mathbf{j} + 2\mathbf{k}$, find the projection of $\mathbf{A} + \mathbf{C}$ in the direction of $\mathbf{B}$.

5.55. A triangle has vertices at $A(2, 3, 1)$, $B(-1, 1, 2)$, $C(1, -2, 3)$. Find (a) the length of the median drawn from B to side AC and (b) the acute angle which this median makes with side BC.

5.56. Prove that the diagonals of a rhombus are perpendicular to each other.

5.57. Prove that the vector $(AB + BA)/(A + B)$ represents the bisector of the angle between **A** and **B**.

THE CROSS OR VECTOR PRODUCT

5.58. If $\mathbf{A} = 2\mathbf{i} - \mathbf{j} + \mathbf{k}$ and $\mathbf{B} = \mathbf{i} + 2\mathbf{j} - 3\mathbf{k}$, find $|(2\mathbf{A} + \mathbf{B}) \times (\mathbf{A} - 2\mathbf{B})|$.

5.59. Find a unit vector perpendicular to the plane of the vectors $\mathbf{A} = 3\mathbf{i} - 2\mathbf{j} + 4\mathbf{k}$ and $\mathbf{B} = \mathbf{i} + \mathbf{j} - 2\mathbf{k}$.

5.60. If $\mathbf{A} \times \mathbf{B} = \mathbf{A} \times \mathbf{C}$, does $\mathbf{B} = \mathbf{C}$ necessarily?

5.61. Find the area of the triangle with vertices $(2, -3, 1)$, $(1, -1, 2)$, $(-1, 2, 3)$.

5.62. Find the shortest distance from the point $(3, 2, 1)$ to the plane determined by $(1, 1, 0)$, $(3, -1, 1)$, $(-1, 0, 2)$.

TRIPLE PRODUCTS

5.63. If $\mathbf{A} = 2\mathbf{i} + \mathbf{j} - 3\mathbf{k}$, $\mathbf{B} = \mathbf{i} - 2\mathbf{j} + \mathbf{k}$, $\mathbf{C} = -\mathbf{i} + \mathbf{j} - 4\mathbf{k}$, find (a) $\mathbf{A} \cdot (\mathbf{B} \times \mathbf{C})$, (b) $\mathbf{C} \cdot (\mathbf{A} \times \mathbf{B})$, (c) $\mathbf{A} \times (\mathbf{B} \times \mathbf{C})$, (d) $(\mathbf{A} \times \mathbf{B}) \times \mathbf{C}$.

5.64. Prove that (a) $\mathbf{A} \cdot (\mathbf{B} \times \mathbf{C}) = \mathbf{B} \cdot (\mathbf{C} \times \mathbf{A}) = \mathbf{C} \cdot (\mathbf{A} \times \mathbf{B})$
(b) $\mathbf{A} \times (\mathbf{B} \times \mathbf{C}) = \mathbf{B}(\mathbf{A} \cdot \mathbf{C}) - \mathbf{C}(\mathbf{A} \cdot \mathbf{B})$.

5.65. Find an equation for the plane passing through $(2, -1, -2)$, $(-1, 2, -3)$, $(4, 1, 0)$.

5.66. Find the volume of the tetrahedron with vertices at $(2, 1, 1)$, $(1, -1, 2)$, $(0, 1, -1)$, $(1, -2, 1)$.

5.67. Prove that $(\mathbf{A} \times \mathbf{B}) \cdot (\mathbf{C} \times \mathbf{D}) + (\mathbf{B} \times \mathbf{C}) \cdot (\mathbf{A} \times \mathbf{D}) + (\mathbf{C} \times \mathbf{A}) \cdot (\mathbf{B} \times \mathbf{D}) = 0$.

DERIVATIVES

5.68. A particle moves along the space curve $\mathbf{r} = e^{-t} \cos t\, \mathbf{i} + e^{-t} \sin t\, \mathbf{j} + e^{-t}\mathbf{k}$. Find the magnitude of the (a) velocity and (b) acceleration at any time t.

5.69. Prove that $\dfrac{d}{du}(\mathbf{A} \times \mathbf{B}) = \mathbf{A} \times \dfrac{d\mathbf{B}}{du} + \dfrac{d\mathbf{A}}{du} \times \mathbf{B}$ where **A** and **B** are differentiable functions of u.

5.70.   Find a unit vector tangent to the space curve $x = t$, $y = t^2$, $z = t^3$ at the point where $t = 1$.

5.71.   If $\mathbf{r} = \mathbf{a} \cos \omega t + \mathbf{b} \sin \omega t$, where $\mathbf{a}$ and $\mathbf{b}$ are any constant noncollinear vectors and ω is a constant scalar, prove that   (a) $\mathbf{r} \times \dfrac{d\mathbf{r}}{dt} = \omega(\mathbf{a} \times \mathbf{b})$, (b) $\dfrac{d^2\mathbf{r}}{dt^2} + \omega^2\mathbf{r} = \mathbf{0}$.

5.72.   If $\mathbf{A} = x^2\mathbf{i} - yj + xz\mathbf{k}$, $\mathbf{B} = y\mathbf{i} + x\mathbf{j} - xyz\mathbf{k}$ and $\mathbf{C} = \mathbf{i} - y\mathbf{j} + x^3z\mathbf{k}$, find (a) $\dfrac{\partial^2}{\partial x\,\partial y}(\mathbf{A} \times \mathbf{B})$ and (b) $d[\mathbf{A} \cdot (\mathbf{B} \times \mathbf{C})]$ at the point $(1, -1, 2)$.

5.73.   If $\mathbf{R} = x^2y\mathbf{i} - 2y^2z\mathbf{j} + xy^2z^2\mathbf{k}$, find $\left| \dfrac{\partial^2\mathbf{R}}{\partial x^2} \times \dfrac{\partial^2\mathbf{R}}{\partial y^2} \right|$ at the point $(2, 1, -2)$.

5.74.   If $\mathbf{A}$ is a differentiable function of u and $|\mathbf{A}(u)| = 1$, prove that $d\mathbf{A}/du$ is perpendicular to $\mathbf{A}$.

5.75.   Let $\mathbf{T}$ and $\mathbf{N}$ denote respectively the unit *tangent vector* and unit *principal normal* vector to a space curve $\mathbf{r} = \mathbf{r}(u)$, where $\mathbf{r}(u)$ is assumed differentiable. Define a vector $\mathbf{B} = \mathbf{T} \times \mathbf{N}$ called the unit *binormal vector* to the space curve. Prove that

$$\frac{d\mathbf{T}}{ds} = \kappa\mathbf{N}, \qquad \frac{d\mathbf{B}}{ds} = -\tau\mathbf{N}, \qquad \frac{d\mathbf{N}}{ds} = \tau\mathbf{B} - \kappa\mathbf{T}$$

These are called the *Frenet-Serret* formulas. In these formulas κ is called the *curvature*, τ is called the *torsion*; and the reciprocals of these, $\rho = 1/\kappa$ and $\sigma = 1/\tau$, are called the *radius of curvature* and *radius of torsion* respectively.

GRADIENT, DIVERGENCE AND CURL

5.76.   If $U, V, \mathbf{A}, \mathbf{B}$ have continuous partial derivatives prove that:
(a) $\nabla(U + V) = \nabla U + \nabla V$, (b) $\nabla \cdot (\mathbf{A} + \mathbf{B}) = \nabla \cdot \mathbf{A} + \nabla \cdot \mathbf{B}$, (c) $\nabla \times (\mathbf{A} + \mathbf{B}) = \nabla \times \mathbf{A} + \nabla \times \mathbf{B}$.

5.77.   If $\phi = xy + yz + zx$ and $\mathbf{A} = x^2y\mathbf{i} + y^2z\mathbf{j} + z^2x\mathbf{k}$, find (a) $\mathbf{A} \cdot \nabla\phi$, (b) $\phi\nabla \cdot \mathbf{A}$ and (c) $(\nabla\phi) \times \mathbf{A}$ at the point $(3, -1, 2)$.

5.78.   Show that $\nabla \times (r^2\mathbf{r}) = \mathbf{0}$ where $\mathbf{r} = x\mathbf{i} + y\mathbf{j} + z\mathbf{k}$ and $r = |\mathbf{r}|$.

5.79.   Prove: (a) $\nabla \times (U\mathbf{A}) = (\nabla U) \times \mathbf{A} + U(\nabla \times \mathbf{A})$, (b) $\nabla \cdot (\mathbf{A} \times \mathbf{B}) = \mathbf{B} \cdot (\nabla \times \mathbf{A}) - \mathbf{A} \cdot (\nabla \times \mathbf{B})$.

5.80.   Prove that $\operatorname{curl} \operatorname{grad} U = \mathbf{0}$, stating appropriate conditions on U.

5.81.   Find a unit normal to the surface $x^2y - 2xz + 2y^2z^4 = 10$ at the point $(2, 1, -1)$.

5.82.   If $\mathbf{A} = 3xz^2\mathbf{i} - yz\mathbf{j} + (x + 2z)\mathbf{k}$, find curl curl $\mathbf{A}$.

5.83.   (a) Prove that $\nabla \times (\nabla \times \mathbf{A}) = -\nabla^2\mathbf{A} + \nabla(\nabla \cdot \mathbf{A})$. (b) Verify the result in (a) if $\mathbf{A}$ is given as in Problem 5.82.

5.84.   Find the equations of the (a) tangent plane and (b) normal line to the surface $x^2 + y^2 = 4z$ at $(2, -4, 5)$.

5.85.   Find the equations of the (a) tangent line and (b) normal plane to the space curve $x = 6 \sin t$, $y = 4 \cos 3t$, $z = 2 \sin 5t$ at the point where $t = \pi/4$.

5.86.   (a) Find the directional derivative of $U = 2xy - z^2$ at $(2, -1, 1)$ in a direction toward $(3, 1, -1)$.
(b) In what direction is the directional derivative a maximum? (c) What is the value of this maximum?

5.87.   Prove that the acute angle γ between the z axis and the normal to the surface $F(x, y, z) = 0$ at any point is given by $\sec \gamma = \sqrt{F_x^2 + F_y^2 + F_z^2}/|F_z|$.

5.88.   (a) Develop a formula for the shortest distance from a point (x_0, y_0, z_0) to a surface. (b) Illustrate the result in (a) by finding the shortest distance from the point $(1, 1, -2)$ to the surface $z = x^2 + y^2$.

5.89. Let **E** and **H** be two vectors assumed to have continuous partial derivatives (of second order at least) with respect to position and time. Suppose further that **E** and **H** satisfy the equations

$$\nabla \cdot \mathbf{E} = 0, \qquad \nabla \cdot \mathbf{H} = 0, \qquad \nabla \times \mathbf{E} = -\frac{1}{c}\frac{\partial \mathbf{H}}{\partial t}, \qquad \nabla \times \mathbf{H} = \frac{1}{c}\frac{\partial \mathbf{E}}{\partial t} \qquad (1)$$

prove that **E** and **H** satisfy the equation

$$\nabla^2 \psi = \frac{1}{c^2}\frac{\partial^2 \psi}{\partial t^2} \qquad\qquad (2)$$

[The vectors **E** and **H** are called *electric* and *magnetic field vectors* in *electromagnetic theory*. Equations (*1*) are a special case of *Maxwell's equations*. The result (*2*) led Maxwell to the conclusion that light was an electromagnetic phenomena. The constant *c* is the velocity of light.]

5.90. Use the relations in Problem 5.89 to show that

$$\frac{\partial}{\partial t}\{\tfrac{1}{2}(E^2 + H^2)\} + c\,\nabla \cdot (\mathbf{E} \times \mathbf{H}) = 0$$

JACOBIANS AND CURVILINEAR COORDINATES

5.91. Prove that $\left|\dfrac{\partial(x, y, z)}{\partial(u_1, u_2, u_3)}\right| = \left|\dfrac{\partial \mathbf{r}}{\partial u_1} \cdot \dfrac{\partial \mathbf{r}}{\partial u_2} \times \dfrac{\partial \mathbf{r}}{\partial u_3}\right|$.

5.92. Express (*a*) grad Φ, (*b*) div **A** in spherical coordinates.

5.93. The transformation from rectangular to *parabolic cylindrical coordinates* is defined by the equations $x = \tfrac{1}{2}(u^2 - v^2)$, $y = uv$, $z = z$. (*a*) Prove that the system is orthogonal. (*b*) Find ds^2 and the scale factors. (*c*) Find the Jacobian of the transformation and the volume element.

5.94. Write (*a*) $\nabla^2 \Phi$ and (*b*) div **A** in parabolic cylindrical coordinates.

5.95. Prove that for orthogonal curvilinear coordinates,

$$\nabla \Phi = \frac{\mathbf{e}_1}{h_1}\frac{\partial \Phi}{\partial u_1} + \frac{\mathbf{e}_2}{h_2}\frac{\partial \Phi}{\partial u_2} + \frac{\mathbf{e}_3}{h_3}\frac{\partial \Phi}{\partial u_3}$$

[*Hint.* Let $\nabla \Phi = a_1\mathbf{e}_1 + a_2\mathbf{e}_2 + a_3\mathbf{e}_3$ and use the fact that $d\Phi = \nabla \Phi \cdot d\mathbf{r}$ must be the same in both rectangular and the curvilinear coordinates.]

5.96. Prove that the acceleration of a particle along a space curve is given respectively in (*a*) cylindrical, (*b*) spherical coordinates by

$$(\ddot{\rho} - \rho\dot{\phi}^2)\mathbf{e}_\rho + (\rho\ddot{\phi} + 2\dot{\rho}\dot{\phi})\mathbf{e}_\phi + \ddot{z}\,\mathbf{e}_z$$

$$(\ddot{r} - r\dot{\theta}^2 - r\dot{\phi}^2\sin^2\theta)\mathbf{e}_r + (r\ddot{\theta} + 2\dot{r}\dot{\theta} - r\dot{\phi}^2\sin\theta\cos\theta)\mathbf{e}_\theta + (2\dot{r}\dot{\phi}\sin\theta + 2r\dot{\theta}\dot{\phi}\cos\theta + r\ddot{\phi}\sin\theta)\mathbf{e}_\phi$$

where dots denote time derivatives and $\mathbf{e}_\rho, \mathbf{e}_\phi, \mathbf{e}_z, \mathbf{e}_r, \mathbf{e}_\theta, \mathbf{e}_\phi$ are unit vectors in the directions of increasing $\rho, \phi, z, r, \theta, \phi$ respectively.

5.97. If $F = x + 3y^2 - z^3$, $G = 2x^2yz$, and $H = 2z^2 - xy$, evaluate $\dfrac{\partial(F, G, H)}{\partial(x, y, z)}$ at $(1, -1, 0)$.

5.98. If $F = xy + yz + zx$, $G = x^2 + y^2 + z^2$, and $H = x + y + z$, determine whether there is a functional relationship connecting F, G, and H, and if so find it.

5.99. If $F(P, V, T) = 0$, prove that (*a*) $\dfrac{\partial P}{\partial T}\Big|_V \dfrac{\partial T}{\partial V}\Big|_P = -\dfrac{\partial P}{\partial V}\Big|_T$, (*b*) $\dfrac{\partial P}{\partial T}\Big|_V \dfrac{\partial T}{\partial V}\Big|_P \dfrac{\partial V}{\partial P}\Big|_T = -1$ where a subscript indicates the variable which is to be held constant. These results are useful in *thermodynamics* where P, V, T correspond to pressure, volume and temperature of a physical system.

5.100. (*a*) If $x = f(u, v, w), y = g(u, v, w)$, and $z = h(u, v, w)$, prove that $\dfrac{\partial(x, y, z)}{\partial(u, v, w)}\dfrac{\partial(u, v, w)}{\partial(x, y, w)} = 1$ provided $\dfrac{\partial(x, y, z)}{\partial(u, v, w)} \neq 0$. (*b*) Give an interpretation of the result of (*a*) in terms of transformations.

Answers to Supplementary Problems

5.45.   33.6 miles, 13.2° north of east

5.50. $(6\mathbf{i} - 2\mathbf{j} + 7\mathbf{k})/\sqrt{89}$

5.51.   24

5.53. $a = -4/3$

5.54.   17/3

5.55.   (a) $\frac{1}{2}\sqrt{26}$,   (b) $\cos^{-1}\sqrt{91}/14$

5.58. $25\sqrt{3}$

5.59. $\pm(2\mathbf{j} + \mathbf{k})/\sqrt{5}$

5.61. $\frac{1}{2}\sqrt{3}$

5.62.   2

5.63.   (a) 20,   (b) 20,   (c) $8\mathbf{i} - 19\mathbf{j} - \mathbf{k}$,   (d) $25\mathbf{i} - 15\mathbf{j} - 10\mathbf{k}$

5.65. $2x + y - 3z = 9$

5.66. $\frac{4}{3}$

5.68.   (a) $\sqrt{3}\, e^{-t}$,   (b) $\sqrt{5}\, e^{-t}$

5.70. $(\mathbf{i} + 2\mathbf{j} + 3\mathbf{k})/\sqrt{14}$

5.72.   (a) $-4\mathbf{i} + 8\mathbf{j}$,   (b) $8\, dx$

5.73. $16\sqrt{5}$

5.77.   (a) 25,   (b) 2,   (c) $56\mathbf{i} - 30\mathbf{j} + 47\mathbf{k}$

5.81. $\pm(3\mathbf{i} + 4\mathbf{j} - 6\mathbf{k})/\sqrt{61}$

5.82. $-6x\mathbf{i} + (6z - 1)\mathbf{k}$

5.84.   (a) $x - 2y - z = 5$,   (b) $\dfrac{x - 2}{1} = \dfrac{y + 4}{-2} = \dfrac{z - 5}{-1}$

5.85.   (a) $\dfrac{x - 3\sqrt{2}}{3} = \dfrac{y + 2\sqrt{2}}{-6} = \dfrac{z + \sqrt{2}}{-5}$ (b) $3x - 6y - 5z = 26\sqrt{2}$

5.86.   (a) 10/3,   (b) $-2\mathbf{i} + 4\mathbf{j} - 2\mathbf{k}$,   (c) $2\sqrt{6}$

5.92.   (a) $\dfrac{\partial \Phi}{\partial r}\mathbf{e}_1 + \dfrac{1}{r}\dfrac{\partial \Phi}{\partial \theta}\mathbf{e}_2 + \dfrac{1}{r \sin \theta}\dfrac{\partial \Phi}{\partial \phi}\mathbf{e}_3$

        (b) $\dfrac{1}{r^2}\dfrac{\partial}{\partial r}(r^2 A_1) + \dfrac{1}{r \sin \theta}\dfrac{\partial}{\partial \theta}(\sin \theta\, A_2) + \dfrac{1}{r \sin \theta}\dfrac{\partial A_3}{\partial \phi}$ where $\mathbf{A} = A_1\mathbf{e}_1 + A_2\mathbf{e}_2 + A_3\mathbf{e}_3$

5.93.   (b) $ds^2 = (u^2 + v^2)\, du^2 + (u^2 + v^2)\, dv^2 + dz^2$, $h_1 = h_2 = \sqrt{u^2 + v^2},\ h_3 = 1$

        (c) $u^2 + v^2$, $(u^2 + v^2)\, du\, dv\, dz$

5.94.   (a) $\nabla^2\Phi = \dfrac{1}{u^2 + v^2}\left(\dfrac{\partial^2 \Phi}{\partial u^2} + \dfrac{\partial^2 \Phi}{\partial v^2}\right) + \dfrac{\partial^2 \Phi}{\partial z^2}$

        (b) $\text{div }\mathbf{A} = \dfrac{1}{u^2 + v^2}\left\{\dfrac{\partial}{\partial u}(\sqrt{u^2 + v^2}\, A_1) + \dfrac{\partial}{\partial v}(\sqrt{u^2 + v^2}\, A_2)\right\} + \dfrac{\partial A_3}{\partial z}$

5.97.   10

5.98. $H^2 - G - 2F = 0$

Chapter 6

Multiple, Line and Surface Integrals and Integral Theorems

DOUBLE INTEGRALS

Let $F(x, y)$ be defined in a closed region $\mathcal{R}$ of the xy plane (see Fig. 6-1). Subdivide $\mathcal{R}$ into n subregions $\Delta\mathcal{R}_k$ of area ΔA_k, $k = 1, 2, \ldots, n$. Let (ξ_k, η_k) be some point of $\Delta\mathcal{R}_k$. Form the sum

$$\sum_{k=1}^{n} F(\xi_k, \eta_k) \, \Delta A_k \tag{1}$$

Consider

$$\lim_{n \to \infty} \sum_{k=1}^{n} F(\xi_k, \eta_k) \, \Delta A_k \tag{2}$$

where the limit is taken so that the number n of subdivisions increases without limit and such that the largest linear dimension of each $\Delta\mathcal{R}_k$ approaches zero. If this limit exists it is denoted by

$$\iint_{\mathcal{R}} F(x, y) \, dA \tag{3}$$

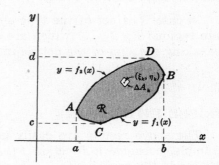

Fig. 6-1

and is called the *double integral* of $F(x, y)$ over the region $\mathcal{R}$.

It can be proved that the limit does exist if $F(x, y)$ is continuous (or piecewise continuous) in $\mathcal{R}$.

ITERATED INTEGRALS

If $\mathcal{R}$ is such that any lines parallel to the y axis meet the boundary of $\mathcal{R}$ in at most two points (as is true in Fig. 6-1), then we can write the equations of the curves ACB and ADB bounding $\mathcal{R}$ as $y = f_1(x)$ and $y = f_2(x)$ respectively, where $f_1(x)$ and $f_2(x)$ are single-valued and continuous in $a \leqq x \leqq b$. In this case we can evaluate the double integral (3) by choosing the regions $\Delta\mathcal{R}_k$ as rectangles formed by constructing a grid of lines parallel to the x and y axes and ΔA_k as the corresponding areas. Then (3) can be written

$$\iint_{\mathcal{R}} F(x, y) \, dx \, dy = \int_{x=a}^{b} \int_{y=f_1(x)}^{f_2(x)} F(x, y) \, dy \, dx \tag{4}$$

$$= \int_{x=a}^{b} \left\{ \int_{y=f_1(x)}^{f_2(x)} F(x, y) \, dy \right\} dx$$

where the integral in braces is to be evaluated first (keeping x constant) and finally integrating with respect to x from a to b. The result (4) indicates how a double integral can be evaluated by expressing it in terms of two single integrals called *iterated integrals*.

If $\mathcal{R}$ is such that any lines parallel to the x axis meet the boundary of $\mathcal{R}$ in at most two points (as in Fig. 6-1), then the equations of curves CAD and CBD can be written $x = g_1(y)$ and $x = g_2(y)$ respectively and we find similarly

$$\iint\limits_{\mathcal{R}} F(x,y)\,dx\,dy \;=\; \int_{y=c}^{d} \int_{x=g_1(y)}^{g_2(y)} F(x,y)\,dx\,dy \tag{5}$$

$$=\; \int_{y=c}^{d} \left\{ \int_{x=g_1(y)}^{g_2(y)} F(x,y)\,dx \right\} dy$$

If the double integral exists, (4) and (5) will in general yield the same value. In writing a double integral, either of the forms (4) or (5), whichever is appropriate, may be used. We call one form an *interchange of the order of integration* with respect to the other form.

In case $\mathcal{R}$ is not of the type shown in the above figure, it can generally be subdivided into regions $\mathcal{R}_1, \mathcal{R}_2, \ldots$ which are of this type. Then the double integral over $\mathcal{R}$ is found by taking the sum of the double integrals over $\mathcal{R}_1, \mathcal{R}_2, \ldots$.

TRIPLE INTEGRALS

The above results are easily generalized to closed regions in three dimensions. For example, consider a function $F(x,y,z)$ defined in a closed three dimensional region $\mathcal{R}$. Subdivide the region into n subregions of volume ΔV_k, $k = 1, 2, \ldots, n$. Letting (ξ_k, η_k, ζ_k) be some point in each subregion, we form

$$\lim_{n \to \infty} \sum_{k=1}^{n} F(\xi_k, \eta_k, \zeta_k)\,\Delta V_k \tag{6}$$

where the number n of subdivisions approaches infinity in such a way that the largest linear dimension of each subregion approaches zero. If this limit exists we denote it by

$$\iiint\limits_{\mathcal{R}} F(x,y,z)\,dV \tag{7}$$

called the *triple integral* of $F(x,y,z)$ over $\mathcal{R}$. The limit does exist if $F(x,y,z)$ is continuous (or piecewise continuous) in $\mathcal{R}$.

If we construct a grid consisting of planes parallel to the xy, yz and xz planes, the region $\mathcal{R}$ is subdivided into subregions which are rectangular parallelepipeds. In such case we can express the triple integral over $\mathcal{R}$ given by (7) as an *iterated integral* of the form

$$\int_{x=a}^{b} \int_{y=g_1(x)}^{g_2(x)} \int_{z=f_1(x,y)}^{f_2(x,y)} F(x,y,z)\,dx\,dy\,dz \;=\; \int_{x=a}^{b} \left[\int_{y=g_1(x)}^{g_2(x)} \left\{ \int_{z=f_1(x,y)}^{f_2(x,y)} F(x,y,z)\,dz \right\} dy \right] dx \tag{8}$$

(where the innermost integral is to be evaluated first) or the sum of such integrals. The integration can also be performed in any other order to give an equivalent result.

Extensions to higher dimensions are also possible.

TRANSFORMATIONS OF MULTIPLE INTEGRALS

In evaluating a multiple integral over a region $\mathcal{R}$, it is often convenient to use coordinates other than rectangular, such as the curvilinear coordinates considered in Chapter 5.

If we let (u, v) be curvilinear coordinates of points in a plane, there will be a set of transformation equations $x = f(u, v)$, $y = g(u, v)$ mapping points (x, y) of the xy plane into points (u, v) of the uv plane. In such case the region $\mathcal{R}$ of the xy plane is mapped into a region $\mathcal{R}'$ of the uv plane. We then have

$$\iint_{\mathcal{R}} F(x, y)\, dx\, dy \;=\; \iint_{\mathcal{R}'} G(u, v) \left| \frac{\partial(x, y)}{\partial(u, v)} \right| du\, dv \tag{9}$$

where $G(u, v) \equiv F\{f(u, v), g(u, v)\}$ and

$$\frac{\partial(x, y)}{\partial(u, v)} \;\equiv\; \begin{vmatrix} \dfrac{\partial x}{\partial u} & \dfrac{\partial x}{\partial v} \\[2mm] \dfrac{\partial y}{\partial u} & \dfrac{\partial y}{\partial v} \end{vmatrix} \tag{10}$$

is the *Jacobian* of x and y with respect to u and v (see Chapter 5).

Similarly if (u, v, w) are curvilinear coordinates in three dimensions, there will be a set of transformation equations $x = f(u, v, w)$, $y = g(u, v, w)$, $z = h(u, v, w)$ and we can write

$$\iiint_{\mathcal{R}} F(x, y, z)\, dx\, dy\, dz \;=\; \iiint_{\mathcal{R}'} G(u, v, w) \left| \frac{\partial(x, y, z)}{\partial(u, v, w)} \right| du\, dv\, dw \tag{11}$$

where $G(u, v, w) \equiv F\{f(u, v, w), g(u, v, w), h(u, v, w)\}$ and

$$\frac{\partial(x, y, z)}{\partial(u, v, w)} \;\equiv\; \begin{vmatrix} \dfrac{\partial x}{\partial u} & \dfrac{\partial x}{\partial v} & \dfrac{\partial x}{\partial w} \\[2mm] \dfrac{\partial y}{\partial u} & \dfrac{\partial y}{\partial v} & \dfrac{\partial y}{\partial w} \\[2mm] \dfrac{\partial z}{\partial u} & \dfrac{\partial z}{\partial v} & \dfrac{\partial z}{\partial w} \end{vmatrix} \tag{12}$$

is the Jacobian of x, y and z with respect to u, v and w.

The results (9) and (11) correspond to change of variables for double and triple integrals.

Generalizations to higher dimensions are easily made.

LINE INTEGRALS

Let C be a curve in the xy plane which connects points $A(a_1, b_1)$ and $B(a_2, b_2)$, (see Fig. 6-2). Let $P(x, y)$ and $Q(x, y)$ be single-valued functions defined at all points of C. Subdivide C into n parts by choosing $(n-1)$ points on it given by $(x_1, y_1), (x_2, y_2), \ldots, (x_{n-1}, y_{n-1})$. Call $\Delta x_k = x_k - x_{k-1}$ and $\Delta y_k = y_k - y_{k-1}$, $k = 1, 2, \ldots, n$ where $(a_1, b_1) \equiv (x_0, y_0)$, $(a_2, b_2) \equiv (x_n, y_n)$ and suppose that points (ξ_k, η_k) are chosen so that they are situated on C between points (x_{k-1}, y_{k-1}) and (x_k, y_k). Form the sum

$$\sum_{k=1}^{n} \{P(\xi_k, \eta_k)\, \Delta x_k \;+\; Q(\xi_k, \eta_k)\, \Delta y_k\} \tag{13}$$

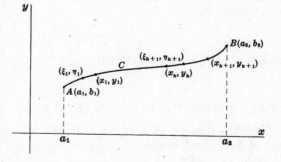

Fig. 6-2

The limit of this sum as $n \to \infty$ in such a way that all the quantities Δx_k, Δy_k approach zero, if such limit exists, is called a *line integral* along C and is denoted by

$$\int_C [P(x, y)\, dx + Q(x, y)\, dy] \quad \text{or} \quad \int_{(a_1, b_1)}^{(a_2, b_2)} [P\, dx + Q\, dy] \tag{14}$$

The limit does exist if P and Q are continuous (or piecewise continuous) at all points of C. The value of the integral depends in general on P, Q, the particular curve C, and on the limits (a_1, b_1) and (a_2, b_2).

In an exactly analogous manner one may define a line integral along a curve C in three dimensional space as

$$\lim_{n \to \infty} \sum_{k=1}^{n} \{A_1(\xi_k, \eta_k, \zeta_k) \, \Delta x_k \ + \ A_2(\xi_k, \eta_k, \zeta_k) \, \Delta y_k \ + \ A_3(\xi_k, \eta_k, \zeta_k) \, \Delta z_k\} \tag{15}$$

$$= \ \int_C [A_1 \, dx \ + \ A_2 \, dy \ + \ A_3 \, dz]$$

where A_1, A_2 and A_3 are functions of x, y and z.

Other types of line integrals, depending on particular curves, can be defined. For example, if Δs_k denotes the arc length along curve C in the above figure between points (x_k, y_k) and (x_{k+1}, y_{k+1}), then

$$\lim_{n \to \infty} \sum_{k=1}^{n} U(\xi_k, \eta_k) \, \Delta s_k \ = \ \int_C U(x, y) \, ds \tag{16}$$

is called the line integral of $U(x, y)$ along curve C. Extensions to three (or higher) dimensions are possible.

VECTOR NOTATION FOR LINE INTEGRALS

It is often convenient to express a line integral in vector form as an aid in physical or geometric understanding as well as for brevity of notation. For example, we can express the line integral (15) in the form

$$\int_C [A_1 \, dx \ + \ A_2 \, dy \ + \ A_3 \, dz] \ = \ \int_C (A_1 \mathbf{i} + A_2 \mathbf{j} + A_3 \mathbf{k}) \cdot (dx \, \mathbf{i} + dy \, \mathbf{j} + dz \, \mathbf{k}) \tag{17}$$

$$= \ \int_C \mathbf{A} \cdot d\mathbf{r}$$

where $\mathbf{A} = A_1 \mathbf{i} + A_2 \mathbf{j} + A_3 \mathbf{k}$ and $d\mathbf{r} = dx \, \mathbf{i} + dy \, \mathbf{j} + dz \, \mathbf{k}$. The line integral (14) is a special case of this with $z = 0$.

If at each point (x, y, z) we associate a force $\mathbf{F}$ acting on an object (i.e. if a *force field* is defined), then

$$\int_C \mathbf{F} \cdot d\mathbf{r} \tag{18}$$

represents physically the total work done in moving the object along the curve C.

EVALUATION OF LINE INTEGRALS

If the equation of a curve C in the plane $z = 0$ is given as $y = f(x)$, the line integral (14) is evaluated by placing $y = f(x)$, $dy = f'(x) \, dx$ in the integrand to obtain the definite integral

$$\int_{a_1}^{a_2} [P\{x, f(x)\} \, dx \ + \ Q\{x, f(x)\} \, f'(x) \, dx] \tag{19}$$

which is then evaluated in the usual manner.

Similarly if C is given as $x = g(y)$, then $dx = g'(y) \, dy$ and the line integral becomes

$$\int_{b_1}^{b_2} [P\{g(y), y\} \, g'(y) \, dy \ + \ Q\{g(y), y\} \, dy] \tag{20}$$

If C is given in parametric form $x = \phi(t)$, $y = \psi(t)$, the line integral becomes

$$\int_{t_1}^{t_2} [P\{\phi(t), \psi(t)\}\, \phi'(t)\, dt + Q\{\phi(t), \psi(t)\}\, \psi'(t)\, dt] \tag{21}$$

where t_1 and t_2 denote the values of t corresponding to points A and B respectively.

Combinations of the above methods may be used in the evaluation.

Similar methods are used for evaluating line integrals along space curves.

PROPERTIES OF LINE INTEGRALS

Line integrals have properties which are analogous to those of ordinary integrals. For example:

1. $\displaystyle\int_C [P(x,y)\, dx + Q(x,y)\, dy] \;=\; \int_C P(x,y)\, dx \;+\; \int_C Q(x,y)\, dy$

2. $\displaystyle\int_{(a_1,b_1)}^{(a_2,b_2)} [P\, dx + Q\, dy] \;=\; -\int_{(a_2,b_2)}^{(a_1,b_1)} [P\, dx + Q\, dy]$

Thus reversal of the path of integration changes the sign of the line integral.

3. $\displaystyle\int_{(a_1,b_1)}^{(a_2,b_2)} [P\, dx + Q\, dy] \;=\; \int_{(a_1,b_1)}^{(a_3,b_3)} [P\, dx + Q\, dy] \;+\; \int_{(a_3,b_3)}^{(a_2,b_2)} [P\, dx + Q\, dy]$

where (a_3, b_3) is another point on C.

Similar properties hold for line integrals in space.

SIMPLE CLOSED CURVES. SIMPLY AND MULTIPLY-CONNECTED REGIONS

A *simple closed curve* is a closed curve which does not intersect itself anywhere. Mathematically, a curve in the xy plane is defined by the parametric equations $x = \phi(t)$, $y = \psi(t)$ where ϕ and ψ are single-valued and continuous in an interval $t_1 \leqq t \leqq t_2$. If $\phi(t_1) = \phi(t_2)$ and $\psi(t_1) = \psi(t_2)$, the curve is said to be *closed*. If $\phi(u) = \phi(v)$ and $\psi(u) = \psi(v)$ only when $u = v$ (except in the special case where $u = t_1$ and $v = t_2$), the curve is closed and does not intersect itself and so is a simple closed curve. We shall also assume, unless otherwise stated, that ϕ and ψ are piecewise differentiable in $t_1 \leqq t \leqq t_2$.

If a plane region has the property that any closed curve in it can be continuously shrunk to a point without leaving the region, then the region is called *simply-connected*, otherwise it is called *multiply-connected* [see Problem 6.19].

As the parameter t varies from t_1 to t_2, the plane curve is described in a certain sense or direction. For curves in the xy plane, we arbitrarily describe this direction as *positive* or *negative* according as a person traversing the curve in this direction with his head pointing in the positive z direction has the region enclosed by the curve always toward his left or right respectively. If we look down upon a simple closed curve in the xy plane, this amounts to saying that traversal of the curve in the counterclockwise direction is taken as positive while traversal in the clockwise direction is taken as negative.

GREEN'S THEOREM IN THE PLANE

Let $P, Q, \partial P/\partial y, \partial Q/\partial x$ be single-valued and continuous in a simply-connected region $\mathcal{R}$ bounded by a simple closed curve C. Then

$$\oint_C [P\, dx + Q\, dy] \;=\; \iint_{\mathcal{R}} \left(\frac{\partial Q}{\partial x} - \frac{\partial P}{\partial y} \right) dx\, dy \tag{22}$$

where $\oint_C$ is used to emphasize that C is closed and that it is described in the positive direction.

This theorem is also true for regions bounded by two or more closed curves (i.e. multiply-connected regions). See Problem 6.19.

CONDITIONS FOR A LINE INTEGRAL TO BE INDEPENDENT OF THE PATH

Theorem 6-1.   A necessary and sufficient condition for $\int_C [P\,dx + Q\,dy]$ to be independent of the path C joining any two given points in a region $\mathcal{R}$ is that in $\mathcal{R}$

$$\partial P/\partial y \;=\; \partial Q/\partial x \tag{23}$$

where it is supposed that these partial derivatives are continuous in $\mathcal{R}$.

The condition (23) is also the condition that $P\,dx + Q\,dy$ is an exact differential, i.e. that there exists a function $\phi(x, y)$ such that $P\,dx + Q\,dy = d\phi$. In such case if the end points of curve C are (x_1, y_1) and (x_2, y_2), the value of the line integral is given by

$$\int_{(x_1,y_1)}^{(x_2,y_2)} [P\,dx + Q\,dy] \;=\; \int_{(x_1,y_1)}^{(x_2,y_2)} d\phi \;=\; \phi(x_2, y_2) - \phi(x_1, y_1) \tag{24}$$

In particular if (23) holds and C is closed, we have $x_1 = x_2$, $y_1 = y_2$ and

$$\oint_C [P\,dx + Q\,dy] \;=\; 0 \tag{25}$$

For proofs and related theorems, see Problems 6.22 and 6.23.

The results in Theorem 6-1 can be extended to line integrals in space. Thus we have

Theorem 6-2.   A necessary and sufficient condition for $\int_C [A_1\,dx + A_2\,dy + A_3\,dz]$ to be independent of the path C joining any two given points in a region $\mathcal{R}$ is that in $\mathcal{R}$

$$\frac{\partial A_1}{\partial y} = \frac{\partial A_2}{\partial x}, \quad \frac{\partial A_3}{\partial x} = \frac{\partial A_1}{\partial z}, \quad \frac{\partial A_2}{\partial z} = \frac{\partial A_3}{\partial y} \tag{26}$$

where it is supposed that these partial derivatives are continuous in $\mathcal{R}$.

The results can be expressed concisely in terms of vectors. If $\mathbf{A} = A_1\mathbf{i} + A_2\mathbf{j} + A_3\mathbf{k}$, the line integral can be written $\int_C \mathbf{A} \cdot d\mathbf{r}$ and condition (26) is equivalent to the condition $\nabla \times \mathbf{A} = 0$. If $\mathbf{A}$ represents a force field $\mathbf{F}$ which acts on an object, the result is equivalent to the statement that the work done in moving the object from one point to another is independent of the path joining the two points if and only if $\nabla \times \mathbf{A} = 0$. Such a force field is often called *conservative*.

The condition (26) [or the equivalent condition $\nabla \times \mathbf{A} = 0$] is also the condition that $A_1\,dx + A_2\,dy + A_3\,dz$ [or $\mathbf{A} \cdot d\mathbf{r}$] is an exact differential, i.e. that there exists a function $\phi(x, y, z)$ such that $A_1\,dx + A_2\,dy + A_3\,dz = d\phi$. In such case if the endpoints of curve C are (x_1, y_1, z_1) and (x_2, y_2, z_2), the value of the line integral is given by

$$\int_{(x_1,y_1,z_1)}^{(x_2,y_2,z_2)} \mathbf{A} \cdot d\mathbf{r} \;=\; \int_{(x_1,y_1,z_1)}^{(x_2,y_2,z_2)} d\phi \;=\; \phi(x_2, y_2, z_2) - \phi(x_1, y_1, z_1) \tag{27}$$

In particular if C is closed and $\nabla \times \mathbf{A} = 0$, we have

$$\oint_C \mathbf{A} \cdot d\mathbf{r} \;=\; 0 \tag{28}$$

SURFACE INTEGRALS

Let S be a two-sided surface having projection $\mathcal{R}$ on the xy plane as in the adjoining Fig. 6-3. Assume that an equation for S is $z = f(x, y)$, where f is single-valued and continuous for all x and y in $\mathcal{R}$. Divide $\mathcal{R}$ into n subregions of area ΔA_p, $p = 1, 2, \ldots, n$, and erect a vertical column on each of these subregions to intersect S in an area ΔS_p.

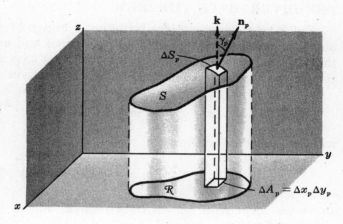

Fig. 6-3

Let $\phi(x, y, z)$ be single-valued and continuous at all points of S. Form the sum

$$\sum_{p=1}^{n} \phi(\xi_p, \eta_p, \zeta_p)\, \Delta S_p \qquad (29)$$

where (ξ_p, η_p, ζ_p) is some point of ΔS_p. If the limit of this sum as $n \to \infty$ in such a way that each $\Delta S_p \to 0$ exists, the resulting limit is called the *surface integral* of $\phi(x, y, z)$ over S and is designated by

$$\iint_S \phi(x, y, z)\, dS \qquad (30)$$

Since $\Delta S_p = |\sec \gamma_p|\, \Delta A_p$ approximately, where γ_p is the angle between the normal line to S and the positive z axis, the limit of the sum (29) can be written

$$\iint_{\mathcal{R}} \phi(x, y, z)\, |\sec \gamma|\, dA \qquad (31)$$

The quantity $|\sec \gamma|$ is given by

$$|\sec \gamma| \;=\; \frac{1}{|\mathbf{n}_p \cdot \mathbf{k}|} \;=\; \sqrt{1 + \left(\frac{\partial z}{\partial x}\right)^2 + \left(\frac{\partial z}{\partial y}\right)^2} \qquad (32)$$

Then assuming that $z = f(x, y)$ has continuous (or sectionally continuous) derivatives in $\mathcal{R}$, (31) can be written in rectangular form as

$$\iint_{\mathcal{R}} \phi(x, y, z)\, \sqrt{1 + \left(\frac{\partial z}{\partial x}\right)^2 + \left(\frac{\partial z}{\partial y}\right)^2}\; dx\, dy \qquad (33)$$

In case the equation for S is given as $F(x, y, z) = 0$, (33) can also be written

$$\iint_S \phi(x, y, z)\, \frac{\sqrt{(F_x)^2 + (F_y)^2 + (F_z)^2}}{|F_z|}\; dx\, dy \qquad (34)$$

The results (33) or (34) can be used to evaluate (30).

In the above we have assumed that S is such that any line parallel to the z axis intersects S in only one point. In case S is not of this type, we can usually subdivide S into surfaces $S_1, S_2, \ldots$ which are of this type. Then the surface integral over S is defined as the sum of the surface integrals over $S_1, S_2, \ldots$.

The results stated hold when S is projected on to a region $\mathcal{R}$ of the xy plane. In some cases it is better to project S on to the yz or xz planes. For such cases (30) can be evaluated by appropriately modifying (33) and (34).

THE DIVERGENCE THEOREM

Let S be a closed surface bounding a region of volume V. Choose the outward drawn normal to the surface as the *positive normal* and assume that α, β, γ are the angles which this normal makes with the positive x, y and z axes respectively. Then if A_1, A_2 and A_3 are continuous and have continuous partial derivatives in the region

$$\iiint\limits_V \left(\frac{\partial A_1}{\partial x} + \frac{\partial A_2}{\partial y} + \frac{\partial A_3}{\partial z} \right) dV \;=\; \iint\limits_S (A_1 \cos \alpha + A_2 \cos \beta + A_3 \cos \gamma)\, dS \qquad (35)$$

which can also be written

$$\iiint\limits_V \left(\frac{\partial A_1}{\partial x} + \frac{\partial A_2}{\partial y} + \frac{\partial A_3}{\partial z} \right) dV \;=\; \iint\limits_S [A_1\, dy\, dz + A_2\, dz\, dx + A_3\, dx\, dy] \qquad (36)$$

In vector form with $\mathbf{A} = A_1\mathbf{i} + A_2\mathbf{j} + A_3\mathbf{k}$ and $\mathbf{n} = \cos \alpha \, \mathbf{i} + \cos \beta \, \mathbf{j} + \cos \gamma \, \mathbf{k}$, these can be simply written as

$$\iiint\limits_V \nabla \cdot \mathbf{A} \, dV \;=\; \iint\limits_S \mathbf{A} \cdot \mathbf{n} \, dS \qquad (37)$$

In words this theorem, called the *divergence theorem* or *Green's theorem in space*, states that the surface integral of the normal component of a vector $\mathbf{A}$ taken over a closed surface is equal to the integral of the divergence of $\mathbf{A}$ taken over the volume enclosed by the surface.

STOKES' THEOREM

Let S be an open, two-sided surface bounded by a closed non-intersecting curve C (simple closed curve). Consider a directed line normal to S as positive if it is on one side of S, and negative if it is on the other side of S. The choice of which side is positive is arbitrary but should be decided upon in advance. Call the direction or sense of C positive if an observer, walking on the boundary of S with his head pointing in the direction of the positive normal, has the surface on his left. Then if A_1, A_2, A_3 are single-valued, continuous, and have continuous first partial derivatives in a region of space including S, we have

$$\int_C [A_1\, dx + A_2\, dy + A_3\, dz] \;=\; \iint\limits_S \left[\left(\frac{\partial A_3}{\partial y} - \frac{\partial A_2}{\partial z} \right) \cos \alpha \right. \qquad (38)$$
$$\left. + \left(\frac{\partial A_1}{\partial z} - \frac{\partial A_3}{\partial x} \right) \cos \beta + \left(\frac{\partial A_2}{\partial x} - \frac{\partial A_1}{\partial y} \right) \cos \gamma \right] dS$$

In vector form with $\mathbf{A} = A_1\mathbf{i} + A_2\mathbf{j} + A_3\mathbf{k}$ and $\mathbf{n} = \cos \alpha \, \mathbf{i} + \cos \beta \, \mathbf{j} + \cos \gamma \, \mathbf{k}$, this is simply expressed as

$$\int_C \mathbf{A} \cdot d\mathbf{r} \;=\; \iint\limits_S (\nabla \times \mathbf{A}) \cdot \mathbf{n} \, dS \qquad (39)$$

In words this theorem, called *Stokes' theorem*, states that the line integral of the tangential component of a vector $\mathbf{A}$ taken around a simple closed curve C is equal to the surface integral of the normal component of the curl of $\mathbf{A}$ taken over any surface S having C as a boundary. Note that if, as a special case $\nabla \times \mathbf{A} = 0$ in (39), we obtain the result (28).

Solved Problems

DOUBLE INTEGRALS

6.1. (a) Sketch the region $\mathcal{R}$ in the xy plane bounded by $y = x^2$, $x = 2$, $y = 1$.

(b) Give a physical interpretation to $\iint\limits_{\mathcal{R}} (x^2 + y^2)\, dx\, dy$.

(c) Evaluate the double integral in (b).

(a) The required region $\mathcal{R}$ is shown shaded in Fig. 6-4 below.

(b) Since $x^2 + y^2$ is the square of the distance from any point (x, y) to $(0, 0)$, we can consider the double integral as representing the *polar moment of inertia* (i.e. moment of inertia with respect to the origin) of the region $\mathcal{R}$ (assuming unit density).

We can also consider the double integral as representing the *mass* of the region $\mathcal{R}$ assuming a density varying as $x^2 + y^2$.

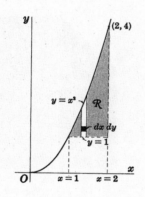

Fig. 6-4

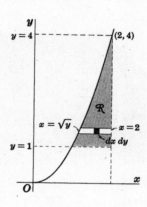

Fig. 6-5

(c) **Method 1.** The double integral can be expressed as the iterated integral

$$\int_{x=1}^{2} \int_{y=1}^{x^2} (x^2 + y^2)\, dy\, dx \;=\; \int_{x=1}^{2} \left\{ \int_{y=1}^{x^2} (x^2 + y^2)\, dy \right\} dx \;=\; \int_{x=1}^{2} \left. x^2 y + \frac{y^3}{3} \right|_{y=1}^{x^2} dx$$

$$=\; \int_{x=1}^{2} \left(x^4 + \frac{x^6}{3} - x^2 - \frac{1}{3} \right) dx \;=\; \frac{1006}{105}$$

The integration with respect to y (keeping x constant) from $y = 1$ to $y = x^2$ corresponds formally to summing in a vertical column (see Fig. 6-4). The subsequent integration with respect to x from $x = 1$ to $x = 2$ corresponds to addition of contributions from all such vertical columns between $x = 1$ and $x = 2$.

Method 2. The double integral can also be expressed as the iterated integral

$$\int_{y=1}^{4} \int_{x=\sqrt{y}}^{2} (x^2 + y^2)\, dx\, dy \;=\; \int_{y=1}^{4} \left\{ \int_{x=\sqrt{y}}^{2} (x^2 + y^2)\, dx \right\} dy \;=\; \int_{y=1}^{4} \left. \frac{x^3}{3} + xy^2 \right|_{x=\sqrt{y}}^{2} dy$$

$$=\; \int_{y=1}^{4} \left(\frac{8}{3} + 2y^2 - \frac{y^{3/2}}{3} - y^{5/2} \right) dy \;=\; \frac{1006}{105}$$

In this case the vertical column of region $\mathcal{R}$ in Fig. 6-4 above is replaced by a horizontal column as in Fig. 6-5 above. Then the integration with respect to x (keeping y constant) from $x = \sqrt{y}$ to $x = 2$ corresponds to summing in this horizontal column. Subsequent integration with respect to y from $y = 1$ to $y = 4$ corresponds to addition of contributions for all such horizontal columns between $y = 1$ and $y = 4$.

6.2. Find the volume of the region common to the intersecting cylinders $x^2 + y^2 = a^2$ and $x^2 + z^2 = a^2$.

Required volume = 8 times volume of region shown in Fig. 6-6

$$= 8 \int_{x=0}^{a} \int_{y=0}^{\sqrt{a^2-x^2}} z \, dy \, dx \;=\; 8 \int_{x=0}^{a} \int_{y=0}^{\sqrt{a^2-x^2}} \sqrt{a^2 - x^2} \, dy \, dx$$

$$= 8 \int_{x=0}^{a} (a^2 - x^2) \, dx \;=\; \frac{16a^3}{3}$$

As an aid in setting up this integral note that $z \, dy \, dx$ corresponds to the volume of a column such as shown darkly shaded in the figure. Keeping x constant and integrating with respect to y from $y = 0$ to $y = \sqrt{a^2 - x^2}$ corresponds to adding the volumes of all such columns in a slab parallel to the yz plane, thus giving the volume of this slab. Finally, integrating with respect to x from $x = 0$ to $x = a$ corresponds to adding the volumes of all such slabs in the region, thus giving the required volume.

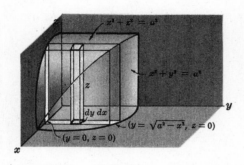

Fig. 6-6

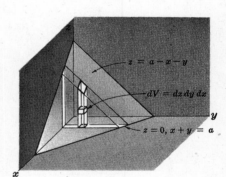

Fig. 6-7

TRIPLE INTEGRALS

6.3. (a) Sketch the 3 dimensional region $\mathcal{R}$ bounded by $x + y + z = a$ $(a > 0)$, $x = 0$, $y = 0$, $z = 0$.

(b) Give a physical interpretation to

$$\iiint_{\mathcal{R}} (x^2 + y^2 + z^2) \, dx \, dy \, dz$$

(c) Evaluate the triple integral in (b).

(a) The required region $\mathcal{R}$ is shown in Fig. 6-7.

(b) Since $x^2 + y^2 + z^2$ is the square of the distance from any point (x, y, z) to $(0, 0, 0)$, we can consider the triple integral as representing the *polar moment of inertia* (i.e. moment of inertia with respect to the origin) of the region $\mathcal{R}$ (assuming unit density).

We can also consider the triple integral as representing the *mass* of the region if the density varies as $x^2 + y^2 + z^2$.

(c) The triple integral can be expressed as the iterated integral

$$\int_{x=0}^{a} \int_{y=0}^{a-x} \int_{z=0}^{a-x-y} (x^2 + y^2 + z^2) \, dz \, dy \, dx$$

$$= \int_{x=0}^{a} \int_{y=0}^{a-x} x^2 z + y^2 z + \frac{z^3}{3} \Big|_{z=0}^{a-x-y} \, dy \, dx$$

$$= \int_{x=0}^{a} \int_{y=0}^{a-x} \left\{ x^2(a-x) - x^2 y + (a-x)y^2 - y^3 + \frac{(a-x-y)^3}{3} \right\} dy\, dx$$

$$= \int_{x=0}^{a} x^2(a-x)y - \frac{x^2 y^2}{2} + \frac{(a-x)y^3}{3} - \frac{y^4}{4} - \frac{(a-x-y)^4}{12} \Bigg|_{y=0}^{a-x} dx$$

$$= \int_{0}^{a} \left\{ x^2(a-x)^2 - \frac{x^2(a-x)^2}{2} + \frac{(a-x)^4}{3} - \frac{(a-x)^4}{4} + \frac{(a-x)^4}{12} \right\} dx$$

$$= \int_{0}^{a} \left\{ \frac{x^2(a-x)^2}{2} + \frac{(a-x)^4}{6} \right\} dx \;=\; \frac{a^5}{20}$$

The integration with respect to z (keeping x and y constant) from $z=0$ to $z=a-x-y$ corresponds to summing the polar moments of inertia (or masses) corresponding to each cube in a vertical column. The subsequent integration with respect to y from $y=0$ to $y=a-x$ (keeping x constant) corresponds to addition of contributions from all vertical columns contained in a slab parallel to the yz plane. Finally, integration with respect to x from $x=0$ to $x=a$ adds up contributions from all slabs parallel to the yz plane.

Although the above integration has been accomplished in the order z, y, x, any other order is clearly possible and the final answer should be the same.

6.4. Find the (a) volume and (b) centroid of the region $\mathcal{R}$ bounded by the parabolic cylinder $z = 4 - x^2$ and the planes $x=0$, $y=0$, $y=6$, $z=0$ assuming the density to be a constant σ.

The region $\mathcal{R}$ is shown in Fig. 6-8.

(a) Required volume $= \iiint\limits_{\mathcal{R}} dx\, dy\, dz$

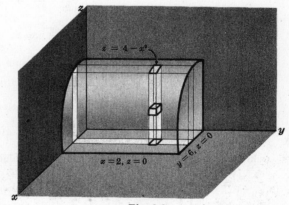

Fig. 6-8

$$= \int_{x=0}^{2} \int_{y=0}^{6} \int_{z=0}^{4-x^2} dz\, dy\, dx$$

$$= \int_{x=0}^{2} \int_{y=0}^{6} (4-x^2)\, dy\, dx$$

$$= \int_{x=0}^{2} (4-x^2)y \Big|_{y=0}^{6} dx$$

$$= \int_{x=0}^{2} (24 - 6x^2)\, dx \;=\; 32$$

(b) Total mass $= \displaystyle\int_{x=0}^{2} \int_{y=0}^{6} \int_{z=0}^{4-x^2} \sigma\, dz\, dy\, dx = 32\sigma$ by part (a), since σ is constant. Then

$$\bar{x} = \frac{\text{Total moment about } yz \text{ plane}}{\text{Total mass}} = \frac{\displaystyle\int_{x=0}^{2} \int_{y=0}^{6} \int_{z=0}^{4-x^2} \sigma x\, dz\, dy\, dx}{\text{Total mass}} = \frac{24\sigma}{32\sigma} = \frac{3}{4}$$

$$\bar{y} = \frac{\text{Total moment about } xz \text{ plane}}{\text{Total mass}} = \frac{\displaystyle\int_{x=0}^{2} \int_{y=0}^{6} \int_{z=0}^{4-x^2} \sigma y\, dz\, dy\, dx}{\text{Total mass}} = \frac{96\sigma}{32\sigma} = 3$$

$$\bar{z} = \frac{\text{Total moment about } xy \text{ plane}}{\text{Total mass}} = \frac{\displaystyle\int_{x=0}^{2} \int_{y=0}^{6} \int_{z=0}^{4-x^2} \sigma z\, dz\, dy\, dx}{\text{Total mass}} = \frac{256\sigma/5}{32\sigma} = \frac{8}{5}$$

Thus the centroid has coordinates $(3/4, 3, 8/5)$.

Note that the value for $\bar{y}$ could have been predicted because of symmetry.

TRANSFORMATION OF DOUBLE INTEGRALS

6.5.   Justify equation *(21)*, page 151, for changing variables in a double integral.

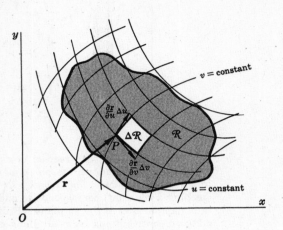

Fig. 6-9

In rectangular coordinates, the double integral
of $F(x, y)$ over the region $\mathcal{R}$ (shaded in Fig. 6-9) is
$\iint\limits_{\mathcal{R}} F(x, y)\, dx\, dy$. We can also evaluate this double

integral by considering a grid formed by a family
of u and v curvilinear coordinate curves constructed
on the region $\mathcal{R}$ as shown in the figure.

Let P be any point with coordinates (x, y) or
(u, v), where $x = f(u, v)$ and $y = g(u, v)$. Then
the vector $\mathbf{r}$ from O to P is given by $\mathbf{r} = x\mathbf{i} + y\mathbf{j} =$
$f(u, v)\mathbf{i} + g(u, v)\mathbf{j}$. The tangent vectors to the coordinate curves $u = c_1$ and $v = c_2$, where c_1 and c_2 are
constants, are $\partial r/\partial v$ and $\partial r/\partial u$ respectively. Then
the area of the region $\Delta\mathcal{R}$ of Fig. 6-9 is given ap-
proximately by $\left|\dfrac{\partial \mathbf{r}}{\partial u} \times \dfrac{\partial \mathbf{r}}{\partial v}\right| \Delta u\, \Delta v$.

But

$$\frac{\partial \mathbf{r}}{\partial u} \times \frac{\partial \mathbf{r}}{\partial v} \;=\; \begin{vmatrix} \mathbf{i} & \mathbf{j} & \mathbf{k} \\[4pt] \dfrac{\partial x}{\partial u} & \dfrac{\partial y}{\partial u} & 0 \\[8pt] \dfrac{\partial x}{\partial v} & \dfrac{\partial y}{\partial v} & 0 \end{vmatrix} \;=\; \begin{vmatrix} \dfrac{\partial x}{\partial u} & \dfrac{\partial y}{\partial u} \\[10pt] \dfrac{\partial x}{\partial v} & \dfrac{\partial y}{\partial v} \end{vmatrix} \mathbf{k} \;=\; \frac{\partial(x, y)}{\partial(u, v)} \mathbf{k}$$

so that

$$\left|\frac{\partial \mathbf{r}}{\partial u} \times \frac{\partial \mathbf{r}}{\partial v}\right| \Delta u\, \Delta v \;=\; \left|\frac{\partial(x, y)}{\partial(u, v)}\right| \Delta u\, \Delta v$$

The double integral is the limit of the sum

$$\sum F\{f(u, v),\, g(u, v)\} \left|\frac{\partial(x, y)}{\partial(u, v)}\right| \Delta u\, \Delta v$$

taken over the entire region $\mathcal{R}$. An investigation reveals that this limit is

$$\iint\limits_{\mathcal{R}'} F\{f(u, v),\, g(u, v)\} \left|\frac{\partial(x, y)}{\partial(u, v)}\right| du\, dv$$

where $\mathcal{R}'$ is the region in the uv plane into which the region $\mathcal{R}$ is mapped under the transformation
$x = f(u, v),\quad y = g(u, v)$.

6.6.   Evaluate $\displaystyle\iint\limits_{\mathcal{R}} \sqrt{x^2 + y^2}\, dx\, dy$, where $\mathcal{R}$ is the region in the xy plane bounded by
$x^2 + y^2 = 4$ and $x^2 + y^2 = 9$.

The presence of $x^2 + y^2$ suggests the use of polar coordinates (ρ, ϕ), where $x = \rho \cos\phi$, $y = \rho \sin\phi$.
Under this transformation the region $\mathcal{R}$ [Fig. 6-10(a)] is mapped into the region $\mathcal{R}'$ [Fig. 6-10(b)].

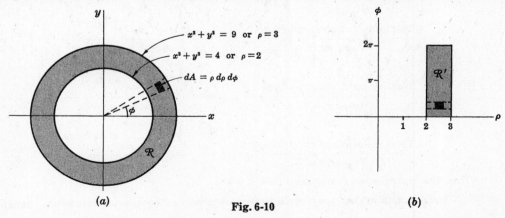

Fig. 6-10

Since $\dfrac{\partial(x,y)}{\partial(\rho,\phi)} = \rho$, it follows that

$$\iint\limits_{\mathcal{R}} \sqrt{x^2+y^2}\,dx\,dy \;=\; \iint\limits_{\mathcal{R}'} \sqrt{x^2+y^2}\left|\frac{\partial(x,y)}{\partial(\rho,\phi)}\right| d\rho\,d\phi \;=\; \iint\limits_{\mathcal{R}'} \rho \cdot \rho\, d\rho\,d\phi$$

$$=\; \int_{\phi=0}^{2\pi}\int_{\rho=2}^{3} \rho^2\,d\rho\,d\phi \;=\; \int_{\phi=0}^{2\pi} \frac{\rho^3}{3}\Big|_{2}^{3}\,d\phi \;=\; \int_{\phi=0}^{2\pi} \frac{19}{3}\,d\phi \;=\; \frac{38\pi}{3}$$

We can also write the integration limits for $\mathcal{R}'$ immediately on observing the region $\mathcal{R}$, since for fixed ϕ, ρ varies from $\rho = 2$ to $\rho = 3$ within the sector shown dashed in Fig. 6-10(a). An integration with respect to ϕ from $\phi = 0$ to $\phi = 2\pi$ then gives the contribution from all sectors. Geometrically $\rho\,d\rho\,d\phi$ represents the area dA as shown in Fig. 6-10(a).

TRANSFORMATION OF TRIPLE INTEGRALS

6.7.   Justify equation (*11*), page 149, for changing variables in a triple integral.

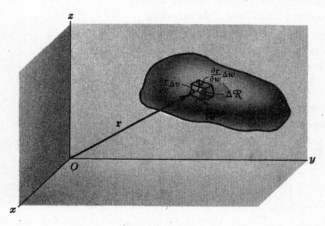

Fig. 6-11

By analogy with Problem 6.5 we construct a grid of curvilinear coordinate surfaces which subdivide the region $\mathcal{R}$ into subregions, a typical one of which is $\Delta\mathcal{R}$ (see Fig. 6-11).

The vector $\mathbf{r}$ from the origin O to point P is

$$r = x\mathbf{i} + y\mathbf{j} + z\mathbf{k} = f(u,v,w)\mathbf{i} + g(u,v,w)\mathbf{j} + h(u,v,w)\mathbf{k}$$

assuming that the transformation equations are $x = f(u,v,w)$, $y = g(u,v,w)$ and $z = h(u,v,w)$.

Tangent vectors to the coordinate curves corresponding to the intersection of pairs of coordinate surfaces are given by $\partial\mathbf{r}/\partial u$, $\partial\mathbf{r}/\partial v$, $\partial\mathbf{r}/\partial w$. Then the volume of the region $\Delta\mathcal{R}$ of Fig. 6-11 is given approximately by

$$\left|\frac{\partial\mathbf{r}}{\partial u}\cdot\frac{\partial\mathbf{r}}{\partial v}\times\frac{\partial\mathbf{r}}{\partial w}\right|\Delta u\,\Delta v\,\Delta w \;=\; \left|\frac{\partial(x,y,z)}{\partial(u,v,w)}\right|\Delta u\,\Delta v\,\Delta w$$

The triple integral of $F(x,y,z)$ over the region is the limit of the sum

$$\sum F\{f(u,v,w),\,g(u,v,w),\,h(u,v,w)\}\left|\frac{\partial(x,y,z)}{\partial(u,v,w)}\right|\Delta u\,\Delta v\,\Delta w$$

An investigation reveals that this limit is

$$\iiint\limits_{\mathcal{R}'} F\{f(u,v,w),\,g(u,v,w),\,h(u,v,w)\}\left|\frac{\partial(x,y,z)}{\partial(u,v,w)}\right| du\,dv\,dw$$

where $\mathcal{R}'$ is the region in the uvw space into which the region $\mathcal{R}$ is mapped under the transformation.

6.8. Express $\displaystyle\iiint\limits_{R} F(x, y, z)\, dx\, dy\, dz$ in cylindrical coordinates.

The transformation equations in cylindrical coordinates are $x = \rho \cos \phi$, $y = \rho \sin \phi$, $z = z$. The Jacobian of the transformation is

$$\frac{\partial(x, y, z)}{\partial(\rho, \phi, z)} \;=\; \begin{vmatrix} \cos \phi & -\rho \sin \phi & 0 \\ \sin \phi & \rho \cos \phi & 0 \\ 0 & 0 & 1 \end{vmatrix} \;=\; \rho$$

Then by Problem 6.7 the triple integral becomes

$$\iiint\limits_{R'} G(\rho, \phi, z)\, \rho\, d\rho\, d\phi\, dz$$

where R' is the region in the ρ, ϕ, z space corresponding to R and where

$$G(\rho, \phi, z) \;=\; F(\rho \cos \phi,\ \rho \sin \phi, z)$$

6.9. Find the volume of the region above the xy plane bounded by the paraboloid $z = x^2 + y^2$ and the cylinder $x^2 + y^2 = a^2$.

The volume is most easily found by using cylindrical coordinates. In these coordinates the equations for the paraboloid and cylinder are respectively $z = \rho^2$ and $\rho = a$. Then

Required volume

$= 4$ times volume shown in Fig. 6-12

$\displaystyle = 4 \int_{\phi=0}^{\pi/2} \int_{\rho=0}^{a} \int_{z=0}^{\rho^2} \rho\, dz\, d\rho\, d\phi$

$\displaystyle = 4 \int_{\phi=0}^{\pi/2} \int_{\rho=0}^{a} \rho^3\, d\rho\, d\phi$

$\displaystyle = 4 \int_{\phi=0}^{\pi/2} \frac{\rho^4}{4} \Big|_{\rho=0}^{a} d\phi = \frac{\pi}{2} a^4$

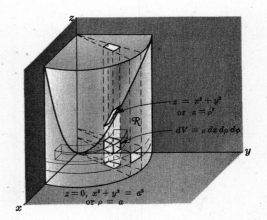

Fig. 6-12

The integration with respect to z (keeping ρ and ϕ constant) from $z = 0$ to $z = \rho^2$ corresponds to summing the cubical volumes (indicated by dV) in a vertical column extending from the xy plane to the paraboloid. The subsequent integration with respect to ρ (keeping ϕ constant) from $\rho = 0$ to $\rho = a$ corresponds to addition of volumes of all columns in the wedge shaped region. Finally, integration with respect to ϕ corresponds to adding volumes of all such wedge shaped regions.

The integration can also be performed in other orders to yield the same result.

We can also set up the integral by determining the region R' in ρ, ϕ, z space into which R is mapped by the cylindrical coordinate transformation.

LINE INTEGRALS

6.10. Evaluate $\displaystyle\int_{(0,1)}^{(1,2)} [(x^2 - y)\, dx + (y^2 + x)\, dy]$ along (a) a straight line from $(0, 1)$ to $(1, 2)$, (b) straight lines from $(0, 1)$ to $(1, 1)$ and then from $(1, 1)$ to $(1, 2)$, (c) the parabola $x = t$, $y = t^2 + 1$.

(a) An equation for the line joining $(0, 1)$ and $(1, 2)$ in the xy plane is $y = x + 1$. Then $dy = dx$ and the line integral equals

$$\int_{x=0}^{1} [\{x^2 - (x+1)\}\, dx + \{(x+1)^2 + x\}\, dx] \;=\; \int_{0}^{1} (2x^2 + 2x)\, dx \;=\; 5/3$$

(b) Along the straight line from $(0, 1)$ to $(1, 1)$, $y = 1$, $dy = 0$ and the line integral equals

$$\int_{x=0}^{1} [(x^2 - 1)\, dx + (1 + x)(0)] \;=\; \int_{0}^{1} (x^2 - 1)\, dx \;=\; -2/3$$

Along the straight line from $(1, 1)$ to $(1, 2)$, $x = 1$, $dx = 0$ and the line integral equals

$$\int_{y=1}^{2} [(1 - y)(0) + (y^2 + 1)\, dy] \;=\; \int_{1}^{2} (y^2 + 1)\, dy \;=\; 10/3$$

Then the required value $= -2/3 + 10/3 = 8/3$.

(c) Since $t = 0$ at $(0, 1)$ and $t = 1$ at $(1, 2)$, the line integral equals

$$\int_{t=0}^{1} [\{t^2 - (t^2 + 1)\}\, dt + \{(t^2 + 1)^2 + t\}\, 2t\, dt] \;=\; \int_{0}^{1} (2t^5 + 4t^3 + 2t^2 + 2t - 1)\, dt \;=\; 2$$

6.11. If $\mathbf{A} = (3x^2 - 6yz)\mathbf{i} + (2y + 3xz)\mathbf{j} + (1 - 4xyz^2)\mathbf{k}$, evaluate $\displaystyle\int_C \mathbf{A} \cdot d\mathbf{r}$ from $(0, 0, 0)$ to $(1, 1, 1)$ along the following paths C:

(a) $x = t$, $y = t^2$, $z = t^3$.

(b) the straight lines from $(0, 0, 0)$ to $(0, 0, 1)$, then to $(0, 1, 1)$, and then to $(1, 1, 1)$.

(c) the straight line joining $(0, 0, 0)$ and $(1, 1, 1)$.

$$\int_C \mathbf{A} \cdot d\mathbf{r} \;=\; \int_C \{(3x^2 - 6yz)\mathbf{i} + (2y + 3xz)\mathbf{j} + (1 - 4xyz^2)\mathbf{k}\} \cdot (dx\,\mathbf{i} + dy\,\mathbf{j} + dz\,\mathbf{k})$$

$$= \int_C [(3x^2 - 6yz)\, dx + (2y + 3xz)\, dy + (1 - 4xyz^2)\, dz]$$

(a) If $x = t,\, y = t^2,\, z = t^3$, points $(0, 0, 0)$ and $(1, 1, 1)$ correspond to $t = 0$ and $t = 1$ respectively. Then

$$\int_C \mathbf{A} \cdot d\mathbf{r} \;=\; \int_{t=0}^{1} [\{3t^2 - 6(t^2)(t^3)\}\, dt + \{2t^2 + 3(t)(t^3)\}\, d(t^2) + \{1 - 4(t)(t^2)(t^3)^2\}\, d(t^3)]$$

$$= \int_{t=0}^{1} [(3t^2 - 6t^5)\, dt + (4t^3 + 6t^5)\, dt + (3t^2 - 12t^{11})\, dt] \;=\; 2$$

Another method.

Along C, $\mathbf{A} = (3t^2 - 6t^5)\mathbf{i} + (2t^2 + 3t^4)\mathbf{j} + (1 - 4t^9)\mathbf{k}$ and $\mathbf{r} = x\mathbf{i} + y\mathbf{j} + z\mathbf{k} = t\mathbf{i} + t^2\mathbf{j} + t^3\mathbf{k}$, $d\mathbf{r} = (\mathbf{i} + 2t\mathbf{j} + 3t^2\mathbf{k})\, dt$. Then

$$\int_C \mathbf{A} \cdot d\mathbf{r} \;=\; \int_{0}^{1} [(3t^2 - 6t^5)\, dt + (4t^3 + 6t^5)\, dt + (3t^2 - 12t^{11})\, dt] \;=\; 2$$

(b) Along the straight line from $(0, 0, 0)$ to $(0, 0, 1)$, $x = 0$, $y = 0$, $dx = 0$, $dy = 0$ while z varies from 0 to 1. Then the integral over this part of the path is

$$\int_{z=0}^{1} [\{3(0)^2 - 6(0)(z)\}0 + \{2(0) + 3(0)(z)\}0 + \{1 - 4(0)(0)(z^2)\}\, dz] \;=\; \int_{z=0}^{1} dz \;=\; 1$$

Along the straight line from $(0, 0, 1)$ to $(0, 1, 1)$, $x = 0$, $z = 1$, $dx = 0$, $dz = 0$ while y varies from 0 to 1. Then the integral over this part of the path is

$$\int_{y=0}^{1} [\{3(0)^2 - 6(y)(1)\}0 + \{2y + 3(0)(1)\}\, dy + \{1 - 4(0)(y)(1)^2\}0] \;=\; \int_{y=0}^{1} 2y\, dy \;=\; 1$$

Along the straight line from $(0, 1, 1)$ to $(1, 1, 1)$, $y = 1$, $z = 1$, $dy = 0$, $dz = 0$ while x varies from 0 to 1. Then the integral over this part of the path is

$$\int_{x=0}^{1} [\{3x^2 - 6(1)(1)\}\, dx + \{2(1) + 3x(1)\}0 + \{1 - 4x(1)(1)^2\}0] \;=\; \int_{x=0}^{1} (3x^2 - 6)\, dx \;=\; -5$$

Adding, $\displaystyle\int_C \mathbf{A} \cdot d\mathbf{r} = 1 + 1 - 5 = -3$.

(c) The straight line joining $(0, 0, 0)$ and $(1, 1, 1)$ is given in parametric form by $x = t$, $y = t$, $z = t$. Then

$$\int_C \mathbf{A} \cdot d\mathbf{r} \;=\; \int_{t=0}^1 [(3t^2 - 6t^2)\, dt \;+\; (2t + 3t^2)\, dt \;+\; (1 - 4t^4)\, dt] \;=\; 6/5$$

6.12. Find the work done in moving a particle once around an ellipse C in the xy plane, if the ellipse has center at the origin with semi-major and semi-minor axes 4 and 3 respectively, as indicated in Fig. 6-13, and if the force field is given by

$$\mathbf{F} \;=\; (3x - 4y + 2z)\mathbf{i} \;+\; (4x + 2y - 3z^2)\mathbf{j} \;+\; (2xz - 4y^2 + z^3)\mathbf{k}$$

In the plane $z = 0$, $\mathbf{F} = (3x - 4y)\mathbf{i} + (4x + 2y)\mathbf{j} - 4y^2\mathbf{k}$ and $d\mathbf{r} = dx\mathbf{i} + dy\mathbf{j}$ so that the work done is

$$\oint_C \mathbf{F} \cdot d\mathbf{r} \;=\; \int_C \{(3x - 4y)\mathbf{i} + (4x + 2y)\mathbf{j} - 4y^2\mathbf{k}\} \cdot (dx\mathbf{i} + dy\mathbf{j})$$

$$=\; \oint_C [(3x - 4y)\, dx + (4x + 2y)\, dy]$$

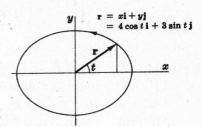

Fig. 6-13

Choose the parametric equations of the ellipse as $x = 4\cos t$, $y = 3\sin t$ where t varies from 0 to 2π (see Fig. 6-13). Then the line integral equals

$$\int_{t=0}^{2\pi} [\{3(4\cos t) - 4(3\sin t)\}\{-4\sin t\}\, dt \;+\; \{4(4\cos t) + 2(3\sin t)\}\{3\cos t\}\, dt]$$

$$=\; \int_{t=0}^{2\pi} (48 - 30\sin t \cos t)\, dt \;=\; (48t - 15\sin^2 t)\Big|_0^{2\pi} \;=\; 96\pi$$

In traversing C we have chosen the counterclockwise direction indicated in Fig. 6-13. We call this the *positive* direction, or say that C has been traversed in the *positive sense*. If C were traversed in the clockwise (negative) direction the value of the integral would be -96π.

6.13. Evaluate $\int_C y\, ds$ along the curve C given by $y = 2\sqrt{x}$ from $x = 3$ to $x = 24$.

Since $ds = \sqrt{dx^2 + dy^2} = \sqrt{1 + (y')^2}\, dx = \sqrt{1 + 1/x}\, dx$, we have

$$\int_C y\, ds \;=\; \int_3^{24} 2\sqrt{x}\sqrt{1 + 1/x}\, dx \;=\; 2\int_3^{24}\sqrt{x + 1}\, dx \;=\; \frac{4}{3}(x + 1)^{3/2}\Big|_3^{24} \;=\; 156$$

GREEN'S THEOREM IN THE PLANE

6.14. Prove Green's theorem in the plane if C is a closed curve which has the property that any straight line parallel to the coordinate axes cuts C in at most two points.

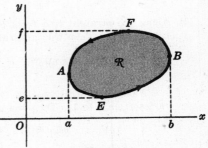

Fig. 6-14

Let the equations of the curves AEB and AFB (see adjoining Fig. 6-14) be $y = Y_1(x)$ and $y = Y_2(x)$ respectively. If $\mathcal{R}$ is the region bounded by C, we have

$$\iint_{\mathcal{R}} \frac{\partial P}{\partial y}\, dx\, dy \;=\; \int_{x=a}^b \left[\int_{y=Y_1(x)}^{Y_2(x)} \frac{\partial P}{\partial y}\, dy \right] dx$$

$$=\; \int_{x=a}^b P(x, y)\Big|_{y=Y_1(x)}^{Y_2(x)}\, dx \;=\; \int_a^b [P(x, Y_2) - P(x, Y_1)]\, dx$$

$$=\; -\int_a^b P(x, Y_1)\, dx \;-\; \int_b^a P(x, Y_2)\, dx \;=\; -\oint_C P\, dx$$

Then
$$\oint_C P\,dx = -\iint_{\mathcal{R}} \frac{\partial P}{\partial y}\,dx\,dy \tag{1}$$

Similarly let the equations of curves EAF and EBF be $x = X_1(y)$ and $x = X_2(y)$ respectively.
Then
$$\iint_{\mathcal{R}} \frac{\partial Q}{\partial x}\,dx\,dy = \int_{y=e}^{f}\left[\int_{x=X_1(y)}^{X_2(y)} \frac{\partial Q}{\partial x}\,dx\right]dy = \int_{e}^{f}[Q(X_2,y) - Q(X_1,y)]\,dy$$

$$= \int_{f}^{e} Q(X_1,y)\,dy + \int_{e}^{f} Q(X_2,y)\,dy = \oint_C Q\,dy$$

Then
$$\oint_C Q\,dy = \iint_{\mathcal{R}} \frac{\partial Q}{\partial x}\,dx\,dy \tag{2}$$

Adding (1) and (2), $\quad\displaystyle\oint_C [P\,dx + Q\,dy] = \iint_{\mathcal{R}}\left(\frac{\partial Q}{\partial x} - \frac{\partial P}{\partial y}\right)dx\,dy$.

6.15. Verify Green's theorem in the plane for

$$\oint_C [(2xy - x^2)\,dx + (x + y^2)\,dy]$$

where C is the closed curve of the region bounded by $y = x^2$ and $y^2 = x$.

The plane curves $y = x^2$ and $y^2 = x$ intersect at $(0,0)$ and $(1,1)$. The positive direction in traversing C is as shown in Fig. 6-15.

Along $y = x^2$, the line integral equals

$$\int_{x=0}^{1} [\{(2x)(x^2) - x^2\}\,dx + \{x + (x^2)^2\}\,d(x^2)] = \int_{0}^{1} (2x^3 + x^2 + 2x^5)\,dx = 7/6$$

Along $y^2 = x$ the line integral equals

$$\int_{y=1}^{0} [\{2(y^2)(y) - (y^2)^2\}\,d(y^2) + \{y^2 + y^2\}\,dy] = \int_{1}^{0} (4y^4 - 2y^5 + 2y^2)\,dy = -17/15$$

Then the required line integral $= 7/6 - 17/15 = 1/30$.

$$\iint_{\mathcal{R}}\left(\frac{\partial Q}{\partial x} - \frac{\partial P}{\partial y}\right)dx\,dy = \iint_{\mathcal{R}}\left\{\frac{\partial}{\partial x}(x+y^2) - \frac{\partial}{\partial y}(2xy - x^2)\right\}dx\,dy$$

$$= \iint_{\mathcal{R}} (1-2x)\,dx\,dy = \int_{x=0}^{1}\int_{y=x^2}^{\sqrt{x}} (1-2x)\,dy\,dx$$

$$= \int_{x=0}^{1} (y - 2xy)\Big|_{y=x^2}^{\sqrt{x}}\,dx = \int_{0}^{1} (x^{1/2} - 2x^{3/2} - x^2 + 2x^3)\,dx = 1/30$$

Hence Green's theorem is verified.

6.16. Extend the proof of Green's theorem in the plane given in Problem 6.14 to the curves C for which lines parallel to the coordinate axes may cut C in more than two points.

Consider a closed curve C such as shown in the adjoining Fig. 6-16, in which lines parallel to the axes may meet C in more than two points. By constructing line ST the region is divided into two regions $\mathcal{R}_1$ and $\mathcal{R}_2$ which are of the type considered in Problem 6.14 and for which Green's theorem applies, i.e.,

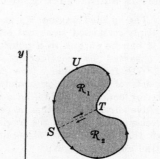

Fig. 6-16

(1) $\displaystyle\int_{STUS} [P\,dx + Q\,dy] = \iint_{\mathcal{R}_1}\left(\frac{\partial Q}{\partial x} - \frac{\partial P}{\partial y}\right) dx\,dy,$ (2) $\displaystyle\int_{SVTS} [P\,dx + Q\,dy] = \iint_{\mathcal{R}_2}\left(\frac{\partial Q}{\partial x} - \frac{\partial P}{\partial y}\right) dx\,dy$

Adding the left hand sides of (1) and (2), we have, omitting the integrand $P\,dx + Q\,dy$ in each case,

$$\int_{STUS} + \int_{SVTS} = \int_{ST} + \int_{TUS} + \int_{SVT} + \int_{TS} = \int_{TUS} + \int_{SVT} = \int_{TUSVT}$$

using the fact that $\displaystyle\int_{ST} = -\int_{TS}$.

Adding the right hand sides of (1) and (2), omitting the integrand, $\displaystyle\iint_{\mathcal{R}_1} + \iint_{\mathcal{R}_2} = \iint_{\mathcal{R}}$ where $\mathcal{R}$ consists of regions $\mathcal{R}_1$ and $\mathcal{R}_2$.

Then $\displaystyle\int_{TUSVT} [P\,dx + Q\,dy] = \iint_{\mathcal{R}}\left(\frac{\partial Q}{\partial x} - \frac{\partial P}{\partial y}\right) dx\,dy$ and the theorem is proved.

A region $\mathcal{R}$ such as considered here for which any closed curve lying in $\mathcal{R}$ can be continuously shrunk to a point without leaving $\mathcal{R}$, is called a *simply-connected region*. A region which is not simply-connected is called *multiply-connected*. We have shown here that Green's theorem in the plane applies to simply-connected regions bounded by closed curves. In Problem 6.19 the theorem is extended to multiply-connected regions.

For more complicated simply-connected regions it may be necessary to construct more lines, such as ST, to establish the theorem.

6.17. Show that the area bounded by a simple closed curve C is given by $\dfrac{1}{2}\displaystyle\oint_C [x\,dy - y\,dx]$.

In Green's theorem, put $P = -y$, $Q = x$. Then

$$\oint_C [x\,dy - y\,dx] = \iint_{\mathcal{R}}\left(\frac{\partial}{\partial x}(x) - \frac{\partial}{\partial y}(-y)\right) dx\,dy = 2\iint_{\mathcal{R}} dx\,dy = 2A$$

where A is the required area. Thus $A = \dfrac{1}{2}\displaystyle\oint_C [x\,dy - y\,dx]$.

6.18. Find the area of the ellipse $x = a\cos\theta$, $y = b\sin\theta$.

$$\text{Area} = \frac{1}{2}\oint_C [x\,dy - y\,dx] = \frac{1}{2}\int_0^{2\pi} [(a\cos\theta)(b\cos\theta)\,d\theta - (b\sin\theta)(-a\sin\theta)\,d\theta]$$

$$= \frac{1}{2}\int_0^{2\pi} ab(\cos^2\theta + \sin^2\theta)\,d\theta = \frac{1}{2}\int_0^{2\pi} ab\,d\theta = \pi ab$$

6.19. Show that Green's theorem in the plane is also valid for a multiply-connected region $\mathcal{R}$ such as shown in Fig. 6-17.

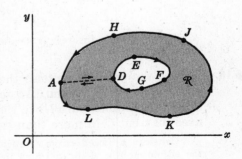

The shaded region $\mathcal{R}$, shown in the figure, is multiply-connected since not every closed curve lying in $\mathcal{R}$ can be shrunk to a point without leaving $\mathcal{R}$, as is observed by considering a curve surrounding $DEFGD$ for example. The boundary of $\mathcal{R}$, which consists of the exterior boundary $AHJKLA$ and the interior boundary $DEFGD$, is to be traversed in the positive direction, so that a person traveling in this direction always has the region on his left. It is seen that the positive directions are those indicated in the adjoining figure.

Fig. 6-17

In order to establish the theorem, construct a line, such as AD, called a *cross-cut*, connecting the exterior and interior boundaries. The region bounded by $ADEFGDALKJHA$ is simply-connected, and so Green's theorem is valid. Then

$$\oint_{ADEFGDALKJHA} [P\,dx + Q\,dy] = \iint_{\mathcal{R}} \left(\frac{\partial Q}{\partial x} - \frac{\partial P}{\partial y} \right) dx\,dy$$

But the integral on the left, leaving out the integrand, is equal to

$$\int_{AD} + \int_{DEFGD} + \int_{DA} + \int_{ALKJHA} = \int_{DEFGD} + \int_{ALKJHA}$$

since $\int_{AD} = -\int_{DA}$. Thus if C_1 is the curve $ALKJHA$, C_2 is the curve $DEFGD$ and C is the boundary of $\mathcal{R}$ consisting of C_1 and C_2 (traversed in the positive directions), then $\int_{C_1} + \int_{C_2} = \int_C$ and so

$$\oint_C [P\,dx + Q\,dy] = \iint_{\mathcal{R}} \left(\frac{\partial Q}{\partial x} - \frac{\partial P}{\partial y} \right) dx\,dy$$

INDEPENDENCE OF THE PATH

6.20. Let $P(x, y)$ and $Q(x, y)$ be continuous and have continuous first partial derivatives at each point of a simply connected region $\mathcal{R}$. Prove that a necessary and sufficient condition that $\oint_C [P\,dx + Q\,dy] = 0$ around every closed path C in $\mathcal{R}$ is that $\partial P/\partial y = \partial Q/\partial x$ identically in $\mathcal{R}$.

Sufficiency. Suppose $\partial P/\partial y = \partial Q/\partial x$. Then by Green's theorem,

$$\oint_C [P\,dx + Q\,dy] = \iint_{\mathcal{R}} \left(\frac{\partial Q}{\partial x} - \frac{\partial P}{\partial y} \right) dx\,dy = 0$$

where $\mathcal{R}$ is the region bounded by C.

Necessity.

Suppose $\oint_C [P\,dx + Q\,dy] = 0$ around every closed path C in $\mathcal{R}$ and that $\partial P/\partial y \neq \partial Q/\partial x$ at some point of $\mathcal{R}$. In particular suppose $\partial P/\partial y - \partial Q/\partial x > 0$ at the point (x_0, y_0).

By hypothesis $\partial P/\partial y$ and $\partial Q/\partial x$ are continuous in $\mathcal{R}$, so that there must be some region τ containing (x_0, y_0) as an interior point for which $\partial P/\partial y - \partial Q/\partial x > 0$. If Γ is the boundary of τ, then by Green's theorem

$$\oint_{\Gamma} [P\,dx + Q\,dy] = \iint_{\tau} \left(\frac{\partial Q}{\partial x} - \frac{\partial P}{\partial y} \right) dx\,dy > 0$$

contradicting the hypothesis that $\oint [P\,dx + Q\,dy] = 0$ for *all* closed curves in $\mathcal{R}$. Thus $\partial Q/\partial x - \partial P/\partial y$ cannot be positive.

Similarly we can show that $\partial Q/\partial x - \partial P/\partial y$ cannot be negative, and it follows that it must be identically zero, i.e. $\partial P/\partial y = \partial Q/\partial x$ identically in $\mathcal{R}$.

6.21. Let P and Q be defined as in Problem 6.20. Prove that a necessary and sufficient condition that $\int_A^B [P\,dx + Q\,dy]$ be independent of the path in $\mathcal{R}$ joining points A and B is that $\partial P/\partial y = \partial Q/\partial x$ identically in $\mathcal{R}$.

Sufficiency.   If $\partial P/\partial y = \partial Q/\partial x$, then by Problem 6.20,

$$\int_{ADBEA} [P\,dx + Q\,dy] = 0$$

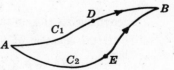

Fig. 6-18

(see Fig. 6-18). From this, omitting for brevity the integrand $P\,dx + Q\,dy$, we have

$$\int_{ADB} + \int_{BEA} = 0, \qquad \int_{ADB} = -\int_{BEA} = \int_{AEB} \qquad \text{and so} \qquad \int_{C_1} = \int_{C_2}$$

i.e. the integral is independent of the path.

6.22. (a) Prove that $\int_{(1,2)}^{(3,4)} [(6xy^2 - y^3)\,dx + (6x^2y - 3xy^2)\,dy]$ is independent of the path joining $(1,2)$ and $(3,4)$. (b) Evaluate the integral in (a).

(a) $P = 6xy^2 - y^3$, $Q = 6x^2y - 3xy^2$. Then $\partial P/\partial y = 12xy - 3y^2 = \partial Q/\partial x$ and by Problem 6.21 the line integral is independent of the path.

(b) **Method 1.**

Since the line integral is independent of the path, choose any path joining $(1,2)$ and $(3,4)$, for example that consisting of lines from $(1,2)$ to $(3,2)$ [along which $y = 2$, $dy = 0$] and then $(3,2)$ to $(3,4)$ [along which $x = 3$, $dx = 0$]. Then the required integral equals

$$\int_{x=1}^{3} (24x - 8)\,dx \;+\; \int_{y=2}^{4} (54y - 9y^2)\,dy \;=\; 80 + 156 \;=\; 236$$

Method 2.

Since $\dfrac{\partial P}{\partial y} = \dfrac{\partial Q}{\partial x}$, we must have (1) $\dfrac{\partial \phi}{\partial x} = 6xy^2 - y^3$, (2) $\dfrac{\partial \phi}{\partial y} = 6x^2y - 3xy^2$.

From (1), $\phi = 3x^2y^2 - xy^3 + f(y)$. From (2), $\phi = 3x^2y^2 - xy^3 + g(x)$. The only way in which these two expressions for ϕ are equal is if $f(y) = g(x) = c$, a constant. Hence $\phi = 3x^2y^2 - xy^3 + c$. Then

$$\int_{(1,2)}^{(3,4)} [(6xy^2 - y^3)\,dx + (6x^2y - 3xy^2)\,dy] \;=\; \int_{(1,2)}^{(3,4)} d(3x^2y^2 - xy^3 + c)$$

$$=\; 3x^2y^2 - xy^3 + c \Big|_{(1,2)}^{(3,4)} \;=\; 236$$

Note that in this evaluation the arbitrary constant c can be omitted.

We could also have noted by inspection that

$$(6xy^2 - y^3)\,dx + (6x^2y - 3xy^2)\,dy \;=\; (6xy^2\,dx + 6x^2y\,dy) - (y^3\,dx + 3xy^2\,dy)$$

$$=\; d(3x^2y^2) - d(xy^3) \;=\; d(3x^2y^2 - xy^3)$$

from which it is clear that $\phi = 3x^2y^2 - xy^3 + c$.

6.23. Evaluate $\oint [(x^2y \cos x + 2xy \sin x - y^2e^x)\,dx + (x^2 \sin x - 2ye^x)\,dy]$ around the hypocycloid $x^{2/3} + y^{2/3} = a^{2/3}$.

$P = x^2y \cos x + 2xy \sin x - y^2e^x$, $Q = x^2 \sin x - 2ye^x$.

Then $\partial P/\partial y = x^2 \cos x + 2x \sin x - 2ye^x = \partial Q/\partial x$, so that by Problem 6.20 the line integral around any closed path, in particular $x^{2/3} + y^{2/3} = a^{2/3}$, is zero.

SURFACE INTEGRALS

6.24. If γ is the angle between the normal line to any point (x, y, z) of a surface S and the positive z axis, prove that

$$|\sec \gamma| \;=\; \sqrt{1 + z_x^2 + z_y^2} \;=\; \frac{\sqrt{F_x^2 + F_y^2 + F_z^2}}{|F_z|}$$

according as the equation for S is $z = f(x, y)$ or $F(x, y, z) = 0$.

If the equation of S is $F(x, y, z) = 0$, a normal to S at (x, y, z) is $\nabla F = F_x\mathbf{i} + F_y\mathbf{j} + F_z\mathbf{k}$. Then

$$\nabla F \cdot \mathbf{k} \;=\; |\nabla F|\,|\mathbf{k}| \cos \gamma \qquad \text{or} \qquad F_z \;=\; \sqrt{F_x^2 + F_y^2 + F_z^2}\,\cos \gamma$$

from which $|\sec \gamma| \;=\; \dfrac{\sqrt{F_x^2 + F_y^2 + F_z^2}}{|F_z|}$ as required.

In case the equation is $z = f(x, y)$, we can write $F(x, y, z) = z - f(x, y) = 0$, from which $F_x = -z_x$, $F_y = -z_y$, $F_z = 1$ and we find $|\sec \gamma| = \sqrt{1 + z_x^2 + z_y^2}$.

6.25. Evaluate $\iint\limits_{S} U(x,y,z)\,dS$ where S is the surface of the paraboloid $z = 2 - (x^2 + y^2)$ above the xy plane and $U(x,y,z)$ is equal to (a) 1, (b) $x^2 + y^2$, (c) $3z$. Give a physical interpretation in each case.

The required integral is equal to

$$\iint\limits_{\mathcal{R}} U(x,y,z)\sqrt{1 + z_x^2 + z_y^2}\,dx\,dy \qquad (1)$$

where $\mathcal{R}$ is the projection of S on the xy plane given by $x^2 + y^2 = 2$, $z = 0$.

Since $z_x = -2x$, $z_y = -2y$, (1) can be written

$$\iint\limits_{\mathcal{R}} U(x,y,z)\sqrt{1 + 4x^2 + 4y^2}\,dx\,dy \qquad (2)$$

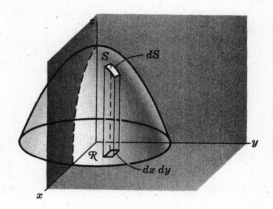

Fig. 6-19

(a) If $U(x,y,z) = 1$, (2) becomes

$$\iint\limits_{\mathcal{R}} \sqrt{1 + 4x^2 + 4y^2}\,dx\,dy$$

To evaluate this, transform to polar coordinates (ρ, ϕ). Then the integral becomes

$$\int_{\phi=0}^{2\pi}\int_{\rho=0}^{\sqrt{2}} \sqrt{1+4\rho^2}\,\rho\,d\rho\,d\phi \;=\; \int_{\phi=0}^{2\pi} \frac{1}{12}(1+4\rho^2)^{3/2}\Big|_{\rho=0}^{\sqrt{2}}\,d\phi \;=\; \frac{13\pi}{3}$$

Physically this could represent the surface area of S, or the mass of S assuming unit density.

(b) If $U(x,y,z) = x^2 + y^2$, (2) becomes $\iint\limits_{\mathcal{R}} (x^2 + y^2)\sqrt{1 + 4x^2 + 4y^2}\,dx\,dy$ or in polar coordinates

$$\int_{\phi=0}^{2\pi}\int_{\rho=0}^{\sqrt{2}} \rho^3\sqrt{1+4\rho^2}\,d\rho\,d\phi \;=\; \frac{149\pi}{30}$$

where the integration with respect to ρ is accomplished by the substitution $\sqrt{1+4\rho^2} = u$.

Physically this could represent the moment of inertia of S about the z axis assuming unit density, or the mass of S assuming a density $= x^2 + y^2$.

(c) If $U(x,y,z) = 3z$, (2) becomes

$$\iint\limits_{\mathcal{R}} 3z\sqrt{1+4x^2+4y^2}\,dx\,dy \;=\; \iint\limits_{\mathcal{R}} 3\{2 - (x^2 + y^2)\}\sqrt{1+4x^2+4y^2}\,dx\,dy$$

or in polar coordinates,

$$\int_{\phi=0}^{2\pi}\int_{\rho=0}^{\sqrt{2}} 3\rho(2-\rho^2)\sqrt{1+4\rho^2}\,d\rho\,d\phi \;=\; \frac{111\pi}{10}$$

Physically this could represent the mass of S assuming a density $= 3z$, or three times the first moment of S about the xy plane.

6.26. Find the surface area of a hemisphere of radius a cut off by a cylinder having this radius as diameter.

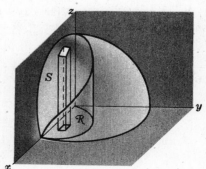

Fig. 6-20

Equations for the hemisphere and cylinder (see Fig. 6-20) are given respectively by $x^2 + y^2 + z^2 = a^2$ (or $z = \sqrt{a^2 - x^2 - y^2}$) and $(x - a/2)^2 + y^2 = a^2/4$ (or $x^2 + y^2 = ax$).

Since

$$z_x = \frac{-x}{\sqrt{a^2 - x^2 - y^2}} \quad \text{and} \quad z_y = \frac{-y}{\sqrt{a^2 - x^2 - y^2}}$$

we have

$$\text{Required surface area} \;=\; 2\iint\limits_{R}\sqrt{1+z_x^2+z_y^2}\,dx\,dy \;=\; 2\iint\limits_{R}\frac{a}{\sqrt{a^2-x^2-y^2}}\,dx\,dy$$

Two methods of evaluation are possible.

Method 1.   Using polar coordinates.

Since $x^2+y^2=ax$ in polar coordinates is $\rho = a\cos\phi$, the integral becomes

$$2\int_{\phi=0}^{\pi/2}\int_{\rho=0}^{a\cos\phi}\frac{a}{\sqrt{a^2-\rho^2}}\,\rho\,d\rho\,d\phi \;=\; 2a\int_{\phi=0}^{\pi/2}-\sqrt{a^2-\rho^2}\,\Big|_{\rho=0}^{a\cos\phi}\,d\phi$$

$$=\; 2a^2\int_0^{\pi/2}(1-\sin\phi)\,d\phi \;=\; (\pi-2)a^2$$

Method 2.   The integral is equal to

$$2\int_{x=0}^{a}\int_{y=0}^{\sqrt{ax-x^2}}\frac{a}{\sqrt{a^2-x^2-y^2}}\,dy\,dx \;=\; 2a\int_{x=0}^{a}\sin^{-1}\frac{y}{\sqrt{ax-x^2}}\,\Big|_{y=0}^{\sqrt{ax-x^2}}dx$$

$$=\; 2a\int_0^a \sin^{-1}\sqrt{\frac{x}{a+x}}\,dx$$

Letting $x = a\tan^2\theta$, this integral becomes

$$4a^2\int_0^{\pi/4}\theta\tan\theta\sec^2\theta\,d\theta \;=\; 4a^2\left\{\tfrac{1}{2}\theta\tan^2\theta\,\Big|_0^{\pi/4}-\tfrac{1}{2}\int_0^{\pi/4}\tan^2\theta\,d\theta\right\}$$

$$=\; 2a^2\left\{\theta\tan^2\theta\,\Big|_0^{\pi/4}-\int_0^{\pi/4}(\sec^2\theta-1)\,d\theta\right\}$$

$$=\; 2a^2\left\{\pi/4-(\tan\theta-\theta)\,\Big|_0^{\pi/4}\right\} \;=\; (\pi-2)a^2$$

Note that the above integrals are *improper* and should actually be treated by appropriate limiting procedures.

6.27. Find the centroid of the surface in Problem 6.25.

By symmetry, $\bar{x}=\bar{y}=0$ and $\quad \bar{z} \;=\; \dfrac{\displaystyle\iint\limits_{S}z\,dS}{\displaystyle\iint\limits_{S}dS} \;=\; \dfrac{\displaystyle\iint\limits_{R}z\sqrt{1+4x^2+4y^2}\,dx\,dy}{\displaystyle\iint\limits_{R}\sqrt{1+4x^2+4y^2}\,dx\,dy}.$

The numerator and denominator can be obtained from the results of Problem 6.25(c) and 6.25(a) respectively, and we thus have $\bar{z} = \dfrac{37\pi/10}{13\pi/3} = \dfrac{111}{130}$.

6.28. Evaluate $\displaystyle\iint\limits_{S}\mathbf{A}\cdot\mathbf{n}\,dS$, where $\mathbf{A} = xy\mathbf{i} - x^2\mathbf{j} + (x+z)\mathbf{k}$, S is that portion of the plane $2x+2y+z=6$ included in the first octant, and $\mathbf{n}$ is a unit normal to S.

A normal to S is $\nabla(2x+2y+z-6) = 2\mathbf{i}+2\mathbf{j}+\mathbf{k}$, and so $\mathbf{n} = \dfrac{2\mathbf{i}+2\mathbf{j}+\mathbf{k}}{\sqrt{2^2+2^2+1^2}} = \dfrac{2\mathbf{i}+2\mathbf{j}+\mathbf{k}}{3}$. Then

$$\mathbf{A}\cdot\mathbf{n} = \{xy\mathbf{i}-x^2\mathbf{j}+(x+z)\mathbf{k}\}\cdot\left(\frac{2\mathbf{i}+2\mathbf{j}+\mathbf{k}}{3}\right)$$

$$= \frac{2xy-2x^2+(x+z)}{3}$$

$$= \frac{2xy-2x^2+(x+6-2x-2y)}{3}$$

$$= \frac{2xy-2x^2-x-2y+6}{3}$$

The required surface integral is therefore

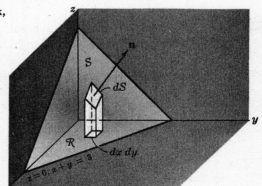

Fig. 6-21

$$\iint_S \left(\frac{2xy - 2x^2 - x - 2y + 6}{3}\right) dS = \iint_{\mathcal{R}} \left(\frac{2xy - 2x^2 - x - 2y + 6}{3}\right) \sqrt{1 + z_x^2 + z_y^2}\, dx\, dy$$

$$= \iint_{\mathcal{R}} \left(\frac{2xy - 2x^2 - x - 2y + 6}{3}\right) \sqrt{1^2 + 2^2 + 2^2}\, dx\, dy$$

$$= \int_{x=0}^{3} \int_{y=0}^{3-x} (2xy - 2x^2 - x - 2y + 6)\, dy\, dx$$

$$= \int_{x=0}^{3} (xy^2 - 2x^2y - xy - y^2 + 6y)\Big|_0^{3-x} dx = 27/4$$

6.29. In dealing with surface integrals we have restricted ourselves to surfaces which are two-sided.   Give an example of a surface which is not two-sided.

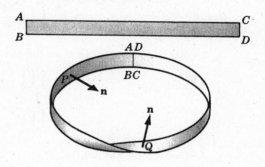

Take a strip of paper such as $ABCD$ as shown in the adjoining Fig. 6-22. Twist the strip so that points A and B fall on D and C respectively, as in the adjoining figure. If **n** is the positive normal at point P of the surface, we find that as **n** moves around the surface it reverses its original direction when it reaches P again. If we tried to color only one side of the surface we would find the whole thing colored. This surface, called a *Moebius strip*, is an example of a one-sided surface. This is sometimes called a *non-orientable* surface. A two-sided surface is *orientable*.

Fig. 6-22

THE DIVERGENCE THEOREM

6.30. Prove the divergence theorem.

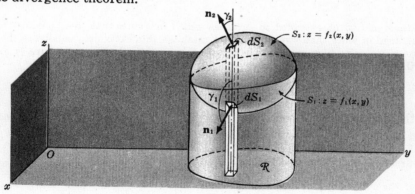

Fig. 6-23

Let S be a closed surface which is such that any line parallel to the coordinate axes cuts S in at most two points. Assume the equations of the lower and upper portions, S_1 and S_2, to be $z = f_1(x, y)$ and $z = f_2(x, y)$ respectively. Denote the projection of the surface on the xy plane by $\mathcal{R}$. Consider

$$\iiint_V \frac{\partial A_3}{\partial z}\, dV = \iiint_V \frac{\partial A_3}{\partial z}\, dz\, dy\, dx = \iint_{\mathcal{R}} \left[\int_{z=f_1(x,y)}^{f_2(x,y)} \frac{\partial A_3}{\partial z}\, dz\right] dy\, dx$$

$$= \iint_{\mathcal{R}} A_3(x, y, z)\Big|_{z=f_1}^{f_2} dy\, dx = \iint_{\mathcal{R}} [A_3(x, y, f_2) - A_3(x, y, f_1)]\, dy\, dx$$

For the upper portion S_2, $dy\, dx = \cos \gamma_2\, dS_2 = \mathbf{k} \cdot \mathbf{n}_2\, dS_2$ since the normal $\mathbf{n}_2$ to S_2 makes an acute angle γ_2 with $\mathbf{k}$.

For the lower portion S_1, $dy\,dx = -\cos\gamma_1\,dS_1 = -\mathbf{k}\cdot\mathbf{n}_1\,dS_1$ since the normal $\mathbf{n}_1$ to S_1 makes an obtuse angle γ_1 with $\mathbf{k}$.

Then

$$\iint_{\mathcal{R}} A_3(x,y,f_2)\,dy\,dx = \iint_{S_2} A_3\mathbf{k}\cdot\mathbf{n}_2\,dS_2$$

$$\iint_{\mathcal{R}} A_3(x,y,f_1)\,dy\,dx = -\iint_{S_1} A_3\mathbf{k}\cdot\mathbf{n}_1\,dS_1$$

and

$$\iint_{\mathcal{R}} A_3(x,y,f_2)\,dy\,dx - \iint_{\mathcal{R}} A_3(x,y,f_1)\,dy\,dx = \iint_{S_2} A_3\mathbf{k}\cdot\mathbf{n}_2\,dS_2 + \iint_{S_1} A_3\mathbf{k}\cdot\mathbf{n}_1\,dS_1$$

$$= \iint_{S} A_3\mathbf{k}\cdot\mathbf{n}\,dS$$

so that

$$\iiint_{V} \frac{\partial A_3}{\partial z}\,dV = \iint_{S} A_3\mathbf{k}\cdot\mathbf{n}\,dS \tag{1}$$

Similarly, by projecting S on the other coordinate planes,

$$\iiint_{V} \frac{\partial A_1}{\partial x}\,dV = \iint_{S} A_1\mathbf{i}\cdot\mathbf{n}\,dS \tag{2}$$

$$\iiint_{V} \frac{\partial A_2}{\partial y}\,dV = \iint_{S} A_2\mathbf{j}\cdot\mathbf{n}\,dS \tag{3}$$

Adding (1), (2) and (3),

$$\iiint_{V} \left(\frac{\partial A_1}{\partial x} + \frac{\partial A_2}{\partial y} + \frac{\partial A_3}{\partial z}\right)dV = \iint_{S} (A_1\mathbf{i} + A_2\mathbf{j} + A_3\mathbf{k})\cdot\mathbf{n}\,dS$$

or

$$\iiint_{V} \nabla\cdot\mathbf{A}\,dV = \iint_{S} \mathbf{A}\cdot\mathbf{n}\,dS$$

The theorem can be extended to surfaces which are such that lines parallel to the coordinate axes meet them in more than two points. To establish this extension, subdivide the region bounded by S into subregions whose surfaces do satisfy this condition. The procedure is analogous to that used in Green's theorem for the plane.

6.31. Verify the divergence theorem for $\mathbf{A} = (2x-z)\mathbf{i} + x^2y\mathbf{j} - xz^2\mathbf{k}$ taken over the region bounded by $x=0$, $x=1$, $y=0$, $y=1$, $z=0$, $z=1$.

We first evaluate $\iint_{S} \mathbf{A}\cdot\mathbf{n}\,dS$ where S is the surface of the cube in Fig. 6-24.

Face DEFG: $\mathbf{n} = \mathbf{i}$, $x = 1$. Then

$$\iint_{DEFG} \mathbf{A}\cdot\mathbf{n}\,dS = \int_0^1\int_0^1 \{(2-z)\mathbf{i} + \mathbf{j} - z^2\mathbf{k}\}\cdot\mathbf{i}\,dy\,dz$$

$$= \int_0^1\int_0^1 (2-z)\,dy\,dz = 3/2$$

Face ABCO: $\mathbf{n} = -\mathbf{i}$, $x = 0$. Then

$$\iint_{ABCO} \mathbf{A}\cdot\mathbf{n}\,dS = \int_0^1\int_0^1 (-z\mathbf{i})\cdot(-\mathbf{i})\,dy\,dz$$

$$= \int_0^1\int_0^1 z\,dy\,dz = 1/2$$

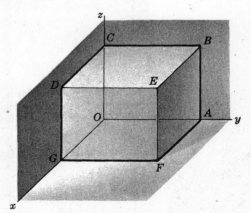

Fig. 6-24

Face ABEF: $\mathbf{n} = \mathbf{j}$, $y = 1$.   Then

$$\iint\limits_{ABEF} \mathbf{A} \cdot \mathbf{n}\, dS \;=\; \int_0^1 \int_0^1 \{(2x - z)\mathbf{i} + x^2\mathbf{j} - xz^2\mathbf{k}\} \cdot \mathbf{j}\, dx\, dz \;=\; \int_0^1 \int_0^1 x^2\, dx\, dz \;=\; 1/3$$

Face OGDC: $\mathbf{n} = -\mathbf{j}$, $y = 0$.   Then

$$\iint\limits_{OGDC} \mathbf{A} \cdot \mathbf{n}\, dS \;=\; \int_0^1 \int_0^1 \{(2x - z)\mathbf{i} - xz^2\mathbf{k}\} \cdot (-\mathbf{j})\, dx\, dz \;=\; 0$$

Face BCDE: $\mathbf{n} = \mathbf{k}$, $z = 1$.   Then

$$\iint\limits_{BCDE} \mathbf{A} \cdot \mathbf{n}\, dS \;=\; \int_0^1 \int_0^1 \{(2x - 1)\mathbf{i} + x^2 y\mathbf{j} - x\mathbf{k}\} \cdot \mathbf{k}\, dx\, dy \;=\; \int_0^1 \int_0^1 -x\, dx\, dy \;=\; -1/2$$

Face AFGO: $\mathbf{n} = -\mathbf{k}$, $z = 0$.   Then

$$\iint\limits_{AFGO} \mathbf{A} \cdot \mathbf{n}\, dS \;=\; \int_0^1 \int_0^1 \{2x\mathbf{i} - x^2 y\mathbf{j}\} \cdot (-\mathbf{k})\, dx\, dy \;=\; 0$$

Adding, $\displaystyle\iint\limits_{S} \mathbf{A} \cdot \mathbf{n}\, dS = \tfrac{3}{2} + \tfrac{1}{2} + \tfrac{1}{3} + 0 - \tfrac{1}{2} + 0 = \tfrac{11}{6}$.   Since

$$\iiint\limits_{V} \nabla \cdot \mathbf{A}\, dV \;=\; \int_0^1 \int_0^1 \int_0^1 (2 + x^2 - 2xz)\, dx\, dy\, dz \;=\; \frac{11}{6}$$

the divergence theorem is verified in this case.

6.32.   Evaluate $\displaystyle\iint\limits_{S} \mathbf{r} \cdot \mathbf{n}\, dS$, where S is a closed surface.

By the divergence theorem,

$$\begin{aligned}
\iint\limits_{S} \mathbf{r} \cdot \mathbf{n}\, dS &= \iiint\limits_{V} \nabla \cdot \mathbf{r}\, dV \\[2mm]
&= \iiint\limits_{V} \left(\frac{\partial}{\partial x}\mathbf{i} + \frac{\partial}{\partial y}\mathbf{j} + \frac{\partial}{\partial z}\mathbf{k} \right) \cdot (x\mathbf{i} + y\mathbf{j} + z\mathbf{k})\, dV \\[2mm]
&= \iiint\limits_{V} \left(\frac{\partial x}{\partial x} + \frac{\partial y}{\partial y} + \frac{\partial z}{\partial z} \right) dV \;=\; 3 \iiint\limits_{V} dV \;=\; 3V
\end{aligned}$$

where V is the volume enclosed by S.

6.33.   Evaluate $\displaystyle\iint\limits_{S} [xz^2\, dy\, dz + (x^2 y - z^3)\, dz\, dx + (2xy + y^2 z)\, dx\, dy]$ where S is the entire surface of the hemispherical region bounded by $z = \sqrt{a^2 - x^2 - y^2}$ and $z = 0$ (a) by the divergence theorem (Green's theorem in space), (b) directly.

(a)   Since $dy\, dz = dS \cos\alpha$, $dz\, dx = dS \cos\beta$, $dx\, dy = dS \cos\gamma$,   the integral can be written

$$\iint\limits_{S} \{xz^2 \cos\alpha + (x^2 y - z^3)\cos\beta + (2xy + y^2 z)\cos\gamma\}\, dS \;=\; \iint\limits_{S} \mathbf{A} \cdot \mathbf{n}\, dS$$

where $\mathbf{A} = xz^2\mathbf{i} + (x^2 y - z^3)\mathbf{j} + (2xy + y^2 z)\mathbf{k}$ and $\mathbf{n} = (\cos\alpha)\mathbf{i} + (\cos\beta)\mathbf{j} + (\cos\gamma)\mathbf{k}$,   the outward drawn unit normal.

Then by the divergence theorem the integral equals

$$\iiint\limits_{V} \nabla \cdot \mathbf{A}\, dV \;=\; \iiint\limits_{V} \left\{ \frac{\partial}{\partial x}(xz^2) + \frac{\partial}{\partial y}(x^2 y - z^3) + \frac{\partial}{\partial z}(2xy + y^2 z) \right\} dV \;=\; \iiint\limits_{V} (x^2 + y^2 + z^2)\, dV$$

where V is the region bounded by the hemisphere and the xy plane.

By use of spherical coordinates, this integral is equal to

$$4 \int_{\phi=0}^{\pi/2} \int_{\theta=0}^{\pi/2} \int_{r=0}^{a} r^2 \cdot r^2 \sin \theta \, dr \, d\theta \, d\phi \;=\; \frac{2\pi a^5}{5}$$

(b) If S_1 is the convex surface of the hemispherical region and S_2 is the base $(z = 0)$, then

$$\iint_{S_1} xz^2 \, dy \, dz = \int_{y=-a}^{a} \int_{z=0}^{\sqrt{a^2-y^2}} z^2 \sqrt{a^2-y^2-z^2} \, dz \, dy \;-\; \int_{y=-a}^{a} \int_{z=0}^{\sqrt{a^2-y^2}} -z^2 \sqrt{a^2-y^2-z^2} \, dz \, dy$$

$$\iint_{S_1} (x^2 y - z^3) \, dz \, dx \;=\; \int_{x=-a}^{a} \int_{z=0}^{\sqrt{a^2-x^2}} \{ x^2 \sqrt{a^2 - x^2 - z^2} - z^3 \} \, dz \, dx$$

$$\qquad\qquad -\; \int_{x=-a}^{a} \int_{z=0}^{\sqrt{a^2-x^2}} \{ -x^2 \sqrt{a^2 - x^2 - z^2} - z^3 \} \, dz \, dx$$

$$\iint_{S_1} (2xy + y^2 z) \, dx \, dy \;=\; \int_{x=-a}^{a} \int_{y=-\sqrt{a^2-x^2}}^{\sqrt{a^2-x^2}} \{ 2xy + y^2 \sqrt{a^2 - x^2 - y^2} \} \, dy \, dx$$

$$\iint_{S_2} xz^2 \, dy \, dz \;=\; 0, \qquad \iint_{S_2} (x^2 y - z^3) \, dz \, dx \;=\; 0$$

$$\iint_{S_2} (2xy + y^2 z) \, dx \, dy \;=\; \iint_{S_2} \{ 2xy + y^2(0) \} \, dx \, dy \;=\; \int_{x=-a}^{a} \int_{y=-\sqrt{a^2-x^2}}^{\sqrt{a^2-x^2}} 2xy \, dy \, dx \;=\; 0$$

By addition of the above, we obtain

$$4 \int_{y=0}^{a} \int_{z=0}^{\sqrt{a^2-y^2}} z^2 \sqrt{a^2-y^2-z^2} \, dz \, dy \;+\; 4 \int_{x=0}^{a} \int_{z=0}^{\sqrt{a^2-x^2}} x^2 \sqrt{a^2-x^2-z^2} \, dz \, dx$$

$$+\; 4 \int_{x=0}^{a} \int_{y=0}^{\sqrt{a^2-x^2}} y^2 \sqrt{a^2-x^2-y^2} \, dy \, dx$$

Since by symmetry all these integrals are equal, the result is, on using polar coordinates,

$$12 \int_{x=0}^{a} \int_{y=0}^{\sqrt{a^2-x^2}} y^2 \sqrt{a^2-x^2-y^2} \, dy \, dx \;=\; 12 \int_{\phi=0}^{\pi/2} \int_{\rho=0}^{a} \rho^2 \sin^2 \phi \, \sqrt{a^2-\rho^2} \, \rho \, d\rho \, d\phi \;=\; \frac{2\pi a^5}{5}$$

STOKES' THEOREM

6.34. Prove Stokes' theorem.

Let S be a surface which is such that its projections on the xy, yz and xz planes are regions bounded by simple closed curves, as indicated in Fig. 6-25. Assume S to have representation $z = f(x, y)$ or $x = g(y, z)$ or $y = h(x, z)$, where f, g, h are single-valued, continuous and differentiable functions. We must show that

$$\iint_S (\nabla \times \mathbf{A}) \cdot \mathbf{n} \, dS$$

$$= \iint_S [\nabla \times (A_1 \mathbf{i} + A_2 \mathbf{j} + A_3 \mathbf{k})] \cdot \mathbf{n} \, dS$$

$$= \int_C \mathbf{A} \cdot d\mathbf{r}$$

where C is the boundary of S.

Consider first $\displaystyle\iint_S [\nabla \times (A_1 \mathbf{i})] \cdot \mathbf{n} \, dS$.

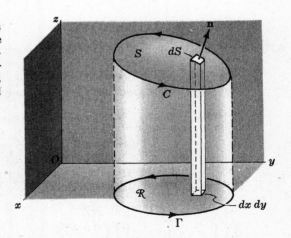

Fig. 6-25

Since $\nabla \times (A_1 \mathbf{i}) = \begin{vmatrix} \mathbf{i} & \mathbf{j} & \mathbf{k} \\ \frac{\partial}{\partial x} & \frac{\partial}{\partial y} & \frac{\partial}{\partial z} \\ A_1 & 0 & 0 \end{vmatrix} = \frac{\partial A_1}{\partial z} \mathbf{j} - \frac{\partial A_1}{\partial y} \mathbf{k},$

$$[\nabla \times (A_1 \mathbf{i})] \cdot \mathbf{n} \, dS = \left(\frac{\partial A_1}{\partial z} \mathbf{n} \cdot \mathbf{j} - \frac{\partial A_1}{\partial y} \mathbf{n} \cdot \mathbf{k} \right) dS \tag{1}$$

If $z = f(x, y)$ is taken as the equation of S, then the position vector to any point of S is $\mathbf{r} = x\mathbf{i} + y\mathbf{j} + z\mathbf{k} = x\mathbf{i} + y\mathbf{j} + f(x, y)\mathbf{k}$ so that $\frac{\partial \mathbf{r}}{\partial y} = \mathbf{j} + \frac{\partial z}{\partial y}\mathbf{k} = \mathbf{j} + \frac{\partial f}{\partial y}\mathbf{k}$. But $\frac{\partial \mathbf{r}}{\partial y}$ is a vector tangent to S and thus perpendicular to $\mathbf{n}$, so that

$$\mathbf{n} \cdot \frac{\partial \mathbf{r}}{\partial y} = \mathbf{n} \cdot \mathbf{j} + \frac{\partial z}{\partial y} \mathbf{n} \cdot \mathbf{k} = 0 \quad \text{or} \quad \mathbf{n} \cdot \mathbf{j} = -\frac{\partial z}{\partial y} \mathbf{n} \cdot \mathbf{k}$$

Substitute in (1) to obtain

$$\left(\frac{\partial A_1}{\partial z} \mathbf{n} \cdot \mathbf{j} - \frac{\partial A_1}{\partial y} \mathbf{n} \cdot \mathbf{k} \right) dS = \left(-\frac{\partial A_1}{\partial z} \frac{\partial z}{\partial y} \mathbf{n} \cdot \mathbf{k} - \frac{\partial A_1}{\partial y} \mathbf{n} \cdot \mathbf{k} \right) dS$$

or $$[\nabla \times (A_1 \mathbf{i})] \cdot \mathbf{n} \, dS = -\left(\frac{\partial A_1}{\partial y} + \frac{\partial A_1}{\partial z} \frac{\partial z}{\partial y} \right) \mathbf{n} \cdot \mathbf{k} \, dS \tag{2}$$

Now on S, $A_1(x, y, z) = A_1[x, y, f(x, y)] = F(x, y)$; hence $\frac{\partial A_1}{\partial y} + \frac{\partial A_1}{\partial z} \frac{\partial z}{\partial y} = \frac{\partial F}{\partial y}$ and (2) becomes

$$[\nabla \times (A_1 \mathbf{i})] \cdot \mathbf{n} \, dS = -\frac{\partial F}{\partial y} \mathbf{n} \cdot \mathbf{k} \, dS = -\frac{\partial F}{\partial y} \, dx \, dy$$

Then $$\iint_S [\nabla \times (A_1 \mathbf{i})] \cdot \mathbf{n} \, dS = \iint_{\mathcal{R}} -\frac{\partial F}{\partial y} \, dx \, dy$$

where $\mathcal{R}$ is the projection of S on the xy plane. By Green's theorem for the plane the last integral equals $\oint_{\Gamma} F \, dx$ where Γ is the boundary of $\mathcal{R}$. Since at each point (x, y) of Γ the value of F is the same as the value of A_1 at each point (x, y, z) of C, and since dx is the same for both curves, we must have

$$\oint_{\Gamma} F \, dx = \oint_C A_1 \, dx$$

or $$\iint_S [\nabla \times (A_1 \mathbf{i})] \cdot \mathbf{n} \, dS = \oint_C A_1 \, dx$$

Similarly, by projections on the other coordinate planes,

$$\iint_S [\nabla \times (A_2 \mathbf{j})] \cdot \mathbf{n} \, dS = \oint_C A_2 \, dy, \qquad \iint_S [\nabla \times (A_3 \mathbf{k})] \cdot \mathbf{n} \, dS = \oint_C A_3 \, dz$$

Thus by addition,

$$\iint_S (\nabla \times \mathbf{A}) \cdot \mathbf{n} \, dS = \oint_C \mathbf{A} \cdot d\mathbf{r}$$

The theorem is also valid for surfaces S which may not satisfy the restrictions imposed above. For assume that S can be subdivided into surfaces $S_1, S_2, \ldots, S_k$ with boundaries $C_1, C_2, \ldots, C_k$ which do satisfy the restrictions. Then Stokes' theorem holds for each such surface. Adding these surface integrals, the total surface integral over S is obtained. Adding the corresponding line integrals over $C_1, C_2, \ldots, C_k$, the line integral over C is obtained.

6.35. Verify Stokes' theorem for $\mathbf{A} = 3y\mathbf{i} - xz\mathbf{j} + yz^2\mathbf{k}$, where S is the surface of the paraboloid $2z = x^2 + y^2$ bounded by $z = 2$ and C is its boundary.

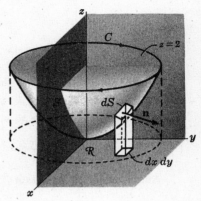

Fig. 6-26

The boundary C of S is a circle with equations $x^2 + y^2 = 4$, $z = 2$ and parametric equations $x = 2\cos t$, $y = 2\sin t$, $z = 2$, where $0 \leqq t < 2\pi$. Then

$$\oint_C \mathbf{A} \cdot d\mathbf{r} = \oint_C [3y\,dx - xz\,dy + yz^2\,dz]$$

$$= \int_{2\pi}^0 [3(2\sin t)(-2\sin t)\,dt - (2\cos t)(2)(2\cos t)\,dt]$$

$$= \int_0^{2\pi} (12\sin^2 t + 8\cos^2 t)\,dt = 20\pi$$

Also,

$$\nabla \times \mathbf{A} = \begin{vmatrix} \mathbf{i} & \mathbf{j} & \mathbf{k} \\ \dfrac{\partial}{\partial x} & \dfrac{\partial}{\partial y} & \dfrac{\partial}{\partial z} \\ 3y & -xz & yz^2 \end{vmatrix} = (z^2 + x)\mathbf{i} - (z + 3)\mathbf{k}$$

and $$\mathbf{n} = \frac{\nabla(x^2 + y^2 - 2z)}{|\nabla(x^2 + y^2 - 2z)|} = \frac{x\mathbf{i} + y\mathbf{j} - \mathbf{k}}{\sqrt{x^2 + y^2 + 1}}$$

Then $$\iint_S (\nabla \times \mathbf{A}) \cdot \mathbf{n}\,dS = \iint_R (\nabla \times A) \cdot \mathbf{n} \frac{dx\,dy}{|\mathbf{n} \cdot \mathbf{k}|} = \iint_R (xz^2 + x^2 + z + 3)\,dx\,dy$$

$$= \iint_R \left\{ x\left(\frac{x^2 + y^2}{2}\right)^2 + x^2 + \frac{x^2 + y^2}{2} + 3 \right\} dx\,dy$$

In polar coordinates this becomes

$$\int_{\phi=0}^{2\pi} \int_{\rho=0}^2 \{(\rho\cos\phi)(\rho^4/2) + \rho^2\cos^2\phi + \rho^2/2 + 3\}\,\rho\,d\rho\,d\phi = 20\pi$$

6.36. Prove that a necessary and sufficient condition that $\oint_C \mathbf{A} \cdot d\mathbf{r} = 0$ for every closed curve C is that $\nabla \times \mathbf{A} = 0$ identically.

Sufficiency.   Suppose $\nabla \times \mathbf{A} = 0$.   Then by Stokes' theorem

$$\oint_C \mathbf{A} \cdot d\mathbf{r} = \iint_S (\nabla \times \mathbf{A}) \cdot \mathbf{n}\,dS = 0$$

Necessity.

Suppose $\oint_C \mathbf{A} \cdot d\mathbf{r} = 0$ around every closed path C, and assume $\nabla \times \mathbf{A} \neq 0$ at some point P. Then assuming $\nabla \times \mathbf{A}$ is continuous there will be a region with P as an interior point, where $\nabla \times \mathbf{A} \neq 0$. Let S be a surface contained in this region whose normal $\mathbf{n}$ at each point has the same direction as $\nabla \times \mathbf{A}$, i.e. $\nabla \times \mathbf{A} = \alpha\mathbf{n}$ where α is a positive constant. Let C be the boundary of S. Then by Stokes' theorem

$$\oint_C \mathbf{A} \cdot d\mathbf{r} = \iint_S (\nabla \times \mathbf{A}) \cdot \mathbf{n}\,dS = \alpha \iint_S \mathbf{n} \cdot \mathbf{n}\,dS > 0$$

which contradicts the hypothesis that $\oint_C \mathbf{A} \cdot d\mathbf{r} = 0$ and shows that $\nabla \times \mathbf{A} = 0$.

It follows that $\nabla \times \mathbf{A} = 0$ is also a necessary and sufficient condition for a line integral $\int_{P_1}^{P_2} \mathbf{A} \cdot d\mathbf{r}$ to be independent of the path joining points P_1 and P_2.

6.37. Prove that a necessary and sufficient condition that $\nabla \times \mathbf{A} = 0$ is that $\mathbf{A} = \nabla \phi$.

Sufficiency. If $\mathbf{A} = \nabla \phi$, then $\nabla \times \mathbf{A} = \nabla \times \nabla \phi = 0$ by Problem 5.80, page 144.

Necessity.

If $\nabla \times \mathbf{A} = 0$, then by Problem 6.36, $\oint \mathbf{A} \cdot d\mathbf{r} = 0$ around every closed path and $\int_C \mathbf{A} \cdot d\mathbf{r}$ is independent of the path joining two points which we take as (a, b, c) and (x, y, z). Let us define

$$\phi(x, y, z) = \int_{(a,b,c)}^{(x,y,z)} \mathbf{A} \cdot d\mathbf{r} = \int_{(a,b,c)}^{(x,y,z)} [A_1\, dx + A_2\, dy + A_3\, dz]$$

Then

$$\phi(x + \Delta x, y, z) - \phi(x, y, z) = \int_{(x,y,z)}^{(x+\Delta x, y, z)} [A_1\, dx + A_2\, dy + A_3\, dz]$$

Since the last integral is independent of the path joining (x, y, z) and $(x + \Delta x, y, z)$, we can choose the path to be a straight line joining these points so that dy and dz are zero. Then

$$\frac{\phi(x + \Delta x, y, z) - \phi(x, y, z)}{\Delta x} = \frac{1}{\Delta x} \int_{(x,y,z)}^{(x+\Delta x, y, z)} A_1\, dx = A_1(x + \theta\, \Delta x, y, z) \qquad 0 < \theta < 1$$

where we have applied the law of the mean for integrals.

Taking the limit of both sides as $\Delta x \to 0$ gives $\partial \phi / \partial x = A_1$.

Similarly we can show that $\partial \phi / \partial y = A_2$, $\partial \phi / \partial z = A_3$.

Thus $\mathbf{A} = A_1 \mathbf{i} + A_2 \mathbf{j} + A_3 \mathbf{k} = \dfrac{\partial \phi}{\partial x} \mathbf{i} + \dfrac{\partial \phi}{\partial y} \mathbf{j} + \dfrac{\partial \phi}{\partial z} \mathbf{k} = \nabla \phi$.

6.38. (a) Prove that a necessary and sufficient condition that $A_1\, dx + A_2\, dy + A_3\, dz = d\phi$, an exact differential, is that $\nabla \times \mathbf{A} = 0$ where $\mathbf{A} = A_1 \mathbf{i} + A_2 \mathbf{j} + A_3 \mathbf{k}$.

(b) Show that in such case,

$$\int_{(x_1, y_1, z_1)}^{(x_2, y_2, z_2)} [A_1\, dx + A_2\, dy + A_3\, dz] = \int_{(x_1, y_1, z_1)}^{(x_2, y_2, z_2)} d\phi = \phi(x_2, y_2, z_2) - \phi(x_1, y_1, z_1)$$

(a) **Necessity.** If $A_1\, dx + A_2\, dy + A_3\, dz = d\phi = \dfrac{\partial \phi}{\partial x} dx + \dfrac{\partial \phi}{\partial y} dy + \dfrac{\partial \phi}{\partial z} dz$, then

$$(1) \quad \frac{\partial \phi}{\partial x} = A_1 \qquad (2) \quad \frac{\partial \phi}{\partial y} = A_2 \qquad (3) \quad \frac{\partial \phi}{\partial z} = A_3$$

Then by differentiating we have, assuming continuity of the partial derivatives,

$$\frac{\partial A_1}{\partial y} = \frac{\partial A_2}{\partial x}, \qquad \frac{\partial A_2}{\partial z} = \frac{\partial A_3}{\partial y}, \qquad \frac{\partial A_1}{\partial z} = \frac{\partial A_3}{\partial x}$$

which is precisely the condition $\nabla \times \mathbf{A} = 0$.

Another method. If $A_1\, dx + A_2\, dy + A_3\, dz = d\phi$, then

$$\mathbf{A} = A_1 \mathbf{i} + A_2 \mathbf{j} + A_3 \mathbf{k} = \frac{\partial \phi}{\partial x} \mathbf{i} + \frac{\partial \phi}{\partial y} \mathbf{j} + \frac{\partial \phi}{\partial z} \mathbf{k} = \nabla \phi$$

from which $\nabla \times \mathbf{A} = \nabla \times \nabla \phi = 0$.

Sufficiency. If $\nabla \times \mathbf{A} = 0$, then by Problem 6.37, $\mathbf{A} = \nabla \phi$ and

$$A_1\, dx + A_2\, dy + A_3\, dz = \mathbf{A} \cdot d\mathbf{r} = \nabla \phi \cdot d\mathbf{r} = \frac{\partial \phi}{\partial x} dx + \frac{\partial \phi}{\partial y} dy + \frac{\partial \phi}{\partial z} dz = d\phi$$

(b) From part (a), $\phi(x, y, z) = \displaystyle\int_{(a,b,c)}^{(x,y,z)} [A_1\, dx + A_2\, dy + A_3\, dz]$.

Then omitting the integrand $A_1\,dx + A_2\,dy + A_3\,dz$, we have

$$\int_{(x_1,y_1,z_1)}^{(x_2,y_2,z_2)} \;=\; \int_{(a,b,c)}^{(x_2,y_2,z_2)} - \int_{(a,b,c)}^{(x_1,y_1,z_1)} \;=\; \phi(x_2,y_2,z_2) \;-\; \phi(x_1,y_1,z_1)$$

6.39. (a) Prove that $\mathbf{F} = (2xz^3 + 6y)\mathbf{i} + (6x - 2yz)\mathbf{j} + (3x^2z^2 - y^2)\mathbf{k}$ is a conservative force field. (b) Evaluate $\int_C \mathbf{F}\cdot d\mathbf{r}$ where C is any path from $(1,-1,1)$ to $(2,1,-1)$. (c) Give a physical interpretation of the results.

(a) A force field $\mathbf{F}$ is conservative if the line integral $\int_C \mathbf{F}\cdot d\mathbf{r}$ is independent of the path C joining any two points. A necessary and sufficient condition that $\mathbf{F}$ be conservative is that $\nabla \times \mathbf{F} = \mathbf{0}$.

Since here $\quad \nabla \times \mathbf{F} = \begin{vmatrix} \mathbf{i} & \mathbf{j} & \mathbf{k} \\[4pt] \dfrac{\partial}{\partial x} & \dfrac{\partial}{\partial y} & \dfrac{\partial}{\partial z} \\[8pt] 2xz^3 + 6y & 6x - 2yz & 3x^2z^2 - y^2 \end{vmatrix} = \mathbf{0}, \quad \mathbf{F}$ is conservative.

(b) Method 1.

By Problem 6.38, $\mathbf{F}\cdot d\mathbf{r} = (2xz^3 + 6y)\,dx + (6x - 2yz)\,dy + (3x^2z^2 - y^2)\,dz$ is an exact differential $d\phi$, where ϕ is such that

$$(1)\quad \frac{\partial \phi}{\partial x} = 2xz^3 + 6y \qquad (2)\quad \frac{\partial \phi}{\partial y} = 6x - 2yz \qquad (3)\quad \frac{\partial \phi}{\partial z} = 3x^2z^2 - y^2$$

From these we obtain respectively

$$\phi = x^2z^3 + 6xy + f_1(y,z) \qquad \phi = 6xy - y^2z + f_2(x,z) \qquad \phi = x^2z^3 - y^2z + f_3(x,y)$$

These are consistent if $f_1(y,z) = -y^2z + c$, $f_2(x,z) = x^2z^3 + c$, $f_3(x,y) = 6xy + c$ in which case $\phi = x^2z^3 + 6xy - y^2z + c$. Thus by Problem 6.38,

$$\int_{(1,-1,1)}^{(2,1,-1)} \mathbf{F}\cdot d\mathbf{r} \;=\; x^2z^3 + 6xy - y^2z + c\,\Big|_{(1,-1,1)}^{(2,1,-1)} \;=\; 15$$

Alternatively we may notice by inspection that

$$\begin{aligned} \mathbf{F}\cdot d\mathbf{r} &= (2xz^3\,dx + 3x^2z^2\,dz) + (6y\,dx + 6x\,dy) - (2yz\,dy + y^2\,dz) \\ &= d(x^2z^3) + d(6xy) - d(y^2z) = d(x^2z^3 + 6xy - y^2z + c) \end{aligned}$$

from which ϕ is determined.

Method 2.

Since the integral is independent of the path, we can choose any path to evaluate it; in particular we can choose the path consisting of straight lines from $(1,-1,1)$ to $(2,-1,1)$, then to $(2,1,1)$ and then to $(2,1,-1)$. The result is

$$\int_{x=1}^{2} (2x-6)\,dx + \int_{y=-1}^{1} (12-2y)\,dy + \int_{z=1}^{-1} (12z^2-1)\,dz \;=\; 15$$

where the first integral is obtained from the line integral by placing $y=-1$, $z=1$, $dy=0$, $dz=0$; the second integral by placing $x=2$, $z=1$, $dx=0$, $dz=0$; and the third integral by placing $x=2$, $y=1$, $dx=0$, $dy=0$.

(c) Physically $\int_C \mathbf{F}\cdot d\mathbf{r}$ represents the work done in moving an object from $(1,-1,1)$ to $(2,1,-1)$ along C. In a conservative force field the work done is independent of the path C joining these points.

Supplementary Problems

DOUBLE INTEGRALS

6.40. (a) Sketch the region $\mathcal{R}$ in the xy plane bounded by $y^2 = 2x$ and $y = x$. (b) Find the area of $\mathcal{R}$. (c) Find the polar moment of inertia of $\mathcal{R}$ assuming constant density σ.

6.41. Find the centroid of the region in the preceding problem.

6.42. Given $\displaystyle\int_{y=0}^{3} \int_{x=1}^{\sqrt{4-y}} (x+y)\, dx\, dy$. (a) Sketch the region and give a possible physical interpretation of the double integral. (b) Interchange the order of integration. (c) Evaluate the double integral.

6.43. Show that $\displaystyle\int_{x=1}^{2} \int_{y=\sqrt{x}}^{x} \sin\frac{\pi x}{2y}\, dy\, dx \;+\; \int_{x=2}^{4} \int_{y=\sqrt{x}}^{2} \sin\frac{\pi x}{2y}\, dy\, dx \;=\; \frac{4(\pi+2)}{\pi^3}$.

6.44. Find the volume of the tetrahedron bounded by $x/a + y/b + z/c = 1$ and the coordinate planes.

6.45. Find the volume of the region bounded by $z = x^2 + y^2$, $z = 0$, $x = -a$, $x = a$, $y = -a$, $y = a$.

6.46. Find (a) the moment of inertia about the z axis and (b) the centroid of the region in Problem 6.45 assuming a constant density σ.

TRIPLE INTEGRALS

6.47. (a) Evaluate $\displaystyle\int_{x=0}^{1} \int_{y=0}^{1} \int_{z=\sqrt{x^2+y^2}}^{2} xyz\, dz\, dy\, dx$. (b) Give a physical interpretation to the integral in (a).

6.48. Find the (a) volume and (b) centroid of the region in the first octant bounded by $x/a + y/b + z/c = 1$, where a, b, c are positive.

6.49. Find the (a) moment of inertia and (b) radius of gyration about the z axis of the region in Problem 6.48.

6.50. Find the mass of the region corresponding to $x^2 + y^2 + z^2 \leqq 4$, $x \geqq 0$, $y \geqq 0$, $z \geqq 0$, if the density is equal to xyz.

6.51. Find the volume of the region bounded by $z = x^2 + y^2$ and $z = 2x$.

TRANSFORMATION OF DOUBLE INTEGRALS

6.52. Evaluate $\displaystyle\iint_{\mathcal{R}} \sqrt{x^2+y^2}\, dx\, dy$, where $\mathcal{R}$ is the region $x^2 + y^2 \leqq a^2$.

6.53. If $\mathcal{R}$ is the region of Problem 6.52, evaluate $\displaystyle\iint_{\mathcal{R}} e^{-(x^2+y^2)}\, dx\, dy$.

6.54. By using the transformation $x + y = u$, $y = uv$, show that

$$\int_{x=0}^{1} \int_{y=0}^{1-x} e^{y/(x+y)}\, dy\, dx \;=\; \frac{e-1}{2}$$

6.55. Find the area of the region bounded by $xy = 4$, $xy = 8$, $xy^3 = 5$, $xy^3 = 15$. [*Hint*. Let $xy = u$, $xy^3 = v$.]

6.56. Show that the volume generated by revolving the region in the first quadrant bounded by the parabolas $y^2 = x$, $y^2 = 8x$, $x^2 = y$, $x^2 = 8y$ about the x axis is $279\pi/2$. [*Hint*. Let $y^2 = ux$, $x^2 = vy$.]

6.57. Find the area of the region in the first quadrant bounded by $y = x^3$, $y = 4x^3$, $x = y^3$, $x = 4y^3$.

6.58. Let $\mathcal{R}$ be the region bounded by $x + y = 1$, $x = 0$, $y = 0$. Show that $\displaystyle\iint_{\mathcal{R}} \cos\left(\frac{x-y}{x+y}\right) dx\, dy = \sin 1$. [*Hint*. Let $x - y = u$, $x + y = v$.]

TRANSFORMATION OF TRIPLE INTEGRALS

6.59. Find the volume of the region bounded by $z = 4 - x^2 - y^2$ and the xy plane.

6.60. Find the centroid of the region in Problem 6.59, assuming constant density σ.

6.61. (a) Evaluate $\iiint\limits_{\mathcal{R}} \sqrt{x^2+y^2+z^2}\,dx\,dy\,dz$, where $\mathcal{R}$ is the region bounded by the plane $z=3$ and

the cone $z=\sqrt{x^2+y^2}$. (b) Give a physical interpretation of the integral in (a). [*Hint.* Perform the integration in cylindrical coordinates in the order ρ, z, ϕ.]

6.62. Show that the volume of the region bounded by the cone $z=\sqrt{x^2+y^2}$ and the paraboloid $z=x^2+y^2$ is $\pi/6$.

6.63. Find the moment of inertia of a right circular cylinder of radius a and height b, about its axis if the density is proportional to the distance from the axis.

6.64. (a) Evaluate $\iiint\limits_{\mathcal{R}} \dfrac{dx\,dy\,dz}{(x^2+y^2+z^2)^{3/2}}$, where $\mathcal{R}$ is the region bounded by the spheres $x^2+y^2+z^2=a^2$

and $x^2+y^2+z^2=b^2$ where $a>b>0$. (b) Give a physical interpretation of the integral in (a).

6.65. (a) Find the volume of the region bounded above by the sphere $r=2a\cos\theta$, and below by the cone $\phi=\alpha$ where $0<\alpha<\pi/2$. (b) Discuss the case $\alpha=\pi/2$.

6.66. Find the centroid of a hemispherical shell having outer radius a and inner radius b if the density (a) is constant, (b) varies as the square of the distance from the base. Discuss the case $a=b$.

LINE INTEGRALS

6.67. Evaluate $\int_{(1,1)}^{(4,2)} [(x+y)\,dx + (y-x)\,dy]$ along (a) the parabola $y^2=x$, (b) a straight line, (c) straight lines from $(1,1)$ to $(1,2)$ and then to $(4,2)$, (d) the curve $x=2t^2+t+1$, $y=t^2+1$.

6.68. Evaluate $\oint [(2x-y+4)\,dx + (5y+3x-6)\,dy]$ around a triangle in the xy plane with vertices at $(0,0)$, $(3,0)$, $(3,2)$ traversed in a counterclockwise direction.

6.69. Evaluate the line integral in the preceding problem around a circle of radius 4 with center at $(0,0)$.

6.70. (a) If $\mathbf{F}=(x^2-y^2)\mathbf{i}+2xy\mathbf{j}$, evaluate $\int_C \mathbf{F}\cdot d\mathbf{r}$ along the curve C in the xy plane given by $y=x^2-x$ from the point $(1,0)$ to $(2,2)$. (b) Interpret physically the result obtained.

6.71. Evaluate $\int_C (2x+y)\,ds$, where C is the curve in the xy plane given by $x^2+y^2=25$ and s is the arc length parameter, from the point $(3,4)$ to $(4,3)$ along the shortest path.

6.72. If $\mathbf{F}=(3x-2y)\mathbf{i}+(y+2z)\mathbf{j}-x^2\mathbf{k}$, evaluate $\int_C \mathbf{F}\cdot d\mathbf{r}$ from $(0,0,0)$ to $(1,1,1)$, where C is a path consisting of (a) the curve $x=t$, $y=t^2$, $z=t^3$, (b) a straight line joining these points, (c) the straight lines from $(0,0,0)$ to $(0,1,0)$, then to $(0,1,1)$ and then to $(1,1,1)$, (d) the curve $x=z^2$, $z=y^2$.

6.73. If $\mathbf{T}$ is the unit tangent vector to a curve C (plane or space curve) and $\mathbf{F}$ is a given force field, prove that under appropriate conditions $\int_C \mathbf{F}\cdot d\mathbf{r} = \int_C \mathbf{F}\cdot\mathbf{T}\,ds$ where s is the arc length parameter. Interpret the result physically and geometrically.

GREEN'S THEOREM IN THE PLANE. INDEPENDENCE OF THE PATH

6.74. Verify Green's theorem in the plane for $\oint_C [(x^2-xy^3)\,dx + (y^2-2xy)\,dy]$ where C is a square with vertices at $(0,0)$, $(2,0)$, $(2,2)$, $(0,2)$.

6.75. Evaluate the line integrals of (a) Problem 6.68 and (b) Problem 6.69 by Green's theorem.

6.76. (a) Let C be any simple closed curve bounding a region having area A. Prove that if a_1, a_2, a_3, b_1, b_2, b_3 are constants,

$$\oint_C [(a_1 x + a_2 y + a_3)\, dx \,+\, (b_1 x + b_2 y + b_3)\, dy] \;=\; (b_1 - a_2)A$$

(b) Under what conditions will the line integral around any path C be zero?

6.77. Find the area bounded by the hypocycloid $x^{2/3} + y^{2/3} = a^{2/3}$.
[*Hint.*   Parametric equations are $x = a \cos^3 t,\ y = a \sin^3 t,\ 0 \le t \le 2\pi$.]

6.78. If $x = \rho \cos \phi,\ y = \rho \sin \phi$, prove that $\frac{1}{2} \oint [x\, dy - y\, dx] = \frac{1}{2} \int \rho^2\, d\phi$ and interpret.

6.79. Verify Green's theorem in the plane for $\oint_C [(x^3 - x^2 y)\, dx + xy^2\, dy]$, where C is the boundary of the region enclosed by the circles $x^2 + y^2 = 4$ and $x^2 + y^2 = 16$.

6.80. (a) Prove that $\int_{(1,0)}^{(2,1)} [(2xy - y^4 + 3)\, dx + (x^2 - 4xy^3)\, dy]$ is independent of the path joining $(1, 0)$ and $(2, 1)$. (b) Evaluate the integral in (a).

6.81. Evaluate $\int_C [(2xy^3 - y^2 \cos x)\, dx \,+\, (1 - 2y \sin x + 3x^2 y^2)\, dy]$ along the parabola $2x = \pi y^2$ from $(0, 0)$ to $(\pi/2, 1)$.

6.82. Evaluate the line integral in the preceding problem around a parallelogram with vertices at $(0, 0)$, $(3, 0)$, $(5, 2)$, $(2, 2)$.

6.83. Prove that if $x = f(u, v),\ y = g(u, v)$ defines a transformation which maps a region $\mathcal{R}$ of the xy plane into a region $\mathcal{R}'$ of the uv plane then

$$\iint_{\mathcal{R}} dx\, dy \;=\; \iint_{\mathcal{R}'} \left| \frac{\partial(x, y)}{\partial(u, v)} \right| du\, dv$$

by using Green's theorem on the integral $\frac{1}{2} \int_C [x\, dy - y\, dx]$ and interpret geometrically.

SURFACE INTEGRALS

6.84. (a) Evaluate $\iint_S (x^2 + y^2)\, dS$, where S is the surface of the cone $z^2 = 3(x^2 + y^2)$ bounded by $z = 0$ and $z = 3$.   (b) Interpret physically the result in (a).

6.85. Determine the surface area of the plane $2x + y + 2z = 16$ cut off by (a) $x = 0,\ y = 0,\ x = 2,\ y = 3$, (b) $x = 0,\ y = 0$ and $x^2 + y^2 = 64$.

6.86. Find the surface area of the paraboloid $2z = x^2 + y^2$ which is outside the cone $z = \sqrt{x^2 + y^2}$.

6.87. Find the area of the surface of the cone $z^2 = 3(x^2 + y^2)$ cut out by the paraboloid $z = x^2 + y^2$.

6.88. Find the surface area of the region common to the intersecting cylinders $x^2 + y^2 = a^2$ and $x^2 + z^2 = a^2$.

6.89. (a) Show that in general the equation $\mathbf{r} = \mathbf{r}(u, v)$ geometrically represents a surface.   (b) Discuss the geometric significance of $u = c_1,\ v = c_2$ where c_1 and c_2 are constants.   (c) Prove that the element of arc length on this surface is given by

$$ds^2 \;=\; E\, du^2 + 2F\, du\, dv + G\, dv^2$$

where $E = \dfrac{\partial \mathbf{r}}{\partial u} \cdot \dfrac{\partial \mathbf{r}}{\partial u},\quad F = \dfrac{\partial \mathbf{r}}{\partial u} \cdot \dfrac{\partial \mathbf{r}}{\partial v},\quad G = \dfrac{\partial \mathbf{r}}{\partial v} \cdot \dfrac{\partial \mathbf{r}}{\partial v}$.

6.90. (a) Referring to Problem 6.89, show that the element of surface area is given by $dS = \sqrt{EG - F^2}\, du\, dv$.

(b) Deduce from (a) that the area of a surface $\mathbf{r} = \mathbf{r}(u, v)$ is $\iint_S \sqrt{EG - F^2}\, du\, dv$.

[*Hint.*   Use the fact that $\left| \dfrac{\partial \mathbf{r}}{\partial u} \times \dfrac{\partial \mathbf{r}}{\partial v} \right| = \sqrt{\left(\dfrac{\partial \mathbf{r}}{\partial u} \times \dfrac{\partial \mathbf{r}}{\partial v} \right) \cdot \left(\dfrac{\partial \mathbf{r}}{\partial u} \times \dfrac{\partial \mathbf{r}}{\partial v} \right)}$ and then use the identity $(\mathbf{A} \times \mathbf{B}) \cdot (\mathbf{C} \times \mathbf{D}) = (\mathbf{A} \cdot \mathbf{C})(\mathbf{B} \cdot \mathbf{D}) - (\mathbf{A} \cdot \mathbf{D})(\mathbf{B} \cdot \mathbf{C})$.

6.91. (a) Prove that $\mathbf{r} = (a \sin u \cos v)\mathbf{i} + a(\sin u \sin v)\mathbf{j} + (a \cos u)\mathbf{k}$, $0 \leqq u \leqq \pi$, $0 \leqq v < 2\pi$ represents a sphere of radius a. (b) Use Problem 6.90 to show that the surface area of this sphere is $4\pi a^2$.

THE DIVERGENCE THEOREM

6.92. Verify the divergence theorem for $\mathbf{A} = (2xy + z)\mathbf{i} + y^2\mathbf{j} - (x + 3y)\mathbf{k}$ taken over the region bounded by $2x + 2y + z = 6$, $x = 0$, $y = 0$, $z = 0$.

6.93. Evaluate $\iint\limits_{S} \mathbf{F} \cdot \mathbf{n} \, dS$, where $\mathbf{F} = (z^2 - x)\mathbf{i} - xy\mathbf{j} + 3z\mathbf{k}$ and S is the surface of the region bounded by $z = 4 - y^2$, $x = 0$, $x = 3$ and the xy plane.

6.94. Evaluate $\iint\limits_{S} \mathbf{A} \cdot \mathbf{n} \, dS$, where $\mathbf{A} = (2x + 3z)\mathbf{i} - (xz + y)\mathbf{j} + (y^2 + 2z)\mathbf{k}$ and S is the surface of the sphere having center at $(3, -1, 2)$ and radius 3.

6.95. Determine the value of $\iint\limits_{S} [x \, dy \, dz + y \, dz \, dx + z \, dx \, dy]$, where S is the surface of the region bounded by the cylinder $x^2 + y^2 = 9$ and the planes $z = 0$ and $z = 3$, (a) by using the divergence theorem, (b) directly.

6.96. Evaluate $\iint\limits_{S} [4xz \, dy \, dz - y^2 \, dz \, dx + yz \, dx \, dy]$, where S is the surface of the cube bounded by $x = 0$, $y = 0$, $z = 0$, $x = 1$, $y = 1$, $z = 1$, (a) directly, (b) by Green's theorem in space (divergence theorem).

6.97. Prove that $\iint\limits_{S} (\nabla \times \mathbf{A}) \cdot \mathbf{n} \, dS = 0$ for any closed surface S.

6.98. Prove that $\iint\limits_{S} \mathbf{n} \, dS = \mathbf{0}$, where n is the outward drawn normal to any closed surface S.

6.99. If $\mathbf{n}$ is the unit outward drawn normal to any closed surface S bounding the region V, prove that

$$\iiint\limits_{V} \operatorname{div} \mathbf{n} \, dV = S$$

STOKES' THEOREM

6.100. Verify Stokes' theorem for $\mathbf{A} = 2y\mathbf{i} + 3x\mathbf{j} - z^2\mathbf{k}$, where S is the upper half surface of the sphere $x^2 + y^2 + z^2 = 9$ and C is its boundary.

6.101. Verify Stokes' theorem for $\mathbf{A} = (y + z)\mathbf{i} - xz\mathbf{j} + y^2\mathbf{k}$, where S is the surface of the region in the first octant bounded by $2x + z = 6$ and $y = 2$ which is not included in the (a) xy plane, (b) plane $y = 2$, (c) plane $2x + z = 6$ and C is the corresponding boundary.

6.102. Evaluate $\iint\limits_{S} (\nabla \times \mathbf{A}) \cdot \mathbf{n} \, dS$, where $\mathbf{A} = (x - z)\mathbf{i} + (x^3 + yz)\mathbf{j} - 3xy^2\mathbf{k}$ and S is the surface of the cone $z = 2 - \sqrt{x^2 + y^2}$ above the xy plane.

6.103. If V is a region bounded by a closed surface S and $\mathbf{B} = \nabla \times \mathbf{A}$, prove that $\iint\limits_{S} \mathbf{B} \cdot \mathbf{n} \, dS = 0$.

6.104. (a) Prove that $\mathbf{F} = (2xy + 3)\mathbf{i} + (x^2 - 4z)\mathbf{j} - 4y\mathbf{k}$ is a conservative force field. (b) Find ϕ such that $\mathbf{F} = \nabla\phi$. (c) Evaluate $\int_{C} \mathbf{F} \cdot d\mathbf{r}$, where C is any path from $(3, -1, 2)$ to $(2, 1, -1)$.

6.105. Let C be any path joining any point on the sphere $x^2 + y^2 + z^2 = a^2$ to any point on the sphere $x^2 + y^2 + z^2 = b^2$. Show that if $\mathbf{F} = 5r^3 \mathbf{r}$, where $\mathbf{r} = x\mathbf{i} + y\mathbf{j} + z\mathbf{k}$, then $\int_{C} \mathbf{F} \cdot d\mathbf{r} = b^5 - a^5$.

6.106. In Problem 6.105 evaluate $\int_{C} \mathbf{F} \cdot d\mathbf{r}$ if $\mathbf{F} = f(r)\mathbf{r}$, where $f(r)$ is assumed to be continuous.

6.107. Determine whether there is a function ϕ such that $\mathbf{F} = \nabla\phi$, where:

    (a) $\mathbf{F} = (xz - y)\mathbf{i} + (x^2y + z^3)\mathbf{j} + (3xz^2 - xy)\mathbf{k}$.

    (b) $\mathbf{F} = 2xe^{-y}\mathbf{i} + (\cos z - x^2e^{-y})\mathbf{j} - (y\sin z)\mathbf{k}$. If so, find it.

6.108. Solve the differential equation $(z^3 - 4xy)\,dx + (6y - 2x^2)\,dy + (3xz^2 + 1)\,dz = 0$.

Answers to Supplementary Problems

6.40. (b) $2/3$; (c) $48\sigma/35 = 72M/35$, where M is the mass of $\mathcal{R}$.

6.41. $\bar{x} = 4/5$, $\bar{y} = 1$

6.42. (b) $\displaystyle\int_{x=1}^{2}\int_{y=0}^{4-x^2}(x+y)\,dy\,dx$, (c) $241/60$

6.44. $abc/6$

6.45. $8a^4/3$

6.46. (a) $\frac{112}{45}a^6\sigma = \frac{14}{15}Ma^2$, where $M = $ mass;

    (b) $\bar{x} = \bar{y} = 0$, $\bar{z} = \frac{7}{15}a^2$

6.47. (a) $3/8$

6.48. (a) $abc/6$; (b) $\bar{x} = a/4$, $\bar{y} = b/4$, $\bar{z} = c/4$

6.49. (a) $M(a^2 + b^2)/10$, (b) $\sqrt{(a^2+b^2)}/10$

6.50. $4/3$

6.51. $\pi/2$

6.52. $\frac{2}{3}\pi a^3$

6.53. $\pi(1 - e^{-a^2})$

6.55. $2\ln 3$

6.57. $1/8$

6.59. 8π

6.60. $\bar{x} = \bar{y} = 0$, $\bar{z} = \frac{4}{3}$

6.61. $27\pi(2\sqrt{2} - 1)/2$

6.63. $\frac{2}{5}Ma^2$

6.64. (a) $4\pi\ln(a/b)$

6.65. $\frac{4}{3}\pi a^3(1 - \cos^4\alpha)$

6.66. Taking the z axis as axis of symmetry:

    (a) $\bar{x} = \bar{y} = 0$, $z = \frac{3}{8}(a^4 - b^4)/(a^3 - b^3)$;

    (b) $\bar{x} = \bar{y} = 0$, $\bar{z} = \frac{5}{8}(a^6 - b^6)/(a^5 - b^5)$

6.67. (a) $34/3$, (b) 11, (c) 14, (d) $32/3$

6.68. 12

6.69. 64π

6.70. (a) $124/15$

6.71. 15

6.72. (a) $23/15$, (b) $5/3$, (c) 0, (d) $13/30$

6.74. Common value $= 8$

6.76. (b) $a_2 = b_1$

6.77. $3\pi a^2/8$

6.79. Common value $= 120\pi$

6.80. (b) 5

6.81. $\pi^2/4$

6.82. 0

6.84. (a) 9π

6.85. (a) 9, (b) 24π

6.86. $\frac{2}{3}\pi(5\sqrt{5} - 1)$

6.87. 6π

6.88. $16a^2$

6.92. Common value $= 27$

6.93. 16

6.94. 108π

6.95. 81π

6.96. $3/2$

6.100. Common value $= 9\pi$

6.101. The common value is

    (a) -6, (b) -9, (c) -18

6.102. 12π

6.104. (b) $\phi = x^2y - 4yz + 3x + $ constant, (c) 6

6.106. $\displaystyle\int_a^b r\,f(r)\,dr$

6.107. (a) ϕ does not exist.

    (b) $\phi = x^2e^{-y} + y\cos z + $ constant

6.108. $xz^3 - 2x^2y + 3y^2 + z = $ constant

Chapter 7

Fourier Series

PERIODIC FUNCTIONS

A function $f(x)$ is said to have a *period* T or to be *periodic* with period T if for all x, $f(x + T) = f(x)$, where T is a positive constant. The least value of $T > 0$ is called the *least period* or simply *the period* of $f(x)$.

    Example 1.   The function $\sin x$ has periods $2\pi, 4\pi, 6\pi, \ldots$, since $\sin(x + 2\pi)$, $\sin(x + 4\pi)$, $\sin(x + 6\pi)$, $\ldots$ all equal $\sin x$. However, 2π is the *least period* or *the period* of $\sin x$.

    Example 2.   The period of $\sin nx$ or $\cos nx$, where n is a positive integer, is $2\pi/n$.

    Example 3.   The period of $\tan x$ is π.

    Example 4.   A constant has any positive number as period.

Other examples of periodic functions are shown in the graphs of Figures 7-1(a), (b) and (c) below.

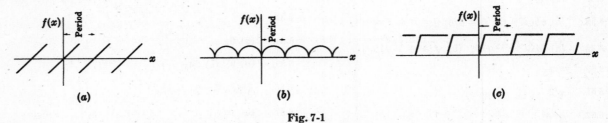

      (a)   **(b)**   **(c)**

Fig. 7-1

FOURIER SERIES

Let $f(x)$ be defined in the interval $(-L, L)$ and outside of this interval by $f(x + 2L) = f(x)$, i.e. assume that $f(x)$ has the period $2L$. The *Fourier series* or *Fourier expansion* corresponding to $f(x)$ is given by

$$\frac{a_0}{2} + \sum_{n=1}^{\infty} \left(a_n \cos \frac{n\pi x}{L} + b_n \sin \frac{n\pi x}{L} \right) \tag{1}$$

where the *Fourier coefficients* a_n and b_n are

$$\begin{cases} a_n = \dfrac{1}{L} \displaystyle\int_{-L}^{L} f(x) \cos \frac{n\pi x}{L}\, dx \\[2mm] b_n = \dfrac{1}{L} \displaystyle\int_{-L}^{L} f(x) \sin \frac{n\pi x}{L}\, dx \end{cases} \qquad n = 0, 1, 2, \ldots \tag{2}$$

If $f(x)$ has the period $2L$, the coefficients a_n and b_n can be determined equivalently from

$$\begin{cases} a_n = \dfrac{1}{L} \displaystyle\int_{c}^{c+2L} f(x) \cos \frac{n\pi x}{L}\, dx \\[2mm] b_n = \dfrac{1}{L} \displaystyle\int_{c}^{c+2L} f(x) \sin \frac{n\pi x}{L}\, dx \end{cases} \tag{3}$$

where c is any real number. In the special case $c = -L$, (3) becomes (2).

To determine a_0 in (1), we use (2) or (3) with $n = 0$. For example, from (2) we see that $a_0 = \dfrac{1}{L} \displaystyle\int_{-L}^{L} f(x)\,dx$. Note that the constant term in (1) is equal to $\dfrac{a_0}{2} = \dfrac{1}{2L} \displaystyle\int_{-L}^{L} f(x)\,dx$, which is the *mean* of $f(x)$ over a period.

If $L = \pi$, the series (1) and the coefficients (2) or (3) are particularly simple. The function in this case has the period 2π.

DIRICHLET CONDITIONS

Theorem 7-1.   Suppose that

(1) $f(x)$ is defined and single-valued except possibly at a finite number of points in $(-L, L)$

(2) $f(x)$ is periodic outside $(-L, L)$ with period $2L$

(3) $f(x)$ and $f'(x)$ are piecewise continuous in $(-L, L)$.

Then the series (1) with coefficients (2) or (3) converges to

(a) $f(x)$ if x is a point of continuity

(b) $\dfrac{f(x + 0) + f(x - 0)}{2}$ if x is a point of discontinuity

In this theorem $f(x + 0)$ and $f(x - 0)$ are the *right* and *left hand limits* of $f(x)$ at x and represent $\lim\limits_{\epsilon \to 0} f(x + \epsilon)$ and $\lim\limits_{\epsilon \to 0} f(x - \epsilon)$ respectively where $\epsilon > 0$. These are often written $\lim\limits_{\epsilon \to 0+} f(x + \epsilon)$ and $\lim\limits_{\epsilon \to 0+} f(x - \epsilon)$ to emphasize that ϵ is approaching zero through positive values. For a proof see Problems 7.18-7.23.

The conditions (1), (2) and (3) imposed on $f(x)$ are *sufficient* but not necessary, and are generally satisfied in practice. There are at present no known necessary and sufficient conditions for convergence of Fourier series. It is of interest that continuity of $f(x)$ does not *alone* insure convergence of a Fourier series.

ODD AND EVEN FUNCTIONS

A function $f(x)$ is called *odd* if $f(-x) = -f(x)$. Thus x^3, $x^5 - 3x^3 + 2x$, $\sin x$, $\tan 3x$ are odd functions.

A function $f(x)$ is called *even* if $f(-x) = f(x)$. Thus x^4, $2x^6 - 4x^2 + 5$, $\cos x$, $e^x + e^{-x}$ are even functions.

The functions portrayed graphically in Figures 7-1(a) and 7-1(b) are odd and even respectively, but that of Fig. 7-1(c) is neither odd nor even.

In the Fourier series corresponding to an odd function, only sine terms can be present. In the Fourier series corresponding to an even function, only cosine terms (and possibly a constant which we shall consider a cosine term) can be present.

HALF RANGE FOURIER SINE OR COSINE SERIES

A half range Fourier sine or cosine series is a series in which only sine terms or only cosine terms are present respectively. When a half range series corresponding to a given function is desired, the function is generally defined in the interval $(0, L)$ [which is half of the interval $(-L, L)$, thus accounting for the name *half range*] and then the function is specified as odd or even, so that it is clearly defined in the other half of the interval, namely $(-L, 0)$. In such case, we have

$$
\begin{cases}
a_n = 0, \quad b_n = \dfrac{2}{L} \int_0^L f(x) \sin \dfrac{n\pi x}{L}\, dx & \text{for } \textit{half range sine series} \\[4mm]
b_n = 0, \quad a_n = \dfrac{2}{L} \int_0^L f(x) \cos \dfrac{n\pi x}{L}\, dx & \text{for } \textit{half range cosine series}
\end{cases}
\tag{4}
$$

PARSEVAL'S IDENTITY states that

$$
\frac{1}{L} \int_{-L}^{L} \{f(x)\}^2\, dx \;=\; \frac{a_0^2}{2} + \sum_{n=1}^{\infty} (a_n^2 + b_n^2)
\tag{5}
$$

if a_n and b_n are the Fourier coefficients corresponding to $f(x)$ and if $f(x)$ satisfies the Dirichlet conditions.

DIFFERENTIATION AND INTEGRATION OF FOURIER SERIES

Differentiation and integration of Fourier series can be justified by using the theorems on page 7 which hold for series in general. It must be emphasized, however, that those theorems provide sufficient conditions and are not necessary. The following theorem for integration is especially useful.

Theorem 7-2. The Fourier series corresponding to $f(x)$ may be integrated term by term from a to x, and the resulting series will converge uniformly to $\int_a^x f(u)\, du$ provided that $f(x)$ is piecewise continuous in $-L \leqq x \leqq L$ and both a and x are in this interval.

COMPLEX NOTATION FOR FOURIER SERIES

Using Euler's identities,

$$
e^{i\theta} \;=\; \cos\theta + i\sin\theta, \qquad e^{-i\theta} \;=\; \cos\theta - i\sin\theta
\tag{6}
$$

where $i = \sqrt{-1}$ [see Problem 1.61, page 30], the Fourier series for $f(x)$ can be written as

$$
f(x) \;=\; \sum_{n=-\infty}^{\infty} c_n e^{in\pi x/L}
\tag{7}
$$

where

$$
c_n \;=\; \frac{1}{2L} \int_{-L}^{L} f(x)\, e^{-in\pi x/L}\, dx
\tag{8}
$$

In writing the equality (7), we are supposing that the Dirichlet conditions are satisfied and further that $f(x)$ is continuous at x. If $f(x)$ is discontinuous at x, the left side of (7) should be replaced by $\dfrac{f(x+0) + f(x-0)}{2}$.

ORTHOGONAL FUNCTIONS

Two vectors $\mathbf{A}$ and $\mathbf{B}$ are called *orthogonal* (perpendicular) if $\mathbf{A} \cdot \mathbf{B} = 0$ or $A_1B_1 + A_2B_2 + A_3B_3 = 0$, where $\mathbf{A} = A_1\mathbf{i} + A_2\mathbf{j} + A_3\mathbf{k}$ and $\mathbf{B} = B_1\mathbf{i} + B_2\mathbf{j} + B_3\mathbf{k}$. Although not geometrically or physically evident, these ideas can be generalized to include vectors with more than three components. In particular we can think of a function, say $A(x)$, as being a vector with an *infinity of components* (i.e. an *infinite dimensional vector*), the value of each component being specified by substituting a particular value of x in some interval (a, b). It is natural in such case to define two functions, $A(x)$ and $B(x)$, as *orthogonal* in (a, b) if

$$\int_a^b A(x)\,B(x)\,dx \;=\; 0 \tag{9}$$

A vector $\mathbf{A}$ is called a *unit vector* or *normalized vector* if its magnitude is unity, i.e. if $\mathbf{A} \cdot \mathbf{A} = A^2 = 1$. Extending the concept, we say that the function $A(x)$ is *normal* or *normalized* in (a, b) if

$$\int_a^b \{A(x)\}^2\,dx \;=\; 1 \tag{10}$$

From the above it is clear that we can consider a set of functions $\{\phi_k(x)\}$, $k = 1, 2, 3, \ldots$, having the properties

$$\int_a^b \phi_m(x)\,\phi_n(x)\,dx \;=\; 0 \qquad m \neq n \tag{11}$$

$$\int_a^b \{\phi_m(x)\}^2\,dx \;=\; 1 \qquad m = 1, 2, 3, \ldots \tag{12}$$

In such case, each member of the set is orthogonal to every other member of the set and is also normalized. We call such a set of functions an *orthonormal set* in (a, b).

The equations (11) and (12) can be summarized by writing

$$\int_a^b \phi_m(x)\,\phi_n(x)\,dx \;=\; \delta_{mn} \tag{13}$$

where δ_{mn}, called *Kronecker's symbol*, is defined as 0 if $m \neq n$ and 1 if $m = n$.

Just as any vector $\mathbf{r}$ in 3 dimensions can be expanded in a set of mutually orthogonal unit vectors $\mathbf{i}, \mathbf{j}, \mathbf{k}$ in the form $\mathbf{r} = c_1\mathbf{i} + c_2\mathbf{j} + c_3\mathbf{k}$, so we consider the possibility of expanding a function $f(x)$ in a set of orthonormal functions, i.e.,

$$f(x) \;=\; \sum_{n=1}^{\infty} c_n\phi_n(x) \qquad a \leq x \leq b \tag{14}$$

Such series, called *orthonormal series,* are generalizations of Fourier series and are of great interest and utility both from theoretical and applied viewpoints.

If
$$\int_a^b w(x)\,\psi_m(x)\,\psi_n(x)\,dx \;=\; \delta_{mn} \tag{15}$$

where $w(x) \geq 0$, we often say that $\psi_m(x)$ and $\psi_n(x)$ are orthonormal with respect to the *density function* or *weight function* $w(x)$. In such case the set of functions $\{\sqrt{w(x)}\,\phi_n(x)\}$ is an orthonormal set in (a, b).

Solved Problems

FOURIER SERIES

7.1. Graph each of the following functions.

(a) $f(x) = \begin{cases} 3 & 0 < x < 5 \\ -3 & -5 < x < 0 \end{cases}$ Period = 10

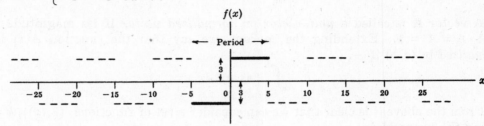

Fig. 7-2

Since the period is 10, that portion of the graph in $-5 < x < 5$ (indicated heavy in Fig. 7-2 above) is extended periodically outside this range (indicated dashed). Note that $f(x)$ is not defined at $x = 0, 5, -5, 10, -10, 15, -15$, etc. These values are the *discontinuities* of $f(x)$.

(b) $f(x) = \begin{cases} \sin x & 0 \leq x \leq \pi \\ 0 & \pi < x < 2\pi \end{cases}$ Period = 2π

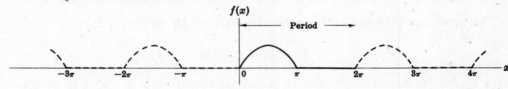

Fig. 7-3

Refer to Fig. 7-3 above. Note that $f(x)$ is defined for all x and is continuous everywhere.

(c) $f(x) = \begin{cases} 0 & 0 \leq x < 2 \\ 1 & 2 \leq x < 4 \\ 0 & 4 \leq x < 6 \end{cases}$ Period = 6

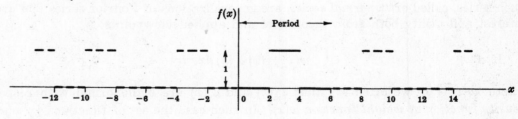

Fig. 7-4

Refer to Fig. 7-4 above. Note that $f(x)$ is defined for all x and is discontinuous at $x = \pm 2, \pm 4, \pm 8, \pm 10, \pm 14, \ldots$.

7.2. Prove $\displaystyle\int_{-L}^{L} \sin\frac{k\pi x}{L}\,dx = \int_{-L}^{L} \cos\frac{k\pi x}{L}\,dx = 0$ if $k = 1, 2, 3, \ldots$.

$$\int_{-L}^{L} \sin\frac{k\pi x}{L}\,dx = -\frac{L}{k\pi}\cos\frac{k\pi x}{L}\bigg|_{-L}^{L} = -\frac{L}{k\pi}\cos k\pi + \frac{L}{k\pi}\cos(-k\pi) = 0$$

$$\int_{-L}^{L} \cos\frac{k\pi x}{L}\,dx = \frac{L}{k\pi}\sin\frac{k\pi x}{L}\bigg|_{-L}^{L} = \frac{L}{k\pi}\sin k\pi - \frac{L}{k\pi}\sin(-k\pi) = 0$$

7.3. Prove (a) $\displaystyle\int_{-L}^{L} \cos\frac{m\pi x}{L}\cos\frac{n\pi x}{L}\,dx \;=\; \int_{-L}^{L}\sin\frac{m\pi x}{L}\sin\frac{n\pi x}{L}\,dx \;=\; \begin{cases} 0 & m \neq n \\ L & m = n \end{cases}$

  (b) $\displaystyle\int_{-L}^{L}\sin\frac{m\pi x}{L}\cos\frac{n\pi x}{L}\,dx \;=\; 0$

where m and n can assume any of the values $1, 2, 3, \ldots$.

(a) From trigonometry: $\cos A \cos B = \frac{1}{2}\{\cos(A-B) + \cos(A+B)\}$, $\sin A \sin B = \frac{1}{2}\{\cos(A-B) - \cos(A+B)\}$.

  Then, if $m \neq n$, we have by Problem 7.2,

$$\int_{-L}^{L}\cos\frac{m\pi x}{L}\cos\frac{n\pi x}{L}\,dx \;=\; \frac{1}{2}\int_{-L}^{L}\left\{\cos\frac{(m-n)\pi x}{L} + \cos\frac{(m+n)\pi x}{L}\right\}dx \;=\; 0$$

  Similarly if $m \neq n$,

$$\int_{-L}^{L}\sin\frac{m\pi x}{L}\sin\frac{n\pi x}{L}\,dx \;=\; \frac{1}{2}\int_{-L}^{L}\left\{\cos\frac{(m-n)\pi x}{L} - \cos\frac{(m+n)\pi x}{L}\right\}dx \;=\; 0$$

  If $m = n$, we have

$$\int_{-L}^{L}\cos\frac{m\pi x}{L}\cos\frac{n\pi x}{L}\,dx \;=\; \frac{1}{2}\int_{-L}^{L}\left(1 + \cos\frac{2n\pi x}{L}\right)dx \;=\; L$$

$$\int_{-L}^{L}\sin\frac{m\pi x}{L}\sin\frac{n\pi x}{L}\,dx \;=\; \frac{1}{2}\int_{-L}^{L}\left(1 - \cos\frac{2n\pi x}{L}\right)dx \;=\; L$$

  Note that if $m = n = 0$ these integrals are equal to $2L$ and 0 respectively.

(b) We have $\sin A \cos B = \frac{1}{2}\{\sin(A-B) + \sin(A+B)\}$. Then by Problem 7.2, if $m \neq n$,

$$\int_{-L}^{L}\sin\frac{m\pi x}{L}\cos\frac{n\pi x}{L}\,dx \;=\; \frac{1}{2}\int_{-L}^{L}\left\{\sin\frac{(m-n)\pi x}{L} + \sin\frac{(m+n)\pi x}{L}\right\}dx \;=\; 0$$

  If $m = n$,

$$\int_{-L}^{L}\sin\frac{m\pi x}{L}\cos\frac{n\pi x}{L}\,dx \;=\; \frac{1}{2}\int_{-L}^{L}\sin\frac{2n\pi x}{L}\,dx \;=\; 0$$

  The results of parts (a) and (b) remain valid even when the limits of integration $-L, L$ are replaced by $c, c + 2L$ respectively.

7.4. If the series $\displaystyle A + \sum_{n=1}^{\infty}\left(a_n\cos\frac{n\pi x}{L} + b_n\sin\frac{n\pi x}{L}\right)$ converges uniformly to $f(x)$ in $(-L, L)$, show that for $n = 1, 2, 3, \ldots$,

(a) $\displaystyle a_n = \frac{1}{L}\int_{-L}^{L}f(x)\cos\frac{n\pi x}{L}\,dx$, (b) $\displaystyle b_n = \frac{1}{L}\int_{-L}^{L}f(x)\sin\frac{n\pi x}{L}\,dx$, (c) $\displaystyle A = \frac{a_0}{2}$.

(a) Multiplying
$$f(x) \;=\; A + \sum_{n=1}^{\infty}\left(a_n\cos\frac{n\pi x}{L} + b_n\sin\frac{n\pi x}{L}\right) \tag{1}$$

  by $\cos\dfrac{m\pi x}{L}$ and integrating from $-L$ to L, using Problem 7.3, we have

$$\int_{-L}^{L}f(x)\cos\frac{m\pi x}{L}\,dx \;=\; A\int_{-L}^{L}\cos\frac{m\pi x}{L}\,dx \tag{2}$$
$$+\; \sum_{n=1}^{\infty}\left\{a_n\int_{-L}^{L}\cos\frac{m\pi x}{L}\cos\frac{n\pi x}{L}\,dx + b_n\int_{-L}^{L}\cos\frac{m\pi x}{L}\sin\frac{n\pi x}{L}\,dx\right\}$$
$$=\; a_m L \quad\text{if } m \neq 0$$

  Thus $\qquad\qquad a_m = \dfrac{1}{L}\displaystyle\int_{-L}^{L}f(x)\cos\frac{m\pi x}{L}\,dx \qquad$ if $m = 1, 2, 3, \ldots$

(b) Multiplying (1) by $\sin\dfrac{m\pi x}{L}$ and integrating from $-L$ to L, using Problem 7.3, we have

$$\int_{-L}^{L} f(x)\sin\frac{m\pi x}{L}\,dx \;=\; A\int_{-L}^{L}\sin\frac{m\pi x}{L}\,dx$$

$$+\sum_{n=1}^{\infty}\left\{a_n\int_{-L}^{L}\sin\frac{m\pi x}{L}\cos\frac{n\pi x}{L}\,dx \;+\; b_n\int_{-L}^{L}\sin\frac{m\pi x}{L}\sin\frac{n\pi x}{L}\,dx\right\}$$

$$=\; b_m L$$

Thus $\qquad b_m \;=\; \dfrac{1}{L}\displaystyle\int_{-L}^{L} f(x)\sin\frac{m\pi x}{L}\,dx \qquad$ if $m = 1, 2, 3, \ldots$

(c) Integration of (1) from $-L$ to L, using Problem 7.2, gives

$$\int_{-L}^{L} f(x)\,dx \;=\; 2AL \qquad\text{or}\qquad A \;=\; \frac{1}{2L}\int_{-L}^{L} f(x)\,dx$$

Putting $m = 0$ in the result of part (a), we find $a_0 = \dfrac{1}{L}\displaystyle\int_{-L}^{L} f(x)\,dx$ and so $A = \dfrac{a_0}{2}$.

The above results also hold when the integration limits $-L, L$ are replaced by $c, c+2L$.

Note that in all parts above, interchange of summation and integration is valid because the series is *assumed* to converge uniformly to $f(x)$ in $(-L, L)$. Even when this assumption is not warranted, the coefficients a_m and b_m as obtained above are called *Fourier coefficients* corresponding to $f(x)$, and the corresponding series with these values of a_m and b_m is called the *Fourier series* corresponding to $f(x)$. An important problem in this case is to investigate conditions under which this series actually converges to $f(x)$. Sufficient conditions for this convergence are the *Dirichlet conditions* established below.

7.5. (a) Find the Fourier coefficients corresponding to the function

$$f(x) \;=\; \begin{cases} 0 & -5 < x < 0 \\ 3 & 0 < x < 5 \end{cases} \qquad \text{Period} = 10$$

(b) Write the corresponding Fourier series.

(c) How should $f(x)$ be defined at $x = -5$, $x = 0$ and $x = 5$ in order that the Fourier series will converge to $f(x)$ for $-5 \leqq x \leqq 5$?

The graph of $f(x)$ is shown in Fig. 7-5 below.

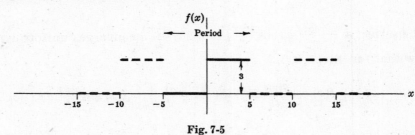

Fig. 7-5

(a) Period $= 2L = 10$ and $L = 5$. Choose the interval c to $c+2L$ as -5 to 5, so that $c = -5$. Then

$$a_n \;=\; \frac{1}{L}\int_{c}^{c+2L} f(x)\cos\frac{n\pi x}{L}\,dx \;=\; \frac{1}{5}\int_{-5}^{5} f(x)\cos\frac{n\pi x}{5}\,dx$$

$$=\; \frac{1}{5}\left\{\int_{-5}^{0}(0)\cos\frac{n\pi x}{5}\,dx + \int_{0}^{5}(3)\cos\frac{n\pi x}{5}\,dx\right\} \;=\; \frac{3}{5}\int_{0}^{5}\cos\frac{n\pi x}{5}\,dx$$

$$=\; \frac{3}{5}\left(\frac{5}{n\pi}\sin\frac{n\pi x}{5}\right)\Bigg|_{0}^{5} \;=\; 0 \qquad \text{if } n \neq 0$$

If $n = 0$, $a_n = a_0 = \dfrac{3}{5}\displaystyle\int_{0}^{5}\cos\frac{0\pi x}{5}\,dx = \dfrac{3}{5}\int_{0}^{5} dx = 3.$

$$b_n = \frac{1}{L}\int_c^{c+2L} f(x)\sin\frac{n\pi x}{L}\,dx = \frac{1}{5}\int_{-5}^5 f(x)\sin\frac{n\pi x}{5}\,dx$$

$$= \frac{1}{5}\left\{\int_{-5}^0 (0)\sin\frac{n\pi x}{5}\,dx + \int_0^5 (3)\sin\frac{n\pi x}{5}\,dx\right\} = \frac{3}{5}\int_0^5 \sin\frac{n\pi x}{5}\,dx$$

$$= \frac{3}{5}\left(-\frac{5}{n\pi}\cos\frac{n\pi x}{5}\right)\Big|_0^5 = \frac{3(1-\cos n\pi)}{n\pi}$$

(b) The corresponding Fourier series is

$$\frac{a_0}{2} + \sum_{n=1}^\infty \left(a_n\cos\frac{n\pi x}{L} + b_n\sin\frac{n\pi x}{L}\right) = \frac{3}{2} + \sum_{n=1}^\infty \frac{3(1-\cos n\pi)}{n\pi}\sin\frac{n\pi x}{5}$$

$$= \frac{3}{2} + \frac{6}{\pi}\left(\sin\frac{\pi x}{5} + \frac{1}{3}\sin\frac{3\pi x}{5} + \frac{1}{5}\sin\frac{5\pi x}{5} + \cdots\right)$$

(c) Since $f(x)$ satisfies the Dirichlet conditions, we can say that the series converges to $f(x)$ at all points of continuity and to $\dfrac{f(x+0)+f(x-0)}{2}$ at points of discontinuity. At $x=-5$, 0 and 5, which are points of discontinuity, the series converges to $(3+0)/2 = 3/2$ as seen from the graph. If we redefine $f(x)$ as follows,

$$f(x) = \begin{cases} 3/2 & x=-5 \\ 0 & -5 < x < 0 \\ 3/2 & x=0 \qquad \text{Period} = 10 \\ 3 & 0 < x < 5 \\ 3/2 & x=5 \end{cases}$$

then the series will converge to $f(x)$ for $-5 \leqq x \leqq 5$.

7.6. Expand $f(x) = x^2$, $0 < x < 2\pi$ in a Fourier series if (a) the period is 2π, (b) the period is not specified.

(a) The graph of $f(x)$ with period 2π is shown in Fig. 7-6 below.

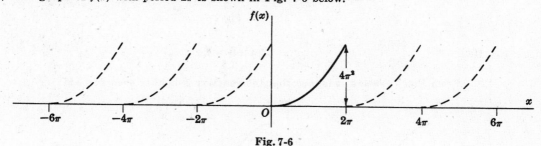

Fig. 7-6

Period $= 2L = 2\pi$ and $L = \pi$. Choosing $c = 0$, we have

$$a_n = \frac{1}{L}\int_c^{c+2L} f(x)\cos\frac{n\pi x}{L}\,dx = \frac{1}{\pi}\int_0^{2\pi} x^2\cos nx\,dx$$

$$= \frac{1}{\pi}\left\{(x^2)\left(\frac{\sin nx}{n}\right) - (2x)\left(\frac{-\cos nx}{n^2}\right) + 2\left(\frac{-\sin nx}{n^3}\right)\right\}\Big|_0^{2\pi} = \frac{4}{n^2}, \quad n \neq 0$$

If $n = 0$, $a_0 = \dfrac{1}{\pi}\displaystyle\int_0^{2\pi} x^2\,dx = \dfrac{8\pi^2}{3}$.

$$b_n = \frac{1}{L}\int_c^{c+2L} f(x)\sin\frac{n\pi x}{L}\,dx = \frac{1}{\pi}\int_0^{2\pi} x^2\sin nx\,dx$$

$$= \frac{1}{\pi}\left\{(x^2)\left(-\frac{\cos nx}{n}\right) - (2x)\left(-\frac{\sin nx}{n^2}\right) + (2)\left(\frac{\cos nx}{n^3}\right)\right\}\Big|_0^{2\pi} = \frac{-4\pi}{n}$$

Then $f(x) = x^2 = \dfrac{4\pi^2}{3} + \displaystyle\sum_{n=1}^\infty \left(\frac{4}{n^2}\cos nx - \frac{4\pi}{n}\sin nx\right)$.

This is valid for $0 < x < 2\pi$. At $x = 0$ and $x = 2\pi$ the series converges to $2\pi^2$.

(b) If the period is not specified, the Fourier series cannot be determined uniquely in general.

7.7. Using the results of Problem 7.6, prove that $\dfrac{1}{1^2} + \dfrac{1}{2^2} + \dfrac{1}{3^2} + \cdots = \dfrac{\pi^2}{6}$.

At $x = 0$ the Fourier series of Problem 7.6 reduces to $\dfrac{4\pi^2}{3} + \displaystyle\sum_{n=1}^{\infty} \dfrac{4}{n^2}$.

By the Dirichlet conditions, the series converges at $x = 0$ to $\frac{1}{2}(0 + 4\pi^2) = 2\pi^2$.

Then $\dfrac{4\pi^2}{3} + \displaystyle\sum_{n=1}^{\infty} \dfrac{4}{n^2} = 2\pi^2$, and so $\displaystyle\sum_{n=1}^{\infty} \dfrac{1}{n^2} = \dfrac{\pi^2}{6}$.

ODD AND EVEN FUNCTIONS. HALF RANGE FOURIER SERIES

7.8. Classify each of the following functions according as they are even, odd, or neither even nor odd.

(a) $f(x) = \begin{cases} 2 & 0 < x < 3 \\ -2 & -3 < x < 0 \end{cases}$ Period = 6

From Fig. 7-7 below it is seen that $f(-x) = -f(x)$, so that the function is odd.

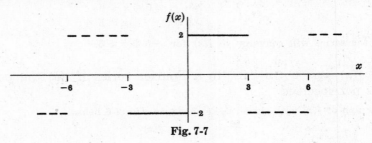

Fig. 7-7

(b) $f(x) = \begin{cases} \cos x & 0 < x < \pi \\ 0 & \pi < x < 2\pi \end{cases}$ Period = 2π

From Fig. 7-8 below it is seen that the function is neither even nor odd.

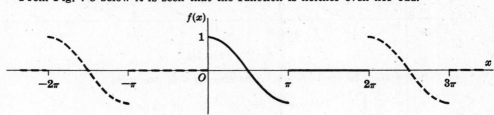

Fig. 7-8

(c) $f(x) = x(10 - x)$, $0 < x < 10$, Period = 10.

From Fig. 7-9 below the function is seen to be even.

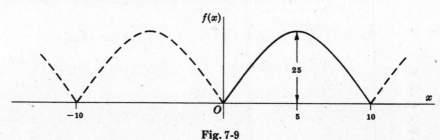

Fig. 7-9

7.9. Show that an even function can have no sine terms in its Fourier expansion.

Method 1.

No sine terms appear if $b_n = 0$, $n = 1, 2, 3, \ldots$. To show this, let us write

$$b_n = \frac{1}{L}\int_{-L}^{L} f(x)\sin\frac{n\pi x}{L}dx = \frac{1}{L}\int_{-L}^{0} f(x)\sin\frac{n\pi x}{L}dx + \frac{1}{L}\int_{0}^{L} f(x)\sin\frac{n\pi x}{L}dx \qquad (1)$$

If we make the transformation $x = -u$ in the first integral on the right of (1), we obtain

$$\frac{1}{L}\int_{-L}^{0} f(x)\sin\frac{n\pi x}{L}dx = \frac{1}{L}\int_{0}^{L} f(-u)\sin\left(-\frac{n\pi u}{L}\right)du = -\frac{1}{L}\int_{0}^{L} f(-u)\sin\frac{n\pi u}{L}du \qquad (2)$$

$$= -\frac{1}{L}\int_{0}^{L} f(u)\sin\frac{n\pi u}{L}du = -\frac{1}{L}\int_{0}^{L} f(x)\sin\frac{n\pi x}{L}dx$$

where we have used the fact that for an even function $f(-u) = f(u)$ and in the last step that the dummy variable of integration u can be replaced by any other symbol, in particular x. Thus from (1), using (2), we have

$$b_n = -\frac{1}{L}\int_{0}^{L} f(x)\sin\frac{n\pi x}{L}dx + \frac{1}{L}\int_{0}^{L} f(x)\sin\frac{n\pi x}{L}dx = 0$$

Method 2.

Assume

$$f(x) = \frac{a_0}{2} + \sum_{n=1}^{\infty}\left(a_n\cos\frac{n\pi x}{L} + b_n\sin\frac{n\pi x}{L}\right)$$

Then

$$f(-x) = \frac{a_0}{2} + \sum_{n=1}^{\infty}\left(a_n\cos\frac{n\pi x}{L} - b_n\sin\frac{n\pi x}{L}\right)$$

If $f(x)$ is even, $f(-x) = f(x)$. Hence

$$\frac{a_0}{2} + \sum_{n=1}^{\infty}\left(a_n\cos\frac{n\pi x}{L} + b_n\sin\frac{n\pi x}{L}\right) = \frac{a_0}{2} + \sum_{n=1}^{\infty}\left(a_n\cos\frac{n\pi x}{L} - b_n\sin\frac{n\pi x}{L}\right)$$

and so

$$\sum_{n=1}^{\infty} b_n\sin\frac{n\pi x}{L} = 0, \quad \text{i.e.} \quad f(x) = \frac{a_0}{2} + \sum_{n=1}^{\infty} a_n\cos\frac{n\pi x}{L}$$

and no sine terms appear.

In a similar manner we can show that an odd function has no cosine terms (or constant term) in its Fourier expansion.

7.10. If $f(x)$ is even, show that (a) $a_n = \dfrac{2}{L}\displaystyle\int_{0}^{L} f(x)\cos\dfrac{n\pi x}{L}dx$, (b) $b_n = 0$.

(a) $$a_n = \frac{1}{L}\int_{-L}^{L} f(x)\cos\frac{n\pi x}{L}dx = \frac{1}{L}\int_{-L}^{0} f(x)\cos\frac{n\pi x}{L}dx + \frac{1}{L}\int_{0}^{L} f(x)\cos\frac{n\pi x}{L}dx$$

Letting $x = -u$,

$$\frac{1}{L}\int_{-L}^{0} f(x)\cos\frac{n\pi x}{L}dx = \frac{1}{L}\int_{0}^{L} f(-u)\cos\left(\frac{-n\pi u}{L}\right)du = \frac{1}{L}\int_{0}^{L} f(u)\cos\frac{n\pi u}{L}du$$

since by definition of an even function $f(-u) = f(u)$. Then

$$a_n = \frac{1}{L}\int_{0}^{L} f(u)\cos\frac{n\pi u}{L}du + \frac{1}{L}\int_{0}^{L} f(x)\cos\frac{n\pi x}{L}dx = \frac{2}{L}\int_{0}^{L} f(x)\cos\frac{n\pi x}{L}dx$$

(b) This follows by Method 1 of Problem 7.9.

7.11. Expand $f(x) = \sin x$, $0 < x < \pi$, in a Fourier cosine series.

A Fourier series consisting of cosine terms alone is obtained only for an even function. Hence we extend the definition of $f(x)$ so that it becomes even (dashed part of Fig. 7-10 below). With this extension, $f(x)$ is then defined in an interval of length 2π. Taking the period as 2π, we have $2L = 2\pi$ so that $L = \pi$.

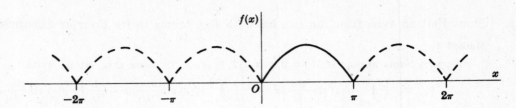

Fig. 7-10

By Problem 7.10, $b_n = 0$ and

$$a_n = \frac{2}{L} \int_0^L f(x) \cos \frac{n\pi x}{L} dx = \frac{2}{\pi} \int_0^\pi \sin x \cos nx \, dx$$

$$= \frac{1}{\pi} \int_0^\pi \{\sin (x + nx) + \sin (x - nx)\} \, dx = \frac{1}{\pi} \left\{ -\frac{\cos (n+1)x}{n+1} + \frac{\cos (n-1)x}{n-1} \right\} \Big|_0^\pi$$

$$= \frac{1}{\pi} \left\{ \frac{1 - \cos (n+1)\pi}{n+1} + \frac{\cos (n-1)\pi - 1}{n-1} \right\} = \frac{1}{\pi} \left\{ -\frac{1 + \cos n\pi}{n+1} - \frac{1 + \cos n\pi}{n-1} \right\}$$

$$= \frac{-2(1 + \cos n\pi)}{\pi(n^2 - 1)} \quad \text{if } n \neq 1$$

For $n = 1$, $\quad a_1 = \frac{2}{\pi} \int_0^\pi \sin x \cos x \, dx = \frac{2}{\pi} \frac{\sin^2 x}{2} \Big|_0^\pi = 0.$

For $n = 0$, $\quad a_0 = \frac{2}{\pi} \int_0^\pi \sin x \, dx = \frac{2}{\pi} (-\cos x) \Big|_0^\pi = \frac{4}{\pi}.$

Then $\qquad f(x) = \frac{2}{\pi} - \frac{2}{\pi} \sum_{n=2}^\infty \frac{(1 + \cos n\pi)}{n^2 - 1} \cos nx$

$$= \frac{2}{\pi} - \frac{4}{\pi} \left(\frac{\cos 2x}{2^2 - 1} + \frac{\cos 4x}{4^2 - 1} + \frac{\cos 6x}{6^2 - 1} + \cdots \right)$$

7.12. Expand $f(x) = x$, $0 < x < 2$, in a half range (a) sine series, (b) cosine series.

(a) Extend the definition of the given function to that of the odd function of period 4 shown in Fig. 7-11 below. This is sometimes called the *odd extension* of $f(x)$. Then $2L = 4$, $L = 2$.

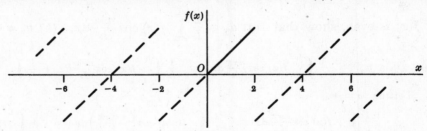

Fig. 7-11

Thus $a_n = 0$ and

$$b_n = \frac{2}{L} \int_0^L f(x) \sin \frac{n\pi x}{L} dx = \frac{2}{2} \int_0^2 x \sin \frac{n\pi x}{2} dx$$

$$= \left\{ (x) \left(\frac{-2}{n\pi} \cos \frac{n\pi x}{2} \right) - (1) \left(\frac{-4}{n^2\pi^2} \sin \frac{n\pi x}{2} \right) \right\} \Big|_0^2 = \frac{-4}{n\pi} \cos n\pi$$

Then $\qquad f(x) = \sum_{n=1}^\infty \frac{-4}{n\pi} \cos n\pi \sin \frac{n\pi x}{2}$

$$= \frac{4}{\pi} \left(\sin \frac{\pi x}{2} - \frac{1}{2} \sin \frac{2\pi x}{2} + \frac{1}{3} \sin \frac{3\pi x}{2} - \cdots \right)$$

(b) Extend the definition of $f(x)$ to that of the even function of period 4 shown in Fig. 7-12 below. This is the *even extension* of $f(x)$.   Then $2L = 4$, $L = 2$.

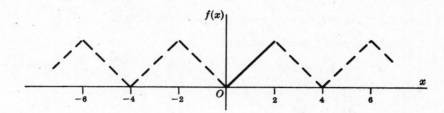

Fig. 7-12

Thus $b_n = 0$,

$$a_n = \frac{2}{L} \int_0^L f(x) \cos \frac{n\pi x}{L} \, dx = \frac{2}{2} \int_0^2 x \cos \frac{n\pi x}{2} \, dx$$

$$= \left\{ (x)\left(\frac{2}{n\pi} \sin \frac{n\pi x}{2} \right) - (1)\left(\frac{-4}{n^2\pi^2} \cos \frac{n\pi x}{2} \right) \right\} \Big|_0^2$$

$$= \frac{4}{n^2\pi^2} (\cos n\pi - 1) \qquad \text{if } n \neq 0$$

If $n = 0$, $a_0 = \int_0^2 x \, dx = 2$.

Then $$f(x) = 1 + \sum_{n=1}^{\infty} \frac{4}{n^2\pi^2} (\cos n\pi - 1) \cos \frac{n\pi x}{2}$$

$$= 1 - \frac{8}{\pi^2} \left(\cos \frac{\pi x}{2} + \frac{1}{3^2} \cos \frac{3\pi x}{2} + \frac{1}{5^2} \cos \frac{5\pi x}{2} + \cdots \right)$$

It should be noted that the given function $f(x) = x$, $0 < x < 2$, is represented *equally well* by the two *different* series in (a) and (b).

PARSEVAL'S IDENTITY

7.13. Assuming that the Fourier series corresponding to $f(x)$ converges uniformly to $f(x)$ in $(-L, L)$, prove Parseval's identity

$$\frac{1}{L} \int_{-L}^{L} \{f(x)\}^2 \, dx = \frac{a_0^2}{2} + \sum (a_n^2 + b_n^2)$$

where the integral is assumed to exist.

If $f(x) = \frac{a_0}{2} + \sum_{n=1}^{\infty} \left(a_n \cos \frac{n\pi x}{L} + b_n \sin \frac{n\pi x}{L} \right)$, then multiplying by $f(x)$ and integrating term by term from $-L$ to L (which is justified since the series is uniformly convergent) we obtain

$$\int_{-L}^{L} \{f(x)\}^2 \, dx = \frac{a_0}{2} \int_{-L}^{L} f(x) \, dx + \sum_{n=1}^{\infty} \left\{ a_n \int_{-L}^{L} f(x) \cos \frac{n\pi x}{L} \, dx + b_n \int_{-L}^{L} f(x) \sin \frac{n\pi x}{L} \, dx \right\}$$

$$= \frac{a_0^2}{2} L + L \sum_{n=1}^{\infty} (a_n^2 + b_n^2) \tag{1}$$

where we have used the results

$$\int_{-L}^{L} f(x) \cos \frac{n\pi x}{L} \, dx = La_n, \qquad \int_{-L}^{L} f(x) \sin \frac{n\pi x}{L} \, dx = Lb_n, \qquad \int_{-L}^{L} f(x) \, dx = La_0 \tag{2}$$

obtained from the Fourier coefficients.

The required result follows on dividing both sides of (1) by L.   Parseval's identity is valid under less restrictive conditions than that imposed here.

7.14. (a) Write Parseval's identity corresponding to the Fourier series of Problem 7.12(b).

(b) Determine from (a) the sum S of the series $\dfrac{1}{1^4} + \dfrac{1}{2^4} + \dfrac{1}{3^4} + \cdots + \dfrac{1}{n^4} + \cdots$.

(a) Here $L = 2$, $a_0 = 2$, $a_n = \dfrac{4}{n^2\pi^2}(\cos n\pi - 1)$, $n \neq 0$, $b_n = 0$.

Then Parseval's identity becomes

$$\frac{1}{2}\int_{-2}^{2} \{f(x)\}^2\, dx \;=\; \frac{1}{2}\int_{-2}^{2} x^2\, dx \;=\; \frac{(2)^2}{2} + \sum_{n=1}^{\infty} \frac{16}{n^4\pi^4}(\cos n\pi - 1)^2$$

or $\dfrac{8}{3} = 2 + \dfrac{64}{\pi^4}\left(\dfrac{1}{1^4} + \dfrac{1}{3^4} + \dfrac{1}{5^4} + \cdots\right)$, i.e. $\dfrac{1}{1^4} + \dfrac{1}{3^4} + \dfrac{1}{5^4} + \cdots = \dfrac{\pi^4}{96}$.

(b) $\begin{aligned} S &= \frac{1}{1^4} + \frac{1}{2^4} + \frac{1}{3^4} + \cdots = \left(\frac{1}{1^4} + \frac{1}{3^4} + \frac{1}{5^4} + \cdots\right) + \left(\frac{1}{2^4} + \frac{1}{4^4} + \frac{1}{6^4} + \cdots\right) \\ &= \left(\frac{1}{1^4} + \frac{1}{3^4} + \frac{1}{5^4} + \cdots\right) + \frac{1}{2^4}\left(\frac{1}{1^4} + \frac{1}{2^4} + \frac{1}{3^4} + \cdots\right) \\ &= \frac{\pi^4}{96} + \frac{S}{16}, \quad \text{from which} \quad S = \frac{\pi^4}{90} \end{aligned}$

7.15. Prove that for all positive integers M,

$$\frac{a_0^2}{2} + \sum_{n=1}^{M}(a_n^2 + b_n^2) \;\leq\; \frac{1}{L}\int_{-L}^{L} \{f(x)\}^2\, dx$$

where a_n and b_n are the Fourier coefficients corresponding to $f(x)$, and $f(x)$ is assumed piecewise continuous in $(-L, L)$.

Let
$$S_M(x) \;=\; \frac{a_0}{2} + \sum_{n=1}^{M}\left(a_n \cos\frac{n\pi x}{L} + b_n \sin\frac{n\pi x}{L}\right) \tag{1}$$

For $M = 1, 2, 3, \ldots$ this is the sequence of partial sums of the Fourier series corresponding to $f(x)$.

We have
$$\int_{-L}^{L} \{f(x) - S_M(x)\}^2\, dx \;\geq\; 0 \tag{2}$$

since the integrand is non-negative. Expanding the integrand, we obtain

$$2\int_{-L}^{L} f(x)\, S_M(x)\, dx - \int_{-L}^{L} S_M^2(x)\, dx \;\leq\; \int_{-L}^{L} \{f(x)\}^2\, dx \tag{3}$$

Multiplying both sides of (1) by $2f(x)$ and integrating from $-L$ to L, using equations (2) of Problem 7.13, gives

$$2\int_{-L}^{L} f(x)\, S_M(x)\, dx \;=\; 2L\left\{\frac{a_0^2}{2} + \sum_{n=1}^{M}(a_n^2 + b_n^2)\right\} \tag{4}$$

Also, squaring (1) and integrating from $-L$ to L, using Problem 7.3, we find

$$\int_{-L}^{L} S_M^2(x)\, dx \;=\; L\left\{\frac{a_0^2}{2} + \sum_{n=1}^{M}(a_n^2 + b_n^2)\right\} \tag{5}$$

Substitution of (4) and (5) into (3) and dividing by L yields the required result.

Taking the limit as $M \to \infty$, we obtain *Bessel's inequality*

$$\frac{a_0^2}{2} + \sum_{n=1}^{\infty}(a_n^2 + b_n^2) \;\leq\; \frac{1}{L}\int_{-L}^{L} \{f(x)\}^2\, dx \tag{6}$$

If the equality holds, we have Parseval's identity (Problem 7.13).

We can think of $S_M(x)$ as representing an *approximation* to $f(x)$, while the left hand side of (2), divided by $2L$, represents the *mean square error* of the approximation. Parseval's identity indicates that as $M \to \infty$ the mean square error approaches zero, while Bessel's inequality indicates the possibility that this mean square error does not approach zero.

The results are connected with the idea of *completeness* of an orthonormal set. If, for example, we were to leave out one or more terms in a Fourier series (say $\cos 4\pi x/L$, for example) we could never get the mean square error to approach zero no matter how many terms we took. For an analogy with 3 dimensional vectors, see Problem 7.46.

DIFFERENTIATION AND INTEGRATION OF FOURIER SERIES

7.16. (a) Find a Fourier series for $f(x) = x^2$, $0 < x < 2$, by integrating the series of Problem 7.12(a). (b) Use (a) to evaluate the series $\sum_{n=1}^{\infty} \frac{(-1)^{n-1}}{n^2}$.

(a) From Problem 7.12(a),

$$x = \frac{4}{\pi}\left(\sin\frac{\pi x}{2} - \frac{1}{2}\sin\frac{2\pi x}{2} + \frac{1}{3}\sin\frac{3\pi x}{2} - \cdots\right) \tag{1}$$

Integrating both sides from 0 to x (applying Theorem 7-2, page 184) and multiplying by 2, we find

$$x^2 = C - \frac{16}{\pi^2}\left(\cos\frac{\pi x}{2} - \frac{1}{2^2}\cos\frac{2\pi x}{2} + \frac{1}{3^2}\cos\frac{3\pi x}{2} - \cdots\right) \tag{2}$$

where $C = \frac{16}{\pi^2}\left(1 - \frac{1}{2^2} + \frac{1}{3^2} - \frac{1}{4^2} + \cdots\right)$.

(b) To determine C in another way, note that (2) represents the Fourier cosine series for x^2 in $0 < x < 2$. Then since $L = 2$ in this case,

$$C = \frac{a_0}{2} = \frac{1}{L}\int_0^L f(x)\,dx = \frac{1}{2}\int_0^2 x^2\,dx = \frac{4}{3}$$

Then from the value of C in (a), we have

$$\sum_{n=1}^{\infty} \frac{(-1)^{n-1}}{n^2} = 1 - \frac{1}{2^2} + \frac{1}{3^2} - \frac{1}{4^2} + \cdots = \frac{\pi^2}{16}\cdot\frac{4}{3} = \frac{\pi^2}{12}$$

7.17. Show that term by term differentiation of the series in Problem 7.12(a) is not valid.

Term by term differentiation yields $2\left(\cos\frac{\pi x}{2} - \cos\frac{2\pi x}{2} + \cos\frac{3\pi x}{2} - \cdots\right)$.

Since the nth term of this series does not approach 0, the series does not converge for any value of x.

CONVERGENCE OF FOURIER SERIES

7.18. Prove that   (a) $\frac{1}{2} + \cos t + \cos 2t + \cdots + \cos Mt = \frac{\sin(M + \frac{1}{2})t}{2\sin\frac{1}{2}t}$

(b) $\frac{1}{\pi}\int_0^\pi \frac{\sin(M + \frac{1}{2})t}{2\sin\frac{1}{2}t}\,dt = \frac{1}{2}$, $\frac{1}{\pi}\int_{-\pi}^0 \frac{\sin(M + \frac{1}{2})t}{2\sin\frac{1}{2}t}\,dt = \frac{1}{2}$.

(a) We have $\cos nt \sin\frac{1}{2}t = \frac{1}{2}\{\sin(n + \frac{1}{2})t - \sin(n - \frac{1}{2})t\}$.

Then summing from $n = 1$ to M,

$$\sin\frac{1}{2}t\{\cos t + \cos 2t + \cdots + \cos Mt\} = (\sin\tfrac{3}{2}t - \sin\tfrac{1}{2}t) + (\sin\tfrac{5}{2}t - \sin\tfrac{3}{2}t)$$
$$+ \cdots + \left(\sin(M + \tfrac{1}{2})t - \sin(M - \tfrac{1}{2})t\right)$$
$$= \tfrac{1}{2}\{\sin(M + \tfrac{1}{2})t - \sin\tfrac{1}{2}t\}$$

On dividing by $\sin\frac{1}{2}t$ and adding $\frac{1}{2}$, the required result follows.

(b) Integrate the result in (a) from $-\pi$ to 0 and 0 to π respectively. This gives the required results, since the integrals of all the cosine terms are zero.

7.19. Prove that $\lim_{n \to \infty} \int_{-\pi}^{\pi} f(x)\sin nx\,dx = \lim_{n \to \infty} \int_{-\pi}^{\pi} f(x)\cos nx\,dx = 0$ if $f(x)$ is piecewise continuous.

This follows at once from Problem 7.15, since if the series $\frac{a_0^2}{2} + \sum_{n=1}^{\infty}(a_n^2 + b_n^2)$ is convergent, $\lim_{n \to \infty} a_n = \lim_{n \to \infty} b_n = 0$.

The result is sometimes called *Riemann's theorem*.

7.20. Prove that $\lim\limits_{M \to \infty} \int_{-\pi}^{\pi} f(x) \sin(M + \tfrac{1}{2})x\, dx = 0$ if $f(x)$ is piecewise continuous.

We have

$$\int_{-\pi}^{\pi} f(x) \sin(M + \tfrac{1}{2})x\, dx = \int_{-\pi}^{\pi} \{f(x) \sin\tfrac{1}{2}x\} \cos Mx\, dx + \int_{-\pi}^{\pi} \{f(x) \cos\tfrac{1}{2}x\} \sin Mx\, dx$$

Then the required result follows at once by using the result of Problem 7.19, with $f(x)$ replaced by $f(x) \sin\tfrac{1}{2}x$ and $f(x) \cos\tfrac{1}{2}x$ respectively which are piecewise continuous if $f(x)$ is.

The result can also be proved when the integration limits are a and b instead of $-\pi$ and π.

7.21. Assuming that $L = \pi$, i.e. that the Fourier series corresponding to $f(x)$ has period $2L = 2\pi$, show that

$$S_M(x) = \frac{a_0}{2} + \sum_{n=1}^{M} (a_n \cos nx + b_n \sin nx) = \frac{1}{\pi} \int_{-\pi}^{\pi} f(t+x) \frac{\sin(M + \tfrac{1}{2})t}{2 \sin\tfrac{1}{2}t}\, dt$$

Using the formulas for the Fourier coefficients with $L = \pi$, we have

$$a_n \cos nx + b_n \sin nx = \left(\frac{1}{\pi} \int_{-\pi}^{\pi} f(u) \cos nu\, du\right) \cos nx + \left(\frac{1}{\pi} \int_{-\pi}^{\pi} f(u) \sin nu\, du\right) \sin nx$$

$$= \frac{1}{\pi} \int_{-\pi}^{\pi} f(u) (\cos nu \cos nx + \sin nu \sin nx)\, du$$

$$= \frac{1}{\pi} \int_{-\pi}^{\pi} f(u) \cos n(u - x)\, du$$

Also,

$$\frac{a_0}{2} = \frac{1}{2\pi} \int_{-\pi}^{\pi} f(u)\, du$$

Then

$$S_M(x) = \frac{a_0}{2} + \sum_{n=1}^{M} (a_n \cos nx + b_n \sin nx)$$

$$= \frac{1}{2\pi} \int_{-\pi}^{\pi} f(u)\, du + \frac{1}{\pi} \sum_{n=1}^{M} \int_{-\pi}^{\pi} f(u) \cos n(u - x)\, du$$

$$= \frac{1}{\pi} \int_{-\pi}^{\pi} f(u) \left\{\frac{1}{2} + \sum_{n=1}^{M} \cos n(u - x)\right\} du$$

$$= \frac{1}{\pi} \int_{-\pi}^{\pi} f(u) \frac{\sin(M + \tfrac{1}{2})(u - x)}{2 \sin\tfrac{1}{2}(u - x)}\, du$$

using Problem 7.18. Letting $u - x = t$, we have

$$S_M(x) = \frac{1}{\pi} \int_{-\pi-x}^{\pi-x} f(t+x) \frac{\sin(M + \tfrac{1}{2})t}{2 \sin\tfrac{1}{2}t}\, dt$$

Since the integrand has period 2π, we can replace the interval $-\pi - x, \pi - x$ by any other interval of length 2π, in particular $-\pi, \pi$. Thus we obtain the required result.

7.22. Prove that

$$S_M(x) - \left(\frac{f(x+0) + f(x-0)}{2}\right) = \frac{1}{\pi} \int_{-\pi}^{0} \left(\frac{f(t+x) - f(x-0)}{2 \sin\tfrac{1}{2}t}\right) \sin(M + \tfrac{1}{2})t\, dt$$

$$+ \frac{1}{\pi} \int_{0}^{\pi} \left(\frac{f(t+x) - f(x+0)}{2 \sin\tfrac{1}{2}t}\right) \sin(M + \tfrac{1}{2})t\, dt$$

From Problem 7.21,

$$S_M(x) = \frac{1}{\pi} \int_{-\pi}^{0} f(t+x) \frac{\sin(M + \tfrac{1}{2})t}{2 \sin\tfrac{1}{2}t}\, dt + \frac{1}{\pi} \int_{0}^{\pi} f(t+x) \frac{\sin(M + \tfrac{1}{2})t}{2 \sin\tfrac{1}{2}t}\, dt \qquad (1)$$

Multiplying the integrals of Problem 7.18(b) by $f(x-0)$ and $f(x+0)$ respectively,

$$\frac{f(x+0) + f(x-0)}{2} = \frac{1}{\pi} \int_{-\pi}^{0} f(x-0) \frac{\sin(M + \tfrac{1}{2})t}{2 \sin\tfrac{1}{2}t}\, dt + \frac{1}{\pi} \int_{0}^{\pi} f(x+0) \frac{\sin(M + \tfrac{1}{2})t}{2 \sin\tfrac{1}{2}t}\, dt \qquad (2)$$

Subtracting (2) from (1) yields the required result.

7.23. If $f(x)$ and $f'(x)$ are piecewise continuous in $(-\pi, \pi)$, prove that

$$\lim_{M \to \infty} S_M(x) = \frac{f(x+0) + f(x-0)}{2}$$

The function $\dfrac{f(t+x) - f(x+0)}{2 \sin \frac{1}{2}t}$ is piecewise continuous in $0 < t \leqq \pi$ because $f(x)$ is piecewise continuous.

Also,

$$\lim_{t \to 0+} \frac{f(t+x) - f(x+0)}{2 \sin \frac{1}{2}t} = \lim_{t \to 0+} \frac{f(t+x) - f(x+0)}{t} \cdot \frac{t}{2 \sin \frac{1}{2}t} = \lim_{t \to 0+} \frac{f(t+x) - f(x+0)}{t}$$

exists, since by hypothesis $f'(x)$ is piecewise continuous so that the right hand derivative of $f(x)$ at each x exists.

Thus $\dfrac{f(t+x) - f(x+0)}{2 \sin \frac{1}{2}t}$ is piecewise continuous in $0 \leqq t \leqq \pi$.

Similarly, $\dfrac{f(t+x) - f(x-0)}{2 \sin \frac{1}{2}t}$ is piecewise continuous in $-\pi \leqq t \leqq 0$.

Then from Problems 7.20 and 7.22, we have

$$\lim_{M \to \infty} S_M(x) - \left\{ \frac{f(x+0) + f(x-0)}{2} \right\} = 0 \quad \text{or} \quad \lim_{M \to \infty} S_M(x) = \frac{f(x+0) + f(x-0)}{2}$$

ORTHOGONAL FUNCTIONS

7.24. (*a*) Show that the set of functions

$$1, \ \sin\frac{\pi x}{L}, \ \cos\frac{\pi x}{L}, \ \sin\frac{2\pi x}{L}, \ \cos\frac{2\pi x}{L}, \ \sin\frac{3\pi x}{L}, \ \cos\frac{3\pi x}{L}, \ldots$$

forms an orthogonal set in the interval $(-L, L)$.

(*b*) Determine the corresponding normalizing constants for the set in (*a*) so that the set is orthonormal in $(-L, L)$.

(*a*) This follows at once from the results of Problems 7.2 and 7.3.

(*b*) By Problem 7.3,

$$\int_{-L}^{L} \sin^2 \frac{m\pi x}{L} \, dx = L, \qquad \int_{-L}^{L} \cos^2 \frac{m\pi x}{L} \, dx = L$$

Then

$$\int_{-L}^{L} \left(\sqrt{\frac{1}{L}} \sin \frac{m\pi x}{L} \right)^2 dx = 1, \qquad \int_{-L}^{L} \left(\sqrt{\frac{1}{L}} \cos \frac{m\pi x}{L} \right)^2 dx = 1$$

Also,

$$\int_{-L}^{L} (1)^2 \, dx = 2L \quad \text{or} \quad \int_{-L}^{L} \left(\frac{1}{\sqrt{2L}} \right)^2 dx = 1$$

Thus the required orthonormal set is given by

$$\frac{1}{\sqrt{2L}}, \ \frac{1}{\sqrt{L}} \sin\frac{\pi x}{L}, \ \frac{1}{\sqrt{L}} \cos\frac{\pi x}{L}, \ \frac{1}{\sqrt{L}} \sin\frac{2\pi x}{L}, \ \frac{1}{\sqrt{L}} \cos\frac{2\pi x}{L}, \ \ldots$$

7.25. Let $\{\phi_n(x)\}$ be a set of functions which are mutually orthonormal in (a, b). Prove that if $\displaystyle\sum_{n=1}^{\infty} c_n \phi_n(x)$ converges uniformly to $f(x)$ in (a, b), then

$$c_n = \int_a^b f(x) \, \phi_n(x) \, dx$$

Multiplying both sides of

$$f(x) = \sum_{n=1}^{\infty} c_n \phi_n(x) \qquad\qquad (1)$$

by $\phi_m(x)$ and integrating from a to b, we have

$$\int_a^b f(x)\,\phi_m(x)\,dx \;=\; \sum_{n=1}^{\infty} c_n \int_a^b \phi_m(x)\,\phi_n(x)\,dx \tag{2}$$

where the interchange of integration and summation is justified by using the fact that the series converges uniformly to $f(x)$. Now since the functions $\{\phi_n(x)\}$ are mutually orthonormal in (a,b), we have

$$\int_a^b \phi_m(x)\,\phi_n(x)\,dx \;=\; \begin{cases} 0 & m \neq n \\ 1 & m = n \end{cases}$$

so that (2) becomes

$$\int_a^b f(x)\,\phi_m(x)\,dx \;=\; c_m \tag{3}$$

as required.

We call the coefficients c_m given by (3) the *generalized Fourier coefficients* corresponding to $f(x)$ even though nothing may be known about the convergence of the series in (1). As in the case of Fourier series, convergence of $\sum_{n=1}^{\infty} c_n \phi_n(x)$ is then investigated using the coefficients (3). The conditions of convergence depend of course on the types of orthonormal functions used.

Supplementary Problems

FOURIER SERIES

7.26. Graph each of the following functions and find their corresponding Fourier series using properties of even and odd functions wherever applicable.

(a) $f(x) = \begin{cases} 8 & 0 < x < 2 \\ -8 & 2 < x < 4 \end{cases}$ Period 4   (b) $f(x) = \begin{cases} -x & -4 \leq x \leq 0 \\ x & 0 \leq x \leq 4 \end{cases}$ Period 8

(c) $f(x) = 4x,\ 0 < x < 10,$ Period 10   (d) $f(x) = \begin{cases} 2x & 0 \leq x < 3 \\ 0 & -3 < x < 0 \end{cases}$ Period 6

7.27. In each part of Problem 7.26, tell where the discontinuities of $f(x)$ are located and to what value the series converges at these discontinuities.

7.28. Expand $f(x) = \begin{cases} 2-x & 0 < x < 4 \\ x-6 & 4 < x < 8 \end{cases}$ in a Fourier series of period 8.

7.29. (a) Expand $f(x) = \cos x,\ 0 < x < \pi$, in a Fourier sine series.

(b) How should $f(x)$ be defined at $x = 0$ and $x = \pi$ so that the series will converge to $f(x)$ for $0 \leq x \leq \pi$?

7.30. (a) Expand in a Fourier series $f(x) = \cos x,\ 0 < x < \pi$ if the period is π; and (b) compare with the result of Problem 7.29, explaining the similarities and differences if any.

7.31. Expand $f(x) = \begin{cases} x & 0 < x < 4 \\ 8-x & 4 < x < 8 \end{cases}$ in a series of (a) sines, (b) cosines.

7.32. Prove that for $0 \leq x \leq \pi$,

(a) $x(\pi - x) \;=\; \dfrac{\pi^2}{6} - \left(\dfrac{\cos 2x}{1^2} + \dfrac{\cos 4x}{2^2} + \dfrac{\cos 6x}{3^2} + \cdots \right)$

(b) $x(\pi - x) \;=\; \dfrac{8}{\pi} \left(\dfrac{\sin x}{1^3} + \dfrac{\sin 3x}{3^3} + \dfrac{\sin 5x}{5^3} + \cdots \right)$

7.33. Use Problem 7.32 to show that

(a) $\displaystyle\sum_{n=1}^{\infty} \dfrac{1}{n^2} = \dfrac{\pi^2}{6},$ (b) $\displaystyle\sum_{n=1}^{\infty} \dfrac{(-1)^{n-1}}{n^2} = \dfrac{\pi^2}{12},$ (c) $\displaystyle\sum_{n=1}^{\infty} \dfrac{(-1)^{n-1}}{(2n-1)^3} = \dfrac{\pi^3}{32}.$

7.34. Show that $\dfrac{1}{1^3} + \dfrac{1}{3^3} - \dfrac{1}{5^3} - \dfrac{1}{7^3} + \dfrac{1}{9^3} + \dfrac{1}{11^3} - \cdots = \dfrac{3\pi^3\sqrt{2}}{128}$.

DIFFERENTIATION AND INTEGRATION OF FOURIER SERIES

7.35. (a) Show that for $-\pi < x < \pi$,

$$x = 2\left(\frac{\sin x}{1} - \frac{\sin 2x}{2} + \frac{\sin 3x}{3} - \cdots\right)$$

(b) By integrating the result of (a), show that for $-\pi \leqq x \leqq \pi$,

$$x^2 = \frac{\pi^2}{3} - 4\left(\frac{\cos x}{1^2} - \frac{\cos 2x}{2^2} + \frac{\cos 3x}{3^2} - \cdots\right)$$

(c) By integrating the result of (b), show that for $-\pi \leqq x \leqq \pi$,

$$x(\pi - x)(\pi + x) = 12\left(\frac{\sin x}{1^3} - \frac{\sin 2x}{2^3} + \frac{\sin 3x}{3^3} - \cdots\right)$$

7.36. (a) Show that for $-\pi < x < \pi$,

$$x\cos x = -\frac{1}{2}\sin x + 2\left(\frac{2}{1\cdot 3}\sin 2x - \frac{3}{2\cdot 4}\sin 3x + \frac{4}{3\cdot 5}\sin 4x - \cdots\right)$$

(b) Use (a) to show that for $-\pi \leqq x \leqq \pi$,

$$x\sin x = 1 - \frac{1}{2}\cos x - 2\left(\frac{\cos 2x}{1\cdot 3} - \frac{\cos 3x}{2\cdot 4} + \frac{\cos 4x}{3\cdot 5} - \cdots\right)$$

7.37. By differentiating the result of Problem 7.32(b), prove that for $0 \leqq x \leqq \pi$,

$$x = \frac{\pi}{2} - \frac{4}{\pi}\left(\frac{\cos x}{1^2} + \frac{\cos 3x}{3^2} + \frac{\cos 5x}{5^2} + \cdots\right)$$

PARSEVAL'S IDENTITY

7.38. By using Problem 7.32 and Parseval's identity, show that

$$(a)\ \sum_{n=1}^{\infty}\frac{1}{n^4} = \frac{\pi^4}{90} \qquad (b)\ \sum_{n=1}^{\infty}\frac{1}{n^6} = \frac{\pi^6}{945}$$

7.39. Show that $\dfrac{1}{1^2\cdot 3^2} + \dfrac{1}{3^2\cdot 5^2} + \dfrac{1}{5^2\cdot 7^2} + \cdots = \dfrac{\pi^2 - 8}{16}$. [*Hint.* Use Problem 7.11.]

7.40. Show that $(a)\ \displaystyle\sum_{n=1}^{\infty}\frac{1}{(2n-1)^4} = \frac{\pi^4}{96}$, $(b)\ \displaystyle\sum_{n=1}^{\infty}\frac{1}{(2n-1)^6} = \frac{\pi^6}{960}$.

7.41. Show that $\dfrac{1}{1^2\cdot 2^2\cdot 3^2} + \dfrac{1}{2^2\cdot 3^2\cdot 4^2} + \dfrac{1}{3^2\cdot 4^2\cdot 5^2} + \cdots = \dfrac{4\pi^2 - 39}{16}$.

ORTHOGONAL FUNCTIONS

7.42. Given the functions a_0, $a_1 + a_2x$, $a_3 + a_4x + a_5x^2$ where $a_0, \ldots, a_5$ are constants. Determine the constants so that these functions are mutually orthonormal in $(-1, 1)$ and thus obtain the functions.

7.43. Generalize Problem 7.42.

7.44. (a) Show that the functions $\dfrac{e^{im\phi}}{\sqrt{2\pi}}$, $m = 0, \pm 1, \pm 2, \ldots$ are mutually orthonormal in $(-\pi, \pi)$. (b) Show how to expand a function $f(x)$ in a series of these functions and explain the connection with Fourier series.

7.45. Let $f(x)$ be approximated by the sum of the first M terms of an orthonormal series

$$\sum_{n=1}^{M} c_n \phi_n(x) = S_M(x)$$

where the functions $\phi_n(x)$ are orthonormal in (a, b). (a) Show that

$$\int_a^b [f(x) - S_M(x)]^2\, dx = \int_a^b [f(x)]^2\, dx - \sum_{n=1}^{M} c_n^2$$

(b) By interpreting

$$\frac{1}{b-a} \int_a^b [f(x) - S_M(x)]^2 \, dx$$

as the *mean square error* of $S_M(x)$ from $f(x)$ [and the square root as the *root mean square* or *r.m.s.* error], show that Parseval's identity is equivalent to the statement that the root mean square error approaches zero as $M \to \infty$.

(c) Show that if the root mean square error may not approach zero as $M \to \infty$, then we still have Bessel's inequality

$$\sum_{n=1}^{\infty} c_n^2 \leq \int_a^b [f(x)]^2 \, dx$$

(d) Discuss the relevance of these results to Fourier series.

7.46. Let **r** be any three dimensional vector. Show that

$$(a) \quad (\mathbf{r} \cdot \mathbf{i})^2 + (\mathbf{r} \cdot \mathbf{j})^2 \leq \mathbf{r}^2 \qquad (b) \quad (\mathbf{r} \cdot \mathbf{i})^2 + (\mathbf{r} \cdot \mathbf{j})^2 + (\mathbf{r} \cdot \mathbf{k})^2 = \mathbf{r}^2$$

and discuss these with reference to Bessel's inequality and Parseval's identity. Compare with Problem 7.15.

7.47. Suppose that one term in any orthonormal series [such as a Fourier series] is omitted. (a) Can we expand a function $f(x)$ into the series? (b) Can Parseval's identity be satisfied? (c) Can Bessel's inequality be satisfied? Justify your answers.

7.48. Let $\{\phi_n(x)\}$, $n = 1, 2, 3, \ldots$, be orthonormal in (a, b). Prove that

$$\int_a^b \left[f(x) - \sum_{n=1}^{M} c_n \phi_n(x) \right]^2 dx$$

is a minimum when

$$c_n = \int_a^b f(x) \, \phi_n(x) \, dx$$

Discuss the connection of this to (a) Fourier series and (b) Problem 7.45.

7.49. (a) Show that the functions $1, 1 - x, 2 - 4x + x^2$ are mutually orthogonal in $(0, \infty)$ with respect to the density function e^{-x}. (b) Obtain a mutually orthonormal set.

7.50. Give a vector interpretation to functions which are orthonormal with respect to a density or weight function.

Answers to Supplementary Problems

7.26. (a) $\dfrac{16}{\pi} \displaystyle\sum_{n=1}^{\infty} \dfrac{(1 - \cos n\pi)}{n} \sin \dfrac{n\pi x}{2}$ $\qquad (b)$ $2 - \dfrac{8}{\pi^2} \displaystyle\sum_{n=1}^{\infty} \dfrac{(1 - \cos n\pi)}{n^2} \cos \dfrac{n\pi x}{4}$

(c) $20 - \dfrac{40}{\pi} \displaystyle\sum_{n=1}^{\infty} \dfrac{1}{n} \sin \dfrac{n\pi x}{5}$ $\qquad (d)$ $\dfrac{3}{2} + \displaystyle\sum_{n=1}^{\infty} \left\{ \dfrac{6(\cos n\pi - 1)}{n^2\pi^2} \cos \dfrac{n\pi x}{3} - \dfrac{6 \cos n\pi}{n\pi} \sin \dfrac{n\pi x}{3} \right\}$

7.27. (a) $x = 0, \pm 2, \pm 4, \ldots; 0$ $\quad (b)$ no discontinuities

(c) $x = 0, \pm 10, \pm 20, \ldots; 20$ $\quad (d)$ $x = \pm 3, \pm 9, \pm 15, \ldots; 3$

7.28. $\dfrac{16}{\pi^2} \left\{ \cos \dfrac{\pi x}{4} + \dfrac{1}{3^2} \cos \dfrac{3\pi x}{4} + \dfrac{1}{5^2} \cos \dfrac{5\pi x}{4} + \cdots \right\}$

7.29. (a) $\dfrac{8}{\pi} \displaystyle\sum_{n=1}^{\infty} \dfrac{n \sin 2nx}{4n^2 - 1}$ $\quad (b)$ $f(0) = f(\pi) = 0$

7.30. Same answer as in Problem 7.29.

7.31. (a) $\dfrac{32}{\pi^2} \displaystyle\sum_{n=1}^{\infty} \dfrac{1}{n^2} \sin \dfrac{n\pi}{2} \sin \dfrac{n\pi x}{8}$ $\qquad (b)$ $2 + \dfrac{16}{\pi^2} \displaystyle\sum_{n=1}^{\infty} \left(\dfrac{2 \cos n\pi/2 - \cos n\pi - 1}{n^2} \right) \cos \dfrac{n\pi x}{8}$

Chapter 8

Fourier Integrals

THE FOURIER INTEGRAL

Let us assume the following conditions on $f(x)$:

1. $f(x)$ satisfies the Dirichlet conditions (page 183) in every finite interval $(-L, L)$.

2. $\int_{-\infty}^{\infty} |f(x)|\, dx$ converges, i.e. $f(x)$ is absolutely integrable in $(-\infty, \infty)$.

Then *Fourier's integral theorem* states that

$$f(x) = \int_0^{\infty} \{A(\alpha) \cos \alpha x + B(\alpha) \sin \alpha x\}\, d\alpha \tag{1}$$

where

$$\begin{cases} A(\alpha) = \dfrac{1}{\pi} \displaystyle\int_{-\infty}^{\infty} f(x) \cos \alpha x\, dx \\[2mm] B(\alpha) = \dfrac{1}{\pi} \displaystyle\int_{-\infty}^{\infty} f(x) \sin \alpha x\, dx \end{cases} \tag{2}$$

The result (1) holds if x is a point of continuity of $f(x)$. If x is a point of discontinuity, we must replace $f(x)$ by $\dfrac{f(x+0) + f(x-0)}{2}$ as in the case of Fourier series. Note that the above conditions are sufficient but not necessary.

The similarity of (1) and (2) with corresponding results for Fourier series is apparent. The right hand side of (1) is sometimes called a *Fourier integral expansion* of $f(x)$.

EQUIVALENT FORMS OF FOURIER'S INTEGRAL THEOREM

Fourier's integral theorem can also be written in the forms

$$f(x) = \frac{1}{\pi} \int_{\alpha=0}^{\infty} \int_{u=-\infty}^{\infty} f(u) \cos \alpha(x-u)\, du\, d\alpha \tag{3}$$

$$f(x) = \frac{1}{2\pi} \int_{-\infty}^{\infty} e^{-i\alpha x}\, d\alpha \int_{-\infty}^{\infty} f(u)\, e^{i\alpha u}\, du \tag{4}$$

$$= \frac{1}{2\pi} \int_{-\infty}^{\infty} \int_{-\infty}^{\infty} f(u)\, e^{i\alpha(u-x)}\, du\, d\alpha$$

where it is understood that if $f(x)$ is not continuous at x the left side must be replaced by $\dfrac{f(x+0) + f(x-0)}{2}$.

These results can be simplified somewhat if $f(x)$ is either an odd or an even function, and we have

$$f(x) = \frac{2}{\pi} \int_0^{\infty} \cos \alpha x\, d\alpha \int_0^{\infty} f(u) \cos \alpha u\, du \qquad \text{if } f(x) \text{ is even} \tag{5}$$

$$f(x) = \frac{2}{\pi} \int_0^{\infty} \sin \alpha x\, d\alpha \int_0^{\infty} f(u) \sin \alpha u\, du \qquad \text{if } f(x) \text{ is odd} \tag{6}$$

FOURIER TRANSFORMS

From (4) it follows that if

$$F(\alpha) = \frac{1}{\sqrt{2\pi}} \int_{-\infty}^{\infty} f(u)\, e^{i\alpha u}\, du \tag{7}$$

then

$$f(x) = \frac{1}{\sqrt{2\pi}} \int_{-\infty}^{\infty} F(\alpha)\, e^{-i\alpha x}\, d\alpha \tag{8}$$

The function $F(\alpha)$ is called the *Fourier transform* of $f(x)$ and is sometimes written $F(\alpha) = \mathcal{F}\{f(x)\}$. The function $f(x)$ is the *inverse Fourier transform* of $F(\alpha)$ and is written $f(x) = \mathcal{F}^{-1}\{F(\alpha)\}$.

Note: The constants preceding the integral signs in (7) and (8) were here taken as equal to $1/\sqrt{2\pi}$. However, they can be any constants different from zero so long as their product is $1/2\pi$. The above is called the *symmetric form*.

If $f(x)$ is an even function, equation (5) yields

$$\begin{cases} F_c(\alpha) = \sqrt{\dfrac{2}{\pi}} \displaystyle\int_0^{\infty} f(u) \cos \alpha u\, du \\[2mm] f(x) = \sqrt{\dfrac{2}{\pi}} \displaystyle\int_0^{\infty} F_c(\alpha) \cos \alpha x\, d\alpha \end{cases} \tag{9}$$

and we call $F_c(\alpha)$ and $f(x)$ *Fourier cosine transforms* of each other.

If $f(x)$ is an odd function, equation (6) yields

$$\begin{cases} F_s(\alpha) = \sqrt{\dfrac{2}{\pi}} \displaystyle\int_0^{\infty} f(u) \sin \alpha u\, du \\[2mm] f(x) = \sqrt{\dfrac{2}{\pi}} \displaystyle\int_0^{\infty} F_s(\alpha) \sin \alpha x\, d\alpha \end{cases} \tag{10}$$

and we call $F_s(\alpha)$ and $f(x)$ *Fourier sine transforms* of each other.

PARSEVAL'S IDENTITIES FOR FOURIER INTEGRALS

If $F_s(\alpha)$ and $G_s(\alpha)$ are Fourier sine transforms of $f(x)$ and $g(x)$ respectively, then

$$\int_0^{\infty} F_s(\alpha)\, G_s(\alpha)\, d\alpha = \int_0^{\infty} f(x)\, g(x)\, dx \tag{11}$$

Similarly if $F_c(\alpha)$ and $G_c(\alpha)$ are Fourier cosine transforms of $f(x)$ and $g(x)$, then

$$\int_0^{\infty} F_c(\alpha)\, G_c(\alpha)\, d\alpha = \int_0^{\infty} f(x)\, g(x)\, dx \tag{12}$$

In the special case where $f(x) = g(x)$, (11) and (12) become respectively

$$\int_0^{\infty} \{F_s(\alpha)\}^2\, d\alpha = \int_0^{\infty} \{f(x)\}^2\, dx \tag{13}$$

$$\int_0^{\infty} \{F_c(\alpha)\}^2\, d\alpha = \int_0^{\infty} \{f(x)\}^2\, dx \tag{14}$$

The above relations are known as *Parseval's identities* for integrals. Similar relations hold for general Fourier transforms. Thus if $F(\alpha)$ and $G(\alpha)$ are Fourier transforms of $f(x)$ and $g(x)$ respectively, we can prove that

$$\int_{-\infty}^{\infty} F(\alpha)\,\overline{G(\alpha)}\,d\alpha \;=\; \int_{-\infty}^{\infty} f(x)\,\overline{g(x)}\,dx \tag{15}$$

where the bar signifies the complex conjugate obtained by replacing i by $-i$. See Problem 8.24.

THE CONVOLUTION THEOREM

If $F(\alpha)$ and $G(\alpha)$ are the Fourier transforms of $f(x)$ and $g(x)$ respectively, then

$$\int_{-\infty}^{\infty} F(\alpha)\,G(\alpha)\,e^{-i\alpha x}\,d\alpha \;=\; \int_{-\infty}^{\infty} f(u)\,g(x-u)\,du \tag{16}$$

If we define the *convolution*, denoted by $f * g$, of the functions f and g to be

$$f * g \;=\; \frac{1}{\sqrt{2\pi}} \int_{-\infty}^{\infty} f(u)\,g(x-u)\,du \tag{17}$$

then (16) can be written

$$\mathcal{F}\{f * g\} \;=\; \mathcal{F}\{f\}\,\mathcal{F}\{g\} \tag{18}$$

or in words, the Fourier transform of the convolution of two functions is equal to the product of their Fourier transforms. This is called the *convolution theorem for Fourier transforms*.

Solved Problems

THE FOURIER INTEGRAL AND FOURIER TRANSFORMS

8.1. (a) Find the Fourier transform of $\quad f(x) = \begin{cases} 1 & |x| < a \\ 0 & |x| > a \end{cases}$.

(b) Graph $f(x)$ and its Fourier transform for $a = 3$.

(a) The Fourier transform of $f(x)$ is

$$F(\alpha) \;=\; \frac{1}{\sqrt{2\pi}} \int_{-\infty}^{\infty} f(u)\,e^{i\alpha u}\,du \;=\; \frac{1}{\sqrt{2\pi}} \int_{-a}^{a} (1)\,e^{i\alpha u}\,du \;=\; \frac{1}{\sqrt{2\pi}}\,\frac{e^{i\alpha u}}{i\alpha}\bigg|_{-a}^{a}$$

$$=\; \frac{1}{\sqrt{2\pi}}\left(\frac{e^{i\alpha a} - e^{-i\alpha a}}{i\alpha}\right) \;=\; \sqrt{\frac{2}{\pi}}\,\frac{\sin \alpha a}{\alpha}, \quad \alpha \neq 0$$

For $\alpha = 0$, we obtain $F(\alpha) = \sqrt{2/\pi}\,a$.

(b) The graphs of $f(x)$ and $F(\alpha)$ for $a = 3$ are shown in Figs. 8-1 and 8-2 respectively.

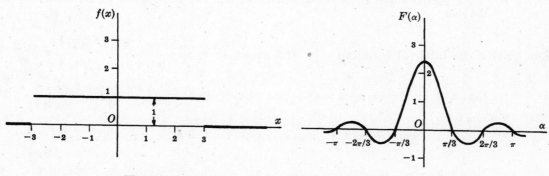

Fig. 8-1   Fig. 8-2

8.2. (a) Use the result of Problem 8.1 to evaluate $\displaystyle\int_{-\infty}^{\infty} \frac{\sin \alpha a \cos \alpha x}{\alpha} \, d\alpha$.

   (b) Deduce the value of $\displaystyle\int_{0}^{\infty} \frac{\sin u}{u} \, du$.

(a) From Fourier's integral theorem, if

$$F(\alpha) = \frac{1}{\sqrt{2\pi}} \int_{-\infty}^{\infty} f(u)\, e^{i\alpha u} \, du \qquad \text{then} \qquad f(x) = \frac{1}{\sqrt{2\pi}} \int_{-\infty}^{\infty} F(\alpha)\, e^{-i\alpha x} \, d\alpha$$

Then from Problem 8.1,

$$\frac{1}{\sqrt{2\pi}} \int_{-\infty}^{\infty} \sqrt{\frac{2}{\pi}} \frac{\sin \alpha a}{\alpha} e^{-i\alpha x} \, d\alpha = \begin{cases} 1 & |x| < a \\ 1/2 & |x| = a \\ 0 & |x| > a \end{cases} \qquad (1)$$

The left side of (1) is equal to

$$\frac{1}{\pi} \int_{-\infty}^{\infty} \frac{\sin \alpha a \cos \alpha x}{\alpha} \, d\alpha - \frac{i}{\pi} \int_{-\infty}^{\infty} \frac{\sin \alpha a \sin \alpha x}{\alpha} \, d\alpha \qquad (2)$$

The integrand in the second integral of (2) is odd and so the integral is zero. Then from (1) and (2), we have

$$\int_{-\infty}^{\infty} \frac{\sin \alpha a \cos \alpha x}{\alpha} \, d\alpha = \begin{cases} \pi & |x| < a \\ \pi/2 & |x| = a \\ 0 & |x| > a \end{cases} \qquad (3)$$

(b) If $x = 0$ and $a = 1$ in the result of (a), we have

$$\int_{-\infty}^{\infty} \frac{\sin \alpha}{\alpha} \, d\alpha = \pi \qquad \text{or} \qquad \int_{0}^{\infty} \frac{\sin \alpha}{\alpha} \, d\alpha = \frac{\pi}{2}$$

since the integrand is even.

8.3. If $f(x)$ is an even function show that:

   (a) $F(\alpha) = \sqrt{\dfrac{2}{\pi}} \displaystyle\int_{0}^{\infty} f(u) \cos \alpha u \, du$, (b) $f(x) = \sqrt{\dfrac{2}{\pi}} \displaystyle\int_{0}^{\infty} F(\alpha) \cos \alpha x \, d\alpha$.

We have

$$F(\alpha) = \frac{1}{\sqrt{2\pi}} \int_{-\infty}^{\infty} f(u)\, e^{i\alpha u} \, du = \frac{1}{\sqrt{2\pi}} \int_{-\infty}^{\infty} f(u) \cos \alpha u \, du + \frac{i}{\sqrt{2\pi}} \int_{-\infty}^{\infty} f(u) \sin \alpha u \, du \qquad (1)$$

(a) If $f(u)$ is even, $f(u) \cos \lambda u$ is even and $f(u) \sin \lambda u$ is odd. Then the second integral on the right of (1) is zero and the result can be written

$$F(\alpha) = \frac{2}{\sqrt{2\pi}} \int_{0}^{\infty} f(u) \cos \alpha u \, du = \sqrt{\frac{2}{\pi}} \int_{0}^{\infty} f(u) \cos \alpha u \, du$$

(b) From (a), $F(-\alpha) = F(\alpha)$ so that $F(\alpha)$ is an even function. Then by using a proof exactly analogous to that in (a), the required result follows.

A similar result holds for odd functions and can be obtained by replacing the cosine by the sine.

8.4. Solve the integral equation $\displaystyle\int_{0}^{\infty} f(x) \cos \alpha x \, dx = \begin{cases} 1 - \alpha & 0 \leqq \alpha \leqq 1 \\ 0 & \alpha > 1 \end{cases}$

Let $\sqrt{\dfrac{2}{\pi}} \displaystyle\int_{0}^{\infty} f(x) \cos \alpha x \, dx = F(\alpha)$ and choose $F(\alpha) = \begin{cases} \sqrt{2/\pi}\,(1-\alpha) & 0 \leqq \alpha \leqq 1 \\ 0 & \alpha > 1 \end{cases}$. Then by Problem 8.3,

$$f(x) = \sqrt{\frac{2}{\pi}} \int_{0}^{\infty} F(\alpha) \cos \alpha x \, d\alpha = \sqrt{\frac{2}{\pi}} \int_{0}^{1} \sqrt{\frac{2}{\pi}} (1-\alpha) \cos \alpha x \, d\alpha$$

$$= \frac{2}{\pi} \int_{0}^{1} (1-\alpha) \cos \alpha x \, d\alpha = \frac{2(1 - \cos x)}{\pi x^2}$$

8.5. Use Problem 8.4 to show that $\displaystyle\int_0^\infty \frac{\sin^2 u}{u^2}\,du = \frac{\pi}{2}$.

As obtained in Problem 8.4,

$$\frac{2}{\pi}\int_0^\infty \frac{1-\cos x}{x^2}\cos \alpha x\,dx \;=\; \begin{cases} 1-\alpha & 0 \leqq \alpha \leqq 1 \\ 0 & \alpha > 1 \end{cases}$$

Taking the limit as $\alpha \to 0+$, we find

$$\int_0^\infty \frac{1-\cos x}{x^2}\,dx \;=\; \frac{\pi}{2}$$

But this integral can be written as $\displaystyle\int_0^\infty \frac{2\sin^2 (x/2)}{x^2}\,dx$ which becomes $\displaystyle\int_0^\infty \frac{\sin^2 u}{u^2}\,du$ on letting $x = 2u$, so that the required result follows.

8.6. Show that $\displaystyle\int_0^\infty \frac{\cos \alpha x}{\alpha^2 + 1}\,d\alpha = \frac{\pi}{2}e^{-x}, \quad x \geqq 0$.

Let $f(x) = e^{-x}$ in the Fourier integral theorem

$$f(x) \;=\; \frac{2}{\pi}\int_0^\infty \cos \alpha x\,d\alpha \int_0^\infty f(u)\cos \lambda u\,du$$

Then

$$\frac{2}{\pi}\int_0^\infty \cos \alpha x\,d\alpha \int_0^\infty e^{-u}\cos \alpha u\,du \;=\; e^{-x}$$

But from the result 16 on page 6 we have $\displaystyle\int_0^\infty e^{-u}\cos \alpha u\,du = \frac{1}{\alpha^2 + 1}$. Then

$$\frac{2}{\pi}\int_0^\infty \frac{\cos \alpha x}{\alpha^2 + 1}\,d\alpha \;=\; e^{-x} \qquad \text{or} \qquad \int_0^\infty \frac{\cos \alpha x}{\alpha^2 + 1}\,d\alpha \;=\; \frac{\pi}{2}e^{-x}$$

PARSEVAL'S IDENTITY

8.7. Verify Parseval's identity for Fourier integrals for the Fourier transforms of Problem 8.1.

We must show that

$$\int_{-\infty}^\infty \{f(x)\}^2\,dx \;=\; \int_{-\infty}^\infty \{F(\alpha)\}^2\,d\alpha$$

where $\quad f(x) = \begin{cases} 1 & |x| < a \\ 0 & |x| > a \end{cases} \quad$ and $\quad F(\alpha) = \sqrt{\dfrac{2}{\pi}}\dfrac{\sin \alpha a}{\alpha}$.

This is equivalent to

$$\int_{-a}^a (1)^2\,dx \;=\; \int_{-\infty}^\infty \frac{2}{\pi}\frac{\sin^2 \alpha a}{\alpha^2}\,d\alpha$$

or

$$\int_{-\infty}^\infty \frac{\sin^2 \alpha a}{\alpha^2}\,d\alpha \;=\; 2\int_0^\infty \frac{\sin^2 \alpha a}{\alpha^2}\,d\alpha \;=\; \pi a$$

i.e.,

$$\int_0^\infty \frac{\sin^2 \alpha a}{\alpha^2}\,d\alpha \;=\; \frac{\pi a}{2}$$

By letting $\alpha a = u$ and using Problem 8.5, it is seen that this is correct. The method can also be used to find $\displaystyle\int_0^\infty \frac{\sin^2 u}{u^2}\,du$ directly.

CONVOLUTION THEOREM

8.8.   Solve the integral equation $y(x) = g(x) + \int_{-\infty}^{\infty} y(u)\,r(x-u)\,du$ where $g(x)$ and $r(x)$ are given.

Suppose that the Fourier transforms of $y(x)$, $g(x)$ and $r(x)$ exist, and denote them by $Y(\alpha)$, $G(\alpha)$ and $R(\alpha)$ respectively. Then taking the Fourier transform of both sides of the given integral equation, we have by the convolution theorem

$$Y(\alpha) = G(\alpha) + \sqrt{2\pi}\,Y(\alpha)\,R(\alpha) \qquad \text{or} \qquad Y(\alpha) = \frac{G(\alpha)}{1 - \sqrt{2\pi}\,R(\alpha)}$$

Then $\qquad y(x) = \mathcal{F}^{-1}\left\{ \dfrac{G(\alpha)}{1 - \sqrt{2\pi}\,R(\alpha)} \right\} = \dfrac{1}{\sqrt{2\pi}} \displaystyle\int_{-\infty}^{\infty} \dfrac{G(\alpha)}{1 - \sqrt{2\pi}\,R(\alpha)}\,e^{-i\alpha x}\,d\alpha$

assuming this integral exists.

8.9.   Solve for $y(x)$ the integral equation

$$\int_{-\infty}^{\infty} \frac{y(u)\,du}{(x-u)^2 + a^2} = \frac{1}{x^2 + b^2} \qquad 0 < a < b$$

We have

$$\mathcal{F}\left\{ \frac{1}{x^2 + b^2} \right\} = \frac{1}{\sqrt{2\pi}} \int_{-\infty}^{\infty} \frac{e^{i\alpha u}}{u^2 + b^2}\,du = \sqrt{\frac{2}{\pi}} \int_{0}^{\infty} \frac{\cos \alpha u}{u^2 + b^2}\,du$$

$$= \frac{1}{b}\sqrt{\frac{2}{\pi}} \int_{0}^{\infty} \frac{\cos b\alpha v}{v^2 + 1}\,dv = \frac{1}{b}\sqrt{\frac{2}{\pi}}\,\frac{\pi}{2}\,e^{-b\alpha} = \frac{1}{b}\sqrt{\frac{\pi}{2}}\,e^{-b\alpha}$$

where we have used the transformation $u = bv$ and the result of Problem 8.6. Then taking the Fourier transform of both sides of the integral equation, we find

$$\sqrt{2\pi}\,\mathcal{F}\{y\}\,\mathcal{F}\left\{ \frac{1}{x^2 + a^2} \right\} = \mathcal{F}\left\{ \frac{1}{x^2 + b^2} \right\}$$

i.e. $\qquad \sqrt{2\pi}\,Y(\alpha)\cdot\dfrac{1}{a}\sqrt{\dfrac{\pi}{2}}\,e^{-a\alpha} = \dfrac{1}{b}\sqrt{\dfrac{\pi}{2}}\,e^{-b\alpha} \qquad \text{or} \qquad Y(\alpha) = \dfrac{1}{\sqrt{2\pi}}\dfrac{a}{b}\,e^{-(b-a)\alpha}$

Thus $\quad y(x) = \dfrac{1}{\sqrt{2\pi}} \displaystyle\int_{-\infty}^{\infty} e^{-i\alpha x}\,Y(\alpha)\,d\alpha = \dfrac{a}{b\pi} \displaystyle\int_{0}^{\infty} e^{-(b-a)\alpha}\cos \alpha x\,d\alpha = \dfrac{a(b-a)}{b\pi[x^2 + (b-a)^2]}$

PROOF OF THE FOURIER INTEGRAL THEOREM

8.10.   Present a heuristic demonstration of Fourier's integral theorem by use of a limiting form of Fourier series.

Let $\qquad f(x) = \dfrac{a_0}{2} + \displaystyle\sum_{n=1}^{\infty} \left(a_n \cos \dfrac{n\pi x}{L} + b_n \sin \dfrac{n\pi x}{L} \right)$ $\qquad\qquad (1)$

where $\quad a_n = \dfrac{1}{L} \displaystyle\int_{-L}^{L} f(u) \cos \dfrac{n\pi u}{L}\,du$ and $b_n = \dfrac{1}{L} \displaystyle\int_{-L}^{L} f(u) \sin \dfrac{n\pi u}{L}\,du$.

Then by substitution of these coefficients into (1) we find

$$f(x) = \frac{1}{2L} \int_{-L}^{L} f(u)\,du + \frac{1}{L} \sum_{n=1}^{\infty} \int_{-L}^{L} f(u) \cos \frac{n\pi}{L}(u-x)\,du \qquad\qquad (2)$$

If we assume that $\displaystyle\int_{-\infty}^{\infty} |f(u)|\,du$ converges, the first term on the right of (2) approaches zero as $L \to \infty$, while the remaining part appears to approach

$$\lim_{L \to \infty} \frac{1}{L} \sum_{n=1}^{\infty} \int_{-\infty}^{\infty} f(u) \cos \frac{n\pi}{L}(u-x)\,du \qquad\qquad (3)$$

This last step is not rigorous and makes the demonstration heuristic.

Calling $\Delta\alpha = \pi/L$, (3) can be written

$$f(x) = \lim_{\Delta\alpha \to 0} \sum_{n=1}^{\infty} \Delta\alpha \, F(n\,\Delta\alpha) \qquad (4)$$

where we have written

$$F(\alpha) = \frac{1}{\pi} \int_{-\infty}^{\infty} f(u) \cos \alpha(u-x) \, du \qquad (5)$$

But the limit (4) is equal to

$$f(x) = \int_0^{\infty} F(\alpha) \, d\alpha = \frac{1}{\pi} \int_0^{\infty} d\alpha \int_{-\infty}^{\infty} f(u) \cos \alpha(u-x) \, du$$

which is Fourier's integral formula.

This demonstration serves only to provide a possible result. To be rigorous, we start with the integral

$$\frac{1}{\pi} \int_0^{\infty} d\alpha \int_{-\infty}^{\infty} f(u) \cos \alpha(u-x) \, dx$$

and examine the convergence. This method is considered in Problems 8.11-8.14.

8.11. Prove that: $(a)\ \lim\limits_{\alpha \to \infty} \int_0^L \dfrac{\sin \alpha v}{v} \, dv = \dfrac{\pi}{2}$, $(b)\ \lim\limits_{\alpha \to \infty} \int_{-L}^0 \dfrac{\sin \alpha v}{v} \, dv = \dfrac{\pi}{2}$.

(a) Let $\alpha v = y$. Then $\lim\limits_{\alpha \to \infty} \int_0^L \dfrac{\sin \alpha v}{v} \, dv = \lim\limits_{\alpha \to \infty} \int_0^{\alpha L} \dfrac{\sin y}{y} \, dy = \int_0^{\infty} \dfrac{\sin y}{y} \, dy = \dfrac{\pi}{2}$ as can be shown by using Problem 8.27.

(b) Let $\alpha v = -y$. Then $\lim\limits_{\alpha \to \infty} \int_{-L}^0 \dfrac{\sin \alpha v}{v} \, dv = \lim\limits_{\alpha \to \infty} \int_0^{\alpha L} \dfrac{\sin y}{y} \, dy = \dfrac{\pi}{2}$.

8.12. Riemann's theorem states that if $F(x)$ is piecewise continuous in (a, b), then

$$\lim_{\alpha \to \infty} \int_a^b F(x) \sin \alpha x \, dx = 0$$

with a similar result for the cosine (see Problem 8.28). Use this to prove that

$$(a)\quad \lim_{\alpha \to \infty} \int_0^L f(x+v) \frac{\sin \alpha v}{v} \, dv = \frac{\pi}{2} f(x+0)$$

$$(b)\quad \lim_{\alpha \to \infty} \int_{-L}^0 f(x+v) \frac{\sin \alpha v}{v} \, dv = \frac{\pi}{2} f(x-0)$$

where $f(x)$ and $f'(x)$ are assumed piecewise continuous in $(0, L)$ and $(-L, 0)$ respectively.

(a) Using Problem 8.11(a), it is seen that a proof of the given result amounts to proving that

$$\lim_{\alpha \to \infty} \int_0^L \{f(x+v) - f(x+0)\} \frac{\sin \alpha v}{v} \, dv = 0$$

This follows at once from Riemann's theorem, because $F(v) = \dfrac{f(x+v) - f(x+0)}{v}$ is piecewise continuous in $(0, L)$ since $\lim\limits_{v \to 0+} F(v)$ exists and $f(x)$ is piecewise continuous.

(b) A proof of this is analogous to that in part (a) if we make use of Problem 8.11(b).

8.13. If $f(x)$ satisfies the additional condition that $\displaystyle\int_{-\infty}^{\infty} |f(x)| \, dx$ converges, prove that

$(a)\ \lim\limits_{\alpha \to \infty} \int_0^{\infty} f(x+v) \dfrac{\sin \alpha v}{v} \, dv = \dfrac{\pi}{2} f(x+0)$, $(b)\ \lim\limits_{\alpha \to \infty} \int_{-\infty}^0 f(x+v) \dfrac{\sin \alpha v}{v} \, dv = \dfrac{\pi}{2} f(x-0)$.

We have

$$\int_0^{\infty} f(x+v) \frac{\sin \alpha v}{v} \, dv = \int_0^L f(x+v) \frac{\sin \alpha v}{v} \, dv + \int_L^{\infty} f(x+v) \frac{\sin \alpha v}{v} \, dv \qquad (1)$$

$$\int_0^{\infty} f(x+0) \frac{\sin \alpha v}{v} \, dv = \int_0^L f(x+0) \frac{\sin \alpha v}{v} \, dv + \int_L^{\infty} f(x+0) \frac{\sin \alpha v}{v} \, dv \qquad (2)$$

Subtracting,

$$\int_0^\infty \{f(x+v) - f(x+0)\} \frac{\sin \alpha v}{v} \, dv \qquad (3)$$

$$= \int_0^L \{f(x+v) - f(x+0)\} \frac{\sin \alpha v}{v} \, dv + \int_L^\infty f(x+v) \frac{\sin \alpha v}{v} \, dv - \int_L^\infty f(x+0) \frac{\sin \alpha v}{v} \, dv$$

Denoting the integrals in (3) by I, I_1, I_2 and I_3 respectively, we have $I = I_1 + I_2 + I_3$ so that

$$|I| \leq |I_1| + |I_2| + |I_3| \qquad (4)$$

Now
$$|I_2| \leq \int_L^\infty \left| f(x+v) \frac{\sin \alpha v}{v} \right| dv \leq \frac{1}{L} \int_L^\infty |f(x+v)| \, dv$$

Also
$$|I_3| \leq |f(x+0)| \left| \int_L^\infty \frac{\sin \alpha v}{v} \, dv \right|$$

Since $\int_0^\infty |f(x)| \, dx$ and $\int_0^\infty \frac{\sin \alpha v}{v} \, dv$ both converge, we can choose L so large that $|I_2| \leq \epsilon/3$, $|I_3| \leq \epsilon/3$. Also, we can choose α so large that $|I_1| \leq \epsilon/3$. Then from (4) we have $|I| < \epsilon$ for α and L sufficiently large, so that the required result follows.

This result follows by reasoning exactly analogous to that in part (a).

8.14. Prove Fourier's integral formula where $f(x)$ satisfies the conditions stated on page 201.

We must prove that $\displaystyle \lim_{L \to \infty} \frac{1}{\pi} \int_{\alpha=0}^L \int_{u=-\infty}^\infty f(u) \cos \alpha(x-u) \, du \, d\alpha = \frac{f(x+0) + f(x-0)}{2}$.

Since $\displaystyle \left| \int_{-\infty}^\infty f(u) \cos \alpha(x-u) \, du \right| \leq \int_{-\infty}^\infty |f(u)| \, du$ which converges, it follows by the Weierstrass test for integrals [see Problem 1.123, page 33] that $\displaystyle \int_{-\infty}^\infty f(u) \cos \alpha(x-u) \, du$ converges absolutely and uniformly for all α. We can show from this that the order of integration can be reversed to obtain

$$\frac{1}{\pi} \int_{\alpha=0}^L d\alpha \int_{u=-\infty}^\infty f(u) \cos \alpha(x-u) \, du = \frac{1}{\pi} \int_{u=-\infty}^\infty f(u) \, du \int_{\alpha=0}^L \cos \alpha(x-u) \, du$$

$$= \frac{1}{\pi} \int_{u=-\infty}^\infty f(u) \frac{\sin L(u-x)}{u-x} \, du$$

$$= \frac{1}{\pi} \int_{v=-\infty}^\infty f(x+v) \frac{\sin Lv}{v} \, dv$$

$$= \frac{1}{\pi} \int_{-\infty}^0 f(x+v) \frac{\sin Lv}{v} \, dv + \frac{1}{\pi} \int_0^\infty f(x+v) \frac{\sin Lv}{v} \, dv$$

where we have let $u = x + v$.

Letting $L \to \infty$, we see by Problem 8.13 that the given integral converges to $\dfrac{f(x+0) + f(x-0)}{2}$ as required.

Supplementary Problems

THE FOURIER INTEGRAL AND FOURIER TRANSFORMS

8.15. (a) Find the Fourier transform of $f(x) = \begin{cases} 1/2\epsilon & |x| \leq \epsilon \\ 0 & |x| > \epsilon \end{cases}$.

(b) Determine the limit of this transform as $\epsilon \to 0+$ and discuss the result.

8.16. (a) Find the Fourier transform of $f(x) = \begin{cases} 1 - x^2 & |x| < 1 \\ 0 & |x| > 1 \end{cases}$.

(b) Evaluate $\displaystyle \int_0^\infty \left(\frac{x \cos x - \sin x}{x^3} \right) \cos \frac{x}{2} \, dx$.

8.17. If $f(x) = \begin{cases} 1 & 0 \leq x < 1 \\ 0 & x \geq 1 \end{cases}$ find the (a) Fourier sine transform, (b) Fourier cosine transform of $f(x)$. In each case obtain the graph of $f(x)$ and its transform.

8.18.   (a) Find the Fourier sine transform of e^{-x}, $x \geqq 0$.

(b) Show that $\displaystyle\int_0^\infty \frac{x \sin mx}{x^2 + 1}\,dx = \frac{\pi}{2} e^{-m}$, $m > 0$ by using the result in (a).

(c) Explain from the viewpoint of Fourier's integral theorem why the result in (b) does not hold for $m = 0$.

8.19.   Solve for $Y(x)$ the integral equation

$$\int_0^\infty Y(x) \sin xt\,dx \;=\; \begin{cases} 1 & 0 \leqq t < 1 \\ 2 & 1 \leqq t < 2 \\ 0 & t \geqq 2 \end{cases}$$

and verify the solution by direct substitution.

8.20   Establish equation (4), page 201, from equation (3), page 201.

PARSEVAL'S IDENTITY

8.21.   Evaluate   (a) $\displaystyle\int_0^\infty \frac{dx}{(x^2 + 1)^2}$,   (b) $\displaystyle\int_0^\infty \frac{x^2\,dx}{(x^2 + 1)^2}$ by use of Parseval's identity.

[*Hint.* Use the Fourier sine and cosine transforms of e^{-x}, $x > 0$.]

8.22.   Use Problem 8.17 to show that   (a) $\displaystyle\int_0^\infty \left(\frac{1 - \cos x}{x}\right)^2 dx = \frac{\pi}{2}$,   (b) $\displaystyle\int_0^\infty \frac{\sin^4 x}{x^2}\,dx = \frac{\pi}{2}$.

8.23.   Show that $\displaystyle\int_0^\infty \frac{(x \cos x - \sin x)^2}{x^6}\,dx = \frac{\pi}{15}$.

8.24.   (a) If $F(\alpha)$ and $G(\alpha)$ are the Fourier transforms of $f(x)$ and $g(x)$ respectively, prove that

$$\int_{-\infty}^\infty F(\alpha)\,\overline{G(\alpha)}\,d\alpha \;=\; \int_{-\infty}^\infty f(x)\,\overline{g(x)}\,dx$$

where the bar signifies the complex conjugate.

(b) From (a) obtain the results (11)-(14), page 202.

CONVOLUTION THEOREM

8.25.   Verify the convolution theorem for the functions $f(x) = g(x) = \begin{cases} 1 & |x| < 1 \\ 0 & |x| > 1 \end{cases}$.

8.26.   Prove the result (18), page 203.

[*Hint.* If $F(\alpha) = \dfrac{1}{\sqrt{2\pi}} \displaystyle\int_{-\infty}^\infty f(u)\,e^{i\alpha u}\,du$ and $G(\alpha) = \dfrac{1}{\sqrt{2\pi}} \displaystyle\int_{-\infty}^\infty g(v)\,e^{i\alpha v}\,dv$, then

$$F(\alpha)\,G(\alpha) \;=\; \frac{1}{2\pi} \int_{-\infty}^\infty \int_{-\infty}^\infty e^{i\alpha(u+v)} f(u)\,g(v)\,du\,dv$$

Now make the transformation $u + v = x$.]

PROOF OF FOURIER INTEGRAL THEOREM

8.27.   By interchanging the order of integration in $\displaystyle\int_{y=0}^\infty \int_{x=0}^\infty e^{-xy} \sin y\,dx\,dy$, prove that

$$\int_0^\infty \frac{\sin y}{y}\,dy = \frac{\pi}{2}$$

and thus complete the proof in Problem 8.11.

8.28.   Prove Riemann's theorem [see Problem 8.12].

Answers to Supplementary Problems

8.15.   (a) $\dfrac{1}{\sqrt{2\pi}} \dfrac{\sin \alpha\epsilon}{\alpha\epsilon}$,   (b) $\dfrac{1}{\sqrt{2\pi}}$ **8.16.**   (a) $2\sqrt{\dfrac{2}{\pi}} \left(\dfrac{\alpha \cos \alpha - \sin \alpha}{\alpha^3}\right)$,   (b) $\dfrac{3\pi}{16}$

8.17.   (a) $\sqrt{\dfrac{2}{\pi}} \left(\dfrac{1 - \cos \alpha}{\alpha}\right)$,   (b) $\sqrt{\dfrac{2}{\pi}} \dfrac{\sin \alpha}{\alpha}$ **8.18.**   (a) $\sqrt{2/\pi}\,[\alpha/(1 + \alpha^2)]$

8.19. $Y(x) = (2 + 2 \cos x - 4 \cos 2x)/\pi x$ **8.21.**   (a) $\pi/4$,   (b) $\pi/4$

Chapter 9

Gamma, Beta and Other Special Functions

THE GAMMA FUNCTION

The *gamma function* denoted by $\Gamma(n)$ is defined by

$$\Gamma(n) = \int_0^\infty x^{n-1} e^{-x}\, dx \qquad (1)$$

which is convergent for $n > 0$.

A *recursion* or *recurrence formula* for the gamma function is

$$\Gamma(n+1) = n\Gamma(n) \qquad (2)$$

where $\Gamma(1) = 1$ (see Problem 9.1). From (2), $\Gamma(n)$ can be determined for all $n > 0$ when the values for $1 \leqq n < 2$ (or any other interval of unit length) are known (see table below). In particular if n is a positive integer, then

$$\Gamma(n+1) = n! \qquad n = 1, 2, 3, \ldots \qquad (3)$$

For this reason $\Gamma(n)$ is sometimes called the *factorial function*.

Examples. $\Gamma(2) = 1! = 1$, $\Gamma(6) = 5! = 120$, $\dfrac{\Gamma(5)}{\Gamma(3)} = \dfrac{4!}{2!} = 12$.

It can be shown (Problem 9.4) that

$$\Gamma(\tfrac{1}{2}) = \sqrt{\pi} \qquad (4)$$

The recurrence relation (2) is a difference equation which has (1) as a solution. By taking (1) as the definition of $\Gamma(n)$ for $n > 0$, we can generalize the gamma function to $n < 0$ by use of (2) in the form

$$\Gamma(n) = \frac{\Gamma(n+1)}{n} \qquad (5)$$

See Problem 9.7, for example. The process is called *analytic continuation*.

TABLE OF VALUES AND GRAPH OF THE GAMMA FUNCTION

n	$\Gamma(n)$
1.00	1.0000
1.10	0.9514
1.20	0.9182
1.30	0.8975
1.40	0.8873
1.50	0.8862
1.60	0.8935
1.70	0.9086
1.80	0.9314
1.90	0.9618
2.00	1.0000

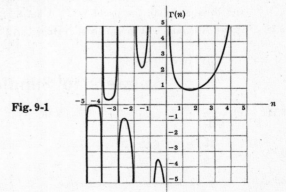

Fig. 9-1

ASYMPTOTIC FORMULA FOR $\Gamma(n)$

If n is large, the computational difficulties inherent in a calculation of $\Gamma(n)$ are apparent. A useful result in such case is supplied by the relation

$$\Gamma(n+1) = \sqrt{2\pi n}\, n^n\, e^{-n}\, e^{\theta/12(n+1)} \qquad 0 < \theta < 1 \tag{6}$$

For most practical purposes the last factor, which is very close to 1 for large n, can be omitted. If n is an integer, we can write

$$n! \sim \sqrt{2\pi n}\, n^n\, e^{-n} \tag{7}$$

where $\sim$ means "is approximately equal to for large n". This is sometimes called *Stirling's factorial approximation or asymptotic formula for n!*.

MISCELLANEOUS RESULTS INVOLVING THE GAMMA FUNCTION

1. $$\Gamma(x)\,\Gamma(1-x) = \frac{\pi}{\sin x\pi} \qquad 0 < x < 1$$

    In particular if $x = \frac{1}{2}$, $\Gamma(\frac{1}{2}) = \sqrt{\pi}$ as in (4).

2. $$2^{2x-1}\,\Gamma(x)\,\Gamma(x+\tfrac{1}{2}) = \sqrt{\pi}\,\Gamma(2x)$$

    This is called the *duplication formula* for the gamma function.

3. $$\Gamma(x)\,\Gamma\left(x + \frac{1}{m}\right)\Gamma\left(x + \frac{2}{m}\right)\cdots\Gamma\left(x + \frac{m-1}{m}\right) = m^{1/2 - mx}\,(2\pi)^{(m-1)/2}\,\Gamma(mx)$$

    The result 2 is a special case of this with $m = 2$.

4. $$\Gamma(x+1) = \sqrt{2\pi x}\, x^x\, e^{-x}\left\{1 + \frac{1}{12x} + \frac{1}{288x^2} - \frac{139}{51{,}840x^3} + \cdots\right\}$$

    This is called *Stirling's asymptotic series* for the gamma function. The series in braces is an asymptotic series (see pages 212 and 219).

5. $$\Gamma'(1) = \int_0^\infty e^{-x}\ln x\, dx = -\gamma$$

    where γ is *Euler's constant* and is defined as

    $$\lim_{M \to \infty}\left(1 + \frac{1}{2} + \frac{1}{3} + \cdots + \frac{1}{M} - \ln M\right) = .577215\ldots$$

6. $$\frac{\Gamma'(p+1)}{\Gamma(p+1)} = 1 + \frac{1}{2} + \frac{1}{3} + \cdots + \frac{1}{p} - \gamma$$

THE BETA FUNCTION

The *beta function*, denoted by $B(m, n)$ is defined by

$$B(m, n) = \int_0^1 x^{m-1}(1-x)^{n-1}\, dx \tag{8}$$

which is convergent for $m > 0$, $n > 0$.

The beta function is connected with the gamma function according to the relation

$$B(m, n) = \frac{\Gamma(m)\,\Gamma(n)}{\Gamma(m+n)} \tag{9}$$

See Problem 9.11.

Many integrals can be evaluated in terms of beta or gamma functions. Two useful results are

$$\int_0^{\pi/2} \sin^{2m-1}\theta\, \cos^{2n-1}\theta\, d\theta = \tfrac{1}{2}B(m, n) = \frac{\Gamma(m)\,\Gamma(n)}{2\,\Gamma(m+n)} \tag{10}$$

valid for $m > 0$ and $n > 0$ [see Problems 9.11 and 9.14] and

$$\int_0^\infty \frac{x^{p-1}}{1+x}\,dx \;=\; \Gamma(p)\,\Gamma(1-p) \;=\; \frac{\pi}{\sin p\pi} \qquad 0 < p < 1 \tag{11}$$

See Problem 9.18.

DIRICHLET INTEGRALS

If V denotes the closed region in the first octant bounded by the surface $\left(\frac{x}{a}\right)^p + \left(\frac{y}{b}\right)^q + \left(\frac{z}{c}\right)^r = 1$ and the coordinate planes, then if all the constants are positive,

$$\iiint_V x^{\alpha-1}\,y^{\beta-1}\,z^{\gamma-1}\,dx\,dy\,dz \;=\; \frac{a^\alpha b^\beta c^\gamma}{pqr}\,\frac{\Gamma\left(\frac{\alpha}{p}\right)\Gamma\left(\frac{\beta}{q}\right)\Gamma\left(\frac{\gamma}{r}\right)}{\Gamma\left(1 + \frac{\alpha}{p} + \frac{\beta}{q} + \frac{\gamma}{r}\right)} \tag{12}$$

Integrals of this type are called *Dirichlet integrals* and are often useful in evaluating multiple integrals (see Problems 9.24 and 9.25).

OTHER SPECIAL FUNCTIONS

Many special functions are of importance in science and engineering. Some of these are given in the following list. Others will be considered in later chapters.

1. **Error function.** $\operatorname{erf}(x) \;=\; \dfrac{2}{\sqrt{\pi}}\displaystyle\int_0^x e^{-u^2}\,du \;=\; 1 - \dfrac{2}{\sqrt{\pi}}\displaystyle\int_x^\infty e^{-u^2}\,du$

2. **Exponential Integral.** $Ei(x) \;=\; \displaystyle\int_x^\infty \dfrac{e^{-u}}{u}\,du$

3. **Sine Integral.** $Si(x) \;=\; \displaystyle\int_0^x \dfrac{\sin u}{u}\,du \;=\; \dfrac{\pi}{2} - \displaystyle\int_x^\infty \dfrac{\sin u}{u}\,du$

4. **Cosine Integral.** $Ci(x) \;=\; \displaystyle\int_x^\infty \dfrac{\cos u}{u}\,du$

5. **Fresnel Sine Integral.** $S(x) \;=\; \sqrt{\dfrac{2}{\pi}}\displaystyle\int_0^x \sin u^2\,du \;=\; 1 - \sqrt{\dfrac{2}{\pi}}\displaystyle\int_0^x \sin u^2\,du$

6. **Fresnel Cosine Integral.** $C(x) \;=\; \sqrt{\dfrac{2}{\pi}}\displaystyle\int_0^x \cos u^2\,du \;=\; 1 - \sqrt{\dfrac{2}{\pi}}\displaystyle\int_x^\infty \cos u^2\,du$

ASYMPTOTIC SERIES OR EXPANSIONS

Consider the series

$$S(x) \;=\; a_0 + \frac{a_1}{x} + \frac{a_2}{x^2} + \cdots + \frac{a_n}{x^n} + \cdots \tag{13}$$

and suppose that

$$S_n(x) \;=\; a_0 + \frac{a_1}{x} + \frac{a_2}{x^2} + \cdots + \frac{a_n}{x^n} \tag{14}$$

are the partial sums of the series.

If $R_n(x) = f(x) - S_n(x)$, where $f(x)$ is given, is such that for every n

$$\lim_{x \to \infty} x^n |R_n(x)| \;=\; 0 \tag{15}$$

then $S(x)$ is called an *asymptotic series* or *expansion* of $f(x)$ and we denote this by writing $f(x) \sim S(x)$.

In practice the series (*13*) diverges. However, by taking the sum of successive terms of the series, stopping just before the terms begin to increase, we may obtain a useful approximation for $f(x)$. The approximation becomes better the larger the value of x.

Various operations with asymptotic series are permissible. For example, asymptotic series may be multiplied or integrated term by term to yield another asymptotic series.

Solved Problems

THE GAMMA FUNCTION

9.1. Prove: (a) $\Gamma(n+1) = n\Gamma(n)$, $n > 0$; (b) $\Gamma(n+1) = n!$, $n = 1, 2, 3, \ldots$.

(a) $\Gamma(n+1) = \displaystyle\int_0^\infty x^n e^{-x}\,dx = \lim_{M\to\infty}\int_0^M x^n e^{-x}\,dx$

$= \displaystyle\lim_{M\to\infty}\left\{ (x^n)(-e^{-x})\Big|_0^M - \int_0^M (-e^{-x})(nx^{n-1})\,dx \right\}$

$= \displaystyle\lim_{M\to\infty}\left\{ -M^n e^{-M} + n\int_0^M x^{n-1} e^{-x}\,dx \right\} = n\,\Gamma(n) \quad \text{if } n > 0$

(b) $\Gamma(1) = \displaystyle\int_0^\infty e^{-x}\,dx = \lim_{M\to\infty}\int_0^M e^{-x}\,dx = \lim_{M\to\infty}(1 - e^{-M}) = 1$

Put $n = 1, 2, 3, \ldots$ in $\Gamma(n+1) = n\,\Gamma(n)$. Then

$\Gamma(2) = 1\,\Gamma(1) = 1, \quad \Gamma(3) = 2\,\Gamma(2) = 2\cdot 1 = 2!, \quad \Gamma(4) = 3\,\Gamma(3) = 3\cdot 2! = 3!$

In general, $\Gamma(n+1) = n!$ if n is a positive integer.

9.2. Evaluate each of the following.

(a) $\dfrac{\Gamma(6)}{2\,\Gamma(3)} = \dfrac{5!}{2\cdot 2!} = \dfrac{5\cdot 4\cdot 3\cdot 2}{2\cdot 2} = 30$

(b) $\dfrac{\Gamma(\frac{5}{2})}{\Gamma(\frac{1}{2})} = \dfrac{\frac{3}{2}\Gamma(\frac{3}{2})}{\Gamma(\frac{1}{2})} = \dfrac{\frac{3}{2}\cdot\frac{1}{2}\Gamma(\frac{1}{2})}{\Gamma(\frac{1}{2})} = \dfrac{3}{4}$

(c) $\dfrac{\Gamma(3)\,\Gamma(2.5)}{\Gamma(5.5)} = \dfrac{2!\,(1.5)(0.5)\,\Gamma(0.5)}{(4.5)(3.5)(2.5)(1.5)(0.5)\,\Gamma(0.5)} = \dfrac{16}{315}$

(d) $\dfrac{6\,\Gamma(\frac{8}{3})}{5\,\Gamma(\frac{2}{3})} = \dfrac{6(\frac{5}{3})(\frac{2}{3})\,\Gamma(\frac{2}{3})}{5\,\Gamma(\frac{2}{3})} = \dfrac{4}{3}$

9.3. Evaluate each integral.

(a) $\displaystyle\int_0^\infty x^3 e^{-x}\,dx = \Gamma(4) = 3! = 6$

(b) $\displaystyle\int_0^\infty x^6 e^{-2x}\,dx$. Let $2x = y$. Then the integral becomes

$\displaystyle\int_0^\infty \left(\frac{y}{2}\right)^6 e^{-y}\,\frac{dy}{2} = \frac{1}{2^7}\int_0^\infty y^6 e^{-y}\,dy = \frac{\Gamma(7)}{2^7} = \frac{6!}{2^7} = \frac{45}{8}$

9.4. Prove that $\Gamma(\tfrac{1}{2}) = \sqrt{\pi}$.

We have $\Gamma(\tfrac{1}{2}) = \displaystyle\int_0^\infty x^{-1/2} e^{-x}\, dx = 2\int_0^\infty e^{-u^2}\, du$ on letting $x = u^2$. It follows that

$$\{\Gamma(\tfrac{1}{2})\}^2 = \left\{2\int_0^\infty e^{-u^2}\, du\right\}\left\{2\int_0^\infty e^{-v^2}\, dv\right\} = 4\int_0^\infty\int_0^\infty e^{-(u^2+v^2)}\, du\, dv$$

Changing to polar coordinates (ρ, ϕ) where $u = \rho\cos\phi$, $v = \rho\sin\phi$, the last integral becomes

$$4\int_{\phi=0}^{\pi/2}\int_{\rho=0}^\infty e^{-\rho^2}\rho\, d\rho\, d\phi = 4\int_{\phi=0}^{\pi/2} -\tfrac{1}{2}e^{-\rho^2}\Big|_{\rho=0}^\infty\, d\phi = \pi$$

and so $\Gamma(\tfrac{1}{2}) = \sqrt{\pi}$.

9.5. Evaluate each integral.

(a) $\displaystyle\int_0^\infty \sqrt{y}\, e^{-y^3}\, dy$. Letting $y^3 = x$, the integral becomes

$$\int_0^\infty \sqrt{x^{1/3}}\, e^{-x}\cdot\tfrac{1}{3} x^{-2/3}\, dx = \tfrac{1}{3}\int_0^\infty x^{-1/2} e^{-x}\, dx = \tfrac{1}{3}\Gamma(\tfrac{1}{2}) = \frac{\sqrt{\pi}}{3}$$

(b) $\displaystyle\int_0^\infty 3^{-4z^2}\, dz = \int_0^\infty (e^{\ln 3})^{(-4z^2)}\, dz = \int_0^\infty e^{-(4\ln 3)z^2}\, dz$. Let $(4\ln 3)z^2 = x$ and the integral becomes

$$\int_0^\infty e^{-x}\, d\left(\frac{x^{1/2}}{\sqrt{4\ln 3}}\right) = \frac{1}{2\sqrt{4\ln 3}}\int_0^\infty x^{-1/2} e^{-x}\, dx = \frac{\Gamma(1/2)}{2\sqrt{4\ln 3}} = \frac{\sqrt{\pi}}{4\sqrt{\ln 3}}$$

(c) $\displaystyle\int_0^1 \frac{dx}{\sqrt{-\ln x}}$. Let $-\ln x = u$. Then $x = e^{-u}$. When $x = 1$, $u = 0$; when $x = 0$, $u = \infty$. The integral becomes

$$\int_0^\infty \frac{e^{-u}}{\sqrt{u}}\, du = \int_0^\infty u^{-1/2} e^{-u}\, du = \Gamma(1/2) = \sqrt{\pi}$$

9.6. Evaluate $\displaystyle\int_0^\infty x^m e^{-ax^n}\, dx$ where m, n, a are positive constants.

Letting $ax^n = y$, the integral becomes

$$\int_0^\infty \left\{\left(\frac{y}{a}\right)^{1/n}\right\}^m e^{-y}\, d\left\{\left(\frac{y}{a}\right)^{1/n}\right\} = \frac{1}{na^{(m+1)/n}}\int_0^\infty y^{(m+1)/n-1} e^{-y}\, dy = \frac{1}{na^{(m+1)/n}}\Gamma\left(\frac{m+1}{n}\right)$$

9.7. Evaluate (a) $\Gamma(-1/2)$, (b) $\Gamma(-5/2)$.

We use the generalization to negative values defined by $\Gamma(n) = \dfrac{\Gamma(n+1)}{n}$.

(a) Letting $n = -\tfrac{1}{2}$, $\Gamma(-1/2) = \dfrac{\Gamma(1/2)}{-1/2} = -2\sqrt{\pi}$.

(b) Letting $n = -3/2$, $\Gamma(-3/2) = \dfrac{\Gamma(-1/2)}{-3/2} = \dfrac{-2\sqrt{\pi}}{-3/2} = \dfrac{4\sqrt{\pi}}{3}$, using (a).

Then $\Gamma(-5/2) = \dfrac{\Gamma(-3/2)}{-5/2} = -\dfrac{8}{15}\sqrt{\pi}$.

9.8. Prove that $\displaystyle\int_0^1 x^m (\ln x)^n\, dx = \frac{(-1)^n n!}{(m+1)^{n+1}}$ where n is a positive integer and $m > -1$.

Letting $x = e^{-y}$, the integral becomes $(-1)^n \displaystyle\int_0^\infty y^n e^{-(m+1)y}\, dy$. If $(m+1)y = u$, this last integral becomes

$$(-1)^n \int_0^\infty \frac{u^n}{(m+1)^n} e^{-u}\, \frac{du}{m+1} = \frac{(-1)^n}{(m+1)^{n+1}}\int_0^\infty u^n e^{-u}\, du = \frac{(-1)^n}{(m+1)^{n+1}}\Gamma(n+1) = \frac{(-1)^n n!}{(m+1)^{n+1}}$$

9.9. Prove that $\displaystyle\int_0^\infty e^{-\alpha\lambda^2}\cos\beta\lambda\,d\lambda = \frac{1}{2}\sqrt{\frac{\pi}{\alpha}}\,e^{-\beta^2/4\alpha}$.

Let $I = I(\alpha,\beta) = \displaystyle\int_0^\infty e^{-\alpha\lambda^2}\cos\beta\lambda\,d\lambda$. Then

$$\frac{\partial I}{\partial \beta} = \int_0^\infty (-\lambda e^{-\alpha\lambda^2})\sin\beta\lambda\,d\lambda$$

$$= \frac{e^{-\alpha\lambda^2}}{2\alpha}\sin\beta\lambda\,\Big|_0^\infty - \frac{\beta}{2\alpha}\int_0^\infty e^{-\alpha\lambda^2}\cos\beta\lambda\,d\lambda = -\frac{\beta}{2\alpha}I$$

Thus
$$\frac{1}{I}\frac{\partial I}{\partial \beta} = -\frac{\beta}{2\alpha} \quad\text{or}\quad \frac{\partial}{\partial\beta}\ln I = -\frac{\beta}{2\alpha} \tag{1}$$

Integration with respect to β yields
$$\ln I = -\frac{\beta^2}{4\alpha} + c_1$$

or
$$I = I(\alpha,\beta) = Ce^{-\beta^2/4\alpha} \tag{2}$$

But $I(\alpha,0) = \displaystyle\int_0^\infty e^{-\alpha\lambda^2}\,d\lambda = \frac{1}{2\sqrt{\alpha}}\int_0^\infty x^{-1/2}e^{-x}\,dx = \frac{\Gamma(\frac{1}{2})}{2\sqrt{\alpha}} = \frac{1}{2}\sqrt{\frac{\pi}{\alpha}}$ on letting $x = \alpha\lambda^2$, so that from (2), $C = \sqrt{\pi}/2\sqrt{\alpha}$. Thus as required,

$$I = \frac{1}{2}\sqrt{\frac{\pi}{\alpha}}\,e^{-\beta^2/4\alpha}$$

9.10. A particle is attracted toward a fixed point O with a force inversely proportional to its instantaneous distance from O. If the particle is released from rest, find the time for it to reach O.

At time $t = 0$ let the particle be located on the x axis at $x = a > 0$ and let O be the origin. Then by Newton's law
$$m\frac{d^2x}{dt^2} = -\frac{k}{x} \tag{1}$$

where m is the mass of the particle and $k > 0$ is a constant of proportionality.

Let $\dfrac{dx}{dt} = v$, the velocity of the particle. Then $\dfrac{d^2x}{dt^2} = \dfrac{dv}{dt} = \dfrac{dv}{dx}\cdot\dfrac{dx}{dt} = v\cdot\dfrac{dv}{dx}$ and (1) becomes

$$mv\frac{dv}{dx} = -\frac{k}{x} \quad\text{or}\quad \frac{mv^2}{2} = -k\ln x + c \tag{2}$$

upon integrating. Since $v = 0$ at $x = a$, we find $c = k\ln a$. Then

$$\frac{mv^2}{2} = k\ln\frac{a}{x} \quad\text{or}\quad v = \frac{dx}{dt} = -\sqrt{\frac{2k}{m}}\sqrt{\ln\frac{a}{x}} \tag{3}$$

where the negative sign is chosen since x is decreasing as t increases. We thus find that the time T taken for the particle to go from $x = a$ to $x = 0$ is given by

$$T = \sqrt{\frac{m}{2k}}\int_0^a \frac{dx}{\sqrt{\ln a/x}} \tag{4}$$

Letting $\ln a/x = u$ or $x = ae^{-u}$, this becomes

$$T = a\sqrt{\frac{m}{2k}}\int_0^\infty u^{-1/2}e^{-u}\,du = a\sqrt{\frac{m}{2k}}\,\Gamma(\tfrac{1}{2}) = a\sqrt{\frac{\pi m}{2k}}$$

THE BETA FUNCTION

9.11. Prove that (a) $\mathrm{B}(m,n) = \mathrm{B}(n,m)$, (b) $\mathrm{B}(m,n) = 2\displaystyle\int_0^{\pi/2}\sin^{2m-1}\theta\,\cos^{2n-1}\theta\,d\theta$.

(a) Using the transformation $x = 1 - y$, we have

$$\mathrm{B}(m,n) = \int_0^1 x^{m-1}(1-x)^{n-1}\,dx = \int_0^1 (1-y)^{m-1}y^{n-1}\,dy = \int_0^1 y^{n-1}(1-y)^{m-1}\,dy = \mathrm{B}(n,m)$$

(b) Using the transformation $x = \sin^2 \theta$, we have

$$B(m, n) = \int_0^1 x^{m-1}(1-x)^{n-1}\, dx = \int_0^{\pi/2} (\sin^2 \theta)^{m-1}(\cos^2 \theta)^{n-1}\, 2\sin\theta\cos\theta\, d\theta$$

$$= 2\int_0^{\pi/2} \sin^{2m-1}\theta\, \cos^{2n-1}\theta\, d\theta$$

9.12. Prove that $B(m, n) = \dfrac{\Gamma(m)\,\Gamma(n)}{\Gamma(m+n)}$ $m, n > 0$.

Letting $z = x^2$, we have $\Gamma(m) = \int_0^\infty z^{m-1} e^{-z}\, dz = 2\int_0^\infty x^{2m-1} e^{-x^2}\, dx$.

Similarly, $\Gamma(n) = 2\int_0^\infty y^{2n-1} e^{-y^2}\, dy$. Then

$$\Gamma(m)\,\Gamma(n) = 4\left(\int_0^\infty x^{2m-1} e^{-x^2}\, dx\right)\left(\int_0^\infty y^{2n-1} e^{-y^2}\, dy\right)$$

$$= 4\int_0^\infty \int_0^\infty x^{2m-1} y^{2n-1} e^{-(x^2+y^2)}\, dx\, dy$$

Transforming to polar coordinates, $x = \rho\cos\phi$, $y = \rho\sin\phi$,

$$\Gamma(m)\,\Gamma(n) = 4\int_{\phi=0}^{\pi/2}\int_{\rho=0}^\infty \rho^{2(m+n)-1} e^{-\rho^2} \cos^{2m-1}\phi\, \sin^{2n-1}\phi\, d\rho\, d\phi$$

$$= 4\left(\int_{\rho=0}^\infty \rho^{2(m+n)-1} e^{-\rho^2}\, d\rho\right)\left(\int_{\phi=0}^{\pi/2} \cos^{2m-1}\phi\, \sin^{2n-1}\phi\, d\phi\right)$$

$$= 2\,\Gamma(m+n)\int_0^{\pi/2} \cos^{2m-1}\phi\, \sin^{2n-1}\phi\, d\phi = \Gamma(m+n)\, B(n, m)$$

$$= \Gamma(m+n)\, B(m, n)$$

using the results of Problem 9.11. Hence the required result follows.

The above argument can be made rigorous by using a limiting procedure.

9.13. Evaluate each of the following integrals.

(a) $\displaystyle\int_0^1 x^4(1-x)^3\, dx = B(5, 4) = \dfrac{\Gamma(5)\,\Gamma(4)}{\Gamma(9)} = \dfrac{4!\,3!}{8!} = \dfrac{1}{280}$

(b) $\displaystyle\int_0^2 \dfrac{x^2\, dx}{\sqrt{2-x}}$. Letting $x = 2v$, the integral becomes

$$4\sqrt{2}\int_0^1 \frac{v^2}{\sqrt{1-v}}\, dv = 4\sqrt{2}\int_0^1 v^2(1-v)^{-1/2}\, dv = 4\sqrt{2}\, B(3, \tfrac{1}{2}) = \frac{4\sqrt{2}\,\Gamma(3)\,\Gamma(1/2)}{\Gamma(7/2)} = \frac{64\sqrt{2}}{15}$$

(c) $\displaystyle\int_0^a y^4\sqrt{a^2-y^2}\, dy$. Letting $y^2 = a^2 x$ or $y = a\sqrt{x}$, the integral becomes

$$\frac{a^6}{2}\int_0^1 x^{3/2}(1-x)^{1/2}\, dx = \frac{a^6}{2}\, B(5/2, 3/2) = \frac{a^6\,\Gamma(5/2)\,\Gamma(3/2)}{2\,\Gamma(4)} = \frac{\pi a^6}{32}$$

9.14. Show that $\displaystyle\int_0^{\pi/2} \sin^{2m-1}\theta\, \cos^{2n-1}\theta\, d\theta = \dfrac{\Gamma(m)\,\Gamma(n)}{2\,\Gamma(m+n)}$ $m, n > 0$.

This follows at once from Problems 9.11 and 9.12.

9.15. Evaluate (a) $\int_0^{\pi/2} \sin^6 \theta \, d\theta$, (b) $\int_0^{\pi/2} \sin^4 \theta \cos^5 \theta \, d\theta$, (c) $\int_0^{\pi} \cos^4 \theta \, d\theta$.

(a) Let $2m - 1 = 6$, $2n - 1 = 0$, i.e. $m = 7/2$, $n = 1/2$, in Problem 9.14.

Then the required integral has the value $\dfrac{\Gamma(7/2)\,\Gamma(1/2)}{2\,\Gamma(4)} = \dfrac{5\pi}{32}$.

(b) Letting $2m - 1 = 4$, $2n - 1 = 5$, the required integral has the value $\dfrac{\Gamma(5/2)\,\Gamma(3)}{2\,\Gamma(11/2)} = \dfrac{8}{315}$.

(c) The given integral $= 2\int_0^{\pi/2} \cos^4 \theta \, d\theta$.

Thus letting $2m - 1 = 0$, $2n - 1 = 4$ in Problem 9.14, the value is $\dfrac{2\,\Gamma(1/2)\,\Gamma(5/2)}{2\,\Gamma(3)} = \dfrac{3\pi}{8}$.

9.16. Prove $\int_0^{\pi/2} \sin^p \theta \, d\theta = \int_0^{\pi/2} \cos^p \theta \, d\theta =$ (a) $\dfrac{1 \cdot 3 \cdot 5 \cdots (p-1)}{2 \cdot 4 \cdot 6 \cdots p} \dfrac{\pi}{2}$ if p is an even positive integer, (b) $\dfrac{2 \cdot 4 \cdot 6 \cdots (p-1)}{1 \cdot 3 \cdot 5 \cdots p}$ if p is an odd positive integer.

From Problem 9.14 with $2m - 1 = p$, $2n - 1 = 0$, we have

$$\int_0^{\pi/2} \sin^p \theta \, d\theta = \frac{\Gamma[\frac{1}{2}(p+1)]\,\Gamma(\frac{1}{2})}{2\,\Gamma[\frac{1}{2}(p+2)]}$$

(a) If $p = 2r$, the integral equals

$$\frac{\Gamma(r + \frac{1}{2})\,\Gamma(\frac{1}{2})}{2\,\Gamma(r+1)} = \frac{(r - \frac{1}{2})(r - \frac{3}{2})\cdots \frac{1}{2}\,\Gamma(\frac{1}{2}) \cdot \Gamma(\frac{1}{2})}{2r(r-1)\cdots 1} = \frac{(2r-1)(2r-3)\cdots 1}{2r(2r-2)\cdots 2}\frac{\pi}{2} = \frac{1 \cdot 3 \cdot 5 \cdots (2r-1)}{2 \cdot 4 \cdot 6 \cdots 2r}\frac{\pi}{2}$$

(b) If $p = 2r + 1$, the integral equals

$$\frac{\Gamma(r + 1)\,\Gamma(\frac{1}{2})}{2\,\Gamma(r + \frac{3}{2})} = \frac{r(r-1)\cdots 1 \cdot \sqrt{\pi}}{2(r + \frac{1}{2})(r - \frac{1}{2})\cdots \frac{1}{2}\sqrt{\pi}} = \frac{2 \cdot 4 \cdot 6 \cdots 2r}{1 \cdot 3 \cdot 5 \cdots (2r+1)}$$

In both cases $\int_0^{\pi/2} \sin^p \theta \, d\theta = \int_0^{\pi/2} \cos^p \theta \, d\theta$, as seen by letting $\theta = \pi/2 - \phi$.

9.17. Evaluate (a) $\int_0^{\pi/2} \cos^6 \theta \, d\theta$, (b) $\int_0^{\pi/2} \sin^3 \theta \cos^2 \theta \, d\theta$, (c) $\int_0^{2\pi} \sin^8 \theta \, d\theta$.

(a) From Problem 9.16 the integral equals $\dfrac{1 \cdot 3 \cdot 5}{2 \cdot 4 \cdot 6}\dfrac{\pi}{2} = \dfrac{5\pi}{32}$ [compare Problem 9.15(a)].

(b) The integral equals

$$\int_0^{\pi/2} \sin^3 \theta (1 - \sin^2 \theta) \, d\theta = \int_0^{\pi/2} \sin^3 \theta \, d\theta - \int_0^{\pi/2} \sin^5 \theta \, d\theta = \frac{2}{1 \cdot 3} - \frac{2 \cdot 4}{1 \cdot 3 \cdot 5} = \frac{2}{15}$$

The method of Problem 9.15(b) can also be used.

(c) The given integral equals $4\int_0^{\pi/2} \sin^8 \theta \, d\theta = 4\left(\dfrac{1 \cdot 3 \cdot 5 \cdot 7}{2 \cdot 4 \cdot 6 \cdot 8}\dfrac{\pi}{2}\right) = \dfrac{35\pi}{64}$.

9.18. Given $\int_0^{\infty} \dfrac{x^{p-1}}{1+x}\,dx = \dfrac{\pi}{\sin p\pi}$, show that $\Gamma(p)\,\Gamma(1-p) = \dfrac{\pi}{\sin p\pi}$ where $0 < p < 1$.

Letting $\dfrac{x}{1+x} = y$ or $x = \dfrac{y}{1-y}$, the given integral becomes

$$\int_0^1 y^{p-1}(1-y)^{-p}\,dy = B(p, 1-p) = \Gamma(p)\,\Gamma(1-p)$$

and the result follows.

9.19. Evaluate $\displaystyle\int_0^\infty \frac{dy}{1+y^4}$.

Let $y^4 = x$. Then the integral becomes $\displaystyle\frac{1}{4}\int_0^\infty \frac{x^{-3/4}}{1+x}\,dx = \frac{\pi}{4\sin(\pi/4)} = \frac{\pi\sqrt{2}}{4}$ by Problem 9.18 with $p = \frac{1}{4}$.

The result can also be obtained by letting $y^2 = \tan\theta$.

9.20. Show that $\displaystyle\int_0^2 x\sqrt[3]{8-x^3}\,dx = \frac{16\pi}{9\sqrt{3}}$.

Letting $x^3 = 8y$ or $x = 2y^{1/3}$, the integral becomes

$$\int_0^1 2y^{1/3}\cdot\sqrt[3]{8(1-y)}\cdot\tfrac{2}{3}y^{-2/3}\,dy \;=\; \frac{8}{3}\int_0^1 y^{-1/3}(1-y)^{1/3}\,dy \;=\; \frac{8}{3}\,B(\tfrac{2}{3},\tfrac{4}{3})$$

$$=\; \frac{8}{3}\frac{\Gamma(\tfrac{2}{3})\,\Gamma(\tfrac{4}{3})}{\Gamma(2)} \;=\; \frac{8}{9}\,\Gamma(\tfrac{1}{3})\,\Gamma(\tfrac{2}{3}) \;=\; \frac{8}{9}\cdot\frac{\pi}{\sin\pi/3} \;=\; \frac{16\pi}{9\sqrt{3}}$$

9.21. Prove the *duplication formula* $2^{2p-1}\,\Gamma(p)\,\Gamma(p+\tfrac{1}{2}) = \sqrt{\pi}\,\Gamma(2p)$.

Let $\displaystyle I = \int_0^{\pi/2} \sin^{2p} x\,dx$, $\displaystyle J = \int_0^{\pi/2} \sin^{2p} 2x\,dx$.

Then $\displaystyle I = \tfrac{1}{2}B(p+\tfrac{1}{2},\tfrac{1}{2}) = \frac{\Gamma(p+\tfrac{1}{2})\sqrt{\pi}}{2\,\Gamma(p+1)}$

Letting $2x = u$, we find

$$J \;=\; \frac{1}{2}\int_0^\pi \sin^{2p} u\,du \;=\; \int_0^{\pi/2} \sin^{2p} u\,du \;=\; I$$

But

$$J \;=\; \int_0^{\pi/2} (2\sin x\cos x)^{2p}\,dx \;=\; 2^{2p}\int_0^{\pi/2} \sin^{2p} x\cos^{2p} x\,dx$$

$$=\; 2^{2p-1}\,B(p+\tfrac{1}{2},\,p+\tfrac{1}{2}) \;=\; \frac{2^{2p-1}\{\Gamma(p+\tfrac{1}{2})\}^2}{\Gamma(2p+1)}$$

Then since $I = J$,

$$\frac{\Gamma(p+\tfrac{1}{2})\sqrt{\pi}}{2p\,\Gamma(p)} \;=\; \frac{2^{2p-1}\{\Gamma(p+\tfrac{1}{2})\}^2}{2p\,\Gamma(2p)}$$

and the required result follows.

9.22. Prove that $\displaystyle\int_0^\infty \frac{\cos x}{x^p}\,dx = \frac{\pi}{2\,\Gamma(p)\cos(p\pi/2)}$, $\quad 0 < p < 1$.

We have $\displaystyle\frac{1}{x^p} = \frac{1}{\Gamma(p)}\int_0^\infty u^{p-1}e^{-xu}\,du$. Then

$$\int_0^\infty \frac{\cos x}{x^p}\,dx \;=\; \frac{1}{\Gamma(p)}\int_0^\infty\int_0^\infty u^{p-1}e^{-xu}\cos x\,du\,dx$$

$$=\; \frac{1}{\Gamma(p)}\int_0^\infty \frac{u^p}{1+u^2}\,du \tag{1}$$

where we have reversed the order of integration and used the integral 16 on page 6.

Letting $u^2 = v$ in the last integral, we have by Problem 9.18

$$\int_0^\infty \frac{u^p}{1+u^2}\,du \;=\; \frac{1}{2}\int_0^\infty \frac{v^{(p-1)/2}}{1+v}\,dv \;=\; \frac{\pi}{2\sin(p+1)\pi/2} \;=\; \frac{\pi}{2\cos p\pi/2} \tag{2}$$

Substitution of (2) in (1) yields the required result.

STIRLING'S FORMULA

9.23. Show that for large n, $n! = \sqrt{2\pi n}\, n^n\, e^{-n}$ approximately.

We have

$$\Gamma(n+1) \;=\; \int_0^\infty x^n\, e^{-x}\, dx \;=\; \int_0^\infty e^{n\ln x - x}\, dx \tag{1}$$

The function $n\ln x - x$ has a relative maximum for $x = n$, as is easily shown by elementary calculus. This leads us to the substitution $x = n + y$. Then (1) becomes

$$\Gamma(n+1) \;=\; e^{-n}\int_{-n}^\infty e^{n\ln(n+y) - y}\, dy \;=\; e^{-n}\int_{-n}^\infty e^{n\ln n + n\ln(1+y/n) - y}\, dy \tag{2}$$

$$=\; n^n\, e^{-n}\int_{-n}^\infty e^{n\ln(1+y/n) - y}\, dy$$

Up to now the analysis is rigorous. The following procedures in which we proceed formally can be made rigorous by suitable limiting procedures, but the proofs become involved and we shall omit them.

In (2) use the result

$$\ln(1+x) \;=\; x - \frac{x^2}{2} + \frac{x^3}{3} - \cdots \tag{3}$$

with $x = y/n$. Then on letting $y = \sqrt{n}\,v$, we find

$$\Gamma(n+1) \;=\; n^n\, e^{-n}\int_{-n}^\infty e^{-y^2/2n + y^3/3n^2 - \cdots}\, dy \;=\; n^n\, e^{-n}\sqrt{n}\int_{-\sqrt{n}}^\infty e^{-v^2/2 + v^3/3\sqrt{n} - \cdots}\, dv \tag{4}$$

When n is large a close approximation is

$$\Gamma(n+1) \;=\; n^n\, e^{-n}\sqrt{n}\int_{-\infty}^\infty e^{-v^2/2}\, dv \;=\; \sqrt{2\pi n}\, n^n\, e^{-n} \tag{5}$$

It is of interest that from (4) we can also obtain the miscellaneous result 4 on page 211 [see Problem 9.38].

DIRICHLET INTEGRALS

9.24. Evaluate $\displaystyle I = \iiint_V x^{\alpha-1}\, y^{\beta-1}\, z^{\gamma-1}\, dx\, dy\, dz$

where V is the region in the first octant bounded by the sphere $x^2 + y^2 + z^2 = 1$ and the coordinate planes.

Let $x^2 = u$, $y^2 = v$, $z^2 = w$. Then

$$I = \iiint_{\mathcal{R}} u^{(\alpha-1)/2}\, v^{(\beta-1)/2}\, w^{(\gamma-1)/2}\, \frac{du}{2\sqrt{u}}\, \frac{dv}{2\sqrt{v}}\, \frac{dw}{2\sqrt{w}}$$

$$= \frac{1}{8}\iiint_{\mathcal{R}} u^{(\alpha/2)-1}\, v^{(\beta/2)-1}\, w^{(\gamma/2)-1}\, du\, dv\, dw \tag{1}$$

where $\mathcal{R}$ is the region in the uvw space bounded by the plane $u + v + w = 1$ and the uv, vw and uw planes as in Fig. 9-2. Thus

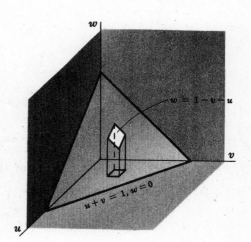

Fig. 9-2

$$I = \frac{1}{8}\int_{u=0}^1 \int_{v=0}^{1-u} \int_{w=0}^{1-u-v} u^{(\alpha/2)-1}\, v^{(\beta/2)-1}\, w^{(\gamma/2)-1}\, du\, dv\, dw \tag{2}$$

$$= \frac{1}{4\gamma}\int_{u=0}^1 \int_{v=0}^{1-u} u^{(\alpha/2)-1}\, v^{(\beta/2)-1}\, (1-u-v)^{\gamma/2}\, du\, dv$$

$$= \frac{1}{4\gamma}\int_{u=0}^1 u^{(\alpha/2)-1}\left\{\int_{v=0}^{1-u} v^{(\beta/2)-1}\, (1-u-v)^{\gamma/2}\, dv\right\} du$$

Letting $v = (1-u)t$, we have

$$\int_{v=0}^{1-u} v^{(\beta/2)-1}(1-u-v)^{\gamma/2}\,dv = (1-u)^{(\beta+\gamma)/2}\int_{t=0}^{1} t^{(\beta/2)-1}(1-t)^{\gamma/2}\,dt$$

$$= (1-u)^{(\beta+\gamma)/2}\frac{\Gamma(\beta/2)\,\Gamma(\gamma/2+1)}{\Gamma[(\beta+\gamma)/2+1]}$$

so that (2) becomes

$$I = \frac{1}{4\gamma}\frac{\Gamma(\beta/2)\,\Gamma(\gamma/2+1)}{\Gamma[(\beta+\gamma)/2+1]}\int_{u=0}^{1} u^{(\alpha/2)-1}(1-u)^{(\beta+\gamma)/2}\,du \qquad (3)$$

$$= \frac{1}{4\gamma}\frac{\Gamma(\beta/2)\,\Gamma(\gamma/2+1)}{\Gamma[(\beta+\gamma)/2+1]}\cdot\frac{\Gamma(\alpha/2)\,\Gamma[(\beta+\gamma)/2+1]}{\Gamma[(\alpha+\beta+\gamma)/2+1]} = \frac{\Gamma(\alpha/2)\,\Gamma(\beta/2)\,\Gamma(\gamma/2)}{8\,\Gamma[(\alpha+\beta+\gamma)/2+1]}$$

where we have used $(\gamma/2)\,\Gamma(\gamma/2) = \Gamma(\gamma/2+1)$.

The integral evaluated here is a special case of the Dirichlet integral (12), page 212. The general case can be evaluated similarly.

9.25. Find the mass of the region bounded by $x^2+y^2+z^2 = a^2$ if the density is $\sigma = x^2y^2z^2$.

The required mass $= 8\iiint\limits_V x^2y^2z^2\,dx\,dy\,dz$, where V is the region in the first octant bounded by the sphere $x^2+y^2+z^2 = a^2$ and the coordinate planes.

In the Dirichlet integral (12), page 212, let $b = c = a$, $p = q = r = 2$ and $\alpha = \beta = \gamma = 3$. Then the required result is

$$8\cdot\frac{a^3\cdot a^3\cdot a^3}{2\cdot 2\cdot 2}\frac{\Gamma(3/2)\,\Gamma(3/2)\,\Gamma(3/2)}{\Gamma(1+3/2+3/2+3/2)} = \frac{4\pi a^9}{945}$$

SPECIAL FUNCTIONS AND ASYMPTOTIC EXPANSIONS

9.26. (a) Prove that if $x > 0$, $p > 0$, then

$$I_p = \int_x^\infty \frac{e^{-u}}{u^p}\,du = S_n(x) + R_n(x)$$

where

$$S_n(x) = e^{-x}\left\{\frac{1}{x^p} - \frac{p}{x^{p+1}} + \frac{p(p+1)}{x^{p+2}} - \cdots (-1)^n\frac{p(p+1)\cdots(p+n)}{x^{p+n}}\right\}$$

$$R_n(x) = (-1)^{n+1}p(p+1)\cdots(p+n)\int_x^\infty \frac{e^{-u}}{u^{p+n+1}}\,du$$

(b) Prove that $\displaystyle\lim_{x\to\infty}\left\{\int_x^\infty\frac{e^{-u}}{u^p}\,du - S_n(x)\right\}x^n = \lim_{x\to\infty}x^n|R_n(x)| = 0$

(c) Explain the significance of the result in (b).

(a) Integrating by parts, we have

$$I_p = \int_x^\infty\frac{e^{-u}}{u^p}\,du = \frac{e^{-x}}{x^p} - p\int_x^\infty\frac{e^{-u}}{u^{p+1}}\,du = \frac{e^{-x}}{x^p} - pI_{p+1}$$

Similarly $I_{p+1} = \dfrac{e^{-x}}{x^{p+1}} - (p+1)I_{p+2}$ so that

$$I_p = \frac{e^{-x}}{x^p} - p\left\{\frac{e^{-x}}{x^{p+1}} - (p+1)I_{p+2}\right\} = \frac{e^{-x}}{x^p} - \frac{pe^{-x}}{x^{p+1}} + p(p+1)I_{p+2}$$

By continuing in this manner the required result follows.

(b) $|R_n(x)| = p(p+1)\cdots(p+n)\displaystyle\int_x^\infty\frac{e^{-u}}{u^{p+n+1}}\,du \leq p(p+1)\cdots(p+n)\int_x^\infty\frac{e^{-u}}{x^{p+n+1}}\,du$

$$\leq \frac{p(p+1)\cdots(p+n)}{x^{p+n+1}}$$

since $\displaystyle\int_x^\infty e^{-u}\,du \leqq \int_0^\infty e^{-u}\,du = 1.$ Thus

$$\lim_{x\to\infty} x^n |R_n(x)| \;\leqq\; \lim_{x\to\infty} \frac{p(p+1)\cdots(p+n)}{x^{p+1}} \;=\; 0$$

(c) Because of the results in (b), we can say that

$$\int_x^\infty \frac{e^{-u}}{u^p}\,du \;\sim\; e^{-x}\left\{\frac{1}{x^p} - \frac{p}{x^{p+1}} + \frac{p(p+1)}{x^{p+2}} - \cdots\right\} \tag{1}$$

i.e. the series on the right is the asymptotic expansion of the function on the left.

9.27. Show that $\displaystyle \operatorname{erf}(x) \;\sim\; 1 - \frac{e^{-x^2}}{\sqrt{\pi}}\left(\frac{1}{x} - \frac{1}{2x^3} + \frac{1\cdot 3}{2^2 x^5} - \frac{1\cdot 3\cdot 5}{2^3 x^7} + \cdots\right)$

We have $\displaystyle \operatorname{erf}(x) \;=\; \frac{2}{\sqrt{\pi}}\int_0^x e^{-v^2}\,dv \;=\; \frac{1}{\sqrt{\pi}}\int_0^{x^2} u^{-1/2}\,e^{-u}\,du$

$$=\; 1 - \frac{1}{\sqrt{\pi}}\int_{x^2}^\infty u^{-1/2}\,e^{-u}\,du$$

Now from equation (1) of Problem 9.26 we have on letting $p = 1/2$ and replacing x by x^2,

$$\int_{x^2}^\infty u^{-1/2}\,e^{-u}\,du \;\sim\; e^{-x^2}\left(\frac{1}{x} - \frac{1}{2x^3} + \frac{1\cdot 3}{2^2 x^5} - \frac{1\cdot 3\cdot 5}{2^3 x^7} + \cdots\right)$$

which gives the required result.

Supplementary Problems

GAMMA FUNCTION

9.28. Evaluate (a) $\dfrac{\Gamma(7)}{2\,\Gamma(4)\,\Gamma(3)}$, (b) $\dfrac{\Gamma(3)\,\Gamma(3/2)}{\Gamma(9/2)}$, (c) $\Gamma(1/2)\,\Gamma(3/2)\,\Gamma(5/2)$.

9.29. Evaluate (a) $\displaystyle\int_0^\infty x^4\,e^{-x}\,dx$, (b) $\displaystyle\int_0^\infty x^6\,e^{-3x}\,dx$, (c) $\displaystyle\int_0^\infty x^2\,e^{-2x^2}\,dx$.

9.30. Find (a) $\displaystyle\int_0^\infty e^{-x^3}\,dx$, (b) $\displaystyle\int_0^\infty \sqrt[4]{x}\,e^{-\sqrt{x}}\,dx$, (c) $\displaystyle\int_0^\infty y^3\,e^{-2y^5}\,dy$.

9.31. Show that $\displaystyle\int_0^\infty \frac{e^{-st}}{\sqrt{t}}\,dt = \sqrt{\frac{\pi}{s}},\ s > 0.$

9.32. Prove that $\displaystyle \Gamma(n) = \int_0^1 \left(\ln\frac{1}{x}\right)^{n-1} dx,\ n > 0.$

9.33. Evaluate (a) $\displaystyle\int_0^1 (\ln x)^4\,dx$, (b) $\displaystyle\int_0^1 (x\ln x)^3\,dx$, (c) $\displaystyle\int_0^1 \sqrt[3]{\ln(1/x)}\,dx$.

9.34. Evaluate (a) $\Gamma(-7/2)$, (b) $\Gamma(-1/3)$.

9.35. Prove that $\displaystyle\lim_{x\to -m} \Gamma(x) = \infty$ where $m = 0, 1, 2, 3, \ldots$

9.36. Prove that if m is a positive integer, $\displaystyle \Gamma(-m + \tfrac{1}{2}) = \frac{(-1)^m\,2^m\,\sqrt{\pi}}{1\cdot 3\cdot 5\cdots(2m-1)}$

9.37. Prove that $\Gamma'(1) = \int_0^\infty e^{-x} \ln x \, dx$ is a negative number (it is equal to $-\gamma$, where $\gamma = 0.577215\ldots$ is called *Euler's constant*).

9.38. Obtain the miscellaneous result 4 on page 211 from the result (4) of Problem 9.23.
[*Hint:* Expand $e^{v^3/(3\sqrt{n})} + \cdots$ in a power series and replace the lower limit of the integral by $-\infty$.]

BETA FUNCTION

9.39. Evaluate   (*a*) $B(3, 5)$,   (*b*) $B(3/2, 2)$,   (*c*) $B(1/3, 2/3)$.

9.40. Find   (*a*) $\int_0^1 x^2(1-x)^3 \, dx$,   (*b*) $\int_0^1 \sqrt{(1-x)/x} \, dx$,   (*c*) $\int_0^2 (4-x^2)^{3/2} \, dx$.

9.41. Evaluate   (*a*) $\int_0^4 u^{3/2}(4-u)^{5/2} \, du$,   (*b*) $\int_0^3 \dfrac{dx}{\sqrt{3x - x^2}}$.

9.42. Prove that $\int_0^a \dfrac{dy}{\sqrt{a^4 - y^4}} = \dfrac{\{\Gamma(1/4)\}^2}{4a\sqrt{2\pi}}$.

9.43. Evaluate   (*a*) $\int_0^{\pi/2} \sin^4\theta \cos^4\theta \, d\theta$,   (*b*) $\int_0^{2\pi} \cos^6\theta \, d\theta$.

9.44. Evaluate   (*a*) $\int_0^\pi \sin^5\theta \, d\theta$,   (*b*) $\int_0^{\pi/2} \cos^5\theta \sin^2\theta \, d\theta$.

9.45. Prove that $\int_0^{\pi/2} \sqrt{\tan\theta} \, d\theta = \pi/\sqrt{2}$.

9.46. Prove that   (*a*) $\int_0^\infty \dfrac{x \, dx}{1 + x^6} = \dfrac{\pi}{3\sqrt{3}}$,   (*b*) $\int_0^\infty \dfrac{y^2 \, dy}{1 + y^4} = \dfrac{\pi}{2\sqrt{2}}$.

9.47. Prove that $\int_{-\infty}^\infty \dfrac{e^{2x}}{ae^{3x} + b} \, dx = \dfrac{2\pi}{3\sqrt{3} \, a^{2/3}b^{1/3}}$ where $a, b > 0$.

9.48. Prove that $\int_{-\infty}^\infty \dfrac{e^{2x}}{(e^{3x} + 1)^2} \, dx = \dfrac{2\pi}{9\sqrt{3}}$.
[*Hint:* Differentiate with respect to b in Problem 9.47.]

DIRICHLET INTEGRALS

9.49. Find the mass of the region in the xy plane bounded by $x + y = 1$, $x = 0$, $y = 0$ if the density is $\sigma = \sqrt{xy}$.

9.50. Find the mass of the region bounded by the ellipsoid $\dfrac{x^2}{a^2} + \dfrac{y^2}{b^2} + \dfrac{z^2}{c^2} = 1$ if the density varies as the square of the distance from its center.

9.51. Find the volume of the region bounded by $x^{2/3} + y^{2/3} + z^{2/3} = 1$.

9.52. Find the centroid of the region in the first octant bounded by $x^{2/3} + y^{2/3} + z^{2/3} = 1$.

9.53. Show that the volume of the region bounded by $x^m + y^m + z^m = a^m$, where $m > 0$, is given by $\dfrac{8 \{\Gamma(1/m)\}^3}{3m^2 \, \Gamma(3/m)} a^3$.

9.54. Show that the centroid of the region in the first octant bounded by $x^m + y^m + z^m = a^m$, where $m > 0$, is given by

$$\bar{x} = \bar{y} = \bar{z} = \frac{3\,\Gamma(2/m)\,\Gamma(3/m)}{4\,\Gamma(1/m)\,\Gamma(4/m)}\,a$$

SPECIAL FUNCTIONS AND ASYMPTOTIC EXPANSIONS

9.55. Show that $\operatorname{erf}(x) = \dfrac{2}{\sqrt{\pi}}\left(x - \dfrac{x^3}{3 \cdot 1!} + \dfrac{x^5}{5 \cdot 2!} - \dfrac{x^7}{7 \cdot 3!} + \cdots\right).$

9.56. Obtain the asymptotic expansion $Ei(x) = \dfrac{e^{-x}}{x}\left(1 - \dfrac{1!}{x} + \dfrac{2!}{x^2} - \dfrac{3!}{x^3} + \cdots\right).$

9.57. Show that　(a) $Si(-x) = -Si(x)$,　(b) $Si(\infty) = \pi/2$.

9.58. Obtain the asymptotic expansions

(a)
$$Si(x) = \frac{\pi}{2} - \frac{\sin x}{x}\left(\frac{1}{x} - \frac{3!}{x^3} + \frac{5!}{x^5} - \cdots\right) - \frac{\cos x}{x}\left(1 - \frac{2!}{x^2} + \frac{4!}{x^4} - \cdots\right)$$

(b)
$$Ci(x) = \frac{\cos x}{x}\left(\frac{1}{x} - \frac{3!}{x^3} + \frac{5!}{x^5} - \cdots\right) - \frac{\sin x}{x}\left(1 - \frac{2!}{x^2} + \frac{4!}{x^4} - \cdots\right)$$

9.59. Show that $\displaystyle\int_0^\infty \frac{\sin x}{x^p}\,dx = \frac{\pi}{2\,\Gamma(p)\,\sin(p\pi/2)}$, $0 < p < 1$, and derive a similar result for $\displaystyle\int_0^\infty \frac{\cos x}{x^p}\,dx$.

9.60. Show that $\displaystyle\int_0^\infty \sin x^2\,dx = \int_0^\infty \cos x^2\,dx = \frac{1}{2}\sqrt{\frac{\pi}{2}}.$

Answers to Supplementary Problems

9.28. (a) 30, (b) 16/105, (c) $\frac{3}{8}\pi^{3/2}$

9.29. (a) 24, (b) $\dfrac{80}{243}$, (c) $\dfrac{\sqrt{2\pi}}{16}$

9.30. (a) $\frac{1}{3}\Gamma(\frac{1}{3})$, (b) $\dfrac{3\sqrt{\pi}}{2}$, (c) $\dfrac{\Gamma(4/5)}{5\sqrt[5]{16}}$

9.33. (a) 24, (b) $-3/128$, (c) $\frac{1}{3}\Gamma(\frac{1}{3})$

9.34. (a) $(16\sqrt{\pi})/105$, (b) $-3\,\Gamma(2/3)$

9.39. (a) 1/105, (b) 4/15, (c) $2\pi/\sqrt{3}$

9.40. (a) 1/60, (b) $\pi/2$, (c) 3π

9.41. (a) 12π, (b) π

9.43. (a) $3\pi/256$, (b) $5\pi/8$

9.44. (a) 16/15, (b) 8/105

9.49. $\pi/24$

9.50. $\dfrac{\pi abck}{30}(a^2 + b^2 + c^2)$, $k = $ constant of proportionality

9.51. $4\pi/35$

9.52. $\bar{x} = \bar{y} = \bar{z} = 21/128$

Chapter 10

Bessel Functions

BESSEL'S DIFFERENTIAL EQUATION

Bessel functions arise as solutions of the differential equation

$$x^2 y'' + xy' + (x^2 - n^2)y = 0 \qquad n \geqq 0 \tag{1}$$

which is called *Bessel's differential equation*. The general solution of *(1)* is given by

$$y = c_1 J_n(x) + c_2 Y_n(x) \tag{2}$$

The solution $J_n(x)$, which has a finite limit as x approaches zero, is called a *Bessel function of the first kind and order n*. The solution $Y_n(x)$ which has no finite limit [i.e. is unbounded] as x approaches zero, is called a *Bessel function of the second kind and order n* or *Neumann function*.

If the independent variable x in *(1)* is changed to λx where λ is a constant, the resulting equation is

$$x^2 y'' + xy' + (\lambda^2 x^2 - n^2)y = 0 \tag{3}$$

with general solution

$$y = c_1 J_n(\lambda x) + c_2 Y_n(\lambda x) \tag{4}$$

BESSEL FUNCTIONS OF THE FIRST KIND

We define the Bessel function of the first kind of order n as

$$J_n(x) = \frac{x^n}{2^n\,\Gamma(n+1)}\left\{1 - \frac{x^2}{2(2n+2)} + \frac{x^4}{2\cdot4(2n+2)(2n+4)} - \cdots\right\} \tag{5}$$

or

$$J_n(x) = \sum_{r=0}^{\infty} \frac{(-1)^r (x/2)^{n+2r}}{r!\,\Gamma(n+r+1)} \tag{6}$$

where $\Gamma(n+1)$ is the *gamma function* [Chapter 9]. If n is a positive integer, $\Gamma(n+1) = n!$, $\Gamma(1) = 1$. For $n = 0$, *(6)* becomes

$$J_0(x) = 1 - \frac{x^2}{2^2} + \frac{x^4}{2^2 4^2} - \frac{x^6}{2^2 4^2 6^2} + \cdots \tag{7}$$

The series *(6)* converges for all x. Graphs of $J_0(x)$ and $J_1(x)$ are shown in Fig. 10-1.

If n is half an odd integer, $J_n(x)$ can be expressed in terms of sines and cosines. See Problems 10.4 and 10.7.

A function $J_{-n}(x)$, $n > 0$, can be defined by replacing n by $-n$ in *(5)* or *(6)*. If n is an integer then we can show that [see Problem 10.3]

$$J_{-n}(x) = (-1)^n J_n(x) \tag{8}$$

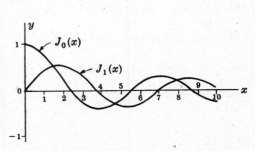

Fig. 10-1

224

If n is not an integer, $J_n(x)$ and $J_{-n}(x)$ are linearly independent, and for this case the general solution of (1) is

$$y = AJ_n(x) + BJ_{-n}(x) \qquad n \neq 0, 1, 2, 3, \ldots \tag{9}$$

BESSEL FUNCTIONS OF THE SECOND KIND

We shall define the Bessel function of the second kind of order n as

$$Y_n(x) = \begin{cases} \dfrac{J_n(x) \cos n\pi - J_{-n}(x)}{\sin n\pi} & n \neq 0, 1, 2, 3, \ldots \\[2ex] \lim_{p \to n} \dfrac{J_p(x) \cos p\pi - J_{-p}(x)}{\sin p\pi} & n = 0, 1, 2, 3, \ldots \end{cases} \tag{10}$$

For the case where $n = 0, 1, 2, 3, \ldots$ we obtain the following series expansion for $Y_n(x)$.

$$Y_n(x) = \frac{2}{\pi} \{\ln(x/2) + \gamma\} J_n(x) - \frac{1}{\pi} \sum_{k=0}^{n-1} (n-k-1)! \, (x/2)^{2k-n}$$

$$- \frac{1}{\pi} \sum_{k=0}^{\infty} (-1)^k \{\Phi(k) + \Phi(n+k)\} \frac{(x/2)^{2k+n}}{k! \, (n+k)!} \tag{11}$$

where $\gamma = .5772156\ldots$ is *Euler's constant* and

$$\Phi(p) = 1 + \frac{1}{2} + \frac{1}{3} + \cdots + \frac{1}{p}, \qquad \Phi(0) = 0 \tag{12}$$

GENERATING FUNCTION FOR $J_n(x)$

The function
$$e^{\frac{x}{2}(t - 1/t)} = \sum_{n=-\infty}^{\infty} J_n(x) t^n \tag{13}$$

is called the *generating function* for Bessel functions of the first kind of integral order. It is very useful in obtaining properties of these functions for integer values of n which can then often be proved for all values of n.

RECURRENCE FORMULAS

The following results are valid for all values of n.

1. $J_{n+1}(x) = \dfrac{2n}{x} J_n(x) - J_{n-1}(x)$

2. $J_n'(x) = \frac{1}{2}[J_{n-1}(x) - J_{n+1}(x)]$

3. $x J_n'(x) = n J_n(x) - x J_{n+1}(x)$

4. $x J_n'(x) = x J_{n-1}(x) - n J_n(x)$

5. $\dfrac{d}{dx}[x^n J_n(x)] = x^n J_{n-1}(x)$

6. $\dfrac{d}{dx}[x^{-n} J_n(x)] = -x^{-n} J_{n+1}(x)$

If n is an integer these can be proved by using the generating function. Note that results 3 and 4 are equivalent respectively to 5 and 6.

The functions $Y_n(x)$ satisfy exactly the same results as those above, where $Y_n(x)$ replaces $J_n(x)$.

FUNCTIONS RELATED TO BESSEL FUNCTIONS

1. Hankel Functions of First and Second Kinds are defined respectively by

$$H_n^{(1)}(x) = J_n(x) + iY_n(x), \quad H_n^{(2)}(x) = J_n(x) - iY_n(x)$$

2. Modified Bessel Functions. The *modified Bessel function of the first kind of order n* is defined as

$$I_n(x) = i^{-n}J_n(ix) = e^{-n\pi i/2}J_n(ix) \tag{14}$$

If n is an integer, $$I_{-n} = I_n(x) \tag{15}$$

but if n is not an integer, $I_n(x)$ and $I_{-n}(x)$ are linearly independent.

The *modified Bessel function of the second kind of order n* is defined as

$$K_n(x) = \begin{cases} \dfrac{\pi}{2}\left[\dfrac{I_{-n}(x) - I_n(x)}{\sin n\pi}\right] & n \neq 0, 1, 2, 3, \ldots \\[3mm] \lim_{p \to n}\dfrac{\pi}{2}\left[\dfrac{I_{-p}(x) - I_p(x)}{\sin p\pi}\right] & n = 0, 1, 2, 3, \ldots \end{cases} \tag{16}$$

These functions satisfy the differential equation

$$x^2y'' + xy' - (x^2 + n^2)y = 0 \tag{17}$$

and the general solution of this equation is

$$y = c_1 I_n(x) + c_2 K_n(x) \tag{18}$$

or if $n \neq 0, 1, 2, 3, \ldots$ $$y = AI_n(x) + BI_{-n}(x) \tag{19}$$

3. Ber, Bei, Ker, Kei Functions. The functions $\text{Ber}_n(x)$ and $\text{Bei}_n(x)$ are the real and imaginary parts of $J_n(i^{3/2}x)$ where $i^{3/2} = e^{3\pi i/4} = (\sqrt{2}/2)(1 - i)$, i.e.

$$J_n(i^{3/2}x) = \text{Ber}_n(x) + i\,\text{Bei}_n(x) \tag{20}$$

The functions $\text{Ker}_n(x)$ and $\text{Kei}_n(x)$ are the real and imaginary parts of $e^{-n\pi i/2}K_n(i^{1/2}x)$ where $i^{1/2} = e^{\pi i/4} = (\sqrt{2}/2)(1 + i)$, i.e.

$$e^{-n\pi i/2}K_n(i^{1/2}x) = \text{Ker}_n(x) + i\,\text{Kei}_n(x) \tag{21}$$

The functions are useful in connection with the equation

$$x^2y'' + xy' - (ix^2 + n^2)y = 0 \tag{22}$$

which arises in electrical engineering and other fields. The general solution of this equation is

$$y = c_1 J_n(i^{3/2}x) + c_2 K_n(i^{1/2}x) \tag{23}$$

EQUATIONS TRANSFORMED INTO BESSEL'S EQUATION

The equation

$$x^2y'' + (2k+1)xy' + (\alpha^2 x^{2r} + \beta^2)y = 0 \tag{24}$$

where k, α, r, β are constants, has the general solution

$$y = x^{-k}[c_1 J_{\kappa/r}(\alpha x^r/r) + c_2 Y_{\kappa/r}(\alpha x^r/r)] \tag{25}$$

where $\kappa = \sqrt{k^2 - \beta^2}$. If $\alpha = 0$ the equation is solvable as an Euler or Cauchy equation [see page 76].

ASYMPTOTIC FORMULAS FOR BESSEL FUNCTIONS

For large values of x we have the following asymptotic formulas

$$J_n(x) \sim \sqrt{\frac{2}{\pi x}} \cos\left(x - \frac{\pi}{4} - \frac{n\pi}{2}\right), \qquad Y_n(x) \sim \sqrt{\frac{2}{\pi x}} \sin\left(x - \frac{\pi}{4} - \frac{n\pi}{2}\right) \qquad (26)$$

ZEROS OF BESSEL FUNCTIONS

We can show that if n is any real number, $J_n(x) = 0$ has an infinite number of roots which are all real. The difference between successive roots approaches π as the roots increase in value. This can be seen from (26). We can also show that the roots of $J_n(x) = 0$ lie between those of $J_{n-1}(x) = 0$ and $J_{n+1}(x) = 0$. Similar remarks can be made for $Y_n(x)$.

ORTHOGONALITY OF BESSEL FUNCTIONS

If λ and μ are two different constants, we can show [see Problem 10.21] that

$$\int_0^1 x J_n(\lambda x) J_n(\mu x)\, dx = \frac{\mu J_n(\lambda) J_n'(\mu) - \lambda J_n(\mu) J_n'(\lambda)}{\lambda^2 - \mu^2} \qquad (27)$$

while [see Problem 10.22]

$$\int_0^1 x J_n^2(\lambda x)\, dx = \frac{1}{2}\left[J_n'^2(\lambda) + \left(1 - \frac{n^2}{\lambda^2}\right) J_n^2(\lambda) \right] \qquad (28)$$

From (27) we can see that if λ and μ are any two different roots of the equation

$$R J_n(x) + S x J_n'(x) = 0 \qquad (29)$$

where R and S are constants, then

$$\int_0^1 x J_n(\lambda x) J_n(\mu x)\, dx = 0 \qquad (30)$$

which states that the functions $\sqrt{x}\, J_n(\lambda x)$ and $\sqrt{x}\, J_n(\mu x)$ are *orthogonal* in $(0, 1)$. Note that as special cases of (29) we see that λ and μ can be any two different roots of $J_n(x) = 0$ or $J_n'(x) = 0$. We can also say that the functions $J_n(\lambda x)$, $J_n(\mu x)$ are orthogonal with respect to the density function x.

SERIES OF BESSEL FUNCTIONS

As in the case of Fourier series, we can show that if $f(x)$ satisfies the Dirichlet conditions [page 183] then at every point of continuity of $f(x)$ in the interval $0 < x < 1$ there will exist a Bessel series expansion having the form

$$f(x) = A_1 J_n(\lambda_1 x) + A_2 J_n(\lambda_2 x) + \cdots = \sum_{p=1}^{\infty} A_p J_n(\lambda_p x) \qquad (31)$$

where $\lambda_1, \lambda_2, \lambda_3, \ldots$ are the positive roots of (29) with $R/S \geqq 0$, $S \neq 0$ and

$$A_p = \frac{2\lambda_p^2}{(\lambda_p^2 - n^2 + R^2/S^2) J_n^2(\lambda_p)} \int_0^1 x J_n(\lambda_p x) f(x)\, dx \qquad (32)$$

At any point of discontinuity the series on the right in (31) converges to $\frac{1}{2}[f(x+0) + f(x-0)]$ which can be used in place of the left side of (31).

In case $S = 0$ so that $\lambda_1, \lambda_2, \ldots$ are the roots of $J_n(x) = 0$,

$$A_p = \frac{2}{J_{n+1}^2(\lambda_p)} \int_0^1 x J_n(\lambda_p x) f(x)\, dx \qquad (33)$$

If $R = 0$ and $n = 0$, then the series (31) starts out with the constant term

$$A_1 = 2 \int_0^1 x f(x)\, dx \qquad (34)$$

Solved Problems

BESSEL'S DIFFERENTIAL EQUATION

10.1. Use the method of Frobenius to find series solutions of Bessel's differential equation $x^2 y'' + x y' + (x^2 - n^2) y = 0$.

Assuming a solution of the form $y = \sum c_k x^{k+\beta}$ where k goes from $-\infty$ to ∞ and $c_k = 0$ for $k < 0$, we have

$$(x^2 - n^2)y = \sum c_k x^{k+\beta+2} - \sum n^2 c_k x^{k+\beta} = \sum c_{k-2} x^{k+\beta} - \sum n^2 c_k x^{k+\beta}$$

$$xy' = \sum (k+\beta) c_k x^{k+\beta}$$

$$x^2 y'' = \sum (k+\beta)(k+\beta-1) c_k x^{k+\beta}$$

Then by addition,

$$\sum [(k+\beta)(k+\beta-1)c_k + (k+\beta)c_k + c_{k-2} - n^2 c_k] x^{k+\beta} = 0$$

and since the coefficients of $x^{k+\beta}$ must be zero, we find

$$[(k+\beta)^2 - n^2] c_k + c_{k-2} = 0 \qquad (1)$$

Letting $k = 0$ in (1) we obtain, since $c_{-2} = 0$, the indicial equation $(\beta^2 - n^2)c_0 = 0$; or assuming $c_0 \neq 0$, $\beta^2 = n^2$. Then there are two cases, given by $\beta = -n$ and $\beta = n$. We shall consider first the case $\beta = n$ and obtain the second case by replacing n by $-n$.

Case 1, $\beta = n$.

In this case (1) becomes

$$k(2n+k)c_k + c_{k-2} = 0 \qquad (2)$$

Putting $k = 1, 2, 3, 4, \ldots$ successively in (2), we have

$$c_1 = 0, \quad c_2 = \frac{-c_0}{2(2n+2)}, \quad c_3 = 0, \quad c_4 = \frac{-c_2}{4(2n+4)} = \frac{c_0}{2 \cdot 4(2n+2)(2n+4)}, \cdots$$

Thus the required series is

$$y = c_0 x^n + c_2 x^{n+2} + c_4 x^{n+4} + \cdots = c_0 x^n \left[1 - \frac{x^2}{2(2n+2)} + \frac{x^4}{2 \cdot 4(2n+2)(2n+4)} - \cdots \right] \qquad (3)$$

Case 2, $\beta = -n$.

Or replacing n by $-n$ in Case 1, we find

$$y = c_0 x^{-n} \left[1 - \frac{x^2}{2(2-2n)} + \frac{x^4}{2 \cdot 4(2-2n)(4-2n)} - \cdots \right] \qquad (4)$$

Now if $n = 0$, both of these series are identical. If $n = 1, 2, \ldots$ the second series fails to exist. However, if $n \neq 0, 1, 2, \ldots$ the two series can be shown to be linearly independent and so for this case the general solution is

$$y = Cx^n \left[1 - \frac{x^2}{2(2n+2)} + \frac{x^4}{2 \cdot 4(2n+2)(2n+4)} - \cdots \right]$$

$$+ Dx^{-n} \left[1 - \frac{x^2}{2(2-2n)} + \frac{x^4}{2 \cdot 4(2-2n)(4-2n)} - \cdots \right] \tag{5}$$

The cases where $n = 0, 1, 2, 3, \ldots$ are treated later [see Problems 10.15 and 10.16].

BESSEL FUNCTIONS OF THE FIRST KIND

10.2. Using the definition (5) of $J_n(x)$ given on page 224, show that if $n \neq 0, 1, 2, \ldots$ then the general solution of Bessel's equation is $y = A J_n(x) + B J_{-n}(x)$.

Note that the definition of $J_n(x)$ on page 224 agrees with the series of Case 1 in Problem 10.1, apart from a constant factor depending only on n. It follows that (5) can be written $y = A J_n(x) + B J_{-n}(x)$ for the cases $n \neq 0, 1, 2, \ldots$.

10.3. (a) Prove that $J_{-n}(x) = (-1)^n J_n(x)$ for $n = 1, 2, 3, \ldots$

(b) Use (a) to explain why $A J_n(x) + B J_{-n}(x)$ is not the general solution of Bessel's equation for integer values of n.

(a) Replacing n by $-n$ in (5) or the equivalent (6) on page 224, we have

$$J_{-n}(x) = \sum_{r=0}^{\infty} \frac{(-1)^r (x/2)^{-n+2r}}{r! \, \Gamma(-n+r+1)}$$

$$= \sum_{r=0}^{n-1} \frac{(-1)^r (x/2)^{-n+2r}}{r! \, \Gamma(-n+r+1)} + \sum_{r=n}^{\infty} \frac{(-1)^r (x/2)^{-n+2r}}{r! \, \Gamma(-n+r+1)}$$

Now since $\Gamma(-n+r+1)$ is infinite for $r = 0, 1, \ldots, n-1$, the first sum on the right is zero. Letting $r = n + k$ in the second sum, it becomes

$$\sum_{k=0}^{\infty} \frac{(-1)^{n+k} (x/2)^{n+2k}}{(n+k)! \, \Gamma(k+1)} = (-1)^n \sum_{k=0}^{\infty} \frac{(-1)^k (x/2)^{n+2k}}{\Gamma(n+k+1)k!} = (-1)^n J_n(x)$$

(b) From (a) it follows that for integer values of n, $J_{-n}(x)$ and $J_n(x)$ are linearly dependent and so $A J_n(x) + B J_{-n}(x)$ cannot be a general solution of Bessel's equation. If n is not an integer, then we can show that $J_{-n}(x)$ and $J_n(x)$ are linearly independent so that $A J_n(x) + B J_{-n}(x)$ is a general solution [see Problem 10.10].

10.4. Prove (a) $J_{1/2}(x) = \sqrt{\dfrac{2}{\pi x}} \sin x$, (b) $J_{-1/2}(x) = \sqrt{\dfrac{2}{\pi x}} \cos x$.

(a)
$$J_{1/2}(x) = \sum_{r=0}^{\infty} \frac{(-1)^r (x/2)^{1/2+2r}}{r! \, \Gamma(r+3/2)} = \frac{(x/2)^{1/2}}{\Gamma(3/2)} - \frac{(x/2)^{5/2}}{1! \, \Gamma(5/2)} + \frac{(x/2)^{9/2}}{2! \, \Gamma(7/2)} - \cdots$$

$$= \frac{(x/2)^{1/2}}{(1/2)\sqrt{\pi}} - \frac{(x/2)^{5/2}}{1! \, (3/2)(1/2)\sqrt{\pi}} + \frac{(x/2)^{7/2}}{2! \, (5/2)(3/2)(1/2)\sqrt{\pi}} - \cdots$$

$$= \frac{(x/2)^{1/2}}{(1/2)\sqrt{\pi}} \left\{ 1 - \frac{x^2}{3!} + \frac{x^4}{5!} - \cdots \right\} = \frac{(x/2)^{1/2}}{(1/2)\sqrt{\pi}} \frac{\sin x}{x} = \sqrt{\frac{2}{\pi x}} \sin x$$

(b)
$$J_{-1/2}(x) = \sum_{r=0}^{\infty} \frac{(-1)^r (x/2)^{-1/2+2r}}{r! \, \Gamma(r+1/2)} = \frac{(x/2)^{-1/2}}{\Gamma(1/2)} - \frac{(x/2)^{3/2}}{1! \, \Gamma(3/2)} + \frac{(x/2)^{7/2}}{2! \, \Gamma(5/2)} - \cdots$$

$$= \frac{(x/2)^{-1/2}}{\sqrt{\pi}} \left\{ 1 - \frac{x^2}{2!} + \frac{x^4}{4!} - \cdots \right\} = \sqrt{\frac{2}{\pi x}} \cos x$$

10.5. Prove (a) $\dfrac{d}{dx}\{x^n J_n(x)\} = x^n J_{n-1}(x)$, (b) $\dfrac{d}{dx}\{x^{-n} J_n(x)\} = -x^{-n} J_{n+1}(x)$ for all n.

(a)
$$\frac{d}{dx}\{x^n J_n(x)\} = \frac{d}{dx}\sum_{r=0}^{\infty} \frac{(-1)^r x^{2n+2r}}{2^{n+2r} r!\,\Gamma(n+r+1)}$$
$$= \sum_{r=0}^{\infty} \frac{(-1)^r x^{2n+2r-1}}{2^{n+2r-1} r!\,\Gamma(n+r)}$$
$$= x^n \sum_{r=0}^{\infty} \frac{(-1)^r x^{(n-1)+2r}}{2^{(n-1)+2r} r!\,\Gamma[(n-1)+r+1]} = x^n J_{n-1}(x)$$

(b)
$$\frac{d}{dx}\{x^{-n} J_n(x)\} = \frac{d}{dx}\sum_{r=0}^{\infty} \frac{(-1)^r x^{2r}}{2^{n+2r} r!\,\Gamma(n+r+1)}$$
$$= x^{-n} \sum_{r=1}^{\infty} \frac{(-1)^r x^{n+2r-1}}{2^{n+2r-1} (r-1)!\,\Gamma(n+r+1)}$$
$$= x^{-n} \sum_{k=0}^{\infty} \frac{(-1)^{k+1} x^{n+2k+1}}{2^{n+2k+1} k!\,\Gamma(n+k+2)} = -x^{-n} J_{n+1}(x)$$

10.6. Prove (a) $J_n'(x) = \tfrac{1}{2}[J_{n-1}(x) - J_{n+1}(x)]$, (b) $J_{n-1}(x) + J_{n+1}(x) = \dfrac{2n}{x} J_n(x)$ for all n.

From Problem 10.5(a), $\;x^n J_n'(x) + nx^{n-1} J_n(x) = x^n J_{n-1}(x)$

or
$$xJ_n'(x) + nJ_n(x) = xJ_{n-1}(x) \tag{1}$$

From Problem 10.5(b), $\;x^{-n} J_n'(x) - nx^{-n-1} J_n(x) = -x^{-n} J_{n+1}(x)$

or
$$xJ_n'(x) - nJ_n(x) = -xJ_{n+1}(x) \tag{2}$$

(a) Adding (1) and (2) and dividing by $2x$ gives
$$J_n'(x) = \tfrac{1}{2}[J_{n-1}(x) - J_{n+1}(x)]$$

(b) Subtracting (2) from (1) and dividing by x gives
$$J_{n-1}(x) + J_{n+1}(x) = \frac{2n}{x} J_n(x)$$

10.7. Show that (a) $J_{3/2}(x) = \sqrt{\dfrac{2}{\pi x}}\left(\dfrac{\sin x - x\cos x}{x}\right)$

                 (b) $J_{-3/2}(x) = -\sqrt{\dfrac{2}{\pi x}}\left(\dfrac{x\sin x + \cos x}{x}\right)$

(a) From Problems 10.6(b) and 10.4 we have on letting $n = 1/2$,
$$J_{3/2}(x) = \frac{1}{x} J_{1/2}(x) - J_{-1/2}(x) = \sqrt{\frac{2}{\pi x}}\left(\frac{\sin x}{x} - \cos x\right) = \sqrt{\frac{2}{\pi x}}\left(\frac{\sin x - x\cos x}{x}\right)$$

(b) From Problems 10.6(b) and 10.4 we have on letting $n = -\tfrac{1}{2}$,
$$J_{-3/2}(x) = -\sqrt{\frac{2}{\pi x}}\left(\frac{x\sin x + \cos x}{x}\right)$$

10.8. Evaluate the integrals (a) $\displaystyle\int x^n J_{n-1}(x)\,dx$, (b) $\displaystyle\int \frac{J_{n+1}(x)}{x^n}\,dx$.

From Problem 10.5,

(a) $\dfrac{d}{dx}\{x^n J_n(x)\} = x^n J_{n-1}(x)$. Then $\displaystyle\int x^n J_{n-1}(x)\,dx = x^n J_n(x) + c$.

(b) $\dfrac{d}{dx}\{x^{-n} J_n(x)\} = -x^{-n} J_{n+1}(x)$. Then $\displaystyle\int \frac{J_{n+1}(x)}{x^n}\,dx = -x^{-n} J_n(x) + c$.

10.9. Evaluate $(a) \int x^4 J_1(x)\,dx$, $(b) \int x^3 J_3(x)\,dx$.

(a) **Method 1.** Integration by parts gives

$$\int x^4 J_1(x)\,dx = \int (x^2)[x^2 J_1(x)\,dx]$$

$$= x^2[x^2 J_2(x)] - \int [x^2 J_2(x)][2x\,dx]$$

$$= x^4 J_2(x) - 2 \int x^3 J_2(x)\,dx$$

$$= x^4 J_2(x) - 2x^3 J_3(x) + c$$

Method 2. We have, using $J_1(x) = -J_0'(x)$ [Problem 10.27(b)],

$$\int x^4 J_1(x)\,dx = -\int x^4 J_0'(x)\,dx = -\left\{x^4 J_0(x) - \int 4x^3 J_0(x)\,dx\right\}$$

$$\int x^3 J_0(x)\,dx = \int x^2[x J_0(x)\,dx] = x^2[x J_1(x)] - \int [x J_1(x)][2x\,dx]$$

$$\int x^2 J_1(x)\,dx = -\int x^2 J_0'(x)\,dx = -\left\{x^2 J_0(x) - \int 2x J_0(x)\,dx\right\}$$

$$= -x^2 J_0(x) + 2x J_1(x)$$

Then

$$\int x^4 J_1(x)\,dx = -x^4 J_0(x) + 4[x^3 J_1(x) - 2\{-x^2 J_0(x) + 2x J_1(x)\}] + c$$

$$= (8x^2 - x^4)J_0(x) + (4x^3 - 16x)J_1(x)$$

(b)

$$\int x^3 J_3(x)\,dx = \int x^5[x^{-2} J_3(x)\,dx]$$

$$= x^5[-x^{-2} J_2(x)] - \int [-x^{-2} J_2(x)]5x^4\,dx$$

$$= -x^3 J_2(x) + 5 \int x^2 J_2(x)\,dx$$

$$\int x^2 J_2(x)\,dx = \int x^3[x^{-1} J_2(x)]\,dx$$

$$= x^3[-x^{-1} J_1(x)] - \int [-x^{-1} J_1(x)]3x^2\,dx$$

$$= -x^2 J_1(x) + 3 \int x J_1(x)\,dx$$

$$\int x J_1(x)\,dx = -\int x J_0'(x)\,dx = -\left[x J_0(x) - \int J_0(x)\,dx\right]$$

$$= -x J_0(x) + \int J_0(x)\,dx$$

Then

$$\int x^3 J_3(x)\,dx = -x^3 J_2(x) + 5\left\{-x^2 J_1(x) + 3\left[-x J_0(x) + \int J_0(x)\,dx\right]\right\}$$

$$= -x^3 J_2(x) - 5x^2 J_1(x) - 15x J_0(x) + 15 \int J_0(x)\,dx$$

The integral $\int J_0(x)\,dx$ cannot be obtained in closed form. In general $\int x^p J_q(x)\,dx$ can be obtained in closed form if $p + q \geqq 0$ and $p + q$ is odd, where p and q are integers. If, however, $p + q$ is even the result can be obtained in terms of $\int J_0(x)\,dx$.

10.10. (a) Prove that $J_n'(x)\, J_{-n}(x) - J_{-n}'(x)\, J_n(x) = \dfrac{2 \sin n\pi}{\pi x}$.

(b) Discuss the significance of the result of (a) from the viewpoint of the linear dependence of $J_n(x)$ and $J_{-n}(x)$.

(a) Since $J_n(x)$, and $J_{-n}(x)$, abbreviated J_n, J_{-n} respectively, satisfy Bessel's equation, we have

$$x^2 J_n'' + x J_n' + (x^2 - n^2) J_n = 0, \qquad x^2 J_{-n}'' + x J_{-n}' + (x^2 - n^2) J_{-n} = 0$$

Multiply the first equation by J_{-n}, the second by J_n and subtract. Then

$$x^2 [J_n'' J_{-n} - J_{-n}'' J_n] + x[J_n' J_{-n} - J_{-n}' J_n] = 0$$

which can be written

$$x \frac{d}{dx} [J_n' J_{-n} - J_{-n}' J_n] + [J_n' J_{-n} - J_{-n}' J_n] = 0$$

or

$$\frac{d}{dx} \{x[J_n' J_{-n} - J_{-n}' J_n]\} = 0$$

Integrating, we find
$$J_n' J_{-n} - J_{-n}' J_n = c/x \qquad (1)$$

To determine c use the series expansions for J_n and J_{-n} to obtain

$$J_n = \frac{x^n}{2^n \Gamma(n+1)} - \cdots, \quad J_n' = \frac{x^{n-1}}{2^n \Gamma(n)} - \cdots, \quad J_{-n} = \frac{x^{-n}}{2^{-n}\, \Gamma(-n+1)} - \cdots,$$

$$J_{-n}' = \frac{x^{-n-1}}{2^{-n}\, \Gamma(-n)} - \cdots$$

and then substitute in (1). We find

$$c = \frac{1}{\Gamma(n)\, \Gamma(1-n)} - \frac{1}{\Gamma(n+1)\, \Gamma(-n)} = \frac{2}{\Gamma(n)\, \Gamma(1-n)} = \frac{2 \sin n\pi}{\pi}$$

using the result 1, page 211. This gives the required result.

(b) The expression $J_n' J_{-n} - J_{-n}' J_n$ in (a) is the *Wronskian* of J_n and J_{-n}. If n is an integer, we see from (a) that the Wronskian is zero so that J_n and J_{-n} are linearly independent as is also clear from Problem 10.3(a). On the other hand if n is not an integer, they are linearly independent since in such case the Wronskian differs from zero.

GENERATING FUNCTION AND MISCELLANEOUS RESULTS

10.11. Prove that $e^{(x/2)(t-1/t)} = \displaystyle\sum_{n=-\infty}^{\infty} J_n(x)\, t^n$.

We have

$$e^{(x/2)(t-1/t)} = e^{xt/2}\, e^{-x/2t} = \left\{ \sum_{r=0}^{\infty} \frac{(xt/2)^r}{r!} \right\} \left\{ \sum_{k=0}^{\infty} \frac{(-x/2t)^k}{k!} \right\} = \sum_{r=0}^{\infty} \sum_{k=0}^{\infty} \frac{(-1)^k (x/2)^{r+k}\, t^{r-k}}{r!\, k!}$$

Let $r - k = n$ so that n varies from $-\infty$ to ∞. Then the sum becomes

$$\sum_{n=-\infty}^{\infty} \sum_{k=0}^{\infty} \frac{(-1)^k (x/2)^{n+2k}\, t^n}{(n+k)!\, k!} = \sum_{n=-\infty}^{\infty} \left\{ \sum_{k=0}^{\infty} \frac{(-1)^k (x/2)^{n+2k}}{k!\, (n+k)!} \right\} t^n = \sum_{n=-\infty}^{\infty} J_n(x)\, t^n$$

10.12. Prove (a) $\cos(x \sin \theta) = J_0(x) + 2J_2(x) \cos 2\theta + 2J_4(x) \cos 4\theta + \cdots$

(b) $\sin(x \sin \theta) = 2J_1(x) \sin \theta + 2J_3(x) \sin 3\theta + 2J_5(x) \sin 5\theta + \cdots$

Let $t = e^{i\theta}$ in Problem 10.11. Then

$$e^{\frac{1}{2}x(e^{i\theta} - e^{-i\theta})} = e^{ix \sin \theta} = \sum_{-\infty}^{\infty} J_n(x)\, e^{in\theta} = \sum_{-\infty}^{\infty} J_n(x)[\cos n\theta + i \sin n\theta]$$

$$= \{J_0(x) + [J_{-1}(x) + J_1(x)] \cos \theta + [J_{-2}(x) + J_2(x)] \cos 2\theta + \cdots\}$$
$$+ i\{[J_1(x) - J_{-1}(x)] \sin \theta + [J_2(x) - J_{-2}(x)] \sin 2\theta + \cdots\}$$

$$= \{J_0(x) + 2J_2(x) \cos 2\theta + \cdots\} + i\{2J_1(x) \sin \theta + 2J_3(x) \sin 3\theta + \cdots\}$$

where we have used Problem 10.3(a). Equating real and imaginary parts gives the required results.

10.13. Prove $\quad J_n(x) = \dfrac{1}{\pi}\displaystyle\int_0^\pi \cos(n\theta - x\sin\theta)\,d\theta \quad n = 0, 1, 2, \ldots$.

Multiply the first and second results of Problem 10.12 by $\cos n\theta$ and $\sin n\theta$ respectively and integrate from 0 to π using

$$\int_0^\pi \cos m\theta\,\cos n\theta\,d\theta = \begin{cases} 0 & m \neq n \\ \pi/2 & m = n \end{cases}, \qquad \int_0^\pi \sin m\theta\,\sin n\theta\,d\theta = \begin{cases} 0 & m \neq n \\ \pi/2 & m = n \neq 0 \end{cases}$$

Then if n is even or zero, we have

$$J_n(x) = \frac{1}{\pi}\int_0^\pi \cos(x\sin\theta)\,\cos n\theta\,d\theta, \qquad 0 = \frac{1}{\pi}\int_0^\pi \sin(x\sin\theta)\,\sin n\theta\,d\theta$$

or on adding,

$$J_n(x) = \frac{1}{\pi}\int_0^\pi [\cos(x\sin\theta)\cos n\theta + \sin(x\sin\theta)\sin n\theta]\,d\theta = \frac{1}{\pi}\int_0^\pi \cos(n\theta - x\sin\theta)\,d\theta$$

Similarly if n is odd,

$$J_n(x) = \frac{1}{\pi}\int_0^\pi \sin(x\sin\theta)\,\sin n\theta\,d\theta, \qquad 0 = \frac{1}{\pi}\int_0^\pi \cos(x\sin\theta)\,\cos n\theta\,d\theta$$

and by adding, $\qquad\qquad J_n(x) = \dfrac{1}{\pi}\displaystyle\int_0^\pi \cos(n\theta - x\sin\theta)\,d\theta$

Thus we have the required result whether n is even or odd, i.e. $n = 0, 1, 2, \ldots$.

10.14. Prove the result of Problem 10.6(b) for integer values of n by using the generating function.

Differentiating both sides of the generating function with respect to t, we have, omitting the limits $-\infty$ to ∞ for n

$$e^{(x/2)(t-1/t)}\,\frac{x}{2}\left(1 + \frac{1}{t^2}\right) = \sum n J_n(x) t^{n-1}$$

or

$$\frac{x}{2}\left(1 + \frac{1}{t^2}\right)\sum J_n(x) t^n = \sum n J_n(x) t^{n-1}$$

i.e.

$$\sum \frac{x}{2}\left(1 + \frac{1}{t^2}\right) J_n(x) t^n = \sum n J_n(x) t^{n-1}$$

This can be written as

$$\sum \frac{x}{2} J_n(x) t^n + \sum \frac{x}{2} J_n(x) t^{n-2} = \sum n J_n(x) t^{n-1}$$

or

$$\sum \frac{x}{2} J_n(x) t^n + \sum \frac{x}{2} J_{n+2}(x) t^n = \sum (n+1) J_{n+1}(x) t^n$$

i.e.

$$\sum \left[\frac{x}{2} J_n(x) + \frac{x}{2} J_{n+2}(x)\right] t^n = \sum (n+1) J_{n+1}(x) t^n$$

Since coefficients of t^n must be equal, we have

$$\frac{x}{2} J_n(x) + \frac{x}{2} J_{n+2}(x) = (n+1) J_n(x)$$

from which the required result is obtained on replacing n by $n-1$.

BESSEL FUNCTIONS OF THE SECOND KIND

10.15. (a) Show that if n is not an integer, the general solution of Bessel's equation is

$$y = E J_n(x) + F\left[\frac{J_n(x)\cos n\pi - J_{-n}(x)}{\sin n\pi}\right]$$

(b) Explain how you can use part (a) to obtain the general solution of Bessel's equation in case n is an integer.

(a) Since J_{-n} and J_n are linearly independent, the general solution of Bessel's equation can be written

$$y = c_1 J_n(x) + c_2 J_{-n}(x)$$

and the required result follows on replacing the arbitrary constants c_1, c_2 by E, F where

$$c_1 = E + \frac{F \cos n\pi}{\sin n\pi}, \qquad c_2 = \frac{-F}{\sin n\pi}$$

Note that we define the Bessel function of the second kind if n is not an integer by

$$Y_n(x) = \frac{J_n(x) \cos n\pi - J_{-n}(x)}{\sin n\pi}$$

(b) The expression

$$\frac{J_n(x) \cos n\pi - J_{-n}(x)}{\sin n\pi}$$

becomes an "indeterminate" of the form 0/0 for the case when n is an integer. This is because for an integer n we have $\cos n\pi = (-1)^n$ and $J_{-n}(x) = (-1)^n J_n(x)$ [see Problem 10.3]. This "indeterminate form" can be evaluated by using L'Hospital's rule, i.e.

$$\lim_{p \to n} \left[\frac{J_p(x) \cos p\pi - J_{-p}(x)}{\sin p\pi} \right]$$

This motivates the definition (10) on page 225.

10.16. Use Problem 10.15 to obtain the general solution of Bessel's equation for $n = 0$.

In this case we must evaluate

$$\lim_{p \to 0} \left[\frac{J_p(x) \cos p\pi - J_{-p}(x)}{\sin p\pi} \right] \tag{1}$$

Using L'Hospital's rule [differentiating the numerator and denominator with respect to p] we find for the limit in (1)

$$\lim_{p \to 0} \left[\frac{(\partial J_p/\partial p) \cos p\pi - (\partial J_{-p}/\partial p)}{\pi \cos p\pi} \right] = \frac{1}{\pi} \left[\frac{\partial J_p}{\partial p} - \frac{\partial J_{-p}}{\partial p} \right]_{p=0}$$

where the notation indicates that we are to take the partial derivatives of $J_p(x)$ and $J_{-p}(x)$ with respect to p and then put $p = 0$. Since $\partial J_{-p}/\partial(-p) = -\partial J_{-p}/\partial p$, the required limit is also equal to

$$\frac{2}{\pi} \frac{\partial J_p}{\partial p} \bigg|_{p=0}$$

To obtain $\partial J_p/\partial p$ we differentiate the series

$$J_p(x) = \sum_{r=0}^{\infty} \frac{(-1)^r (x/2)^{p+2r}}{r!\, \Gamma(p+r+1)}$$

with respect to p and obtain

$$\frac{\partial J_p}{\partial p} = \sum_{r=0}^{\infty} \frac{(-1)^r}{r!} \frac{\partial}{\partial p} \left\{ \frac{(x/2)^{p+2r}}{\Gamma(p+r+1)} \right\} \tag{2}$$

Now if we let $\dfrac{(x/2)^{p+2r}}{\Gamma(p+r+1)} = G$, then

$$\ln G = (p+2r) \ln (x/2) - \ln \Gamma(p+r+1)$$

so that differentiation with respect to p gives

$$\frac{1}{G} \frac{\partial G}{\partial p} = \ln (x/2) - \frac{\Gamma'(p+r+1)}{\Gamma(p+r+1)}$$

Then for $p = 0$, we have

$$\frac{\partial G}{\partial p} \bigg|_{p=0} = \frac{(x/2)^{2r}}{\Gamma(r+1)} \left[\ln (x/2) - \frac{\Gamma'(r+1)}{\Gamma(r+1)} \right] \tag{3}$$

Using (2) and (3), we have

$$\frac{2}{\pi}\frac{\partial J_p}{\partial p}\bigg|_{p=0} = \frac{2}{\pi}\sum_{r=0}^{\infty}\frac{(-1)^r(x/2)^{2r}}{r!\,\Gamma(r+1)}\left[\ln(x/2) - \frac{\Gamma'(r+1)}{\Gamma(r+1)}\right]$$

$$= \frac{2}{\pi}\{\ln(x/2)+\gamma\}J_0(x) + \frac{2}{\pi}\left[\frac{x^2}{2^2} - \frac{x^4}{2^24^2}(1+\tfrac{1}{2}) + \frac{x^6}{2^24^26^2}(1+\tfrac{1}{2}+\tfrac{1}{3}) - \cdots\right]$$

where the last series is obtained on using the result 6 on page 224. This last series is the series for $Y_0(x)$. We can in a similar manner obtain the series (11), page 225, for $Y_n(x)$ where n is an integer. The general solution if n is an integer is then given by $y = c_1 J_n(x) + c_2 Y_n(x)$.

FUNCTIONS RELATED TO BESSEL FUNCTIONS

10.17. Prove that the recurrence formula for the modified Bessel function of the first kind $I_n(x)$ is given by

$$I_{n+1}(x) = I_{n-1}(x) - \frac{2n}{x}I_n(x)$$

From Problem 10.6(b) we have

$$J_{n+1}(x) = \frac{2n}{x}J_n(x) - J_{n-1}(x) \tag{1}$$

Replace x by ix to obtain

$$J_{n+1}(ix) = -\frac{2in}{x}J_n(ix) - J_{n-1}(ix) \tag{2}$$

Now by definition $I_n(x) = i^{-n}J_n(ix)$ or $J_n(ix) = i^n I_n(x)$ so that (2) becomes

$$i^{n+1}I_{n+1}(x) = -\frac{2in}{x}i^n I_n(x) - i^{n-1}I_n(x)$$

Dividing by i^{n+1} then gives the required result.

10.18. If n is not an integer, show that

(a) $H_n^{(1)}(x) = \dfrac{J_{-n}(x) - e^{-in\pi}J_n(x)}{i\sin n\pi}$, (b) $H_n^{(2)}(x) = \dfrac{e^{in\pi}J_n(x) - J_{-n}(x)}{i\sin n\pi}$

(a) By definition of $H_n^{(1)}(x)$ and $Y_n(x)$ [see pages 226 and 225 respectively] we have

$$H_n^{(1)}(x) = J_n(x) + iY_n(x) = J_n(x) + i\left[\frac{J_n(x)\cos n\pi - J_{-n}(x)}{\sin n\pi}\right]$$

$$= \frac{J_n(x)\sin n\pi + iJ_n(x)\cos n\pi - iJ_{-n}(x)}{\sin n\pi}$$

$$= i\left[\frac{J_n(x)(\cos n\pi - i\sin n\pi) - J_{-n}(x)}{\sin n\pi}\right]$$

$$= i\left[\frac{J_n(x)e^{-in\pi} - J_{-n}(x)}{\sin n\pi}\right] = \frac{J_{-n}(x) - e^{-in\pi}J_n(x)}{i\sin n\pi}$$

(b) Since $H_n^{(2)}(x) = J_n(x) - iY_n(x)$, we find on replacing i by $-i$ in the result of part (a),

$$H_n^{(2)}(x) = \frac{J_{-n}(x) - e^{in\pi}J_n(x)}{-i\sin n\pi} = \frac{e^{in\pi}J_n(x) - J_{-n}(x)}{i\sin n\pi}$$

10.19. Show that   (a) $\text{Ber}_0(x) = 1 - \dfrac{x^4}{2^24^2} + \dfrac{x^8}{2^24^26^28^2} - \cdots$

(b) $\text{Bei}_0(x) = \dfrac{x^2}{2^2} - \dfrac{x^6}{2^24^26^2} + \dfrac{x^{10}}{2^24^26^28^210^2} - \cdots$

We have

$$J_0(i^{3/2}x) = 1 - \frac{(i^{3/2}x)^2}{2^2} + \frac{(i^{3/2}x)^4}{2^24^2} - \frac{(i^{3/2}x)^6}{2^24^26^2} + \frac{(i^{3/2}x)^8}{2^24^26^28^2} - \cdots$$

$$= 1 - \frac{i^3x^2}{2^2} + \frac{i^6x^4}{2^24^2} - \frac{i^9x^6}{2^24^26^2} + \frac{i^{12}x^8}{2^24^26^28^2} - \cdots$$

$$= 1 + \frac{ix^2}{2^2} - \frac{x^4}{2^24^2} - \frac{ix^6}{2^24^26^2} + \frac{x^8}{2^24^26^28^2} - \cdots$$

$$= \left(1 - \frac{x^4}{2^24^2} + \frac{x^8}{2^24^26^28^2} - \cdots\right) + i\left(\frac{x^2}{2^2} - \frac{x^6}{2^24^26^2} + \cdots\right)$$

and the required result follows on noting that $J_0(i^{3/2}x) = \text{Ber}_0(x) + i\,\text{Bei}_0(x)$ and equating real and imaginary parts. Note that sometimes the subscript zero is omitted in $\text{Ber}_0(x)$ and $\text{Bei}_0(x)$.

EQUATIONS TRANSFORMED INTO BESSEL'S EQUATION

10.20. Find the general solution of the equation $xy'' + y' + ay = 0$.

The equation can be written as $x^2y'' + xy' + axy = 0$ and is a special case of equation (24), page 226, where $k = 0$, $\alpha = \sqrt{a}$, $r = 1/2$, $\beta = 0$. Then the solution as given by (25), page 226, is

$$y = c_1 J_0(2\sqrt{ax}) + c_2 Y_0(2\sqrt{ax})$$

ORTHOGONALITY OF BESSEL FUNCTIONS

10.21. Prove that $\displaystyle\int_0^1 xJ_n(\lambda x)\,J_n(\mu x)\,dx = \frac{\mu J_n(\lambda)\,J_n'(\mu) - \lambda J_n(\mu)\,J_n'(\lambda)}{\lambda^2 - \mu^2}$ if $\lambda \neq \mu$.

From (3) and (4), page 224, we see that $y_1 = J_n(\lambda x)$ and $y_2 = J_n(\mu x)$ are solutions of the equations

$$x^2y_1'' + xy_1' + (\lambda^2x^2 - n^2)y_1 = 0, \qquad x^2y_2'' + xy_2' + (\mu^2x^2 - n^2)y_2 = 0$$

Multiplying the first equation by y_2, the second by y_1 and subtracting, we find

$$x^2[y_2y_1'' - y_1y_2''] + x[y_2y_1' - y_1y_2'] = (\mu^2 - \lambda^2)x^2y_1y_2$$

which on division by x can be written as

$$x\frac{d}{dx}[y_2y_1' - y_1y_2'] + [y_2y_1' - y_1y_2'] = (\mu^2 - \lambda^2)xy_1y_2$$

or

$$\frac{d}{dx}\{x[y_2y_1' - y_1y_2']\} = (\mu^2 - \lambda^2)xy_1y_2$$

Then by integrating and omitting the constant of integration,

$$(\mu^2 - \lambda^2)\int xy_1y_2\,dx = x[y_2y_1' - y_1y_2']$$

or using $y_1 = J_n(\lambda x)$, $y_2 = J_n(\mu x)$ and dividing by $\mu^2 - \lambda^2 \neq 0$,

$$\int xJ_n(\lambda x)\,J_n(\mu x)\,dx = \frac{x[\lambda J_n(\mu x)\,J_n'(\lambda x) - \mu J_n(\lambda x)\,J_n'(\mu x)]}{\mu^2 - \lambda^2}$$

Thus

$$\int_0^1 xJ_n(\lambda x)\,J_n(\mu x)\,dx = \frac{\lambda J_n(\mu)\,J_n'(\lambda) - \mu J_n(\lambda)\,J_n'(\mu)}{\mu^2 - \lambda^2}$$

which is equivalent to the required result.

10.22. Prove that $\displaystyle\int_0^1 xJ_n^2(\lambda x)\,dx = \frac{1}{2}\left[J_n'^2(\lambda) + \left(1 - \frac{n^2}{\lambda^2}\right)J_n^2(\lambda)\right]$.

Let $\mu \to \lambda$ in the result of Problem 10.21. Then using L'Hospital's rule, we find

$$\int_0^1 xJ_n^2(\lambda x)\,dx = \lim_{\mu \to \lambda}\frac{\lambda J_n'(\mu)\,J_n'(\lambda) - J_n(\lambda)\,J_n'(\mu) - \mu J_n(\lambda)\,J_n''(\mu)}{2\mu}$$

$$= \frac{\lambda J_n'^2(\lambda) - J_n(\lambda)\,J_n'(\lambda) - \lambda J_n(\lambda)\,J_n''(\lambda)}{2\lambda}$$

But since $\lambda^2 J_n''(\lambda) + \lambda J_n'(\lambda) + (\lambda^2 - n^2) J_n(\lambda) = 0$, we find on solving for $J_n''(\lambda)$ and substituting,

$$\int_0^1 x J_n^2(\lambda x)\, dx = \frac{1}{2}\left[J_n'^2(\lambda) + \left(1 - \frac{n^2}{\lambda^2}\right) J_n^2(x) \right]$$

10.23. Prove that if λ and μ are any two different roots of the equation $R J_n(x) + S x J_n'(x) = 0$ where R and S are constants, then

$$\int_0^1 x J_n(\lambda x)\, J_n(\mu x)\, dx = 0$$

i.e. $\sqrt{x}\, J_n(\lambda x)$ and $\sqrt{x}\, J_n(\mu x)$ are orthogonal in $(0, 1)$.

Since λ and μ are roots of $R J_n(x) + S x J_n'(x) = 0$, we have

$$R J_n(\lambda) + S\lambda J_n'(\lambda) = 0, \qquad R J_n(\mu) + S\mu J_n'(\mu) = 0 \qquad (1)$$

Then if $R \neq 0$, $S \neq 0$ we find from (1),

$$\mu J_n(\lambda)\, J_n'(\mu) - \lambda J_n(\mu)\, J_n'(\lambda) = 0$$

and so from Problem 10.21 we have the required result

$$\int_0^1 x J_n(\lambda x)\, J_n(\lambda x)\, dx = 0$$

In case $R = 0$, $S \neq 0$ or $R \neq 0$, $S = 0$, the result is also easily proved.

SERIES OF BESSEL FUNCTIONS

10.24. If $f(x) = \sum_{p=1}^{\infty} A_p J_n(\lambda_p x)$, $0 < x < 1$, where λ_p, $p = 1, 2, 3, \ldots$, are the positive roots of $J_n(x) = 0$, show that

$$A_p = \frac{2}{J_{n+1}^2(\lambda_p)} \int_0^1 x J_n(\lambda_p x)\, f(x)\, dx$$

Multiply the series for $f(x)$ by $x J_n(\lambda_k x)$ and integrate term by term from 0 to 1. Then

$$\int_0^1 x J_n(\lambda_k x)\, f(x)\, dx = \sum_{p=1}^{\infty} A_p \int_0^1 x J_n(\lambda_k x)\, J_n(\lambda_p x)\, dx$$

$$= A_k \int_0^1 x J_n^2(\lambda_k x)\, dx$$

$$= \tfrac{1}{2} A_k J_n'^2(\lambda_k)$$

where we have used Problems 10.22 and 10.23 together with the fact that $J_n(\lambda_k) = 0$. It follows that

$$A_k = \frac{2}{J_n'^2(\lambda_k)} \int_0^1 x J_n(\lambda_k x)\, f(x)\, dx$$

To obtain the required result from this, we note that from the recurrence formula 3, page 225, which is equivalent to the formula 6 on that page, we have

$$\lambda_k J_n'(\lambda_k) = n J_n(\lambda_k) - \lambda_k J_{n+1}(\lambda_k)$$

or since $J_n(\lambda_k) = 0$, $\qquad\qquad J_n'(\lambda_k) = -J_{n+1}(\lambda_k)$

10.25. Expand $f(x) = 1$ in a series of the form

$$\sum_{p=1}^{\infty} A_p J_0(\lambda_p x)$$

for $0 < x < 1$, if λ_p, $p = 1, 2, 3, \ldots$, are the positive roots of $J_0(x) = 0$.

From Problem 10.24 we have

$$A_p = \frac{2}{J_1^2(\lambda_p)} \int_0^1 x J_0(\lambda_p x)\, dx = \frac{2}{\lambda_p^2 J_1^2(\lambda_p)} \int_0^{\lambda_p} v J_0(v)\, dv$$

$$= \frac{2}{\lambda_p^2 J_1^2(\lambda_p)} v J_1(v)\Big|_0^{\lambda_p} = \frac{2}{\lambda_p J_1(\lambda_p)}$$

where we have made the substitution $v = \lambda_p x$ in the integral and used the result of Problem 10.8(a) with $n = 1$.

Thus we have the required series

$$f(x) = 1 = \sum_{p=1}^{\infty} \frac{2}{\lambda_k J_1(\lambda_p)} J_0(\lambda_p x)$$

which can be written $\dfrac{J_0(\lambda_1 x)}{\lambda_1 J_1(\lambda_1)} + \dfrac{J_0(\lambda_2 x)}{\lambda_2 J_2(\lambda_2)} + \cdots = \dfrac{1}{2}$

Supplementary Problems

BESSEL'S DIFFERENTIAL EQUATION

10.26. Show that if x is replaced by λx where λ is a constant, then Bessel's equation $x^2 y'' + xy' + (x^2 - n^2)y = 0$ is transformed into $x^2 y'' + xy' + (\lambda^2 x^2 - n^2)y = 0$.

BESSEL FUNCTIONS OF THE FIRST KIND

10.27. (a) Show that $J_1(x) = \dfrac{x}{2} - \dfrac{x^3}{2^2 4} + \dfrac{x^5}{2^2 4^2 6} - \dfrac{x^7}{2^2 4^2 6^2 8} + \cdots$ and verify that the interval of convergence is $-\infty < x < \infty$.

(b) Show that $J_0'(x) = -J_1(x)$.

(c) Show that $\dfrac{d}{dx}[x J_1(x)] = x J_0(x)$.

10.28. Evaluate (a) $J_{5/2}(x)$ and (b) $J_{-5/2}(x)$ in terms of sines and cosines.

10.29. Find $J_3(x)$ in terms of $J_0(x)$ and $J_1(x)$.

10.30. Prove that (a) $J_n''(x) = \frac{1}{4}[J_{n-2}(x) - 2J_n(x) + J_{n+2}(x)]$,

(b) $J_n'''(x) = \frac{1}{8}[J_{n-3}(x) - 3J_{n-1}(x) + 3J_{n+1}(x) - J_{n+3}(x)]$

and generalize these results.

10.31. Evaluate (a) $\displaystyle\int x^3 J_2(x)\, dx$, (b) $\displaystyle\int_0^1 x^3 J_0(x)\, dx$, (c) $\displaystyle\int x^2 J_0(x)\, dx$.

10.32. Evaluate (a) $\displaystyle\int J_1(\sqrt[3]{x})\, dx$, (b) $\displaystyle\int \frac{J_2(x)}{x^2}\, dx$.

10.33. Evaluate $\displaystyle\int J_0(x) \sin x\, dx$.

10.34. Verify directly the result $J_n'(x) J_{-n}(x) - J_{-n}'(x) J_n(x) = \dfrac{2 \sin n\pi}{|\pi x|}$ for $n = \frac{1}{2}$.

GENERATING FUNCTION AND MISCELLANEOUS RESULTS

10.35. Use the generating function to prove that $J'_n(x) = \frac{1}{2}[J_{n-1}(x) + J_{n+1}(x)]$ for the case where n is an integer.

10.36. Use the generating function to work Problem 10.30 for the case where n is an integer.

10.37. Show that   (a) $1 = J_0(x) + 2J_2(x) + 2J_4(x) + \cdots$

(b) $J_1(x) - J_3(x) + J_5(x) - J_7(x) + \cdots = \sin x$.

10.38. Show that $\frac{x}{4} J_1(x) = J_2(x) - 2J_4(x) + 3J_6(x) - \cdots$.

10.39. Show that $J_0(x) = \frac{2}{\pi} \int_0^{\pi/2} \cos(x \sin \theta) \, d\theta$.

10.40. Show that   (a) $\displaystyle\int_0^{\pi/2} J_1(x \cos \theta) \, d\theta = \frac{1 - \cos x}{x}$

(b) $\displaystyle\int_0^{\pi/2} J_0(x \sin \theta) \cos \theta \sin \theta \, d\theta = \frac{J_1(x)}{x}$.

10.41. Show that $\displaystyle\int_0^x J_0(t) \, dt = 2 \sum_{k=0}^{\infty} J_{2k+1}(x)$.

10.42. Show that $\displaystyle\int_0^{\infty} e^{-ax} J_0(bx) \, dx = \frac{1}{\sqrt{a^2 + b^2}}$ and thus find

(a) $\mathcal{L}\{J_0(bt)\}$,   (b) $\mathcal{L}\{J_1(bt)\}$,   (c) $\mathcal{L}\{J_n(bt)\}$, $n = 0, 1, 2, \ldots$

10.43. Show that $\displaystyle\int_0^{\infty} J_0(x) \, dx = 1$.

10.44. Prove that $|J_n(x)| \leqq 1$ for all integers n. Is the result true if n is not an integer?

BESSEL FUNCTIONS OF THE SECOND KIND

10.45. Show that   (a) $Y_{n+1}(x) = \frac{2n}{x} Y_n(x) - Y_{n-1}(x)$,   (b) $Y'_n(x) = \frac{1}{2}[Y_{n-1}(x) - Y_{n+1}(x)]$.

10.46. Explain why the recurrence formulas for $J_n(x)$ on page 225 hold if $J_n(x)$ is replaced by $Y_n(x)$.

10.47. Prove that $Y'_0(x) = -Y_1(x)$.

10.48. Evaluate   (a) $Y_{1/2}(x)$,   (b) $Y_{-1/2}(x)$.

10.49. Prove that $J_n(x) Y'_n(x) - J'_n(x) Y_n(x) = 2/\pi x$.

10.50. Evaluate   (a) $\displaystyle\int x^3 Y_2(x) \, dx$,   (b) $\displaystyle\int Y_3(x) \, dx$,   (c) $\displaystyle\int \frac{Y_3(x)}{x^3} \, dx$.

10.51. Prove the result (11), page 225.

FUNCTIONS RELATED TO BESSEL FUNCTIONS

10.52. Show that $I_0(x) = 1 + \frac{x^2}{2^2} + \frac{x^4}{2^2 4^2} + \frac{x^6}{2^2 4^2 6^2} + \cdots$.

10.53. Show that   (a) $I'_n(x) = \frac{1}{2}\{I_{n-1}(x) + I_{n+1}(x)\}$,   (b) $xI'_n(x) = xI_{n-1}(x) - nI_n(x)$.

10.54. Show that $e^{(x/2)(t + 1/t)} = \sum_{-\infty}^{\infty} I_n(x) t^n$ is the generating function for $I_n(x)$.

10.55. Show that $I_0(x) = \frac{2}{\pi} \int_0^{\pi/2} \cosh(x \sin \theta) \, d\theta$.

10.56. Show that   (a) $\sinh x = 2[I_1(x) + I_3(x) + \cdots]$

(b) $\cosh x = I_0(x) + 2[I_2(x) + I_4(x) + \cdots]$.

10.57. Show that   (a) $I_{3/2}(x) = \sqrt{\dfrac{2}{\pi x}} \left(\cosh x - \dfrac{\sinh x}{x} \right)$,   (b) $I_{-3/2}(x) = \sqrt{\dfrac{2}{\pi x}} \left(\sinh x - \dfrac{\cosh x}{x} \right)$.

10.58.   (a) Show that $K_{n+1}(x) = K_{n-1}(x) + \dfrac{2n}{x} K_n(x)$.   (b) Explain why the functions $K_n(x)$ satisfy recurrence formulas which are the same as those for $I_n(x)$ with $I_n(x)$ replaced by $K_n(x)$.

10.59. Give asymptotic formulas for   (a) $H_n^{(1)}(x)$,   (b) $H_n^{(2)}(x)$.

10.60. Show that   (a) $\mathrm{Ber}_n(x) = \displaystyle\sum_{k=0}^{\infty} \frac{(x/2)^{2k+n}}{k!\,\Gamma(n+k+1)} \cos \frac{(3n+2k)\pi}{4}$

                (b) $\mathrm{Bei}_n(x) = \displaystyle\sum_{k=0}^{\infty} \frac{(x/2)^{2k+n}}{k!\,\Gamma(n+k+1)} \sin \left(\frac{3n+2k}{4} \right) \pi$.

10.61. Show that
$$\mathrm{Ker}_0(x) = -\{\ln(x/2) + \gamma\}\,\mathrm{Ber}_0(x) + \frac{\pi}{4}\mathrm{Bei}_0(x) + 1 - \frac{(x/2)^4}{2!^2}\left(1 + \tfrac{1}{2}\right) + \frac{(x/2)^8}{4!^2}\left(1 + \tfrac{1}{2} + \tfrac{1}{3} + \tfrac{1}{4}\right) - \cdots$$

EQUATIONS TRANSFORMED INTO BESSEL'S EQUATION

10.62. Prove that (25), page 226, is a solution of (24).

10.63. Solve $4xy'' + 4y' + y = 0$.

10.64. Solve   (a) $xy'' + 2y' + xy = 0$,   (b) $y'' + x^2 y = 0$.

10.65. Solve $y'' + e^{2x}y = 0$.   [*Hint.* Let $e^x = u$.]

10.66.   (a) Show by direct substitution that $y = J_0(2\sqrt{x})$ is a solution of $xy'' + y' + y = 0$ and (b) write the general solution.

10.67.   (a) Show by direct substitution that $y = \sqrt{x}\,J_{1/3}(\tfrac{2}{3}x^{3/2})$ is a solution of $y'' + xy = 0$ and (b) write the general solution.

10.68.   (a) Show that Bessel's equation $x^2 y'' + xy' + (x^2 - n^2)y = 0$ can be transformed into
$$\frac{d^2 u}{dx^2} + \left(1 - \frac{n^2 - 1/4}{x^2} \right) u = 0$$
where $y = u/\sqrt{x}$.   (b) Discuss the case where $n = \pm 1/2$.

   (b) Discuss the case where x is large and explain the connection with the asymptotic formulas on page 227.

ORTHOGONAL SERIES OF BESSEL FUNCTIONS

10.69. Complete Problem 10.23, page 237, for the cases   (a) $R \neq 0$, $S = 0$,   (b) $R = 0$, $S \neq 0$.

10.70. Show that $\displaystyle\int x J_n^2(\lambda x)\,dx = \frac{x^2}{2}[J_n^2(\lambda x) + J_{n+1}^2(\lambda x)] - \frac{nx}{\lambda} J_n(\lambda x)\,J_{n+1}(\lambda x) + c$

10.71. Prove the results (31) and (32), page 227.

10.72. Show that $\dfrac{1 - x^2}{8} = \displaystyle\sum_{p=1}^{\infty} \frac{J_0(\lambda_p x)}{\lambda_p^3 J_1(\lambda_p)} \qquad 0 < x < 1$

where λ_p are the positive roots of $J_0(\lambda) = 0$.

10.73. Show that $x = 2 \displaystyle\sum_{p=1}^{\infty} \frac{J_1(\lambda_p x)}{\lambda J_2(\lambda_p)} \qquad -1 < x < 1$

where λ_p are the positive roots of $J_1(\lambda) = 0$.

10.74. Show that $x^3 = \displaystyle\sum_{p=1}^{\infty} \frac{2(8 - \lambda_p^2)\,J_1(\lambda_p x)}{\lambda_p^3 J_1'(\lambda_p)} \qquad 0 \leq x < 1$

where λ_p are the positive roots of $J_1(\lambda) = 0$.

10.75. If $f(x) = \sum_{p=1}^{\infty} A_p J_0(\lambda_p x)$ where $J_0(\lambda_p) = 0$, $p = 1, 2, 3, \ldots$, show that

$$\int_0^1 x[f(x)]^2 \, dx = \sum_{p=1}^{\infty} A_p^2 J_1^2(\lambda_p)$$

Compare with *Parseval's identity* for Fourier series.

10.76. Use Problems 10.73 and 10.75 to show that

$$\sum_{p=1}^{\infty} \frac{1}{\lambda_p^2} = \frac{1}{4}$$

where λ_p are the positive roots of $J_0(\lambda) = 0$.

Answers to Supplementary Problems

10.28. (a) $\sqrt{\dfrac{2}{\pi x}} \left[\dfrac{(3 - x^2) \sin x - 3x \cos x}{x^2} \right]$ (b) $\sqrt{\dfrac{2}{\pi x}} \left[\dfrac{3x \sin x + (3 - x^2) \cos x}{x^2} \right]$

10.29. $\left(\dfrac{8 - x^2}{x^2} \right) J_1(x) - \dfrac{4}{x} J_0(x)$

10.31. (a) $x^3 J_3(x) + c$ (b) $2J_0(1) - 3J_1(1)$ (c) $x^2 J_1(x) + x J_0(x) - \displaystyle\int J_0(x) \, dx$

10.32. (a) $6\sqrt[3]{x} \, J_1(\sqrt[3]{x}) - 3\sqrt[3]{x^2} \, J_0(\sqrt[3]{x}) + c$

   (b) $-\dfrac{J_2(x)}{3x} - \dfrac{J_1(x)}{3} + \dfrac{1}{3} \displaystyle\int J_0(x) \, dx$

10.33. $x J_0(x) \sin x - x J_1(x) \cos x + c$

10.42. (a) $\dfrac{1}{\sqrt{s^2 + b^2}}$ (b) $\dfrac{\sqrt{s^2 + b^2} - s}{b\sqrt{s^2 + b^2}}$ (c) $\dfrac{(\sqrt{s^2 + b^2} - s)^n}{b^n \sqrt{s^2 + b^2}}$

10.48. (a) $-\sqrt{\dfrac{2}{\pi x}} \cos x$ (b) $\sqrt{\dfrac{2}{\pi x}} \sin x$

10.50. (a) $x^3 Y_3(x) + c$

   (b) $-Y_2(x) - 2Y_1(x)/x + c$ (c) $-\dfrac{1}{15} Y_1(x) - \dfrac{1}{15x} Y_2(x) - \dfrac{1}{5x^2} Y_3(x) + \dfrac{1}{15} \displaystyle\int Y_0(x) \, dx$

10.59. (a) $\sqrt{\dfrac{2}{\pi x}} \, e^{i(x - \frac{1}{4}\pi - \frac{1}{2}n\pi)}$ (b) $\sqrt{\dfrac{2}{\pi x}} \, e^{-i(x - \frac{1}{4}\pi - \frac{1}{2}n\pi)}$

10.63. $y = A J_0(\sqrt{x}) + B Y_0(\sqrt{x})$

10.64. (a) $y = \dfrac{A \sin x + B \cos x}{x}$ (b) $y = \sqrt{x} \, [A J_{1/4}(\tfrac{1}{2}x^2) + B J_{-1/4}(\tfrac{1}{2}x^2)]$

10.65. $y = A J_0(e^x) + B Y_0(e^x)$

Chapter 11

Legendre Functions and Other Orthogonal Functions

LEGENDRE'S DIFFERENTIAL EQUATION

Legendre functions arise as solutions of the differential equation

$$(1-x^2)y'' - 2xy' + n(n+1)y = 0 \tag{1}$$

which is called *Legendre's differential equation*. The general solution of (1) in the case where $n = 0, 1, 2, 3, \ldots$ is given by

$$y = c_1 P_n(x) + c_2 Q_n(x)$$

where $P_n(x)$ are polynomials called *Legendre polynomials* and $Q_n(x)$ are called *Legendre functions of the second kind* which are unbounded at $x = \pm 1$.

LEGENDRE POLYNOMIALS

The Legendre polynomials are defined by

$$P_n(x) = \frac{(2n-1)(2n-3)\cdots 1}{n!}\left\{ x^n - \frac{n(n-1)}{2(n-1)}x^{n-2} + \frac{n(n-1)(n-2)(n-3)}{2\cdot 4(2n-1)(2n-3)}x^{n-4} - \cdots \right\} \tag{2}$$

Note that $P_n(x)$ is a polynomial of degree n. The first few Legendre polynomials are as follows:

1. $P_0(x) = 1$
2. $P_1(x) = x$
3. $P_2(x) = \frac{1}{2}(3x^2 - 1)$
4. $P_3(x) = \frac{1}{2}(5x^3 - 3x)$
5. $P_4(x) = \frac{1}{8}(35x^4 - 30x^2 + 3)$
6. $P_5(x) = \frac{1}{8}(63x^5 - 70x^3 + 15x)$

In all cases $P_n(1) = 1$, $P_n(-1) = (-1)^n$.

The Legendre polynomials can also be expressed by *Rodrigue's formula* given by

$$P_n(x) = \frac{1}{2^n n!}\frac{d^n}{dx^n}(x^2 - 1)^n \tag{3}$$

GENERATING FUNCTION FOR LEGENDRE POLYNOMIALS

The function

$$\frac{1}{\sqrt{1 - 2xt + t^2}} = \sum_{n=0}^{\infty} P_n(x)t^n \tag{4}$$

is called the *generating function* for Legendre polynomials and is useful in obtaining their properties.

RECURRENCE FORMULAS

1. $P_{n+1}(x) = \frac{2n+1}{n+1}xP_n(x) - \frac{n}{n+1}P_{n-1}(x)$

2. $P'_{n+1}(x) - P'_{n-1}(x) = (2n+1)P_n(x)$

LEGENDRE FUNCTIONS OF THE SECOND KIND

If $|x| < 1$, the Legendre functions of the second kind are given by the following according as n is even or odd respectively:

$$Q_n(x) = \frac{(-1)^{n/2} 2^n [(n/2)!]^2}{n!} \left\{ x - \frac{(n-1)(n+2)}{3!} x^3 + \frac{(n-1)(n-3)(n+2)(n+4)}{5!} x^5 - \cdots \right\} \tag{5}$$

$$Q_n(x) = \frac{(-1)^{(n+1)/2} 2^{n-1} [(n-1)/2]!^2}{1 \cdot 3 \cdot 5 \cdots n} \left\{ 1 - \frac{n(n+1)}{2!} x^2 + \frac{n(n-2)(n+1)(n+3)}{4!} x^4 - \cdots \right\} \tag{6}$$

For $n > 1$, these coefficients are taken so that the recurrence formulas for $P_n(x)$ above apply also to $Q_n(x)$.

ORTHOGONALITY OF LEGENDRE POLYNOMIALS

The following results are fundamental:

$$\int_{-1}^{1} P_m(x) P_n(x) \, dx = 0 \qquad \text{if } m \neq n \tag{7}$$

$$\int_{-1}^{1} P_n^2(x) \, dx = \frac{2}{2n+1} \tag{8}$$

The first shows that any two different Legendre polynomials are orthogonal in the interval $-1 < x < 1$.

SERIES OF LEGENDRE POLYNOMIALS

If $f(x)$ satisfies the Dirichlet conditions [Page 183], then at every point of continuity of $f(x)$ in the interval $-1 < x < 1$ there will exist a Legendre series expansion having the form

$$f(x) = A_0 P_0(x) + A_1 P_1(x) + A_2 P_2(x) + \cdots = \sum_{k=0}^{\infty} A_k P_k(x) \tag{9}$$

where

$$A_k = \frac{2k+1}{2} \int_{-1}^{1} f(x) P_k(x) \, dx \tag{10}$$

At any point of discontinuity the series on the right in (9) converges to $\frac{1}{2}[f(x+0) + f(x-0)]$ which can be used to replace the left side of (9).

ASSOCIATED LEGENDRE FUNCTIONS

The differential equation

$$(1 - x^2) y'' - 2xy' + \left[n(n+1) - \frac{m^2}{1 - x^2} \right] y = 0 \tag{11}$$

is called *Legendre's associated differential equation*. If $m = 0$ this reduces to Legendre's equation (1). Solutions to (11) are called *associated Legendre functions*. We consider the case where m and n are non-negative integers. In this case the general solution of (11) is given by

$$y = c_1 P_n^m(x) + c_2 Q_n^m(x) \tag{12}$$

where $P_n^m(x)$ and $Q_n^m(x)$ are called *associated Legendre functions of the first and second kinds* respectively. They are given in terms of the ordinary Legendre functions by

$$P_n^m(x) = (1 - x^2)^{m/2} \frac{d^m}{dx^m} P_n(x) \tag{13}$$

$$Q_n^m(x) = (1 - x^2)^{m/2} \frac{d^m}{dx^m} Q_n(x) \tag{14}$$

Note that if $m > n$, $P_n^m(x) = 0$.

As in the case of Legendre polynomials, the Legendre functions $P_n^m(x)$ are orthogonal in $-1 < x < 1$, i.e.

$$\int_{-1}^{1} P_n^m(x) P_k^m(x) \, dx \;=\; 0 \qquad n \neq k \tag{15}$$

We also have

$$\int_{-1}^{1} [P_n^m(x)]^2 \, dx \;=\; \frac{2}{2n+1} \frac{(n+m)!}{(n-m)!} \tag{16}$$

Using these, we can expand a function $f(x)$ in a series of the form

$$f(x) \;=\; \sum_{k=0}^{\infty} A_k P_k^m(x) \tag{17}$$

OTHER SPECIAL FUNCTIONS

The following special functions are some of the important ones arising in science and engineering.

1. **Hermite polynomials.** These polynomials, denoted by $H_n(x)$, are solutions of *Hermite's differential equation*

$$y'' - 2xy' + 2ny \;=\; 0 \tag{18}$$

The polynomials are given by a corresponding *Rodrigue's formula*

$$H_n(x) \;=\; (-1)^n e^{x^2} \frac{d^n}{dx^n}(e^{-x^2}) \tag{19}$$

Their generating function is given by

$$e^{2tx - t^2} \;=\; \sum_{n=0}^{\infty} \frac{H_n(x)}{n!} t^n \tag{20}$$

and they satisfy the *recursion formulas*

$$H_{n+1}(x) \;=\; 2xH_n(x) - 2nH_{n-1}(x) \tag{21}$$

$$H_n'(x) \;=\; 2nH_{n-1}(x) \tag{22}$$

The important results

$$\int_{-\infty}^{\infty} e^{-x^2} H_m(x) H_n(x) \, dx \;=\; 0 \qquad m \neq n \tag{23}$$

$$\int_{-\infty}^{\infty} e^{-x^2} H_n^2(x) \, dx \;=\; 2^n n! \sqrt{\pi} \tag{24}$$

enable us to expand a function into a *Hermite series* of the form

$$f(x) \;=\; \sum_{k=0}^{\infty} A_k H_k(x) \tag{25}$$

where

$$A_k \;=\; \frac{1}{2^k k! \sqrt{\pi}} \int_{-\infty}^{\infty} e^{-x^2} f(x) H_k(x) \, dx \tag{26}$$

2. **Laguerre polynomials.** These polynomials, denoted by $L_n(x)$, are solutions of *Laguerre's differential equation*

$$xy'' + (1-x)y' + ny \;=\; 0 \tag{27}$$

The polynomials are given by the *Rodrigue's formula*

$$L_n(x) \;=\; e^x \frac{d^n}{dx^n}(x^n e^{-x}) \tag{28}$$

Their generating function is given by

$$\frac{e^{-xt/(1-t)}}{1-t} = \sum_{n=0}^{\infty} \frac{L_n(x)}{n!} t^n \tag{29}$$

and they satisfy the *recursion formulas*

$$L_{n+1}(x) = (2n+1-x)L_n(x) - n^2 L_{n-1}(x) \tag{30}$$

$$nL_{n-1}(x) = nL'_{n-1}(x) - L'_n(x) \tag{31}$$

The important results

$$\int_0^\infty e^{-x} L_m(x) L_n(x) \, dx = 0 \qquad m \neq n \tag{32}$$

$$\int_0^\infty e^{-x} L_n^2(x) \, dx = (n!)^2 \tag{33}$$

enable us to expand a function into a *Laguerre series* of the form

$$f(x) = \sum_{k=0}^{\infty} A_k L_k(x) \tag{34}$$

where

$$A_k = \frac{1}{(k!)^2} \int_0^\infty e^{-x} f(x) L_k(x) \, dx \tag{35}$$

STURM-LIOUVILLE SYSTEMS

A boundary-value problem having the form

$$\left.\begin{aligned}
\frac{d}{dx}\left[p(x)\frac{dy}{dx} \right] + [q(x) + \lambda r(x)]y &= 0 \qquad a \leqq x \leqq b \\
a_1 y(a) + a_2 y'(a) = 0, \quad b_1 y(b) + b_2 y'(b) &= 0
\end{aligned}\right\} \tag{36}$$

where a_1, a_2, b_1, b_2 are given constants; $p(x), q(x), r(x)$ are given functions which we shall assume to be differentiable and λ is an unspecified parameter independent of x, is called a *Sturm-Liouville boundary-value problem* or *Sturm-Liouville system*.

A non-trivial solution of this system, i.e. one which is not identically zero, exists in general only for a particular set of values of the parameter λ. These values are called the *characteristic values*, or more often *eigenvalues*, of the system. The corresponding solutions are called *characteristic functions* or *eigenfunctions* of the system. In general to each eigenvalue there is one eigenfunction, although exceptions can occur.

If $p(x), q(x)$ are real, then the eigenvalues are real. Also the eigenfunctions form an orthogonal set with respect to the *density function* $r(x)$ which is generally taken as non-negative, i.e. $r(x) \geqq 0$. It follows that by suitable normalization the set of functions can be made an orthonormal set with respect to $r(x)$ in $a \leqq x \leqq b$.

Solved Problems

LEGENDRE'S DIFFERENTIAL EQUATION

11.1. Use the method of Frobenius to find series solutions of Legendre's differential equation $(1 - x^2)y'' - 2xy' + n(n+1)y = 0$.

Assuming a solution of the form $y = \sum c_k x^{k+\beta}$ where the summation index k goes from $-\infty$ to ∞ and $c_k = 0$ for $k < 0$, we have

$$n(n+1)y = \sum n(n+1)c_k x^{k+\beta}$$

$$-2xy' = \sum -2(k+\beta)c_k x^{k+\beta}$$

$$(1-x^2)y'' = \sum (k+\beta)(k+\beta-1)c_k x^{k+\beta-2} - \sum (k+\beta)(k+\beta-1)c_k x^{k+\beta}$$

$$= \sum (k+\beta+2)(k+\beta+1)c_{k+2} x^{k+\beta} - \sum (k+\beta)(k+\beta-1)c_k x^{k+\beta}$$

Then by addition,

$$\sum [(k+\beta+2)(k+\beta+1)c_{k+2} - (k+\beta)(k+\beta-1)c_k - 2(k+\beta)c_k + n(n+1)c_k]x^{k+\beta} = 0$$

and since the coefficient of $x^{k+\beta}$ must be zero, we find

$$(k+\beta+2)(k+\beta+1)c_{k+2} + [n(n+1) - (k+\beta)(k+\beta+1)]c_k = 0 \qquad (1)$$

Letting $k = -2$ we obtain, since $c_{-2} = 0$, the indicial equation $\beta(\beta-1)c_0 = 0$ or, assuming $c_0 \neq 0$, $\beta = 0$ or 1.

Case 1, $\beta = 0$.

In this case (1) becomes

$$(k+2)(k+1)c_{k+2} + [n(n+1) - k(k+1)]c_k = 0 \qquad (2)$$

Putting $k = -1, 0, 1, 2, 3, \ldots$ in succession, we find that c_1 is arbitrary while

$$c_2 = -\frac{n(n+1)}{2 \cdot 1}, \quad c_3 = \frac{1 \cdot 2 - n(n+1)}{3 \cdot 2} c_1, \quad c_4 = \frac{[2 \cdot 3 - n(n+1)]}{4 \cdot 3} c_2, \quad \ldots$$

and so we obtain

$$y = c_0 \left[1 - \frac{n(n+1)}{2!} x^2 + \frac{n(n-2)(n+1)(n+3)}{4!} x^4 - \cdots \right]$$

$$+ c_1 \left[x - \frac{(n-1)(n+2)}{3!} x^3 + \frac{(n-1)(n-3)(n+2)(n+4)}{5!} x^5 - \cdots \right] \qquad (3)$$

Since this leads to a solution with two arbitrary constants, we need not consider Case 2, $\beta = 1$.

For an even integer $n \geqq 0$, the first of the above series terminates and gives a polynomial solution. For an odd integer $n > 0$, the second series terminates and gives a polynomial solution. Thus for any integer $n \geqq 0$ the equation has polynomial solutions. If $n = 0, 1, 2, 3$, for example, we obtain from (3) the polynomial

$$c_0, \quad c_1 x, \quad c_0(1 - 3x^2), \quad c_1 \left(\frac{3x - 5x^3}{2} \right)$$

which are, apart from a multiplicative constant, Legendre polynomials.

LEGENDRE POLYNOMIALS

11.2. Derive formula (2), page 242, for the Legendre polynomials.

From equation (2) of Problem 11.1 we see that if $k = n$ then $c_{n+2} = 0$ and thus $c_{n+4} = 0$, $c_{n+6} = 0, \ldots$. Then letting $k = n-2, n-4, \ldots$ we find from the equation (2) of Problem 11.1,

$$c_{n-2} = -\frac{n(n-1)}{2(2n-1)} c_n, \quad c_{n-4} = -\frac{(n-2)(n-3)}{4(2n-3)} c_{n-2} = \frac{n(n-1)(n-2)(n-3)}{2 \cdot 4(2n-1)(2n-3)} c_n, \quad \ldots$$

This leads to the polynomial solutions

$$y \;=\; c_n\left[\,x^n \;-\; \frac{n(n-1)}{2(2n-1)}\,x^{n-2} \;+\; \frac{n(n-1)(n-2)(n-3)}{2\cdot 4(2n-1)(2n-3)}\,x^{n-4} \;-\; \cdots\right]$$

The Legendre polynomials $P_n(x)$ are defined by choosing

$$c_n \;=\; \frac{(2n-1)(2n-3)\cdots 3\cdot 1}{n!}$$

This choice is made in order that $P_n(1) = 1$.

11.3. Derive Rodrigue's formula $P_n(x) = \dfrac{1}{2^n n!}\dfrac{d^n}{dx^n}(x^2-1)^n$.

By Problem 11.2 the Legendre polynomials are given by

$$P_n(x) \;=\; \frac{(2n-1)(2n-3)\cdots 3\cdot 1}{n!}\left\{x^n \;-\; \frac{n(n-1)}{2(2n-1)}\,x^{n-2} \;+\; \frac{n(n-1)(n-2)(n-3)}{2\cdot 4(2n-1)(2n-3)}\,x^{n-4} \;-\; \cdots\right\}$$

Now integrating this n times from 0 to x, we obtain

$$\frac{(2n-1)(2n-3)\cdots 3\cdot 1}{(2n)!}\left\{x^{2n} \;-\; n x^{2n-2} \;+\; \frac{n(n-1)}{2!}\,x^{2n-4} \;-\; \cdots\right\}$$

which can be written

$$\frac{(2n-1)(2n-3)\cdots 3\cdot 1}{(2n)(2n-1)(2n-2)\cdots 2\cdot 1}\,(x^2-1)^n \quad\text{or}\quad \frac{1}{2^n n!}\,(x^2-1)^n$$

which proves that

$$P_n(x) \;=\; \frac{1}{2^n n!}\frac{d^n}{dx^n}(x^2-1)^n$$

GENERATING FUNCTION

11.4. Prove that $\dfrac{1}{\sqrt{1-2xt+t^2}} = \displaystyle\sum_{n=0}^{\infty} P_n(x)t^n$.

Using the binomial theorem

$$(1+v)^p \;=\; 1 \;+\; pv \;+\; \frac{p(p-1)}{2!}\,v^2 \;+\; \frac{p(p-1)(p-2)}{3!}\,v^3 \;+\; \cdots$$

we have

$$\frac{1}{\sqrt{1-2xt+t^2}} \;=\; [1 - t(2x-t)]^{-1/2}$$

$$=\; 1 \;+\; \frac{1}{2}\,t(2x-t) \;+\; \frac{1\cdot 3}{2\cdot 4}\,t^2(2x-t)^2 \;+\; \frac{1\cdot 3\cdot 5}{2\cdot 4\cdot 6}\,t^3(2x-t)^3 \;+\; \cdots$$

and the coefficient of t^n in this expansion is

$$\frac{1\cdot 3\cdot 5\cdots(2n-1)}{2\cdot 4\cdot 6\cdots 2n}\,(2x)^n \;-\; \frac{1\cdot 3\cdot 5\cdots(2n-3)}{2\cdot 4\cdot 6\cdots(2n-2)}\cdot\frac{(n-1)}{1!}\,(2x)^{n-2}$$

$$+\; \frac{1\cdot 3\cdot 5\cdots 2n-5}{2\cdot 4\cdot 6\cdots 2n-4}\cdot\frac{(n-2)(n-3)}{2!}\,(2x)^{n-4} \;-\; \cdots$$

which can be written as

$$\frac{1\cdot 3\cdot 5\cdots(2n-1)}{n!}\left\{x^n \;-\; \frac{n(n-1)}{2(2n-1)}\,x^{n-2} \;+\; \frac{n(n-1)(n-2)(n-3)}{2\cdot 4(2n-1)(2n-3)}\,x^{n-4} \;-\; \cdots\right\}$$

i.e. $P_n(x)$. The required result thus follows.

RECURRENCE FORMULAS FOR LEGENDRE POLYNOMIALS

11.5. Prove that $P_{n+1}(x) = \dfrac{2n+1}{n+1} x P_n(x) - \dfrac{n}{n+1} P_{n-1}(x).$

From the generating function of Problem 11.4 we have

$$\frac{1}{\sqrt{1-2xt+t^2}} = \sum_{n=0}^{\infty} P_n(x) t^n \qquad (1)$$

Differentiating with respect to t,

$$\frac{x-t}{(1-2xt+t^2)^{3/2}} = \sum_{n=0}^{\infty} n P_n(x) t^{n-1}$$

Multiplying by $1 - 2xt + t^2$,

$$\frac{x-t}{\sqrt{1-2xt+t^2}} = \sum_{n=0}^{\infty} (1-2xt+t^2) n P_n(x) t^{n-1} \qquad (2)$$

Now the left side of (2) can be written in terms of (1) and we have

$$\sum_{n=0}^{\infty} (x-t) P_n(x) t^n = \sum_{n=0}^{\infty} (1-2xt+t^2) n P_n(x) t^{n-1}$$

i.e. $\displaystyle\sum_{n=0}^{\infty} x P_n(x) t^n - \sum_{n=0}^{\infty} P_n(x) t^{n+1} = \sum_{n=0}^{\infty} n P_n(x) t^{n-1} - \sum_{n=0}^{\infty} 2nx P_n(x) t^n + \sum_{n=0}^{\infty} n P_n(x) t^{n+1}$

Equating the coefficients of t^n on each side, we find

$$x P_n(x) - P_{n-1}(x) = (n+1) P_{n+1}(x) - 2nx P_n(x) + (n-1) P_{n-1}(x)$$

which yields the required result.

11.6. Given that $P_0(x) = 1$, $P_1(x) = x$,   find   (a) $P_2(x)$ and   (b) $P_3(x)$.

Using the recurrence formula of Problem 11.5, we have on letting $n = 1$,

$$P_2(x) = \frac{3}{2} x P_1(x) - \frac{1}{2} P_0(x) = \frac{3}{2} x^2 - \frac{1}{2} = \frac{1}{2}(3x^2 - 1)$$

Similarly letting $n = 2$,

$$P_3(x) = \frac{5}{3} x P_2(x) - \frac{2}{3} P_1(x) = \frac{5}{3} x \left(\frac{3x^2 - 1}{2} \right) - \frac{2}{3} x = \frac{1}{2}(5x^3 - 3x)$$

LEGENDRE FUNCTIONS OF THE SECOND KIND

11.7. Obtain the results (5) and (6), page 243, for the Legendre functions of the second kind in the case where n is a non-negative integer.

The Legendre functions of the second kind are the series solutions of Legendre's equation which do not terminate. From Problem 11.1, equation (3), we see that if n is even the series which does not terminate is

$$x - \frac{(n-1)(n+2)}{3!} x^3 + \frac{(n-1)(n-3)(n+2)(n+4)}{5!} x^5 - \cdots$$

while if n is odd the series which does not terminate is

$$1 - \frac{n(n+1)}{2!} x^2 + \frac{n(n-2)(n+1)(n+3)}{4!} x^4 - \cdots$$

These series solutions, apart from multiplicative constants, provide definitions for Legendre functions of the second kind and are given by (5) and (6) on page 243.

11.8. Obtain the Legendre functions of the second kind (a) $Q_0(x)$, (b) $Q_1(x)$ and (c) $Q_2(x)$.

(a) From equation (5), page 243, we have if $n = 0$,

$$Q_0(x) = x + \frac{2}{3!}x^3 + \frac{1\cdot3\cdot2\cdot4}{5!}x^5 + \frac{1\cdot3\cdot5\cdot2\cdot4\cdot6}{6!}x^7 + \cdots$$

$$= x + \frac{x^3}{3} + \frac{x^5}{5} + \frac{x^7}{7} + \cdots = \frac{1}{2}\ln\left(\frac{1+x}{1-x}\right)$$

where we have used the expansion $\ln(1+u) = u - u^2/2 + u^3/3 - u^4/4 + \cdots$.

(b) From equation (6), page 243, we have if $n = 1$,

$$Q_1(x) = -\left\{1 - \frac{(1)(2)}{2!}x^2 + \frac{(1)(-1)(2)(4)}{4!}x^4 - \frac{(1)(-1)(-3)(2)(4)(6)}{6!}x^6 + \cdots\right\}$$

$$= x\left\{x + \frac{x^3}{3} + \frac{x^5}{5} + \cdots\right\} - 1 = \frac{x}{2}\ln\left(\frac{1+x}{1-x}\right) - 1$$

(c) The recurrence formulas for $Q_n(x)$ are identical with those of $P_n(x)$. Then from Problem 11.5,

$$Q_{n+1}(x) = \frac{2n+1}{n+1}xQ_n(x) - \frac{n}{n+1}Q_{n-1}(x)$$

Putting $n = 1$, we have on using parts (a) and (b),

$$Q_2(x) = \frac{3}{2}xQ_1(x) - \frac{1}{2}Q_0(x) = \left(\frac{3x^2-1}{4}\right)\ln\left(\frac{1+x}{1-x}\right) - \frac{3x}{2}$$

ORTHOGONALITY OF LEGENDRE POLYNOMIALS

11.9. Prove that $\displaystyle\int_{-1}^{1} P_m(x)P_n(x)\,dx = 0$ if $m \neq n$.

Since $P_m(x)$, $P_n(x)$ satisfy Legendre's equation,

$$(1-x^2)P_m'' - 2xP_m' + m(m+1)P_m = 0$$
$$(1-x^2)P_n'' - 2xP_n' + n(n+1)P_n = 0$$

Then multiplying the first equation by P_n, the second equation by P_m and subtracting, we find

$$(1-x^2)[P_nP_m'' - P_mP_n''] - 2x[P_nP_m' - P_mP_n'] = [n(n+1)-m(m+1)]P_mP_n$$

which can be written

$$(1-x^2)\frac{d}{dx}[P_nP_m' - P_mP_n'] - 2x[P_nP_m' - P_mP_n'] = [n(n+1)-m(m+1)]P_mP_n$$

or

$$\frac{d}{dx}\{(1-x^2)[P_nP_m' - P_mP_n']\} = [n(n+1)-m(m+1)]P_mP_n$$

Thus by integrating we have

$$[n(n+1)-m(m+1)]\int_{-1}^{1} P_m(x)P_n(x)\,dx = (1-x^2)[P_nP_m' - P_mP_n']\Big|_{-1}^{1} = 0$$

Then since $m \neq n$, $\displaystyle\int_{-1}^{1} P_m(x)P_n(x)\,dx = 0$

11.10. Prove that $\displaystyle\int_{-1}^{1} P_n^2(x)\,dx = \frac{2}{2n+1}$.

From the generating function

$$\frac{1}{\sqrt{1-2tx+t^2}} = \sum_{n=0}^{\infty} P_n(x)t^n$$

we have on squaring both sides,

$$\frac{1}{1-2tx+t^2} = \sum_{m=0}^{\infty}\sum_{n=0}^{\infty} P_m(x)P_n(x)t^{m+n}$$

Then by integrating from -1 to 1 we have

$$\int_{-1}^{1} \frac{dx}{1 - 2tx + t^2} = \sum_{m=0}^{\infty} \sum_{n=0}^{\infty} \left\{ \int_{-1}^{1} P_m(x) P_n(x) \, dx \right\} t^{m+n}$$

Using the result of Problem 11.9 on the right side and performing the integration on the left side,

$$-\frac{1}{2t} \ln (1 - 2tx + t^2) \Big|_{-1}^{1} = \sum_{n=0}^{\infty} \left\{ \int_{-1}^{1} P_n^2(x) \, dx \right\} t^{2n}$$

or

$$\frac{1}{t} \ln \left(\frac{1 + t}{1 - t} \right) = \sum_{n=0}^{\infty} \left\{ \int_{-1}^{1} P_n^2(x) \, dx \right\} t^{2n}$$

i.e.

$$\sum_{n=0}^{\infty} \frac{2t^{2n}}{2n + 1} = \sum_{n=0}^{\infty} \left\{ \int_{-1}^{1} P_n^2(x) \, dx \right\} t^{2n}$$

Equating coefficients of t^{2n} we have as required

$$\int_{-1}^{1} P_n^2(x) \, dx = \frac{2}{2n + 1}$$

SERIES OF LEGENDRE POLYNOMIALS

11.11. If $f(x) = \sum_{k=0}^{\infty} A_k P_k(x)$, $-1 < x < 1$, show that

$$A_k = \frac{2k + 1}{2} \int_{-1}^{1} P_k(x) f(x) \, dx$$

Multiplying the given series by $P_m(x)$ and integrating from -1 to 1, we have on using Problems 11.9 and 11.10,

$$\int_{-1}^{1} P_m(x) f(x) \, dx = \sum_{k=0}^{\infty} A_k \int_{-1}^{1} P_m(x) P_k(x) \, dx$$

$$= A_m \int_{-1}^{1} P_m^2(x) \, dx = \frac{2A_m}{2m + 1}$$

Then as required,

$$A_m = \frac{2m + 1}{2} \int_{-1}^{1} P_m(x) f(x) \, dx$$

11.12. Expand the function $f(x) = \begin{cases} 1 & 0 < x < 1 \\ 0 & -1 < x < 0 \end{cases}$ in a series of the form $\sum_{k=0}^{\infty} A_k P_k(x)$.

By Problem 11.11,

$$A_k = \frac{2k + 1}{2} \int_{-1}^{1} P_k(x) f(x) \, dx = \frac{2k + 1}{2} \int_{-1}^{0} P_k(x)[0] \, dx + \frac{2k + 1}{2} \int_{0}^{1} P_k(x)[1] \, dx$$

$$= \frac{2k + 1}{2} \int_{0}^{1} P_k(x) \, dx$$

Then

$$A_0 = \frac{1}{2} \int_{0}^{1} P_0(x) \, dx = \frac{1}{2} \int_{0}^{1} (1) \, dx = \frac{1}{2}$$

$$A_1 = \frac{3}{2} \int_{0}^{1} P_1(x) \, dx = \frac{3}{2} \int_{0}^{1} x \, dx = \frac{3}{4}$$

$$A_2 = \frac{5}{2} \int_{0}^{1} P_2(x) \, dx = \frac{5}{2} \int_{0}^{1} \frac{3x^2 - 1}{2} \, dx = 0$$

$$A_3 = \frac{7}{2} \int_{0}^{1} P_3(x) \, dx = \frac{7}{2} \int_{0}^{1} \frac{5x^3 - 3x}{2} \, dx = -\frac{7}{16}$$

$$A_4 = \frac{9}{2}\int_0^1 P_4(x)\,dx = \frac{9}{2}\int_0^1 \frac{35x^4 - 30x^2 + 3}{8}\,dx = 0$$

$$A_5 = \frac{11}{2}\int_0^1 P_5(x)\,dx = \frac{11}{2}\int_0^1 \frac{63x^5 - 70x^3 + 15x}{8}\,dx = \frac{11}{32}$$

etc. Thus $\qquad f(x) = \frac{1}{2}P_0(x) + \frac{3}{4}P_1(x) - \frac{7}{16}P_3(x) + \frac{11}{32}P_5(x) - \cdots$

The general term for the coefficients in this series can be obtained by using the recurrence formula 2 on page 242 and the results of Problem 11.29. We find

$$A_n = \frac{2n+1}{2}\int_0^1 P_n(x)\,dx = \frac{1}{2}\int_0^1 [P'_{n+1}(x) - P'_{n-1}(x)]\,dx = \frac{1}{2}[P_{n-1}(0) - P_{n+1}(0)]$$

For n even $A_n = 0$, while for n odd we can use Problem 11.29(c).

ASSOCIATED LEGENDRE FUNCTIONS

11.13. Obtain the associated Legendre functions (a) $P_2^1(x)$, (b) $P_3^2(x)$, (c) $P_2^3(x)$.

(a) $P_2^1(x) = (1-x^2)^{1/2}\dfrac{d}{dx}P_2(x) = (1-x^2)^{1/2}\dfrac{d}{dx}\left(\dfrac{3x^2-1}{2}\right) = 3x(1-x^2)^{1/2}$

(b) $P_3^2(x) = (1-x^2)^{2/2}\dfrac{d^2}{dx^2}P_3(x) = (1-x^2)\dfrac{d^2}{dx^2}\left(\dfrac{5x^3-3x}{2}\right) = 15x - 15x^3$

(c) $P_2^3(x) = (1-x^2)^{3/2}\dfrac{d^3}{dx^3}P_2(x) = 0$

11.14. Verify that $P_3^2(x)$ is a solution of Legendre's associated equation (11), page 243, for $m = 2$, $n = 3$.

By Problem 11.13, $P_3^2(x) = 15x - 15x^3$. Substituting this in the equation

$$(1-x^2)y'' - 2xy' + \left[3\cdot4 - \frac{4}{1-x^2}\right]y = 0$$

we find after simplifying,

$$(1-x^2)(-90x) - 2x(15 - 45x^2) + \left[12 - \frac{4}{1-x^2}\right][15x - 15x^3] = 0$$

and so $P_3^2(x)$ is a solution.

11.15. Verify the result (15), page 244, for the functions $P_2^1(x)$ and $P_3^1(x)$.

We have from Problem 11.13(a), $P_2^1(x) = 3x(1-x^2)^{1/2}$. Also,

$$P_3^1(x) = (1-x^2)^{1/2}\frac{d}{dx}P_3(x) = (1-x^2)^{1/2}\frac{d}{dx}\left(\frac{5x^3 - 3x}{2}\right) = (1-x^2)^{1/2}\frac{15x^2 - 3}{2}$$

Then $\qquad \displaystyle\int_{-1}^1 P_2^1(x)\,P_3^1(x)\,dx = \int_{-1}^1 3x(1-x^2)\frac{15x^2 - 3}{2}\,dx = 0 \qquad$ (odd function)

11.16. Verify the result (16), page 244, for the function $P_2^1(x)$.

Since $P_2^1(x) = 3x(1-x^2)^{1/2}$,

$$\int_{-1}^1 [P_2^1(x)]^2\,dx = 9\int_{-1}^1 x^2(1-x^2)\,dx = 9\left[\frac{x^3}{3} - \frac{x^5}{5}\right]\Bigg|_{-1}^1 = \frac{36}{15} = \frac{12}{5}$$

Now according to (16), page 244, the required result should be

$$\frac{2}{2(2)+1}\frac{(2+1)!}{(2-1)!} = \frac{2}{5}\cdot\frac{3!}{1!} = \frac{12}{5}$$

so that the verification is achieved.

HERMITE POLYNOMIALS

11.17. Use the generating function for the Hermite polynomials to find (a) $H_0(x)$, (b) $H_1(x)$, (c) $H_2(x)$, (d) $H_3(x)$.

We have

$$e^{2tx-t^2} = \sum_{n=0}^{\infty} \frac{H_n(x)t^n}{n!} = H_0(x) + H_1(x)t + \frac{H_2(x)}{2!}t^2 + \frac{H_3(x)}{3!}t^3 + \cdots$$

Now

$$e^{2tx-t^2} = 1 + (2tx - t^2) + \frac{(2tx - t^2)^2}{2!} + \frac{(2tx - t^2)^3}{3!} + \cdots$$

$$= 1 + (2x)t + (2x^2 - 1)t^2 + \left(\frac{4x^3 - 6x}{3}\right)t^3 + \cdots$$

Comparing the two series, we have

$$H_0(x) = 1, \quad H_1(x) = 2x, \quad H_2(x) = 4x^2 - 2, \quad H_3(x) = 8x^3 - 12x$$

11.18. Prove that $H_n'(x) = 2nH_{n-1}(x)$.

Differentiating $e^{2tx-t^2} = \sum_{n=0}^{\infty} \frac{H_n(x)}{n!} t^n$ with respect to x,

$$2te^{2tx-t^2} = \sum_{n=0}^{\infty} \frac{H_n'(x)}{n!} t^n$$

or

$$\sum_{n=0}^{\infty} \frac{2H_n(x)}{n!} t^{n+1} = \sum_{n=0}^{\infty} \frac{H_n'(x)}{n!} t^n$$

Equating coefficients of t^n on both sides,

$$\frac{2H_{n-1}(x)}{(n-1)!} = \frac{H_n'(x)}{n!} \quad \text{or} \quad H_n'(x) = 2nH_{n-1}(x)$$

11.19. Prove that $H_n(x) = (-1)^n e^{x^2} \dfrac{d^n}{dx^n}(e^{-x^2})$.

We have

$$e^{2tx-t^2} = e^{x^2 - (t-x)^2} = \sum_{n=0}^{\infty} \frac{H_n(x)}{n!} t^n$$

Then

$$\frac{\partial^n}{\partial t^n}(e^{2tx-t^2})\bigg|_{t=0} = H_n(x)$$

But

$$\frac{\partial^n}{\partial t^n}(e^{2tx-t^2})\bigg|_{t=0} = e^{x^2}\frac{\partial^n}{\partial t^n}[e^{-(t-x)^2}]\bigg|_{t=0}$$

$$= e^{x^2}\frac{\partial^n}{\partial(-x)^n}[e^{-(t-x)^2}]\bigg|_{t=0} = (-1)^n\frac{d^n}{dx^n}(e^{-x^2})$$

11.20. Prove that $\displaystyle\int_{-\infty}^{\infty} e^{-x^2} H_m(x) H_n(x)\, dx = \begin{cases} 0 & m \neq n \\ 2^n n! \sqrt{\pi} & m = n \end{cases}$.

We have

$$e^{2tx-t^2} = \sum_{n=0}^{\infty} \frac{H_n(x)t^n}{n!}, \quad e^{2sx-s^2} = \sum_{m=0}^{\infty} \frac{H_m(x)s^m}{m!}$$

Multiplying these,

$$e^{2tx-t^2+2sx-s^2} = \sum_{m=0}^{\infty}\sum_{n=0}^{\infty} \frac{H_m(x)H_n(x)s^m t^n}{m!\, n!}$$

Multiplying by e^{-x^2} and integrating from $-\infty$ to ∞,

$$\int_{-\infty}^{\infty} e^{-[(x+s+t)^2 - 2st]}\, dx = \sum_{m=0}^{\infty}\sum_{n=0}^{\infty} \frac{s^m t^n}{m!\, n!}\int_{-\infty}^{\infty} e^{-x^2} H_m(x) H_n(x)\, dx$$

Now the left side is equal to

$$e^{2st} \int_{-\infty}^{\infty} e^{-(x+s+t)^2} \, dx \;=\; e^{2st} \int_{-\infty}^{\infty} e^{-u^2} \, du \;=\; e^{2st} \sqrt{\pi} \;=\; \sqrt{\pi} \sum_{m=0}^{\infty} \frac{2^m s^m t^m}{m!}$$

By equating coefficients the required result follows.

The result

$$\int_{-\infty}^{\infty} e^{-x^2} H_m(x) \, H_n(x) \, dx \;=\; 0 \qquad m \neq n$$

can also be proved by using a method similar to that of Problem 11.9.

LAGUERRE POLYNOMIALS

11.21. Determine the Laguerre polynomials (a) $L_0(x)$, (b) $L_1(x)$, (c) $L_2(x)$, (d) $L_3(x)$.

We have $L_n(x) = e^x \dfrac{d^n}{dx^n}(x^n e^{-x})$. Then

(a) $L_0(x) = 1$

(b) $L_1(x) = e^x \dfrac{d}{dx}(xe^{-x}) = 1 - x$

(c) $L_2(x) = e^x \dfrac{d^2}{dx^2}(x^2 e^{-x}) = 2 - 4x + x^2$

(d) $L_3(x) = e^x \dfrac{d^3}{dx^3}(x^3 e^{-x}) = 6 - 18x + 9x^2 - x^3$

11.22. Prove that the Laguerre polynomials $L_n(x)$ are orthogonal in $(0, \infty)$ with respect to the weight function e^{-x}.

From Laguerre's differential equation we have for any two Laguerre polynomials $L_m(x)$ and $L_n(x)$,

$$xL_m'' + (1-x)L_m' + mL_m \;=\; 0$$

$$xL_n'' + (1-x)L_n' + nL_n \;=\; 0$$

Multiplying these equations by L_n and L_m respectively and subtracting, we find

$$x[L_n L_m'' - L_m L_n''] + (1-x)[L_n L_m' - L_m L_n'] \;=\; (n-m)L_m L_n$$

or

$$\frac{d}{dx}[L_n L_m' - L_m L_n'] + \frac{1-x}{x}[L_n L_m' - L_m L_n'] \;=\; \frac{(n-m)L_m L_n}{x}$$

Multiplying by the integrating factor

$$e^{\int (1-x)/x \, dx} \;=\; e^{\ln x - x} \;=\; xe^{-x}$$

this can be written as

$$\frac{d}{dx}\{xe^{-x}[L_n L_m' - L_m L_n']\} \;=\; (n-m)e^{-x} L_m L_n$$

so that by integrating from 0 to ∞,

$$(n-m) \int_0^{\infty} e^{-x} L_m(x) \, L_n(x) \, dx \;=\; xe^{-x}[L_n L_m' - L_m L_n']\Big|_0^{\infty} \;=\; 0$$

Thus if $m \neq n$,

$$\int_0^{\infty} e^{-x} L_m(x) \, L_n(x) \, dx \;=\; 0$$

which proves the required result.

STURM-LIOUVILLE SYSTEMS

11.23. (a) Verify that the system $y'' + \lambda y = 0$, $y(0) = 0$, $y(1) = 0$ is a Sturm-Liouville system. (b) Find the eigenvalues and eigenfunctions of the system. (c) Prove that the eigenfunctions are orthogonal in $(0, 1)$. (d) Find the corresponding set of normalized eigenfunctions. (e) Expand $f(x) = 1$ in a series of these orthonormal functions.

(a) The system is a special case of (36), page 245, with $p(x) = 1$, $q(x) = 0$, $r(x) = 1$, $a = 0$, $b = 1$, $\alpha_1 = 1$, $\alpha_2 = 0$, $\beta_1 = 1$, $\beta_2 = 0$ and thus is a Sturm-Liouville system.

(b) The general solution of $y'' + \lambda y = 0$ is $y = A \cos \sqrt{\lambda}\, x + B \sin \sqrt{\lambda}\, x$. From the boundary condition $y(0) = 0$ we have $A = 0$, i.e. $y = B \sin \sqrt{\lambda}\, x$. From the boundary condition $y(1) = 0$ we have $B \sin \sqrt{\lambda} = 0$, so that since B cannot be zero [otherwise the solution will be identically zero, i.e. trivial] we must have $\sin \sqrt{\lambda} = 0$. Then $\sqrt{\lambda} = m\pi$, $\lambda = m^2\pi^2$ where $m = 1, 2, 3, \ldots$ are the required eigenvalues.

The eigenfunctions belonging to the eigenvalues $\lambda = m^2\pi^2$ can be designated by $B_m \sin m\pi x$, $m = 1, 2, 3, \ldots$.

Note that we exclude the value $m = 0$ or $\lambda = 0$ as eigenvalue since the corresponding eigenfunction is zero.

(c) The eigenfunctions are orthogonal since

$$\int_0^1 [B_m \sin m\pi x][B_n \sin n\pi x]\, dx \;=\; B_m B_n \int_0^1 \sin m\pi x \sin n\pi x\, dx$$

$$= B_m B_n \int_0^1 [\cos (m-n)\pi x - \cos (m+n)\pi x]\, dx$$

$$= B_m B_n \left[\frac{\sin (m-n)\pi x}{(m-n)\pi} - \frac{\sin (m+n)\pi x}{(m+n)\pi} \right] \Bigg|_0^1 \;=\; 0, \quad m \neq n$$

(d) The eigenfunctions will be orthonormal if

$$\int_0^1 [B_m \sin m\pi x]^2\, dx \;=\; 1$$

i.e. if $B_m^2 \displaystyle\int_0^1 \sin^2 n\pi x\, dx = \dfrac{B_m^2}{2} \int_0^1 (1 - \cos 2n\pi x)\, dx = \dfrac{B_m^2}{2} = 1$ or $B_m = \sqrt{2}$ taking the positive square root. Thus the set $\sqrt{2} \sin m\pi x$, $m = 1, 2, \ldots$, is an orthonormal set.

(e) We must find constants $c_1, c_2, \ldots$ such that

$$f(x) \;=\; \sum_{m=1}^{\infty} c_m \phi_m(x)$$

where $f(x) = 1$, $\phi_m(x) = \sqrt{2} \sin m\pi x$. By the methods of Chapter 8,

$$c_m \;=\; \int_0^1 f(x)\, \phi_m(x)\, dx \;=\; \sqrt{2} \int_0^1 \sin m\pi x\, dx \;=\; \frac{\sqrt{2}\,(1 - \cos m\pi)}{m\pi}$$

Then the required series [Fourier series] is

$$1 \;=\; \sum_{m=1}^{\infty} \frac{2(1 - \cos m\pi)}{m\pi} \sin m\pi x$$

11.24. Show that the eigenvalues of a Sturm-Liouville system are real.

We have
$$\frac{d}{dx} \left[p(x) \frac{dy}{dx} \right] + [q(x) + \lambda r(x)] y \;=\; 0 \tag{1}$$

$$\alpha_1 y(a) + \alpha_2 y'(a) \;=\; 0, \qquad \beta_1 y(b) + \beta_2 y'(b) \;=\; 0 \tag{2}$$

Then assuming $p(x)$, $q(x)$, $r(x)$, α_1, α_2, β_1, β_2 are real while λ and y may be complex, we have on taking the complex conjugate,

$$\frac{d}{dx} \left[p(x) \frac{d\bar{y}}{dx} \right] + [q(x) + \bar{\lambda} r(x)] y \;=\; 0 \tag{3}$$

$$\alpha_1 \bar{y}(a) + \alpha_2 \bar{y}'(a) \;=\; 0, \qquad \beta_1 \bar{y}(b) + \beta_2 \bar{y}'(b) \;=\; 0 \tag{4}$$

Multiplying equation (1) by $\bar{y}$, (3) by y and subtracting, we find after simplifying,

$$\frac{d}{dx}[p(x)(y\bar{y}' - \bar{y}y')] = (\lambda - \bar{\lambda})r(x)y\bar{y}\,dx$$

Then integrating from a to b, we have

$$(\lambda - \bar{\lambda})\int_a^b r(x)|y|^2\,dx = p(x)(y\bar{y}' - \bar{y}y')\Big|_a^b = 0 \tag{5}$$

on using the conditions (2) and (4).   Since $r(x) \geqq 0$ and is not identically zero in (a, b), the integral on the left of (5) is positive and so $\lambda - \bar{\lambda} = 0$ or $\lambda = \bar{\lambda}$ so that λ is real.

11.25. Show that the eigenfunctions belonging to two different eigenvalues are orthogonal with respect to $r(x)$ in (a, b).

If y_1 and y_2 are eigenfunctions belonging to the eigenvalues λ_1 and λ_2 respectively,

$$\frac{d}{dx}\left[p(x)\frac{dy_1}{dx}\right] + [q(x) + \lambda_1 r(x)]y_1 = 0 \tag{1}$$

$$\alpha_1 y_1(a) + \alpha_2 y_1'(a) = 0, \qquad \beta_1 y_1(b) + \beta_2 y_1'(b) = 0 \tag{2}$$

$$\frac{d}{dx}\left[p(x)\frac{dy_2}{dx}\right] + [q(x) + \lambda_2 r(x)]y_2 = 0 \tag{3}$$

$$\alpha_1 y_2(a) + \alpha_2 y_2'(a) = 0, \qquad \beta_1 y_2(b) + \beta_2 y_2'(b) = 0 \tag{4}$$

Then multiplying (1) by y_2, (3) by y_1 and subtracting, we find as in Problem 11.24,

$$\frac{d}{dx}[p(x)(y_1 y_2' - y_2 y_1')] = (\lambda_1 - \lambda_2)r(x)y_1 y_2$$

Integrating from a to b, we have on using (2) and (4),

$$(\lambda_1 - \lambda_2)\int_a^b r(x)y_1 y_2\,dx = p(x)(y_1 y_2' - y_2 y_1')\Big|_a^b = 0$$

and since $\lambda_1 \neq \lambda_2$ we have the required result

$$\int_a^b r(x)y_1 y_2\,dx = 0$$

Supplementary Problems

LEGENDRE POLYNOMIALS

11.26.   Use Rodrigue's formula (3), page 242, to verify the formulas for $P_0(x)$, $P_1(x)$, ..., $P_6(x)$ on page 242.

11.27.   Obtain the formulas for $P_4(x)$ and $P_5(x)$ using the recursion formula.

11.28.   Evaluate   (a) $\displaystyle\int_0^1 xP_5(x)\,dx$,   (b) $\displaystyle\int_{-1}^1 [P_2(x)]^2\,dx$,   (c) $\displaystyle\int_{-1}^1 P_2(x)P_4(x)\,dx$.

11.29.   Show that   (a) $P_n(1) = 1$ (c) $P_{2n-1}(0) = 0$

                 (b) $P_n(-1) = (-1)^n$ (d) $P_{2n}(0) = (-1)^n \dfrac{1 \cdot 3 \cdot 5 \cdots (2n-1)}{2 \cdot 4 \cdot 6 \cdots (2n)}$

      for $n = 1, 2, 3, \ldots$ using (2), page 242.

11.30.   Use the generating function to prove that $P_{n+1}'(x) - P_{n-1}'(x) = (2n+1)P_n(x)$.

11.31.   Prove that   (a) $P_{n+1}'(x) - xP_n'(x) = (n+1)P_n(x)$,   (b) $xP_n'(x) - P_{n-1}'(x) = nP_n(x)$.

11.32.   Show that $\displaystyle\sum_{n=0}^\infty P_n(\cos\theta) = \frac{1}{2}\csc\frac{\theta}{2}$.

11.33.   Show that   (a) $P_2(\cos\theta) = \frac{1}{4}(1 + 3\cos 2\theta)$,   (b) $P_3(\cos\theta) = \frac{1}{8}(3\cos\theta + 5\cos 3\theta)$.

LEGENDRE FUNCTIONS OF THE SECOND KIND

11.34. Prove that the series (5) and (6) on page 243 which are non-terminating are convergent for $-1 < x < 1$.

11.35. Find $Q_3(x)$.

11.36. Write the general solution of $(1-x^2)y'' - 2xy' + 2y = 0$.

SERIES OF LEGENDRE POLYNOMIALS

11.37. Expand $x^4 - 3x^2 + x$ in a series of the form $\sum_{k=0}^{\infty} A_k P_k(x)$.

11.38. Expand $f(x) = \begin{cases} 2x+1 & 0 < x \leqq 1 \\ 0 & -1 \leqq x < 0 \end{cases}$ in a series of the form $\sum_{k=0}^{\infty} A_k P_k(x)$, writing the first four nonzero terms.

11.39. If $f(x) = \sum_{k=0}^{\infty} A_k P_k(x)$, obtain *Parseval's identity*

$$\int_{-1}^{1} [f(x)]^2\, dx = \sum_{k=0}^{\infty} \frac{A_k^2}{2k+1}$$

and illustrate by using the function of Problem 11.37.

ASSOCIATED LEGENDRE FUNCTIONS

11.40. Find   (a) $P_2^2(x)$,   (b) $P_4^2(x)$,   (c) $P_4^3(x)$.

11.41. Find   (a) $Q_1^1(x)$,   (b) $Q_1^2(x)$.

11.42. Verify that the expressions for $P_2^1(x)$ and $Q_2^1(x)$ are solutions of the corresponding differential equation and thus write the general solution.

11.43. Verify formulas (15) and (16), page 244, for the case where   (a) $m=1$, $n=1$, $l=2$,   (b) $m=1$, $n=1$, $l=1$.

11.44. Obtain a generating function for $P_n^m(x)$.

11.45. Use the generating function to obtain results (15) and (16) on page 244.

11.46. Show how to expand $f(x)$ in a series of the form $\sum_{k=0}^{\infty} A_k P_k^m(x)$ and illustrate by using the cases (a) $f(x) = x^2$, $m=2$ and   (b) $f(x) = x(1-x)$, $m=1$. Verify the corresponding Parseval identity in each case.

HERMITE POLYNOMIALS

11.47. Use Rodrigue's formula (19), page 244, to obtain the Hermite polynomials $H_0(x), H_1(x), H_2(x), H_3(x)$.

11.48. Use the generating function to obtain the recurrence formula (21) on page 244 and obtain $H_2(x), H_3(x)$ given that $H_0(x) = 1$, $H_1(x) = 2x$.

11.49. Show directly that   (a) $\displaystyle\int_{-\infty}^{\infty} e^{-x^2} H_2(x)\, H_3(x)\, dx = 0$,   (b) $\displaystyle\int_{-\infty}^{\infty} e^{-x^2} [H_2(x)]^2\, dx = 8\sqrt{\pi}$.

11.50. Evaluate $\displaystyle\int_{-\infty}^{\infty} x^2 e^{-x^2} H_n(x)\, dx$.

11.51. Show that $H_{2n}(0) = \dfrac{(-1)^n (2n)!}{n!}$.

11.52. (a) Show how to expand $f(x)$ in a series of the form $\sum_{k=0}^{\infty} A_k H_k(x)$ and   (b) illustrate by using $f(x) = x^3 - 3x^2 + 2x$.

11.53. If $f(x) = \sum_{k=0}^{\infty} A_k H_k(x)$, obtain *Parseval's identity*

$$\int_{-\infty}^{\infty} e^{-x^2} [f(x)]^2\, dx = \sqrt{\pi} \sum_{k=0}^{\infty} 2^k k!\, c_k^2$$

and illustrate by using the function of Problem 11.52.

11.54. Find the general solution of Hermite's differential equation if (a) $n = 0$, (b) $n = 1$.

LAGUERRE POLYNOMIALS

11.55. Find $L_4(x)$ and show that it satisfies Laguerre's equation (27), page 244, for $n = 4$.

11.56. Use the generating function to obtain the recursion formula (30) on page 245.

11.57. Use formula (30) to determine $L_2(x)$, $L_3(x)$ and $L_4(x)$ if we define $L_n(x) = 0$ when $n = -1$ and 1 when $n = 0$.

11.58. Show that $nL_{n-1}(x) = nL'_{n-1}(x) - L'_n(x)$.

11.59. Prove the result (33) on page 245.

11.60. Expand $f(x) = x^3 - 3x^2 + 2x$ in a series of the form $\sum\limits_{k=0}^{\infty} A_k L_k(x)$.

11.61. Illustrate Parseval's identity for Problem 11.60.

11.62. Find the general solution of Laguerre's equation if (a) $n = 0$, (b) $n = 1$.

STURM-LIOUVILLE SYSTEMS

11.63. (a) Verify that the system $y'' + \lambda y = 0$, $y'(0) = 0$, $y(1) = 0$ is a Sturm-Liouville system.
(b) Find the eigenvalues and eigenfunctions of the system.
(c) Prove that the eigenfunctions are orthogonal and determine the corresponding orthonormal functions.

11.64. Show how to write (a) Legendre's equation, (b) Hermite's equation, (c) Bessel's equation and (d) Laguerre's equation in Sturm-Liouville form and discuss the significance.

Answers to Supplementary Problems

11.28. (a) 0   (b) 2/5   (c) 0

11.36. $y = Ax + B\left[1 + \dfrac{x}{2}\ln\left(\dfrac{1-x}{1+x}\right)\right]$

11.37. $\dfrac{-4}{7}P_0(x) + P_1(x) - \dfrac{10}{7}P_2(x) + \dfrac{8}{35}P_4(x)$

11.38. $P_0(x) + \dfrac{7}{4}P_1(x) + \dfrac{5}{8}P_2(x) - \dfrac{7}{16}P_3(x) + \cdots$

11.40. (a) $3(1 - x^2)$ (b) $-\dfrac{5}{2}(3 - 24x^2 + 21x^4)$ (c) $105x(1 - x^2)^{3/2}$

11.50. $\dfrac{1}{2}\sqrt{\pi}$ if $n = 0$, $2\sqrt{\pi}$ if $n = 2$,   0 otherwise

11.52. (b) $-\dfrac{3}{2}H_0(x) + \dfrac{7}{4}H_1(x) - \dfrac{3}{4}H_2(x) + \dfrac{1}{8}H_3(x)$

11.54. (a) $y = c_1 + c_2 \displaystyle\int e^{x^2}\, dx$ (b) $y = c_1 x + c_2 x \displaystyle\int \dfrac{e^{x^2}\, dx}{x^2}$

11.55. $L_4(x) = 24 - 96x + 72x^2 - 16x^3 + x^4$

11.60. $2L_0(x) - 8L_1(x) + 6L_2(x) - L_3(x)$

11.62. (a) $y = c_1 + c_2 \displaystyle\int \dfrac{e^x}{x}\, dx$ (b) $y = c_1(1 - x) + c_2(1 - x)\displaystyle\int \dfrac{e^x}{x(1-x)^2}\, dx$

11.63. $\dfrac{1}{4}[2m - 1]^2\pi^2$, $\cos(m - \dfrac{1}{2})\pi x$ $m = 1, 2, 3, \ldots$

11.64. (a) $\dfrac{d}{dx}[(1 - x^2)y'] + n(n+1)y = 0$ (c) $\dfrac{d}{dx}[xy'] + \left(x - \dfrac{n^2}{x}\right)y = 0$

(b) $\dfrac{d}{dx}[e^{-x^2}y'] + 2ne^{-x^2}y = 0$ (d) $\dfrac{d}{dx}[xe^{-x}y'] + ne^{-x}y = 0$

Chapter 12

Partial Differential Equations

SOME DEFINITIONS INVOLVING PARTIAL DIFFERENTIAL EQUATIONS

A *partial differential equation* is an equation containing an unknown function of two or more variables and its partial derivatives with respect to these variables.

The *order* of a partial differential equation is that of the highest ordered derivative present.

Example 1. $\dfrac{\partial^2 u}{\partial x\,\partial y} = 2x - y$ is a partial differential equation of order two, or a second order partial differential equation.

A *solution* of a partial differential equation is any function which satisfies the equation identically.

The *general solution* is a solution which contains a number of arbitrary independent functions equal to the order of the equation.

A *particular solution* is one which can be obtained from the general solution by particular choice of the arbitrary functions.

Example 2. As seen by substitution, $u = x^2 y - \frac{1}{2}xy^2 + F(x) + G(y)$ is a *solution* of the partial differential equation of Example 1. Because it contains two arbitrary independent functions $F(x)$ and $G(y)$, it is the *general solution*. If in particular $F(x) = 2\sin x$, $G(y) = 3y^4 - 5$, we obtain the *particular solution* $u = x^2 y - \frac{1}{2}xy^2 + 2\sin x + 3y^4 - 5$.

A *singular solution* is one which cannot be obtained from the general solution by particular choice of the arbitrary functions.

A *boundary-value problem* involving a partial differential equation seeks all solutions of a partial differential equation which satisfy conditions called *boundary conditions*. Theorems relating to the existence and uniqueness of such solutions are called *existence and uniqueness theorems*.

LINEAR PARTIAL DIFFERENTIAL EQUATIONS

The general *linear partial differential equation* of order two in two independent variables has the form

$$A\frac{\partial^2 u}{\partial x^2} + B\frac{\partial^2 u}{\partial x\,\partial y} + C\frac{\partial^2 u}{\partial y^2} + D\frac{\partial u}{\partial x} + E\frac{\partial u}{\partial y} + Fu = G \tag{1}$$

where $A, B, \ldots, G$ may depend on x and y but not on u. A second order equation with independent variables x and y which does not have the form (1) is called *nonlinear*.

If $G = 0$ the equation is called *homogeneous*, while if $G \neq 0$ it is called *non-homogeneous*. Generalizations to higher order equations are easily made.

Because of the nature of the solutions of (1) the equation is often classified as *elliptic*, *hyperbolic* or *parabolic* according as $B^2 - 4AC$ is less than, greater than or equal to zero respectively.

SOME IMPORTANT PARTIAL DIFFERENTIAL EQUATIONS

1. Heat Conduction Equation
$$\frac{\partial u}{\partial t} = \kappa \nabla^2 u$$

Here $u(x, y, z, t)$ is the temperature in a solid at position (x, y, z) at time t. The constant κ, called the *diffusivity*, is equal to $K/\sigma\tau$ where the *thermal conductivity K*, the *specific heat* σ and the density (mass per unit volume) τ are assumed constant.

In case u does not depend on y and z, the equation reduces to $\frac{\partial u}{\partial t} = \kappa \frac{\partial^2 u}{\partial x^2}$ called the *one-dimensional heat conduction equation*.

2. Vibrating String Equation
$$\frac{\partial^2 y}{\partial t^2} = a^2 \frac{\partial^2 y}{\partial x^2}$$

This equation is applicable to the small transverse vibrations of a taut, flexible string, such as a violin string, initially located on the x axis and set into motion [see Fig. 12-1]. The function $y(x, t)$ is the displacement of any point x of the string at time t. The constant $a^2 = T/\mu$, where T is the (constant) tension in the string and μ is the (constant) mass per unit length of the string. It is assumed that no external forces act on the string but that it vibrates only due to its elasticity.

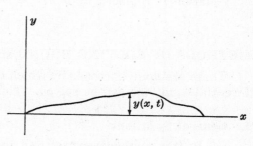

Fig. 12-1

The equation can easily be generalized to higher dimensions as for example the vibrations of a membrane or drum head in two dimensions. In two dimensions, for example, the equation is
$$\frac{\partial^2 z}{\partial t^2} = a^2 \left(\frac{\partial^2 z}{\partial x^2} + \frac{\partial^2 z}{\partial y^2} \right)$$

3. Laplace's Equation
$$\nabla^2 v = 0$$

This equation occurs in many fields. In the theory of heat conduction, for example, v is the *steady-state temperature*, i.e. the temperature after a long time has elapsed, and is equivalent to putting $\partial u/\partial t = 0$ in the heat conduction equation above. In the theory of gravitation or electricity v represents the *gravitational* or *electric potential* respectively. For this reason the equation is often called the *potential equation*.

4. Longitudinal Vibrations of a Beam
$$\frac{\partial^2 u}{\partial t^2} = c^2 \frac{\partial^2 u}{\partial x^2}$$

This equation describes the motion of a beam [Fig. 12-2] which can vibrate longitudinally [i.e. in the x direction]. The variable $u(x, t)$ is the longitudinal displacement from the equilibrium position of the cross section at x. The constant $c^2 = gE/\tau$ where g is the acceleration due to gravity, E is the modulus of elasticity [stress divided by strain] and depends on the properties of the beam, τ is the density [mass per unit volume].

Note that this equation is the same as that for a vibrating string.

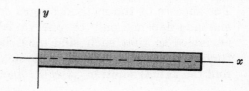

Fig. 12-2

5. Transverse Vibrations of a Beam $\boxed{\dfrac{\partial^2 y}{\partial t^2} + b^2 \dfrac{\partial^4 y}{\partial x^4} = 0}$

This equation describes the motion of a beam [initially located on the x axis, see Fig. 12-3] which is vibrating transversely (i.e. perpendicular to the x direction). In this case $y(x, t)$ is the transverse displacement or deflection at any time t of any point x. The constant $b^2 = EIg/\mu$ where E is the modulus of elasticity, I is the moment of inertia of any cross section about the x axis, g is the acceleration due to gravity and μ is the mass per unit length. In case an external transverse force $F(x, t)$ is applied, the right hand side of the equation is replaced by $b^2 F(x, t)/EI$.

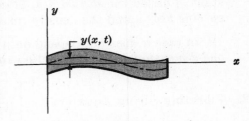

Fig. 12-3

METHODS OF SOLVING BOUNDARY-VALUE PROBLEMS

There are many methods by which boundary-value problems involving linear partial differential equations can be solved. The following are among the most important.

1. General Solutions.

In this method we first find the general solution and then that particular solution which satisfies the boundary conditions. The following theorems are of fundamental importance.

Theorem 12-1 **[Superposition principle].** If $u_1, u_2, \ldots, u_n$ are solutions of a linear homogeneous partial differential equation, then $c_1 u_1 + c_2 u_2 + \cdots + c_n u_n$ where $c_1, c_2, \ldots, c_n$ are constants is also a solution.

Theorem 12-2. The general solution of a linear non-homogeneous partial differential equation is obtained by adding a particular solution of the non-homogeneous equation to the general solution of the homogeneous equation.

We can sometimes find general solutions by using the methods of ordinary differential equations. See Problems 12.8 and 12.9.

If $A, B, \ldots, F$ in (1) are constants, then the general solution of the homogeneous equation can be found by assuming that $u = e^{ax+by}$ where a and b are constants to be determined. See Problems 12.10-12.13.

2. Separation of Variables.

In this method it is assumed that a solution can be expressed as a product of unknown functions each of which depends on only one of the independent variables. The success of the method hinges on being able to write the resulting equation so that one side depends only on one variable while the other side depends on the remaining variables so that each side must be a constant. By repetition of this the unknown functions can then be determined. Superposition of these solutions can then be used to find the actual solution. See Problems 12.15-12.17.

The method often makes use of Fourier series, Fourier integrals, Bessel series and Legendre series. For illustrations see Problems 12.18-12.28.

3. Laplace Transform Methods.

In this method the Laplace transform of the partial differential equation and associated boundary conditions are first obtained with respect to one of the independent variables. We then solve the resulting equation for the Laplace transform of the required solution which is then found by taking the inverse Laplace transform. In cases where Laplace inversion is of some difficulty the complex inversion formula of Chapter 14 can be used. For illustrations see Problems 12.29 and 14.15-14.19.

4. Complex Variable Methods. [See Chapter 13.]

Solved Problems

CLASSIFICATION OF PARTIAL DIFFERENTIAL EQUATIONS

12.1. Determine whether each of the following partial differential equations are linear or nonlinear, state the order of each equation, and name the dependent and independent variables.

(a) $\dfrac{\partial u}{\partial t} = 4\dfrac{\partial^2 u}{\partial x^2}$ linear, order 2, dep. var. u, ind. var. x, t

(b) $x^2\dfrac{\partial^3 R}{\partial y^3} = y^3\dfrac{\partial^2 R}{\partial x^2}$ linear, order 3, dep. var. R, ind. var. x, y

(c) $W\dfrac{\partial^2 W}{\partial r^2} = rst$ nonlinear, order 2, dep. var. W, ind. var. r, s, t

(d) $\dfrac{\partial^2 \phi}{\partial x^2} + \dfrac{\partial^2 \phi}{\partial y^2} + \dfrac{\partial^2 \phi}{\partial z^2} = 0$ linear, order 2, dep. var. ϕ, ind. var. x, y, z

(e) $\left(\dfrac{\partial z}{\partial u}\right)^2 + \left(\dfrac{\partial z}{\partial v}\right)^2 = 1$ nonlinear, order 1, dep. var. z, ind. var. u, v

12.2. Classify each of the following equations as elliptic, hyperbolic or parabolic.

(a) $\dfrac{\partial^2 \phi}{\partial x^2} + \dfrac{\partial^2 \phi}{\partial y^2} = 0$ $u = \phi,\ A = 1,\ B = 0,\ C = 1$

Then $B^2 - 4AC = -4 < 0$ and the equation is *elliptic*.

(b) $\dfrac{\partial u}{\partial t} = \kappa\dfrac{\partial^2 u}{\partial x^2}$ $y = t,\ A = \kappa,\ B = 0,\ C = 0$

Then $B^2 - 4AC = 0$ and the equation is *parabolic*.

(c) $\dfrac{\partial^2 y}{\partial t^2} = a^2\dfrac{\partial^2 y}{\partial x^2}$ $y = t,\ u = Y,\ A = a^2,\ B = 0,\ C = -1$

Then $B^2 - 4AC = 4a^2 > 0$ and the equation is *hyperbolic*.

(d) $\dfrac{\partial^2 u}{\partial x^2} + 3\dfrac{\partial^2 u}{\partial x\,\partial y} + 4\dfrac{\partial^2 u}{\partial y^2} + 5\dfrac{\partial u}{\partial x} - 2\dfrac{\partial u}{\partial y} + 4u = 2x - 3y$ $A = 1,\ B = 3,\ C = 4.$

Then $B^2 - 4AC = -7 < 0$ and the equation is *elliptic*.

(e) $x\dfrac{\partial^2 u}{\partial x^2} + y\dfrac{\partial^2 u}{\partial y^2} + 3y^2\dfrac{\partial u}{\partial x} = 0$ $A = x,\ B = 0,\ C = y$

Then $B^2 - 4AC = -4xy.$

In the region $xy > 0$ the equation is *elliptic*.

In the region $xy < 0$ the equation is *hyperbolic*.

If $xy = 0$, the equation is *parabolic*.

SOLUTIONS OF PARTIAL DIFFERENTIAL EQUATIONS

12.3. Show that $u(x, t) = e^{-8t} \sin 2x$ is a solution to the boundary-value problem

$$\frac{\partial u}{\partial t} = 2\frac{\partial^2 u}{\partial x^2}, \quad u(0, t) = u(\pi, t) = 0, \quad u(x, 0) = \sin 2x$$

From $u(x, t) = e^{-8t} \sin 2x$ we have

$$u(0, t) = e^{-8t} \sin 0 = 0, \quad u(\pi, t) = e^{-8t} \sin 2\pi = 0, \quad u(x, 0) = e^{-0} \sin 2x = \sin 2x$$

and the boundary conditions are satisfied.

Also $\quad\quad \dfrac{\partial u}{\partial t} = -8e^{-8t} \sin 2x, \quad \dfrac{\partial u}{\partial x} = 2e^{-8t} \cos 2x, \quad \dfrac{\partial^2 u}{\partial x^2} = -4e^{-8t} \sin 2x$

Then substituting into the differential equation, we have

$$-8e^{-8t} \sin 2x = 2(-4e^{-8t} \sin 2x)$$

which is an identity.

12.4. (a) Show that $v = F(y - 3x)$, where F is an arbitrary differentiable function, is a general solution of the equation

$$\frac{\partial v}{\partial x} + 3\frac{\partial v}{\partial y} = 0$$

(b) Find the particular solution which satisfies the condition $v(0, y) = 4 \sin y$.

(a) Let $y - 3x = u$. Then $v = F(u)$ and

$$\frac{\partial v}{\partial x} = \frac{\partial v}{\partial u}\frac{\partial u}{\partial x} = F'(u)(-3) = -3F'(u)$$

$$\frac{\partial v}{\partial y} = \frac{\partial v}{\partial u}\frac{\partial u}{\partial y} = F'(u)(1) = F'(u)$$

Thus $\quad\quad\quad\quad\quad\quad \dfrac{\partial v}{\partial x} + 3\dfrac{\partial v}{\partial y} = 0$

Since the equation is of order one, the solution $v = F(u) = F(y - 3x)$ which involves only one arbitrary function is a general solution.

(b) $v(x, y) = F(y - 3x)$. Then $v(0, y) = F(y) = 4 \sin y$.

If $F(y) = 4 \sin y$, then $v(x, y) = F(y - 3x) = 4 \sin (y - 3x)$ is the required solution.

12.5. (a) Show that $y(x, t) = F(2x + 5t) + G(2x - 5t)$ is a general solution of

$$4\frac{\partial^2 y}{\partial t^2} = 25\frac{\partial^2 y}{\partial x^2}$$

(b) Find a particular solution satisfying the conditions

$$y(0, t) = y(\pi, t) = 0, \quad y(x, 0) = \sin 2x, \quad y_t(x, 0) = 0$$

(a) Let $2x + 5t = u$, $2x - 5t = v$. Then $y = F(u) + G(v)$.

$$\frac{\partial y}{\partial t} = \frac{\partial F}{\partial u}\frac{\partial u}{\partial t} + \frac{\partial G}{\partial v}\frac{\partial v}{\partial t} = F'(u)(5) + G'(v)(-5) = 5F'(u) - 5G'(v) \tag{1}$$

$$\frac{\partial^2 y}{\partial t^2} = \frac{\partial}{\partial t}(5F'(u) - 5G'(v)) = 5\frac{\partial F'}{\partial u}\frac{\partial u}{\partial t} - 5\frac{\partial G'}{\partial u}\frac{\partial u}{\partial t} = 25F''(u) + 25G''(v) \tag{2}$$

$$\frac{\partial y}{\partial x} = \frac{\partial F}{\partial u}\frac{\partial u}{\partial x} + \frac{\partial G}{\partial v}\frac{\partial v}{\partial x} = F'(u)(2) + G'(v)(2) = 2F'(u) + 2G'(v) \tag{3}$$

$$\frac{\partial^2 y}{\partial x^2} = \frac{\partial}{\partial x}[2F'(u) + 2G'(v)] = 2\frac{\partial F'}{\partial u}\frac{\partial u}{\partial x} + 2\frac{\partial G'}{\partial u}\frac{\partial u}{\partial x} = 4F''(u) + 4G''(v) \tag{4}$$

From (2) and (4), $4\dfrac{\partial^2 y}{\partial t^2} = 25\dfrac{\partial^2 y}{\partial x^2}$ and the equation is satisfied. Since the equation is of order 2 and the solution involves two arbitrary functions, it is a general solution.

(b) We have from $y(x, t) = F(2x + 5t) + G(2x - 5t)$,

$$y(x, 0) = F(2x) + G(2x) = \sin 2x \tag{5}$$

Also $\qquad y_t(x, t) = \partial y/\partial t = 5F'(2x + 5t) - 5G'(2x - 5t)$

so that $\qquad\qquad y_t(x, 0) = 5F'(2x) - 5G'(2x) = 0 \tag{6}$

Differentiating (5), $\qquad 2F'(2x) + 2G'(2x) = 2\cos 2x$

From (6), $\qquad\qquad\qquad F'(2x) = G'(2x)$

Then $\qquad\qquad\qquad F'(2x) = G'(2x) = \tfrac{1}{2}\cos 2x$

from which $\qquad F(2x) = \tfrac{1}{2}\sin 2x + c_1, \qquad G(2x) = \tfrac{1}{2}\sin 2x + c_2$

i.e. $\qquad\qquad y(x, t) = \tfrac{1}{2}\sin(2x + 5t) + \tfrac{1}{2}\sin(2x - 5t) + c_1 + c_2$

Using $\;y(0, t) = 0$ or $Y(n, t) = 0$, $\;c_1 + c_2 = 0\;$ so that

$$y(x, t) = \tfrac{1}{2}\sin(2x + 5t) + \tfrac{1}{2}\sin(2x - 5t) = \sin 2x \cos 5t$$

which can be checked as the required solution.

SOME IMPORTANT PARTIAL DIFFERENTIAL EQUATIONS

12.6. If the temperature at any point (x, y, z) of a solid at time t is $u(x, y, z, t)$ and if κ, σ and τ are respectively the thermal conductivity, specific heat and density of the solid, assumed constant, show that

$$\frac{\partial u}{\partial t} = \kappa \nabla^2 u \qquad \text{where } \kappa = K/\sigma\tau$$

Let V be an arbitrary volume lying within the solid, and let S denote its surface. The total flux of heat across S, or the quantity of heat leaving S per unit time, is

$$\iint_S (-K\nabla u) \cdot \mathbf{n}\, dS$$

Thus the quantity of heat entering S per unit time is

$$\iint_S (K\nabla u) \cdot \mathbf{n}\, dS = \iiint_V \nabla \cdot (K\nabla u)\, dV \tag{1}$$

by the divergence theorem. The heat contained in a volume V is given by

$$\iiint_V \sigma\tau u\, dV$$

Then the time rate of increase of heat is

$$\frac{\partial}{\partial t} \iiint_V \sigma\tau u\, dV = \iiint_V \sigma\tau \frac{\partial u}{\partial t}\, dV \tag{2}$$

Equating the right hand sides of (1) and (2),

$$\iiint_V \left[\sigma\tau \frac{\partial u}{\partial t} - \nabla \cdot (K\nabla u) \right] dV = 0$$

and since V is arbitrary, the integrand, assumed continuous, must be identically zero so that

$$\sigma\tau \frac{\partial u}{\partial t} = \nabla \cdot (K\nabla u)$$

or if K, σ, τ are constants, $\qquad \dfrac{\partial u}{\partial t} = \dfrac{K}{\sigma\tau} \nabla \cdot \nabla u = \kappa \nabla^2 u$

The quantity κ is called the *diffusivity*. For steady-state heat flow (i.e. $\partial u/\partial t = 0$ or u is independent of time) the equation reduces to Laplace's equation $\nabla^2 u = 0$.

12.7. Derive the vibrating string equation on page 259.

Referring to Fig. 12-4 assume that Δs represents an element of arc of the string. Since the tension is assumed constant the net upward vertical force acting on Δs is given by

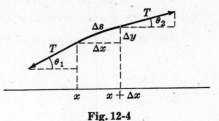

$$T \sin \theta_2 - T \sin \theta_1 \tag{1}$$

Since $\sin \theta = \tan \theta$ approximately for small angles, this force is

$$T \frac{\partial y}{\partial x}\bigg|_{x+\Delta x} - T \frac{\partial y}{\partial x}\bigg|_{x} \tag{2}$$

Fig. 12-4

using the fact that the slope is $\tan \theta = \partial y / \partial x$. By Newton's law this net force is equal to the mass of the string ($\mu \Delta s$) times the acceleration of Δs which is given by $\frac{\partial^2 y}{\partial t^2} + \epsilon$ where $\epsilon \to 0$ as $\Delta s \to 0$. Thus we have approximately

$$T\left[\frac{\partial y}{\partial x}\bigg|_{x+\Delta x} - \frac{\partial y}{\partial x}\bigg|_{x}\right] = [\mu \Delta s]\left[\frac{\partial^2 y}{\partial t^2} + \epsilon\right] \tag{3}$$

If the vibrations are small then $\Delta s = \Delta x$ approximately so that (3) becomes on division by $T \Delta x$,

$$\frac{T}{\mu} \frac{\frac{\partial y}{\partial x}\big|_{x+\Delta x} - \frac{\partial y}{\partial x}\big|_{x}}{\Delta x} = \frac{\partial^2 y}{\partial t^2} + \epsilon$$

Taking the limit as $\Delta x \to 0$ [in which case $\epsilon \to 0$ also], we have

$$\frac{T}{\mu} \frac{\partial}{\partial x}\left(\frac{\partial y}{\partial x}\right) = \frac{\partial^2 y}{\partial t^2} \quad \text{or} \quad \frac{\partial^2 y}{\partial t^2} = a^2 \frac{\partial^2 y}{\partial x^2} \quad \text{where} \quad a^2 = T/\mu$$

METHODS OF FINDING SOLUTIONS OF PARTIAL DIFFERENTIAL EQUATIONS

12.8. (a) Solve the equation $\dfrac{\partial^2 z}{\partial x \, \partial y} = x^2 y$.

(b) Find the particular solution for which $z(x, 0) = x^2$, $z(1, y) = \cos y$.

(a) Write the equation as $\dfrac{\partial}{\partial x}\left(\dfrac{\partial z}{\partial y}\right) = x^2 y$. Then integrating with respect to x, we find

$$\partial z / \partial y = \tfrac{1}{3}x^3 y + F(y) \tag{1}$$

where $F(y)$ is arbitrary.

Integrating (1) with respect to y,

$$z = \tfrac{1}{6}x^3 y^2 + \int F(y)\, dy + G(x) \tag{2}$$

where $G(x)$ is arbitrary.

The result (2) can be written

$$z = z(x, y) = \tfrac{1}{6}x^3 y^2 + H(y) + G(x) \tag{3}$$

which has two arbitrary (essential) functions and is therefore a general solution.

(b) Since $z(x, 0) = x^2$, we have from (3)

$$x^2 = H(0) + G(x) \quad \text{or} \quad G(x) = x^2 - H(0) \tag{4}$$

Thus

$$z = \tfrac{1}{6}x^3 y^2 + H(y) + x^2 - H(0) \tag{5}$$

Since $z(1, y) = \cos y$, we have from (5)

$$\cos y = \tfrac{1}{6}y^2 + H(y) + 1 - H(0)$$

or

$$H(y) = \cos y - \tfrac{1}{6}y^2 - 1 + H(0) \tag{6}$$

Thus using (6) in (5) we find the required solution

$$z = \tfrac{1}{6}x^3 y^2 + \cos y - \tfrac{1}{6}y^2 + x^2 - 1$$

12.9. Solve $t \dfrac{\partial^2 u}{\partial x\,\partial t} + 2\dfrac{\partial u}{\partial x} = x^2$.

Write the equation as $\dfrac{\partial}{\partial x}\left[t\dfrac{\partial u}{\partial t} + 2u \right] = x^2$. Integrating with respect to x,

$$t\dfrac{\partial u}{\partial t} + 2u = \tfrac{1}{3}x^3 + F(t)$$

or

$$\dfrac{\partial u}{\partial t} + \dfrac{2}{t}u = \dfrac{1}{3}\dfrac{x^3}{t} + \dfrac{F(t)}{t}$$

This is a linear equation having integrating factor $e^{\int (2/t)\,dt} = e^{2\ln t} = e^{\ln t^2} = t^2$. Then

$$\dfrac{\partial}{\partial t}(t^2 u) = \tfrac{1}{3}tx^3 + tF(t)$$

Integrating, $t^2 u = \tfrac{1}{6}t^2 x^3 + \displaystyle\int tF(t)\,dt + H(x) = \tfrac{1}{6}t^2 x^3 + G(t) + H(x)$

the required general solution.

12.10. Find solutions of $\dfrac{\partial^2 u}{\partial x^2} + 3\dfrac{\partial^2 u}{\partial x\,\partial y} + 2\dfrac{\partial^2 u}{\partial y^2} = 0$.

Assume $u = e^{ax+by}$. Substituting in the given equation, we find

$$(a^2 + 3ab + 2b^2)e^{ax+by} = 0 \quad \text{or} \quad a^2 + 3ab + 2b^2 = 0$$

Then $(a+b)(a+2b) = 0$ and $a = -b$, $a = -2b$. If $a = -b$, $e^{-bx+by} = e^{b(y-x)}$ is a solution for any value of b. If $a = -2b$, $e^{-2bx+by} = e^{b(y-2x)}$ is a solution for any value of b.

Since the equation is linear and homogeneous, sums of these solutions are solutions. For example, $3e^{2(y-x)} - 2e^{3(y-x)} + 5e^{\pi(y-x)}$ is a solution (among many others) and one is thus led to $F(y-x)$ where F is arbitrary, which can be verified as a solution. Similarly $G(y-2x)$ where G is arbitrary is a solution. The general solution found by addition is then given by

$$u = F(y-x) + G(y-2x)$$

12.11. Find a general solution of (a) $2\dfrac{\partial u}{\partial x} + 3\dfrac{\partial u}{\partial y} = 2u$, (b) $4\dfrac{\partial^2 u}{\partial x^2} - 4\dfrac{\partial^2 u}{\partial x\,\partial y} + \dfrac{\partial^2 u}{\partial y^2} = 0$.

(a) Let $u = e^{ax+by}$. Then $2a + 3b = 2$, $a = \dfrac{2-3b}{2}$, and $e^{[(2-3b)/2]x + by} = e^x e^{(b/2)(2y-3x)}$ is a solution.

Thus $u = e^x F(2y - 3x)$ is a general solution.

(b) Let $u = e^{ax+by}$. Then $4a^2 - 4ab + b^2 = 0$ and $b = 2a, 2a$. From this $U = e^{a(x+2y)}$ and so $F(x + 2y)$ is a solution.

By analogy with repeated roots for ordinary differential equations we might be led to believe $xG(x+2y)$ or $yG(x+2y)$ to be another solution, and that this is in fact true is easy to verify. Thus the general solution is

$$u = F(x+2y) + xG(x+2y) \quad \text{or} \quad u = F(x+2y) + yG(x+2y)$$

12.12. Solve $\dfrac{\partial^2 u}{\partial x^2} + \dfrac{\partial^2 u}{\partial y^2} = 10e^{2x+y}$.

The homogeneous equation $\dfrac{\partial^2 u}{\partial x^2} + \dfrac{\partial^2 u}{\partial y^2} = 0$ has general solution $u = F(x+iy) + G(x-iy)$ by Problem 12.42(c).

To find a particular solution of the given equation assume $u = \alpha e^{2x+y}$ where α is an unknown constant. This is the *method of undetermined coefficients* as in ordinary differential equations. We find $\alpha = 2$ so that the required general solution is

$$u = F(x+iy) + G(x-iy) + 2e^{2x+y}$$

12.13. Solve $\dfrac{\partial^2 u}{\partial x^2} - 4\dfrac{\partial^2 u}{\partial y^2} = e^{2x+y}$.

The homogeneous equation has general solution

$$u = F(2x+y) + G(2x-y)$$

To find a particular solution, we would normally assume $u = \alpha e^{2x+y}$ as in Problem 12.12 but this assumed solution is already included in $F(2x+y)$. Hence we assume as in ordinary differential equations that $u = \alpha x e^{2x+y}$ (or $u = \alpha y e^{2x+y}$). Substituting, we find $\alpha = \frac{1}{4}$.

Then the general solution is

$$u = F(2x+y) + G(2x-y) + \tfrac{1}{4}x e^{2x+y}$$

SEPARATION OF VARIABLES

12.14. Solve the boundary-value problem

$$\frac{\partial u}{\partial x} = 4\frac{\partial u}{\partial y}, \qquad u(0,y) = 8e^{-3y}$$

by the method of separation of variables.

Let $u = XY$ in the given equation. Then

$$X'Y = 4XY' \quad \text{or} \quad X'/4X = Y'/Y$$

Since X depends only on x and Y depends only on y and since x and y are independent variables, each side must be a constant, say c.

Then $X' - 4cX = 0$, $Y' - cY = 0$ whose solutions are $X = Ae^{4cx}$, $Y = Be^{cy}$.

A solution is thus given by

$$u(x,y) = XY = ABe^{c(4x+y)} = Ke^{c(4x+y)}$$

From the boundary condition,

$$u(0,y) = Ke^{cy} = 8e^{-3y}$$

which is possible if and only if $K = 8$ and $c = -3$.

Then $u(x,y) = 8e^{-3(4x+y)} = 8e^{-12x-3y}$ is the required solution.

12.15. Solve Problem 12.14 if $u(0,y) = 8e^{-3y} + 4e^{-5y}$.

As before a solution is $Ke^{c(4x+y)}$.

Then $K_1 e^{c_1(4x+y)}$ and $K_2 e^{c_2(4x+y)}$ are solutions and by the principle of superposition so also is their sum, i.e. a solution is

$$u(x,y) = K_1 e^{c_1(4x+y)} + K_2 e^{c_2(4x+y)}$$

From the boundary condition,

$$u(0,y) = K_1 e^{c_1 y} + K_2 e^{c_2 y} = 8e^{-3y} + 4e^{-5y}$$

which is possible if and only if $K_1 = 8$, $K_2 = 4$, $c_1 = -3$, $c_2 = -5$.

Then $u(x,y) = 8e^{-3(4x+y)} + 4e^{-5(4x+y)} = 8e^{-12x-3y} + 4e^{-20x-5y}$ is the required solution.

12.16. Solve $\dfrac{\partial u}{\partial t} = 2\dfrac{\partial^2 u}{\partial x^2}$, $0 < x < 3$, $t > 0$, given that $u(0,t) = u(3,t) = 0$,

$$u(x,0) = 5\sin 4\pi x - 3\sin 8\pi x + 2\sin 10\pi x, \qquad |u(x,t)| < M$$

where the last condition states that u is bounded for $0 < x < 3$, $t > 0$.

Let $u = XT$. Then

$$XT' = 2X''T \quad \text{and} \quad X''/X = T'/2T$$

Each side must be a constant which we call $-\lambda^2$. [If we use $+\lambda^2$, the resulting solution obtained does not satisfy the boundedness condition for real values of λ.] Then

$$X'' + \lambda^2 X \;=\; 0, \qquad T' + 2\lambda^2 T \;=\; 0$$

with solutions
$$X \;=\; A_1 \cos \lambda x \,+\, B_1 \sin \lambda x, \qquad T \;=\; c_1 e^{-2\lambda^2 t}$$

A solution of the partial differential equation is thus given by

$$u(x,t) \;=\; XT \;=\; c_1 e^{-2\lambda^2 t}(A_1 \cos \lambda x + B_1 \sin \lambda x) \;=\; e^{-2\lambda^2 t}(A \cos \lambda x + B \sin \lambda x)$$

Since $u(0,t) = 0$, $e^{-2\lambda^2 t}(A) = 0$ or $A = 0$. Then

$$u(x,t) \;=\; Be^{-2\lambda^2 t} \sin \lambda x$$

Since $u(3,t) = 0$, $Be^{-2\lambda^2 t} \sin 3\lambda = 0$. If $B = 0$, the solution is identically zero, so we must have $\sin 3\lambda = 0$ or $3\lambda = m\pi$, $\lambda = m\pi/3$ where $m = 0, \pm 1, \pm 2, \ldots$. Thus

$$u(x,t) \;=\; Be^{-2m^2\pi^2 t/9} \sin \frac{m\pi x}{3}$$

is a solution.

Also, by the principle of superposition,

$$u(x,t) \;=\; B_1 e^{-2m_1^2\pi^2 t/9} \sin \frac{m_1\pi x}{3} + B_2 e^{-2m_2^2\pi^2 t/9} \sin \frac{m_2\pi x}{3} + B_3 e^{-2m_3^2\pi^2 t/9} \sin \frac{m_3\pi x}{3} \qquad (1)$$

is a solution.

By the last boundary condition,

$$u(x,0) \;=\; B_1 \sin \frac{m_1\pi x}{3} + B_2 \sin \frac{m_2\pi x}{3} + B_3 \sin \frac{m_3\pi x}{3}$$
$$=\; 5 \sin 4\pi x - 3 \sin 8\pi x + 2 \sin 10\pi x$$

This is possible if and only if $B_1 = 5$, $m_1 = 12$, $B_2 = -3$, $m_2 = 24$, $B_3 = 2$, $m_3 = 30$.

Substituting these in (1), the required solution is

$$u(x,t) \;=\; 5e^{-32\pi^2 t} \sin 4\pi x - 3e^{-128\pi^2 t} \sin 8\pi x + 2e^{-200\pi^2 t} \sin 10\pi x \qquad (2)$$

This boundary-value problem has the following heat flow interpretation. A bar whose surface is insulated [Fig. 12-5] has a length of 3 units and a diffusivity of 2 units. If its ends are kept at temperature zero units and its initial temperature $u(x,0) = 5 \sin 4\pi x - 3 \sin 8\pi x + 2 \sin 10\pi x$, find the temperature at position x at time t, i.e. find $u(x,t)$. We shall assume that c.g.s. units are used and that temperature is in degrees centigrade (°C). However, other units could of course be used.

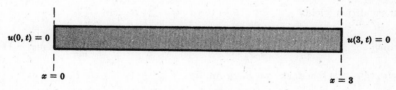

$u(0,t) = 0$ $u(3,t) = 0$

$x = 0$ $x = 3$

Fig. 12-5

SOLUTIONS USING FOURIER SERIES

12.17. Find the temperature of the bar in Problem 12.16 if the initial temperature is 25°C.

This problem is identical with Problem 12.16 except that to satisfy the initial condition $u(x,0) = 25$ it is necessary to superimpose an *infinite number of solutions*, i.e. we must replace equation (1) of the last problem by

$$u(x,t) \;=\; \sum_{m=1}^{\infty} B_m e^{-m^2\pi^2 t/9} \sin \frac{m\pi x}{3}$$

which for $t = 0$ yields
$$25 \;=\; \sum_{m=1}^{\infty} B_m \sin \frac{m\pi x}{3} \qquad 0 < x < 3$$

This amounts to the same thing as expanding 25 in a *Fourier sine series*. By the methods of Chapter 7 we then find

$$B_m \;=\; \frac{2}{L} \int_0^L f(x) \sin \frac{m\pi x}{L}\, dx \;=\; \frac{2}{3} \int_0^3 25 \sin \frac{m\pi x}{3}\, dx \;=\; \frac{50(1 - \cos m\pi)}{m\pi}$$

The result can be written

$$u(x, t) = \sum_{m=1}^{\infty} \frac{50(1 - \cos m\pi)}{m\pi} e^{-m^2\pi^2 t/9} \sin \frac{m\pi x}{3}$$

$$= \frac{100}{\pi} \left\{ e^{-\pi^2 t/9} \sin \frac{\pi x}{3} + \frac{1}{3} e^{-\pi^2 t} \sin \pi x + \cdots \right\}$$

which can be verified as the required solution.

This problem illustrates the importance of Fourier series (and orthogonal series in general) in solving boundary-value problems.

12.18. Solve the boundary-value problem

$$\frac{\partial u}{\partial t} = 2 \frac{\partial^2 u}{\partial x^2}, \quad u(0, t) = 10, \quad u(3, t) = 40, \quad u(x, 0) = 25, \quad |u(x, t)| < M$$

This is the same as Problem 12.17 except that the ends of the bar are at temperatures 10°C and 40°C instead of 0°C. As far as the solution goes this makes quite a difference since we can no longer conclude that $A = 0$ and $\lambda = m\pi/3$ as in that problem.

To solve the present problem assume that $u(x, t) = v(x, t) + \psi(x)$ where $\psi(x)$ is to be suitably determined. In terms of $v(x, t)$ the boundary-value problem becomes

$$\frac{\partial v}{\partial t} = 2 \frac{\partial^2 v}{\partial x^2} + 2\psi''(x), \quad v(0, t) + \psi(0) = 10, \quad v(3, t) + \psi(3) = 40, \quad v(x, 0) + \psi(x) = 25, \quad |v(x, t)| < M$$

This can be simplified by choosing

$$\psi''(x) = 0, \quad \psi(0) = 10, \quad \psi(3) = 40$$

from which we find $\psi(x) = 10x + 10$ so that the resulting boundary-value problem is

$$\frac{\partial v}{\partial t} = 2 \frac{\partial^2 v}{\partial x^2}, \quad v(0, t) = 0, \quad v(3, t) = 0, \quad v(x, 0) = 15 - 10x$$

As in Problem 12.17 we now find from the first three of these,

$$v(x, t) = \sum_{m=1}^{\infty} B_m e^{-m^2\pi^2 t/9} \sin \frac{m\pi x}{3}$$

The last condition yields

$$15 - 10x = \sum_{m=1}^{\infty} B_m \sin \frac{m\pi x}{3}$$

from which

$$B_m = \frac{2}{3} \int_0^3 (15 - 10x) \sin \frac{m\pi x}{3} dx = \frac{30}{m\pi} (\cos m\pi - 1)$$

Since $u(x, t) = v(x, t) + \psi(x)$, we have finally

$$u(x, t) = 10x + 10 + \sum_{m=1}^{\infty} \frac{30}{m\pi} (\cos m\pi - 1) e^{-m^2\pi^2 t/9} \sin \frac{m\pi x}{3}$$

as the required solution.

The term $10x + 10$ is the *steady-state temperature*, i.e. the temperature after a long time has elapsed.

12.19. A string of length L is stretched between points $(0, 0)$ and $(L, 0)$ on the x axis. At time $t = 0$ it has a shape given by $f(x)$, $0 < x < L$ and it is released from rest. Find the displacement of the string at any later time.

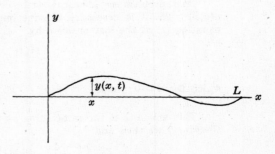

The equation of the vibrating string is

$$\frac{\partial^2 y}{\partial t^2} = a^2 \frac{\partial^2 y}{\partial x^2} \quad 0 < x < L, \ t > 0$$

where $y(x, t) =$ displacement from x axis at time t [Fig. 12-6].

Fig. 12-6

Since the ends of the string are fixed at $x = 0$ and $x = L$,

$$y(0, t) = y(L, t) = 0 \qquad t > 0$$

Since the initial shape of the string is given by $f(x)$,

$$y(x, 0) = f(x) \qquad 0 < x < L$$

Since the initial velocity of the string is zero,

$$y_t(x, 0) = 0 \qquad 0 < x < L$$

To solve this boundary-value problem, let $y = XT$ as usual.

Then $\qquad\qquad XT'' = a^2 X''T \qquad$ or $\qquad T''/a^2T = X''/X$

Calling the separation constant $-\lambda^2$, we have

$$T'' + \lambda^2 a^2 T = 0, \qquad X'' + \lambda^2 X = 0$$

and $\qquad\qquad T = A_1 \sin \lambda at + B_1 \cos \lambda at, \qquad X = A_2 \sin \lambda x + B_2 \cos \lambda x$

A solution is thus given by

$$y(x, t) = XT = (A_2 \sin \lambda x + B_2 \cos \lambda x)(A_1 \sin \lambda at + B_1 \cos \lambda at)$$

From $y(0, t) = 0$, $A_2 = 0$, Then

$$y(x, t) = B_2 \sin \lambda x(A_1 \sin \lambda at + B_1 \cos \lambda at) = \sin \lambda x(A \sin \lambda at + B \cos \lambda at)$$

From $y(L, t) = 0$, we have $\qquad \sin \lambda L(A \sin \lambda at + B \cos \lambda at) = 0$

so that $\sin \lambda L = 0$, $\lambda L = m\pi$ or $\lambda = m\pi/L$ since the second factor must not be equal to zero. Now

$$y_t(x, t) = \sin \lambda x(A\lambda a \cos \lambda at - B\lambda a \sin \lambda at)$$

and $y_t(x, 0) = (\sin \lambda x)(A\lambda a) = 0$ from which $A = 0$. Thus

$$y(x, t) = B \sin \frac{m\pi x}{L} \cos \frac{m\pi at}{L}$$

To satisfy the condition $y(x, 0) = f(x)$, it will be necessary to superimpose solutions. This yields

$$y(x, t) = \sum_{m=1}^{\infty} B_m \sin \frac{m\pi x}{L} \cos \frac{m\pi at}{L}$$

Then $\qquad\qquad y(x, 0) = f(x) = \sum_{m=1}^{\infty} B_m \sin \frac{m\pi x}{L}$

and from the theory of Fourier series,

$$B_m = \frac{2}{L} \int_0^L f(x) \sin \frac{m\pi x}{L} \, dx$$

The result is $\qquad y(x, t) = \sum_{m=1}^{\infty} \left(\frac{2}{L} \int_0^L f(x) \sin \frac{m\pi x}{L} \, dx \right) \sin \frac{m\pi x}{L} \cos \frac{m\pi at}{L}$

which can be verified as the solution.

The terms in this series represent the *natural* or *normal modes of vibration*. The frequency of the mth normal mode f_m is obtained from the term involving $\cos \dfrac{m\pi at}{L}$ and is given by

$$2\pi f_m = \frac{m\pi a}{L} \qquad \text{or} \qquad f_m = \frac{ma}{2L} = \frac{m}{2L} \sqrt{\frac{T}{\mu}}$$

Since all the frequencies are integer multiples of the lowest frequency f_1, the vibrations of the string will yield a musical tone as in the case of a violin or piano string.

12.20. A circular plate of unit radius, whose faces are insulated, has half of its boundary kept at constant temperature u_1 and the other half at constant temperature u_2 [see Fig. 12-7]. Find the steady-state temperature of the plate.

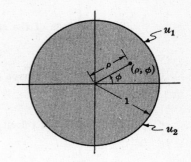

Fig. 12-7

In polar coordinates (ρ, ϕ) the partial differential equation for steady-state heat flow is

$$\frac{\partial^2 u}{\partial \rho^2} + \frac{1}{\rho}\frac{\partial u}{\partial \rho} + \frac{1}{\rho^2}\frac{\partial^2 u}{\partial \phi^2} \;=\; 0 \tag{1}$$

The boundary conditions are

$$u(1, \phi) \;=\; \begin{cases} u_1 & 0 < \phi < \pi \\ u_2 & \pi < \phi < 2\pi \end{cases} \tag{2}$$

$$|u(\rho, \phi)| \;<\; M, \quad \text{i.e. } u \text{ is bounded in the region} \tag{3}$$

Let $u(\rho, \phi) = P\Phi$ where P is a function of ρ and Φ is a function of ϕ. Then equation (1) becomes

$$P''\Phi + \frac{1}{\rho}P'\Phi + \frac{1}{\rho^2}P\Phi'' \;=\; 0$$

Dividing by $P\Phi$, multiplying by ρ^2 and rearranging terms,

$$\frac{\rho^2 P''}{P} + \frac{\rho P'}{P} \;=\; -\frac{\Phi''}{\Phi}$$

Setting each side equal to λ^2,

$$\Phi'' + \lambda^2 \Phi \;=\; 0, \qquad \rho^2 P'' + \rho P' - \lambda^2 P \;=\; 0$$

which have solutions

$$\Phi \;=\; A_1 \cos \lambda\phi + B_1 \sin \lambda\phi, \qquad P \;=\; A_2 \rho^\lambda + B_2/\rho^\lambda$$

[See Problem 3.70(a), page 94.]

Since $u(\rho, \phi)$ must have period 2π in ϕ, we must have $\lambda = m = 0, 1, 2, 3, \ldots$.

Also, since u must be bounded at $\rho = 0$, we must have $B_2 = 0$. Thus

$$u \;=\; P\Phi \;=\; A_2\rho^m(A_1 \cos m\phi + B_1 \sin m\phi) \;=\; \rho^m(A \cos m\phi + B \sin m\phi)$$

By superposition, a solution is

$$u(\rho, \phi) \;=\; \frac{A_0}{2} + \sum_{m=1}^{\infty} \rho^m(A_m \cos m\phi + B_m \sin m\phi)$$

from which
$$u(1, \phi) \;=\; \frac{A_0}{2} + \sum_{m=1}^{\infty} (A_m \cos m\phi + B_m \sin m\phi)$$

Then from the theory of Fourier series,

$$A_m \;=\; \frac{1}{\pi}\int_0^{2\pi} u(1, \phi) \cos m\phi \, d\phi$$

$$\;=\; \frac{1}{\pi}\int_0^{\pi} u_1 \cos m\phi \, J\phi + \frac{1}{\pi}\int_\pi^{2\pi} u_2 \cos m\phi \, d\phi \;=\; \begin{cases} 0 & \text{if } m > 0 \\ u_1 + u_2 & \text{if } m = 0 \end{cases}$$

$$B_m \;=\; \frac{1}{\pi}\int_0^{2\pi} u(1, \phi) \sin m\phi \, d\phi$$

$$\;=\; \frac{1}{\pi}\int_0^{\pi} u_1 \sin m\phi \, d\phi + \frac{1}{\pi}\int_\pi^{2\pi} u_2 \sin m\phi \, d\phi \;=\; \frac{(u_1 - u_2)}{m\pi}(1 - \cos m\pi)$$

Then $\quad u(\rho, \phi) \;=\; \dfrac{u_1 + u_2}{2} + \displaystyle\sum_{m=1}^{\infty} \dfrac{(u_1 - u_2)(1 - \cos m\pi)}{m\pi}\, \rho^m \sin m\phi$

$\qquad\qquad\qquad =\; \dfrac{u_1 + u_2}{2} + \dfrac{2(u_1 - u_2)}{\pi} \{\rho \sin \phi + \tfrac{1}{3}\rho^3 \sin 3\phi + \tfrac{1}{5}\rho^5 \sin 5\phi + \cdots\}$

$\qquad\qquad\qquad =\; \dfrac{u_1 + u_2}{2} + \dfrac{(u_1 - u_2)}{\pi} \tan^{-1}\left(\dfrac{2\rho \sin \phi}{1 - \rho^2}\right)$

by Problem 12.52.

SOLUTIONS USING FOURIER INTEGRALS

12.21. A semi-infinite thin bar $x \geqq 0$ whose surface is insulated has an initial temperature equal to $f(x)$. A temperature of zero is suddenly applied to the end $x = 0$ and maintained. (a) Set up the boundary-value problem for the temperature $u(x, t)$ at any point x at time t. (b) Show that

$$u(x, t) \;=\; \frac{1}{\pi} \int_0^\infty \int_0^\infty f(v) e^{-\kappa\lambda^2 t} \sin \lambda v \sin \lambda x \, d\lambda \, dv$$

(a) The boundary-value problem is

$$\frac{\partial u}{\partial t} \;=\; \kappa \frac{\partial^2 u}{\partial x^2} \qquad x > 0, \; t > 0 \tag{1}$$

$$u(x, 0) \;=\; f(x), \qquad u(0, t) \;=\; 0, \qquad |u(x, t)| < M \tag{2}$$

where the last condition is used since the temperature must be bounded for physical reasons.

(b) A solution of (1) obtained by separation of variables is

$$u(x, t) \;=\; e^{-\kappa\lambda^2 t}(A \cos \lambda x + B \sin \lambda x)$$

From the second of boundary conditions (2) we find $A = 0$ so that

$$u(x, t) \;=\; B e^{-\kappa\lambda^2 t} \sin \lambda x \tag{3}$$

Now since there is no restriction on λ we can replace B in (3) by a function $B(\lambda)$ and still have a solution. Furthermore we can integrate over λ from 0 to ∞ and still have a solution. This is the analog of the superposition theorem for discrete values of λ used in connection with Fourier series. We thus arrive at the possible solution

$$u(x, t) \;=\; \int_0^\infty B(\lambda) e^{-\kappa\lambda^2 t} \sin \lambda x \, d\lambda \tag{4}$$

From the first of boundary conditions (2) we find

$$f(x) \;=\; \int_0^\infty B(\lambda) \sin \lambda x \, d\lambda$$

which is an integral equation for the determination of $B(\lambda)$. From page 201 we see that since $f(x)$ must be an odd function we have

$$B(\lambda) \;=\; \frac{2}{\pi} \int_0^\infty f(x) \sin \lambda x \, dx \;=\; \frac{2}{\pi} \int_0^\infty f(v) \sin \lambda v \, dv$$

Using this in (4) we find

$$u(x, t) \;=\; \frac{2}{\pi} \int_0^\infty \int_0^\infty f(v) e^{-\kappa\lambda^2 t} \sin \lambda v \sin \lambda x \, d\lambda \, dv$$

12.22. Show that the result of Problem 12.21 can be written

$$u(x, t) \;=\; \frac{1}{\sqrt{\pi}}\left[\int_{-x/2\sqrt{\kappa t}}^{\infty} e^{-w^2} f(2w\sqrt{\kappa t} + x)\, dw - \int_{x/2\sqrt{\kappa t}}^{\infty} e^{-w^2} f(2w\sqrt{\kappa t} - x)\, dw\right]$$

Using $\quad \sin \lambda v \sin \lambda x = \tfrac{1}{2}[\cos \lambda(v - x) - \cos \lambda(v + x)] \quad$ the result of Problem 12.21 can be written

$$u(x, t) = \frac{1}{\pi} \int_0^\infty \int_0^\infty f(v)e^{-\kappa\lambda^2 t} \left[\cos\lambda(v - x) - \cos\lambda(v + x)\right] d\lambda \, dv$$

$$= \frac{1}{\pi} \int_0^\infty f(v) \left[\int_0^\infty e^{-\kappa\lambda^2 t} \cos\lambda(v - x) \, d\lambda - \int_0^\infty e^{-\kappa\lambda^2 t} \cos\lambda(v + x) \, d\lambda\right] dv$$

From the integral $$\int_0^\infty e^{-\alpha\lambda^2} \cos\beta\lambda \, d\lambda = \frac{1}{2}\sqrt{\frac{\pi}{\alpha}} e^{-\beta^2/4\alpha}$$

[see Problem 9.9, page 215] we find

$$u(x, t) = \frac{1}{2\sqrt{\pi\kappa t}} \left[\int_0^\infty f(v)e^{-(v-x)^2/4\kappa t} \, dv - \int_0^\infty f(v)e^{-(v+x)^2/4\kappa t} \, dv\right]$$

Letting $(v - x)/2\sqrt{\kappa t} = w$ in the first integral and $(v + x)/2\sqrt{\kappa t} = w$ in the second integral, we find that

$$u(x, t) = \frac{1}{\sqrt{\pi}} \left[\int_{-x/2\sqrt{\kappa t}}^\infty e^{-w^2} f(2w\sqrt{\kappa t} + x) \, dw - \int_{x/2\sqrt{\kappa t}}^\infty e^{-w^2} f(2w\sqrt{\kappa t} - x) \, dw\right]$$

12.23. In case the initial temperature $f(x)$ in Problem 12.21 is the constant u_0, show that

$$u(x, t) = \frac{2u_0}{\sqrt{\pi}} \int_0^{x/2\sqrt{\kappa t}} e^{-w^2} \, dw = u_0 \, \mathrm{erf}\,(x/2\sqrt{\kappa t})$$

where $\mathrm{erf}\,(x/2\sqrt{\kappa t})$ is the error function [see page 212].

If $f(x) = u_0$, we obtain from Problem 12.22

$$u(x, t) = \frac{u_0}{\sqrt{\pi}} \left[\int_{-x/2\sqrt{\kappa t}}^\infty e^{-w^2} \, dw - \int_{x/2\sqrt{\kappa t}}^\infty e^{-w^2} \, dw\right]$$

$$= \frac{u_0}{\sqrt{\pi}} \int_{-x/2\sqrt{\kappa t}}^{x/2\sqrt{\kappa t}} e^{-w^2} \, dw = \frac{2u_0}{\sqrt{\pi}} \int_0^{x/2\sqrt{\kappa t}} e^{-w^2} \, dw = u_0 \, \mathrm{erf}\,(x/2\sqrt{\kappa t})$$

SOLUTIONS USING BESSEL FUNCTIONS

12.24. A circular plate of unit radius [as in Fig. 12-7, page 270] has its plane faces insulated. If the initial temperature is $F(\rho)$ and if the rim is kept at temperature zero, find the temperature of the plate at any time.

Since the temperature is independent of ϕ, the boundary value problem for determining $u(\rho, t)$ is

$$\frac{\partial u}{\partial t} = \kappa\left(\frac{\partial^2 u}{\partial\rho^2} + \frac{1}{\rho}\frac{\partial u}{\partial\rho}\right) \tag{1}$$

$$u(1, t) = 0, \quad u(\rho, 0) = F(\rho), \quad |u(\rho, t)| < M$$

Let $u = P(\rho)T(t) = PT$ in equation (1). Then

$$PT' = \kappa\left(P''T + \frac{1}{\rho}P'T\right)$$

or dividing by κPT,

$$\frac{T'}{\kappa T} = \frac{P''}{P} + \frac{1}{\rho}\frac{P'}{P} = -\lambda^2$$

from which $$T' + \kappa\lambda^2 T = 0, \quad P'' + \frac{1}{\rho}P' + \lambda^2 P = 0$$

These have general solutions [see Chapter 10]

$$T = c_1 e^{-\kappa\lambda^2 t}, \quad P = A_1 J_0(\lambda\rho) + B_1 Y_0(\lambda\rho)$$

Since $u = PT$ is bounded at $\rho = 0$, $B_1 = 0$. Then

$$u(\rho, t) = Ae^{-\kappa\lambda^2 t} J_0(\lambda\rho)$$

where $A = A_1 c_1$.

From the first boundary condition,

$$u(1, t) = A e^{-\kappa \lambda^2 t} J_0(\lambda) = 0$$

from which $J_0(\lambda) = 0$ and $\lambda = \lambda_1, \lambda_2, \ldots$ are the positive roots.

Thus a solution is

$$u(\rho, t) = A e^{-\kappa \lambda_m^2 t} J_0(\lambda_m \rho) \qquad m = 1, 2, 3, \ldots$$

By superposition, a solution is

$$u(\rho, t) = \sum_{m=1}^{\infty} A_m e^{-\kappa \lambda_m^2 t} J_0(\lambda_m \rho)$$

From the second boundary condition,

$$u(\rho, 0) = F(\rho) = \sum_{m=1}^{\infty} A_m J_0(\lambda_m \rho)$$

Then by the methods of Chapter 10,

$$A_m = \frac{2}{J_1^2(\lambda_m)} \int_0^1 \rho F(\rho) J_0(\lambda_m \rho) \, d\rho$$

and so

$$u(\rho, t) = \sum_{m=1}^{\infty} \left\{ \left[\frac{2}{J_1^2(\lambda_m)} \int_0^1 \rho F(\rho) J_0(\lambda_m \rho) \, d\rho \right] e^{-\kappa \lambda_m^2 t} J_0(\lambda_m \rho) \right\} \qquad (2)$$

which can be established as the required solution.

Note that this solution also gives the temperature of an infinitely long solid cylinder whose convex surface is kept at temperature zero and whose initial temperature is $F(\rho)$.

12.25. A drum consists of a stretched circular membrane of unit radius whose rim represented by the circle of Fig. 12-7, page 270, is fixed. If the membrane is struck so that its initial displacement is $F(\rho, \phi)$ and is then released, find the displacement at any time.

The boundary-value problem for the displacement $z(\rho, \phi, t)$ from the equilibrium or rest position (the xy plane) is

$$\frac{\partial^2 z}{\partial t^2} = a^2 \left(\frac{\partial^2 z}{\partial \rho^2} + \frac{1}{\rho} \frac{\partial z}{\partial \rho} + \frac{1}{\rho^2} \frac{\partial^2 z}{\partial \phi^2} \right)$$

$$z(1, \phi, t) = 0, \quad z(\rho, \phi, 0) = 0, \quad z_t(\rho, \phi, 0) = 0, \quad z(\rho, \phi, 0) = F(\rho, \phi)$$

Let $z = P(\rho) \, \Phi(\phi) \, T(t) = P\Phi T$. Then

$$P\Phi T'' = a^2 \left(P''\Phi T + \frac{1}{\rho} P'\Phi T + \frac{1}{\rho^2} P\Phi'' T \right)$$

Dividing by $a^2 P\Phi T$,

$$\frac{T''}{a^2 T} = \frac{P''}{P} + \frac{1}{\rho} \frac{P'}{P} + \frac{1}{\rho^2} \frac{\Phi''}{\Phi} = -\lambda^2$$

and so

$$T'' + \lambda^2 a^2 T = 0 \qquad (1)$$

$$\frac{P''}{P} + \frac{1}{\rho} \frac{P'}{P} + \frac{1}{\rho^2} \frac{\Phi''}{\Phi} = -\lambda^2 \qquad (2)$$

Multiplying equation (2) by ρ^2, the variables can be separated to yield

$$\frac{\rho^2 P''}{P} + \frac{\rho P'}{P} + \lambda^2 \rho^2 = -\frac{\Phi''}{\Phi} = \mu^2$$

so that

$$\phi'' + \mu^2 \Phi = 0 \qquad (3)$$

$$\rho^2 P'' + \rho P' + (\lambda^2 \rho^2 - \mu^2) P = 0 \qquad (4)$$

General solutions of (1), (3) and (4) are

$$T = A_1 \cos \lambda a t + B_1 \sin \lambda a t \qquad (5)$$

$$\Phi = A_2 \cos \mu \phi + B_2 \sin \mu \phi \qquad (6)$$

$$P = A_3 J_\mu(\lambda \rho) + B_3 Y_\mu(\lambda \rho) \qquad (7)$$

A solution $z(\rho, \phi, t)$ is given by the product of these.

Since z must have period 2π in the variable ϕ, we must have $\mu = m$ where $m = 0, 1, 2, 3, \ldots$ from equation (6).

Also, since z is bounded at $\rho = 0$ we must take $B_3 = 0$.

Furthermore, to satisfy $z_t(\rho, \phi, 0) = 0$ we must choose $B_1 = 0$.

Then a solution is

$$u(\rho, \phi, t) = J_m(\lambda\rho) \cos \lambda at \, (A \cos m\phi + B \sin m\phi)$$

Since $z(1, \phi, t) = 0$, $J_m(\lambda) = 0$ so that $\lambda = \lambda_{mk}$, $k = 1, 2, 3, \ldots$, are the positive roots.

By superposition (summing over both m and k),

$$z(\rho, \phi, t) = \sum_{m=0}^{\infty} \sum_{k=1}^{\infty} J_m(\lambda_{mk}\rho) \cos(\lambda_{mk}at)(A_{mk} \cos m\phi + B_{mk} \sin m\phi)$$

$$= \sum_{m=0}^{\infty} \left\{ \left[\sum_{k=1}^{\infty} A_{mk} J_m(\lambda_{mk}\rho) \right] \cos m\phi + \left[\sum_{k=1}^{\infty} B_{mk} J_m(\lambda_{mk}\rho) \right] \sin m\phi \right\} \cos \lambda_{mk}at \qquad (8)$$

Putting $t = 0$, we have

$$z(\rho, \phi, 0) = F(\rho, \phi) = \sum_{m=0}^{\infty} \{C_m \cos m\phi + D_m \sin m\phi\} \qquad (9)$$

where

$$C_m = \sum_{k=1}^{\infty} A_{mk} J_m(\lambda_{mk}\rho)$$

$$D_m = \sum_{k=1}^{\infty} B_{mk} J_m(\lambda_{mk}\rho) \qquad (10)$$

But (9) is simply a Fourier series and we can determine C_m and D_m by the usual methods. We find

$$C_m = \begin{cases} \dfrac{1}{\pi} \displaystyle\int_0^{2\pi} F(\rho, \phi) \cos m\phi \, d\phi & m = 1, 2, 3, \ldots \\[3mm] \dfrac{1}{2\pi} \displaystyle\int_0^{2\pi} F(\rho, \phi) \, d\phi & m = 0 \end{cases}$$

$$D_m = \dfrac{1}{\pi} \int_0^{2\pi} F(\rho, \phi) \sin m\phi \, d\phi \qquad m = 0, 1, 2, 3, \ldots$$

From (10), using the results of Bessel series expansions, we have

$$A_{mk} = \frac{2}{[J_{m+1}(\lambda_{mk})]^2} \int_0^1 \rho \, J_m(\lambda_{mk}\rho) C_m \, d\rho$$

$$= \begin{cases} \dfrac{2}{\pi[J_{m+1}(\lambda_{mk})]^2} \displaystyle\int_0^1 \int_0^{2\pi} \rho F(\rho, \phi) J_m(\lambda_{mk}\rho) \cos m\phi \, d\rho \, d\phi & \text{if } m = 1, 2, 3, \ldots \\[3mm] \dfrac{1}{\pi[J_1(\lambda_{0k})]^2} \displaystyle\int_0^1 \int_0^{2\pi} \rho F(\rho, \phi) J_0(\lambda_{0k}\rho) \, d\rho \, d\phi & \text{if } m = 0 \end{cases}$$

$$B_{mk} = \frac{2}{[J_{m+1}(\lambda_{mk})]^2} \int_0^1 \rho \, J_m(\lambda_{mk}\rho) D_m \, d\rho$$

$$= \frac{2}{\pi[J_{m+1}(\lambda_{mk})]^2} \int_0^1 \int_0^{2\pi} \rho F(\rho, \phi) J_m(\lambda_{mk}\rho) \sin m\phi \, d\rho \, d\phi \qquad \text{if } m = 0, 1, 2, \ldots$$

Using these values of A_{mk} and B_{mk} in (8) yields the required solution.

Note that the various modes of vibration of the drum are obtained by specifying particular values of m and k. The frequency of vibration is then given by

$$f_{mk} = \frac{\lambda_{mk}}{2\pi} a$$

Because these are not integer multiples of the lowest frequency, we would expect *noise* rather than a *musical tone*.

SOLUTIONS USING LEGENDRE FUNCTIONS

12.26. Find solutions to Laplace's equation in spherical coordinates which are independent of ϕ.

Laplace's equation $\nabla^2 v = 0$ in spherical coordinates if there is no ϕ dependence can be written

$$r \frac{\partial}{\partial r}\left(r \frac{\partial v}{\partial r} \right) + \frac{1}{\sin \theta} \frac{\partial}{\partial \theta}\left(\sin \theta \frac{\partial v}{\partial \theta} \right) = 0 \tag{1}$$

Let $v = R(r)\, \Theta(\theta) = R\Theta$ in (1). Then after dividing by $R\Theta$, the equation becomes separable as follows

$$\frac{r}{R} \frac{d}{dr}\left(r \frac{dR}{dr} \right) = -\frac{1}{\Theta \sin \theta} \frac{d}{d\theta}\left(\sin \theta \frac{d\Theta}{d\theta} \right) = -\lambda^2$$

and we find

$$r \frac{d}{dr}\left(r \frac{dR}{dr} \right) + \lambda^2 R = 0 \tag{2}$$

$$\frac{d}{d\theta}\left(\sin \theta \frac{d\Theta}{d\theta} \right) - \lambda^2 \sin \theta \, \Theta = 0 \tag{3}$$

Equation (1) can be written $\qquad r^2 R'' + 2r R' + \lambda^2 R = 0 \tag{4}$

a Cauchy equation [see Problem 3.70(e), page 94] having solution

$$R = A_1 r^n + \frac{B_1}{r^{n+1}} \tag{5}$$

where we have placed $n = -\frac{1}{2} + \sqrt{\frac{1}{4} - \lambda^2}$ so that $\lambda^2 = -n(n+1)$.

Equation (3) with $\lambda^2 = -n(n+1)$ can be written

$$\frac{d}{d\theta}\left(\sin \theta \frac{d\Theta}{d\theta} \right) + n(n+1) \sin \theta \, \Theta = 0 \tag{6}$$

Now if we let $\cos \theta = x$, then

$$\frac{d\Theta}{dx} = \frac{d\Theta/d\theta}{dx/d\theta} = -\frac{1}{\sin \theta} \frac{d\Theta}{d\theta} \quad \text{or} \quad \sin \theta \frac{d\Theta}{d\theta} = -\sin^2 \theta \frac{d\Theta}{dx} = -(1-x^2)\frac{d\Theta}{dx}$$

so that

$$\frac{d}{d\theta}\left(\sin \theta \frac{d\Theta}{d\theta} \right) = \frac{d}{d\theta}\left[-(1-x^2)\frac{d\Theta}{dx} \right] = \frac{d}{dx}\left[-(1-x^2)\frac{d\Theta}{dx} \right] \frac{dx}{d\theta}$$

$$= \frac{d}{dx}\left[(1-x^2)\frac{d\Theta}{dx} \right] \sin \theta$$

Then (6) becomes

$$\frac{d}{dx}\left[(1-x^2)\frac{d\Theta}{dx} \right] + n(n+1)\Theta = 0$$

or

$$(1-x^2)\frac{d^2\Theta}{dx^2} - 2x\frac{d\Theta}{dx} + n(n+1)\Theta = 0 \tag{7}$$

which is Legendre's differential equation having solution

$$\Theta = A_2 P_n(x) + B_2 Q_n(x) \tag{8}$$

[see Chapter 11]. Thus the solution of (6) is

$$\Theta = A_2 P_n(\cos \theta) + B_2 Q_n(\cos \theta) \tag{9}$$

Using (5) and (8) or (9), we then see that a solution of (1) is

$$v = R\Theta = \left[A_1 r^n + \frac{B_1}{r^{n+1}} \right][A_2 P_n(x) + B_2 Q_n(x)] \tag{10}$$

where $x = \cos \theta$.

12.27. Find the potential v (a) interior to and (b) exterior to a hollow sphere of unit radius if half of its surface is charged to potential v_0 and the other half to potential zero.

Choose the sphere in the position shown in Fig. 12-8. Then v is independent of ϕ and we can use the results of Problem 12.26. A solution is

$$v(r, \theta) = \left(A_1 r^n + \frac{B_1}{r^{n+1}}\right)(A_2 P_n(x) + B_2 Q_n(x))$$

where $x = \cos \theta$. Since v must be bounded at $\theta = 0$ and π, i.e. $x = \pm 1$, we must choose $B_2 = 0$. Then

$$v(r, \theta) = \left(A r^n + \frac{B}{r^{n+1}}\right) P_n(x)$$

The boundary conditions are

$$v(1, \theta) = \begin{cases} v_0 & \text{if } 0 < \theta < \frac{\pi}{2} \quad \text{i.e.} \quad 0 < x < 1 \\ 0 & \text{if } \frac{\pi}{2} < \theta < \pi \quad \text{i.e.} \quad -1 < x < 0 \end{cases}$$

and v is bounded.

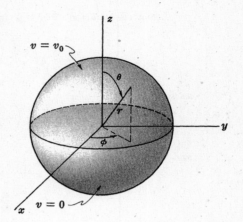

Fig. 12-8

(a) **Interior Potential,** $0 \leqq r < 1$.

Since v is bounded at $r = 0$, choose $B = 0$. Then a solution is

$$A r^n P_n(x) = A r^n P_n(\cos \theta)$$

By superposition, $$v(r, \theta) = \sum_{n=0}^{\infty} A_n r^n P_n(\cos \theta) = \sum_{n=0}^{\infty} A_n r^n P_n(x)$$

When $r = 1$, $$v(1, \theta) = \sum_{n=0}^{\infty} A_n P_n(x)$$

Then as in Problem 8.16,

$$A_n = \frac{2n+1}{2} \int_{-1}^{1} v(1, \theta) P_n(x)\, dx = \left(\frac{2n+1}{2}\right) v_0 \int_{0}^{1} P_n(x)\, dx$$

from which

$$A_0 = \tfrac{1}{2} v_0, \quad A_1 = \tfrac{3}{4} v_0, \quad A_2 = 0, \quad A_3 = -\tfrac{7}{16} v_0, \quad A_4 = 0, \quad A_5 = \tfrac{11}{32} v_0$$

Thus $$V(r, \theta) = \frac{v_0}{2}[1 + \tfrac{3}{2} r P_1(\cos \theta) - \tfrac{7}{8} r^3 P_3(\cos \theta) + \tfrac{11}{16} P_5(\cos \theta) + \cdots]$$

(b) **Exterior Potential,** $1 < r < \infty$.

Since v is bounded as $r \to \infty$, choose $A = 0$. Then a solution is

$$\frac{B}{r^{n+1}} P_n(x) = \frac{B}{r^{n+1}} P_n(\cos \theta)$$

By superposition, $$v(r, \theta) = \sum_{n=0}^{\infty} \frac{B_n}{r^{n+1}} P_n(x)$$

When $r = 1$, $$v(1, \theta) = \sum_{n=0}^{\infty} B_n P_n(x)$$

Then $B_n = A_n$ of part (a) and so

$$v(r, \theta) = \frac{v_0}{2r}\left[1 + \frac{3}{2r} P_1(\cos \theta) - \frac{7}{8r^3} P_3(\cos \theta) + \frac{11}{16r^5} P_5(\cos \theta) + \cdots\right]$$

SOLUTIONS USING LAPLACE TRANSFORMS

12.28. Solve by Laplace transforms the boundary-value problem

$$\frac{\partial u}{\partial t} = 4\frac{\partial^2 u}{\partial x^2}$$

$$u(0, t) = 0, \quad u(3, t) = 0, \quad u(x, 0) = 10 \sin 2\pi x - 6 \sin 4\pi x$$

Taking the Laplace transform of the given differential equation with respect to t, we have

$$\int_0^\infty e^{-st}\left(\frac{\partial u}{\partial t}\right)dt = \int_0^\infty e^{-st}\left(4\frac{\partial^2 u}{\partial x^2}\right)dt$$

which can be written as

$$s\int_0^\infty e^{-st}u\,dt - u(x,0) = 4\frac{d^2}{dx^2}\int_0^\infty e^{-st}u\,dt$$

or

$$sU - u(x,0) = 4\frac{d^2U}{dx^2} \tag{1}$$

where

$$U = U(x,s) = \mathcal{L}\{u(x,t)\} = \int_0^\infty e^{-st}u\,dt$$

Using the given condition $u(x, 0) = 10 \sin 2\pi x - 6 \sin 4\pi x$, (1) becomes

$$4\frac{d^2U}{dx^2} - sU = 6 \sin 4\pi x - 10 \sin 2\pi x \tag{2}$$

Taking the Laplace transform of the conditions $U(0, t) = 0$, $U(3, t) = 0$, we have

$$\mathcal{L}\{u(0,t)\} = 0, \quad \mathcal{L}\{u(3,t)\} = 0$$

or

$$U(0, s) = 0, \quad U(3, s) = 0 \tag{3}$$

Solving the ordinary differential equation (2) subject to conditions (3) by the usual elementary methods, we find

$$U(x,s) = \frac{10 \sin 2\pi x}{s + 16\pi^2} - \frac{6 \sin 4\pi x}{s + 64\pi^2}$$

Thus taking the inverse Laplace transform we find

$$u(x,t) = \mathcal{L}^{-1}\{U(x,s)\} = \mathcal{L}^{-1}\left\{\frac{10}{s + 16\pi^2}\right\}\sin 2\pi x - \mathcal{L}^{-1}\left\{\frac{6}{s + 64\pi^2}\right\}\sin 4\pi x$$

$$= 10e^{-16\pi^2 t}\sin 2\pi x - 6e^{-64\pi^2 t}\sin 4\pi x$$

which is the required solution.

Note that theoretically we could have taken the Laplace transform of the given differential equation with respect to x rather than t. However, this would lead to various difficulties as is evident upon carrying out the procedure. In practice we take Laplace transforms with respect to the various independent variables and then choose that variable which leads to the greatest simplification.

Supplementary Problems

CLASSIFICATION OF PARTIAL DIFFERENTIAL EQUATIONS

12.29. Determine whether each of the following partial differential equations is linear or nonlinear, state the order of each equation, and name the dependent and independent variables.

(a) $\dfrac{\partial^2 u}{\partial x^2} + 2\dfrac{\partial^2 U}{\partial x\,\partial y} + \dfrac{\partial^2 U}{\partial y^2} = 0$ (c) $\phi\dfrac{\partial\phi}{\partial x} = \dfrac{\partial^3\phi}{\partial y^3}$ (e) $\dfrac{\partial z}{\partial r} + \dfrac{\partial z}{\partial s} = \dfrac{1}{z^2}$

(b) $(x^2+y^2)\dfrac{\partial^4 T}{\partial z^4} = \dfrac{\partial^2 T}{\partial x^2} + \dfrac{\partial^2 T}{\partial y^2}$ (d) $\dfrac{\partial^2 y}{\partial t^2} - 4\dfrac{\partial^2 y}{\partial x^2} = x^2$

12.30. Classify each of the following equations as elliptic, hyperbolic or parabolic.

(a) $\dfrac{\partial^2 \phi}{\partial x^2} - \dfrac{\partial^2 \phi}{\partial y^2} = 0$

(b) $\dfrac{\partial u}{\partial x} + \dfrac{\partial^2 u}{\partial x\,\partial y} = 4$

(c) $\dfrac{\partial^2 z}{\partial x^2} - 2\dfrac{\partial^2 z}{\partial x\,\partial y} + 2\dfrac{\partial^2 z}{\partial y^2} = x + 3y$

(d) $x^2\dfrac{\partial^2 u}{\partial x^2} + 2xy\dfrac{\partial^2 u}{\partial x\,\partial y} + y^2\dfrac{\partial^2 u}{\partial y^2} = 0$

(e) $(x^2 - 1)\dfrac{\partial^2 u}{\partial x^2} + 2xy\dfrac{\partial^2 u}{\partial x\,\partial y} + (y^2 - 1)\dfrac{\partial^2 U}{\partial y^2} = x\dfrac{\partial U}{\partial x} + y\dfrac{\partial U}{\partial y}$

SOLUTIONS OF PARTIAL DIFFERENTIAL EQUATIONS

12.31. Show that $z(x, y) = 4e^{-3x}\cos 3y$ is a solution to the boundary-value problem

$$\frac{\partial^2 z}{\partial x^2} + \frac{\partial^2 z}{\partial y^2} = 0, \qquad z(x, \pi/2) = 0, \qquad z(x, 0) = 4e^{-3x}$$

12.32. (a) Show that $v(x, y) = xF(2x + y)$ is a general solution of $x\dfrac{\partial v}{\partial x} - 2x\dfrac{\partial v}{\partial y} = v$.

(b) Find a particular solution satisfying $v(1, y) = y^2$.

12.33. Find a partial differential equation having general solution $u = F(x - 3y) + G(2x + y)$.

12.34. Find a partial differential equation having general solution

(a) $z = e^x f(2y - 3x)$,   (b) $z = f(2x + y) + g(x - 2y)$

SOME IMPORTANT PARTIAL DIFFERENTIAL EQUATIONS

12.35. If a taut, horizontal string with fixed ends vibrates in a vertical plane under the influence of gravity, show that its equation is

$$\frac{\partial^2 y}{\partial t^2} = a^2\frac{\partial^2 y}{\partial x^2} - g$$

where g is the acceleration due to gravity.

12.36. A thin bar located on the x axis has its ends at $x = 0$ and $x = L$. The initial temperature of the bar is $f(x)$, $0 < x < L$, and the ends $x = 0$, $x = L$ are maintained at constant temperatures T_1, T_2 respectively. Assuming the surrounding medium is at temperature u_0 and that Newton's law of cooling applies, show that the partial differential equation for the temperature of the bar at any point at any time is given by

$$\frac{\partial u}{\partial t} = k\frac{\partial^2 u}{\partial x^2} - \beta(u - u_0)$$

and write the corresponding boundary conditions.

12.37. Write the boundary conditions in Problem 12.36 if the ends $x = 0$ and $x = L$ are insulated.

12.38. The gravitational potential v at any point (x, y, z) outside of a mass m located at the point (X, Y, Z) is defined as the mass m divided by the distance of the point (x, y, z) from (X, Y, Z). Show that v satisfies Laplace's equation $\nabla^2 v = 0$.

12.39. Work Problem 12.38 for a solid.

12.40. Assuming that the tension of a vibrating string is variable, show that its equation is

$$\frac{\partial}{\partial x}\left(T\frac{\partial y}{\partial x}\right) = \mu\frac{\partial^2 y}{\partial t^2}$$

METHODS OF FINDING SOLUTIONS OF PARTIAL DIFFERENTIAL EQUATIONS

12.41. (a) Solve $x\dfrac{\partial^2 z}{\partial x\,\partial y} + \dfrac{\partial z}{\partial y} = 0$.   (b) Find the particular solution for which

$$z(x, 0) = x^5 + x, \qquad z(2, y) = 3y^4$$

12.42. Find general solutions of each of the following.

(a) $\dfrac{\partial^2 u}{\partial x^2} = \dfrac{\partial^2 u}{\partial y^2}$

(d) $\dfrac{\partial^2 z}{\partial x^2} - 2\dfrac{\partial^2 z}{\partial x \, \partial y} - 3\dfrac{\partial^2 z}{\partial y^2} = 0$

(b) $\dfrac{\partial u}{\partial x} + 2\dfrac{\partial u}{\partial y} = 3u$

(e) $\dfrac{\partial^2 z}{\partial x^2} - 2\dfrac{\partial^2 z}{\partial x \, \partial y} + \dfrac{\partial^2 z}{\partial y^2} = 0$

(c) $\dfrac{\partial^2 u}{\partial x^2} + \dfrac{\partial^2 u}{\partial y^2} = 0$

12.43. Find general solutions of each of the following.

(a) $\dfrac{\partial u}{\partial x} + 2\dfrac{\partial u}{\partial y} = x$

(c) $\dfrac{\partial^4 u}{\partial x^4} + 2\dfrac{\partial^4 u}{\partial x^3 \, \partial y} = 4$

(b) $\dfrac{\partial^2 y}{\partial x^2} = \dfrac{\partial^2 y}{\partial t^2} + 12t^2$

(d) $\dfrac{\partial^2 z}{\partial x^2} - 3\dfrac{\partial^2 z}{\partial x \, \partial y} + 2\dfrac{\partial^2 z}{\partial y^2} = x \sin y$

12.44. Solve $\dfrac{\partial^4 u}{\partial x^4} + 2\dfrac{\partial^4 u}{\partial x^2 \, \partial y^2} + \dfrac{\partial^4 u}{\partial y^4} = 16$.

12.45. Show that the general solution of $\dfrac{\partial^2 v}{\partial r^2} + \dfrac{2}{r}\dfrac{\partial v}{\partial r} = \dfrac{1}{c^2}\dfrac{\partial^2 v}{\partial t^2}$ is

$$v = \frac{F(r - ct) + G(r + ct)}{r}$$

SEPARATION OF VARIABLES

12.46. Solve each of the following boundary-value problems by the method of separation of variables.

(a) $3\dfrac{\partial u}{\partial x} + 2\dfrac{\partial u}{\partial y} = 0, \qquad u(x, 0) = 4e^{-x}$

(b) $\dfrac{\partial u}{\partial x} = 2\dfrac{\partial u}{\partial y} + u, \qquad u(x, 0) = 3e^{-5x} + 2e^{-3x}$

(c) $\dfrac{\partial u}{\partial t} = 4\dfrac{\partial^2 u}{\partial x^2}, \qquad u(0, t) = 0, \qquad u(\pi, t) = 0, \qquad u(x, 0) = 2 \sin 3x - 4 \sin 5x$

(d) $\dfrac{\partial u}{\partial t} = \dfrac{\partial^2 u}{\partial x^2}, \qquad u_x(0, t) = 0, \qquad u(2, t) = 0, \qquad u(x, 0) = 8 \cos \dfrac{3\pi x}{4} - 6 \cos \dfrac{9\pi x}{4}$

12.47. Solve the boundary-value problem

$$\frac{\partial^2 y}{\partial t^2} = 4\frac{\partial^2 y}{\partial x^2}, \qquad y(0, t) = y(5, t) = 0, \qquad y(x, 0) = 0, \qquad y_t(x, 0) = f(x)$$

if   (a) $f(x) = 5 \sin \pi x$,   (b) $f(x) = 3 \sin 2\pi x - 2 \sin 5\pi x$.

SOLUTIONS USING FOURIER SERIES

12.48. (a) Solve the boundary-value problem

$$\frac{\partial u}{\partial t} = 2\frac{\partial^2 u}{\partial x^2}, \qquad u(0, t) = u(4, t) = 0, \qquad u(x, 0) = 25x$$

where $0 < x < 4, \ t > 0$.

(b) Interpret physically the boundary-value problem in (a).

12.49. (a) Show that the solution of the boundary-value problem

$$\frac{\partial u}{\partial t} = \frac{\partial^2 u}{\partial x^2}, \qquad u_x(0, t) = u_x(\pi, t) = 0, \qquad u(x, 0) = f(x)$$

where $0 < x < \pi, \ t > 0$ is given by

$$u(x, t) = \frac{1}{\pi}\int_0^\pi f(x) \, dx + \frac{2}{\pi}\sum_{m=1}^{\infty} e^{-m^2 t} \cos mx \int_0^\pi f(x) \cos mx \, dx$$

(b) Interpret physically the boundary-value problem in (a).

12.50. Find the steady-state temperature in a bar whose ends are located at $x = 0$ and $x = 10$ if these ends are kept at 150°C and 100°C respectively.

12.51. A circular plate of unit radius [see Fig. 12-7, page 270] whose faces are insulated has its boundary kept at temperature $120 + 60 \cos 2\phi$. Find the steady-state temperature of the plate.

12.52. Show that
$$\rho \sin \phi + \tfrac{1}{3}\rho^3 \sin 3\phi + \tfrac{1}{5}\rho^5 \sin 5\phi + \cdots = \tfrac{1}{2} \tan^{-1}\left(\frac{2\rho \sin \phi}{1 - \rho^2}\right)$$
and thus complete Problem 12.20.

12.53. A string 2 ft long is stretched between two fixed points $x = 0$ and $x = 2$. If the displacement of the string from the x axis at $t = 0$ is given by $f(x) = .03x(2 - x)$ and if the initial velocity is zero, find the displacement at any later time.

12.54. A square plate of side a has one side maintained at temperature $f(x)$ and the others at zero, as indicated in Fig. 12-9. Show that the steady-state temperature at any point of the plate is given by
$$u(x, y) = \sum_{k=1}^{\infty} \left[\frac{2}{a \sinh (k\pi)} \int_0^a f(x) \sin \frac{k\pi x}{a}\, dx\right] \sin \frac{k\pi x}{a} \sinh \frac{k\pi y}{a}$$
Obtain the result for $f(x) = u_0$, a constant.

12.55. Work Problem 12.54 if the sides are maintained at temperatures $f_1(x), g_1(y), f_2(x), g_2(y)$ respectively. [*Hint*. Use the principle of superposition and the result of Problem 12.54.]

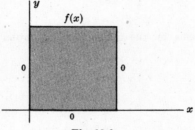

Fig. 12-9

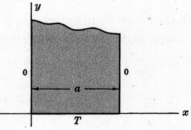

Fig. 12-10

12.56. An infinitely long plate of width a indicated by the shaded region of Fig. 12-10 has its two parallel sides maintained at temperature 0 and its other side at constant temperature u_0. (*a*) Show that the steady-state temperature is given by
$$u(x, y) = \frac{4u_0}{\pi}\left(e^{-y} \sin \frac{\pi x}{a} + \frac{1}{3} e^{-3y} \sin \frac{3\pi x}{a} + \frac{1}{5} e^{-5y} \sin \frac{5\pi x}{a} + \cdots\right)$$

(*b*) Use Problem 12.52 to show that
$$u(x, y) = \frac{2u_0}{\pi} \tan^{-1}\left(\frac{\sin (\pi x/a)}{\sinh y}\right)$$

12.57. Solve Problem 12.35 if the string has its ends fixed at $x = 0$ and $x = l$ and if its initial displacement and velocity are given by $f(x)$ and $g(x)$ respectively.

12.58. A square plate [Fig. 12-10] having sides of unit length has its edges fixed in the xy plane and is set into transverse vibration.

(*a*) Show that the transverse displacement $z(x, y, t)$ of any point (x, y) at time t is given by
$$\frac{\partial^2 z}{\partial t^2} = a^2 \left(\frac{\partial^2 z}{\partial x^2} + \frac{\partial^2 z}{\partial y^2}\right)$$
where a^2 is a constant.

(*b*) Show that if the plate is given an initial shape $f(x, y)$ and released with velocity $g(x, y)$, then the displacement is given by

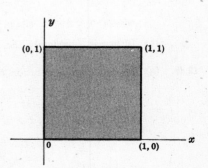

Fig. 12-11

$$z(x, y, t) = \sum_{m=1}^{\infty} \sum_{n=1}^{\infty} [A_{mn} \cos \lambda_{mn} at + B_{mn} \sin \lambda_{mn} at] \sin m\pi x \sin n\pi y$$

where

$$A_{mn} = 4 \int_0^1 \int_0^1 f(x, y) \sin m\pi x \sin n\pi y \, dx \, dy$$

$$B_{mn} = \frac{4}{a\lambda_{mn}} \int_0^1 \int_0^1 g(x, y) \sin m\pi x \sin n\pi y \, dx \, dy$$

and $\lambda_{mn} = \pi \sqrt{m^2 + n^2}$.

12.59. Show that the natural frequencies for the plate of Problem 12.57 are given by $f_{mn} = a\sqrt{m^2 + n^2}/2\pi$. Would you expect to get a musical tone from the vibration? Explain.

12.60. A beam has its ends hinged at $x = 0$ and $x = l$. At time $t = 0$, a concentrated transverse load of constant magnitude w is suddenly applied at the midpoint. Show that the resulting transverse displacement of any point x of the beam at any time $t > 0$ is

$$y(x, t) = \frac{wx}{12EI}(\tfrac{3}{4}l^2 - x^2) - \frac{2wl^3}{\pi^4 EI}\left\{ \frac{\sin(\pi x/l)}{1^4} + \frac{\sin(3\pi x/l)}{3^4} + \frac{\sin(5\pi x/l)}{5^4} + \cdots \right\}$$

if $0 < x < l/2$, while the corresponding result for $l/2 < x < l$ is obtained by symmetry.

12.61. Solve the boundary-value problem

$$\frac{\partial u}{\partial t} = \frac{\partial^2 u}{\partial x^2} - \alpha^2 u \qquad 0 < x < l, \ t > 0$$

$$u(0, t) = u_1, \qquad u(l, t) = u_2, \qquad u(x, 0) = 0$$

where α and l are constants, and interpret physically.

12.62. Show that the solution of the boundary-value problem

$$\frac{\partial u}{\partial t} = k\frac{\partial^2 u}{\partial x^2} \qquad 0 < x < l, \ t > 0$$

$$u_x(0, t) = hu(0, t), \qquad u_x(l, t) = -hu(l, t), \qquad u(x, 0) = f(x)$$

where k, h and l are constants is

$$u(x, t) = \sum_{n=1}^{\infty} e^{-k\lambda_n^2 t} \frac{\lambda_n \cos \lambda_n x + h \sin \lambda_n x}{(\lambda_n^2 + h^2)l + 2h} \int_0^l f(x)[\lambda_n \cos \lambda_n x + h \sin \lambda_n x] \, dx$$

where λ_n are solutions of the equation $\tan \lambda l = \dfrac{2h\lambda}{\lambda^2 - h^2}$. Give a physical interpretation.

SOLUTIONS USING FOURIER INTEGRALS

12.63. An infinite thin bar $(-\infty < x < \infty)$ whose surface is insulated has an initial temperature equal to $f(x)$. Show that the temperature at any point x at any time t is given by

$$u(x, t) = \frac{1}{\sqrt{\pi}} \int_{-\infty}^{\infty} e^{-w^2} f(x + 2\sqrt{kt} \, w) dw$$

12.64. If $f(x) = u_0$ in Problem 12.63, show that $u(x, t) = u_0 \, \text{erf}\,(x/2\sqrt{kt})$ and explain the connection with Problem 12.23.

12.65. Solve the boundary-value problem

$$\frac{\partial^2 u}{\partial x^2} + \frac{\partial^2 u}{\partial y^2} = 0 \qquad y > 0$$

$$u(x, 0) = f(x) \qquad |u(x, y)| < M$$

showing that

$$u(x, y) = \frac{1}{\pi} \int_{\lambda=0}^{\infty} \int_{v=-\infty}^{\infty} f(v) e^{-\lambda y} \cos \lambda(v - x) \, d\lambda \, dv$$

and give a physical interpretation.

12.66. Show that if $f(x) = \begin{cases} 0 & x < 0 \\ u_0 & x > 0 \end{cases}$ in Problem 12.65, then

$$u(x, y) = \frac{1}{\pi} \tan^{-1} \frac{x}{y}$$

12.67. The region bounded by $x > 0$, $y > 0$ has one edge $x = 0$ kept at potential zero and the other edge $y = 0$ kept at potential $f(x)$. (*a*) Show that the potential at any point (x, y) is given by

$$v(x, y) = \frac{1}{\pi} \int_0^\infty y f(v) \left[\frac{1}{(v - x)^2 + y^2} - \frac{1}{(v + x)^2 + y^2} \right] dv$$

(*b*) If $f(x) = 1$, show that $v(x, y) = \frac{2}{\pi} \tan^{-1} \frac{x}{y}$.

SOLUTIONS USING BESSEL FUNCTIONS

12.68. The temperature of a long solid circular cylinder of unit radius is initially zero. At $t = 0$ the surface is given a constant temperature u_0 which is then maintained. Show that the temperature of the cylinder is given by

$$u(\rho, t) = u_0 \left\{ 1 - 2 \sum_{n=1}^\infty \frac{J_0(\lambda_n \rho)}{\lambda_n J_1(\lambda_n)} e^{-k\lambda_n^2 t} \right\}$$

where λ_n, $n = 1, 2, 3, \ldots$, are the positive roots of $J_0(\lambda) = 0$ and k is the diffusivity.

12.69. Show that if $F(\rho) = u_0(1 - \rho^2)$, then the temperature of the cylinder of Problem 12.23 is given by

$$u(\rho, t) = 4ku_0 \sum_{n=1}^\infty \frac{J_0(\lambda_n \rho) J_2(\lambda_n)}{\lambda_n^2 J_1^2(\lambda_n)} e^{-k\lambda_n^2 t}$$

12.70. A cylinder $0 < \rho < a$, $0 < z < l$ has the end $z = 0$ at temperature $f(\rho)$ while the other surfaces are kept at temperature zero. Show that the steady-state temperature at any point is given by

$$u(\rho, z) = \frac{2}{a^2} \sum_{n=1}^\infty \frac{J_0(\lambda_n \rho) \sinh \lambda_n(l - z)}{J_1^2(\lambda_n a) \sinh \lambda_n l} \int_0^a \rho f(\rho) J_0(\lambda_n \rho) \, d\rho$$

where $J_0(\lambda_n a) = 0$, $n = 1, 2, 3, \ldots$.

12.71. A circular membrane of unit radius lies in the xy plane with its center at the origin. Its edge $\rho = 1$ is fixed in the xy plane and it is set into vibration by displacing it an amount $f(\rho)$ and then releasing it. Show that the displacement is given by

$$z(\rho, t) = \sum_{n=1}^\infty \frac{J_0(\lambda_n \rho) \cos \lambda_n a t}{2J_1^2(\lambda_n)} \int_0^1 \rho f(\rho) J_0(\lambda_n \rho) \, d\rho$$

where λ_n are the roots of $J_0(\lambda) = 0$.

12.72. (*a*) Solve the boundary-value problem

$$\frac{\partial^2 u}{\partial t^2} = \frac{\partial^2 u}{\partial \rho^2} + \frac{1}{\rho} \frac{\partial u}{\partial \rho} + \frac{1}{\rho^2} \frac{\partial^2 u}{\partial \phi^2}$$

where $0 < \rho < 1$, $0 < \phi < 2\pi$, $t > 0$ if u is bounded, and

$$u(1, \phi, t) = 0, \qquad u(\rho, \phi, 0) = \rho \cos 3\phi, \qquad u_t(\rho, \phi, 0) = 0$$

(*b*) Give a physical interpretation to the solution.

12.73. Solve the boundary-value problem

$$\frac{\partial}{\partial x} \left(x \frac{\partial y}{\partial x} \right) = \frac{\partial^2 y}{\partial t^2}$$

given that $y(x, 0) = f(x)$, $y_t(x, 0) = 0$, $y(1, t) = 0$ and $y(x, t)$ is bounded for $0 \leqq x \leqq 1$, $t > 0$.

12.74. A chain of constant mass per unit length and length l is suspended vertically from one end O as indicated in Fig. 12-12. If the chain is displaced slightly at time $t = 0$ so that its shape is given by $f(x)$, $0 < x < l$, and then released, show that the displacement of any point x at time t is given by

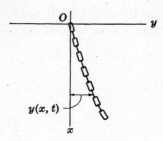

$$y(x, t) = \sum_{n=1}^{\infty} A_n J_0\left(2\lambda_n\sqrt{\frac{l-x}{g}}\right)\cos\lambda_n t$$

where λ_n are the roots of $J_0(2\lambda\sqrt{l/g}) = 0$ and

$$A_n = \frac{2}{J_1^2(\lambda_n)}\int_0^1 v J_0(\lambda_n v)\, f(l - \tfrac{1}{4}gv^2)\, dv$$

Fig. 12-12

12.75. Determine the frequencies of the normal modes and indicate whether you would expect music or noise from the vibrations.

12.76. A solid circular cylinder $0 < \rho < a$, $0 < z < l$ has its bases kept at temperature zero and the convex surface at constant temperature u_0. Show that the steady-state temperature at any point of the cylinder is

$$u(\rho, \phi, z) = \frac{4u_0}{\pi}\sum_{n=1}^{\infty}\frac{I_0[(2n-1)\pi\rho/l]\,\sin[(2n-1)\pi z/l]}{(2n-1)I_0[(2n-1)\pi a/l]}$$

where I_0 is the modified Bessel function of order zero.

12.77. A membrane in the form of a circular ring $a \leqq \rho \leqq b$ is set into vibration. Show that the frequencies of the various modes of vibration are given by

$$f_{mn} = \frac{\lambda_{mn}}{2\pi}\sqrt{\frac{\tau}{\sigma}}$$

where τ is the tension per unit length, σ is the mass per unit area, and λ_{mn} are roots of the equation

$$J_m(\lambda a)\,Y_m(\lambda b) - J_m(\lambda b)\,Y_m(\lambda a) = 0$$

12.78. Explain how you would determine the displacement of the membrane in Problem 12.77.

12.79. The surface $\rho = 1$ of an infinite cylinder is kept at temperature $f(z)$. Show that the steady-state temperature everywhere in the cylinder is given by

$$u(\rho, z) = \frac{1}{\pi}\int_{\lambda=0}^{\infty}\int_{v=-\infty}^{\infty}\frac{f(v)\cos\lambda(v-z)I_0(\lambda\rho)}{I_0(\lambda)}\, d\lambda\, dv$$

SOLUTIONS USING LEGENDRE FUNCTIONS

12.80. Find the potential v (a) interior and (b) exterior to a hollow sphere of unit radius with center at the origin if the surface is charged to potential $v_0(1 + 3\cos\theta)$ where v_0 is constant.

12.81. Solve Problem 12.80 if the surface potential is $v_0\sin^2\theta$.

12.82. Find the steady-state temperature within the region bounded by two concentric spheres of radius a and $2a$ if the temperatures of the outer and inner spheres are u_0 and 0 respectively.

12.83. Find the gravitational potential at any point outside a solid uniform sphere of radius a of mass m.

12.84. Is there a solution to Problem 12.82, if the point is inside the sphere? Explain.

12.85. Find the potential at any point due to a thin circular ring of radius a and mass m lying in the xy plane with center at the origin. [*Hint.* First find the potential at any point on the axis.]

12.86. Work Problem 12.85 for a solid circular disc.

12.87. Show that a solution of Laplace's equation $\nabla^2 v = 0$ in spherical coordinates is given by

$$v = \sum_{m=0}^{\infty}\sum_{n=0}^{\infty}\left[A_n r^n + \frac{B_n}{r^{n+1}}\right][C_m\cos m\theta + D_m\sin m\theta][EP_n^m(\cos\phi) + FQ_n^m(\cos\phi)]$$

where P_n^m and Q_n^m are the associated Legendre functions.

SOLUTIONS USING LAPLACE TRANSFORMS

12.88. Solve each of the following boundary-value problems by using Laplace transforms.

    (a) $\dfrac{\partial u}{\partial t} = 3\dfrac{\partial u}{\partial x}$, $u(x, 0) = 4e^{-2x}$

    (b) $\dfrac{\partial u}{\partial t} = \dfrac{\partial u}{\partial x} - 2u$, $u(x, 0) = 10e^{-x} - 6e^{-4x}$

    (c) $\dfrac{\partial u}{\partial t} = \dfrac{\partial^2 u}{\partial x^2}$, $u(0, t) = 0$, $u(4, t) = 0$, $u(x, 0) = 6 \sin(\pi x/2) + 3 \sin \pi x$

    (d) $\dfrac{\partial u}{\partial t} = \dfrac{\partial^2 u}{\partial x^2}$, $u_x(0, t) = 0$, $u_x(2, t) = 0$, $u(x, 0) = 4 \cos \pi x - 2 \cos 3\pi x$

12.89. Work (a) Problem 12.46(d), (b) Problem 12.47, by use of Laplace transforms.

12.90. Show how you would attempt to solve Problem 12.17 by Laplace transforms. Explain what difficulties arise. [Problems of this type can best be solved by using the methods of Chapter 14.]

Answers to Supplementary Problems

12.29. (a) linear, dep. var. u, ind. var. x, y, order 2   (d) linear, dep. var. y, ind. var. x, t, order 2

       (b) linear, dep. var. T, ind. var. x, y, z, order 4   (e) nonlinear, dep. var. z. ind. var. r, s, order 1

       (c) nonlinear, dep. var. ϕ, ind. var. x, y, order 3

12.30. (a) hyperbolic   (b) hyperbolic   (c) elliptic   (d) parabolic

       (e) elliptic if $x^2 + y^2 < 1$, hyperbolic if $x^2 + y^2 > 1$, parabolic if $x^2 + y^2 = 1$

12.32. (b) $x(2x + y - 2)^2$

12.33. $3\dfrac{\partial^2 u}{\partial x^2} - 5\dfrac{\partial^2 u}{\partial x\,\partial y} - 2\dfrac{\partial^2 u}{\partial y^2} = 0$

12.34. (a) $2\dfrac{\partial z}{\partial x} + 3\dfrac{\partial z}{\partial y} = 2z$ (b) $2\dfrac{\partial^2 z}{\partial x^2} - 3\dfrac{\partial^2 z}{\partial x\,\partial y} - 2\dfrac{\partial^2 z}{\partial y^2} = 0$

12.37. $u_x(0, t) = 0$, $u_x(L, t) = 0$

12.41. (a) $xz = F(x) + G(y)$ (b) $xz = x^6 + x^2 + 6y^4 - 68$

12.42. (a) $u = F(x + y) + G(x - y)$ (d) $z = F(3x + y) + G(y - x)$

       (b) $u = e^{3x}F(y - 2x)$ (e) $z = F(x + y) + xG(x + y)$

       (c) $u = F(x + iy) + G(x - iy)$

12.43. (a) $u = F(y - 2x) + \dfrac{x^2}{2}$ (c) $u = F(y) + xG(y) + x^2H(y) + I(y - 2x) + \tfrac{2}{3}x^3$

       (b) $y = F(x - t) + G(x + t) - t^4$ (d) $z = F(x + y) + G(2x + y) - \dfrac{x}{2}\sin y + \tfrac{3}{4}\cos y$

12.44. $u = F(x + iy) + G(x - iy) + xH(x + iy) + xJ(x - iy) + 4(x^2 + y^2)^2$

12.46. (a) $u = 4e^{(3y - 2x)/2}$

       (b) $u = 3e^{-5x - 3y} + 2e^{-3x - 2y}$

       (c) $u = 2e^{-36t} \sin 3x - 4e^{-100t} \sin 5x$

       (d) $u = 8e^{-9\pi^2 t/16} \cos \dfrac{3\pi x}{4} - 6e^{-81\pi^2 t/16} \cos \dfrac{9\pi x}{4}$

12.47. (a) $y = \dfrac{5}{2\pi} \sin \pi x \sin 2\pi t$ (b) $y = \dfrac{3}{4\pi} \sin 2\pi x \sin 4\pi t - \dfrac{1}{5\pi} \sin 5\pi x \sin 10\pi t$

12.48. (a) $u(x, t) = -\dfrac{200}{\pi} \displaystyle\sum_{m=1}^{\infty} \dfrac{e^{-m^2\pi^2 t/8} \cos m\pi}{m} \sin \dfrac{m\pi x}{4}$

12.50. $150 - 5x$

12.51. $u(\rho, \phi) = 120 + 60\rho^2 \cos 2\phi$

12.53. $y(x, t) = \dfrac{.96}{\pi^3} \displaystyle\sum_{n=1}^{\infty} \dfrac{1}{(2n-1)^3} \sin \dfrac{(2n-1)\pi x}{2} \cos \dfrac{(2n-1)\pi c t}{2}$

12.72. $u(\rho, \phi, t) = \displaystyle\sum_{n=1}^{\infty} A_n J_3(\lambda_n \rho) \cos 3\phi \cos \lambda_n t$ where λ_n are the positive roots of $J_3(\lambda) = 0$ and

$A_n = \dfrac{2[(\lambda_n^2 - 8) J_0(\lambda_n) - 6\lambda_n J_1(\lambda_n) + 8]}{\lambda_n^3 J_4^2(\lambda_n)}$.

12.73. $y(x, t) = \displaystyle\sum_{n=1}^{\infty} \dfrac{J_0(\lambda_n \sqrt{x})}{J_1^2(\lambda_n)} \cos(\tfrac{1}{2}\lambda_n t) \int_0^1 f(x) J_0(\lambda_n \sqrt{x})\, dx$ where $J_0(\lambda_n) = 0$, $n = 1, 2, \ldots$

12.80. (a) $v_0(1 + 3r \cos \theta)$ (b) $\dfrac{v_0}{r}\left(1 + \dfrac{3}{r} \cos \theta\right)$

12.81. (a) $\tfrac{2}{3} v_0 [1 - r^2 P_2(\cos \theta)]$ (b) $\dfrac{2v_0}{3r}\left[1 - \dfrac{P_2(\cos \theta)}{r^2}\right]$

12.82. $2u_0\left(1 - \dfrac{a}{r}\right)$

12.83. m/r where $r > a$ is the distance from the center of the sphere

12.85. $v = \begin{cases} \dfrac{m}{a}\left[P_0(\cos \phi) - \dfrac{1}{2}\dfrac{r^2}{a^2} P_2(\cos \phi) + \dfrac{1 \cdot 3}{2 \cdot 4}\dfrac{r^4}{a^4} P_4(\cos \phi) - \cdots\right] & \text{if } r < a \\[3mm] \dfrac{m}{a}\left[\dfrac{a}{r}P_0(\cos \phi) - \dfrac{1}{2}\dfrac{a^3}{r^3} P_2(\cos \phi) + \dfrac{1 \cdot 3}{2 \cdot 4}\dfrac{a^5}{r^5} P_4(\cos \phi) - \cdots\right] & \text{if } r > a \end{cases}$

12.86. $v = \begin{cases} \dfrac{2m}{a}\Big[P_0(\cos \phi) - \dfrac{r}{a} P_1(\cos \phi) + \dfrac{1}{2}\dfrac{r^2}{a^2} P_2(\cos \phi) - \dfrac{1}{2 \cdot 4}\dfrac{r^4}{a^4} P_4(\cos \phi) \\[3mm] \qquad\qquad\qquad + \dfrac{1 \cdot 3}{2 \cdot 4 \cdot 6}\dfrac{r^6}{a^6} P_6(\cos \phi) - \cdots\Big] & \text{if } r < a \\[3mm] \dfrac{2m}{a}\left[\dfrac{1}{2}\dfrac{a}{r}P_0(\cos \phi) - \dfrac{1}{2 \cdot 4}\dfrac{a^3}{r^3} P_2(\cos \phi) + \dfrac{1 \cdot 3}{2 \cdot 4 \cdot 6}\dfrac{a^5}{r^5} P_4(\cos \phi) - \cdots\right] & \text{if } r > a \end{cases}$

12.88. (a) $u = 4e^{-2x - 6t}$

(b) $u = 10e^{-x - 3t} - 6e^{-4x - 6t}$

(c) $u = 6e^{-\pi^2 t/4} \sin(\pi x/2) + 3e^{-\pi^2 t} \sin \pi x$

(d) $u = 4e^{-\pi^2 t} \cos \pi x - 2e^{-9\pi^2 t/4} \cos 3\pi x$

Chapter 13

FUNCTIONS

If to each of a set of complex numbers which a variable z may assume there corresponds one or more values of a variable w, then w is called a *function of the complex variable z*, written $w = f(z)$. The fundamental operations with complex numbers have already been considered in Chapter 1.

A function is *single-valued* if for each value of z there corresponds only one value of w; otherwise it is *multiple-valued* or *many-valued*. In general we can write $w = f(z) = u(x, y) + iv(x, y)$, where u and v are real functions of x and y.

> **Example 1.** $w = z^2 = (x + iy)^2 = x^2 - y^2 + 2ixy = u + iv$ so that $u(x, y) = x^2 - y^2$, $v(x, y) = 2xy$. These are called the *real* and *imaginary parts* of $w = z^2$ respectively.

Unless otherwise specified we shall assume that $f(z)$ is single-valued. A function which is multiple-valued can be considered as a collection of single-valued functions.

LIMITS AND CONTINUITY

Definitions of limits and continuity for functions of a complex variable are analogous to those for a real variable. Thus $f(z)$ is said to have the *limit l* as z approaches z_0 if, given any $\epsilon > 0$, there exists a $\delta > 0$ such that $|f(z) - l| < \epsilon$ whenever $0 < |z - z_0| < \delta$.

Similarly, $f(z)$ is said to be *continuous* at z_0 if, given any $\epsilon > 0$, there exists a $\delta > 0$ such that $|f(z) - f(z_0)| < \epsilon$ whenever $|z - z_0| < \delta$. Alternatively, $f(z)$ is continuous at z_0 if $\lim_{z \to z_0} f(z) = f(z_0)$.

DERIVATIVES

If $f(z)$ is single-valued in some region of the z plane the *derivative* of $f(z)$, denoted by $f'(z)$, is defined as

$$\lim_{\Delta z \to 0} \frac{f(z + \Delta z) - f(z)}{\Delta z} \tag{1}$$

provided the limit exists independent of the manner in which $\Delta z \to 0$. If the limit (1) exists for $z = z_0$, then $f(z)$ is called *analytic* at z_0. If the limit exists for all z in a region $\mathcal{R}$, then $f(z)$ is called *analytic in $\mathcal{R}$*. In order to be analytic, $f(z)$ must be single-valued and continuous. The converse, however, is not necessarily true.

We define elementary functions of a complex variable by a natural extension of the corresponding functions of a real variable. Where series expansions for real functions $f(x)$ exist, we can use as definition the series with x replaced by z.

> **Example 2.** We define $e^z = 1 + z + \frac{z^2}{2!} + \frac{z^3}{3!} + \cdots$, $\sin z = z - \frac{z^3}{3!} + \frac{z^5}{5!} - \frac{z^7}{7!} + \cdots$, $\cos z = 1 - \frac{z^2}{2!} + \frac{z^4}{4!} - \frac{z^6}{6!} + \cdots$. From these we can show that $e^z = e^{x+iy} = e^x(\cos y + i \sin y)$, as well as numerous other relations.

Example 3. We define a^b as $e^{b \ln a}$ even when a and b are complex numbers. Since $e^{2k\pi i} = 1$, it follows that $e^{i\phi} = e^{i(\phi + 2k\pi)}$ and we define $\ln z = \ln (\rho e^{i\phi}) = \ln \rho + i(\phi + 2k\pi)$. Thus $\ln z$ is a many-valued function. The various single-valued functions of which this many-valued function is composed are called its *branches*.

Rules for differentiating functions of a complex variable are much the same as for those of real variables. Thus $\dfrac{d}{dz}(z^n) = nz^{n-1}$, $\dfrac{d}{dz}(\sin z) = \cos z$, etc.

CAUCHY-RIEMANN EQUATIONS

A necessary condition that $w = f(z) = u(x, y) + iv(x, y)$ be analytic in a region $\mathcal{R}$ is that u and v satisfy the *Cauchy-Riemann equations*

$$\frac{\partial u}{\partial x} = \frac{\partial v}{\partial y}, \quad \frac{\partial u}{\partial y} = -\frac{\partial v}{\partial x} \tag{2}$$

(see Problem 13.7). If the partial derivatives in (2) are continuous in $\mathcal{R}$, the equations are sufficient conditions that $f(z)$ be analytic in $\mathcal{R}$.

If the second derivatives of u and v with respect to x and y exist and are continuous, we find by differentiating (2) that

$$\frac{\partial^2 u}{\partial x^2} + \frac{\partial^2 u}{\partial y^2} = 0, \quad \frac{\partial^2 v}{\partial x^2} + \frac{\partial^2 v}{\partial y^2} = 0 \tag{3}$$

Thus the real and imaginary parts satisfy Laplace's equation in two dimensions. Functions satisfying Laplace's equation are called *harmonic functions*.

INTEGRALS

If $f(z)$ is defined, single-valued and continuous in a region $\mathcal{R}$, we define the *integral* of $f(z)$ along some path C in $\mathcal{R}$ from point z_1 to point z_2, where $z_1 = x_1 + iy_1$, $z_2 = x_2 + iy_2$, as

$$\int_C f(z)\, dz = \int_{(x_1, y_1)}^{(x_2, y_2)} (u + iv)(dx + i\, dy) = \int_{(x_1, y_1)}^{(x_2, y_2)} u\, dx - v\, dy + i \int_{(x_1, y_1)}^{(x_2, y_2)} v\, dx + u\, dy$$

with this definition the integral of a function of a complex variable can be made to depend on line integrals for real functions already considered in Chapter 6. An alternative definition based on the limit of a sum, as for functions of a real variable, can also be formulated and turns out to be equivalent to the one above.

The rules for complex integration are similar to those for real integrals. An important result is

$$\left| \int_C f(z)\, dz \right| \leq \int_C |f(z)|\, |dz| \leq M \int_C ds = ML \tag{4}$$

where M is an upper bound of $|f(z)|$ on C, i.e. $|f(z)| \leq M$, and L is the length of the path C.

CAUCHY'S THEOREM

Let C be a simple closed curve. If $f(z)$ is analytic within the region bounded by C as well as on C, then we have *Cauchy's theorem* that

$$\int_C f(z)\, dz = \oint_C f(z)\, dz = 0 \tag{5}$$

where the second integral emphasizes the fact that C is a simple closed curve.

Expressed in another way, (5) is equivalent to the statement that $\int_{z_1}^{z_2} f(z)\,dz$ has a value *independent of the path* joining z_1 and z_2. Such integrals can be evaluated as $F(z_2) - F(z_1)$ where $F'(z) = f(z)$. These results are similar to corresponding results for line integrals developed in Chapter 6.

Example 4. Since $f(z) = 2z$ is analytic everywhere, we have for any simple closed curve C

$$\oint_C 2z\,dz = 0$$

Also, $\int_{2i}^{1+i} 2z\,dz = z^2 \Big|_{2i}^{1+i} = (1+i)^2 - (2i)^2 = 2i + 4$

CAUCHY'S INTEGRAL FORMULAS

If $f(z)$ is analytic within and on a simple closed curve C and a is any point interior to C, then

$$f(a) = \frac{1}{2\pi i} \oint_C \frac{f(z)}{z-a}\,dz \tag{6}$$

where C is traversed in the positive (counterclockwise) sense.

Also, the nth derivative of $f(z)$ at $z = a$ is given by

$$f^{(n)}(a) = \frac{n!}{2\pi i} \oint_C \frac{f(z)}{(z-a)^{n+1}}\,dz \tag{7}$$

These are called *Cauchy's integral formulas*. They are quite remarkable because they show that if the function $f(z)$ is known *on* the closed curve C then it is also known *within* C, and the various derivatives at points within C can be calculated. Thus if a function of a complex variable has a first derivative, it has all higher derivatives as well. This of course is not necessarily true for functions of real variables.

TAYLOR'S SERIES

Let $f(z)$ be analytic inside and on a circle having its center at $z = a$. Then for all points z in the circle we have the *Taylor series* representation of $f(z)$ given by

$$f(z) = f(a) + f'(a)(z-a) + \frac{f''(a)}{2!}(z-a)^2 + \frac{f'''(a)}{3!}(z-a)^3 + \cdots \tag{8}$$

See Problem 13.21.

SINGULAR POINTS

A singular point of a function $f(z)$ is a value of z at which $f(z)$ fails to be analytic. If $f(z)$ is analytic everywhere in some region except at an interior point $z = a$, we call $z = a$ an *isolated singularity* of $f(z)$.

Example 5. If $f(z) = \dfrac{1}{(z-3)^2}$, then $z = 3$ is an isolated singularity of $f(z)$.

POLES

If $f(z) = \dfrac{\phi(z)}{(z-a)^n}$, $\phi(a) \neq 0$, where $\phi(z)$ is analytic everywhere in a region including $z = a$, and if n is a positive integer, then $f(z)$ has an isolated singularity at $z = a$ which is called a *pole of order n*. If $n = 1$, the pole is often called a *simple pole*; if $n = 2$ it is called a *double pole*, etc.

Example 6. $f(z) = \dfrac{z}{(z-3)^2(z+1)}$ has two singularities: a pole of order 2 or double pole at $z = 3$, and a pole of order 1 or simple pole at $z = -1$.

Example 7. $f(z) = \dfrac{3z-1}{z^2+4} = \dfrac{3z-1}{(z+2i)(z-2i)}$ has two simple poles at $z = \pm 2i$.

A function can have other types of singularities besides poles. For example, $f(z) = \sqrt{z}$ has a *branch point* at $z = 0$ (see Problem 13.36). The function $f(z) = \dfrac{\sin z}{z}$ has a singularity at $z = 0$. However, due to the fact that $\lim\limits_{z \to 0} \dfrac{\sin z}{z}$ is finite, we call such a singularity a *removable singularity*.

LAURENT'S SERIES

If $f(z)$ has a pole of order n at $z = a$ but is analytic at every other point inside and on a circle C with center at a, then $(z-a)^n f(z)$ is analytic at all points inside and on C and has a Taylor series about $z = a$ so that

$$f(z) = \frac{a_{-n}}{(z-a)^n} + \frac{a_{-n+1}}{(z-a)^{n-1}} + \cdots + \frac{a_{-1}}{z-a} + a_0 + a_1(z-a) + a_2(z-a)^2 + \cdots \qquad (9)$$

This is called a *Laurent series* for $f(z)$. The part $a_0 + a_1(z-a) + a_2(z-a)^2 + \cdots$ is called the *analytic part*, while the remainder consisting of inverse powers of $z - a$ is called the *principal part*. More generally, we refer to the series $\sum\limits_{k=-\infty}^{\infty} a_k(z-a)^k$ as a Laurent series where the terms with $k < 0$ constitute the principal part. A function which is analytic in a region bounded by two concentric circles having center at $z = a$ can always be expanded into such a Laurent series (see Problem 13.82).

It is possible to define various types of singularities of a function $f(z)$ from its Laurent series. For example, when the principal part of a Laurent series has a finite number of terms and $a_{-n} \neq 0$ while $a_{-n-1}, a_{-n-2}, \ldots$ are all zero, then $z = a$ is a pole of order n. If the principal part has infinitely many terms, $z = a$ is called an *essential singularity* or sometimes a *pole of infinite order*.

Example 8. The function $e^{1/z} = 1 + \dfrac{1}{z} + \dfrac{1}{2! \, z^2} + \cdots$ has an essential singularity at $z = 0$.

RESIDUES

The coefficients in (9) can be obtained in the customary manner by writing the coefficients for the Taylor series corresponding to $(z-a)^n f(z)$. In further developments, the coefficient a_{-1}, called the *residue* of $f(z)$ at the pole $z = a$, is of considerable importance. It can be found from the formula

$$a_{-1} = \lim_{z \to a} \frac{1}{(n-1)!} \frac{d^{n-1}}{dz^{n-1}} \{(z-a)^n f(z)\} \qquad (10)$$

where n is the order of the pole. For simple poles the calculation of the residue is of particular simplicity since it reduces to

$$a_{-1} = \lim_{z \to a} (z-a) f(z) \qquad (11)$$

RESIDUE THEOREM

If $f(z)$ is analytic in a region $\mathcal{R}$ except for a pole of order n at $z = a$ and if C is any simple closed curve in $\mathcal{R}$ containing $z = a$, then $f(z)$ has the form (9). Integrating (9), using the fact that

$$\oint_C \frac{dz}{(z-a)^n} = \begin{cases} 0 & \text{if } n \neq 1 \\ 2\pi i & \text{if } n = 1 \end{cases} \tag{12}$$

(see Problem 13.13), it follows that

$$\oint_C f(z)\, dz = 2\pi i a_{-1} \tag{13}$$

i.e. the integral of $f(z)$ around a closed path enclosing a single pole of $f(z)$ is $2\pi i$ times the residue at the pole.

More generally, we have the following important

Theorem 13.1. If $f(z)$ is analytic within and on the boundary C of a region $\mathcal{R}$ except at a finite number of poles $a, b, c, \ldots$ within $\mathcal{R}$, having residues $a_{-1}, b_{-1}, c_{-1}, \ldots$ respectively, then

$$\oint_C f(z)\, dz = 2\pi i(a_{-1} + b_{-1} + c_{-1} + \cdots) \tag{14}$$

i.e. the integral of $f(z)$ is $2\pi i$ times the sum of the residues of $f(z)$ at the poles enclosed by C.

Cauchy's theorem and integral formulas are special cases of this result which we call the *residue theorem*.

EVALUATION OF DEFINITE INTEGRALS

The evaluation of various definite integrals can often be achieved by using the residue theorem together with a suitable function $f(z)$ and a suitable path or *contour* C, the choice of which may require great ingenuity. The following types are most common in practice.

1. $\int_0^\infty F(x)\, dx$, $F(x)$ is an even function.

   Consider $\oint_C F(z)\, dz$ along a contour C consisting of the line along the x axis from $-R$ to $+R$ and the semi-circle above the x axis having this line as diameter. Then let $R \to \infty$. See Problems 13.29, 13.30.

2. $\int_0^{2\pi} G(\sin\theta, \cos\theta)\, d\theta$, G is a rational function of $\sin\theta$ and $\cos\theta$.

   Let $z = e^{i\theta}$ Then $\sin\theta = \dfrac{z - z^{-1}}{2i}$, $\cos\theta = \dfrac{z + z^{-1}}{2}$ and $dz = ie^{i\theta}\, d\theta$ or $d\theta = dz/iz$. The given integral is equivalent to $\oint_C F(z)\, dz$ where C is the unit circle with center at the origin. See Problems 13.31, 13.32.

3. $\int_{-\infty}^\infty F(x) \begin{Bmatrix} \cos mx \\ \sin mx \end{Bmatrix} dx$, $F(x)$ is a rational function.

   Here we consider $\oint_C F(z)\, e^{imz}\, dz$ where C is the same contour as that in Type 1. See Problem 13.34.

4. Miscellaneous integrals involving particular contours. See Problems 13.35, 13.37.

CONFORMAL MAPPING

The analytic function $w = f(z) = u(x,y) + iv(x,y)$ defines a transformation $u = u(x,y)$, $v = v(x,y)$ which establishes a correspondence between points of the uv and xy planes.

Suppose that under this transformation point (x_0, y_0) of the xy plane is mapped into point (u_0, v_0) of the uv plane [see Figs. 13-1 and 13-2] while curves C_1 and C_2 [intersecting at (x_0, y_0)] are mapped respectively into curves C_1' and C_2' [intersecting at (u_0, v_0)]. Then if the transformation is such that the angle at (x_0, y_0) between C_1 and C_2 is equal to the angle at (u_0, v_0) between C_1' and C_2' both in magnitude and sense, the transformation or mapping is said to be *conformal* at (x_0, y_0). A mapping which preserves the magnitudes of angles but not necessarily the sense is called *isogonal*.

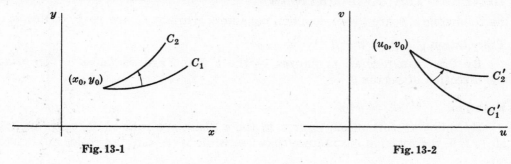

Fig. 13-1   Fig. 13-2

The following theorem is fundamental.

Theorem 13.2.   If $f(z)$ is analytic and $f'(z) \neq 0$ in a region $\mathcal{R}$, then the mapping $w = f(z)$ is conformal at all points of $\mathcal{R}$.

For conformal mappings or transformations, small figures in the neighborhood of a point z_0 in the z plane map into similar small figures in the w plane and are magnified [or reduced] by an amount given approximately by $|f'(z_0)|^2$, called the *area magnification factor* or simply *magnification factor*. Short distances in the z plane in the neighborhood of z_0 are magnified [or reduced] in the w plane by an amount given approximately by $|f'(z_0)|$, called the *linear magnification factor*. Large figures in the z plane usually map into figures in the w plane which are far from similar.

RIEMANN'S MAPPING THEOREM

Let C [Fig. 13-3] be a simple closed curve in the z plane forming the boundary of a region $\mathcal{R}$. Let C' [Fig. 13-4] be a circle of radius one and center at the origin [the *unit circle*] forming the boundary of region $\mathcal{R}'$ in the w plane. The region $\mathcal{R}'$ is sometimes called the *unit disk*. Then *Riemann's mapping theorem* states that there exists a function $w = f(z)$, analytic in $\mathcal{R}$, which maps each point of $\mathcal{R}$ into a corresponding point of $\mathcal{R}'$ and each point of C into a corresponding point of C', the correspondence being one to one.

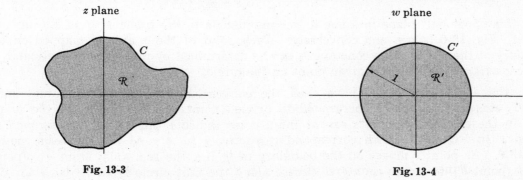

Fig. 13-3   Fig. 13-4

This function $f(z)$ contains three arbitrary real constants which can be determined by making the center of C' correspond to some given point in $\mathcal{R}$, while some point on C' corresponds to a given point on C. It should be noted that while Riemann's mapping theorem demonstrates the *existence* of a mapping function, it does not actually produce this function.

It is possible to extend Riemann's mapping theorem to the case where a region bounded by two simple closed curves, one inside the other, is mapped into a region bounded by two concentric circles.

SOME GENERAL TRANSFORMATIONS

In the following α, β are given complex constants while a, θ_0 are real constants.

1. Translation. $w = z + \beta$

By this transformation, figures in the z plane are *displaced* or *translated* in the direction of vector β.

2. Rotation. $w = e^{i\theta_0}z$

By this transformation, figures in the z plane are rotated through an angle θ_0. If $\theta_0 > 0$ the rotation is counterclockwise, while if $\theta_0 < 0$ the rotation is clockwise.

3. Stretching. $w = az$

By this transformation, figures in the z plane are stretched (or contracted) in the direction z if $a > 1$ (or $0 < a < 1$). We consider contraction as a special case of stretching.

4. Inversion. $w = 1/z$

5. Linear Transformation. $w = \alpha z + \beta$

This is a combination of the transformations of translation, rotation and stretching.

6. Bilinear or Fractional Transformation. $w = \dfrac{\alpha z + \beta}{\gamma z + \delta}, \ \alpha\delta - \beta\gamma \neq 0$

This is a combination of the transformations of translation, rotation, stretching and inversion.

MAPPING OF A HALF PLANE ON TO A CIRCLE

Let z_0 be any point P in the upper half of the z plane denoted by $\mathcal{R}$ in Fig. 13-5 below. Then the transformation

$$w = e^{i\theta_0}\left(\frac{z - z_0}{z - \bar{z}_0}\right) \tag{15}$$

maps this upper half plane in a one to one manner on to the interior $\mathcal{R}'$ of the unit circle $|w| = 1$, Fig. 13-6 below, and conversely. Each point of the x axis is mapped on to the boundary of the circle. The constant θ_0 can be determined by making one particular point of the x axis correspond to a given point on the circle.

In the above figures we have used the convention that unprimed points such as A, B, C, etc., in the z plane correspond to primed points A', B', C', etc., in the w plane. Also, in the case where points are at infinity we indicate this by an arrow such as at A and F in Fig. 13-5 which correspond respectively to A' and F' (the same point) in Fig. 13-6. As point z moves on the boundary of $\mathcal{R}$ [i.e. the real axis] from $-\infty$ (point A) to $+\infty$ (point F), w moves counterclockwise along the unit circle from A' back to A'.

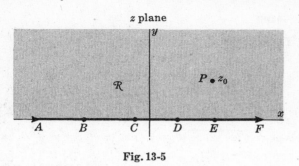

Fig. 13-5

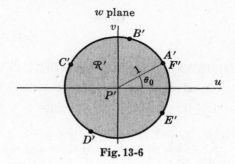

Fig. 13-6

THE SCHWARZ-CHRISTOFFEL TRANSFORMATION

Consider a polygon [Fig. 13-7] in the w plane having vertices at $w_1, w_2, \ldots, w_n$ with corresponding interior angles $\alpha_1, \alpha_2, \ldots \alpha_n$ respectively. Let the points $w_1, w_2, \ldots, w_n$ map respectively into points $x_1, x_2, \ldots, x_n$ on the real axis of the z plane [Fig. 13-8].

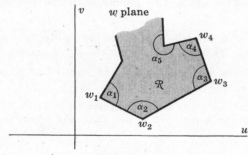

Fig. 13-7

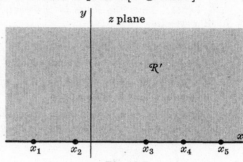

Fig. 13-8

A transformation which maps the interior $\mathcal{R}$ of the polygon of the w plane on to the upper half $\mathcal{R}'$ of the z plane and the boundary of the polygon on to the real axis is given by

$$dw/dz \;=\; A\,(z-x_1)^{\alpha_1/\pi-1}(z-x_2)^{\alpha_2/\pi-1}\cdots(z-x_n)^{\alpha_n/\pi-1} \tag{16}$$

or

$$w \;=\; A\int (z-x_1)^{\alpha_1/\pi-1}(z-x_2)^{\alpha_2/\pi-1}\cdots(z-x_n)^{\alpha_n/\pi-1}\,dz \;+\; B \tag{17}$$

where A and B are complex constants.

The following facts should be noted:

1. Any three of the points $x_1, x_2, \ldots, x_n$ can be chosen at will.
2. The constants A and B determine the size, orientation and position of the polygon.
3. It is convenient to choose one point, say x_n, at infinity in which case the last factor of (16) and (17) involving x_n is not present.
4. Infinite open polygons can be considered as limiting cases of closed polygons.

SOLUTIONS OF LAPLACE'S EQUATION BY CONFORMAL MAPPING

The problem of determining a function which is harmonic, i.e. satisfies Laplace's equation, in some region $\mathcal{R}$ [Fig. 13-7] and which takes prescribed values on the boundary C is often called a *Dirichlet problem*. It can often be solved by mapping $\mathcal{R}$ into the unit circle or upper half plane. In such case C is mapped into a corresponding boundary C' and boundary conditions on C are transformed into boundary conditions for C'. Since Laplace's equation in $\mathcal{R}$ is transformed into Laplace's equation in $\mathcal{R}'$ [see Problem 13.49], the problem is thus reduced to solving Laplace's equation in $\mathcal{R}'$ with boundary conditions on C', which is generally simple to do. By then transforming back we obtain the required solution.

For illustrations of the procedure see Problems 13.51-13.54.

Solved Problems

FUNCTIONS, LIMITS, CONTINUITY

13.1. Determine the locus represented by

    (a) $|z-2| = 3$,   (b) $|z-2| = |z+4|$,   (c) $|z-3| + |z+3| = 10$.

(a) **Method 1.** $|z-2| = |x+iy-2| = |x-2+iy| = \sqrt{(x-2)^2+y^2} = 3$ or $(x-2)^2+y^2 = 9$, a circle with center at $(2,0)$ and radius 3.

    Method 2. $|z-2|$ is the distance between the complex numbers $z = x+iy$ and $2+0i$. If this distance is always 3, the locus is a circle of radius 3 with center at $2+0i$ or $(2,0)$.

(b) **Method 1.** $|x+iy-2| = |x+iy+4|$ or $\sqrt{(x-2)^2+y^2} = \sqrt{(x+4)^2+y^2}$. Squaring, we find $x = -1$, a straight line.

    Method 2. The locus is such that the distances from any point on it to $(2,0)$ and $(-4,0)$ are equal. Thus the locus is the perpendicular bisector of the line joining $(2,0)$ and $(-4,0)$, or $x = -1$.

(c) **Method 1.** The locus is given by $\sqrt{(x-3)^2+y^2} + \sqrt{(x+3)^2+y^2} = 10$ or $\sqrt{(x-3)^2+y^2} = 10 - \sqrt{(x+3)^2+y^2}$. Squaring and simplifying, $25 + 3x = 5\sqrt{(x+3)^2+y^2}$. Squaring and simplifying again yields $\dfrac{x^2}{25} + \dfrac{y^2}{16} = 1$, an ellipse with semi-major and semi-minor axes of lengths 5 and 4 respectively.

    Method 2. The locus is such that the sum of the distances from any point on it to $(3,0)$ and $(-3,0)$ is 10. Thus the locus is an ellipse whose foci are at $(-3,0)$ and $(3,0)$ and whose major axis has length 10.

13.2. Determine the region in the z plane represented by each of the following.

(a) $|z| < 1$.

    Interior of a circle of radius 1. See Fig. 13-9.

(b) $1 < |z+2i| \leqq 2$.

    $|z+2i|$ is the distance from z to $-2i$, so that $|z+2i| = 1$ is a circle of radius 1 with center at $-2i$, i.e. $(0,-2)$; and $|z+2i| = 2$ is a circle of radius 2 with center at $-2i$. Then $1 < |z+2i| \leqq 2$ represents the region *exterior* to $|z+2i| = 1$ but *interior* to or *on* $|z+2i| = 2$. See Fig. 13-10.

(c) $\pi/3 \leqq \arg z \leqq \pi/2$.

    Note that $\arg z = \phi$, where $z = \rho e^{i\phi}$. The required region is the infinite region bounded by the lines $\phi = \pi/3$ and $\phi = \pi/2$, including these lines. See Fig. 13-11.

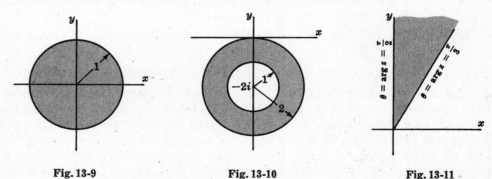

      Fig. 13-9   **Fig. 13-10**   **Fig. 13-11**

13.3. Express each function in the form $u(x,y) + iv(x,y)$, where u and v are real:

    (a) z^3,   (b) $1/(1-z)$,   (c) e^{3z},   (d) $\ln z$.

(a) $w = z^3 = (x + iy)^3 = x^3 + 3x^2(iy) + 3x(iy)^2 + (iy)^3 = x^3 + 3ix^2y - 3xy^2 - iy^3$
$= x^3 - 3xy^2 + i(3x^2y - y^3)$

Then $u(x, y) = x^3 - 3xy^2, \quad v(x, y) = 3x^2y - y^3.$

(b) $w = \dfrac{1}{1-z} = \dfrac{1}{1 - (x + iy)} = \dfrac{1}{1 - x - iy} \cdot \dfrac{1 - x + iy}{1 - x + iy} = \dfrac{1 - x + iy}{(1-x)^2 + y^2}$

Then $u(x, y) = \dfrac{1 - x}{(1-x)^2 + y^2}, \quad v(x, y) = \dfrac{y}{(1-x)^2 + y^2}.$

(c) $e^{3z} = e^{3(x+iy)} = e^{3x} e^{3iy} = e^{3x}(\cos 3y + i \sin 3y)$ and $u = e^{3x} \cos 3y, \quad v = e^{3x} \sin 3y$

(d) $\ln z = \ln (\rho e^{i\phi}) = \ln \rho + i\phi = \ln \sqrt{x^2 + y^2} + i \tan^{-1} y/x$ and

$$u = \tfrac{1}{2} \ln (x^2 + y^2), \quad v = \tan^{-1} y/x$$

Note that $\ln z$ is a multiple-valued function (in this case it is *infinitely* many-valued) since ϕ can be increased by any multiple of 2π. The *principal value* of the logarithm is defined as that value for which $0 \leqq \phi < 2\pi$ and is called the *principal branch* of $\ln z$.

13.4. Prove (a) $\sin (x + iy) = \sin x \cosh y + i \cos x \sinh y$
(b) $\cos (x + iy) = \cos x \cosh y - i \sin x \sinh y$

We use the relations $e^{iz} = \cos z + i \sin z, \quad e^{-iz} = \cos z - i \sin z,$ from which

$$\sin z = \frac{e^{iz} - e^{-iz}}{2i}, \qquad \cos z = \frac{e^{iz} + e^{-iz}}{2}$$

Then $\sin z = \sin (x + iy) = \dfrac{e^{i(x+iy)} - e^{-i(x+iy)}}{2i} = \dfrac{e^{ix-y} - e^{-ix+y}}{2i}$

$= \tfrac{1}{2i}\{e^{-y}(\cos x + i \sin x) - e^{y}(\cos x - i \sin x)\}$

$= (\sin x)\left(\dfrac{e^y + e^{-y}}{2}\right) + i(\cos x)\left(\dfrac{e^y - e^{-y}}{2}\right) = \sin x \cosh y + i \cos x \sinh y$

Similarly, $\cos z = \cos (x + iy) = \dfrac{e^{i(x+iy)} + e^{-i(x+iy)}}{2}$

$= \tfrac{1}{2}\{e^{ix-y} + e^{-ix+y}\} = \tfrac{1}{2}\{e^{-y}(\cos x + i \sin x) + e^{y}(\cos x - i \sin x)\}$

$= (\cos x)\left(\dfrac{e^y + e^{-y}}{2}\right) - i(\sin x)\left(\dfrac{e^y - e^{-y}}{2}\right) = \cos x \cosh y - i \sin x \sinh y$

DERIVATIVES. CAUCHY-RIEMANN EQUATIONS

13.5. Prove that $\dfrac{d}{dz} \bar{z}$, where $\bar{z}$ is the conjugate of z, does not exist anywhere.

By definition, $\dfrac{d}{dz} f(z) = \lim\limits_{\Delta z \to 0} \dfrac{f(z + \Delta z) - f(z)}{\Delta z}$ if this limit exists independent of the manner in which $\Delta z = \Delta x + i \Delta y$ approaches zero. Then

$$\frac{d}{dz} \bar{z} = \lim_{\Delta z \to 0} \frac{\overline{z + \Delta z} - \bar{z}}{\Delta z} = \lim_{\substack{\Delta x \to 0 \\ \Delta y \to 0}} \frac{\overline{x + iy + \Delta x + i\Delta y} - \overline{x + iy}}{\Delta x + i \Delta y}$$

$$= \lim_{\substack{\Delta x \to 0 \\ \Delta y \to 0}} \frac{x - iy + \Delta x - i\Delta y - (x - iy)}{\Delta x + i \Delta y} = \lim_{\substack{\Delta x \to 0 \\ \Delta y \to 0}} \frac{\Delta x - i\Delta y}{\Delta x + i \Delta y}$$

If $\Delta y = 0$, the required limit is $\lim\limits_{\Delta x \to 0} \dfrac{\Delta x}{\Delta x} = 1.$

If $\Delta x = 0$, the required limit is $\lim\limits_{\Delta y \to 0} \dfrac{-i \Delta y}{i \Delta y} = -1.$

These two possible approaches show that the limit depends on the manner in which $\Delta z \to 0$, so that the derivative does not exist; i.e. $\bar{z}$ is *non-analytic* anywhere.

13.6. (a) If $w = f(z) = \dfrac{1+z}{1-z}$, find $\dfrac{dw}{dz}$. (b) Determine where w is non-analytic.

(a) **Method 1.** $\dfrac{dw}{dz} = \lim\limits_{\Delta z \to 0} \dfrac{\dfrac{1+(z+\Delta z)}{1-(z+\Delta z)} - \dfrac{1+z}{1-z}}{\Delta z} = \lim\limits_{\Delta z \to 0} \dfrac{2}{(1-z-\Delta z)(1-z)}$

$= \dfrac{2}{(1-z)^2}$ provided $z \neq 1$, independent of the manner in which $\Delta z \to 0$.

Method 2. The usual rules of differentiation apply provided $z \neq 1$. Thus by the quotient rule for differentiation,

$$\frac{d}{dz}\left(\frac{1+z}{1-z}\right) = \frac{(1-z)\dfrac{d}{dz}(1+z) - (1+z)\dfrac{d}{dz}(1-z)}{(1-z)^2} = \frac{(1-z)(1) - (1+z)(-1)}{(1-z)^2} = \frac{2}{(1-z)^2}$$

(b) The function is analytic everywhere except at $z = 1$, where the derivative does not exist; i.e. the function is non-analytic at $z = 1$.

13.7. Prove that a necessary condition for $w = f(z) = u(x, y) + iv(x, y)$ to be analytic in a region is that the Cauchy-Riemann equations $\dfrac{\partial u}{\partial x} = \dfrac{\partial v}{\partial y}$, $\dfrac{\partial u}{\partial y} = -\dfrac{\partial v}{\partial x}$ be satisfied in the region.

Since $f(z) = f(x + iy) = u(x, y) + iv(x, y)$, we have

$$f(z + \Delta z) = f[x + \Delta x + i(y + \Delta y)] = u(x + \Delta x, y + \Delta y) + iv(x + \Delta x, y + \Delta y)$$

Then

$$\lim_{\Delta z \to 0} \frac{f(z + \Delta z) - f(z)}{\Delta z} = \lim_{\substack{\Delta x \to 0 \\ \Delta y \to 0}} \frac{u(x + \Delta x, y + \Delta y) - u(x, y) + i\{v(x + \Delta x, y + \Delta y) - v(x, y)\}}{\Delta x + i\Delta y}$$

If $\Delta y = 0$, the required limit is

$$\lim_{\Delta x \to 0} \frac{u(x + \Delta x, y) - u(x, y)}{\Delta x} + i\left\{\frac{v(x + \Delta x, y) - v(x, y)}{\Delta x}\right\} = \frac{\partial u}{\partial x} + i\frac{\partial v}{\partial x}$$

If $\Delta x = 0$, the required limit is

$$\lim_{\Delta y \to 0} \frac{u(x, y + \Delta y) - u(x, y)}{i\,\Delta y} + \left\{\frac{v(x, y + \Delta y) - v(x, y)}{\Delta y}\right\} = \frac{1}{i}\frac{\partial u}{\partial y} + \frac{\partial v}{\partial y}$$

If the derivative is to exist, these two special limits must be equal, i.e.,

$$\frac{\partial u}{\partial x} + i\frac{\partial v}{\partial x} = \frac{1}{i}\frac{\partial u}{\partial y} + \frac{\partial v}{\partial y} = -i\frac{\partial u}{\partial y} + \frac{\partial v}{\partial y}$$

so that we must have $\dfrac{\partial u}{\partial x} = \dfrac{\partial v}{\partial y}$ and $\dfrac{\partial v}{\partial x} = -\dfrac{\partial u}{\partial y}$.

Conversely, we can prove that if the first partial derivatives of u and v with respect to x and y are continuous in a region, then the Cauchy-Riemann equations provide sufficient conditions for $f(z)$ to be analytic.

13.8. (a) If $f(z) = u(x, y) + iv(x, y)$ is analytic in a region $\mathcal{R}$, prove that the one parameter families of curves $u(x, y) = C_1$ and $v(x, y) = C_2$ are orthogonal families. (b) Illustrate by using $f(z) = z^2$.

(a) Consider any two particular members of these families $u(x, y) = u_0$, $v(x, y) = v_0$ which intersect at the point (x_0, y_0).

Since $du = u_x\,dx + u_y\,dy = 0$, we have $\dfrac{dy}{dx} = -\dfrac{u_x}{u_y}$.

Also since $dv = v_x\,dx + v_y\,dy = 0$, $\dfrac{dy}{dx} = -\dfrac{v_x}{v_y}$.

When evaluated at (x_0, y_0), these represent respectively the slopes of the two curves at this point of intersection.

By the Cauchy-Riemann equations, $u_x = v_y$, $u_y = -v_x$, we have the product of the slopes at the point (x_0, y_0) equal to

$$\left(-\frac{u_x}{u_y}\right)\left(-\frac{v_x}{v_y}\right) = -1$$

so that any two members of the respective families are orthogonal, and thus the two families are orthogonal.

(b) If $f(z) = z^2$, then $u = x^2 - y^2$, $v = 2xy$. The graphs of several members of $x^2 - y^2 = C_1$, $2xy = C_2$ are shown in Fig. 13-12.

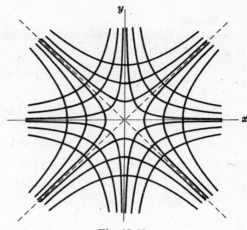

Fig. 13-12

13.9. In aerodynamics and fluid mechanics, the functions ϕ and ψ in $f(z) = \phi + i\psi$, where $f(z)$ is analytic, are called the *velocity potential* and *stream function* respectively. If $\phi = x^2 + 4x - y^2 + 2y$, (a) find ψ and (b) find $f(z)$.

(a) By the Cauchy-Riemann equations, $\dfrac{\partial \phi}{\partial x} = \dfrac{\partial \psi}{\partial y}$, $\dfrac{\partial \psi}{\partial x} = -\dfrac{\partial \phi}{\partial y}$. Then

$$(1)\quad \frac{\partial \psi}{\partial y} = 2x + 4 \qquad (2)\quad \frac{\partial \psi}{\partial x} = 2y - 2$$

Method 1. Integrating (1), $\psi = 2xy + 4y + F(x)$.

Integrating (2), $\psi = 2xy - 2x + G(y)$.

These are identical if $F(x) = -2x + c$, $G(y) = 4y + c$ where c is any real constant. Thus $\psi = 2xy + 4y - 2x + c$.

Method 2.

Integrating (1), $\psi = 2xy + 4y + F(x)$. Then substituting in (2), $2y + F'(x) = 2y - 2$ or $F'(x) = -2$ and $F(x) = -2x + c$. Hence $\psi = 2xy + 4y - 2x + c$.

(b) From (a), $f(z) = \phi + i\psi = x^2 + 4x - y^2 + 2y + i(2xy + 4y - 2x + c)$

$= (x^2 - y^2 + 2ixy) + 4(x + iy) - 2i(x + iy) + ic = z^2 + 4z - 2iz + c_1$

where c_1 is a pure imaginary constant.

This can also be accomplished by noting that $z = x + iy$, $\bar{z} = x - iy$ so that $x = \dfrac{z + \bar{z}}{2}$, $y = \dfrac{z - \bar{z}}{2i}$. The result is then obtained by substitution; the terms involving $\bar{z}$ drop out.

INTEGRALS, CAUCHY'S THEOREM, CAUCHY'S INTEGRAL FORMULAS

13.10. Evaluate $\displaystyle\int_{1+i}^{2+4i} z^2\, dz$

(a) along the parabola $x = t$, $y = t^2$ where $1 \leqq t \leqq 2$,

(b) along the straight line joining $1 + i$ and $2 + 4i$,

(c) along straight lines from $1 + i$ to $2 + i$ and then to $2 + 4i$.

We have

$$\int_{1+i}^{2+4i} z^2\, dz = \int_{(1,1)}^{(2,4)} (x+iy)^2\,(dx + i\,dy) = \int_{(1,1)}^{(2,4)} (x^2 - y^2 + 2ixy)(dx + i\,dy)$$

$$= \int_{(1,1)}^{(2,4)} (x^2 - y^2)\,dx - 2xy\,dy + i\int_{(1,1)}^{(2,4)} 2xy\,dx + (x^2 - y^2)\,dy$$

Method 1.

(a) The points $(1, 1)$ and $(2, 4)$ correspond to $t = 1$ and $t = 2$ respectively. Then the above line integrals become

$$\int_{t=1}^{2} \{(t^2 - t^4)\, dt - 2(t)(t^2)2t\, dt\} + i \int_{t=1}^{2} \{2(t)(t^2) + (t^2 - t^4)(2t)\, dt\} = -\frac{86}{3} - 6i$$

(b) The line joining $(1, 1)$ and $(2, 4)$ has the equation $y - 1 = \frac{4 - 1}{2 - 1}(x - 1)$ or $y = 3x - 2$. Then we find

$$\int_{x=1}^{2} \{[x^2 - (3x - 2)^2]\, dx - 2x(3x - 2)3\, dx\}$$
$$+ i \int_{x=1}^{2} \{2x(3x - 2)\, dx + [x^2 - (3x - 2)^2]3\, dx\} = -\frac{86}{3} - 6i$$

(c) From $1 + i$ to $2 + i$ [or $(1, 1)$ to $(2, 1)$], $y = 1$, $dy = 0$ and we have

$$\int_{x=1}^{2} (x^2 - 1)\, dx + i \int_{x=1}^{2} 2x\, dx = \frac{4}{3} + 3i$$

From $2 + i$ to $2 + 4i$ [or $(2, 1)$ to $(2, 4)$], $x = 2$, $dx = 0$ and we have

$$\int_{y=1}^{4} -4y\, dy + i \int_{y=1}^{4} (4 - y^2)\, dy = -30 - 9i$$

Adding, $(\frac{4}{3} + 3i) + (-30 - 9i) = -\frac{86}{3} - 6i$.

Method 2.

By the methods of Chapter 6 it is seen that the line integrals are independent of the path, thus accounting for the same values obtained in (a), (b) and (c) above. In such case the integral can be evaluated directly, as for real variables, as follows:

$$\int_{1+i}^{2+4i} z^2\, dz = \frac{z^3}{3}\Big|_{1+i}^{2+4i} = \frac{(2 + 4i)^3}{3} - \frac{(1 + i)^3}{3} = -\frac{86}{3} - 6i$$

13.11. (a) Prove Cauchy's theorem: If $f(z)$ is analytic inside and on a simple closed curve C, then $\oint_C f(z)\, dz = 0$.

(b) Under these conditions prove that $\int_{P_1}^{P_2} f(z)\, dz$ is independent of the path joining P_1 and P_2.

(a) $$\oint_C f(z)\, dz = \oint_C (u + iv)(dx + i\, dy) = \oint_C (u\, dx - v\, dy) + i \oint_C (v\, dx + u\, dy)$$

By Green's theorem (Chapter 6),

$$\oint_C (u\, dx - v\, dy) = \iint_{\mathcal{R}} \left(-\frac{\partial v}{\partial x} - \frac{\partial u}{\partial y}\right) dx\, dy, \qquad \oint_C (v\, dx + u\, dy) = \iint_{\mathcal{R}} \left(\frac{\partial u}{\partial x} - \frac{\partial v}{\partial y}\right) dx\, dy$$

where $\mathcal{R}$ is the region (simply-connected) bounded by C.

Since $f(z)$ is analytic, $\frac{\partial u}{\partial x} = \frac{\partial v}{\partial y}$, $\frac{\partial v}{\partial x} = -\frac{\partial u}{\partial y}$ (Problem 13.7), and so the above integrals are zero. Then $\oint_C f(z)\, dz = 0$, assuming $f'(z)$ [and thus the partial derivatives] to be continuous.

(b) Consider any two paths joining points P_1 and P_2 (see Fig. 13-13). By Cauchy's theorem,

$$\int_{P_1 A P_2 B P_1} f(z)\, dz = 0$$

Then $$\int_{P_1 A P_2} f(z)\, dz + \int_{P_2 B P_1} f(z)\, dz = 0$$

or $$\int_{P_1 A P_2} f(z)\, dz = -\int_{P_2 B P_1} f(z)\, dz = \int_{P_1 B P_2} f(z)\, dz$$

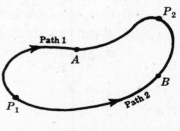

Fig. 13-13

i.e. the integral along $P_1 A P_2$ (path 1) = integral along $P_1 B P_2$ (path 2), and so the integral is independent of the path joining P_1 and P_2.

This explains the results of Problem 13.10, since $f(z) = z^2$ is analytic.

13.12. If $f(z)$ is analytic within and on the boundary of a region bounded by two closed curves C_1 and C_2 (see Fig. 13-14), prove that

$$\oint_{C_1} f(z)\,dz \;=\; \oint_{C_2} f(z)\,dz$$

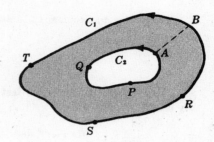

As in Fig. 13-14, construct line AB (called a *cross-cut*) connecting any point on C_2 and a point on C_1. By Cauchy's theorem (Problem 13.11),

$$\int_{AQPABRSTBA} f(z)\,dz \;=\; 0$$

since $f(z)$ is analytic within the region shaded and also on the boundary.   Then

$$\int_{AQPA} f(z)\,dz \;+\; \int_{AB} f(z)\,dz \;+\; \int_{BRSTB} f(z)\,dz \;+\; \int_{BA} f(z)\,dz \;=\; 0 \qquad (1)$$

Fig. 13-14

But $\displaystyle\int_{AB} f(z)\,dz = -\int_{BA} f(z)\,dz$.   Hence (1) gives

$$\int_{AQPA} f(z)\,dz \;=\; -\int_{BRSTB} f(z)\,dz \;=\; \int_{BTSRB} f(z)\,dz$$

i.e.

$$\oint_{C_1} f(z)\,dz \;=\; \oint_{C_2} f(z)\,dz$$

Note that $f(z)$ need not be analytic *within* curve C_2.

13.13. (a) Prove that $\displaystyle\oint_C \frac{dz}{(z-a)^n} = \begin{cases} 2\pi i & \text{if } n = 1 \\ 0 & \text{if } n = 2, 3, 4, \ldots \end{cases}$ where C is a simple closed curve bounding a region having $z = a$ as interior point.

(b) What is the value of the integral if $n = 0, -1, -2, -3, \ldots$?

(a) Let C_1 be a circle of radius ϵ having center at $z = a$ (see Fig. 13-15).   Since $(z-a)^{-n}$ is analytic within and on the boundary of the region bounded by C and C_1, we have by Problem 13.12,

$$\oint_C \frac{dz}{(z-a)^n} \;=\; \oint_{C_1} \frac{dz}{(z-a)^n}$$

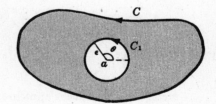

To evaluate this last integral, note that on C_1, $|z - a| = \epsilon$ or $z - a = \epsilon e^{i\theta}$ and $dz = i\epsilon e^{i\theta}\,d\theta$. The integral equals

Fig. 13-15

$$\int_0^{2\pi} \frac{i\epsilon e^{i\theta}\,d\theta}{\epsilon^n e^{in\theta}} \;=\; \frac{i}{\epsilon^{n-1}}\int_0^{2\pi} e^{(1-n)i\theta}\,d\theta \;=\; \frac{i}{\epsilon^{n-1}}\,\frac{e^{(1-n)i\theta}}{(1-n)i}\Big|_0^{2\pi} \;=\; 0 \qquad \text{if } n \neq 1$$

If $n = 1$, the integral equals $\displaystyle i\int_0^{2\pi} d\theta = 2\pi i$.

(b) For $n = 0, -1, -2, \ldots$ the integrand is $1, (z-a), (z-a)^2, \ldots$ and is analytic everywhere inside C_1, including $z = a$.   Hence by Cauchy's theorem the integral is zero.

13.14. Evaluate $\displaystyle\oint_C \frac{dz}{z-3}$ where C is (a) the circle $|z| = 1$, (b) the circle $|z+i| = 4$.

(a) Since $z = 3$ is not interior to $|z| = 1$, the integral equals zero (Problem 13.11).

(b) Since $z = 3$ is interior to $|z+i| = 4$, the integral equals $2\pi i$ (Problem 13.13).

13.15. If $f(z)$ is analytic inside and on a simple closed curve C, and a is any point within C, prove that

$$f(a) \;=\; \frac{1}{2\pi i}\oint_C \frac{f(z)}{z-a}\,dz$$

Referring to Problem 13.12 and the figure of Problem 13.13, we have

$$\oint_C \frac{f(z)}{z-a}\,dz \;=\; \oint_{C_1}\frac{f(z)}{z-a}\,dz$$

Letting $z-a=\epsilon e^{i\theta}$, the last integral becomes $i\displaystyle\int_0^{2\pi} f(a+\epsilon e^{i\theta})\,d\theta$. But since $f(z)$ is analytic, it is continuous. Hence

$$\lim_{\epsilon\to 0} i\int_0^{2\pi} f(a+\epsilon e^{i\theta})\,d\theta \;=\; i\int_0^{2\pi}\lim_{\epsilon\to 0} f(a+\epsilon e^{i\theta})\,d\theta \;=\; i\int_0^{2\pi} f(a)\,d\theta \;=\; 2\pi i\,f(a)$$

and the required result follows.

13.16. Evaluate $(a)\ \displaystyle\oint_C \frac{\cos z}{z-\pi}\,dz$, $(b)\ \displaystyle\oint_C \frac{e^z}{z(z+1)}\,dz$ where C is the circle $|z-1|=3$.

(a) Since $z=\pi$ lies within C, $\dfrac{1}{2\pi i}\displaystyle\oint_C \frac{\cos z}{z-\pi}\,dz \;=\; \cos\pi \;=\; -1$ by Problem 13.15 with $f(z)=\cos z$, $a=\pi$. Then $\displaystyle\oint_C \frac{\cos z}{z-\pi}\,dz \;=\; -2\pi i$.

(b)
$$\oint_C \frac{e^z}{z(z+1)}\,dz \;=\; \oint_C e^z\left(\frac{1}{z}-\frac{1}{z+1}\right)dz \;=\; \oint_C \frac{e^z}{z}\,dz - \oint_C \frac{e^z}{z+1}\,dz$$
$$=\; 2\pi i e^0 - 2\pi i e^{-1} \;=\; 2\pi i(1-e^{-1})$$

by Problem 13.15, since $z=0$ and $z=-1$ are both interior to C.

13.17. Evaluate $\displaystyle\oint_C \frac{5z^2-3z+2}{(z-1)^3}\,dz$ where C is any simple closed curve enclosing $z=1$.

Method 1. By Cauchy's integral formula, $f^{(n)}(a) = \dfrac{n!}{2\pi i}\displaystyle\oint_C \frac{f(z)}{(z-a)^{n+1}}\,dz$.

If $n=2$ and $f(z)=5z^2-3z+2$, then $f''(1)=10$. Hence

$$10 \;=\; \frac{2!}{2\pi i}\oint_C \frac{5z^2-3z+2}{(z-1)^3}\,dz \quad\text{or}\quad \oint_C \frac{5z^2-3z+2}{(z-1)^3}\,dz \;=\; 10\pi i$$

Method 2. $5z^2-3z+2 = 5(z-1)^2 + 7(z-1) + 4$. Then

$$\oint_C \frac{5z^2-3z+2}{(z-1)^3}\,dz \;=\; \oint_C \frac{5(z-1)^2+7(z-1)+4}{(z-1)^3}\,dz$$
$$=\; 5\oint_C \frac{dz}{z-1} + 7\oint_C \frac{dz}{(z-1)^2} + 4\oint_C \frac{dz}{(z-1)^3} \;=\; 5(2\pi i) + 7(0) + 4(0)$$
$$=\; 10\pi i$$

by Problem 13.13.

SERIES AND SINGULARITIES

13.18. For what values of z does each series converge?

(a) $\displaystyle\sum_{n=1}^{\infty} \frac{z^n}{n^2\, 2^n}$. The nth term $= u_n = \dfrac{z^n}{n^2\, 2^n}$. Then

$$\lim_{n\to\infty} \left|\frac{u_{n+1}}{u_n}\right| \;=\; \lim_{n\to\infty} \left|\frac{z^{n+1}}{(n+1)^2\, 2^{n+1}} \cdot \frac{n^2\, 2^n}{z^n}\right| \;=\; \frac{|z|}{2}$$

By the ratio test the series converges if $|z| < 2$ and diverges if $|z| > 2$. If $|z| = 2$ the ratio test fails.

However, the series of absolute values $\displaystyle\sum_{n=1}^{\infty}\left|\frac{z^n}{n^2\, 2^n}\right| = \sum_{n=1}^{\infty}\frac{|z|^n}{n^2\, 2^n}$ converges if $|z| = 2$, since $\displaystyle\sum_{n=1}^{\infty}\frac{1}{n^2}$ converges.

Thus the series converges (absolutely) for $|z| \leqq 2$, i.e. at all points inside and on the circle $|z| = 2$.

(b) $\displaystyle\sum_{n=1}^{\infty} \frac{(-1)^{n-1}\, z^{2n-1}}{(2n-1)!} = z - \frac{z^3}{3!} + \frac{z^5}{5!} - \cdots$. We have

$$\lim_{n\to\infty} \left|\frac{u_{n+1}}{u_n}\right| \;=\; \lim_{n\to\infty} \left|\frac{(-1)^n z^{2n+1}}{(2n+1)!} \cdot \frac{(2n-1)!}{(-1)^{n-1} z^{2n-1}}\right| \;=\; \lim_{n\to\infty} \left|\frac{-z^2}{2n(2n+1)}\right| \;=\; 0$$

Then the series, which represents $\sin z$, converges for all values of z.

(c) $\displaystyle\sum_{n=1}^{\infty} \frac{(z-i)^n}{3^n}$. We have $\displaystyle\lim_{n\to\infty}\left|\frac{u_{n+1}}{u_n}\right| = \lim_{n\to\infty}\left|\frac{(z-i)^{n+1}}{3^{n+1}}\cdot\frac{3^n}{(z-i)^n}\right| = \frac{|z-i|}{3}$.

The series converges if $|z - i| < 3$, and diverges if $|z - i| > 3$.

If $|z - i| = 3$, then $z - i = 3e^{i\theta}$ and the series becomes $\displaystyle\sum_{n=1}^{\infty} e^{in\theta}$. This series diverges since the nth term does not approach zero as $n \to \infty$.

Thus the series converges within the circle $|z - i| = 3$ but not on the boundary.

13.19. If $\displaystyle\sum_{n=0}^{\infty} a_n z^n$ is absolutely convergent for $|z| \leqq R$, show that it is uniformly convergent for these values of z.

The definitions, theorems and proofs for series of complex numbers are analogous to those for real series.

In this case we have $|a_n z^n| \leqq |a_n|\, R^n = M_n$. Since by hypothesis $\displaystyle\sum_{n=1}^{\infty} M_n$ converges, it follows by the Weierstrass M test that $\displaystyle\sum_{n=0}^{\infty} a_n z^n$ converges uniformly for $|z| \leqq R$.

13.20. Locate in the finite z plane all the singularities, if any, of each function and name them.

(a) $\dfrac{z^2}{(z+1)^3}$. $z = -1$ is a pole of order 3.

(b) $\dfrac{2z^3 - z + 1}{(z-4)^2(z-i)(z-1+2i)}$. $z = 4$ is a pole of order 2 (double pole); $z = i$ and $z = 1 - 2i$ are poles of order 1 (simple poles).

(c) $\dfrac{\sin mz}{z^2 + 2z + 2}$, $m \neq 0$. Since $z^2 + 2z + 2 = 0$ when $z = \dfrac{-2 \pm \sqrt{4-8}}{2} = \dfrac{-2 \pm 2i}{2} = -1 \pm i$, we can write $z^2 + 2z + 2 = \{z - (-1+i)\}\{z - (-1-i)\} = (z+1-i)(z+1+i)$.

The function has the two simple poles: $z = -1 + i$ and $z = -1 - i$.

(d) $\dfrac{1 - \cos z}{z}$. $z = 0$ appears to be a singularity. However, since $\lim\limits_{z \to 0} \dfrac{1 - \cos z}{z} = 0$, it is a removable singularity.

Another method.

Since $\dfrac{1 - \cos z}{z} = \dfrac{1}{z}\left\{1 - \left(1 - \dfrac{z^2}{2!} + \dfrac{z^4}{4!} - \dfrac{z^6}{6!} + \cdots\right)\right\} = \dfrac{z}{2!} - \dfrac{z^3}{4!} + \cdots$, we see that $z = 0$ is a removable singularity.

(e) $e^{-1/(z-1)^2} = 1 - \dfrac{1}{(z-1)^2} + \dfrac{1}{2!\,(z-1)^4} - \cdots$.

This is a Laurent series where the principal part has an infinite number of non-zero terms. Then $z = 1$ is an *essential singularity*.

(f) e^z.

This function has no finite singularity. However, letting $z = 1/u$, we obtain $e^{1/u}$ which has an essential singularity at $u = 0$. We conclude that $z = \infty$ is an essential singularity of e^z.

In general, to determine the nature of a possible singularity of $f(z)$ at $z = \infty$, we let $z = 1/u$ and then examine the behavior of the new function at $u = 0$.

13.21. If $f(z)$ is analytic at all points inside and on a circle of radius R with center at a, and if $a + h$ is any point inside C, prove *Taylor's theorem* that

$$f(a + h) = f(a) + h\,f'(a) + \dfrac{h^2}{2!}f''(a) + \dfrac{h^3}{3!}f'''(a) + \cdots$$

By Cauchy's integral formula (Problem 13.15), we have

$$f(a + h) = \dfrac{1}{2\pi i}\oint_C \dfrac{f(z)\,dz}{z - a - h} \tag{1}$$

By division,

$$\dfrac{1}{z - a - h} = \dfrac{1}{(z-a)[1 - h/(z-a)]}$$

$$= \dfrac{1}{(z-a)}\left\{1 + \dfrac{h}{(z-a)} + \dfrac{h^2}{(z-a)^2} + \cdots + \dfrac{h^n}{(z-a)^n} + \dfrac{h^{n+1}}{(z-a)^n(z-a-h)}\right\} \tag{2}$$

Substituting (2) in (1) and using Cauchy's integral formulas, we have

$$f(a + h) = \dfrac{1}{2\pi i}\oint_C \dfrac{f(z)\,dz}{z - a} + \dfrac{h}{2\pi i}\oint_C \dfrac{f(z)\,dz}{(z-a)^2} + \cdots + \dfrac{h^n}{2\pi i}\oint_C \dfrac{f(z)\,dz}{(z-a)^{n+1}} + R_n$$

$$= f(a) + h\,f'(a) + \dfrac{h^2}{2!}f''(a) + \cdots + \dfrac{h^n}{n!}f^{(n)}(a) + R_n$$

where

$$R_n = \dfrac{h^{n+1}}{2\pi i}\oint_C \dfrac{f(z)\,dz}{(z-a)^{n+1}(z-a-h)}$$

Now when z is on C, $\left|\dfrac{f(z)}{z - a - h}\right| \leqq M$ and $|z - a| = R$, so that by (4), page 287, we have, since $2\pi R$ is the length of C,

$$|R_n| \leqq \dfrac{|h|^{n+1}M}{2\pi\,R^{n+1}} \cdot 2\pi R$$

As $n \to \infty$, $|R_n| \to 0$. Then $R_n \to 0$ and the required result follows.

If $f(z)$ is analytic in an annular region $r_1 \leqq |z - a| \leqq r_2$, we can generalize the Taylor series to a Laurent series (see Problem 13.82). In some cases, as shown in Problem 13.22, the Laurent series can be obtained by use of known Taylor series.

13.22. Find Laurent series about the indicated singularity for each of the following functions. Name the singularity in each case and give the region of convergence of each series.

(a) $\dfrac{e^z}{(z-1)^2}$; $z=1$. Let $z-1=u$. Then $z=1+u$ and

$$\frac{e^z}{(z-1)^2} = \frac{e^{1+u}}{u^2} = e\cdot\frac{e^u}{u^2} = \frac{e}{u^2}\left\{1+u+\frac{u^2}{2!}+\frac{u^3}{3!}+\frac{u^4}{4!}+\cdots\right\}$$

$$= \frac{e}{(z-1)^2}+\frac{e}{z-1}+\frac{e}{2!}+\frac{e(z-1)}{3!}+\frac{e(z-1)^2}{4!}+\cdots$$

$z=1$ is a *pole of order 2*, or *double pole*.

The series converges for all values of $z\neq 1$.

(b) $z\cos\dfrac{1}{z}$; $z=0$.

$$z\cos\frac{1}{z} = z\left(1-\frac{1}{2!\,z^2}+\frac{1}{4!\,z^4}-\frac{1}{6!\,z^6}+\cdots\right) = z-\frac{1}{2!\,z}+\frac{1}{4!\,z^3}-\frac{1}{6!\,z^5}+\cdots$$

$z=0$ is an *essential singularity*.

The series converges for all values of $z\neq 0$.

(c) $\dfrac{\sin z}{z-\pi}$; $z=\pi$. Let $z-\pi=u$. Then $z=\pi+u$ and

$$\frac{\sin z}{z-\pi} = \frac{\sin(u+\pi)}{u} = -\frac{\sin u}{u} = -\frac{1}{u}\left(u-\frac{u^3}{3!}+\frac{u^5}{5!}-\cdots\right)$$

$$= -1+\frac{u^2}{3!}-\frac{u^4}{5!}+\cdots = -1+\frac{(z-\pi)^2}{3!}-\frac{(z-\pi)^4}{5!}+\cdots$$

$z=\pi$ is a *removable singularity*.

The series converges for all values of z.

(d) $\dfrac{z}{(z+1)(z+2)}$; $z=-1$. Let $z+1=u$. Then

$$\frac{z}{(z+1)(z+2)} = \frac{u-1}{u(u+1)} = \frac{u-1}{u}(1-u+u^2-u^3+u^4-\cdots)$$

$$= -\frac{1}{u}+2-2u+2u^2-2u^3+\cdots$$

$$= -\frac{1}{z+1}+2-2(z+1)+2(z+1)^2-\cdots$$

$z=-1$ is a *pole of order 1*, or *simple pole*.

The series converges for values of z such that $0<|z+1|<1$.

(e) $\dfrac{1}{z(z+2)^3}$; $z=0,-2$.

Case 1, $z=0$. Using the binomial theorem,

$$\frac{1}{z(z+2)^3} = \frac{1}{8z(1+z/2)^3} = \frac{1}{8z}\left\{1+(-3)\left(\frac{z}{2}\right)+\frac{(-3)(-4)}{2!}\left(\frac{z}{2}\right)^2+\frac{(-3)(-4)(-5)}{3!}\left(\frac{z}{2}\right)^3+\cdots\right\}$$

$$= \frac{1}{8z}-\frac{3}{16}+\frac{3}{16}z-\frac{5}{32}z^2+\cdots$$

$z=0$ is a *pole of order 1*, or *simple pole*.

The series converges for $0<|z|<2$.

Case 2, $z = -2$. Let $z + 2 = u$. Then

$$\frac{1}{z(z+2)^3} = \frac{1}{(u-2)u^3} = \frac{1}{-2u^3(1-u/2)} = -\frac{1}{2u^3}\left\{1 + \frac{u}{2} + \left(\frac{u}{2}\right)^2 + \left(\frac{u}{2}\right)^3 + \left(\frac{u}{2}\right)^4 + \cdots\right\}$$

$$= -\frac{1}{2u^3} - \frac{1}{4u^2} - \frac{1}{8u} - \frac{1}{16} - \frac{1}{32}u - \cdots$$

$$= -\frac{1}{2(z+2)^3} - \frac{1}{4(z+2)^2} - \frac{1}{8(z+2)} - \frac{1}{16} - \frac{1}{32}(z+2) - \cdots$$

$z = -2$ is a *pole of order 3*.

The series converges for $0 < |z+2| < 2$.

RESIDUES AND THE RESIDUE THEOREM

13.23. If $f(z)$ is analytic everywhere inside and on a simple closed curve C except at $z = a$ which is a pole of order n so that

$$f(z) = \frac{a_{-n}}{(z-a)^n} + \frac{a_{-n+1}}{(z-a)^{n-1}} + \cdots + a_0 + a_1(z-a) + a_2(z-a)^2 + \cdots$$

where $a_{-n} \ne 0$, prove that

(a) $\oint_C f(z)\,dz = 2\pi i a_{-1}$

(b) $a_{-1} = \lim\limits_{z \to a} \dfrac{1}{(n-1)!} \dfrac{d^{n-1}}{dz^{n-1}}\{(z-a)^n f(z)\}$.

(a) By integration, we have on using Problem 13.13

$$\oint_C f(z)\,dz = \oint_C \frac{a_{-n}}{(z-a)^n}\,dz + \cdots + \oint_C \frac{a_{-1}}{z-a}\,dz + \oint_C \{a_0 + a_1(z-a) + a_2(z-a)^2 + \cdots\}\,dz$$

$$= 2\pi i a_{-1}$$

Since only the term involving a_{-1} remains, we call a_{-1} the *residue* of $f(z)$ at the pole $z = a$.

(b) Multiplication by $(z-a)^n$ gives the Taylor series

$$(z-a)^n f(z) = a_{-n} + a_{-n+1}(z-a) + \cdots + a_{-1}(z-a)^{n-1} + \cdots$$

Taking the $(n-1)$st derivative of both sides and letting $z \to a$, we find

$$(n-1)!\,a_{-1} = \lim_{z \to a} \frac{d^{n-1}}{dz^{n-1}}\{(z-a)^n f(z)\}$$

from which the required result follows.

13.24. Determine the residues of each function at the indicated poles.

(a) $\dfrac{z^2}{(z-2)(z^2+1)}$; $z = 2, i, -i$. These are simple poles. Then:

Residue at $z = 2$ is $\quad \lim\limits_{z \to 2}\,(z-2)\left\{\dfrac{z^2}{(z-2)(z^2+1)}\right\} = \dfrac{4}{5}$.

Residue at $z = i$ is $\quad \lim\limits_{z \to i}\,(z-i)\left\{\dfrac{z^2}{(z-2)(z-i)(z+i)}\right\} = \dfrac{i^2}{(i-2)(2i)} = \dfrac{1-2i}{10}$.

Residue at $z = -i$ is $\quad \lim\limits_{z \to -i}\,(z+i)\left\{\dfrac{z^2}{(z-2)(z-i)(z+i)}\right\} = \dfrac{i^2}{(-i-2)(-2i)} = \dfrac{1+2i}{10}$.

(b) $\dfrac{1}{z(z+2)^3}$; $z = 0, -2$. $z = 0$ is a simple pole, $z = -2$ is a pole of order 3.   Then:

Residue at $z = 0$ is $\lim\limits_{z \to 0} z \cdot \dfrac{1}{z(z+2)^3} = \dfrac{1}{8}$.

Residue at $z = -2$ is $\lim\limits_{z \to -2} \dfrac{1}{2!} \dfrac{d^2}{dz^2} \left\{ (z+2)^3 \cdot \dfrac{1}{z(z+2)^3} \right\}$

$$= \lim\limits_{z \to -2} \dfrac{1}{2} \dfrac{d^2}{dz^2} \left(\dfrac{1}{z} \right) = \lim\limits_{z \to -2} \dfrac{1}{2} \left(\dfrac{2}{z^3} \right) = -\dfrac{1}{8}.$$

Note that these residues can also be obtained from the coefficients of $1/z$ and $1/(z+2)$ in the respective Laurent series [see Problem 13.22(e)].

(c) $\dfrac{ze^{zt}}{(z-3)^2}$; $z = 3$, a pole of order 2 or double pole.   Then:

Residue is $\lim\limits_{z \to 3} \dfrac{d}{dz} \left\{ (z-3)^2 \cdot \dfrac{ze^{zt}}{(z-3)^2} \right\} = \lim\limits_{z \to 3} \dfrac{d}{dz} (ze^{zt}) = \lim\limits_{z \to 3} (e^{zt} + zte^{zt})$

$$= e^{3t} + 3te^{3t}$$

(d) $\cot z$; $z = 5\pi$, a pole of order 1.   Then:

Residue is $\lim\limits_{z \to 5\pi} (z - 5\pi) \cdot \dfrac{\cos z}{\sin z} = \left(\lim\limits_{z \to 5\pi} \dfrac{z - 5\pi}{\sin z} \right) \left(\lim\limits_{z \to 5\pi} \cos z \right) = \left(\lim\limits_{z \to 5\pi} \dfrac{1}{\cos z} \right) (-1)$

where we have used L'Hospital's rule, which can be shown applicable for functions of a complex variable.

13.25. If $f(z)$ is analytic within and on a simple closed curve C except at a number of poles $a, b, c, \dots$ interior to C, prove that

$$\oint_C f(z)\, dz = 2\pi i \{\text{sum of residues of } f(z) \text{ at poles } a, b, c, \text{ etc.}\}$$

Refer to Fig. 13-16.

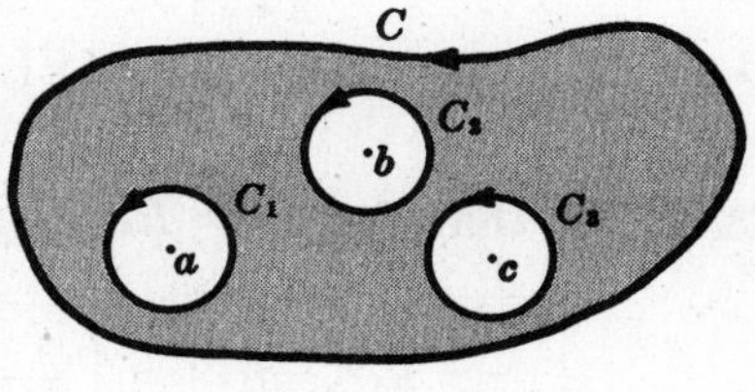

By reasoning similar to that of Problem 13.12 (i.e. by constructing cross cuts from C to C_1, C_2, C_3, etc.), we have

$$\oint_C f(z)\, dz = \oint_{C_1} f(z)\, dz + \oint_{C_2} f(z)\, dz + \cdots$$

For pole a,

$$f(z) = \dfrac{a_{-m}}{(z-a)^m} + \cdots + \dfrac{a_{-1}}{(z-a)} + a_0 + a_1(z-a) + \cdots$$

hence, as in Problem 13.23, $\oint_{C_1} f(z)\, dz = 2\pi i a_{-1}$.

Similarly for pole b, $f(z) = \dfrac{b_{-n}}{(z-b)^n} + \cdots + \dfrac{b_{-1}}{(z-b)} + b_0 + b_1(z-b) + \cdots$

so that

$$\oint_{C_2} f(z)\, dz = 2\pi i b_{-1}$$

Continuing in this manner, we see that

$$\oint_C f(z)\, dz = 2\pi i (a_{-1} + b_{-1} + \cdots) = 2\pi i \,(\text{sum of residues})$$

Fig. 13-16

13.26. Evaluate $\displaystyle\oint_C \frac{e^z\,dz}{(z-1)(z+3)^2}$ where C is given by (a) $|z| = 3/2$, (b) $|z| = 10$.

Residue at simple pole $z = 1$ is $\displaystyle\lim_{z \to 1}\left\{(z-1)\frac{e^z}{(z-1)(z+3)^2}\right\} = \frac{e}{16}$

Residue at double pole $z = -3$ is

$$\lim_{z \to -3}\frac{d}{dz}\left\{(z+3)^2\frac{e^z}{(z-1)(z+3)^2}\right\} = \lim_{z \to -3}\frac{(z-1)e^z - e^z}{(z-1)^2} = \frac{-5e^{-3}}{16}$$

(a) Since $|z| = 3/2$ encloses only the pole $z = 1$,

the required integral $= 2\pi i\left(\dfrac{e}{16}\right) = \dfrac{\pi i e}{8}$

(b) Since $|z| = 10$ encloses both poles $z = 1$ and $z = -3$,

the required integral $= 2\pi i\left(\dfrac{e}{16} - \dfrac{5e^{-3}}{16}\right) = \dfrac{\pi i(e - 5e^{-3})}{8}$

EVALUATION OF DEFINITE INTEGRALS

13.27. If $|f(z)| \leqq \dfrac{M}{R^k}$ for $z = Re^{i\theta}$, where $k > 1$ and M are constants, prove that $\displaystyle\lim_{R \to \infty}\int_\Gamma f(z)\,dz = 0$ where Γ is the semi-circular arc of radius R shown in Fig. 13-17.

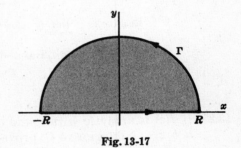

Fig. 13-17

By the result (4), page 287, we have

$$\left|\int_\Gamma f(z)\,dz\right| \leqq \int_\Gamma |f(z)|\,|dz| \leqq \frac{M}{R^k}\cdot\pi R = \frac{\pi M}{R^{k-1}}$$

since the length of arc $L = \pi R$. Then

$$\lim_{R \to \infty}\left|\int_\Gamma f(z)\,dz\right| = 0 \qquad \text{and so} \qquad \lim_{R \to \infty}\int_\Gamma f(z)\,dz = 0$$

13.28. Show that for $z = Re^{i\theta}$, $|f(z)| \leqq \dfrac{M}{R^k}$, $k > 1$ if $f(z) = \dfrac{1}{1 + z^4}$.

If $z = Re^{i\theta}$, $|f(z)| = \left|\dfrac{1}{1 + R^4e^{4i\theta}}\right| \leqq \dfrac{1}{|R^4e^{4i\theta}| - 1} = \dfrac{1}{R^4 - 1} \leqq \dfrac{2}{R^4}$ if R is large enough (say $R > 2$, for example) so that $M = 2$, $k = 4$.

Note that we have made use of the inequality $|z_1 + z_2| \geqq |z_1| - |z_2|$ with $z_1 = R^4e^{4i\theta}$ and $z_2 = 1$.

13.29. Evaluate $\displaystyle\int_0^\infty \frac{dx}{x^4 + 1}$.

Consider $\displaystyle\oint_C \frac{dz}{z^4 + 1}$, where C is the closed contour of Problem 13.27 consisting of the line from $-R$ to R and the semi-circle Γ, traversed in the positive (counterclockwise) sense.

Since $z^4 + 1 = 0$ when $z = e^{\pi i/4}, e^{3\pi i/4}, e^{5\pi i/4}, e^{7\pi i/4}$, these are simple poles of $1/(z^4 + 1)$. Only the poles $e^{\pi i/4}$ and $e^{3\pi i/4}$ lie within C. Then using L'Hospital's rule,

$$\text{Residue at } e^{\pi i/4} \;=\; \lim_{z \to e^{\pi i/4}} \left\{ (z - e^{\pi i/4}) \frac{1}{z^4 + 1} \right\}$$

$$=\; \lim_{z \to e^{\pi i/4}} \frac{1}{4z^3} \;=\; \frac{1}{4} e^{-3\pi i/4}$$

$$\text{Residue at } e^{3\pi i/4} \;=\; \lim_{z \to e^{3\pi i/4}} \left\{ (z - e^{3\pi i/4}) \frac{1}{z^4 + 1} \right\}$$

$$=\; \lim_{z \to e^{3\pi i/4}} \frac{1}{4z^3} \;=\; \frac{1}{4} e^{-9\pi i/4}$$

Thus
$$\oint_C \frac{dz}{z^4 + 1} \;=\; 2\pi i \{ \tfrac{1}{4} e^{-3\pi i/4} + \tfrac{1}{4} e^{-9\pi i/4} \} \;=\; \frac{\pi \sqrt{2}}{2} \tag{1}$$

i.e.
$$\int_{-R}^{R} \frac{dx}{x^4 + 1} + \int_{\Gamma} \frac{dz}{z^4 + 1} \;=\; \frac{\pi \sqrt{2}}{2} \tag{2}$$

Taking the limit of both sides of (2) as $R \to \infty$ and using the results of Problem 13.28, we have

$$\lim_{R \to \infty} \int_{-R}^{R} \frac{dx}{x^4 + 1} \;=\; \int_{-\infty}^{\infty} \frac{dx}{x^4 + 1} \;=\; \frac{\pi \sqrt{2}}{2}$$

Since $\displaystyle \int_{-\infty}^{\infty} \frac{dx}{x^4 + 1} = 2 \int_0^{\infty} \frac{dx}{x^4 + 1}$, the required integral has the value $\dfrac{\pi \sqrt{2}}{4}$.

13.30. Show that $\displaystyle \int_{-\infty}^{\infty} \frac{x^2 \, dx}{(x^2 + 1)^2 (x^2 + 2x + 2)} = \frac{7\pi}{50}$.

The poles of $\dfrac{z^2}{(z^2 + 1)^2 (z^2 + 2z + 2)}$ enclosed by the contour C of Problem 13.27 are $z = i$ of order 2 and $z = -1 + i$ of order 1.

Residue at $z = i$ is $\displaystyle \lim_{z \to i} \frac{d}{dz} \left\{ (z - i)^2 \frac{z^2}{(z + i)^2 (z - i)^2 (z^2 + 2z + 2)} \right\} = \frac{9i - 12}{100}$.

Residue at $z = -1 + i$ is $\displaystyle \lim_{z \to -1 + i} (z + 1 - i) \frac{z^2}{(z^2 + 1)^2 (z + 1 - i)(z + 1 + i)} = \frac{3 - 4i}{25}$.

Then
$$\oint_C \frac{z^2 \, dz}{(z^2 + 1)^2 (z^2 + 2z + 2)} \;=\; 2\pi i \left\{ \frac{9i - 12}{100} + \frac{3 - 4i}{25} \right\} \;=\; \frac{7\pi}{50}$$

or
$$\int_{-R}^{R} \frac{x^2 \, dx}{(x^2 + 1)^2 (x^2 + 2x + 2)} + \int_{\Gamma} \frac{z^2 \, dz}{(z^2 + 1)^2 (z^2 + 2z + 2)} \;=\; \frac{7\pi}{50}$$

Taking the limit as $R \to \infty$ and noting that the second integral approaches zero by Problem 13.27, we obtain the required result.

13.31. Evaluate $\displaystyle \int_0^{2\pi} \frac{d\theta}{5 + 3 \sin \theta}$.

Let $z = e^{i\theta}$. Then $\sin \theta = \dfrac{e^{i\theta} - e^{-i\theta}}{2i} = \dfrac{z - z^{-1}}{2i}$, $dz = i e^{i\theta} \, d\theta = iz \, d\theta$ so that

$$\int_0^{2\pi} \frac{d\theta}{5 + 3 \sin \theta} \;=\; \oint_C \frac{dz/iz}{5 + 3 \left(\dfrac{z - z^{-1}}{2i} \right)} \;=\; \oint_C \frac{2 \, dz}{3z^2 + 10iz - 3}$$

where C is the circle of unit radius with center at the origin, as shown in Fig. 13-18 below.

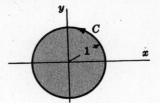

Fig. 13-18

The poles of $\dfrac{2}{3z^2 + 10iz - 3}$ are the simple poles

$$z = \frac{-10i \pm \sqrt{-100 + 36}}{6}$$

$$= \frac{-10i \pm 8i}{6}$$

$$= -3i,\ -i/3$$

Only $-i/3$ lies inside C.

Residue at $-i/3 = \lim\limits_{z \to -i/3} \left(z + \dfrac{i}{3} \right)\left(\dfrac{2}{3z^2 + 10iz - 3} \right) = \lim\limits_{z \to -i/3} \dfrac{2}{6z + 10i} = \dfrac{1}{4i}$ by L'Hospital's rule.

Then $\displaystyle\oint_C \dfrac{2\,dz}{3z^2 + 10iz - 3} = 2\pi i\left(\dfrac{1}{4i}\right) = \dfrac{\pi}{2}$, the required value.

13.32. Show that $\displaystyle\int_0^{2\pi} \dfrac{\cos 3\theta}{5 - 4\cos\theta}\,d\theta = \dfrac{\pi}{12}$.

If $z = e^{i\theta}$, $\cos\theta = \dfrac{z + z^{-1}}{2}$, $\cos 3\theta = \dfrac{e^{3i\theta} + e^{-3i\theta}}{2} = \dfrac{z^3 + z^{-3}}{2}$, $dz = iz\,d\theta$.

Then $\displaystyle\int_0^{2\pi} \dfrac{\cos 3\theta}{5 - 4\cos\theta}\,d\theta = \oint_C \dfrac{(z^3 + z^{-3})/2}{5 - 4\left(\dfrac{z + z^{-1}}{2}\right)} \dfrac{dz}{iz}$

$$= -\frac{1}{2i}\oint_C \frac{z^6 + 1}{z^3(2z - 1)(z - 2)}\,dz$$

where C is the contour of Problem 13.31.

The integrand has a pole of order 3 at $z = 0$ and a simple pole $z = \frac{1}{2}$ within C.

Residue at $z = 0$ is $\lim\limits_{z \to 0} \dfrac{1}{2!}\dfrac{d^2}{dz^2}\left\{ z^3 \cdot \dfrac{z^6 + 1}{z^3(2z - 1)(z - 2)} \right\} = \dfrac{21}{8}$.

Residue at $z = \frac{1}{2}$ is $\lim\limits_{z \to 1/2}\left\{ (z - \tfrac{1}{2}) \cdot \dfrac{z^6 + 1}{z^3(2z - 1)(z - 2)} \right\} = -\dfrac{65}{24}$.

Then $-\dfrac{1}{2i}\displaystyle\oint_C \dfrac{z^6 + 1}{z^3(2z - 1)(z - 2)}\,dz = -\dfrac{1}{2i}(2\pi i)\left\{ \dfrac{21}{8} - \dfrac{65}{24} \right\} = \dfrac{\pi}{12}$ as required.

13.33. If $|f(z)| \leqq \dfrac{M}{R^k}$ for $z = Re^{i\theta}$, where $k > 0$ and M are constants, prove that

$$\lim_{R \to \infty} \int_\Gamma e^{imz} f(z)\,dz = 0$$

where Γ is the semi-circular arc of the contour in Problem 13.27 and m is a positive constant.

If $z = Re^{i\theta}$, $\displaystyle\int_\Gamma e^{imz} f(z)\,dz = \int_0^\pi e^{imRe^{i\theta}} f(Re^{i\theta})\,iRe^{i\theta}\,d\theta$.

Then $\left| \displaystyle\int_0^\pi e^{imRe^{i\theta}} f(Re^{i\theta})\,iRe^{i\theta}\,d\theta \right| \leqq \displaystyle\int_0^\pi |e^{imRe^{i\theta}} f(Re^{i\theta})\,iRe^{i\theta}|\,d\theta$

$$= \int_0^\pi |e^{imR\cos\theta\,-\,mR\sin\theta}\,f(Re^{i\theta})\,iRe^{i\theta}|\,d\theta$$

$$= \int_0^\pi e^{-mR\sin\theta}\,|f(Re^{i\theta})|\,R\,d\theta$$

$$\leqq \frac{M}{R^{k-1}}\int_0^\pi e^{-mR\sin\theta}\,d\theta = \frac{2M}{R^{k-1}}\int_0^{\pi/2} e^{-mR\sin\theta}\,d\theta$$

Now $\sin \theta \geqq 2\theta/\pi$ for $0 \leqq \theta \leqq \pi/2$ [see Problem 14.3, page 327]. Then the last integral is less than or equal to

$$\frac{2M}{R^{k-1}} \int_0^{\pi/2} e^{-2mR\theta/\pi} d\theta = \frac{\pi M}{mR^k}(1 - e^{-mR})$$

As $R \to \infty$ this approaches zero, since m and k are positive, and the required result is proved.

13.34. Show that $\displaystyle \int_0^\infty \frac{\cos mx}{x^2 + 1} dx = \frac{\pi}{2} e^{-m}, \ m > 0.$

Consider $\displaystyle \oint_C \frac{e^{imz}}{z^2 + 1} dz$ where C is the contour of Problem 13.27.

The integrand has simple poles at $z = \pm i$, but only $z = i$ lies within C.

Residue at $z = i$ is $\displaystyle \lim_{z \to i} \left\{ (z - i) \frac{e^{imz}}{(z-i)(z+i)} \right\} = \frac{e^{-m}}{2i}.$

Then

$$\oint_C \frac{e^{imz}}{z^2 + 1} dz = 2\pi i \left(\frac{e^{-m}}{2i} \right) = \pi e^{-m}$$

or

$$\int_{-R}^R \frac{e^{imx}}{x^2 + 1} dx + \int_\Gamma \frac{e^{imz}}{z^2 + 1} dz = \pi e^{-m}$$

i.e.

$$\int_{-R}^R \frac{\cos mx}{x^2 + 1} dx + i \int_{-R}^R \frac{\sin mx}{x^2 + 1} dx + \int_\Gamma \frac{e^{imz}}{z^2 + 1} dz = \pi e^{-m}$$

and so

$$2 \int_0^R \frac{\cos mx}{x^2 + 1} dx + \int_\Gamma \frac{e^{imz}}{z^2 + 1} dz = \pi e^{-m}$$

Taking the limit as $R \to \infty$ and using Problem 13.33 to show that the integral around Γ approaches zero, we obtain the required result.

13.35. Show that $\displaystyle \int_0^\infty \frac{\sin x}{x} dx = \frac{\pi}{2}.$

The method of Problem 13.34 leads us to consider the integral of e^{iz}/z around the contour of Problem 13.27. However, since $z = 0$ lies on this path of integration and since we cannot integrate through a singularity, we modify that contour by indenting the path at $z = 0$, as shown in Fig. 13-19, which we call contour C' or $ABDEFGHJA$.

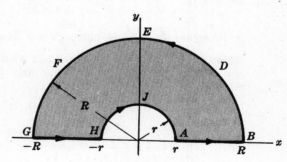

Fig. 13-19

Since $z = 0$ is outside C', we have

$$\int_{C'} \frac{e^{iz}}{z} dz = 0$$

or

$$\int_{-R}^{-r} \frac{e^{ix}}{x} dx + \int_{HJA} \frac{e^{iz}}{z} dz + \int_r^R \frac{e^{ix}}{x} dx + \int_{BDEFG} \frac{e^{iz}}{z} dz = 0$$

Replacing x by $-x$ in the first integral and combining with the third integral, we find,

$$\int_r^R \frac{e^{ix} - e^{-ix}}{x} dx + \int_{HJA} \frac{e^{iz}}{z} dz + \int_{BDEFG} \frac{e^{iz}}{z} dz = 0$$

or

$$2i \int_r^R \frac{\sin x}{x} dx = - \int_{HJA} \frac{e^{iz}}{z} dz - \int_{BDEFG} \frac{e^{iz}}{z} dz$$

13.36. Let $w = \sqrt{z}$ define a transformation from the z plane to the w plane. A point moves counterclockwise along the circle $|z| = 1$. Show that when it has returned to its starting position for the first time its image point has not yet returned, but that when it has returned for the second time its image point returns for the first time.

Let $z = e^{i\theta}$. Then $w = \sqrt{z} = e^{i\theta/2}$. Let $\theta = 0$ correspond to the starting position. Then $z = 1$ and $w = 1$ [corresponding to A and P in Figs. 13-20 and 13-21].

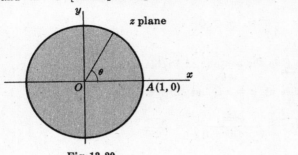
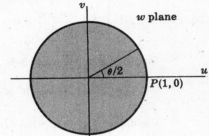

Fig. 13-20   Fig. 13-21

When one complete revolution in the z plane has been made, $\theta = 2\pi$, $z = 1$ but $w = e^{i\theta/2} = e^{i\pi} = -1$ so the image point has not yet returned to its starting position.

However, after two complete revolutions in the z plane have been made, $\theta = 4\pi$, $z = 1$ and $w = e^{i\theta/2} = e^{2\pi i} = 1$ so the image point has returned for the first time.

It follows from the above that w is not a single-valued function of z but is a *double-valued function* of z; i.e. given z, there are two values of w. If we wish to consider it a single-valued function, we must restrict θ. We can, for example, choose $0 \leqq \theta < 2\pi$, although other possibilities exist. This represents one branch of the double-valued function $w = \sqrt{z}$. In continuing beyond this interval we are on the second branch, e.g. $2\pi \leqq \theta < 4\pi$. The point $z = 0$ about which the rotation is taking place is called a *branch point*. Equivalently, we can insure that $f(z) = \sqrt{z}$ will be single-valued by agreeing not to cross the line Ox, called a *branch line*.

13.37. Show that $\displaystyle \int_0^\infty \frac{x^{p-1}}{1+x}\, dx = \frac{\pi}{\sin p\pi}$, $\quad 0 < p < 1$.

Consider $\displaystyle \oint_C \frac{z^{p-1}}{1+z}\, dz$. Since $z = 0$ is a branch point, choose C as the contour of Fig. 13-22 where AB and GH are actually coincident with the x axis but are shown separated for visual purposes.

The integrand has the pole $z = -1$ lying within C.

Residue at $z = -1 = e^{\pi i}$ is

$$\lim_{z \to -1} (z+1)\frac{z^{p-1}}{1+z} = (e^{\pi i})^{p-1} = e^{(p-1)\pi i}$$

Then $\displaystyle \oint_C \frac{z^{p-1}}{1+z}\, dz = 2\pi i e^{(p-1)\pi i}$

or, omitting the integrand,

$$\int_{AB} + \int_{BDEFG} + \int_{GH} + \int_{HJA} = 2\pi i e^{(p-1)\pi i}$$

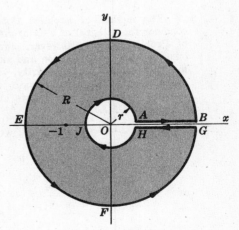

Fig. 13-22

We thus have

$$\int_r^R \frac{x^{p-1}}{1+x}\, dx + \int_0^{2\pi} \frac{(Re^{i\theta})^{p-1} iRe^{i\theta}\, d\theta}{1 + Re^{i\theta}} + \int_R^r \frac{(xe^{2\pi i})^{p-1}}{1 + xe^{2\pi i}}\, dx + \int_{2\pi}^0 \frac{(re^{i\theta})^{p-1} ire^{i\theta}\, d\theta}{1 + re^{i\theta}} = 2\pi i e^{(p-1)\pi i}$$

where we have to use $z = xe^{2\pi i}$ for the integral along GH, since the argument of z is increased by 2π in going around the circle $BDEFG$.

Taking the limit as $r \to 0$ and $R \to \infty$ and noting that the second and fourth integrals approach zero, we find

$$\int_0^\infty \frac{x^{p-1}}{1+x}\,dx + \int_\infty^0 \frac{e^{2\pi i(p-1)}x^{p-1}}{1+x}\,dx = 2\pi e^{(p-1)\pi i}$$

or

$$(1 - e^{2\pi i(p-1)}) \int_0^\infty \frac{x^{p-1}}{1+x}\,dx = 2\pi i e^{(p-1)\pi i}$$

so that

$$\int_0^\infty \frac{x^{p-1}}{1+x}\,dx = \frac{2\pi i e^{(p-1)\pi i}}{1 - e^{2\pi i(p-1)}} = \frac{2\pi i}{e^{p\pi i} - e^{-p\pi i}} = \frac{\pi}{\sin p\pi}$$

CONFORMAL MAPPING

13.38. Consider the transformation $w = f(z)$ where $f(z)$ is analytic at z_0 and $f'(z_0) \neq 0$. Prove that under this transformation the tangent at z_0 to any curve C in the z plane passing through z_0 [Fig. 13-23] is rotated through the angle $\arg f'(z_0)$.

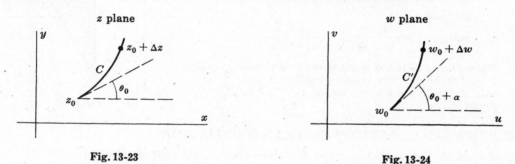

Fig. 13-23   Fig. 13-24

As a point moves from z_0 to $z_0 + \Delta z$ along C [Fig. 13-23] the image point moves along C' in the w plane [Fig. 13-24] from w_0 to $w_0 + \Delta w$. If the parameter used to describe the curve is t, then corresponding to the path $z = z(t)$ [or $x = x(t)$, $y = y(t)$] in the z plane, we have the path $w = w(t)$ [or $u = u(t)$, $v = v(t)$] in the w plane.

The derivatives dz/dt and dw/dt represent tangent vectors to corresponding points on C and C'.

Now $\dfrac{dw}{dt} = \dfrac{dw}{dz} \cdot \dfrac{dz}{dt} = f'(z) \dfrac{dz}{dt}$ and, in particular at z_0 and w_0,

$$\frac{dw}{dt}\Big|_{w=w_0} = f'(z_0) \frac{dz}{dt}\Big|_{z=z_0} \tag{1}$$

provided $f(z)$ is analytic at $z = z_0$. Writing $\dfrac{dw}{dt}\Big|_{w=w_0} = \rho_0 e^{i\phi_0}$, $f'(z) = Re^{i\alpha}$, $\dfrac{dz}{dt}\Big|_{z=z_0} = r_0 e^{i\theta_0}$, we have from (1)

$$\rho_0 e^{i\phi_0} = R r_0 e^{i(\theta_0 + \alpha)} \tag{2}$$

so that, as required, $$\phi_0 = \theta_0 + \alpha = \theta_0 + \arg f'(z_0) \tag{3}$$

Note that if $f'(z_0) = 0$, then α is indeterminate. Points where $f'(z) = 0$ are called *critical points*.

13.39. Prove that the angle between two curves C_1 and C_2 passing through the point z_0 in the z plane [see Figs. 13-1 and 13-2, page 291] is preserved [in magnitude and sense] under the transformation $w = f(z)$, i.e. the mapping is conformal, if $f(z)$ is analytic at z_0 and $f'(z_0) \neq 0$.

By Problem 13.38 each curve is rotated through the angle $\arg f'(z_0)$. Hence the angle between the curves must be preserved, both in magnitude and sense, in the mapping.

13.40. If $w = f(z) = u + iv$ is analytic in a region $\mathcal{R}$, prove that $\dfrac{\partial(u, v)}{\partial(x, y)} = |f'(z)|^2$.

If $f(z)$ is analytic in $\mathcal{R}$, then the Cauchy-Riemann equations

$$\frac{\partial u}{\partial x} = \frac{\partial v}{\partial y}, \quad \frac{\partial v}{\partial x} = -\frac{\partial u}{\partial y}$$

are satisfied in $\mathcal{R}$.   Hence

$$\frac{\partial(u, v)}{\partial(x, y)} = \begin{vmatrix} \dfrac{\partial u}{\partial x} & \dfrac{\partial u}{\partial y} \\ \dfrac{\partial v}{\partial x} & \dfrac{\partial v}{\partial y} \end{vmatrix} = \begin{vmatrix} \dfrac{\partial u}{\partial x} & \dfrac{\partial u}{\partial y} \\ -\dfrac{\partial u}{\partial y} & \dfrac{\partial u}{\partial x} \end{vmatrix} = \left(\frac{\partial u}{\partial x}\right)^2 + \left(\frac{\partial u}{\partial y}\right)^2 = \left| \frac{\partial u}{\partial x} + i\frac{\partial u}{\partial y} \right|^2 = |f'(z)|^2$$

13.41. If z_0 is in the upper half of the z plane, show that the bilinear transformation $w = e^{i\theta_0}\left(\dfrac{z - z_0}{z - \bar{z}_0}\right)$ maps the upper half of the z plane into the interior of the unit circle in the w plane, i.e. $|w| \leqq 1$.

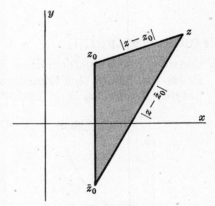

Fig. 13-25

We have

$$|w| = \left| e^{i\theta_0}\left(\frac{z - z_0}{z - \bar{z}_0}\right) \right| = \left| \frac{z - z_0}{z - \bar{z}_0} \right|$$

From Fig. 13-25 if z is in the upper half plane, $|z - z_0| \leqq |z - \bar{z}_0|$, the equality holding if and only if z is on the x axis. Hence $|w| \leqq 1$, as required.

THE SCHWARZ-CHRISTOFFEL TRANSFORMATION

13.42. Establish the validity of the Schwarz-Christoffel transformation.

We must show that the mapping function obtained from

$$\frac{dw}{dz} = A(z - x_1)^{\alpha_1/\pi - 1}(z - x_2)^{\alpha_2/\pi - 1} \cdots (z - x_n)^{\alpha_n/\pi - 1} \tag{1}$$

maps a given polygon of the w plane [Fig. 13-26] into the real axis of the z plane [Fig. 13-27].

To show this observe that from (1) we have

$$\arg dw = \arg dz + \arg A + \left(\frac{\alpha_1}{\pi} - 1\right)\arg(z - x_1) + \left(\frac{\alpha_2}{\pi} - 1\right)\arg(z - x_2)$$

$$+ \cdots + \left(\frac{\alpha_n}{\pi} - 1\right)\arg(z - x_n) \tag{2}$$

As z moves along the real axis from the left toward x_1, let us assume that w moves along a side of the polygon toward w_1. When z crosses from the left of x_1 to the right of x_1, $\theta_1 = \arg(z - x_1)$ changes from π to 0 while all other terms in (2) stay constant. Hence $\arg dw$ decreases by $(\alpha_1/\pi - 1)\arg(z - x_1) = (\alpha_1/\pi - 1)\pi = \alpha_1 - \pi$ or, what is the same thing, increases by $\pi - \alpha_1$ [an increase being in the counterclockwise direction].

w plane

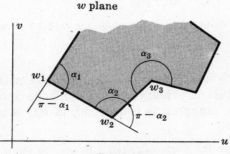

Fig. 13-26

z plane

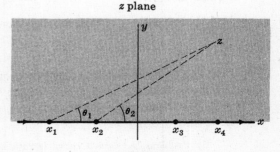

Fig. 13-27

It follows from this that the direction through w_1 turns through the angle $\pi - \alpha_1$, and thus w now moves along the side w_1w_2 of the polygon.

When z moves through x_2, $\theta_1 = \arg(z - x_1)$ and $\theta_2 = \arg(z - x_2)$ change from π to 0 while all other terms stay constant. Hence another turn through angle $\pi - \alpha_2$ in the w plane is made. By continuing the process we see that as z traverses the x axis, w traverses the polygon, and conversely.

We can prove that the interior of the polygon (if it is closed) is mapped on to the upper half plane by (1).

13.43. Prove that for closed polygons the sum of the exponents $\dfrac{\alpha_1}{\pi} - 1$, $\dfrac{\alpha_2}{\pi} - 1$, $\ldots$, $\dfrac{\alpha_n}{\pi} - 1$ in the Schwarz-Christoffel transformation (16) or (17), page 293, is equal to -2.

The sum of the exterior angles of any closed polygon is 2π. Then

$$(\pi - \alpha_1) + (\pi - \alpha_2) + \cdots + (\pi - \alpha_n) = 2\pi$$

and dividing by $-\pi$, we obtain as required,

$$\left(\frac{\alpha_1}{\pi} - 1\right) + \left(\frac{\alpha_2}{\pi} - 1\right) + \cdots + \left(\frac{\alpha_n}{\pi} - 1\right) = -2$$

13.44. If in the Schwarz-Christoffel transformation (16) or (17), page 293, one point, say x_n, is chosen at infinity, show that the last factor is not present.

In (16), page 293, let $A = K/(-x_n)^{\alpha_n/\pi - 1}$ where K is a constant. Then the right side of (16) can be written

$$K(z - x_1)^{\alpha_1/\pi - 1}(z - x_2)^{\alpha_2/\pi - 1} \cdots (z - x_{n-1})^{\alpha_{n-1}/\pi - 1}\left(\frac{x_n - z}{x_n}\right)^{\alpha_n/\pi - 1}$$

As $x_n \to \infty$, this last factor approaches 1; this is equivalent to removal of the factor.

13.45. Determine a function which maps the region of Fig. 13-28 in the w plane on to the upper half of the z plane of Fig. 13-29.

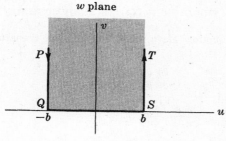

Fig. 13-28

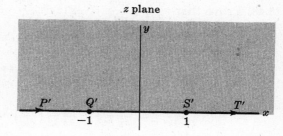

Fig. 13-29

Let points P, Q, S and T [Fig. 13-28] map respectively into P', Q', S' and T' [Fig. 13-29]. We can consider $PQST$ as a limiting case of a polygon (a triangle) with two vertices at Q and S and the third vertex P or T at infinity.

By the Schwarz-Christoffel transformation, since the angles at Q and S are equal to $\pi/2$, we have

$$\frac{dw}{dz} = A(z+1)^{\frac{\pi/2}{\pi} - 1}(z-1)^{\frac{\pi/2}{\pi} - 1} = \frac{A}{\sqrt{z^2 - 1}} = \frac{K}{\sqrt{1 - z^2}}$$

Integrating, $$w = K\int \frac{dz}{\sqrt{1 - z^2}} + B = K\sin^{-1}z + B$$

When $z = 1$, $w = b$. Hence   (1) $b = K\sin^{-1}(1) + B = K\pi/2 + B$.

When $z = -1$, $w = -b$. Hence,   (2) $-b = K\sin^{-1}(-1) + B = -K\pi/2 + B$.

Solving (1) and (2) simultaneously, we find $B = 0$, $K = 2b/\pi$. Then

$$w = \frac{2b}{\pi}\sin^{-1}z \qquad \text{or} \qquad z = \sin\frac{\pi w}{2b}$$

SOLUTIONS OF LAPLACE'S EQUATION BY CONFORMAL MAPPING

13.46. Show that the functions (a) $x^2 - y^2 + 2y$ and (b) $\sin x \cosh y$ are harmonic in any finite region $\mathcal{R}$ of the z plane.

(a) If $\Phi = x^2 - y^2 + 2y$, we have $\dfrac{\partial^2 \Phi}{\partial x^2} = 2$, $\dfrac{\partial^2 \Phi}{\partial y^2} = -2$. Then $\dfrac{\partial^2 \Phi}{\partial x^2} + \dfrac{\partial^2 \Phi}{\partial y^2} = 0$ and Φ is harmonic in $\mathcal{R}$.

(b) If $\Phi = \sin x \cosh y$, we have $\dfrac{\partial^2 \Phi}{\partial x^2} = -\sin x \cosh y$, $\dfrac{\partial^2 \Phi}{\partial y^2} = \sin x \cosh y$. Then $\dfrac{\partial^2 \Phi}{\partial x^2} + \dfrac{\partial^2 \Phi}{\partial y^2} = 0$ and Φ is harmonic in $\mathcal{R}$.

13.47. Show that the functions of Problem 13.46 are harmonic in the w plane under the transformation $z = w^3$.

If $z = w^3$, then $x + iy = (u + iv)^3 = u^3 - 3uv^2 + i(3u^2v - v^3)$ and $x = u^3 - 3uv^2$, $y = 3u^2v - v^3$.

(a) $\Phi = x^2 - y^2 + 2y = (u^3 - 3uv^2)^2 - (3u^2v - v^3)^2 + 2(3u^2v - v^3)$

$\qquad = u^6 - 15u^4v^2 + 15u^2v^4 - v^6 + 6u^2v - 2v^3$

Then $\dfrac{\partial^2 \Phi}{\partial u^2} = 30u^4 - 180u^2v^2 + 30v^4 + 12v$, $\dfrac{\partial^2 \Phi}{\partial v^2} = -30u^4 + 180u^2v^2 - 30v^4 - 12v$

and $\dfrac{\partial^2 \Phi}{\partial u^2} + \dfrac{\partial^2 \Phi}{\partial v^2} = 0$ as required.

(b) We must show that $\Phi = \sin(u^3 - 3uv^2) \cosh(3u^2v - v^3)$ satisfies $\dfrac{\partial^2 \Phi}{\partial u^2} + \dfrac{\partial^2 \Phi}{\partial v^2} = 0$. This can readily be established by straightforward but tedious differentiation.

This problem illustrates a general result proved in Problem 13.49.

13.48. Prove that $\dfrac{\partial^2 \Phi}{\partial x^2} + \dfrac{\partial^2 \Phi}{\partial y^2} = |f'(z)|^2 \left(\dfrac{\partial^2 \Phi}{\partial u^2} + \dfrac{\partial^2 \Phi}{\partial v^2} \right)$ where $w = f(z)$ is analytic and $f'(z) \neq 0$.

The function $\Phi(x, y)$ is transformed into a function $\Phi[x(u, v), y(u, v)]$ by the transformation. By differentiation we have

$$\frac{\partial \Phi}{\partial x} = \frac{\partial \Phi}{\partial u} \frac{\partial u}{\partial x} + \frac{\partial \Phi}{\partial v} \frac{\partial v}{\partial x}, \qquad \frac{\partial \Phi}{\partial y} = \frac{\partial \Phi}{\partial u} \frac{\partial u}{\partial y} + \frac{\partial \Phi}{\partial v} \frac{\partial v}{\partial y}$$

$$\frac{\partial^2 \Phi}{\partial x^2} = \frac{\partial \Phi}{\partial u} \frac{\partial^2 u}{\partial x^2} + \frac{\partial u}{\partial x} \frac{\partial}{\partial x} \left(\frac{\partial \Phi}{\partial u} \right) + \frac{\partial \Phi}{\partial v} \frac{\partial^2 v}{\partial x^2} + \frac{\partial v}{\partial x} \frac{\partial}{\partial x} \left(\frac{\partial \Phi}{\partial v} \right)$$

$$= \frac{\partial \Phi}{\partial u} \frac{\partial^2 u}{\partial x^2} + \frac{\partial u}{\partial x} \left[\frac{\partial}{\partial u} \left(\frac{\partial \Phi}{\partial u} \right) \frac{\partial u}{\partial x} + \frac{\partial}{\partial v} \left(\frac{\partial \Phi}{\partial u} \right) \frac{\partial v}{\partial x} \right]$$

$$\qquad + \frac{\partial \Phi}{\partial v} \frac{\partial^2 v}{\partial x^2} + \frac{\partial v}{\partial x} \left[\frac{\partial}{\partial u} \left(\frac{\partial \Phi}{\partial v} \right) \frac{\partial u}{\partial x} + \frac{\partial}{\partial v} \left(\frac{\partial \Phi}{\partial v} \right) \frac{\partial v}{\partial x} \right]$$

$$= \frac{\partial \Phi}{\partial u} \frac{\partial^2 u}{\partial x^2} + \frac{\partial u}{\partial x} \left[\frac{\partial^2 \Phi}{\partial u^2} \frac{\partial u}{\partial x} + \frac{\partial^2 \Phi}{\partial v \, \partial u} \frac{\partial v}{\partial x} \right] + \frac{\partial \Phi}{\partial v} \frac{\partial^2 v}{\partial x^2} + \frac{\partial v}{\partial x} \left[\frac{\partial^2 \Phi}{\partial u \, \partial v} \frac{\partial u}{\partial x} + \frac{\partial^2 \Phi}{\partial v^2} \frac{\partial v}{\partial x} \right]$$

Similarly,

$$\frac{\partial^2 \Phi}{\partial y^2} = \frac{\partial \Phi}{\partial u} \frac{\partial^2 u}{\partial y^2} + \frac{\partial u}{\partial y} \left[\frac{\partial^2 \Phi}{\partial u^2} \frac{\partial u}{\partial y} + \frac{\partial^2 \Phi}{\partial v \, \partial u} \frac{\partial v}{\partial y} \right] + \frac{\partial \Phi}{\partial v} \frac{\partial^2 v}{\partial y^2} + \frac{\partial v}{\partial y} \left[\frac{\partial^2 \Phi}{\partial u \, \partial v} \frac{\partial u}{\partial y} + \frac{\partial^2 \Phi}{\partial v^2} \frac{\partial v}{\partial y} \right]$$

Adding,

$$\frac{\partial^2 \Phi}{\partial x^2} + \frac{\partial^2 \Phi}{\partial y^2} = \frac{\partial \Phi}{\partial u} \left(\frac{\partial^2 u}{\partial x^2} + \frac{\partial^2 u}{\partial y^2} \right) + \frac{\partial \Phi}{\partial v} \left(\frac{\partial^2 v}{\partial x^2} + \frac{\partial^2 v}{\partial y^2} \right) + \frac{\partial^2 \Phi}{\partial u^2} \left[\left(\frac{\partial u}{\partial x} \right)^2 + \left(\frac{\partial u}{\partial y} \right)^2 \right]$$

$$\qquad + 2 \frac{\partial^2 \Phi}{\partial u \, \partial v} \left[\frac{\partial u}{\partial x} \frac{\partial v}{\partial x} + \frac{\partial u}{\partial y} \frac{\partial v}{\partial y} \right] + \frac{\partial^2 \Phi}{\partial v^2} \left[\left(\frac{\partial v}{\partial x} \right)^2 + \left(\frac{\partial v}{\partial y} \right)^2 \right]$$

(1)

Since u and v are harmonic, $\dfrac{\partial^2 u}{\partial x^2} + \dfrac{\partial^2 u}{\partial y^2} = 0$, $\dfrac{\partial^2 v}{\partial x^2} + \dfrac{\partial^2 v}{\partial y^2} = 0$. Also, by the Cauchy-Riemann equations, $\dfrac{\partial u}{\partial x} = \dfrac{\partial v}{\partial y}$, $\dfrac{\partial v}{\partial x} = -\dfrac{\partial u}{\partial y}$. Then

$$\left(\frac{\partial u}{\partial x}\right)^2 + \left(\frac{\partial u}{\partial y}\right)^2 \;=\; \left(\frac{\partial v}{\partial x}\right)^2 + \left(\frac{\partial v}{\partial y}\right)^2 \;=\; \left(\frac{\partial u}{\partial x}\right)^2 + \left(\frac{\partial v}{\partial x}\right)^2 \;=\; \left|\frac{\partial u}{\partial x} + i\frac{\partial v}{\partial x}\right|^2 \;=\; |f'(z)|^2$$

$$\frac{\partial u}{\partial x}\frac{\partial v}{\partial x} + \frac{\partial u}{\partial y}\frac{\partial v}{\partial y} \;=\; 0$$

Hence (1) becomes
$$\frac{\partial^2 \Phi}{\partial x^2} + \frac{\partial^2 \Phi}{\partial y^2} \;=\; |f'(z)|^2\left(\frac{\partial^2 \Phi}{\partial u^2} + \frac{\partial^2 \Phi}{\partial v^2}\right)$$

13.49. Prove that a harmonic function $\Phi(x, y)$ remains harmonic under the transformation $w = f(z)$ where $f(z)$ is analytic and $f'(z) \neq 0$.

This follows at once from Problem 13.48, since if $\dfrac{\partial^2 \Phi}{\partial x^2} + \dfrac{\partial^2 \Phi}{\partial y^2} = 0$ and $f'(z) \neq 0$, then

$$\frac{\partial^2 \Phi}{\partial u^2} + \frac{\partial^2 \Phi}{\partial v^2} \;=\; 0$$

13.50. If a is real, show that the real and imaginary parts of $w = \ln(z - a)$ are harmonic functions in any region $\mathcal{R}$ not containing $z = a$.

Method 1. If $\mathcal{R}$ does not contain a, then $w = \ln(z - a)$ is analytic in $\mathcal{R}$. Hence the real and imaginary parts are harmonic in $\mathcal{R}$.

Method 2. Let $z - a = re^{i\theta}$. Then if principal values are used for θ, $w = u + iv = \ln(z - a) = \ln r + i\theta$ so that $u = \ln r$, $v = \theta$.

In the polar coordinates (r, θ), Laplace's equation is $\dfrac{\partial^2 \Phi}{\partial r^2} + \dfrac{1}{r}\dfrac{\partial \Phi}{\partial r} + \dfrac{1}{r^2}\dfrac{\partial^2 \Phi}{\partial \theta^2} = 0$ and by direct substitution we find that $u = \ln r$ and $v = \theta$ are solutions if $\mathcal{R}$ does not contain $r = 0$, i.e. $z = a$.

Method 3. If $z - a = re^{i\theta}$, then $x - a = r\cos\theta$, $y = r\sin\theta$ and
$$r \;=\; \sqrt{(x-a)^2 + y^2}, \qquad \theta \;=\; \tan^{-1}\{y/(x-a)\}$$
Then $w = u + iv = \tfrac{1}{2}\ln\{(x-a)^2 + y^2\} + i\tan^{-1}\{y/(x-a)\}$ and $u = \tfrac{1}{2}\ln\{(x-a)^2 + y^2\}$, $v = \tan^{-1}\{y/(x-a)\}$. Substituting these into Laplace's equation $\dfrac{\partial^2 \Phi}{\partial x^2} + \dfrac{\partial^2 \Phi}{\partial y^2} = 0$, we find after straightforward differentiation that u and v are solutions if $z \neq a$.

13.51. Find a function harmonic in the upper half of the z plane, $\text{Im}\{z\} > 0$, which takes the prescribed values on the x axis given by $G(x) = \begin{cases} 1 & x > 0 \\ 0 & x < 0 \end{cases}$.

We must solve for $\Phi(x, y)$ the boundary-value problem
$$\frac{\partial^2 \Phi}{\partial x^2} + \frac{\partial^2 \Phi}{\partial y^2} = 0, \quad y > 0; \qquad \lim_{y \to 0+} \Phi(x, y) = G(x) = \begin{cases} 1 & x > 0 \\ 0 & x < 0 \end{cases}$$
This is a Dirichlet problem for the upper half plane [see Fig. 13-30].

The function $A\theta + B$, where A and B are real constants, is harmonic since it is the imaginary part of $A\ln z + Bi$.

To determine A and B note that the boundary conditions are $\Phi = 1$ for $x > 0$, i.e. $\theta = 0$ and $\Phi = 0$ for $x < 0$, i.e. $\theta = \pi$. Thus

(1) $1 = A(0) + B$, (2) $0 = A(\pi) + B$

from which $A = -1/\pi$, $B = 1$.

Then the required solution is
$$\Phi = A\theta + B = 1 - \frac{\theta}{\pi} = 1 - \frac{1}{\pi}\tan^{-1}\left(\frac{y}{x}\right)$$

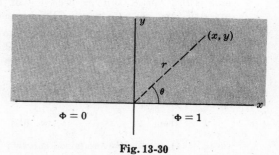

Fig. 13-30

13.52. Solve the boundary-value problem

$$\frac{\partial^2 \Phi}{\partial x^2} + \frac{\partial^2 \Phi}{\partial y^2} = 0, \quad y > 0$$

$$\lim_{y \to 0+} \Phi(x,y) = G(x) = \begin{cases} T_0 & x < -1 \\ T_1 & -1 < x < 1 \\ T_2 & x > 1 \end{cases}$$

where T_0, T_1, T_2 are constants.

This is a Dirichlet problem for the upper half plane [see Fig. 13-31].

The function $A\theta_1 + B\theta_2 + C$ where A, B and C are real constants, is harmonic since it is the imaginary part of $A \ln(z+1) + B \ln(z-1) + Ci$.

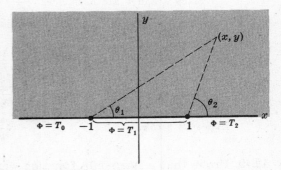

Fig. 13-31

To determine A, B, C note that the boundary conditions are: $\Phi = T_2$ for $x > 1$, i.e. $\theta_1 = \theta_2 = 0$; $\Phi = T_1$ for $-1 < x < 1$, i.e. $\theta_1 = 0$, $\theta_2 = \pi$; $\Phi = T_0$ for $x < -1$, i.e. $\theta_1 = \pi$, $\theta_2 = \pi$. Thus

$(1) \quad T_2 = A(0) + B(0) + C \qquad (2) \quad T_1 = A(0) + B(\pi) + C \qquad (3) \quad T_0 = A(\pi) + B(\pi) + C$

from which $C = T_2$, $B = (T_1 - T_2)/\pi$, $A = (T_0 - T_1)/\pi$.

Then the required solution is

$$\Phi = A\theta_1 + B\theta_2 + C = \frac{T_0 - T_1}{\pi} \tan^{-1}\left(\frac{y}{x+1}\right) + \frac{T_1 - T_2}{\pi} \tan^{-1}\left(\frac{y}{x-1}\right) + T_2$$

13.53. Find a function harmonic inside the unit circle $|z| = 1$ and taking the prescribed values given by $F(\theta) = \begin{cases} 1 & 0 < \theta < \pi \\ 0 & \pi < \theta < 2\pi \end{cases}$ on its circumference.

This is a Dirichlet problem for the unit circle [Fig. 13-32] in which we seek a function satisfying Laplace's equation inside $|z| = 1$ and taking the values 0 on arc ABC and 1 on arc CDE.

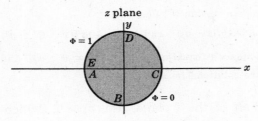

Fig. 13-32

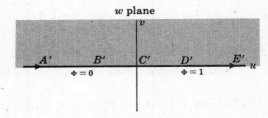

Fig. 13-33

We map the interior of the circle $|z| = 1$ on to the upper half of the w plane [Fig. 13-33] by using the mapping function $z = \frac{i-w}{i+w}$ or $w = i\left(\frac{1-z}{1+z}\right)$ obtained by using (15), page 292, with w and z interchanged.

Under this transformation, arcs ABC and CDE are mapped on to the negative and positive real axis $A'B'C'$ and $C'D'E'$ respectively of the w plane. Then the boundary conditions $\Phi = 0$ on arc ABC and $\Phi = 1$ on arc CDE become respectively $\Phi = 0$ on $A'B'C'$ and $\Phi = 1$ on $C'D'E'$.

Thus we have reduced the problem to finding a function Φ harmonic in the upper half w plane and taking the values 0 for $u < 0$ and 1 for $u > 0$. But this problem has already been solved in Problem 13.51 and the solution (replacing x by u and y by v) is given by

$$\Phi = 1 - \frac{1}{\pi} \tan^{-1}\left(\frac{v}{u}\right) \tag{1}$$

Now from $w = i\left(\frac{1-z}{1+z}\right)$, we find $u = \frac{2y}{(1+x)^2 + y^2}$, $v = \frac{1-(x^2+y^2)}{(1+x)^2 + y^2}$. Then substituting these in (1), we find the required solution

$$\Phi = 1 - \frac{1}{\pi} \tan^{-1}\left(\frac{2y}{1 - [x^2 + y^2]}\right) \tag{2}$$

or in polar coordinates (r, θ), where $x = r \cos \theta$, $y = r \sin \theta$,

$$\Phi = 1 - \frac{1}{\pi} \tan^{-1}\left(\frac{2r \sin \theta}{1 - r^2}\right) \tag{3}$$

13.54. A semi-infinite slab (shaded in Fig. 13-34) has its boundaries maintained at the indicated temperatures where T is constant.   Find the steady-state temperature.

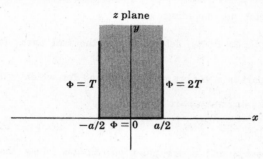

Fig. 13-34

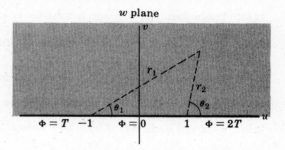

Fig. 13-35

We can solve this problem by methods of conformal mapping since the steady-state temperature satisfies Laplace's equation.

The shaded region of the z plane is mapped into the upper half of the w plane [Fig. 13-35] by the mapping function $w = \sin(\pi z/a)$ which is equivalent to $u = \sin(\pi x/a) \cosh(\pi y/a)$, $v = \cos(\pi x/a) \sinh(\pi y/a)$.

We must now solve the equivalent problem in the w plane.   We use the method of Problem 13.52 to find that the solution in the w plane is

$$\Phi = \frac{T}{\pi} \tan^{-1}\left(\frac{v}{u+1}\right) - \frac{2T}{\pi} \tan^{-1}\left(\frac{v}{u-1}\right) + 2T$$

and the required solution to the problem in the z plane is therefore

$$\Phi = \frac{T}{\pi} \tan^{-1}\left\{\frac{\cos(\pi x/a) \sinh(\pi y/a)}{\sin(\pi x/a) \cosh(\pi y/a) + 1}\right\} - \frac{2T}{\pi} \tan^{-1}\left\{\frac{\cos(\pi x/a) \sinh(\pi y/a)}{\sin(\pi x/a) \cosh(\pi y/a) - 1}\right\} + 2T$$

Supplementary Problems

FUNCTIONS, LIMITS, CONTINUITY

13.55. Describe the locus represented by (a) $|z + 2 - 3i| = 5$, (b) $|z + 2| = 2|z - 1|$, (c) $|z + 5| - |z - 5| = 6$. Construct a figure in each case.

13.56. Determine the region in the z plane represented by each of the following:
(a) $|z - 2 + i| \geqq 4$,   (b) $|z| \leqq 3$, $0 \leqq \arg z \leqq \frac{\pi}{4}$,   (c) $|z - 3| + |z + 3| < 10$.
Construct a figure in each case.

13.57. Express each function in the form $u(x, y) + iv(x, y)$, where u and v are real.
(a) $z^3 + 2iz$,   (b) $z/(3 + z)$,   (c) e^{z^2},   (d) $\ln(1 + z)$.

13.58. Prove that (a) $\lim\limits_{z \to z_0} z^2 = z_0^2$, (b) $f(z) = z^2$ is continuous at $z = z_0$ directly from the definition.

13.59. (a) If $z = \omega$ is any root of $z^5 = 1$ different from 1, prove that all the roots are $1, \omega, \omega^2, \omega^3, \omega^4$.
(b) Show that $1 + \omega + \omega^2 + \omega^3 + \omega^4 = 0$.
(c) Generalize the results in (a) and (b) to the equation $z^n = 1$.

DERIVATIVES, CAUCHY-RIEMANN EQUATIONS

13.60. (a) If $w = f(z) = z + \dfrac{1}{z}$, find $\dfrac{dw}{dz}$ directly from the definition.

(b) For what finite values of z is $f(z)$ non-analytic?

13.61. Given the function $w = z^4$. (a) Find real functions u and v such that $w = u + iv$. (b) Show that the Cauchy-Riemann equations hold at all points in the finite z plane. (c) Prove that u and v are harmonic functions. (d) Determine dw/dz.

13.62. Prove that $f(z) = z|z|$ is not analytic anywhere.

13.63. Prove that $f(z) = \dfrac{1}{z-2}$ is analytic in any region not including $z = 2$.

13.64. If the imaginary part of an analytic function is $2x(1-y)$, determine (a) the real part, (b) the function.

13.65. Construct an analytic function $f(z)$ whose real part is $e^{-x}(x \cos y + y \sin y)$ and for which $f(0) = 1$.

13.66. Prove that there is no analytic function whose imaginary part is $x^2 - 2y$.

13.67. Find $f(z)$ such that $f'(z) = 4z - 3$ and $f(1 + i) = -3i$.

13.68. If $z = \rho e^{i\phi}$ and $f(z) = u(\rho, \phi) + iv(\rho, \phi)$, where ρ and ϕ are polar coordinates, show that the Cauchy-Riemann equations are

$$\frac{\partial u}{\partial \rho} = \frac{1}{\rho}\frac{\partial v}{\partial \phi}, \qquad \frac{\partial v}{\partial \rho} = -\frac{1}{\rho}\frac{\partial u}{\partial \phi}$$

INTEGRALS, CAUCHY'S THEOREM, CAUCHY'S INTEGRAL FORMULAS

13.69. Evaluate $\displaystyle\int_{1-2i}^{3+i} (2z + 3)\, dz$:

(a) along the path $x = 2t + 1$, $y = 4t^2 - t - 2 \quad 0 \le t \le 1$,

(b) along the straight line joining $1 - 2i$ and $3 + i$,

(c) along straight lines from $1 - 2i$ to $1 + i$ and then to $3 + i$.

13.70. Evaluate $\displaystyle\int_C (z^2 - z + 2)\, dz$, where C is the upper half of the circle $|z| = 1$ traversed in the positive direction.

13.71. Evaluate $\displaystyle\oint_C \frac{z\, dz}{2z - 5}$, where C is the circle (a) $|z| = 2$, (b) $|z - 3| = 2$.

13.72. Evaluate $\displaystyle\oint_C \frac{z^2}{(z+2)(z-1)}\, dz$, where C is: (a) a square with vertices at $-1 - i, -1 + i, -3 + i$, $-3 - i$; (b) the circle $|z + i| = 3$; (c) the circle $|z| = \sqrt{2}$.

13.73. Evaluate (a) $\displaystyle\oint_C \frac{\cos \pi z}{z - 1}\, dz$, (b) $\displaystyle\oint_C \frac{e^z + z}{(z-1)^4}\, dz$ where C is any simple closed curve enclosing $z = 1$.

13.74. Prove Cauchy's integral formulas.
[*Hint.* Use the definition of derivative and then apply mathematical induction.]

13.75. If $f(z)$ is analytic inside and on the circle $|z - a| = R$, prove *Cauchy's inequality*, namely,

$$|f^{(n)}(a)| \le \frac{n!\, M}{R^n}$$

where $|f(z) \le M|$ on the circle. [*Hint:* Use Cauchy's integral formulas.]

SERIES AND SINGULARITIES

13.76. For what values of z does each series converge?

(a) $\displaystyle\sum_{n=1}^{\infty} \frac{(z+2)^n}{n!}$, (b) $\displaystyle\sum_{n=1}^{\infty} \frac{n(z-i)^n}{n+1}$, (c) $\displaystyle\sum_{n=1}^{\infty} (-1)^n (z^2 + 2z + 2)^{2n}$

13.77. Prove that the series $\displaystyle\sum_{n=1}^{\infty}\frac{z^n}{n(n+1)}$ is (a) absolutely convergent, (b) uniformly convergent for $|z| \leqq 1$.

13.78. Prove that the series $\displaystyle\sum_{n=0}^{\infty}\frac{(z+i)^n}{2^n}$ converges uniformly within any circle of radius R such that $|z+i| < R < 2$.

13.79. Locate in the finite z plane all the singularities, if any, of each function and name them:

(a) $\dfrac{z-2}{(2z+1)^4}$, (b) $\dfrac{z}{(z-1)(z+2)^2}$, (c) $\dfrac{z^2+1}{z^2+2z+2}$, (d) $\cos\dfrac{1}{z}$, (e) $\dfrac{\sin(z-\pi/3)}{3z-\pi}$, (f) $\dfrac{\cos z}{(z^2+4)^2}$.

13.80. Find Laurent series about the indicated singularity for each of the following functions, naming the singularity in each case. Indicate the region of convergence of each series.

(a) $\dfrac{\cos z}{z-\pi}$; $z=\pi$ (b) $z^2 e^{-1/z}$; $z=0$ (c) $\dfrac{z^2}{(z-1)^2(z+3)}$; $z=1$

13.81. Find a Laurent series expansion for the function $f(z)=\dfrac{z}{(z+1)(z+2)}$ which converges for $1<|z|<2$ and diverges elsewhere.

$$\left[\text{Hint: Write}\quad \frac{z}{(z+1)(z+2)} = \frac{-1}{z+1} + \frac{2}{z+2} = \frac{-1}{z(1+1/z)} + \frac{1}{1+z/2}.\right]$$

13.82. If $a+h$ is any point in the annular region bounded by C_1 and C_2, and $f(z)$ is analytic in this region, prove *Laurent's theorem* that

$$f(a+h) = \sum_{-\infty}^{\infty} a_n h^n$$

where

$$a_n = \frac{1}{2\pi i}\oint_C \frac{f(z)\,dz}{(z-a)^{n+1}}$$

C being any closed curve in the angular region surrounding C_1.

$$\left[\text{Hint. Write}\quad f(a+h) = \frac{1}{2\pi i}\oint_{C_2}\frac{f(z)\,dz}{z-(a+h)} - \frac{1}{2\pi i}\oint_{C_1}\frac{f(z)\,dz}{z-(a+h)}\quad\text{and expand}\quad \frac{1}{z-a-h}\text{ in two different ways.}\right]$$

RESIDUES AND THE RESIDUE THEOREM

13.83. Determine the residues of each function at its poles:

(a) $\dfrac{2z+3}{z^2-4}$, (b) $\dfrac{z-3}{z^3+5z^2}$, (c) $\dfrac{e^{zt}}{(z-2)^3}$, (d) $\dfrac{z}{(z^2+1)^2}$.

13.84. Find the residue of $e^{zt}\tan z$ at the simple pole $z=3\pi/2$.

13.85. Evaluate $\displaystyle\oint_C \frac{z^2\,dz}{(z+1)(z+3)}$, where C is a simple closed curve enclosing all the poles.

13.86. If C is a simple closed curve enclosing $z=\pm i$, show that

$$\oint_C \frac{ze^{zt}}{(z^2+1)^2}\,dz = \tfrac{1}{2}t\sin t$$

13.87. If $f(z)=P(z)/Q(z)$, where $P(z)$ and $Q(z)$ are polynomials such that the degree of $P(z)$ is at least two less than the degree of $Q(z)$, prove that $\displaystyle\oint_C f(z)\,dz = 0$, where C encloses all the poles of $f(z)$.

EVALUATION OF DEFINITE INTEGRALS

Use contour integration to verify each of the following

13.88. $\displaystyle\int_0^{\infty}\frac{x^2\,dx}{x^4+1} = \frac{\pi}{2\sqrt{2}}$

13.90. $\displaystyle\int_0^{\infty}\frac{dx}{(x^2+4)^2} = \frac{\pi}{32}$

13.89. $\displaystyle\int_{-\infty}^{\infty}\frac{dx}{x^6+a^6} = \frac{2\pi}{3a^5}$, $a>0$

13.91. $\displaystyle\int_0^{\infty}\frac{\sqrt{x}}{x^3+1}\,dx = \frac{\pi}{3}$

13.92. $\displaystyle\int_0^\infty \frac{dx}{(x^4+a^4)^2} = \frac{3\pi}{8\sqrt{2}} a^{-7}, \quad a > 0$

13.95. $\displaystyle\int_0^{2\pi} \frac{d\theta}{(2+\cos\theta)^2} = \frac{4\pi\sqrt{3}}{9}$

13.93. $\displaystyle\int_{-\infty}^\infty \frac{dx}{(x^2+1)^2(x^2+4)} = \frac{\pi}{9}$

13.96. $\displaystyle\int_0^\pi \frac{\sin^2\theta}{5-4\cos\theta} d\theta = \frac{\pi}{8}$

13.94. $\displaystyle\int_0^{2\pi} \frac{d\theta}{2-\cos\theta} = \frac{2\pi}{\sqrt{3}}$

13.97. $\displaystyle\int_0^{2\pi} \frac{d\theta}{(1+\sin^2\theta)^2} = \frac{3\pi}{2\sqrt{2}}$

13.98. $\displaystyle\int_0^{2\pi} \frac{\cos n\theta \, d\theta}{1-2a\cos\theta+a^2} = \frac{2\pi a^n}{1-a^2}, \quad n = 0,1,2,3,\ldots, \quad 0 < a < 1$

13.99. $\displaystyle\int_0^{2\pi} \frac{d\theta}{(a+b\cos\theta)^3} = \frac{(2a^2+b^2)\pi}{(a^2-b^2)^{5/2}}, \quad a > |b|$

13.100. $\displaystyle\int_0^\infty \frac{x\sin 2x}{x^2+4} dx = \frac{\pi e^{-4}}{4}$

13.103. $\displaystyle\int_0^\infty \frac{\sin x}{x(x^2+1)^2} dx = \frac{\pi(2e-3)}{4e}$

13.101. $\displaystyle\int_0^\infty \frac{\cos 2\pi x}{x^4+4} dx = \frac{\pi e^{-\pi}}{8}$

13.104. $\displaystyle\int_0^\infty \frac{\sin^2 x}{x^2} dx = \frac{\pi}{2}$

13.102. $\displaystyle\int_0^\infty \frac{x\sin\pi x}{(x^2+1)^2} dx = \frac{\pi^2 e^{-\pi}}{4}$

13.105. $\displaystyle\int_0^\infty \frac{\sin^3 x}{x^3} dx = \frac{3\pi}{8}$

13.106. $\displaystyle\int_0^\infty \frac{\cos x}{\cosh x} dx = \frac{\pi}{2\cosh(\pi/2)}$.   [*Hint.* Consider $\displaystyle\oint_C \frac{e^{iz}}{\cosh z} dz$, where C is a rectangle with vertices at $(-R,0)$, $(R,0)$, (R,π), $(-R,\pi)$.   Then let $R \to \infty$.]

CONFORMAL MAPPING

13.107. Prove that (*a*) $w = z + \beta$ represents a translation, (*b*) $w = e^{i\theta_0}z$ represents a rotation, (*c*) $w = az + b$ represents a stretching [or contraction].

13.108. Prove that (*a*) $w = \alpha z + \beta$ represents a combination of translation, rotation and stretching.

13.109. Prove that $w = \dfrac{\alpha z + \beta}{\gamma z + \delta}$ represents a combination of translation, rotation, stretching and inversion if $\alpha\delta - \beta\gamma \neq 0$.   Discuss the case $\alpha\delta - \beta\gamma = 0$.

13.110. (*a*) Prove that under the transformation $w = (z-i)/(iz-1)$ the region $\text{Im}\{z\} \leqq 0$ is mapped into the region $|w| \leqq 1$. (*b*) Into what region is $\text{Im}\{z\} \leqq 0$ mapped under the transformation?

13.111. Determine the equation of the curve in the w plane into which the straight line $x + y = 1$ is mapped under the transformations (*a*) $w = z^2$, (*b*) $w = 1/z$.

13.112. Show that $w = \left(\dfrac{1+z}{1-z}\right)^{2/3}$ maps the unit circle on to a wedge-shaped region and illustrate graphically.

THE SCHWARZ-CHRISTOFFEL TRANSFORMATION

13.113. Use the Schwarz-Christoffel transformation to determine a function which maps each of the indicated regions in the w plane on to the upper half of the z plane.

(*a*) z plane w plane

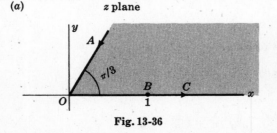

Fig. 13-36

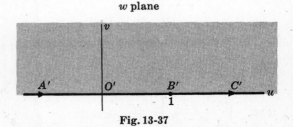

Fig. 13-37

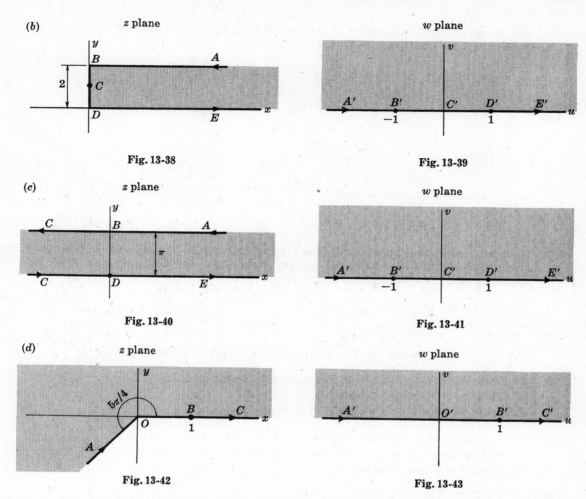

(b)　　z plane

Fig. 13-38

w plane

Fig. 13-39

(c)　　z plane

Fig. 13-40

w plane

Fig. 13-41

(d)　　z plane

Fig. 13-42

w plane

Fig. 13-43

13.114. Find a function which maps the upper half plane on to the interior of a triangle with vertices at $w = 0, 1, i$ corresponding to $z = 0, 1, \infty$ respectively.

13.115. Show that the functions (a) $2xy + y^3 - 3x^2y$, (b) $e^{-x} \sin y$ are harmonic.

13.116. Show that the functions of Problem 13.115 remain harmonic under the transformations (a) $z = w^2$, (b) $z = \sin w$.

13.117. Find a function harmonic in the upper half z plane $\text{Im}\{z\} > 0$ which takes the prescribed values on the x axis given by $G(x) = \begin{cases} 1 & x > 0 \\ -1 & x < 0 \end{cases}$.

13.118. Work Problem 13.117 if $G(x) = \begin{cases} 1 & x < -1 \\ 0 & -1 < x < 1 \\ -1 & x > 1 \end{cases}$.

13.119. Find a function harmonic inside the circle $|z| = 1$ and taking the values $F(\theta) = \begin{cases} T & 0 < \theta < \pi \\ -T & \pi < \theta < 2\pi \end{cases}$ on its circumference.

13.120. Work Problem 13.119 if $F(\theta) = \begin{cases} T & 0 < \theta < \pi/2 \\ 0 & \pi/2 < \theta < 3\pi/2 \\ -T & 3\pi/2 < \theta < 2\pi \end{cases}$.

13.121. Find the steady-state temperature at the point $(5, 2)$ in the shaded region of Fig. 13-44 below if the temperatures are maintained as shown.

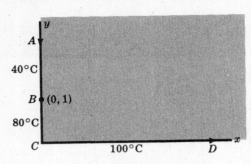

Fig. 13-44   Fig. 13-45

13.122. An infinite conducting plate has in it a circular hole $ABCD$ of unit radius [Fig. 13-45 above]. Temperatures of 20°C and 80°C are applied to arcs ABC and ADC and maintained indefinitely. Find the steady-state temperature at any point of the plate.

Answers to Supplementary Problems

13.55. (a) Circle $(x + 2)^2 + (y - 3)^2 = 25$, center $(-2, 3)$, radius 5.

(b) Circle $(x - 2)^2 + y^2 = 4$, center $(2, 0)$, radius 2.

(c) Branch of hyperbola $x^2/9 - y^2/16 = 1$, where $x \geqq 3$.

13.56. (a) Boundary and exterior of circle $(x - 2)^2 + (y + 1)^2 = 16$.

(b) Region in the first quadrant bounded by $x^2 + y^2 = 9$ the x axis and the line $y = x$.

(c) Interior of ellipse $x^2/25 + y^2/16 = 1$.

13.57. (a) $u = x^3 - 3xy^2 - 2y$, $v = 3x^2y - y^3 + 2x$

(b) $u = \dfrac{x^2 + 3x + y^2}{x^2 + 6x + y^2 + 9}$, $v = \dfrac{3y}{x^2 + 6x + y^2 + 9}$

(c) $u = e^{x^2 - y^2} \cos 2xy$, $v = e^{x^2 - y^2} \sin 2xy$

(d) $u = \frac{1}{2} \ln \{(1 + x)^2 + y^2\}$, $v = \tan^{-1} \dfrac{y}{1 + x} + 2k\pi$, $k = 0, \pm 1, \pm 2, \ldots$

13.60. (a) $1 - 1/z^2$, (b) $z = 0$

13.61. (a) $u = x^4 - 6x^2y^2 + y^4$, $v = 4x^3y - 4xy^3$ (d) $4z^3$

13.64. (a) $y^2 - x^2 - 2y + c$, (b) $2iz - z^2 + c$, where c is real

13.65. $ze^{-z} + 1$

13.67. $f(z) = 2z^2 - 3z + 3 - 4i$

13.69. $17 + 19i$ in all cases

13.71. (a) 0, (b) $5\pi i/2$

13.72. (a) $-8\pi i/3$ (b) $-2\pi i$ (c) $2\pi i/3$

13.73. (a) $-2\pi i$ (b) $\pi i e/3$

13.76. (a) all z (b) $|z - i| < 1$ (c) $z = -1 \pm i$

13.79. *(a)* $z = -\frac{1}{2}$, pole of order 4   *(d)* $z = 0$, essential singularity

(b) $z = 1$, simple pole; $z = -2$, double pole   *(e)* $z = \pi/3$, removable singularity

(c) Simple poles $z = -1 \pm i$ *(f)* $z = \pm 2i$, double poles

13.80. *(a)* $-\dfrac{1}{z-\pi} + \dfrac{z-\pi}{2!} - \dfrac{(z-\pi)^3}{4!} + \dfrac{(z-\pi)^5}{6!} - \cdots$,   simple pole, all $z \neq \pi$

(b) $z^2 - z + \dfrac{1}{2!} - \dfrac{1}{3!\,z} + \dfrac{1}{4!\,z^2} - \dfrac{1}{5!\,z^3} + \cdots$,   essential singularity, all $z \neq 0$

(c) $\dfrac{1}{4(z-1)^2} + \dfrac{7}{16(z-1)} + \dfrac{9}{64} - \dfrac{9(z-1)}{256} + \cdots$,   double pole, $0 < |z-1| < 4$

13.81. $\cdots - \dfrac{1}{z^5} + \dfrac{1}{z^4} - \dfrac{1}{z^3} + \dfrac{1}{z^2} - \dfrac{1}{z} + 1 - \dfrac{z}{2} + \dfrac{z^2}{4} - \dfrac{z^3}{8} + \cdots$

13.83. *(a)* $z = 2$; $7/4$, $z = -2$; $1/4$ *(c)* $z = 2$; $\frac{1}{2}t^2 e^{2t}$

(b) $z = 0$; $8/25$, $z = -5$; $-8/25$ *(d)* $z = i$; 0, $z = -i$; 0

13.84. $-e^{3\pi t/2}$

13.85. $-8\pi i$

13.111. *(a)* $u^2 + 2v = 1$,   *(b)* $u^2 + 2uv + 2v^2 = u + v$

13.113. *(a)* $w = z^3$,   *(b)* $w = \cosh(\pi z/2)$,   *(c)* $w = e^z$,   *(d)* $w = z^{4/5}$

13.114. $w = \dfrac{\Gamma(3/4)}{\sqrt{\pi}\,\Gamma(1/4)} \displaystyle\int_0^z t^{-1/2}(1-t)^{-3/4}\,dt$

13.117. $1 - (2/\pi)\tan^{-1}(y/x)$

13.118. $1 - \dfrac{1}{\pi}\tan^{-1}\!\left(\dfrac{y}{x-1}\right) - \dfrac{1}{\pi}\tan^{-1}\!\left(\dfrac{y}{x+1}\right)$

13.119. $T\left\{1 - \dfrac{2}{\pi}\tan^{-1}\!\left(\dfrac{2r\sin\theta}{1-r^2}\right)\right\}$

13.121. $45.9°$ C

Chapter 14

Complex Inversion Formula for Laplace Transforms

THE COMPLEX INVERSION FORMULA

If $F(s) = \mathcal{L}\{f(t)\}$, then $\mathcal{L}^{-1}\{F(s)\}$ is given by

$$f(t) = \frac{1}{2\pi i} \int_{\gamma - i\infty}^{\gamma + i\infty} e^{st} F(s)\, ds, \qquad t > 0 \tag{1}$$

and $f(t) = 0$ for $t < 0$. This result is called the *complex inversion integral* or *formula*. It is also known as *Bromwich's integral formula*. The result provides a direct means for obtaining the inverse Laplace transform of a given function $F(s)$.

The integration in *(1)* is to be performed along a line $s = \gamma + iy$ in the complex plane where $s = x + iy$. The real number γ is chosen so that $s = \gamma$ lies to the right of all the singularities (poles, branch points or essential singularities) but is otherwise arbitrary.

THE BROMWICH CONTOUR

In practice, the integral in *(1)* is evaluated by considering the contour integral

$$\frac{1}{2\pi i} \oint_C e^{st} F(s)\, ds \tag{2}$$

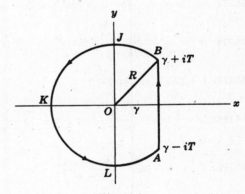

Fig. 14-1

where C is the contour of Fig. 14-1. This contour, sometimes called the *Bromwich contour*, is composed of line AB and the arc $BJKLA$ of a circle of radius R with center at the origin O.

If we represent arc $BJKLA$ by Γ, it follows from *(1)* that since $T = \sqrt{R^2 - \gamma^2}$,

$$f(t) = \lim_{R \to \infty} \frac{1}{2\pi i} \int_{\gamma - iT}^{\gamma + iT} e^{st} F(s)\, ds$$

$$= \lim_{R \to \infty} \left\{ \frac{1}{2\pi i} \oint_C e^{st} F(s)\, ds - \frac{1}{2\pi i} \int_\Gamma e^{st} F(s)\, ds \right\} \tag{3}$$

USE OF RESIDUE THEOREM IN FINDING INVERSE LAPLACE TRANSFORMS

Suppose that the only singularities of $F(s)$ are poles all of which lie to the left of the line $s = \gamma$ for some real constant γ. Suppose further that the integral around Γ in *(3)* approaches zero as $R \to \infty$. Then by the residue theorem we can write *(3)* as

$$f(t) = \text{sum of residues of } e^{st} F(s) \text{ at poles of } F(s)$$

$$= \sum \text{residues of } e^{st} F(s) \text{ at poles of } F(s) \tag{4}$$

A SUFFICIENT CONDITION FOR THE INTEGRAL AROUND Γ TO APPROACH ZERO

The validity of the result (4) hinges on the assumption that the integral around Γ in (3) approaches zero as $R \to \infty$. A sufficient condition under which this assumption is correct is supplied in the following

Theorem 14-1.   If we can find constants $M > 0$, $k > 0$ such that on Γ (where $s = Re^{i\theta}$),

$$|F(s)| \; < \; \frac{M}{R^k} \tag{5}$$

then the integral around Γ of $e^{st}F(s)$ approaches zero as $R \to \infty$, i.e.,

$$\lim_{R \to \infty} \int_{\Gamma} e^{st} F(s) \, ds \; = \; 0 \tag{6}$$

The condition (5) always holds if $F(s) = P(s)/Q(s)$ where $P(s)$ and $Q(s)$ are polynomials and the degree of $P(s)$ is less than the degree of $Q(s)$.

The result is valid even if $F(s)$ has other singularities besides poles.

MODIFICATION OF BROMWICH CONTOUR IN CASE OF BRANCH POINTS

If $F(s)$ has branch points, extensions of the above results can be made provided that the Bromwich contour is suitably modified. For example, if $F(s)$ has only one branch point at $s = 0$, then we can use the contour of Fig. 14-2. In this figure, BDE and LNA represent arcs of a circle of radius R with center at origin O, while HJK is the arc of a circle of radius ϵ with center at O. For details of evaluating inverse Laplace transforms in such cases see Prob. 14.9.

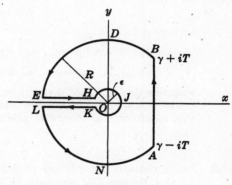

Fig. 14-2

CASE OF INFINITELY MANY SINGULARITIES

If we wish to find the inverse Laplace transform of functions which have infinitely many isolated singularities, the above methods can be applied. In such case the curved portion of the Bromwich contour is chosen to be of such radius R_m so as to enclose only a finite number of the singularities and so as not to pass through any singularity. The required inverse Laplace transform is then found by taking an appropriate limit as $m \to \infty$. See Problems 14.13 and 14.14.

APPLICATIONS TO BOUNDARY-VALUE PROBLEMS

The method of Laplace transforms combined with the complex inversion formula provide powerful tools in the solution of various boundary-value problems arising in science and engineering. See Problems 14.15-14.19.

Solved Problems

THE COMPLEX INVERSION FORMULA

14.1. Establish the validity of the complex inversion formula.

We have, by definition, $F(s) = \int_0^\infty e^{-su} f(u)\, du$. Then

$$\lim_{T \to \infty} \frac{1}{2\pi i} \int_{\gamma-iT}^{\gamma+iT} e^{st} F(s)\, ds = \lim_{T \to \infty} \frac{1}{2\pi i} \int_{\gamma-iT}^{\gamma+iT} \int_0^\infty e^{st-su} f(u)\, du\, ds$$

Letting $s = \gamma + iy$, $ds = i\, dy$, this becomes

$$\lim_{T \to \infty} \frac{1}{2\pi} e^{\gamma t} \int_{-T}^{T} e^{iyt}\, dy \int_0^\infty e^{-iyu} [e^{-\gamma u} f(u)]\, du = \frac{1}{2\pi} e^{\gamma t} \begin{cases} 2\pi e^{-\gamma t} f(t) & t > 0 \\ 0 & t < 0 \end{cases}$$

$$= \begin{cases} f(t) & t > 0 \\ 0 & t < 0 \end{cases}$$

by Fourier's integral theorem [see Chapter 8]. Thus we find

$$f(t) = \frac{1}{2\pi i} \int_{\gamma-i\infty}^{\gamma+i\infty} e^{st} F(s)\, ds \qquad t > 0$$

as required.

In the above proof, we assume that $e^{-\gamma u} f(u)$ is absolutely integrable in $(0, \infty)$, i.e. $\int_0^\infty e^{-\gamma u} |f(u)|\, du$ converges, so that Fourier's integral theorem can be applied. To insure this condition it is sufficient that $f(t)$ be of exponential order γ where the real number γ is chosen so that the line $x = \gamma$ in the complex plane lies to the right of all the singularities of $F(s)$. Except for this condition, γ is otherwise arbitrary.

14.2. Let Γ denote the curved portion $BJPKQLA$ of the Bromwich contour [Fig. 14-3] with equation $s = Re^{i\theta}$, $\theta_0 \leqq \theta \leqq 2\pi - \theta_0$, i.e. Γ is the arc of a circle of radius R with center at O. Suppose that on Γ we have

$$|F(s)| < \frac{M}{R^k}$$

where $k > 0$ and M are constants. Show that

$$\lim_{R \to \infty} \int_\Gamma e^{st} F(s)\, ds = 0$$

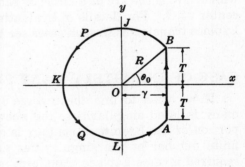

Fig. 14-3

If Γ_1, Γ_2, Γ_3 and Γ_4 represent arcs BJ, JPK, KQL and LA respectively, we have

$$\int_\Gamma e^{st} F(s)\, ds = \int_{\Gamma_1} e^{st} F(s)\, ds + \int_{\Gamma_2} e^{st} F(s)\, ds + \int_{\Gamma_3} e^{st} F(s)\, ds + \int_{\Gamma_4} e^{st} F(s)\, ds$$

Then if we can show that each of the integrals on the right approach zero as $R \to \infty$ we will have proved the required result. To do this we consider these four integrals.

Case 1. Integral over Γ_1 or BJ.

Along Γ_1 we have, since $s = Re^{i\theta}$, $\theta_0 \leqq \theta \leqq \pi/2$,

$$I_1 = \int_{\Gamma_1} e^{st} F(s)\, ds = \int_{\theta_0}^{\pi/2} e^{Re^{i\theta} t} F(Re^{i\theta})\, iRe^{i\theta}\, d\theta$$

Then
$$|I_1| \;\leqq\; \int_{\theta_0}^{\pi/2} |e^{(R\cos\theta)t}| \, |e^{i(R\sin\theta)t}| \, |F(Re^{i\theta})| \, |iRe^{i\theta}| \, d\theta$$

$$\leqq\; \int_{\theta_0}^{\pi/2} e^{(R\cos\theta)t} \, |F(Re^{i\theta})| \, R \, d\theta$$

$$\leqq\; \frac{M}{R^{k-1}} \int_{\theta_0}^{\pi/2} e^{(R\cos\theta)t} \, d\theta \;=\; \frac{M}{R^{k-1}} \int_{0}^{\phi_0} e^{(R\sin\phi)t} \, d\phi$$

where we have used the given condition $|F(s)| \leqq M/R^k$ on Γ_1 and the transformation $\theta = \pi/2 - \phi$ where $\phi_0 = \pi/2 - \theta_0 = \sin^{-1}(\gamma/R)$.

Since $\sin\phi \leqq \sin\phi_0 \leqq \cos\theta_0 = \gamma/R$, this last integral is less than or equal to

$$\frac{M}{R^{k-1}} \int_{0}^{\phi_0} e^{\gamma t} \, d\phi \;=\; \frac{M \, e^{\gamma t} \, \phi_0}{R^{k-1}} \;=\; \frac{M \, e^{\gamma t}}{R^{k-1}} \sin^{-1}\frac{\gamma}{R}$$

But as $R \to \infty$, this last quantity approaches zero [as can be seen by noting, for example, that $\sin^{-1}(\gamma/R) \approx \gamma/R$ for large R]. Thus $\lim_{R \to \infty} I_1 = 0$.

Case 2.   Integral over Γ_2 or JPK.

Along Γ_2 we have, since $s = Re^{i\theta}$, $\pi/2 \leqq \theta \leqq \pi$,

$$I_2 \;=\; \int_{\Gamma_2} e^{st} F(s) \, ds \;=\; \int_{\pi/2}^{\pi} e^{Re^{i\theta}t} F(Re^{i\theta}) \, iRe^{i\theta} \, d\theta$$

Then, as in Case 1, we have

$$|I_2| \;\leqq\; \frac{M}{R^{k-1}} \int_{\pi/2}^{\pi} e^{(R\cos\theta)t} \, d\theta \;\leqq\; \frac{M}{R^{k-1}} \int_{0}^{\pi/2} e^{-(R\sin\phi)t} \, d\phi$$

upon letting $\theta = \pi/2 + \phi$.

Now $\sin\phi \geqq 2\phi/\pi$ for $0 \leqq \phi \leqq \pi/2$ [see Problem 14.3], so that the last integral is less than or equal to

$$\frac{M}{R^{k-1}} \int_{0}^{\pi/2} e^{-2R\phi t/\pi} \, d\phi \;=\; \frac{\pi M}{2tR^k}(1 - e^{-Rt})$$

which approaches zero as $R \to \infty$. Thus $\lim_{R \to \infty} I_2 = 0$.

Case 3.   Integral over Γ_3 or KQL.

This case can be treated in a manner similar to Case 2 [see Problem 14.28(a)].

Case 4.   Integral over Γ_4 or LA.

This case can be treated in a manner similar to Case 1 [see Problem 14.28(b)].

14.3.   Show that $\sin\phi \geqq 2\phi/\pi$ for $0 \leqq \phi \leqq \pi/2$.

Method 1.   Geometrical proof.

From Fig. 14-4, in which curve OPQ represents an arc of the sine curve $y = \sin\phi$ and $y = 2\phi/\pi$ represents line OP, it is geometrically evident that $\sin\phi \geqq 2\phi/\pi$ for $0 \leqq \phi \leqq \pi/2$.

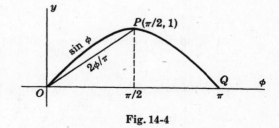

Fig. 14-4

Method 2.   Analytical proof.

Consider $G(\phi) = \dfrac{\sin\phi}{\phi}$. We have

$$\frac{dG}{d\phi} \;=\; G'(\phi) \;=\; \frac{\phi\cos\phi - \sin\phi}{\phi^2} \tag{1}$$

If $H(\phi) = \phi \cos \phi - \sin \phi$, then

$$\frac{dH}{d\phi} = H'(\phi) = -\phi \sin \phi \qquad (2)$$

Thus for $0 \leq \phi < \pi/2$, $H'(\phi) \leq 0$ and $H(\phi)$ is a decreasing function. Since $H(0) = 0$, it follows that $H(\phi) \leq 0$. Then from (1) we see that $G'(\phi) \leq 0$, or $G(\phi)$ is a decreasing function. Defining $G(0) = \lim\limits_{\phi \to 0} G(\phi) = 1$, we see that $G(\phi)$ decreases from 1 to $2/\pi$ as ϕ goes from 0 to $\pi/2$. Thus

$$1 \geq \frac{\sin \phi}{\phi} \geq \frac{2}{\pi}$$

from which the required result follows.

USE OF RESIDUE THEOREM IN FINDING INVERSE LAPLACE TRANSFORMS

14.4. Suppose that the only singularities of $F(s)$ are poles which all lie to the left of the line $x = \gamma$ for some real constant γ. Suppose further that $F(s)$ satisfies the condition given in Problem 14.2. Prove that the inverse Laplace transform of $F(s)$ is given by

$$f(t) \quad = \quad \text{sum of residues of } e^{st} F(s) \text{ at all the poles of } F(s)$$

We have $\qquad \dfrac{1}{2\pi i} \oint_C e^{st} F(s)\, ds \quad = \quad \dfrac{1}{2\pi i} \int_{\gamma - iT}^{\gamma + iT} e^{st} F(s)\, ds + \dfrac{1}{2\pi i} \int_\Gamma e^{st} F(s)\, ds$

where C is the Bromwich contour of Problem 14.2 and Γ is the circular arc *BJPKQLA* of Fig. 14-3. By the residue theorem,

$$\frac{1}{2\pi i} \oint_C e^{st} F(s)\, ds \quad = \quad \text{sum of residues of } e^{st} F(s) \text{ at all poles of } F(s) \text{ inside } C$$

$$= \quad \Sigma \text{ residues inside } C$$

Thus $\qquad \dfrac{1}{2\pi i} \int_{\gamma - iT}^{\gamma + iT} e^{st} F(s)\, ds \quad = \quad \Sigma \text{ residues inside } C - \dfrac{1}{2\pi i} \int_\Gamma e^{st} F(s)\, ds$

Taking the limit as $R \to \infty$, we find by Problem 14.2,

$$f(t) \quad = \quad \text{sum of residues of } e^{st} F(s) \text{ at all the poles of } F(s)$$

14.5. (a) Show that $F(s) = \dfrac{1}{s-2}$ satisfies the condition in Problem 14.2.

(b) Find the residue of $\dfrac{e^{st}}{s-2}$ at the pole $s = 2$.

(c) Evaluate $\mathcal{L}^{-1}\left\{\dfrac{1}{s-2}\right\}$ by using the complex inversion formula.

(a) For $s = Re^{i\theta}$, we have

$$\left|\frac{1}{s-2}\right| \quad = \quad \left|\frac{1}{Re^{i\theta} - 2}\right| \quad \leq \quad \frac{1}{|Re^{i\theta}| - 2} \quad = \quad \frac{1}{R-2} \quad < \quad \frac{2}{R}$$

for large enough R (e.g. $R > 4$). Thus the condition in Problem 14.2 is satisfied when $k = 1$, $M = 2$. Note that in establishing the above we have used the result $|z_1 - z_2| \geqq |z_1| - |z_2|$ which follows from the result $|a + b| \leqq |a| + |b|$ if $a = z_1 - z_2$, $b = z_2$.

(b) The residue at the simple pole $s = 2$ is

$$\lim_{s \to 2} (s - 2)\left(\frac{e^{st}}{s - 2}\right) = e^{2t}$$

(c) By Problem 14.4 and the results of parts (a) and (b), we see that

$$\mathcal{L}^{-1}\left\{\frac{1}{s - 2}\right\} = \text{ sum of residues of } e^{st} f(s) = e^{2t}$$

Note that the Bromwich contour in this case is chosen so that γ is any real number greater than 2 and the contour encloses the pole $s = 2$.

14.6. Evaluate $\mathcal{L}^{-1}\left\{\dfrac{1}{(s + 1)(s - 2)^2}\right\}$ by using the method of residues.

Since the function whose Laplace inverse is sought satisfies condition (5) of the theorem on page 325 [this can be established as in Problem 14.5], we have

$$\mathcal{L}^{-1}\left\{\frac{1}{(s + 1)(s - 2)^2}\right\} = \frac{1}{2\pi i} \int_{\gamma - i\infty}^{\gamma + i\infty} \frac{e^{st}\, ds}{(s + 1)(s - 2)^2}$$

$$= \frac{1}{2\pi i} \oint_C \frac{e^{st}\, ds}{(s + 1)(s - 2)^2}$$

$$= \sum \text{ residues of } \frac{e^{st}}{(s + 1)(s - 2)^2} \text{ at poles } s = -1 \text{ and } s = 2$$

Now the residue at simple pole $s = -1$ is

$$\lim_{s \to -1} (s + 1)\left\{\frac{e^{st}}{(s + 1)(s - 2)^2}\right\} = \frac{1}{9} e^{-t}$$

and the residue at double pole $s = 2$ is

$$\lim_{s \to 2} \frac{1}{1!} \frac{d}{ds}\left[(s - 2)^2 \left\{\frac{e^{st}}{(s + 1)(s - 2)^2}\right\}\right] = \lim_{s \to 2} \frac{d}{ds}\left[\frac{e^{st}}{s + 1}\right]$$

$$= \lim_{s \to 2} \frac{(s + 1)te^{st} - e^{st}}{(s + 1)^2} = \frac{1}{3} te^{2t} - \frac{1}{9} e^{2t}$$

Then

$$\mathcal{L}^{-1}\left\{\frac{1}{(s + 1)(s - 2)^2}\right\} = \sum \text{ residues } = \frac{1}{9} e^{-t} + \frac{1}{3} te^{2t} - \frac{1}{9} e^{2t}$$

14.7. Evaluate $\mathcal{L}^{-1}\left\{\dfrac{s}{(s + 1)^3(s - 1)^2}\right\}$.

As in Problem 14.6, the required inverse is the sum of the residues of

$$\frac{se^{st}}{(s + 1)^3(s - 1)^2}$$

at the poles $s = -1$ and $s = 1$ which are of orders three and two respectively.

Now the residue at $s = -1$ is

$$\lim_{s \to -1} \frac{1}{2!} \frac{d^2}{ds^2} \left[(s+1)^3 \frac{se^{st}}{(s+1)^3(s-1)^2} \right] = \lim_{s \to -1} \frac{1}{2} \frac{d^2}{ds^2} \left[\frac{se^{st}}{(s-1)^2} \right] = \frac{1}{16} e^{-t}(1-2t^2)$$

and the residue at $s = 1$ is

$$\lim_{s \to 1} \frac{1}{1!} \frac{d}{ds} \left[(s-1)^2 \frac{se^{st}}{(s+1)^3(s-1)^2} \right] = \lim_{s \to 1} \frac{d}{ds} \left[\frac{se^{st}}{(s-1)^2} \right] = \frac{1}{16} e^{t}(2t-1)$$

Then $\mathcal{L}^{-1} \left\{ \dfrac{s}{(s+1)^3(s-1)^2} \right\} = \sum \text{residues} = \dfrac{1}{16} e^{-t}(1-2t^2) + \dfrac{1}{16} e^{t}(2t-1)$

14.8.   Evaluate $\mathcal{L}^{-1} \left\{ \dfrac{1}{(s^2+1)^2} \right\}$.

We have $\dfrac{1}{(s^2+1)^2} = \dfrac{1}{[(s+i)(s-i)]^2} = \dfrac{1}{(s+i)^2(s-i)^2}$

The required inverse is the sum of the residues of

$$\frac{e^{st}}{(s+i)^2(s-i)^2}$$

at the poles $s = i$ and $s = -i$ which are of order two each.

Now the residue at $s = i$ is

$$\lim_{s \to i} \frac{d}{ds} \left[(s-i)^2 \frac{e^{st}}{(s+i)^2(s-i)^2} \right] = -\frac{1}{4} te^{it} - \frac{1}{4} ie^{it}$$

and the residue at $s = -i$ is

$$\lim_{s \to -i} \frac{d}{ds} \left[(s+i)^2 \frac{e^{st}}{(s+i)^2(s-i)^2} \right] = -\frac{1}{4} te^{-it} + \frac{1}{4} ie^{-it}$$

which can also be obtained from the residue at $s = i$ by replacing i by $-i$.   Then

$$\sum \text{residues} = -\frac{1}{4} t(e^{it} + e^{-it}) - \frac{1}{4} i(e^{it} - e^{-it})$$

$$= -\frac{1}{2} t \cos t + \frac{1}{2} \sin t = \frac{1}{2}(\sin t - t \cos t)$$

Compare with Problem 4.37, page 113.

INVERSE LAPLACE TRANSFORMS OF FUNCTIONS WITH BRANCH POINTS

14.9.   Find $\mathcal{L}^{-1} \left\{ \dfrac{e^{-a\sqrt{s}}}{s} \right\}$ by use of the complex inversion formula.

By the complex inversion formula, the required inverse Laplace transform is given by

$$f(t) = \frac{1}{2\pi i} \int_{\gamma - i\infty}^{\gamma + i\infty} \frac{e^{st - a\sqrt{s}}}{s} ds \qquad (1)$$

Since $s = 0$ is a branch point of the integrand, we consider

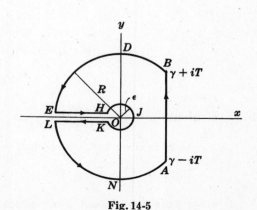

Fig. 14-5

$$\frac{1}{2\pi i}\oint_C \frac{e^{st-a\sqrt{s}}}{s}ds = 2\pi i\int_{AB}\frac{e^{st-a\sqrt{s}}}{s}ds + \frac{1}{2\pi i}\int_{BDE}\frac{e^{st-a\sqrt{s}}}{s}ds$$

$$+ \frac{1}{2\pi i}\int_{EH}\frac{e^{st-a\sqrt{s}}}{s}ds + \frac{1}{2\pi i}\int_{HJK}\frac{e^{st-a\sqrt{s}}}{s}ds$$

$$+ \frac{1}{2\pi i}\int_{KL}\frac{e^{st-a\sqrt{s}}}{s}ds + \frac{1}{2\pi i}\int_{LNA}\frac{e^{st-a\sqrt{s}}}{s}ds$$

where C is the contour of Fig. 14-5 consisting of the line AB ($s = \gamma + iy$), the arcs BDE and LNA of a circle of radius R and center at origin O, and the arc HJK of a circle of radius ϵ with center at O.

Since the only singularity $s = 0$ of the integrand is not inside C, the integral on the left is zero by Cauchy's theorem. Also, the integrand satisfies the condition of Problem 14.2 so that on taking the limit as $R \to \infty$ the integrals along BDE and LNA approach zero. It follows that

$$f(t) = \lim_{\substack{R\to\infty\\\epsilon\to 0}}\frac{1}{2\pi i}\int_{AB}\frac{e^{st-a\sqrt{s}}}{s}ds = \frac{1}{2\pi i}\int_{\gamma-i\infty}^{\gamma+i\infty}\frac{e^{st-a\sqrt{s}}}{s}ds$$

$$= -\lim_{\substack{R\to\infty\\\epsilon\to 0}}\frac{1}{2\pi i}\left\{\int_{EH}\frac{e^{st-a\sqrt{s}}}{s}ds + \int_{HJK}\frac{e^{st-a\sqrt{s}}}{s}ds + \int_{KL}\frac{e^{st-a\sqrt{s}}}{s}ds\right\} \quad (2)$$

Along EH, $s = xe^{\pi i}$, $\sqrt{s} = \sqrt{x}\,e^{\pi i/2} = i\sqrt{x}$ and as s goes from $-R$ to $-\epsilon$, x goes from R to ϵ. Hence we have

$$\int_{EH}\frac{e^{st-a\sqrt{s}}}{s}ds = \int_{-R}^{-\epsilon}\frac{e^{st-a\sqrt{s}}}{s}ds = \int_R^\epsilon \frac{e^{-xt-ai\sqrt{x}}}{x}dx$$

Similarly, along KL, $s = xe^{-\pi i}$, $\sqrt{s} = \sqrt{x}\,e^{-\pi i/2} = -i\sqrt{x}$ and as s goes from $-\epsilon$ to $-R$, x goes from ϵ to R. Then

$$\int_{KL}\frac{e^{st-a\sqrt{s}}}{s}ds = \int_{-\epsilon}^{-R}\frac{e^{st-a\sqrt{s}}}{s}ds = \int_\epsilon^R \frac{e^{-xt+ai\sqrt{x}}}{x}dx$$

Along HJK, $s = \epsilon e^{i\theta}$ and we have

$$\int_{HJK}\frac{e^{st-a\sqrt{s}}}{s}ds = \int_\pi^{-\pi}\frac{e^{\epsilon e^{i\theta}t - a\sqrt{\epsilon}e^{i\theta/2}}}{\epsilon e^{i\theta}}i\epsilon e^{i\theta}d\theta$$

$$= i\int_\pi^{-\pi}e^{\epsilon e^{i\theta}t - a\sqrt{\epsilon}e^{i\theta/2}}d\theta$$

Thus (2) becomes

$$f(t) = -\lim_{\substack{R\to\infty\\\epsilon\to 0}}\frac{1}{2\pi i}\left\{\int_R^\epsilon\frac{e^{-xt-ai\sqrt{x}}}{x}dx + \int_\epsilon^R\frac{e^{-xt+ai\sqrt{x}}}{x}dx + i\int_\pi^{-\pi}e^{\epsilon e^{i\theta}t - a\sqrt{\epsilon}e^{i\theta/2}}d\theta\right\}$$

$$= -\lim_{\substack{R\to\infty\\\epsilon\to 0}}\frac{1}{2\pi i}\left\{\int_\epsilon^R\frac{e^{-xt}(e^{ai\sqrt{x}}-e^{-ai\sqrt{x}})}{x}dx + i\int_\pi^{-\pi}e^{\epsilon e^{i\theta}t - a\sqrt{\epsilon}e^{i\theta/2}}d\theta\right\}$$

$$= -\lim_{\substack{R\to\infty\\\epsilon\to 0}}\frac{1}{2\pi i}\left\{2i\int_\epsilon^R\frac{e^{-xt}\sin a\sqrt{x}}{x}dx + i\int_\pi^{-\pi}e^{\epsilon e^{i\theta}t - a\sqrt{\epsilon}e^{i\theta/2}}d\theta\right\}$$

Since the limit can be taken underneath the integral sign, we have

$$\lim_{\epsilon\to 0}\int_\pi^{-\pi}e^{\epsilon e^{i\theta}t - a\sqrt{\epsilon}e^{i\theta/2}}d\theta = \int_\pi^{-\pi}1\,d\theta = -2\pi$$

and so we find

$$f(t) = 1 - \frac{1}{\pi}\int_0^\infty\frac{e^{-xt}\sin a\sqrt{x}}{x}dx \quad (3)$$

This can be written (see Problem 14.10) as

$$f(t) = 1 - \mathrm{erf}\,(a/2\sqrt{t}) = \mathrm{erfc}\,(a/2\sqrt{t}) \quad (4)$$

14.10. Prove that $\dfrac{1}{\pi} \displaystyle\int_0^\infty \dfrac{e^{-xt} \sin a\sqrt{x}}{x} \, dx = \operatorname{erf}(a/2\sqrt{t})$ and thus establish the final result (4) of Problem 14.9.

Letting $x = u^2$, the required integral becomes

$$I = \frac{2}{\pi} \int_0^\infty \frac{e^{-u^2 t} \sin au}{u} \, du$$

Then differentiating with respect to a and using Problem 9, we have

$$\frac{\partial I}{\partial a} = \frac{2}{\pi} \int_0^\infty e^{-u^2 t} \cos au \, du = \frac{2}{\pi}\left(\frac{\sqrt{\pi}}{2\sqrt{t}} e^{-a^2/4t}\right) = \frac{1}{\sqrt{\pi t}} e^{-a^2/4t}$$

Hence, using the fact that $I = 0$ when $a = 0$,

$$I = \int_0^a \frac{1}{\sqrt{\pi t}} e^{-p^2/4t} \, dp = \frac{2}{\sqrt{\pi}} \int_0^{a/2\sqrt{t}} e^{-u^2} \, du = \operatorname{erf}(a/2\sqrt{t})$$

and the required result is established.

14.11. Find $\mathcal{L}^{-1}\{e^{-a\sqrt{s}}\}$.

If $\mathcal{L}\{f(t)\} = F(s)$, then we have $\mathcal{L}\{f'(t)\} = sF(s) - f(0) = sf(s)$ if $F(0) = 0$. Thus if $\mathcal{L}^{-1}\{F(s)\} = f(t)$ and $f(0) = 0$, then $\mathcal{L}^{-1}\{sF(s)\} = f'(t)$.

By Problems 14.9 and 14.10, we have

$$f(t) = \operatorname{erfc}(a/2\sqrt{t}) = 1 - \frac{2}{\sqrt{\pi}} \int_0^{a/2\sqrt{t}} e^{-u^2} \, du$$

so that $f(0) = 0$ and

$$F(s) = \mathcal{L}\{f(t)\} = \frac{e^{-a\sqrt{s}}}{s}$$

Then it follows that

$$\mathcal{L}^{-1}\{e^{-a\sqrt{s}}\} = f'(t) = \frac{d}{dt}\left\{1 - \frac{2}{\sqrt{\pi}} \int_0^{a/2\sqrt{t}} e^{-u^2} \, du\right\}$$

$$= \frac{a}{2\sqrt{\pi}} t^{-3/2} e^{-a^2/4t}$$

INVERSE LAPLACE TRANSFORMS OF FUNCTIONS WITH INFINITELY MANY SINGULARITIES

14.12. Find all the singularities of $F(s) = \dfrac{\cosh x\sqrt{s}}{s \cosh \sqrt{s}}$ where $0 < x < 1$.

Because of the presence of $\sqrt{s}$, it would appear that $s = 0$ is a branch point. That this is not so, however, can be seen by noting that

$$F(s) = \frac{\cosh x\sqrt{s}}{s \cosh \sqrt{s}} = \frac{1 + (x\sqrt{s})^2/2! + (x\sqrt{s})^4/4! + \cdots}{s\{1 + (\sqrt{s})^2/2! + (\sqrt{s})^4/4! + \cdots\}}$$

$$= \frac{1 + x^2 s/2! + x^4 s^2/4! + \cdots}{s\{1 + s/2! + s^2/4! + \cdots\}}$$

from which it is evident that there is no branch point at $s = 0$. However, there is a simple pole at $s = 0$.

The function $F(s)$ also has infinitely many poles given by the roots of the equation

$$\cosh \sqrt{s} = \frac{e^{\sqrt{s}} + e^{-\sqrt{s}}}{2} = 0$$

These occur where $\qquad e^{2\sqrt{s}} = -1 = e^{\pi i + 2k\pi i} \qquad k = 0, \pm 1, \pm 2, \ldots$

from which $\qquad \sqrt{s} = (k + \tfrac{1}{2})\pi i \qquad$ or $\qquad s = -(k + \tfrac{1}{2})^2 \pi^2$

These are simple poles.

Thus $F(s)$ has simple poles at

$$s = 0 \quad\text{and}\quad s = s_n \quad\text{where}\quad s_n = -(n - \tfrac{1}{2})^2\pi^2, \ n = 1, 2, 3, \ldots$$

14.13. Find $\mathcal{L}^{-1}\left\{\dfrac{\cosh x\sqrt{s}}{s \cosh \sqrt{s}}\right\}$ where $0 < x < 1$.

The required inverse can be found by using the Bromwich contour of Fig. 14-6. The line AB is chosen so as to lie to the right of all the poles which, as seen in Problem 14.12, are given by

$$s = 0 \quad\text{and}\quad s = s_n = -(n - \tfrac{1}{2})^2\pi^2, \ n = 1, 2, 3, \ldots$$

We choose the Bromwich contour so that the curved portion $BDEFGHA$ is an arc of a circle Γ_m with center at the origin and radius

$$R_m = m^2\pi^2$$

where m is a positive integer. This choice insures that the contour does not pass through any of the poles.

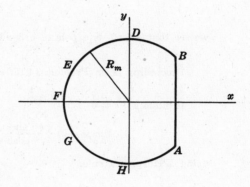

Fig. 14-6

We now find the residues of $\qquad \dfrac{e^{st}\cosh x\sqrt{s}}{s \cosh \sqrt{s}}$

at the poles. We have:

Residue at $s = 0$ is $\qquad \displaystyle\lim_{s\to 0}(s - 0)\left\{\frac{e^{st}\cosh x\sqrt{s}}{s \cosh \sqrt{s}}\right\} = 1$

Residue at $s = -(n - \tfrac{1}{2})^2\pi^2, \ n = 1, 2, 3, \ldots$ is

$$\lim_{s\to s_n}(s - s_n)\left\{\frac{e^{st}\cosh x\sqrt{s}}{s \cosh \sqrt{s}}\right\} = \lim_{s\to s_n}\left\{\frac{s - s_n}{\cosh \sqrt{s}}\right\}\lim_{s\to s_n}\left\{\frac{e^{st}\cosh x\sqrt{s}}{s}\right\}$$

$$= \lim_{s\to s_n}\left\{\frac{1}{(\sinh \sqrt{s})(1/2\sqrt{s})}\right\}\lim_{s\to s_n}\left\{\frac{e^{st}\cosh x\sqrt{s}}{s}\right\}$$

$$= \frac{4(-1)^n}{\pi(2n - 1)}\, e^{-(n-\frac{1}{2})^2\pi^2 t}\cos(n - \tfrac{1}{2})\pi x$$

If C_m is the contour of Fig. 14-6, then

$$\frac{1}{2\pi i}\oint_{C_m}\frac{e^{st}\cosh x\sqrt{s}}{s \cosh \sqrt{s}}\, ds = 1 + \frac{4}{\pi}\sum_{n=1}^{m}\frac{(-1)^n}{2n - 1}\, e^{-(n-\frac{1}{2})^2\pi^2 t}\cos(n - \tfrac{1}{2})\pi x$$

Taking the limit as $m \to \infty$ and noting that the integral around Γ_m approaches zero, we find

$$\mathcal{L}^{-1}\left\{\frac{\cosh x\sqrt{s}}{s \cosh \sqrt{s}}\right\} = 1 + \frac{4}{\pi}\sum_{n=1}^{\infty}\frac{(-1)^n}{2n - 1}\, e^{-(n-\frac{1}{2})^2\pi^2 t}\cos(n - \tfrac{1}{2})\pi x$$

$$= 1 + \frac{4}{\pi}\sum_{n=1}^{\infty}\frac{(-1)^n}{2n - 1}\, e^{-(2n-1)^2\pi^2 t/4}\cos\frac{(2n - 1)\pi x}{2}$$

14.14. Find $\mathcal{L}^{-1}\left\{\dfrac{\sinh sx}{s^2 \cosh sa}\right\}$ where $0 < x < a$.

The function $F(s) = \dfrac{\sinh sx}{s^2 \cosh sa}$ has poles at $s = 0$ and at values of s for which $\cosh sa = 0$, i.e.,

$$s = s_k = (k + \tfrac{1}{2})\pi i/a \qquad k = 0, \pm 1, \pm 2, \ldots$$

Because of the presence of s^2, it would appear that $s = 0$ is a pole of order two. However, by observing that near $s = 0$,

$$\frac{\sinh sx}{s^2 \cosh sa} = \frac{sx + (sx)^3/3! + (sx)^5/5! + \cdots}{s^2\{1 + (sa)^2/2! + (sa)^4/4! + \cdots\}}$$

$$= \frac{x + s^2 x^3/3! + s^4 x^5/5!}{s\{1 + s^2 a^2/2! + s^4 a^4/4! + \cdots\}}$$

we see that $s = 0$ is a pole of order one, i.e. a simple pole. The poles s_k are also simple poles.

Proceeding as in Problem 14.13, we obtain the residues of $e^{st} f(s)$ at these poles.

Residue at $s = 0$ is

$$\lim_{s \to 0} (s - 0)\left\{\frac{e^{st} \sinh sx}{s^2 \cosh sa}\right\} = \left\{\lim_{s \to 0} \frac{\sinh sx}{s}\right\}\left\{\lim_{s \to 0} \frac{e^{st}}{\cosh sa}\right\} = x$$

using L'Hospital's rule.

Residue at $s = s_k$ is

$$\lim_{s \to s_k} (s - s_k)\left\{\frac{e^{st} \sinh sx}{s^2 \cosh sa}\right\}$$

$$= \left\{\lim_{s \to s_k} \frac{s - s_k}{\cosh sa}\right\}\left\{\lim_{s \to s_k} \frac{e^{st} \sinh sx}{s^2}\right\}$$

$$= \left\{\lim_{s \to s_k} \frac{1}{a \sinh sa}\right\}\left\{\lim_{s \to s_k} \frac{e^{st} \sinh sx}{s^2}\right\}$$

$$= \frac{1}{ai \sin (k + \tfrac{1}{2})\pi} \cdot \frac{e^{(k + \frac{1}{2})\pi it/a}\, i \sin (k + \tfrac{1}{2})\pi x/a}{-(k + \tfrac{1}{2})^2 \pi^2/a^2}$$

$$= -\frac{a(-1)^k e^{(k + \frac{1}{2})\pi it/a} \sin (k + \tfrac{1}{2})\pi x/a}{\pi^2 (k + \tfrac{1}{2})^2}$$

By an appropriate limiting procedure similar to that used in Problem 14.13, we find on taking the sum of the residues the required result,

$$\mathcal{L}^{-1}\left\{\frac{\sinh sx}{s^2 \cosh sa}\right\} = x - \frac{a}{\pi^2} \sum_{k = -\infty}^{\infty} \frac{(-1)^k e^{(k + \frac{1}{2})\pi it/a} \sin (k + \tfrac{1}{2})\pi x/a}{(k + \tfrac{1}{2})^2}$$

$$= x + \frac{2a}{\pi^2} \sum_{n = 1}^{\infty} \frac{(-1)^n \cos (n - \tfrac{1}{2})\pi t/a \sin (n - \tfrac{1}{2})\pi x/a}{(n - \tfrac{1}{2})^2}$$

$$= x + \frac{8a}{\pi^2} \sum_{n = 1}^{\infty} \frac{(-1)^n}{(2n - 1)^2} \sin \frac{(2n - 1)\pi x}{2a} \cos \frac{(2n - 1)\pi t}{2a}$$

APPLICATIONS TO BOUNDARY-VALUE PROBLEMS

14.15. A semi-infinite solid $x > 0$ [see Fig. 14-7] is initially at temperature zero. At time $t = 0$, a constant temperature $u_0 > 0$ is applied and maintained at the face $x = 0$. Find the temperature at any point of the solid at any later time $t > 0$.

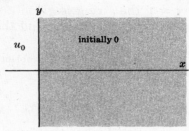

Fig. 14-7

The boundary-value problem for the determination of the temperature $u(x, t)$ at any point x and any time t is

$$\frac{\partial u}{\partial t} = \kappa \frac{\partial^2 u}{\partial x^2} \qquad x > 0,\ t > 0$$

$$u(x, 0) = 0, \qquad u(0, t) = u_0, \qquad |u(x, t)| < M$$

where the last condition expresses the requirement that the temperature is bounded for all x and t.

Taking Laplace transforms, we find

$$sU - u(x, 0) = \kappa \frac{d^2U}{dx^2} \qquad \text{or} \qquad \frac{d^2U}{dx^2} - \frac{s}{\kappa} U = 0 \tag{1}$$

where

$$U(0, s) = \mathcal{L}\{u(0, t)\} = \frac{u_0}{s} \tag{2}$$

and $U = U(x, s)$ is required to be bounded.

Solving (1), we find

$$U(x, s) = c_1 e^{\sqrt{s/\kappa}\, x} + c_2 e^{-\sqrt{s/\kappa}\, x}$$

Then we choose $c_1 = 0$ so that u is bounded as $x \to \infty$, and we have

$$U(x, s) = c_2 e^{-\sqrt{s/\kappa}\, x} \tag{3}$$

From (2) we have $c_2 = u_0/s$, so that

$$U(x, s) = \frac{u_0}{s} e^{-\sqrt{s/\kappa}\, x}$$

Hence by Problems 14.9 and 14.10 we find

$$u(x, t) = u_0 \operatorname{erfc}(x/2\sqrt{\kappa t}) = u_0 \left\{ 1 - \frac{2}{\sqrt{\pi}} \int_0^{x/2\sqrt{\kappa t}} e^{-u^2}\, du \right\}$$

14.16. Work Problem 15 if at $t = 0$ the temperature applied is given by $g(t)$, $t > 0$.

The boundary-value problem in this case is the same as in the preceding problem except that the boundary condition $u(0, t) = u_0$ is replaced by $u(0, t) = g(t)$. Then if the Laplace transform of $g(t)$ is $G(s)$, we find from (3) of Problem 14.15 that $c_2 = G(s)$ and so

$$U(x, s) = G(s) e^{-\sqrt{s/\kappa}\, x}$$

Now by Problem 14.11,

$$\mathcal{L}^{-1}\{e^{-\sqrt{s/\kappa}\, x}\} = \frac{x}{2\sqrt{\pi\kappa}} t^{-3/2} e^{-x^2/4\kappa t}$$

Hence by the convolution theorem,

$$u(x, t) = \int_0^t \frac{x}{2\sqrt{\pi\kappa}} u^{-3/2} e^{-x^2/4\kappa u}\, g(t - u)\, du$$

$$= \frac{2}{\sqrt{\pi}} \int_{x/2\sqrt{\kappa t}}^{\infty} e^{-v^2}\, g\left(t - \frac{x^2}{4\kappa v^2}\right) dv$$

on letting $v = x^2/4\kappa u$.

14.17. A tightly stretched flexible string has its ends fixed at $x = 0$ and $x = l$. At time $t = 0$ the string is given a shape defined by $f(x) = \mu x(l - x)$, where μ is a constant, and then released. Find the displacement of any point x of the string at any time $t > 0$.

The boundary-value problem is

$$\frac{\partial^2 y}{\partial t^2} = a^2 \frac{\partial^2 y}{\partial x^2} \qquad 0 < x < l,\ t > 0$$

$$y(0, t) = 0, \qquad y(l, t) = 0, \qquad y(x, 0) = \mu x(l - x), \qquad y_t(x, 0) = 0$$

Taking Laplace transforms, we find, if $Y(x, s) = \mathcal{L}\{y(x, t)\}$,

$$s^2 Y - s\, y(x, 0) - y_t(x, 0) = a^2 \frac{d^2 Y}{dx^2}$$

or

$$\frac{d^2 Y}{dx^2} - \frac{s^2}{a^2} Y = -\frac{\mu s x(l - x)}{a^2} \tag{1}$$

where

$$Y(0, s) = 0, \qquad Y(l, s) = 0 \tag{2}$$

The general solution of (1) is

$$Y = c_1 \cosh \frac{sx}{a} + c_2 \sinh \frac{sx}{a} + \frac{\mu x(l - x)}{s} - \frac{2a^2 \mu}{s^3} \tag{3}$$

Then from conditions (2) we find

$$c_1 = \frac{2a^2 \mu}{s^3}, \qquad c_2 = \frac{2a^2 \mu}{s^3}\left(\frac{1 - \cosh sl/a}{\sinh sl/a}\right) = -\frac{2a^2 \mu}{s^3}\tanh sl/2a \tag{4}$$

so that (3) becomes $\qquad Y = \dfrac{2a^2 \mu}{s^3} \dfrac{\cosh s(2x - l)/2a}{\cosh sl/2a} + \dfrac{\mu x(l - x)}{s} - \dfrac{2a^2 \mu}{s^3}$

By using residues we find

$$y(x, t) = a^2 \mu \left\{ t^2 + \left(\frac{2x - l}{2a}\right)^2 - \left(\frac{l}{2a}\right)^2 \right\}$$

$$- \frac{32 a^2 \mu}{\pi^3}\left(\frac{l}{2a}\right)^2 \sum_{n=1}^{\infty} \frac{(-1)^n}{(2n - 1)^3} \cos \frac{(2n - 1)\pi(2x - l)}{2l} \cos \frac{(2n - 1)\pi at}{l}$$

$$+ \mu x(l - x) - a^2 \mu t^2$$

or

$$y(x, t) = \frac{8\mu l^2}{\pi^3} \sum_{n=1}^{\infty} \frac{1}{(2n - 1)^3} \sin \frac{(2n - 1)\pi x}{l} \cos \frac{(2n - 1)\pi at}{l}$$

14.18. A semi-infinite beam which is initially at rest on the x axis is at time $t = 0$ given a transverse displacement h at its end $x = 0$. Determine the transverse displacement $y(x, t)$ at any position $x > 0$ and at any time $t > 0$.

The boundary-value problem is

$$\frac{\partial^2 y}{\partial t^2} + b^2 \frac{\partial^4 y}{\partial x^4} = 0 \qquad x > 0,\ t > 0 \tag{1}$$

$$y(x, 0) = 0, \quad y_t(x, 0) = 0, \quad y(0, t) = h, \quad y_{xx}(0, t) = 0, \quad |y(x, t)| < M \tag{2}$$

Taking Laplace transforms, we find

$$s^2 Y(x, s) - s\, y(x, 0) - y_t(x, 0) + b^2 \frac{d^4 Y}{dx^4} = 0 \qquad \text{or} \qquad \frac{d^4 Y}{dx^4} + \frac{s^2}{b^2} Y = 0$$

$$Y(0, s) = h/s, \qquad Y_{xx}(0, s) = 0, \qquad Y(x, s)\text{ is bounded} \tag{3}$$

The general solution of the differential equation is

$$Y(x, s) = e^{\sqrt{s/2b}\,x}(c_1 \cos \sqrt{s/2b}\,x + c_2 \sin \sqrt{s/2b}\,x) + e^{-\sqrt{s/2b}\,x}(c_3 \cos \sqrt{s/2b}\,x + c_4 \sin \sqrt{s/2b}\,x)$$

From the boundedness condition we require $c_1 = c_2 = 0$ so that

$$Y(x, s) = e^{-\sqrt{s/2b}\,x}(c_3 \cos \sqrt{s/2b}\,x + c_4 \sin \sqrt{s/2b}\,x)$$

From the first and second boundary conditions in (3), we find $c_4 = 0$ and $c_3 = h/s$ so that

$$Y(x, s) = \frac{h}{s} e^{-\sqrt{s/2b}\, x} \cos \sqrt{s/2b}\, x$$

The inverse Laplace transform is, by the complex inversion formula,

$$y(x, t) = \frac{1}{2\pi i} \int_{\gamma-i\infty}^{\gamma+i\infty} \frac{h e^{st - \sqrt{s/2b}\, x} \cos \sqrt{s/2b}\, x}{s} \, ds$$

To evaluate this we use the contour of Fig. 14-8 since $s = 0$ is a branch point. Proceeding as in Problem 14.9, we find, omitting the integrand for the sake of brevity, that

$$y(x, t) = -\lim_{\substack{R \to \infty \\ \epsilon \to 0}} \frac{1}{2\pi i} \left\{ \int_{EH} + \int_{HJK} + \int_{KL} \right\} \quad (4)$$

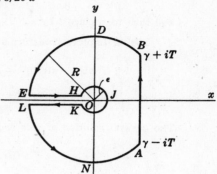

Fig. 14-8

Along EH, $s = ue^{\pi i}$, $\sqrt{s} = i\sqrt{u}$ and we find

$$\int_{EH} = \int_{R}^{\epsilon} \frac{h e^{-ut - i\sqrt{u/2b}\, x} \cosh \sqrt{u/2b}\, x}{u} \, du$$

Along KL, $s = ue^{-\pi i}$, $\sqrt{s} = -i\sqrt{u}$ and we find

$$\int_{KL} = \int_{\epsilon}^{R} \frac{h e^{-ut + i\sqrt{u/2b}\, x} \cosh \sqrt{u/2b}\, x}{u} \, du$$

Along HJK, $s = \epsilon e^{i\theta}$ and we find

$$\int_{HJK} = \int_{\pi}^{-\pi} h e^{\epsilon e^{i\theta} t - \sqrt{\epsilon e^{i\theta}/2b}\, x} \cos \sqrt{\epsilon e^{i\theta}/2b}\, x \, d\theta$$

Then (4) becomes

$$y(x, t) = h \left\{ 1 - \frac{1}{\pi} \int_0^\infty \frac{e^{-ut} \sin \sqrt{u/2b}\, x \cosh \sqrt{u/2b}\, x}{u} \, du \right\}$$

Letting $u/2b = v^2$, this can be written

$$y(x, t) = h \left\{ 1 - \frac{2}{\pi} \int_0^\infty \frac{e^{-2bv^2 t} \sin vx \cosh vx}{v} \, dv \right\}$$

The result can also be written in terms of *Fresnel integrals* as

$$Y(x, t) = h \left\{ 1 - \sqrt{\frac{2}{\pi}} \int_0^{x/\sqrt{bt}} (\cos w^2 + \sin w^2) \, dw \right\}$$

14.19. An infinitely long circular cylinder of unit radius has a constant initial temperature u_0. At $t = 0$ a temperature of $0°C$ is applied to the surface and is maintained. Find the temperature at any point of the cylinder at any later time t.

If (r, ϕ, z) are cylindrical coordinates of any point of the cylinder and the cylinder has its axis coincident with the z axis [see Fig. 14-9], it is clear that the temperature is independent of ϕ and z and can thus be denoted by $u(r, t)$. The boundary-value problem is

$$\frac{\partial u}{\partial t} = \kappa \left(\frac{\partial^2 u}{\partial r^2} + \frac{1}{r} \frac{\partial u}{\partial r} \right) \quad 0 < r < 1,\ t > 0 \quad (1)$$

$$u(1, t) = 0, \quad u(r, 0) = u_0, \quad |u(r, t)| < M \quad (2)$$

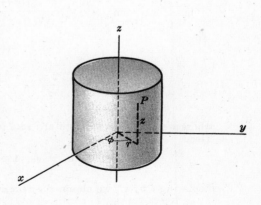

Fig. 14-9

It is convenient to consider instead of (1) the equation

$$\frac{\partial u}{\partial t} = \frac{\partial^2 u}{\partial r^2} + \frac{1}{r}\frac{\partial u}{\partial r}$$

and then to replace t by κt.

Taking Laplace transforms, we find

$$sU - u(r,0) = \frac{d^2 U}{dr^2} + \frac{1}{r}\frac{dU}{dr} \quad\text{or}\quad \frac{d^2 U}{dr^2} + \frac{1}{r}\frac{dU}{dr} - sU = -u_0$$

$$U(1,s) = 0, \qquad U(r,s) \text{ is bounded}$$

The general solution of this equation is given in terms of Bessel functions as

$$U(r,s) = c_1 J_0(i\sqrt{s}\,r) + c_2 Y_0(i\sqrt{s}\,r) + \frac{u_0}{s}$$

Since $Y_0(i\sqrt{s}\,r)$ is unbounded as $r \to 0$, we must choose $c_2 = 0$. Then

$$U(r,s) = c_1 J_0(i\sqrt{s}\,r) + \frac{u_0}{s}$$

From $U(1,s) = 0$, we find

$$c_1 J_0(i\sqrt{s}) + \frac{u_0}{s} = 0 \quad\text{or}\quad c_1 = -\frac{u_0}{s\,J_0(i\sqrt{s})}$$

Thus

$$U(r,s) = \frac{u_0}{s} - \frac{u_0 J_0(i\sqrt{s}\,r)}{s\,J_0(i\sqrt{s})}$$

By the inversion formula,

$$u(r,t) = u_0 - \frac{u_0}{2\pi i}\int_{\gamma-i\infty}^{\gamma+i\infty} \frac{e^{st} J_0(i\sqrt{s}\,r)}{s\,J_0(i\sqrt{s})}\,ds$$

Now $J_0(i\sqrt{s})$ has simple zeros where $i\sqrt{s} = \lambda_1, \lambda_2, \ldots \lambda_n, \ldots$. Thus the integrand has simple poles at $s = -\lambda_n^2$, $n = 1, 2, 3, \ldots$ and also at $s = 0$. Furthermore it can be shown that the integrand satisfies the conditions of Problem 14.2 so that the method of residues can be used.

We have:

Residue of integrand at $s = 0$ is

$$\lim_{s\to 0} s\,\frac{e^{st} J_0(i\sqrt{s}\,r)}{s\,J_0(i\sqrt{s})} = 1$$

Residue of integrand at $s = -\lambda_n^2$ is

$$\lim_{s\to -\lambda_n^2} (s+\lambda_n^2)\frac{e^{st} J_0(i\sqrt{s}\,r)}{s\,J_0(i\sqrt{s})} = \left\{\lim_{s\to -\lambda_n^2}\frac{(s+\lambda_n^2)}{J_0(i\sqrt{s})}\right\}\left\{\lim_{s\to -\lambda_n^2}\frac{e^{st} J_0(i\sqrt{s}\,r)}{s}\right\}$$

$$= \left\{\lim_{s\to -\lambda_n^2}\frac{1}{J_0'(i\sqrt{s})\,i/2\sqrt{s}}\right\}\left\{\frac{e^{-\lambda_n^2 t} J_0(\lambda_n r)}{-\lambda_n^2}\right\}$$

$$= -\frac{2e^{-\lambda_n^2 t} J_0(\lambda_n r)}{\lambda_n J_1(\lambda_n)}$$

where we have used L'Hospital's rule in evaluating the limit and also the fact that $J_0'(u) = -J_1(u)$. Then

$$u(r,t) = u_0 - u_0\left\{1 - \sum_{n=1}^{\infty}\frac{2e^{-\lambda_n^2 t} J_0(\lambda_n r)}{\lambda_n J_1(\lambda_n)}\right\} = 2u_0\sum_{n=1}^{\infty}\frac{e^{-\lambda_n^2 t} J_0(\lambda_n r)}{\lambda_n J_1(\lambda_n)}$$

Replacing t by κt, we obtain the required solution

$$u(r,t) = 2u_0\sum_{n=1}^{\infty}\frac{e^{-\kappa\lambda_n^2 t} J_0(\lambda_n r)}{\lambda_n J_1(\lambda_n)}$$

Supplementary Problems

THE COMPLEX INVERSION FORMULA AND USE OF RESIDUE THEOREM

14.20. Use the complex inversion formula to evaluate

(a) $\mathcal{L}^{-1}\left\{\dfrac{s}{s^2+a^2}\right\}$ (b) $\mathcal{L}^{-1}\left\{\dfrac{1}{s^2+a^2}\right\}$ (c) $\mathcal{L}^{-1}\left\{\dfrac{1}{(s+1)(s^2+1)}\right\}$

14.21. Find the inverse Laplace transform of each of the following using the complex inversion formula:
(a) $1/(s+1)^2$,   (b) $1/s^3(s^2+1)$.

14.22. (a) Show that $F(s) = \dfrac{1}{s^2-3s+2}$ satisfies the conditions of the inversion formula.   (b) Find $\mathcal{L}^{-1}\{F(s)\}$.

14.23. Evaluate $\mathcal{L}^{-1}\left\{\dfrac{s^2}{(s^2+4)^2}\right\}$ justifying all steps.

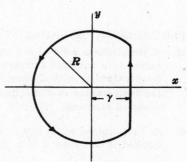

Fig. 14-10

14.24. (a) Evaluate $\mathcal{L}^{-1}\left\{\dfrac{s}{(s^2+1)^3}\right\}$ justifying all steps and

(b) check your answer.

14.25. (a) Evaluate $\dfrac{1}{2\pi i}\displaystyle\oint_C \dfrac{se^{st}}{(s^2-1)^2}\,ds$ around the contour C

shown in Fig. 14-10 where $R \geqq 3$ and $\gamma > 1$.

(b) Give an interpretation of your answer as far as Laplace transform theory is concerned.

14.26. Use the inversion formula to evaluate $\mathcal{L}^{-1}\left\{\dfrac{s}{(s+a)(s-b)^2}\right\}$ where a and b are any positive constants.

14.27. Use the inversion formula to work: (a) Problem 4.39, page 113, (b) Problem 4.40, page 113, (c) Problem 4.41, page 114, (d) Problem 4.78(c), page 118.

14.28. Complete the proofs of (a) Case 3 and (b) Case 4 of Problem 14.2.

INVERSE LAPLACE TRANSFORMS OF FUNCTIONS WITH BRANCH POINTS

14.29. Find $\mathcal{L}^{-1}\{e^{-\sqrt{s}}\}$ using the complex inversion formula.

14.30. Find $\mathcal{L}^{-1}\left\{\dfrac{1}{\sqrt{s}}\right\}$ by the inversion formula.

14.31. Show that $\mathcal{L}^{-1}\left\{\dfrac{1}{s\sqrt{s+1}}\right\} = \operatorname{erf}\sqrt{t}$ by using the inversion formula.

14.32. Find $\mathcal{L}^{-1}\left\{\dfrac{\sqrt{s}}{s-1}\right\}$ by using the complex inversion formula.

14.33. (a) Use the complex inversion formula to evaluate $\mathcal{L}^{-1}\{s^{-1/3}\}$ and (b) check your result by another method.

14.34. Evaluate $\mathcal{L}^{-1}\{\ln(1+1/s)\}$ by using the inversion formula.

14.35. Evaluate $\mathcal{L}^{-1}\{\ln(1+1/s^2)\}$ by the inversion formula.

INVERSE LAPLACE TRANSFORMS OF FUNCTIONS WITH INFINITELY MANY SINGULARITIES

14.36. Find $\mathcal{L}^{-1}\left\{\dfrac{1}{s(e^s+1)}\right\}$ using the complex inversion formula.

14.37. Prove that $\mathcal{L}^{-1}\left\{\dfrac{1}{s\cosh s}\right\} = 1 - \dfrac{4}{\pi}\left\{\cos\dfrac{\pi t}{2} - \dfrac{1}{3}\cos\dfrac{3\pi t}{2} + \dfrac{1}{5}\cos\dfrac{5\pi t}{2} - \cdots\right\}$.

14.38. Find $\mathcal{L}^{-1}\left\{\dfrac{1}{s^2\sinh s}\right\}$.

14.39. By using the complex inversion formula, prove that

$$\mathcal{L}^{-1}\left\{\frac{1}{s^3 \sinh as}\right\} = \frac{t(t^2 - a^2)}{6a} - \frac{2a^2}{\pi^3}\sum_{n=1}^{\infty}\frac{(-1)^n}{n^3}\sin\frac{n\pi t}{a}$$

14.40. Show that $\mathcal{L}^{-1}\left\{\dfrac{1}{(s^2 + \omega^2)(1 + e^{-2as})}\right\} = \dfrac{\sin\omega(t + a)}{2\omega} + \dfrac{1}{a}\sum_{n=1}^{\infty}\dfrac{\cos(2n-1)\pi t/2a}{\omega^2 - (2n-1)^2\pi^2/4a^2}$.

14.41. Show that for $0 < x < a$,

$$\mathcal{L}^{-1}\left\{\frac{\sinh\sqrt{s}\,(a - x)}{\sinh\sqrt{s}\,a}\right\} = \frac{a - x}{a} - \frac{2}{\pi}\sum_{n=1}^{\infty}\frac{e^{-n^2\pi^2 t/a^2}}{n}\sin\frac{n\pi x}{a}$$

APPLICATIONS TO BOUNDARY-VALUE PROBLEMS

14.42. A semi-infinite solid $x \geqq 0$ is initially at temperature zero. At $t = 0$ the face $x = 0$ is suddenly raised to a constant temperature u_0 and kept at this temperature for a time t_0, after which the temperature is immediately reduced to zero. Show that after an additional time t_0 has elapsed, the temperature is a maximum at a distance given by $x = 2\sqrt{\kappa t_0 \ln 2}$ where κ is the diffusivity, assumed constant.

14.43. A semi-infinite solid $x > 0$ has its initial temperature equal to zero. A constant heat flux A is applied at the face $x = 0$ so that $-Ku_x(0, t) = A$. Show that the temperature at the face after time t is $\dfrac{A}{K}\sqrt{\dfrac{\kappa t}{\pi}}$.

14.44. Find the temperature at any point $x > 0$ of the solid in Problem 14.43.

14.45. A solid $0 \leqq x \leqq l$ is insulated at both ends $x = 0$ and $x = l$. If the initial temperature is equal to $ax(l - x)$ where a is a constant, find the temperature at any point x and at any time t.

14.46. A tightly stretched flexible string has its ends fixed at $x = 0$ and $x = l$. At $t = 0$ its midpoint is displaced a distance h and released. Find the resulting displacement at any time $t > 0$.

14.47. Solve the boundary-value problem

$$y_{tt} = y_{xx} + g \qquad 0 < x < \pi,\ t > 0$$

$$y(0, t) = 0,\quad y(\pi, t) = 0,\quad y(x, 0) = \mu\,x(\pi - x),\quad y_t(x, 0) = 0$$

and interpret physically.

14.48. At $t = 0$, a semi-infinite solid $x > 0$ which is at temperature zero has a sinusoidal heat flux applied to the face $x = 0$ so that $-Ku_x(0, t) = A + B\sin\omega t,\ t > 0$. Show that the temperature of the face at any later time is given by

$$\frac{2\sqrt{\kappa}\,A}{K\sqrt{\pi}}\,t^{1/2} + \frac{2B\sqrt{\kappa}\,\omega}{K}\left\{\left(\int_0^{\sqrt{t}}\cos\omega v^2\,dv\right)\sin\omega t - \left(\int_0^{\sqrt{t}}\sin\omega v^2\,dv\right)\cos\omega t\right\}$$

14.49. A solid, $0 < x < l$, is initially at constant temperature u_0 while the ends $x = 0$ and $x = l$ are maintained at temperature zero. Show that the temperature at any position x at any time t is given by

$$u(x, t) = u_0\,\mathrm{erf}\left(\frac{x}{2\sqrt{\kappa t}}\right) + u_0\sum_{n=1}^{\infty}(-1)^n\left\{\mathrm{erf}\left(\frac{nl - x}{2\sqrt{\kappa t}}\right) - \mathrm{erf}\left(\frac{nl + x}{2\sqrt{\kappa t}}\right)\right\}$$

14.50. A beam has its ends hinged at $x = 0$ and $x = l$. At time $t = 0$, a concentrated transverse load of magnitude w is suddenly applied at the midpoint. Show that the resulting transverse displacement of any point x of the beam at any time $t > 0$ is

$$y(x, t) = \frac{wx}{12EI}\left(\tfrac{3}{4}l^2 - x^2\right) - \frac{2wl^3}{\pi^4 EI}\left\{\frac{\sin\pi x/l}{1^4} + \frac{\sin 3\pi x/l}{3^4} + \frac{\sin 5\pi x/l}{5^4} + \cdots\right\}$$

if $0 < x < l/2$, while the corresponding result for $l/2 < x < l$ is obtained by symmetry.

14.51. An infinite circular cylinder of unit radius has its initial temperature zero. A constant flux A is applied to the convex surface. Show that the temperature at points distant r from the axis at any time t is given by

$$u(r, t) = \frac{A}{4\kappa}\{1 - 8\kappa t - 2r^2\} + \frac{2A}{\kappa}\sum_{n=1}^{\infty} e^{-\kappa\lambda_n^2 t}\frac{J_0(\lambda_n r)}{\lambda_n^2 J_0(\lambda_n)}$$

where λ_n are the positive roots of $J_0(\lambda) = 0$.

14.52. A cylinder of unit radius and height has its circular ends maintained at temperature zero while its convex surface is maintained at constant temperature u_0. Assuming that the cylinder has its axis coincident with the z axis, show that the steady-state temperature at any distance r from the axis and z from one end is

$$u(r, z) = \frac{4u_0}{\pi}\sum_{n=1}^{\infty}\frac{\sin (2n-1)\pi z}{2n-1}\frac{I_0\{(2n-1)\pi r\}}{I_0\{(2n-1)\pi\}}$$

Answers to Supplementary Problems

14.20. (a) $\cos at$, (b) $(\sin at)/a$, (c) $\frac{1}{2}(\sin t - \cos t + e^{-t})$

14.21. (a) te^{-t}, (b) $\frac{1}{2}t^2 + \cos t - 1$

14.22. (b) $e^{2t} - e^t$

14.23. $\frac{1}{4}\sin 2t + \frac{1}{2}t\cos 2t$

14.34. $(1 - e^{-t})/t$

14.35. $2(1 - \cos t)/t$

14.38. $\frac{1}{2}t^2 + \frac{2}{\pi^2}\sum_{n=1}^{\infty}\frac{(-1)^n}{n^2}(1 - \cos n\pi t)$

14.44. $\frac{A}{K}\{\sqrt{\kappa t/\pi}\, e^{-x^2/4\kappa t} - \frac{1}{2}x\, \text{erfc}\,(x/2\sqrt{\kappa t}\,)\}$

14.45. $\frac{al^2}{6} - \frac{al^2}{\pi^2}\sum_{n=1}^{\infty}\frac{e^{-4\kappa n^2\pi^2 t/l^2}}{n^2}\cos\frac{2n\pi x}{l}$

14.46. $y(x, t) = \frac{8h}{\pi^2}\sum_{n=1}^{\infty}\frac{(-1)^{n-1}}{(2n-1)^2}\sin\frac{(2n-1)\pi x}{l}\cos\frac{(2n-1)\pi at}{l}$

14.47. $y(x, t) = \frac{1}{2}gx(\pi - x) + \frac{4(2\mu x - g)}{\pi}\sum_{n=1}^{\infty}\frac{1}{(2n-1)^3}\sin (2n-1)x\cos (2n-1)t$

Chapter 15

DEFINITION OF A MATRIX

A *matrix of order* $m \times n$, or *m by n matrix*, is a rectangular array of numbers having m rows and n columns. It can be written in the form

$$A = \begin{pmatrix} a_{11} & a_{12} & a_{13} & \ldots & a_{1n} \\ a_{21} & a_{22} & a_{23} & \ldots & a_{2n} \\ \cdots\cdots\cdots\cdots\cdots\cdots\cdots \\ a_{m1} & a_{m2} & a_{m3} & \ldots & a_{mn} \end{pmatrix} \tag{1}$$

Each number a_{jk} in this matrix is called an *element*. The subscripts j and k indicate respectively the row and column of the matrix in which the element appears.

We shall often denote a matrix by a letter, such as A in (1), or by the symbol (a_{jk}) which shows a representative element.

A matrix having only one row is called a *row matrix* [or *row vector*] while a matrix having only one column is called a *column matrix* [or *column vector*]. If the number of rows m and columns n are equal the matrix is called a *square matrix* of order $n \times n$ or briefly n. A matrix is said to be a *real matrix* or *complex matrix* according as its elements are real or complex numbers.

SOME SPECIAL DEFINITIONS AND OPERATIONS INVOLVING MATRICES

1. **Equality of Matrices.** Two matrices $A = (a_{jk})$ and $B = (b_{jk})$ of the same order [i.e. equal numbers of rows and columns] are *equal* if and only if $a_{jk} = b_{jk}$.

2. **Addition of Matrices.** If $A = (a_{jk})$ and $B = (b_{jk})$ have the same order we define the *sum* of A and B as $A + B = (a_{jk} + b_{jk})$.

   Example 1. If $A = \begin{pmatrix} 2 & 1 & 4 \\ -3 & 0 & 2 \end{pmatrix}$, $B = \begin{pmatrix} 3 & -5 & 1 \\ 2 & 1 & 3 \end{pmatrix}$ then

   $$A + B = \begin{pmatrix} 2+3 & 1-5 & 4+1 \\ -3+2 & 0+1 & 2+3 \end{pmatrix} = \begin{pmatrix} 5 & -4 & 5 \\ -1 & 1 & 5 \end{pmatrix}$$

   Note that the commutative and associative laws for addition are satisfied by matrices, i.e. for any matrices A, B, C of the same order

   $$A + B = B + A, \quad A + (B + C) = (A + B) + C \tag{2}$$

3. **Subtraction of Matrices.** If $A = (a_{jk})$, $B = (b_{jk})$ have the same order, we define the *difference* of A and B as $A - B = (a_{jk} - b_{jk})$.

   Example 2. If A and B are the matrices of Example 1, then

   $$A - B = \begin{pmatrix} 2-3 & 1+5 & 4-1 \\ -3-2 & 0-1 & 2-3 \end{pmatrix} = \begin{pmatrix} -1 & 6 & 3 \\ -5 & -1 & -1 \end{pmatrix}$$

4. **Multiplication of a Matrix by a Number.** If $A = (a_{jk})$ and λ is any number [or *scalar*], we define the *product* of A by λ as $\lambda A = A\lambda = (\lambda a_{jk})$.

    Example 3. If A is the matrix of Example 1 and $\lambda = 4$, then

$$\lambda A = 4 \begin{pmatrix} 2 & 1 & 4 \\ -3 & 0 & 2 \end{pmatrix} = \begin{pmatrix} 8 & 4 & 16 \\ -12 & 0 & 8 \end{pmatrix}$$

5. **Multiplication of Matrices.** If $A = (a_{jk})$ is an $m \times n$ matrix while $B = (b_{jk})$ is an $n \times p$ matrix, then we define the *product* $A \cdot B$ or AB of A and B as the matrix $C = (c_{jk})$ where

$$c_{jk} = \sum_{l=1}^{n} a_{jl} b_{lk} \tag{3}$$

and where C is of order $m \times p$.

    Note that matrix multiplication is defined if and only if the number of columns of A is the same as the number of rows of B. Such matrices are sometimes called *conformable*.

    Example 4. Let $A = \begin{pmatrix} 2 & 1 & 4 \\ -3 & 0 & 2 \end{pmatrix}$, $D = \begin{pmatrix} 3 & 5 \\ 2 & -1 \\ 4 & 2 \end{pmatrix}$. Then

$$AD = \begin{pmatrix} (2)(3) + (1)(2) + (4)(4) & (2)(5) + (1)(-1) + (4)(2) \\ (-3)(3) + (0)(2) + (2)(4) & (-3)(5) + (0)(-1) + (2)(2) \end{pmatrix} = \begin{pmatrix} 24 & 17 \\ -1 & -11 \end{pmatrix}$$

    Note that in general $AB \neq BA$, i.e. the commutative law for multiplication of matrices is not satisfied in general. However, the associative and distributive laws are satisfied, i.e.

$$A(BC) = (AB)C, \quad A(B+C) = AB + AC, \quad (B+C)A = BA + CA \tag{4}$$

    A matrix A can be multiplied by itself if and only if it is a square matrix. The product $A \cdot A$ can in such case be written A^2. Similarly we define powers of a square matrix, i.e. $A^3 = A \cdot A^2$, $A^4 = A \cdot A^3$, etc.

6. **Transpose of a Matrix.** If we interchange rows and columns of a matrix A, the resulting matrix is called the *transpose* of A and is denoted by A^T. In symbols, if $A = (a_{jk})$ then $A^T = (a_{kj})$.

    Example 5. The transpose of $A = \begin{pmatrix} 2 & 1 & 4 \\ -3 & 0 & 2 \end{pmatrix}$ is

$$A^T = \begin{pmatrix} 2 & -3 \\ 1 & 0 \\ 4 & 2 \end{pmatrix}$$

We can prove that

$$(A+B)^T = A^T + B^T, \quad (AB)^T = B^T A^T, \quad (A^T)^T = A \tag{5}$$

7. **Symmetric and Skew-Symmetric Matrices.** A square matrix A is called *symmetric* if $A^T = A$ and *skew-symmetric* if $A^T = -A$.

    Example 6. The matrix $E = \begin{pmatrix} 2 & -4 \\ -4 & 3 \end{pmatrix}$ is symmetric while $F = \begin{pmatrix} 0 & -2 \\ 2 & 0 \end{pmatrix}$ is skew-symmetric.

    Any real square matrix [i.e. one having only real elements] can always be expressed as the sum of a real symmetric matrix and a real skew-symmetric matrix.

8. **Complex Conjugate of a Matrix.** If all elements a_{jk} of a matrix A are replaced by their complex conjugates $\bar{a}_{jk}$, the matrix obtained is called the *complex conjugate* of A and is denoted by $\bar{A}$.

9. **Hermitian and Skew-Hermitian Matrices.** A square matrix A which is the same as the complex conjugate of its transpose, i.e. if $A = \bar{A}^T$, is called *Hermitian*. If $A = -\bar{A}^T$, then A is called *skew-Hermitian*. If A is real these reduce to symmetric and skew-symmetric matrices respectively.

10. **Principal Diagonal and Trace of a Matrix.** If $A = (a_{jk})$ is a square matrix, then the diagonal which contains all elements a_{jk} for which $j = k$ is called the *principal* or *main diagonal* and the sum of all such elements is called the *trace* of A.

> **Example 7.** The principal or main diagonal of the matrix
>
> $$\begin{pmatrix} 5 & 2 & 0 \\ 3 & 1 & -2 \\ -1 & 4 & 2 \end{pmatrix}$$
>
> is indicated by the shading, and the trace of the matrix is $5 + 1 + 2 = 8$.

A matrix for which $a_{jk} = 0$ when $j \neq k$ is called a *diagonal matrix*.

11. **Unit Matrix.** A square matrix in which all elements of the principal diagonal are equal to 1 while all other elements are zero is called the *unit matrix* and is denoted by I. An important property of I is that

$$AI = IA = A, \quad I^n = I, \quad n = 1, 2, 3, \ldots \tag{6}$$

The unit matrix plays a role in matrix algebra similar to that played by the number one in ordinary algebra.

12. **Zero or Null Matrix.** A matrix whose elements are all equal to zero is called the *null* or *zero matrix* and is often denoted by O or simply 0. For any matrix A having the same order as 0 we have

$$A + 0 = 0 + A = A \tag{7}$$

Also if A and 0 are square matrices, then

$$A0 = 0A = 0 \tag{8}$$

The zero matrix plays a role in matrix algebra similar to that played by the number zero of ordinary algebra.

DETERMINANTS

If the matrix A in (1) is a square matrix, then we associate with A a number denoted by

$$\Delta = \begin{vmatrix} a_{11} & a_{12} & \ldots & a_{1n} \\ a_{21} & a_{22} & \ldots & a_{2n} \\ \hdotsfor{4} \\ a_{n1} & a_{n2} & \ldots & a_{nn} \end{vmatrix} \tag{9}$$

called the *determinant* of A of *order n*, written $\det(A)$. In order to define the value of a determinant, we introduce the following concepts.

1. **Minor.** Given any element a_{jk} of Δ we associate a new determinant of order $(n-1)$ obtained by removing all elements of the jth row and kth column called the minor of a_{jk}.

Example 8.   The minor corresponding to the element 5 in the 2nd row and 3rd column of the fourth order determinant

$$\begin{vmatrix} 2 & -1 & 1 & 3 \\ -3 & 2 & 5 & 0 \\ 1 & 0 & -2 & 2 \\ 4 & -2 & 3 & 1 \end{vmatrix} \qquad \text{is} \qquad \begin{vmatrix} 2 & -1 & 3 \\ 1 & 0 & 2 \\ 4 & -2 & 1 \end{vmatrix}$$

which is obtained by removing the elements shown shaded.

2.   Cofactor.   If we multiply the minor of a_{jk} by $(-1)^{j+k}$, the result is called the *cofactor* of a_{jk} and is denoted by A_{jk}.

Example 9.   The cofactor corresponding to the element 5 in the determinant of Example 8 is $(-1)^{2+3}$ times its minor, or

$$-\begin{vmatrix} 2 & -1 & 3 \\ 1 & 0 & 2 \\ 4 & -2 & 1 \end{vmatrix}$$

The value of a determinant is then defined as the sum of the products of the elements in any row [or column] by their corresponding cofactors and is called the *Laplace expansion*. In symbols,

$$\det A \;=\; \sum_{k=1}^{n} a_{jk} A_{jk} \tag{10}$$

We can show that this value is independent of the row [or column] used [see Problem 15.7].

THEOREMS ON DETERMINANTS

Theorem 15-1.   The value of a determinant remains the same if rows and columns are interchanged.   In symbols, $\det(A) = \det(A^T)$.

Theorem 15-2.   If all elements of any row [or column] are zero except for one element, then the value of the determinant is equal to the product of that element by its cofactor.   In particular, if all elements of a row [or column] are zero the determinant is zero.

Theorem 15-3.   An interchange of any two rows [or columns] changes the sign of the determinant.

Theorem 15-4.   If all elements in any row [or column] are multiplied by a number, the determinant is also multiplied by this number.

Theorem 15-5.   If any two rows [or columns] are the same or proportional, the determinant is zero.

Theorem 15-6.   If we express the elements of each row [or column] as the sum of two terms, then the determinant can be expressed as the sum of two determinants having the same order.

Theorem 15-7.   If we multiply the elements of any row [or column] by a given number and add to corresponding elements of any other row [or column], then the value of the determinant remains the same.

Theorem 15-8.   If A and B are square matrices of the same order, then

$$\det(AB) \;=\; \det(A)\det(B) \tag{11}$$

Theorem 15-9.   The sum of the products of the elements of any row [or column] by the cofactors of another row [or column] is zero.   In symbols,

$$\sum_{k=1}^{n} a_{qk} A_{pk} = 0 \quad \text{or} \quad \sum_{k=1}^{n} a_{kq} A_{kp} = 0 \quad \text{if } p \neq q \tag{12}$$

If $p = q$, the sum is $\det(A)$ by (10).

Theorem 15-10.   Let $v_1, v_2, \ldots, v_n$ represent row vectors [or column vectors] of a square matrix A of order n.   Then $\det(A) = 0$ if and only if there exist constants [scalars] $\lambda_1, \lambda_2, \ldots, \lambda_n$ not all zero such that

$$\lambda_1 v_1 + \lambda_2 v_2 + \cdots + \lambda_n v_n = O \tag{13}$$

where O is the null or zero row matrix.   If condition (13) is satisfied we say that the vectors $v_1, v_2, \ldots, v_n$ are *linearly dependent*.   Otherwise they are *linearly independent*.   A matrix A such that $\det(A) = 0$ is called a *singular matrix*.   If $\det(A) \neq 0$, then A is a *non-singular matrix*.

In practice we evaluate a determinant of order n by using Theorem 15-7 successively to replace all but one of the elements in a row or column by zeros and then using Theorem 15-2 to obtain a new determinant of order $n-1$.   We continue in this manner, arriving ultimately at determinants of orders 2 or 3 which are easily evaluated.

INVERSE OF A MATRIX

If for a given square matrix A there exists a matrix B such that $AB = I$, then B is called an *inverse* of A and is denoted by A^{-1}.   The following theorem is fundamental.

Theorem 15-11.   If A is a non-singular square matrix of order n [i.e. $\det(A) \neq 0$], then there exists a unique inverse A^{-1} such that $AA^{-1} = A^{-1}A = I$ and we can express A^{-1} in the following form

$$A^{-1} = \frac{(A_{jk})^T}{\det(A)} \tag{14}$$

where (A_{jk}) is the matrix of cofactors A_{jk} and $(A_{jk})^T = (A_{kj})$ is its transpose.

The following express some properties of the inverse:

$$(AB)^{-1} = B^{-1}A^{-1}, \quad (A^{-1})^{-1} = A \tag{15}$$

ORTHOGONAL AND UNITARY MATRICES

A real matrix A is called an *orthogonal matrix* if its transpose is the same as its inverse, i.e. if $A^T = A^{-1}$ or $A^T A = I$.

A complex matrix A is called a *unitary matrix* if its complex conjugate transpose is the same as its inverse, i.e. if $\bar{A}^T = A^{-1}$ or $\bar{A}^T A = I$.   It should be noted that a real unitary matrix is an orthogonal matrix.

ORTHOGONAL VECTORS

In Chapter 5 we found that the scalar or dot product of two vectors $a_1\mathbf{i} + a_2\mathbf{j} + a_3\mathbf{k}$ and $b_1\mathbf{i} + b_2\mathbf{j} + b_3\mathbf{k}$ is $a_1 b_1 + a_2 b_2 + a_3 b_3$ and that the vectors are perpendicular or orthogonal if $a_1 b_1 + a_2 b_2 + a_3 b_3 = 0$.   From the point of view of matrices we can consider these vectors as column vectors

$$A = \begin{pmatrix} a_1 \\ a_2 \\ a_3 \end{pmatrix}, \quad B = \begin{pmatrix} b_1 \\ b_2 \\ b_3 \end{pmatrix}$$

from which it follows that $A^T B = a_1 b_1 + a_2 b_2 + a_3 b_3$

This leads us to define the *scalar product of real column vectors A* and *B* as $A^T B$ and to define *A* and *B* to be *orthogonal* if $A^T B = 0$.

It is convenient to generalize this to cases where the vectors can have complex components and we adopt the following definition:

Definition 1.   Two column vectors *A* and *B* are called *orthogonal* if $\bar{A}^T B = 0$, and $\bar{A}^T B$ is called the *scalar product* of *A* and *B*.

It should be noted also that if *A* is a unitary matrix then $\bar{A}^T A = 1$, which means that the scalar product of *A* with itself is 1 or equivalently *A* is a *unit vector*, i.e. having length 1. Thus a unitary column vector is a unit vector. Because of these remarks we have the following

Definition 2.   A set of vectors $X_1, X_2, \ldots$ for which

$$\bar{X}_j^T X_k = \begin{cases} 0 & j \neq k \\ 1 & j = k \end{cases}$$

is called a *unitary set or system of vectors* or, in the case where the vectors are real, an *orthonormal set* or an *orthogonal set of unit vectors*.

SYSTEMS OF LINEAR EQUATIONS

A set of equations having the form

$$\left. \begin{aligned} a_{11}x_1 + a_{12}x_2 + \cdots + a_{1n}x_n &= r_1 \\ a_{21}x_1 + a_{22}x_2 + \cdots + a_{2n}x_n &= r_2 \\ \cdots\cdots\cdots\cdots\cdots\cdots\cdots\cdots\cdots\cdots \\ a_{m1}x_1 + a_{m2}x_2 + \cdots + a_{mn}x_n &= r_n \end{aligned} \right\} \tag{16}$$

is called a *system of m linear equations in the n unknowns* $x_1, x_2, \ldots, x_n$. If $r_1, r_2, \ldots, r_n$ are all zero the system is called *homogeneous*. If they are not all zero it is called *nonhomogeneous*. Any set of numbers $x_1, x_2, \ldots, x_n$ which satisfies (*16*) is called a *solution* of the system.

In matrix form (*16*) can be written

$$\begin{pmatrix} a_{11} & a_{12} & \ldots & a_{1n} \\ a_{21} & a_{22} & \ldots & a_{2n} \\ \cdots\cdots\cdots\cdots\cdots\cdots \\ a_{m1} & a_{m2} & \ldots & a_{mn} \end{pmatrix} \begin{pmatrix} x_1 \\ x_2 \\ \vdots \\ x_n \end{pmatrix} = \begin{pmatrix} r_1 \\ r_2 \\ \vdots \\ r_n \end{pmatrix} \tag{17}$$

or more briefly $AX = R$ (*18*)

where A, X, R represent the corresponding matrices in (*17*).

SYSTEMS OF n EQUATIONS IN n UNKNOWNS.   CRAMER'S RULE

If $m = n$ and if *A* is a non-singular matrix so that A^{-1} exists, we can solve (*17*) or (*18*) by writing

$$X = A^{-1} R \tag{19}$$

and the system has a unique solution.

Alternatively we can express the unknowns $x_1, x_2, \ldots, x_n$ as

$$x_1 = \frac{\Delta_1}{\Delta}, \quad x_2 = \frac{\Delta_2}{\Delta}, \quad \ldots, \quad x_n = \frac{\Delta_n}{\Delta} \tag{20}$$

where $\Delta = \det(A)$, called the *determinant of the system*, is given by (*9*) and Δ_k, $k = 1, 2, \ldots, n$ is the determinant obtained from Δ by removing the kth column and replacing it by the column vector R. The rule expressed in (*20*) is called *Cramer's rule*.

The following four cases can arise.

Case 1, $\Delta \neq 0$, $R \neq 0$.   In this case there will be a unique solution where not all x_k will be zero.

Case 2, $\Delta \neq 0$, $R = 0$.   In this case the only solution will be $x_1 = 0$, $x_2 = 0$, $\ldots$, $x_n = 0$, i.e. $X = 0$. This is often called the *trivial solution*.

Case 3, $\Delta = 0$, $R = 0$.   In this case there will be infinitely many solutions other than the trivial solution. This means that at least one of the equations can be obtained from the others, i.e. the equations are linearly dependent.

Case 4, $\Delta = 0$, $R \neq 0$.   In this case infinitely many solutions will exist if and only if all of the determinants Δ_k in (*20*) are zero. Otherwise there will be no solution.

The cases where $m \neq n$ are considered in Problems 15.93-15.96.

EIGENVALUES AND EIGENVECTORS

Let $A = (a_{jk})$ be an $n \times n$ matrix and X a column vector. The equation

$$AX = \lambda X \tag{21}$$

where λ is a number can be written as

$$\begin{pmatrix} a_{11} & a_{12} & \ldots & a_{1n} \\ a_{21} & a_{22} & \ldots & a_{2n} \\ \cdots & \cdots & \cdots & \cdots \\ a_{n1} & a_{n2} & \ldots & a_{nn} \end{pmatrix} \begin{pmatrix} x_1 \\ x_2 \\ \vdots \\ x_n \end{pmatrix} = \lambda \begin{pmatrix} x_1 \\ x_2 \\ \vdots \\ x_n \end{pmatrix} \tag{22}$$

or

$$\left. \begin{array}{l} (a_{11} - \lambda)x_1 + a_{12}x_2 + \cdots + a_{1n}x_n = 0 \\ a_{21}x_1 + (a_{22} - \lambda)x_2 + \cdots + a_{2n}x_n = 0 \\ \cdots\cdots\cdots\cdots\cdots\cdots\cdots\cdots\cdots\cdots\cdots \\ a_{n1}x_1 + a_{n2}x_2 + \cdots + (a_{nn} - \lambda)x_n = 0 \end{array} \right\} \tag{23}$$

The equation (*23*) will have non-trivial solutions if and only if

$$\begin{vmatrix} a_{11} - \lambda & a_{12} & \ldots & a_{1n} \\ a_{21} & a_{22} - \lambda & \ldots & a_{2n} \\ \cdots\cdots & \cdots\cdots & \cdots & \cdots\cdots \\ a_{n1} & a_{n2} & \ldots & a_{nn} - \lambda \end{vmatrix} = 0 \tag{24}$$

which is a polynomial equation of degree n in λ. The roots of this polynomial equation are called *eigenvalues* or *characteristic values* of the matrix A. Corresponding to each eigenvalue there will be a solution $X \neq 0$, i.e. a non-trivial solution, which is called an *eigenvector* or *characteristic vector* belonging to the eigenvalue. The equation (*24*) can also be written

$$\det(A - \lambda I) = 0 \tag{25}$$

and the equation in λ is often called the *characteristic equation*.

THEOREMS ON EIGENVALUES AND EIGENVECTORS

Theorem 15-12.   The eigenvalues of a Hermitian matrix [or symmetric real matrix] are real.   The eigenvalues of a skew-Hermitian matrix [or skew-symmetric real matrix] are zero or pure imaginary.   The eigenvalues of a unitary [or real orthogonal matrix] all have absolute value equal to 1.

Theorem 15-13.   The eigenvectors belonging to different eigenvalues of a Hermitian matrix [or symmetric real matrix] are orthogonal.

Theorem 15-14   [Cayley-Hamilton].   A matrix satisfies its own characteristic equation [see Problem 15.40].

Theorem 15-15   [Reduction of matrix to diagonal form].   If a non-singular matrix A has distinct eigenvalues $\lambda_1, \lambda_2, \lambda_3, \ldots$ with corresponding eigenvectors written as columns in the matrix

$$B = \begin{pmatrix} b_{11} & b_{12} & b_{13} & \ldots \\ b_{21} & b_{22} & b_{23} & \ldots \\ \cdots\cdots\cdots\cdots\cdots \end{pmatrix}$$

then

$$B^{-1}AB = \begin{pmatrix} \lambda_1 & 0 & 0 & \ldots \\ 0 & \lambda_2 & 0 & \ldots \\ 0 & 0 & \lambda_3 & \ldots \\ \cdots\cdots\cdots\cdots\cdots \end{pmatrix}$$

i.e. $B^{-1}AB$, called the *transform* of A by B, is a diagonal matrix containing the eigenvalues of A in the main diagonal and zeros elsewhere.   We say that A has been *transformed* or *reduced to diagonal form*.   See Problem 15.41.

Theorem 15-16   [Reduction of quadratic form to canonical form].

Let A be a symmetric real matrix, for example,

$$A = \begin{pmatrix} a_{11} & a_{12} & a_{13} \\ a_{21} & a_{22} & a_{23} \\ a_{31} & a_{32} & a_{33} \end{pmatrix} \qquad a_{12} = a_{21},\ a_{13} = a_{31},\ a_{23} = a_{32}$$

Then if $X = \begin{pmatrix} x_1 \\ x_2 \\ x_3 \end{pmatrix}$, we obtain the *quadratic form*

$$X^T A X = a_{11}x_1^2 + a_{22}x_2^2 + a_{33}^2 x_3^2 + 2a_{12}x_1x_2 + 2a_{13}x_1x_3 + 2a_{23}x_2x_3$$

The cross product terms of this quadratic form can be removed by letting $X = BU$ where U is the column vector with elements u_1, u_2, u_3 and B is an orthogonal matrix which diagonalizes A.   The new quadratic form in u_1, u_2, u_3 with no cross product terms is called the *canonical form*.   See Problem 15.43.   A generalization can be made to Hermitian quadratic forms [see Problem 15.114].

OPERATOR INTERPRETATION OF MATRICES

If A is an $n \times n$ matrix, we can think of it as an *operator* or *transformation* acting on a column vector X to produce AX which is another column vector.   With this interpretation equation (21) asks for those vectors X which are transformed by A into constant multiples of themselves [or equivalently into vectors which have the same direction but possibly different magnitude].

If case A is an orthogonal matrix, the transformation is a *rotation* and explains why the absolute value of all the eigenvalues in such case are equal to one [Theorem 15-12], since an ordinary rotation of a vector would not change its magnitude.

The ideas of transformation are very convenient in giving interpretations to many properties of matrices.

Solved Problems

OPERATIONS WITH MATRICES

15.1. If $A = \begin{pmatrix} 2 & -1 \\ 4 & 3 \end{pmatrix}$, $B = \begin{pmatrix} -1 & 1 \\ 2 & -4 \end{pmatrix}$, $C = \begin{pmatrix} 1 & 4 \\ -2 & -1 \end{pmatrix}$ find (a) $A + B$, (b) $A - B$, (c) $2A - 3C$, (d) $3A + 2B - 4C$, (e) AB, (f) BA, (g) $(AB)C$, (h) $A(BC)$, (i) $A^T + B^T$, (j) $B^T A^T$.

(a) $A + B = \begin{pmatrix} 2 & -1 \\ 4 & 3 \end{pmatrix} + \begin{pmatrix} -1 & 1 \\ 2 & -4 \end{pmatrix} = \begin{pmatrix} 1 & 0 \\ 6 & -1 \end{pmatrix}$

(b) $A - B = \begin{pmatrix} 2 & -1 \\ 4 & 3 \end{pmatrix} - \begin{pmatrix} -1 & 1 \\ 2 & -4 \end{pmatrix} = \begin{pmatrix} 2 & -1 \\ 4 & 3 \end{pmatrix} + \begin{pmatrix} 1 & -1 \\ -2 & 4 \end{pmatrix} = \begin{pmatrix} 3 & -2 \\ 2 & 7 \end{pmatrix}$

(c) $2A - 3C = 2\begin{pmatrix} 2 & -1 \\ 4 & 3 \end{pmatrix} - 3\begin{pmatrix} 1 & 4 \\ -2 & -1 \end{pmatrix} = \begin{pmatrix} 4 & -2 \\ 8 & 6 \end{pmatrix} + \begin{pmatrix} -3 & -12 \\ 6 & 3 \end{pmatrix} = \begin{pmatrix} 1 & -14 \\ 14 & 9 \end{pmatrix}$

(d) $3A + 2B - 4C = 3\begin{pmatrix} 2 & -1 \\ 4 & 3 \end{pmatrix} + 2\begin{pmatrix} -1 & 1 \\ 2 & -4 \end{pmatrix} - 4\begin{pmatrix} 1 & 4 \\ -2 & -1 \end{pmatrix}$

$\qquad = \begin{pmatrix} 6 & -3 \\ 12 & 9 \end{pmatrix} + \begin{pmatrix} -2 & 2 \\ 4 & -8 \end{pmatrix} + \begin{pmatrix} -4 & -16 \\ 8 & 4 \end{pmatrix} = \begin{pmatrix} 0 & -17 \\ 24 & 5 \end{pmatrix}$

(e) $AB = \begin{pmatrix} 2 & -1 \\ 4 & 3 \end{pmatrix}\begin{pmatrix} -1 & 1 \\ 2 & -4 \end{pmatrix} = \begin{pmatrix} (2)(-1) + (-1)(2) & (2)(1) + (-1)(-4) \\ (4)(-1) + (3)(2) & (4)(1) + (3)(-4) \end{pmatrix} = \begin{pmatrix} -4 & 6 \\ 2 & -8 \end{pmatrix}$

(f) $BA = \begin{pmatrix} -1 & 1 \\ 2 & -4 \end{pmatrix}\begin{pmatrix} 2 & -1 \\ 4 & 3 \end{pmatrix} = \begin{pmatrix} (-1)(2) + (1)(4) & (-1)(-1) + (1)(3) \\ (2)(2) + (-4)(4) & (2)(-1) + (-4)(3) \end{pmatrix} = \begin{pmatrix} 2 & 4 \\ -12 & -14 \end{pmatrix}$

Note that $AB \neq BA$ using (e), illustrating the fact that the commutative law for products does not hold in general.

(g) $(AB)C = \begin{pmatrix} -4 & 6 \\ 2 & -8 \end{pmatrix}\begin{pmatrix} 1 & 4 \\ -2 & -1 \end{pmatrix} = \begin{pmatrix} -16 & -22 \\ 18 & 16 \end{pmatrix}$

(h) $A(BC) = \begin{pmatrix} 2 & -1 \\ 4 & 3 \end{pmatrix}\left[\begin{pmatrix} -1 & 1 \\ 2 & -4 \end{pmatrix}\begin{pmatrix} 1 & 4 \\ -2 & -1 \end{pmatrix}\right] = \begin{pmatrix} 2 & -1 \\ 4 & 3 \end{pmatrix}\begin{pmatrix} -3 & -5 \\ 10 & 12 \end{pmatrix} = \begin{pmatrix} -16 & -22 \\ 18 & 16 \end{pmatrix}$

Note that $(AB)C = A(BC)$ using (g), illustrating the fact that the associative law for products holds.

(i) $A^T + B^T = \begin{pmatrix} 2 & -1 \\ 4 & 3 \end{pmatrix}^T + \begin{pmatrix} -1 & 1 \\ 2 & -4 \end{pmatrix}^T = \begin{pmatrix} 2 & 4 \\ -1 & 3 \end{pmatrix} + \begin{pmatrix} -1 & 2 \\ 1 & -4 \end{pmatrix} = \begin{pmatrix} 1 & 6 \\ 0 & -1 \end{pmatrix}$

Note that $A^T + B^T = (A + B)^T$ using (a).

(j) $B^T A^T = \begin{pmatrix} -1 & 1 \\ 2 & -4 \end{pmatrix}^T\begin{pmatrix} 2 & -1 \\ 4 & 3 \end{pmatrix}^T = \begin{pmatrix} -1 & 2 \\ 1 & -4 \end{pmatrix}\begin{pmatrix} 2 & 4 \\ -1 & 3 \end{pmatrix} = \begin{pmatrix} -4 & 2 \\ 6 & -8 \end{pmatrix}$

Note that $B^T A^T = (AB)^T$ using (e).

15.2. If $A = \begin{pmatrix} 2 & 1 & -1 \\ 1 & -2 & 3 \\ -2 & 1 & 2 \end{pmatrix}$, $B = \begin{pmatrix} 1 & -1 & 2 \\ -2 & 1 & 3 \\ 2 & -1 & 1 \end{pmatrix}$ show that

$$(A+B)^2 = A^2 + AB + BA + B^2$$

We have
$$A + B = \begin{pmatrix} 3 & 0 & 1 \\ -1 & -1 & 6 \\ 0 & 0 & 3 \end{pmatrix}$$

Then
$$(A+B)^2 = (A+B)(A+B) = \begin{pmatrix} 3 & 0 & 1 \\ -1 & -1 & 6 \\ 0 & 0 & 3 \end{pmatrix}\begin{pmatrix} 3 & 0 & 1 \\ -1 & -1 & 6 \\ 0 & 0 & 3 \end{pmatrix}$$

$$= \begin{pmatrix} (3)(3) + (0)(-1) + (1)(0) & (3)(0) + (0)(-1) + (1)(0) & (3)(1) + (0)(6) + (1)(3) \\ (-1)(3) + (-1)(-1) + (6)(0) & (-1)(0) + (-1)(-1) + (6)(0) & (-1)(1) + (-1)(6) + (6)(3) \\ (0)(3) + (0)(-1) + (3)(0) & (0)(0) + (0)(-1) + (3)(0) & (0)(1) + (0)(6) + (3)(3) \end{pmatrix}$$

$$= \begin{pmatrix} 9 & 0 & 6 \\ -2 & 1 & 11 \\ 0 & 0 & 9 \end{pmatrix}$$

Now
$$A^2 = \begin{pmatrix} 2 & 1 & -1 \\ 1 & -2 & 3 \\ -2 & 1 & 2 \end{pmatrix}\begin{pmatrix} 2 & 1 & -1 \\ 1 & -2 & 3 \\ -2 & 1 & 2 \end{pmatrix} = \begin{pmatrix} 7 & -1 & -1 \\ -6 & 8 & -1 \\ -7 & -2 & 9 \end{pmatrix}$$

$$AB = \begin{pmatrix} 2 & 1 & -1 \\ 1 & -2 & 3 \\ -2 & 1 & 2 \end{pmatrix}\begin{pmatrix} 1 & -1 & 2 \\ -2 & 1 & 3 \\ 2 & -1 & 1 \end{pmatrix} = \begin{pmatrix} -2 & 0 & 6 \\ 11 & -6 & -1 \\ 0 & 1 & 1 \end{pmatrix}$$

$$BA = \begin{pmatrix} 1 & -1 & 2 \\ -2 & 1 & 3 \\ 2 & -1 & 1 \end{pmatrix}\begin{pmatrix} 2 & 1 & -1 \\ 1 & -2 & 3 \\ -2 & 1 & 2 \end{pmatrix} = \begin{pmatrix} -3 & 5 & 0 \\ -9 & -1 & 11 \\ 1 & 5 & -3 \end{pmatrix}$$

$$B^2 = \begin{pmatrix} 1 & -1 & 2 \\ -2 & 1 & 3 \\ 2 & -1 & 1 \end{pmatrix}\begin{pmatrix} 1 & -1 & 2 \\ -2 & 1 & 3 \\ 2 & -1 & 1 \end{pmatrix} = \begin{pmatrix} 7 & -4 & 1 \\ 2 & 0 & 2 \\ 6 & -4 & 2 \end{pmatrix}$$

Thus
$$A^2 + AB + BA + B^2 = \begin{pmatrix} 9 & 0 & 6 \\ -2 & 1 & 11 \\ 0 & 0 & 9 \end{pmatrix} = (A+B)^2$$

15.3. Prove that any real square matrix can always be expressed as the sum of a real symmetric matrix and a real skew-symmetric matrix.

If A is any real square matrix, then

$$A = \tfrac{1}{2}(A + A^T) + \tfrac{1}{2}(A - A^T)$$

But since $(A + A^T)^T = A^T + A = A + A^T$, it follows that $\tfrac{1}{2}(A + A^T)$ is symmetric. Also, since $(A - A^T)^T = A^T - A = -(A - A^T)$, it follows that $\tfrac{1}{2}(A - A^T)$ is skew-symmetric. The required result is thus proved.

15.4. Show that the matrix $A = \begin{pmatrix} 0 & -i \\ i & 0 \end{pmatrix}$ is Hermitian.

We have $A^T = \begin{pmatrix} 0 & i \\ -i & 0 \end{pmatrix}$ and $\overline{A^T} = \begin{pmatrix} 0 & -i \\ i & 0 \end{pmatrix} = A$. Thus A is Hermitian.

15.5. Prove that a unit matrix I of order n commutes with any square matrix A of order n and the resulting product is A.

We illustrate the proof for $n = 3$. In such case

$$I = \begin{pmatrix} 1 & 0 & 0 \\ 0 & 1 & 0 \\ 0 & 0 & 1 \end{pmatrix}, \qquad A = \begin{pmatrix} a_{11} & a_{12} & a_{13} \\ a_{21} & a_{22} & a_{23} \\ a_{31} & a_{32} & a_{33} \end{pmatrix}$$

Then $\quad IA = \begin{pmatrix} 1 & 0 & 0 \\ 0 & 1 & 0 \\ 0 & 0 & 1 \end{pmatrix}\begin{pmatrix} a_{11} & a_{12} & a_{13} \\ a_{21} & a_{22} & a_{23} \\ a_{31} & a_{32} & a_{33} \end{pmatrix} = \begin{pmatrix} a_{11} & a_{12} & a_{13} \\ a_{21} & a_{22} & a_{23} \\ a_{31} & a_{32} & a_{33} \end{pmatrix} = A$

$$AI = \begin{pmatrix} a_{11} & a_{12} & a_{13} \\ a_{21} & a_{22} & a_{23} \\ a_{31} & a_{32} & a_{33} \end{pmatrix}\begin{pmatrix} 1 & 0 & 0 \\ 0 & 1 & 0 \\ 0 & 0 & 1 \end{pmatrix} = \begin{pmatrix} a_{11} & a_{12} & a_{13} \\ a_{21} & a_{22} & a_{23} \\ a_{31} & a_{32} & a_{33} \end{pmatrix} = A$$

i.e. $IA = AI = A$.

Extensions are easily made for $n > 3$.

DETERMINANTS

15.6. Use the definition of a determinant [Laplace expansion] as given on page 345 to evaluate a determinant of (a) order 2, (b) order 3.

(a) Let the determinant be $\begin{vmatrix} a_{11} & a_{12} \\ a_{21} & a_{22} \end{vmatrix}$. Use the elements of the first row. The corresponding cofactors are

$$A_{11} = (-1)^{1+1} a_{22} = a_{22}, \qquad A_{12} = (-1)^{2+1} a_{21} = -a_{21}$$

Then by the Laplace expansion the determinant has the value

$$a_{11}A_{11} + a_{12}A_{12} = a_{11}a_{22} - a_{12}a_{21}$$

The same value is obtained by using the elements of the second row [or first and second columns].

(b) Let the determinant be $\begin{vmatrix} a_{11} & a_{12} & a_{13} \\ a_{21} & a_{22} & a_{23} \\ a_{31} & a_{32} & a_{33} \end{vmatrix}$. The cofactors of the elements in the first row are

$$A_{11} = (-1)^{1+1} \begin{vmatrix} a_{22} & a_{23} \\ a_{32} & a_{33} \end{vmatrix} = a_{22}a_{33} - a_{23}a_{32}$$

$$A_{12} = (-1)^{1+2} \begin{vmatrix} a_{21} & a_{23} \\ a_{31} & a_{33} \end{vmatrix} = a_{23}a_{31} - a_{21}a_{33}$$

$$A_{13} = (-1)^{1+3} \begin{vmatrix} a_{21} & a_{22} \\ a_{31} & a_{32} \end{vmatrix} = a_{21}a_{32} - a_{22}a_{31}$$

Then the value of the determinant is

$$a_{11}A_{11} + a_{12}A_{12} + a_{13}A_{13} = a_{11}(a_{22}a_{33} - a_{23}a_{32})$$
$$+ a_{12}(a_{23}a_{31} - a_{21}a_{33})$$
$$+ a_{13}(a_{21}a_{32} - a_{22}a_{31})$$

$$= a_{11}a_{22}a_{33} + a_{12}a_{23}a_{31} + a_{13}a_{21}a_{32}$$
$$- a_{11}a_{23}a_{32} - a_{12}a_{21}a_{33} - a_{13}a_{22}a_{31}$$

The same value is obtained by using elements of the second or third rows [or first, second and third columns].

15.7. Prove that the value of a determinant remains the same regardless of which row [or column] is taken for the Laplace expansion.

Consider the determinant $\Delta = (a_{jk})$ of order n. The result is true for $n = 2$ by Problem 15.6. We use proof by induction, i.e. assuming it to be true for order $n - 1$ we shall prove it true for order n. The plan will be to expand Δ using two different rows p and q and show that the expansions are the same.

Let us first expand Δ by elements in the pth row. Then a typical term in the expansion is

$$a_{pk}A_{pk} \;=\; a_{pk}(-1)^{p+k}M_{pk} \tag{1}$$

where M_{pk} is the minor corresponding to the cofactor A_{pk} of a_{pk}. Since this minor is of order $n - 1$, any row can be used in its expansion.

We shall use the qth row where we assume that $q > p$ since a similar argument holds if $q < p$. This row consists of elements a_{qr} where $r \neq k$ and corresponds to the $(q-1)$st row of M_{pk}.

Now if $r < k$, a_{qr} is located in the rth column of M_{pk} so that in the expansion the term corresponding to a_{qr} is

$$a_{qr}(-1)^{(q-1)+r}M_{pkqr} \tag{2}$$

where M_{pkqr} is the minor corresponding to the element a_{qr} in M_{pk}. From (1) and (2) it follows that a typical term in the expansion of Δ is

$$a_{pk}(-1)^{p+k}a_{qr}(-1)^{q-1+r}M_{pkqr} \;=\; a_{pk}a_{qr}(-1)^{p+k+q+r-1}M_{pkqr} \tag{3}$$

If $r > k$ then a_{qr} is located in the $(r-1)$st column and so there is an additional minus sign in (3).

If we now expand Δ by elements in the qth row, a typical term is

$$a_{qr}A_{qr} \;=\; a_{qr}(-1)^{q+r}M_{qr} \tag{4}$$

We can expand M_{qr} by elements in the pth row where $p > q$. As before if $k > r$, a typical term in the expansion of M_{qr} is

$$a_{pk}(-1)^{p+(k-1)}M_{pkqr} \tag{5}$$

From (4) and (5) we see that a typical term in the expansion of Δ is

$$a_{qr}(-1)^{q+r}a_{pk}(-1)^{p+k-1}M_{pkqr} \;=\; a_{pk}a_{qr}(-1)^{p+k+q+r-1}M_{pkqr} \tag{6}$$

which is the same as (3). If $k < r$ an additional minus sign appears in (6), agreeing with the case corresponding to $r > k$ using the first expansion. Thus the required result is proved.

In a similar manner we can prove that expansion by columns is the same and gives the same result as the expansion by rows [Theorem 15-1, page 345].

15.8. Evaluate by the Laplace expansion the determinant $\begin{vmatrix} 3 & -2 & 2 \\ 1 & 2 & -3 \\ 4 & 1 & 2 \end{vmatrix}$ (a) using elements in the first row and (b) using elements in the second row.

(a) Using elements in the first row, the expansion is

$$(3)\begin{vmatrix} 2 & -3 \\ 1 & 2 \end{vmatrix} - (-2)\begin{vmatrix} 1 & -3 \\ 4 & 2 \end{vmatrix} + (2)\begin{vmatrix} 1 & 2 \\ 4 & 1 \end{vmatrix}$$
$$= (3)(7) - (-2)(14) + (2)(-7) = 35$$

(b) Using elements in the second row, the expansion is

$$-(1)\begin{vmatrix} -2 & 2 \\ 1 & 2 \end{vmatrix} + (2)\begin{vmatrix} 3 & 2 \\ 4 & 2 \end{vmatrix} - (-3)\begin{vmatrix} 3 & -2 \\ 4 & 1 \end{vmatrix}$$
$$= -(1)(-6) + (2)(-2) - (-3)(11) = 35$$

15.9. Prove Theorem 15-4, page 345.

Let the determinant be

$$\Delta = \begin{vmatrix} a_{11} & a_{12} & \cdots & a_{1n} \\ \cdots\cdots\cdots\cdots\cdots\cdots \\ a_{k1} & a_{k2} & \cdots & a_{kn} \\ \cdots\cdots\cdots\cdots\cdots\cdots \\ a_{n1} & a_{n2} & \cdots & a_{nn} \end{vmatrix} \qquad (1)$$

and suppose that the elements in the kth row are multiplied by λ to give the determinant

$$\Delta_1 = \begin{vmatrix} a_{11} & a_{12} & \cdots & a_{1n} \\ \cdots\cdots\cdots\cdots\cdots\cdots \\ \lambda a_{k1} & \lambda a_{k2} & \cdots & \lambda a_{kn} \\ \cdots\cdots\cdots\cdots\cdots\cdots \\ a_{n1} & a_{n2} & \cdots & a_{nn} \end{vmatrix} \qquad (2)$$

Expanding (1) and (2) according to elements in the kth row, we find respectively

$$\Delta = a_{k1}A_{k1} + a_{k2}A_{k2} + \cdots + a_{kn}A_{kn} \qquad (3)$$

$$\Delta_1 = (\lambda a_{k1})A_{k1} + (\lambda a_{k2})A_{k2} + \cdots + (\lambda a_{kn})A_{kn} \qquad (4)$$

from which $\Delta_1 = \lambda\Delta$ as required.

15.10. Prove Theorem 15-5, page 345.

(a) If two rows have the same elements, then the value of the determinant will not change if the rows are interchanged. However, according to Theorem 15-3, page 345, the sign must change. Thus we have $\Delta = -\Delta$ or $\Delta = 0$.

(b) If the two rows have proportional elements, then they can be made the same by factoring out the proportionality constants and thus the determinant must be zero by (a).

15.11. Prove Theorem 15-6, page 345.

Write the determinant as

$$\Delta = \begin{vmatrix} a_{11}+b_1 & a_{12}+b_2 & \cdots & a_{1n}+b_n \\ a_{21} & a_{22} & \cdots & a_{2n} \\ a_{31} & a_{32} & \cdots & a_{3n} \\ \cdots\cdots\cdots\cdots\cdots\cdots\cdots \\ a_{n1} & a_{n2} & \cdots & a_{nn} \end{vmatrix}$$

in which the first row has each element expressed as the sum of two terms. Then by the Laplace expansion we have

$$\Delta = (a_{11}+b_1)A_{11} + (a_{12}+b_2)A_{12} + \cdots + (a_{1n}+b_n)A_{1n} \qquad (1)$$

where $A_{11}, A_{12}, \ldots, A_{1n}$ are the cofactors of the corresponding elements in the first row. But (1) can be written as

$$\Delta = (a_{11}A_{11} + a_{12}A_{12} + \cdots + a_{1n}A_{1n}) + (b_1A_{11} + \cdots + b_nA_{1n})$$

$$= \begin{vmatrix} a_{11} & a_{12} & \cdots & a_{1n} \\ a_{21} & a_{22} & \cdots & a_{2n} \\ \cdots\cdots\cdots\cdots\cdots \\ a_{n1} & a_{n2} & \cdots & a_{nn} \end{vmatrix} + \begin{vmatrix} b_1 & b_2 & \cdots & b_n \\ a_{21} & a_{22} & \cdots & a_{2n} \\ \cdots\cdots\cdots\cdots\cdots \\ a_{n1} & a_{n2} & \cdots & a_{nn} \end{vmatrix}$$

as required. A similar procedure proves the result if any other row [or column] is chosen.

15.12. Prove Theorem 15-7, page 345.

Suppose we multiply the elements of the second row of $\Delta = (a_{jk})$ by λ and add to the elements of the first row [a similar proof can be used for any other rows or columns]. Then the determinant can be written as

$$\begin{vmatrix} a_{11} + \lambda a_{21} & a_{12} + \lambda a_{22} & \dots & a_{1n} + \lambda a_{2n} \\ a_{21} & a_{22} & \dots & a_{2n} \\ \hdotsfor{4} \\ a_{n1} & a_{n2} & \dots & a_{nn} \end{vmatrix}$$

But by Problem 15.11 this can be written as

$$\begin{vmatrix} a_{11} & a_{12} & \dots & a_{1n} \\ a_{21} & a_{22} & \dots & a_{2n} \\ \hdotsfor{4} \\ a_{n1} & a_{n2} & \dots & a_{nn} \end{vmatrix} + \begin{vmatrix} \lambda a_{21} & \lambda a_{22} & \dots & \lambda a_{2n} \\ a_{21} & a_{22} & \dots & a_{2n} \\ \hdotsfor{4} \\ a_{n1} & a_{n2} & \dots & a_{nn} \end{vmatrix}$$

Then the required result follows since the second determinant is zero because the elements of its first and second rows are proportional [Theorem 15-5].

15.13. Evaluate $\begin{vmatrix} 2 & 1 & -1 & 4 \\ -2 & 3 & 2 & -5 \\ 1 & -2 & -3 & 2 \\ -4 & -3 & 2 & -2 \end{vmatrix}$.

Multiplying the elements of the first row by $-3, 2, 3$ and adding to the elements of the second, third and fourth rows respectively, we find

$$\begin{vmatrix} 2 & 1 & -1 & 4 \\ -8 & 0 & 5 & -17 \\ 5 & 0 & -5 & 10 \\ 2 & 0 & -1 & 10 \end{vmatrix}$$

which by Theorem 15-7 has a value equal to that of the given determinant. Note that this new determinant has three zeros in the 2nd column, which was precisely our intention in choosing the numbers $-3, 2, 3$ in the first place.

Multiplying each element in the second column by its cofactor, we see that the value of the determinant is

$$- \begin{vmatrix} -8 & 5 & -17 \\ 5 & -5 & 10 \\ 2 & -1 & 10 \end{vmatrix} = -5 \begin{vmatrix} -8 & 5 & -17 \\ 1 & -1 & 2 \\ 2 & -1 & 10 \end{vmatrix}$$

on removing the factor 5 from the second row, using Theorem 15-4.

Now multiplying the elements in the second row by 5 and -1 and adding to the elements of the first and third rows respectively, we find

$$-5 \begin{vmatrix} -3 & 0 & -7 \\ 1 & -1 & 2 \\ 1 & 0 & 8 \end{vmatrix}$$

which on expanding by the elements in the second column gives

$$(-5)(-1) \begin{vmatrix} -3 & -7 \\ 1 & 8 \end{vmatrix} = -85$$

15.14. Verify Theorem 15-8 if $A = \begin{pmatrix} 2 & -1 \\ 3 & 2 \end{pmatrix}$, $B = \begin{pmatrix} 7 & 2 \\ -3 & 4 \end{pmatrix}$.

The theorem states that $\det(AB) = \det(A)\det(B)$. Then since

$$AB = \begin{pmatrix} 2 & -1 \\ 3 & 2 \end{pmatrix}\begin{pmatrix} 7 & 2 \\ -3 & 4 \end{pmatrix} = \begin{pmatrix} 17 & 0 \\ 15 & 14 \end{pmatrix}$$

it states that
$$\begin{vmatrix} 2 & -1 \\ 3 & 2 \end{vmatrix}\begin{vmatrix} 7 & 2 \\ -3 & 4 \end{vmatrix} = \begin{vmatrix} 17 & 0 \\ 15 & 14 \end{vmatrix}$$

or
$$(7)(34) = (17)(14)$$

But since this is correct, the theorem is verified for this case.

15.15. Let $v_1 = (2 \ -1 \ 3)$, $v_2 = (1 \ 2 \ -1)$, $v_3 = (-3 \ 4 \ -7)$. (a) Show that v_1, v_2, v_3 are linearly dependent. (b) Illustrate Theorem 15-10, page 346, by showing that

$$\begin{vmatrix} 2 & -1 & 3 \\ 1 & 2 & -1 \\ -3 & 4 & -7 \end{vmatrix} = 0$$

(a) We must show that there exist constants $\lambda_1, \lambda_2, \lambda_3$ not all zero such that $\lambda_1 v_1 + \lambda_2 v_2 + \lambda_3 v_3 = 0 = (0 \ 0 \ 0)$. Now

$$\lambda_1(2 \ -1 \ 3) + \lambda_2(1 \ 2 \ -1) + \lambda_3(-3 \ 4 \ -7) = (0 \ 0 \ 0)$$

when
$$2\lambda_1 + \lambda_2 - 3\lambda_3 = 0$$
$$-\lambda_1 + 2\lambda_2 + 4\lambda_3 = 0$$
$$3\lambda_1 - \lambda_2 - 7\lambda_3 = 0$$

Assuming that $\lambda_3 = 1$, for example, the equations become $2\lambda_1 + \lambda_2 = 3$, $\lambda_1 - 2\lambda_2 = 4$, $3\lambda_1 - \lambda_2 = 7$. Solving any two of these simultaneously, we find $\lambda_1 = 2$, $\lambda_2 = -1$. Thus $\lambda_1 = 2$, $\lambda_2 = -1$, $\lambda_3 = 1$ provide the required constants.

(b) Multiplying the elements of the second row by $-2, 3$ and adding to the first and third rows respectively, the given determinant equals

$$\begin{vmatrix} 0 & -5 & 5 \\ 1 & 2 & -1 \\ 0 & 10 & -10 \end{vmatrix} = -(1)\begin{vmatrix} -5 & 5 \\ 10 & -10 \end{vmatrix} = 0$$

15.16. Prove Theorem 15-9, page 346.

By definition the determinant

$$A = \begin{vmatrix} a_{11} & a_{12} & \dots & a_{1n} \\ \dots\dots\dots\dots\dots\dots \\ a_{p1} & a_{p2} & \dots & a_{pn} \\ \dots\dots\dots\dots\dots\dots \\ a_{n1} & a_{n2} & \dots & a_{nn} \end{vmatrix}$$

when expanded according to the elements of the pth row has the value

$$\det(A) = a_{p1}A_{p1} + a_{p2}A_{p2} + \cdots + a_{pn}A_{pn} = \sum_{k=1}^{n} a_{pk}A_{pk} \qquad (1)$$

Let us now replace the elements a_{pk} in the pth row of A by corresponding elements a_{qk} of the qth row where $p \neq q$. Then two rows will be identical and the new determinant thus obtained will be zero by Theorem 15-5. Since $a_{pk} = a_{qk}$, (1) is replaced by

$$0 = a_{q1}A_{p1} + a_{q2}A_{p2} + \cdots + a_{qn}A_{pn} = \sum_{k=1}^{n} a_{qk}A_{pk}$$

i.e.
$$\sum_{k=1}^{n} a_{qk}A_{pk} = 0 \qquad p \neq q \qquad (2)$$

Similarly by using columns rather than rows we can show that

$$\sum_{k=1}^{n} a_{kq}A_{kp} \;=\; 0 \qquad p \neq q \tag{3}$$

If $p = q$, then (2) and (3) become respectively

$$\sum_{k=1}^{n} a_{pk}A_{pk} \;=\; \det(A) \tag{4}$$

$$\sum_{k=1}^{n} a_{kp}A_{kp} \;=\; \det(A) \tag{5}$$

INVERSE OF A MATRIX

15.17. Prove that $\quad A^{-1} = \dfrac{(A_{jk})^T}{\det(A)} = \dfrac{(A_{kj})}{\det(A)}$.

We must show that $AA^{-1} = I$, the unit matrix. To do this consider the product

$$A(A_{jk})^T \;=\; \begin{vmatrix} a_{11} & a_{12} & \ldots & a_{1n} \\ a_{21} & a_{22} & \ldots & a_{2n} \\ \hdotsfor{4} \\ a_{n1} & a_{n2} & \ldots & a_{nn} \end{vmatrix} \begin{vmatrix} A_{11} & A_{21} & \ldots & A_{n1} \\ A_{12} & A_{22} & \ldots & A_{n2} \\ \hdotsfor{4} \\ A_{1n} & A_{2n} & \ldots & A_{nn} \end{vmatrix}$$

Now by the rule for multiplying determinants [which is the same as that for multiplying matrices], the element c_{pq} in the resulting determinant is found by taking the sum of the products of elements in the qth row of the first determinant and the pth column of the second determinant. We thus have

$$c_{pq} \;=\; a_{q1}A_{p1} + a_{q2}A_{p2} + \cdots + a_{qn}A_{pn} \;=\; \sum_{k=1}^{n} a_{qk}A_{pk}$$

But by the results of Problem 15.16,

$$c_{pq} \;=\; \begin{cases} 0 & p \neq q \\ \det(A) & p = q \end{cases}$$

It follows that

$$A(A_{jk})^T \;=\; \begin{vmatrix} \det(A) & 0 & \ldots & 0 \\ 0 & \det(A) & \ldots & 0 \\ \hdotsfor{4} \\ 0 & 0 & \ldots & \det(A) \end{vmatrix}$$

Then if $\det(A) \neq 0$, this can be written

$$\frac{A(A_{jk})^T}{\det(A)} \;=\; \begin{vmatrix} 1 & 0 & \ldots & 0 \\ 0 & 1 & \ldots & 0 \\ \hdotsfor{4} \\ 0 & 0 & \ldots & 1 \end{vmatrix} \;=\; I$$

and it thus follows that $AB = I$ where

$$B \;=\; A^{-1} \;=\; \frac{(A_{jk})^T}{\det(A)}$$

15.18. (a) Find the inverse of the matrix $A = \begin{pmatrix} 3 & -2 & 2 \\ 1 & 2 & -3 \\ 4 & 1 & 2 \end{pmatrix}$ and (b) check the answer by direct multiplication.

(a) The matrix of cofactors of A is given by

$$(A_{jk}) \;=\; \begin{pmatrix} 7 & -14 & -7 \\ 6 & -2 & -11 \\ 2 & 11 & 8 \end{pmatrix}$$

The transpose of this matrix is

$$(A_{jk})^T = (A_{kj}) = \begin{pmatrix} 7 & 6 & 2 \\ -14 & -2 & 11 \\ -7 & -11 & 8 \end{pmatrix}$$

Since $\det(A) = 35$ [see Problem 15.8], we have

$$A^{-1} = \frac{(A_{jk})^T}{\det(A)} = \frac{1}{35}\begin{pmatrix} 7 & 6 & 2 \\ -14 & -2 & 11 \\ -7 & -11 & 8 \end{pmatrix} = \begin{pmatrix} \frac{1}{5} & \frac{6}{35} & \frac{2}{35} \\ -\frac{2}{5} & -\frac{2}{35} & \frac{11}{35} \\ -\frac{1}{5} & -\frac{11}{35} & \frac{8}{35} \end{pmatrix}$$

(b)
$$AA^{-1} = \begin{pmatrix} 3 & -2 & 2 \\ 1 & 2 & -3 \\ 4 & 1 & 2 \end{pmatrix}\begin{pmatrix} \frac{1}{5} & \frac{6}{35} & \frac{2}{35} \\ -\frac{2}{5} & -\frac{2}{35} & \frac{11}{35} \\ -\frac{1}{5} & -\frac{11}{35} & \frac{8}{35} \end{pmatrix} = \begin{pmatrix} 1 & 0 & 0 \\ 0 & 1 & 0 \\ 0 & 0 & 1 \end{pmatrix} = I$$

We can also show that $A^{-1}A = I$. This supplies the required check.

15.19. Prove that $(AB)^{-1} = B^{-1}A^{-1}$.

Let $X = (AB)^{-1}$. Then $(AB)X = I$ where I is the unit matrix. By the associative law this becomes $A(BX) = I$. Multiplying by A^{-1}, we have $A^{-1}[A(BX)] = A^{-1}I = A^{-1}$ which again using the associative law becomes $(A^{-1}A)(BX) = A^{-1}$ or $I(BX) = A^{-1}$, i.e. $BX = A^{-1}$. Multiplying by B^{-1} and using the associative law once more, we have $B^{-1}(BX) = B^{-1}A^{-1}$, $(B^{-1}B)X = B^{-1}A^{-1}$, $IX = B^{-1}A^{-1}$, i.e. $X = B^{-1}A^{-1}$, as required.

15.20. Prove that if A is a non-singular matrix, then $\det(A^{-1}) = \dfrac{1}{\det A}$.

Since $AA^{-1} = I$, $\det(AA^{-1}) = \det(I) = 1$. But by Theorem 15-8, $\det(AA^{-1}) = \det(A)\det(A^{-1})$. Thus $\det(A^{-1})\det(A) = 1$ and the required result follows.

ORTHOGONAL AND UNITARY MATRICES. ORTHOGONAL VECTORS

15.21. Show that $A = \begin{pmatrix} \cos\theta & -\sin\theta \\ \sin\theta & \cos\theta \end{pmatrix}$ is an orthogonal matrix.

We have, using the fact that A is real,

$$A^TA = \begin{pmatrix} \cos\theta & \sin\theta \\ -\sin\theta & \cos\theta \end{pmatrix}\begin{pmatrix} \cos\theta & -\sin\theta \\ \sin\theta & \cos\theta \end{pmatrix} = \begin{pmatrix} 1 & 0 \\ 0 & 1 \end{pmatrix} = I$$

since $\cos^2\theta + \sin^2\theta = 1$. Thus A is an orthogonal matrix.

15.22. Show that $A = \begin{pmatrix} \sqrt{2}/2 & -i\sqrt{2}/2 & 0 \\ i\sqrt{2}/2 & -\sqrt{2}/2 & 0 \\ 0 & 0 & 1 \end{pmatrix}$ is a unitary matrix.

Since A is complex, we must show that $\bar{A}^TA = I$. We have

$$\bar{A}^TA = \begin{pmatrix} \sqrt{2}/2 & -i\sqrt{2}/2 & 0 \\ i\sqrt{2}/2 & -\sqrt{2}/2 & 0 \\ 0 & 0 & 1 \end{pmatrix}\begin{pmatrix} \sqrt{2}/2 & -i\sqrt{2}/2 & 0 \\ i\sqrt{2}/2 & -\sqrt{2}/2 & 0 \\ 0 & 0 & 1 \end{pmatrix} = \begin{pmatrix} 1 & 0 & 0 \\ 0 & 1 & 0 \\ 0 & 0 & 1 \end{pmatrix} = I$$

so that A is a unitary matrix.

15.23. If A is an orthogonal matrix, prove that $\det(A) = \pm 1$.

If A is orthogonal, then $A^T A = I$ so that by Theorem 15-8, page 345,

$$\det(A^T A) = \det(A^T)\det(A) = \det I = 1 \tag{1}$$

But $\det(A^T) = \det(A)$ so that (1) becomes

$$[\det(A)]^2 = 1 \quad \text{or} \quad \det(A) = \pm 1$$

15.24. Show that the vectors

$$A_1 = \begin{pmatrix} \cos\theta \\ \sin\theta \\ 0 \end{pmatrix}, \quad A_2 = \begin{pmatrix} -\sin\theta \\ \cos\theta \\ 0 \end{pmatrix}, \quad A_3 = \begin{pmatrix} 0 \\ 0 \\ 1 \end{pmatrix}$$

form an orthonormal set or system of vectors.

Since the vectors are real, we must show that

$$A_j^T A_k = \begin{cases} 1 & \text{if } j = k \\ 0 & \text{if } j \neq k \end{cases}$$

If $j = k = 1$, we have

$$A_1^T A_1 = (\cos\theta \;\; \sin\theta \;\; 0) \begin{pmatrix} \cos\theta \\ \sin\theta \\ 0 \end{pmatrix} = \cos^2\theta + \sin^2\theta = 1$$

Similarly we find if $j = k = 2$ and $j = k = 3$, $A_2^T A_2 = 1$, $A_3^T A_3 = 1$. Thus A_1, A_2, A_3 are unit vectors.

To show the orthogonality of any two of the vectors consider, for example, $j = 1$, $k = 2$. Then we have

$$A_1^T A_2 = (\cos\theta \;\; \sin\theta \;\; 0) \begin{pmatrix} -\sin\theta \\ \cos\theta \\ 0 \end{pmatrix} = 0$$

Similarly $A_1^T A_3 = 0$, $A_2^T A_3 = 0$ and so the vectors are mutually orthogonal. Thus the vectors form an orthonormal system.

SYSTEMS OF LINEAR EQUATIONS

15.25. Prove Cramer's rule (20), page 348, for solving the system of equations (16), page 347, in the case where $m = n$.

The system of equations can be written

$$\sum_{q=1}^{n} a_{kq} x_q = r_k \qquad k = 1, \ldots, n$$

Multiplying by A_{kp} and adding from $k = 1$ to n, we have

$$A_{kp} \sum_{q=1}^{n} a_{kq} x_q = r_k A_{kp}$$

and

$$\sum_{k=1}^{n} A_{kp} \sum_{q=1}^{n} a_{kq} x_q = \sum_{k=1}^{n} r_k A_{kp}$$

This can be written

$$\sum_{q=1}^{n} \left\{ \sum_{k=1}^{n} A_{kp} a_{kq} \right\} x_q = \sum_{k=1}^{n} r_k A_{kp} \tag{1}$$

Now by equations (3) and (5) of Problem 15.16, we have

$$\sum_{k=1}^{n} A_{kp} a_{kq} = \begin{cases} 0 & q \neq p \\ \det(A) & q = p \end{cases}$$

Thus (1) becomes

$$\det(A)x_p = \sum_{k=1}^{n} r_k A_{kp}$$

so that if $\Delta = \det(A)$,

$$x_p = \frac{\sum_{k=1}^{n} r_k A_{kp}}{\Delta} \qquad p = 1, \ldots, n \tag{2}$$

Now the numerator of (2) is a determinant in which the pth column is replaced by the column vector $(r_1 \ r_2 \ \ldots \ r_n)^T$, and so Cramer's rule follows.

15.26. Work Problem 15.25 by using the inverse matrix.

As on page 347, we solve the system (17) or (18) in the form (19), i.e.

$$X = A^{-1}R$$

Now

$$A^{-1} = \frac{(A_{kj})}{\det(A)} = \frac{(A_{kj})}{\Delta}, \qquad R = \begin{pmatrix} r_1 \\ \vdots \\ r_n \end{pmatrix}$$

Thus we have

$$X = \begin{pmatrix} x_1 \\ \vdots \\ x_n \end{pmatrix} = A^{-1}R = \frac{1}{\Delta} \begin{pmatrix} A_{11} & A_{21} & \ldots & A_{n1} \\ A_{12} & A_{22} & \ldots & A_{n2} \\ \cdots\cdots\cdots\cdots\cdots\cdots\cdots \\ A_{1n} & A_{2n} & \ldots & A_{nn} \end{pmatrix} \begin{pmatrix} r_1 \\ r_2 \\ \vdots \\ r_n \end{pmatrix}$$

$$= \frac{1}{\Delta} \begin{pmatrix} A_{11}r_1 + A_{21}r_2 + \cdots + A_{n1}r_n \\ \cdots\cdots\cdots\cdots\cdots\cdots\cdots\cdots \\ A_{1n}r_1 + A_{2n}r_2 + \cdots + A_{nn}r_n \end{pmatrix}$$

from which it follows that

$$x_p = \frac{A_{1p}r_1 + A_{2p}r_2 + \cdots + A_{np}r_n}{\Delta}$$

agreeing with (2) of Problem 15.25.

15.27. Solve the system of equations

$$\begin{cases} 3x_1 - 2x_2 + 2x_3 = 10 \\ x_1 + 2x_2 - 3x_3 = -1 \\ 4x_1 + x_2 + 2x_3 = 3 \end{cases}$$

(a) by Cramer's rule and (b) by using inverse matrices.

(a) By Cramer's rule,

$$x_1 = \frac{\begin{vmatrix} 10 & -2 & 2 \\ -1 & 2 & -3 \\ 3 & 1 & 2 \end{vmatrix}}{\Delta}, \qquad x_2 = \frac{\begin{vmatrix} 3 & 10 & 2 \\ 1 & -1 & -3 \\ 4 & 3 & 2 \end{vmatrix}}{\Delta}, \qquad x_3 = \frac{\begin{vmatrix} 3 & -2 & 10 \\ 1 & 2 & -1 \\ 4 & 1 & 3 \end{vmatrix}}{\Delta}$$

where the determinant of the coefficients is

$$\Delta = \begin{vmatrix} 3 & -2 & 2 \\ 1 & 2 & -3 \\ 4 & 1 & 2 \end{vmatrix} = 35$$

See Problem 15.8. Evaluation of the other determinants yields the solution $x_1 = 2$, $x_2 = -3$, $x_3 = -1$.

(b) The system can be written in matrix form as

$$\begin{pmatrix} 3 & -2 & 2 \\ 1 & 2 & -3 \\ 4 & 1 & 2 \end{pmatrix} \begin{pmatrix} x_1 \\ x_2 \\ x_3 \end{pmatrix} = \begin{pmatrix} 10 \\ -1 \\ 3 \end{pmatrix} \quad \text{or} \quad AX = R \qquad (1)$$

Now the inverse of the first matrix A in (1) was found in Problem 15.18, so that multiplying both sides of (1) by this matrix we have on using the fact that $A^{-1}A = I$,

$$\begin{pmatrix} x_1 \\ x_2 \\ x_3 \end{pmatrix} = \begin{pmatrix} \frac{1}{5} & \frac{6}{35} & \frac{2}{35} \\ -\frac{2}{5} & -\frac{2}{35} & \frac{11}{35} \\ -\frac{1}{5} & -\frac{11}{35} & \frac{8}{35} \end{pmatrix} \begin{pmatrix} 10 \\ -1 \\ 3 \end{pmatrix} = \begin{pmatrix} 2 \\ -3 \\ -1 \end{pmatrix}$$

Thus $x_1 = 2$, $x_2 = -3$, $x_3 = -1$.

Geometrically the equations with $x_1 = x$, $x_2 = y$, $x_3 = z$ represent three planes intersecting in the point $(2, -3, -1)$.

15.28. Solve $\begin{cases} 2x_1 + 5x_2 - 3x_3 = 3 \\ x_1 - 2x_2 + x_3 = 2 \\ 7x_1 + 4x_2 - 3x_3 = -4 \end{cases}$.

Cramer's rule gives

$$x_1 = \frac{\begin{vmatrix} 3 & 5 & -3 \\ 2 & -2 & 1 \\ -4 & 4 & -3 \end{vmatrix}}{\Delta}, \quad x_2 = \frac{\begin{vmatrix} 2 & 3 & -3 \\ 1 & 2 & 1 \\ 7 & -4 & -3 \end{vmatrix}}{\Delta}, \quad x_3 = \frac{\begin{vmatrix} 2 & 5 & 3 \\ 1 & -2 & 2 \\ 7 & 4 & -4 \end{vmatrix}}{\Delta}$$

where

$$\Delta = \begin{vmatrix} 2 & 5 & -3 \\ 1 & -2 & 1 \\ 7 & 4 & -3 \end{vmatrix}$$

Evaluation of the determinants gives formally

$$x_1 = \frac{16}{0}, \quad x_2 = \frac{80}{0}, \quad x_3 = \frac{144}{0} \qquad (1)$$

illustrating the fact that the system has no solution.

On multiplying the first of the given equations by 2, the second by 3, and adding, we obtain $7x_1 + 4x_2 - 3x_3 = 12$ which is not consistent with the third equation given, i.e. $7x_1 + 4x_2 - 3x_3 = -4$. Thus the system of equations is *inconsistent*.

Geometrically the first two equations represent two planes which intersect in a line. The third equation represents a plane which is parallel to this line. Theoretically the planes meet at a point at infinity, which is a possible interpretation of (1).

15.29. Solve $\begin{cases} 2x_1 + 5x_2 - 3x_3 = 3 \\ x_1 - 2x_2 + x_3 = 2 \\ 7x_1 + 4x_2 - 3x_3 = 12 \end{cases}$.

In this case a formal application of Cramer's rule gives

$$x_1 = \frac{0}{0}, \quad x_2 = \frac{0}{0}, \quad x_3 = \frac{0}{0}$$

since theoretically 0/0 can represent any number, our result illustrates the fact that the system has infinitely many solutions.

On multiplying the first equation by 2, the second by 3 and adding, we obtain the third equation. Thus the third equation can be obtained from the first two and so is not needed. We call the system of equations *dependent* or more precisely *linearly dependent*.

Geometrically the planes represented by the first two equations intersect in a line. The plane represented by the third equation passes through the line.

To obtain possible solutions, assign different values to x_3 for example. Thus if $x_3 = 1$, then $x_1 = \frac{17}{9}$, $x_2 = \frac{4}{9}$ and we have a point on the line whose coordinates are $(\frac{17}{9}, \frac{4}{9}, 1)$. Other solutions can be obtained similarly.

15.30. Solve $\begin{cases} 3x_1 - 2x_2 + 2x_3 = 0 \\ x_1 + 2x_2 - 3x_3 = 0 \\ 4x_1 + x_2 + 2x_3 = 0 \end{cases}$.

Cramer's rule gives the solution [see Problem 15.8]

$$x_1 = \frac{0}{35} = 0, \quad x_2 = \frac{0}{35} = 0, \quad x_3 = \frac{0}{35} = 0$$

so that the only solution is the trivial solution.

Geometrically the equations represent three planes which intersect in the point $(0, 0, 0)$.

15.31. Solve $\begin{cases} 2x_1 + 5x_2 - 3x_3 = 0 \\ x_1 - 2x_2 + x_3 = 0 \\ 7x_1 + 4x_2 - 3x_3 = 0 \end{cases}$.

Formal application of Cramer's rule gives

$$x_1 = \frac{0}{0}, \quad y_1 = \frac{0}{0}, \quad z_1 = \frac{0}{0}$$

illustrating the fact that there are infinitely many solutions besides the trivial and obvious one $x_1 = 0$, $x_2 = 0$, $x_3 = 0$. Such solutions can be found by assigning different values to x_3 as in Problem 15.29. Note that the third equation is obtained by adding twice the first equation to three times the second equation, so the equations are dependent.

15.32. For what values of k will the system

$$\begin{cases} 2x + ky + z = 0 \\ (k-1)x - y + 2z = 0 \\ 4x + y + 4z = 0 \end{cases}$$

have non-trivial solutions?

Solving formally by Cramer's rule, we would have

$$x = \frac{0}{\Delta}, \quad y = \frac{0}{\Delta}, \quad z = \frac{0}{\Delta}$$

Then if $\Delta \neq 0$, the system would have the trivial solution $x = 0$, $y = 0$, $z = 0$. In order that the system have non-trivial solutions, we must then have $\Delta = 0$, i.e.

$$\begin{vmatrix} 2 & k & 1 \\ k-1 & -1 & 2 \\ 4 & 1 & 4 \end{vmatrix} = 0 \quad \text{or} \quad 9k - 9 - 4k(k-1) = 0$$

Solving, we find that $k = 1, 9/4$.

EIGENVALUES AND EIGENVECTORS

15.33. Find the eigenvalues of the matrix $\quad A = \begin{pmatrix} 5 & 7 & -5 \\ 0 & 4 & -1 \\ 2 & 8 & -3 \end{pmatrix}$.

Method 1.

If $\quad X = \begin{pmatrix} x_1 \\ x_2 \\ x_3 \end{pmatrix} \quad$ we must consider the equation $AX = \lambda X$, i.e.

$$\begin{pmatrix} 5 & 7 & -5 \\ 0 & 4 & -1 \\ 2 & 8 & -3 \end{pmatrix} \begin{pmatrix} x_1 \\ x_2 \\ x_3 \end{pmatrix} = \lambda \begin{pmatrix} x_1 \\ x_2 \\ x_3 \end{pmatrix}$$

or $\qquad \begin{pmatrix} 5x_1 + 7x_2 - 5x_3 \\ 4x_2 - x_3 \\ 2x_1 + 8x_2 - 3x_3 \end{pmatrix} = \begin{pmatrix} \lambda x_1 \\ \lambda x_2 \\ \lambda x_3 \end{pmatrix}$

Equating corresponding elements of these matrices, we find

$$\begin{aligned} (5 - \lambda)x_1 + 7x_2 - 5x_3 &= 0 \\ (4 - \lambda)x_2 - x_3 &= 0 \\ 2x_1 + 8x_2 - (3 + \lambda)x_3 &= 0 \end{aligned} \tag{1}$$

This system will have non-trivial solutions if

$$\begin{vmatrix} 5 - \lambda & 7 & -5 \\ 0 & 4 - \lambda & -1 \\ 2 & 8 & -3 - \lambda \end{vmatrix} = 0 \tag{2}$$

Expansion of this determinant yields

$$\lambda^3 - 6\lambda^2 + 11\lambda - 6 = 0 \quad \text{or} \quad (\lambda - 1)(\lambda - 2)(\lambda - 3) = 0$$

Then the eigenvalues are $\lambda = 1, 2, 3$.

Method 2.

We can write $AX = \lambda X$ as $AX = \lambda IX$ or $(A - \lambda I)X = 0$ where I and 0 are the unit and zero matrix and

$$A - \lambda I = \begin{pmatrix} 5 - \lambda & 7 & -5 \\ 0 & 4 - \lambda & -1 \\ 2 & 8 & -3 - \lambda \end{pmatrix}$$

Then non-trivial solutions will exist if $\det(A - \lambda I) = 0$ and we can then proceed as in method 1. Note that equation (2) can be written at once by subtracting λ from each of the diagonal elements of A.

15.34. (a) Find eigenvectors corresponding to the eigenvalues of the matrix A in Problem 15.33 and (b) determine a set of unit eigenvectors.

(a) Corresponding to $\lambda = 1$, equations (1) of Problem 15.33 become

$$\begin{aligned} 4x_1 + 7x_2 - 5x_3 &= 0 \\ 3x_2 - x_3 &= 0 \\ 2x_1 + 8x_2 - 4x_3 &= 0 \end{aligned}$$

Solving for x_1 and x_3 in terms of x_2 we find $x_3 = 3x_2$, $x_1 = 2x_2$. Then an eigenvector is

$$\begin{pmatrix} x_1 \\ x_2 \\ x_3 \end{pmatrix} = \begin{pmatrix} 2x_2 \\ x_2 \\ 3x_2 \end{pmatrix} = x_2 \begin{pmatrix} 2 \\ 1 \\ 3 \end{pmatrix} \quad \text{or simply} \quad \begin{pmatrix} 2 \\ 1 \\ 3 \end{pmatrix}$$

since any eigenvector is a scalar (constant) multiple of this.

Similarly, corresponding to $\lambda = 2$ equations (1) of Problem 15.33 lead to $x_3 = 2x_2$, $x_1 = x_2$ which in turn leads to the eigenvector

$$\begin{pmatrix} x_1 \\ x_2 \\ x_3 \end{pmatrix} = \begin{pmatrix} x_2 \\ x_2 \\ 2x_2 \end{pmatrix} = x_2 \begin{pmatrix} 1 \\ 1 \\ 2 \end{pmatrix} \quad \text{or simply} \quad \begin{pmatrix} 1 \\ 1 \\ 2 \end{pmatrix}$$

Finally if $\lambda = 3$ we obtain $x_3 = x_2$, $x_1 = -x_2$, giving the eigenvector

$$\begin{pmatrix} x_1 \\ x_2 \\ x_3 \end{pmatrix} = \begin{pmatrix} -x_2 \\ x_2 \\ x_2 \end{pmatrix} = x_2 \begin{pmatrix} -1 \\ 1 \\ 1 \end{pmatrix} \quad \text{or simply} \quad \begin{pmatrix} -1 \\ 1 \\ 1 \end{pmatrix}$$

(b) The unit eigenvectors have the property that they have length 1, i.e. the sum of the squares of their components $= 1$. To obtain such eigenvectors we divide each vector by the square root of the sum of the squares of the components. Thus the above become respectively

$$\begin{pmatrix} 2/\sqrt{14} \\ 1/\sqrt{14} \\ 3/\sqrt{14} \end{pmatrix}, \quad \begin{pmatrix} 1/\sqrt{5} \\ 1/\sqrt{5} \\ 2/\sqrt{5} \end{pmatrix}, \quad \begin{pmatrix} -1/\sqrt{3} \\ 1/\sqrt{3} \\ 1/\sqrt{3} \end{pmatrix}$$

15.35. Find the (a) eigenvalues and (b) eigenvectors of $A = \begin{pmatrix} 2 & 0 & -2 \\ 0 & 4 & 0 \\ -2 & 0 & 5 \end{pmatrix}$.

(a) The eigenvalues are solutions of $\begin{vmatrix} 2-\lambda & 0 & -2 \\ 0 & 4-\lambda & 0 \\ -2 & 0 & 5-\lambda \end{vmatrix} = 0$ which gives $\lambda = 1, 4, 6$.

(b) From the equations $(A - \lambda I)X = 0$ we obtain

$$(2-\lambda)x_1 - 2x_3 = 0$$
$$(4-\lambda)x_2 = 0$$
$$-2x_1 + (5-\lambda)x_3 = 0$$

Then corresponding to $\lambda = 1$ we find the eigenvector $\begin{pmatrix} 2 \\ 0 \\ 1 \end{pmatrix}$.

Corresponding to $\lambda = 4$ we find the eigenvector $\begin{pmatrix} 0 \\ 1 \\ 0 \end{pmatrix}$.

Corresponding to $\lambda = 6$ we find the eigenvector $\begin{pmatrix} 1 \\ 0 \\ -2 \end{pmatrix}$.

15.36. Find the (a) eigenvalues and (b) eigenvectors of $A = \begin{pmatrix} \cos\theta & -\sin\theta \\ \sin\theta & \cos\theta \end{pmatrix}$.

(a) By the usual procedure the eigenvalues are solutions of

$$\begin{vmatrix} \cos\theta - \lambda & -\sin\theta \\ \sin\theta & \cos\theta - \lambda \end{vmatrix} = 0 \quad \text{or} \quad \lambda^2 - 2\lambda\cos\theta + 1 = 0$$

Then $\lambda = \dfrac{2\cos\theta \pm \sqrt{4\cos^2\theta - 4}}{2} = \cos\theta \pm i\sin\theta = e^{\pm i\theta}$

(b) The equations for determining the eigenvectors are found from

$$\begin{pmatrix} \cos\theta - \lambda & -\sin\theta \\ \sin\theta & \cos\theta - \lambda \end{pmatrix}\begin{pmatrix} x_1 \\ x_2 \end{pmatrix} = 0$$

i.e.
$$\left.\begin{array}{c} (\cos\theta - \lambda)x_1 - (\sin\theta)x_2 = 0 \\ (\sin\theta)x_1 + (\cos\theta - \lambda)x_2 = 0 \end{array}\right\} \tag{1}$$

Using $\lambda = e^{i\theta} = \cos\theta + i\sin\theta$ we find from (1), $x_2 = -ix$, so that a corresponding eigenvector is

$$\begin{pmatrix} x_1 \\ x_2 \end{pmatrix} = \begin{pmatrix} x_1 \\ -ix_1 \end{pmatrix} = x_1\begin{pmatrix} 1 \\ -i \end{pmatrix} \qquad \text{or simply} \qquad \begin{pmatrix} 1 \\ -i \end{pmatrix}$$

Using $\lambda = e^{-i\theta} = \cos\theta - i\sin\theta$ we find $x_2 = ix_1$ so that a corresponding eigenvector is

$$\begin{pmatrix} x_1 \\ x_2 \end{pmatrix} = \begin{pmatrix} x_1 \\ ix_1 \end{pmatrix} = x_1\begin{pmatrix} 1 \\ i \end{pmatrix} \qquad \text{or simply} \qquad \begin{pmatrix} 1 \\ i \end{pmatrix}$$

THEOREMS ON EIGENVALUES AND EIGENVECTORS

15.37. Prove that the eigenvalues of a Hermitian matrix [or symmetric real matrix] are real.

Let A be a Hermitian matrix and λ an eigenvalue. Then by definition there is a non-trivial eigenvector X such that

$$AX = \lambda X$$

Multiplying by $\bar{X}^T$, $$\bar{X}^T A X = \lambda \bar{X}^T X \tag{1}$$

Taking the conjugate, $$X^T \bar{A} \bar{X} = \bar{\lambda} X^T \bar{X} \tag{2}$$

Taking the transpose, using the second and third equations in (5), page 343, we find

$$\bar{X}^T \bar{A}^T X = \bar{\lambda} \bar{X}^T X \tag{3}$$

Now since A is Hermitian, $\bar{A}^T = A$ so that (3) becomes

$$\bar{X}^T A X = \bar{\lambda} \bar{X}^T X \tag{4}$$

Subtracting (4) from (1) we thus obtain

$$(\lambda - \bar{\lambda})\bar{X}^T X = 0$$

Then since $\bar{X}^T X$ cannot be zero, it follows that $\lambda = \bar{\lambda}$ or that λ must be real.

15.38. Prove that the eigenvectors of a Hermitian matrix [or symmetric real matrix] belonging to different eigenvalues are orthogonal.

Let X_1 and X_2 be eigenvectors belonging to eigenvalues λ_1, λ_2. Then denoting the matrix by A, we have

$$AX_1 = \lambda_1 X_1, \quad AX_2 = \lambda_2 X_2 \tag{1}$$

Multiplying these by $\bar{X}_2^T$ and $\bar{X}_1^T$ respectively, we find

$$\bar{X}_2^T A X_1 = \lambda_1 \bar{X}_2^T X_1, \quad \bar{X}_1^T A X_2 = \lambda_2 \bar{X}_1^T X_2 \tag{2}$$

Taking the conjugate of the first equation in (2), we find since λ_1 is real,

$$X_2^T \bar{A} \bar{X}_1 = \lambda_1 X_2^T \bar{X}_1 \tag{3}$$

Now taking the transpose of (3),

$$\bar{X}_1^T \bar{A}^T X_2 = \lambda_1 \bar{X}_1^T X_2 \tag{4}$$

Since A is Hermitian, i.e. $\bar{A}^T = A$, (4) becomes

$$\bar{X}_1^T A X_2 = \lambda_1 \bar{X}_1^T X_2$$

Subtracting this from the second of equations (2),

$$(\lambda_1 - \lambda_2)\bar{X}_1^T X_2 = 0$$

Then since $\lambda_1 \neq \lambda_2$, we have $\bar{X}_1^T X_2 = 0$ or X_1 and X_2 are orthogonal.

15.39. (a) Illustrate by an example the results of Problems 15.37 and 15.38.

(b) If a matrix has real eigenvalues, must it be Hermitian? Explain.

(a) The matrix A of Problem 15.35 is real and symmetric and thus Hermitian. As shown in that problem, the eigenvalues are all real. Also the eigenvectors

$$\begin{pmatrix} 2 \\ 0 \\ 1 \end{pmatrix}, \quad \begin{pmatrix} 0 \\ 1 \\ 0 \end{pmatrix}, \quad \begin{pmatrix} 1 \\ 0 \\ -2 \end{pmatrix}$$

are mutually orthogonal as is easily verified.

(b) A matrix can have real eigenvalues without being Hermitian. See, for example, the matrix of Problem 15.33.

15.40. Verify the Cayley-Hamilton Theorem 15-14, page 349, for the matrix $A = \begin{pmatrix} 3 & 2 \\ -1 & 4 \end{pmatrix}$.

The characteristic equation is

$$\begin{vmatrix} 3 - \lambda & 2 \\ -1 & 4 - \lambda \end{vmatrix} = 0 \quad \text{or} \quad \lambda^2 - 7\lambda + 14 = 0$$

To verify the theorem we must show that the matrix A satisfies

$$A^2 - 7A + 14I = 0$$

where λ in the characteristic equation is replaced by A, the constant term [in this case 14] is replaced by $14I$ and 0 is replaced by O.

We have

$$A^2 - 7A + 14I = \begin{pmatrix} 3 & 2 \\ -1 & 4 \end{pmatrix}\begin{pmatrix} 3 & 2 \\ -1 & 4 \end{pmatrix} - 7\begin{pmatrix} 3 & 2 \\ -1 & 4 \end{pmatrix} + 14\begin{pmatrix} 1 & 0 \\ 0 & 1 \end{pmatrix}$$

$$= \begin{pmatrix} 7 & 14 \\ -7 & 14 \end{pmatrix} + \begin{pmatrix} -21 & -14 \\ 7 & -28 \end{pmatrix} + \begin{pmatrix} 14 & 0 \\ 0 & 14 \end{pmatrix} = \begin{pmatrix} 0 & 0 \\ 0 & 0 \end{pmatrix} = O$$

as required.

15.41. Verify Theorem 15-15, page 349, by transforming the matrix of Problem 15.33 into diagonal form.

The eigenvectors for the matrix of Problem 15.33 are the columns of

$$B = \begin{pmatrix} 2 & 1 & -1 \\ 1 & 1 & 1 \\ 3 & 2 & 1 \end{pmatrix}$$

as shown in Problem 15.34. The inverse of B is then given by

$$B^{-1} = \begin{pmatrix} -1 & -3 & 2 \\ 2 & 5 & -3 \\ -1 & -1 & 1 \end{pmatrix}$$

Thus $B^{-1}AB = \begin{pmatrix} -1 & -3 & 2 \\ 2 & 5 & -3 \\ -1 & -1 & 1 \end{pmatrix} \begin{pmatrix} 5 & 7 & -5 \\ 0 & 4 & -1 \\ 2 & 8 & -3 \end{pmatrix} \begin{pmatrix} 2 & 1 & -1 \\ 1 & 1 & 1 \\ 3 & 2 & 1 \end{pmatrix} = \begin{pmatrix} 1 & 0 & 0 \\ 0 & 2 & 0 \\ 0 & 0 & 3 \end{pmatrix}$

15.42. Prove Theorem 15-15.

We prove the theorem for the case of a third order matrix, since the proof for any square matrix is exactly analogous. Denote the eigenvectors of A by the columns in

$$B = \begin{pmatrix} b_{11} & b_{12} & b_{13} \\ b_{21} & b_{22} & b_{23} \\ b_{31} & b_{32} & b_{33} \end{pmatrix}$$

and the corresponding distinct eigenvalues by $\lambda_1, \lambda_2, \lambda_3$. Then by definition,

$$A\begin{pmatrix} b_{11} \\ b_{21} \\ b_{31} \end{pmatrix} = \lambda_1 \begin{pmatrix} b_{11} \\ b_{21} \\ b_{31} \end{pmatrix}, \quad A\begin{pmatrix} b_{12} \\ b_{22} \\ b_{32} \end{pmatrix} = \lambda_2 \begin{pmatrix} b_{12} \\ b_{22} \\ b_{32} \end{pmatrix}, \quad A\begin{pmatrix} b_{13} \\ b_{23} \\ b_{33} \end{pmatrix} = \lambda_3 \begin{pmatrix} b_{13} \\ b_{23} \\ b_{33} \end{pmatrix}$$

from which $AB = A\begin{pmatrix} b_{11} & b_{12} & b_{13} \\ b_{21} & b_{22} & b_{23} \\ b_{31} & b_{32} & b_{33} \end{pmatrix} = \begin{pmatrix} \lambda_1 b_{11} & \lambda_2 b_{12} & \lambda_3 b_{13} \\ \lambda_1 b_{21} & \lambda_2 b_{22} & \lambda_3 b_{23} \\ \lambda_1 b_{31} & \lambda_2 b_{32} & \lambda_3 b_{33} \end{pmatrix}$

$$= \begin{pmatrix} b_{11} & b_{12} & b_{13} \\ b_{12} & b_{22} & b_{23} \\ b_{13} & b_{32} & b_{33} \end{pmatrix} \begin{pmatrix} \lambda_1 & 0 & 0 \\ 0 & \lambda_2 & 0 \\ 0 & 0 & \lambda_3 \end{pmatrix}$$

$$= B\begin{pmatrix} \lambda_1 & 0 & 0 \\ 0 & \lambda_2 & 0 \\ 0 & 0 & \lambda_3 \end{pmatrix}$$

Thus multiplying by B^{-1} we have, as required,

$$B^{-1}AB = \begin{pmatrix} \lambda_1 & 0 & 0 \\ 0 & \lambda_2 & 0 \\ 0 & 0 & \lambda_3 \end{pmatrix}$$

15.43. (a) Show that the quadratic form $2x_1^2 + 4x_2^2 + 5x_3^2 - 4x_1x_3 = X^TAX$ where

$$X = \begin{pmatrix} x_1 \\ x_2 \\ x_3 \end{pmatrix}, \quad A = \begin{pmatrix} 2 & 0 & -2 \\ 0 & 4 & 0 \\ -2 & 0 & 5 \end{pmatrix}$$

(b) Find a linear transformation from x_1, x_2, x_3 to u_1, u_2, u_3 which will remove the cross product term in the quadratic form of (a) and thus write the resulting quadratic form in u_1, u_2, u_3.

(a) We have $X^TAX = (x_1 \ x_2 \ x_3) \begin{pmatrix} 2 & 0 & -2 \\ 0 & 4 & 0 \\ -2 & 0 & 5 \end{pmatrix} \begin{pmatrix} x_1 \\ x_2 \\ x_3 \end{pmatrix}$

$$= (x_1 \ x_2 \ x_3) \begin{pmatrix} 2x_1 - 2x_3 \\ 4x_2 \\ -2x_1 + 5x_3 \end{pmatrix}$$

$$= 2x_1^2 + 4x_2^2 + 5x_3^2 - 4x_1x_3$$

Note that the coefficients of x_1^2, x_2^2, x_3^2 namely 2, 4, 5 appear in the main diagonal while half the coefficients of $x_j x_k$, $j \neq k$ appear as elements in the jth row and kth column.

(b) A linear transformation from x_1, x_2, x_3 to u_1, u_2, u_3 can be written as $X = BU$ where

$$X = \begin{pmatrix} x_1 \\ x_2 \\ x_3 \end{pmatrix}, \quad U = \begin{pmatrix} u_1 \\ u_2 \\ u_3 \end{pmatrix} \quad \text{and } B \text{ is a } 3 \times 3 \text{ matrix. Then we have}$$

$$X^T A X = (BU)^T A (BU) = U^T (B^T A B) U \tag{1}$$

Now the right side of (1) will not have cross product terms if $B^T A B$ is a diagonal matrix. Thus we see that if $B^T = B^{-1}$ [i.e. if B is an orthogonal matrix] the problem becomes one of finding the eigenvalues and eigenvectors of A. This has already been done in Problem 15.35. We choose B as the matrix of unit eigenvectors, i.e.

$$B = \begin{pmatrix} 2/\sqrt{5} & 0 & 1/\sqrt{5} \\ 0 & 1 & 0 \\ 1/\sqrt{5} & 0 & -2/\sqrt{5} \end{pmatrix}$$

from which we easily find that $B^T = B^{-1}$ so that B is orthogonal and we have

$$B^{-1} A B = \begin{pmatrix} 2/\sqrt{5} & 0 & 1/\sqrt{5} \\ 0 & 1 & 0 \\ 1/\sqrt{5} & 0 & -2/\sqrt{5} \end{pmatrix} \begin{pmatrix} 2 & 0 & -2 \\ 0 & 4 & 0 \\ -2 & 0 & 5 \end{pmatrix} \begin{pmatrix} 2/\sqrt{5} & 0 & 1/\sqrt{5} \\ 0 & 1 & 0 \\ 1/\sqrt{5} & 0 & -2/\sqrt{5} \end{pmatrix} = \begin{pmatrix} 1 & 0 & 0 \\ 0 & 4 & 0 \\ 0 & 0 & 6 \end{pmatrix}$$

as required. Then (1) becomes

$$X^T A X = U^T (B^{-1} A B) U = (u_1 \; u_2 \; u_3) \begin{pmatrix} 1 & 0 & 0 \\ 0 & 4 & 0 \\ 0 & 0 & 6 \end{pmatrix} \begin{pmatrix} u_1 \\ u_2 \\ u_3 \end{pmatrix} = u_1^2 + 4u_2^2 + 6u_3^2$$

which is the required quadratic form, called the *canonical form*. The transformation from X to U is $X = BU$, from which we find

$$x_1 = \frac{2u_1 + u_3}{\sqrt{5}}, \quad x_2 = u_2, \quad x_3 = \frac{u_1 - 2u_3}{\sqrt{5}}$$

Supplementary Problems

OPERATIONS WITH MATRICES

15.44. (a) If $A = \begin{pmatrix} -2 & 4 \\ 1 & 5 \end{pmatrix}$, $B = \begin{pmatrix} 1 & -4 \\ 2 & 3 \end{pmatrix}$, $C = \begin{pmatrix} 3 & 2 \\ -1 & 1 \end{pmatrix}$ verify that $A(B + C) = AB + AC$, $(A + B)(A - B) = A^2 - B^2 + BA - AB$, and $(ABC)^T = C^T B^T A^T$.

(b) Find $2A - 3B - C$ and $(A - 2B)(C + 3B)$ where A, B, C are the matrices in (a).

15.45. If $A = \begin{pmatrix} 3 & 2 \\ -2 & 4 \end{pmatrix}$, $B = (4 \; -2)$, $C = \begin{pmatrix} -1 \\ 5 \end{pmatrix}$ find (a) AC, (b) CA, (c) BC, (d) $C^T B^T$, (e) $A(B^T + C)$, (f) BB^T.

15.46. If $A = \begin{pmatrix} 3 & 1 & 2 \\ -2 & 4 & -2 \end{pmatrix}$, $B = \begin{pmatrix} 2 & -1 \\ 3 & 4 \end{pmatrix}$, $C = \begin{pmatrix} -1 \\ 2 \end{pmatrix}$ find (a) $C^T A$, (b) $A^T C$, (c) $AA^T BC$.

15.47. If $A = \begin{pmatrix} 3 & 2 & -2 \\ 4 & -1 & 2 \\ -2 & 0 & -1 \end{pmatrix}$, $B = \begin{pmatrix} 1 & -2 & 0 \\ 2 & 1 & -1 \\ 1 & -3 & 2 \end{pmatrix}$ find (a) $(A - B)(A + B)$, (b) $A^2 - B^2$, (c) $AB - BA$, (d) $A^T B + B^T A$.

15.48. Prove that for any $m \times n$ matrices (a) $A + B = B + A$, (b) $A + (B + C) = (A + B) + C$,
(c) $\lambda(A + B) = \lambda A + \lambda B$ where λ is any scalar.

15.49. Find x and y such that $\begin{pmatrix} 2 & -1 \\ -3 & 4 \end{pmatrix} \begin{pmatrix} x \\ y \end{pmatrix} + \begin{pmatrix} 8 \\ 1 \end{pmatrix} = \begin{pmatrix} 0 \\ 0 \end{pmatrix}$.

15.50. If A and B are square matrices such that $AB = 0$, prove that we can have $A \neq 0$, $B \neq 0$. Is the result true for non-square matrices?

15.51. If $AB = AC$, is it true that $B = C$? Explain.

15.52. If A, B and C are any square matrices of the same order, prove that (a) $A(BC) = (AB)C$,
(b) $A(B + C) = AB + AC$, (c) $(ABC)^T = C^T B^T A^T$ and generalize these results.

15.53. A *linear transformation* from an (x_1, x_2) to a (y_1, y_2) coordinate system is defined as $y_1 = a_{11}x_1 + a_{12}x_2$, $y_2 = a_{21}x_1 + a_{22}x_2$. (a) If $Y = \begin{pmatrix} y_1 \\ y_2 \end{pmatrix}$, $X = \begin{pmatrix} x_1 \\ x_2 \end{pmatrix}$, $A = \begin{pmatrix} a_{11} & a_{12} \\ a_{21} & a_{22} \end{pmatrix}$ show that the transformation can be written $Y = AX$. (b) If $X = BU$ where $X = \begin{pmatrix} x_1 \\ x_2 \end{pmatrix}$, $U = \begin{pmatrix} u_1 \\ u_2 \end{pmatrix}$, $B = \begin{pmatrix} b_{11} & b_{12} \\ b_{21} & b_{22} \end{pmatrix}$ show that $x_1 = b_{11}u_1 + b_{12}u_2$, $x_2 = b_{21}u_1 + b_{22}u_2$. Thus obtain y_1, y_2 in terms of u_1, u_2 and explain how you can use the approach to motivate a definition of AB.

15.54. Generalize the ideas of Problem 15.53 to 3 or more dimensions.

15.55. Let P [Fig. 15-1] have coordinates $(x_1 y)$ relative to an xy coordinate system and (x', y') relative to an $x'y'$ coordinate system which is rotated through angle θ relative to the xy coordinate system. (a) Prove that the relationship between the coordinates or transformation from (x, y) to (x', y') is given by

$$\begin{pmatrix} x' \\ y' \end{pmatrix} = \begin{pmatrix} \cos\theta & -\sin\theta \\ \sin\theta & \cos\theta \end{pmatrix} \begin{pmatrix} x \\ y \end{pmatrix}$$

(b) Show that the square matrix in (a) is skew-symmetric.

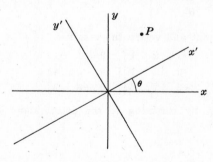

Fig. 15-1

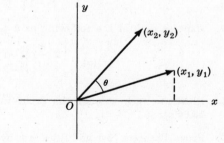

Fig. 15-2

15.56. A vector with components (x_1, y_1) in an xy coordinate system [Fig. 15-2] is rotated through angle θ so that its new components are (x_2, y_2). Show that

$$\begin{pmatrix} x_2 \\ y_2 \end{pmatrix} = \begin{pmatrix} \cos\theta & -\sin\theta \\ \sin\theta & \cos\theta \end{pmatrix} \begin{pmatrix} x_1 \\ y_1 \end{pmatrix}$$

and explain the relationship with Problem 15.55.

15.57. Let $A(\theta)$ denote the square matrix in Problems 15.55 or 15.56. Show that (a) $A(\theta_1 + \theta_2) = A(\theta_1)A(\theta_2)$, (b) $A(\theta_1 + \theta_2 + \cdots + \theta_n) = A(\theta_1)A(\theta_2)\cdots A(\theta_n)$, (c) $[A(\theta)]^n = A(n\theta)$ and discuss the significance of these results in terms of the transformations in Problems 15.55 and 15.56.

15.58. Let $X = \begin{pmatrix} x_1 \\ x_2 \\ \vdots \\ x_n \end{pmatrix}$, $A = \begin{pmatrix} a_{11} & a_{12} & \cdots & a_{1n} \\ a_{21} & a_{22} & \cdots & a_{2n} \\ \cdots\cdots\cdots\cdots\cdots\cdots \\ a_{n1} & a_{n2} & \cdots & a_{nn} \end{pmatrix}$.   (*a*) Show that $X^T A X = \sum_{j=1}^{n} \sum_{k=1}^{n} a_{jk} x_j x_k$, which is called a *quadratic form* in $x_1, \ldots, x_n$.   (*b*) Show that if A is a real symmetric matrix, i.e. $a_{jk} = a_{kj}$, then $X^T A X = a_{11}x_1^2 + a_{22}x_2^2 + \cdots + a_{nn}x_n^2 + 2a_{12}x_1x_2 + 2a_{13}x_1x_3 + \cdots$, which is called a *symmetric quadratic form*.   (*c*) What does the quadratic form become if A is skew-symmetric?

15.59. Write the quadratic forms   (*a*) $4x_1^2 - 6x_1x_2 + 3x_2^2$,   (*b*) $x_1^2 - 2x_2^2 + 4x_3^2 + 2x_1x_2 - 6x_1x_3 + 4x_2x_3$ in terms of matrices.

15.60. If A is Hermitian or skew-Hermitian, the quadratic form $\bar{X}^T A X$ is called a Hermitian or skew-Hermitian form respectively.   Prove that for every choice of X (*a*) the value of a Hermitian form is always real,   (*b*) the value of a skew-Hermitian form is zero or pure imaginary.

15.61. Prove that every square matrix C can be written as $A + B$ where A is Hermitian and B is skew-Hermitian.

15.62. The concept of a matrix whose elements are real or complex numbers can be extended to one whose elements themselves are matrices.   In such case the elements are called *submatrices*.   Rules for addition, multiplication, etc., analogous to those on pages 312-313 can be made.   Prove that if, for example, $A = \begin{pmatrix} A_{11} & A_{12} \\ A_{21} & A_{22} \end{pmatrix}$, $B = \begin{pmatrix} B_{11} & B_{12} \\ B_{21} & B_{22} \end{pmatrix}$ where the elements are matrices, then

(*a*) $A + B = \begin{pmatrix} A_{11} + B_{11} & A_{12} + B_{12} \\ A_{21} + B_{21} & A_{22} + B_{22} \end{pmatrix}$ (*b*) $AB = \begin{pmatrix} A_{11}B_{11} + A_{12}B_{21} & A_{11}B_{12} + A_{12}B_{22} \\ A_{21}B_{11} + A_{22}B_{21} & A_{21}B_{12} + A_{22}B_{22} \end{pmatrix}$

assuming the submatrices are conformable.

DETERMINANTS

15.63. Evaluate the determinants   (*a*) $\begin{vmatrix} 3 & -2 \\ -4 & -5 \end{vmatrix}$,   (*b*) $\begin{vmatrix} \cos\theta & -\sin\theta \\ \sin\theta & \cos\theta \end{vmatrix}$,   (*c*) $\begin{vmatrix} 3 & -2 & 4 \\ 2 & 1 & -1 \\ -1 & 2 & 2 \end{vmatrix}$,   (*d*) $\begin{vmatrix} a & b & 1 \\ 1 & a & b \\ b & 1 & a \end{vmatrix}$.

15.64. Evaluate   (*a*) $\begin{vmatrix} 2 & 1 & -1 & 1 \\ -2 & 1 & 3 & -2 \\ 1 & -2 & 1 & 1 \\ -3 & 1 & 2 & -1 \end{vmatrix}$,   (*b*) $\begin{vmatrix} 1 & a & a^2 & a^3 \\ 1 & b & b^2 & b^3 \\ 1 & c & c^2 & c^3 \\ 1 & d & d^2 & d^3 \end{vmatrix}$.

15.65. Express each of the following as a single determinant and verify the result:

(*a*) $\begin{vmatrix} 2 & -1 \\ 3 & -4 \end{vmatrix} \begin{vmatrix} -1 & 5 \\ -3 & -2 \end{vmatrix}$ (*b*) $\begin{vmatrix} -1 & 2 & 1 \\ 0 & 4 & -1 \\ -3 & 1 & 4 \end{vmatrix} \begin{vmatrix} 2 & 1 & -1 \\ -3 & 2 & 4 \\ 0 & -4 & 2 \end{vmatrix}$

15.66. Illustrate each step of the proof in Problem 15.7 by referring to the determinant of Problem 15.64(*a*).

15.67. Prove Theorem 15-1 and illustrate by an example.

15.68. A triangular matrix A has all elements above [or below] the main diagonal equal to zero.   Prove that $\det(A) = 0$.

15.69. If A and B are the matrices of Problem 15.50, prove that at least one of them is singular.

15.70. Prove Theorem 15-2, page 345, and illustrate by an example.

15.71. Prove Theorem 15-3, page 345, and illustrate by an example.   [*Hint:* Use Problem 15.7.]

15.72. (*a*) Show that $\begin{vmatrix} a_{11} & a_{12} & 0 & 0 \\ a_{21} & a_{22} & 0 & 0 \\ -1 & 0 & b_{11} & b_{12} \\ 0 & -1 & b_{21} & b_{22} \end{vmatrix} = \begin{vmatrix} a_{11} & a_{12} \\ a_{21} & a_{22} \end{vmatrix} \begin{vmatrix} b_{11} & b_{12} \\ b_{21} & b_{22} \end{vmatrix}$

(b) By transforming the fourth order determinant in (a) using Theorem 15-7, page 345, show that it is equal to

$$\begin{vmatrix} 0 & 0 & a_{11}b_{11} + a_{12}b_{21} & a_{11}b_{12} + a_{12}b_{22} \\ 0 & 0 & a_{21}b_{11} + a_{22}b_{21} & a_{21}b_{12} + a_{22}b_{22} \\ -1 & 0 & b_{11} & b_{12} \\ 0 & -1 & b_{21} & b_{22} \end{vmatrix} = \begin{vmatrix} a_{11}b_{11} + a_{12}b_{21} & a_{11}b_{12} + a_{12}b_{22} \\ a_{21}b_{11} + a_{22}b_{21} & a_{21}b_{12} + a_{22}b_{22} \end{vmatrix}$$

Thus prove Theorem 15-8, page 345, for 2×2 matrices.

15.73. Generalize Problem 15.72 and thus prove Theorem 15-8 for square matrices of any order.

15.74. Illustrate Theorem 15-9, page 346, by an example.

15.75. Prove Theorem 15-10, page 346.

INVERSE OF A MATRIX

15.76. Find the inverse of each of the following matrices and check results.

(a) $\begin{pmatrix} 2 & -1 \\ -1 & 1 \end{pmatrix}$ (b) $\begin{pmatrix} \cos\theta & -\sin\theta \\ \sin\theta & \cos\theta \end{pmatrix}$ (c) $\begin{pmatrix} 1 & 0 & -1 \\ 0 & 1 & 1 \\ 1 & -1 & 1 \end{pmatrix}$ (d) $\begin{pmatrix} 2 & 1 & 0 \\ 1 & 1 & 0 \\ 0 & 0 & 1 \end{pmatrix}$ (e) $\begin{pmatrix} 1 & 1 & 2 \\ 1 & 2 & 2 \\ 2 & 2 & 3 \end{pmatrix}$

15.77. (a) Prove that if $AB = I$, then $B = A^{-1}$. Use this result to find directly the inverses of

(a) $\begin{pmatrix} 3 & 2 \\ -2 & 4 \end{pmatrix}$ and (b) $\begin{pmatrix} 2 & 1 & -1 \\ 1 & 1 & 1 \\ 3 & 2 & 1 \end{pmatrix}$. [*Hint:* In (a) assume $B = \begin{pmatrix} b_{11} & b_{12} \\ b_{21} & b_{22} \end{pmatrix}$ and find $b_{11}, b_{12}, b_{21}, b_{22}$.]

15.78. Prove that $(A^{-1})^{-1} = A$ where A is a non-singular matrix and illustrate by an example.

15.79. Is it true that if $\det(A) \neq 0$ (a) $(A^{-1})^2 = (A^2)^{-1}$, (b) $(A^m)^n = (A^n)^m = A^{mn}$? Justify your statements.

15.80. Prove that $(ABC)^{-1} = C^{-1}B^{-1}A^{-1}$ and generalize.

15.81. Discuss the significance of the inverse of a matrix with special reference to (a) Problem 15.53, (b) Problems 15.55 and 15.56.

ORTHOGONAL AND UNITARY MATRICES. ORTHOGONAL VECTORS

15.82. Show that (a) $\begin{pmatrix} \frac{3}{5} & -\frac{4}{5} \\ \frac{4}{5} & \frac{3}{5} \end{pmatrix}$ (b) $\begin{pmatrix} 2/\sqrt{5} & 0 & 1/\sqrt{5} \\ 0 & 1 & 0 \\ 1 & 0 & 0 \end{pmatrix}$ are orthogonal matrices.

15.83. Show that (a) $\begin{pmatrix} 0 & -i \\ i & 0 \end{pmatrix}$ (b) $\begin{pmatrix} \cos\phi & i\sin\phi \\ i\sin\phi & \cos\phi \end{pmatrix}$ (c) $\begin{pmatrix} \cos\phi & -i\sin\phi & 0 \\ -i\sin\phi & \cos\phi & 0 \\ 0 & 0 & 1 \end{pmatrix}$ are unitary matrices.

15.84. Determine the form of the most general second order unitary square matrix of (a) order 2, (b) order 3.

15.85. If A is a unitary matrix, prove that $\det(A) = e^{i\alpha}$ for some constant α. Illustrate by an example.

15.86. (a) Show that the vectors $\begin{pmatrix} 2 \\ -1 \\ -2 \end{pmatrix}$, $\begin{pmatrix} 1 \\ 0 \\ 1 \end{pmatrix}$, $\begin{pmatrix} 1 \\ 4 \\ 1 \end{pmatrix}$ are mutually orthogonal. (b) From the vectors in (a) determine a set of mutually orthonormal vectors.

15.87. Find a unit vector which is orthogonal to each of the vectors $\begin{pmatrix} 2 \\ -3 \\ 1 \end{pmatrix}$ and $\begin{pmatrix} -2 \\ -1 \\ 1 \end{pmatrix}$.

15.88. If $\begin{pmatrix} a_1 \\ a_2 \\ a_3 \end{pmatrix}$, $\begin{pmatrix} b_1 \\ b_2 \\ b_3 \end{pmatrix}$, $\begin{pmatrix} c_1 \\ c_2 \\ c_3 \end{pmatrix}$ are mutually orthogonal, prove that $\begin{pmatrix} a_1 & b_1 & c_1 \\ a_2 & b_2 & c_2 \\ a_3 & b_3 & c_3 \end{pmatrix}$ is an orthogonal

matrix. Can you determine an analogous result for a unitary matrix?

SYSTEMS OF LINEAR EQUATIONS

15.89. Solve the systems of equations

(a) $\begin{cases} 3x_1 - 2x_2 = 17 \\ 5x + 3x_2 = 3 \end{cases}$
(b) $\begin{cases} 2x_1 + 3x_2 - 4x_3 = -3 \\ 3x_1 - 2x_2 + 5x_3 = 24 \\ x_1 + 4x_2 - 3x_3 = -6 \end{cases}$
(c) $\begin{cases} x_1 + 2x_3 = -8 \\ x_2 - 4x_3 = 16 \\ x_3 - 3x_1 = 3 \end{cases}$

15.90. The currents I_1, I_2, I_3 and I_4 in an electric network satisfy the system of equations

$$\begin{cases} 3I_1 + 2I_3 - I_4 = 60 \\ 2I_1 - I_2 + 4I_3 = 160 \\ 4I_2 + I_3 - 2I_4 = 20 \\ 5I_1 - I_2 - 2I_3 + I_4 = 0 \end{cases}$$

Find I_3.

15.91. Classify each of the following systems of equations according as they (i) have a unique solution, (ii) are inconsistent (iii) are dependent. Determine solutions where they exist.

(a) $\begin{cases} 2x_1 - x_2 + 2x_3 = 4 \\ -2x_1 + 2x_2 - 5x_3 = -2 \\ 4x_1 + x_2 + x_3 = 2 \end{cases}$
(c) $\begin{cases} x_1 + x_2 = 0 \\ 2x_1 - 3x_3 = 0 \\ x_2 + 5x_3 = 0 \end{cases}$
(e) $\begin{cases} x_1 - 2x_3 = 0 \\ 2x_2 + 3x_4 = 0 \\ x_3 - 2x_2 = 0 \\ 2x_1 - 7x_3 - 9x_4 = 0 \end{cases}$

(b) $\begin{cases} 3x_1 + 2x_2 - 5x_3 = 21 \\ x_1 - 4x_2 + 2x_3 = -17 \\ 2x_1 + x_2 - x_3 = 4 \end{cases}$
(d) $\begin{cases} x_1 + 3x_2 - 2x_3 = 4 \\ 2x_1 - x_2 + x_3 = 3 \\ 4x_1 - 9x_2 + 7x_3 = 1 \end{cases}$

15.92. For what value or values of k will the system of equation

$$\begin{cases} 2x_1 - kx_2 + x_3 = 0 \\ x_2 + x_3 - 2x_4 = 0 \\ kx_1 - x_2 - x_4 = 0 \\ x_1 + 3x_4 = 0 \end{cases}$$

have solutions other than the trivial one? Determine some of these solutions.

15.93. Given the system of equations $\begin{cases} 3x_1 + 2x_2 - 4x_3 = 2 \\ 2x_1 - 3x_2 + x_3 = -1 \end{cases}$ show that any two of x_1, x_2, x_3 can be solved in terms of the remaining one and thus that there are infinitely many solutions.

15.94. Given the system of equations $\begin{cases} 2x_1 + x_2 - 4x_3 = 4 \\ 3x_1 - 2x_2 + 8x_3 = 6 \end{cases}$ determine whether any two of x_1, x_2, x_3 can be solved in terms of the remaining one.

15.95. Investigate each of the following systems for possible solutions.

(a) $\begin{cases} 3x_1 - 2x_2 + x_3 + x_4 = 5 \\ 2x_1 + x_2 - x_3 - 2x_4 = 2 \\ x_1 + 4x_2 - 3x_3 + x_4 = -11 \end{cases}$
(d) $\begin{cases} x_1 + 2x_2 = 4 \\ 2x_1 - x_2 = 3 \\ 3x_1 - 4x_2 = 2 \end{cases}$

(b) $\begin{cases} 2x_1 - x_2 + x_3 = 0 \\ x_1 + x_2 - 2x_3 = 0 \end{cases}$

(e) $\begin{cases} 2x_1 - x_3 = 1 \\ x_2 + 2x_3 = -2 \\ x_1 + x_3 = 2 \\ 5x_1 + x_2 + x_3 = 1 \end{cases}$

(c) $\begin{cases} 3x_1 + x_2 + 2x_3 - x_4 = 0 \\ 2x_1 - x_2 - 3x_3 + 2x_4 = 0 \\ 2x_1 + 3x_2 + x_3 - 3x_4 = 0 \end{cases}$

EIGENVALUES AND EIGENVECTORS

15.96. Find eigenvalues and corresponding eigenvectors for each of the following matrices.

(a) $\begin{pmatrix} 2 & 2 \\ -1 & 5 \end{pmatrix}$ (b) $\begin{pmatrix} 1 & -2 \\ -2 & 4 \end{pmatrix}$ (c) $\begin{pmatrix} 3 & -4 & 0 \\ 4 & 3 & 0 \\ 0 & 0 & 5 \end{pmatrix}$ (d) $\begin{pmatrix} -2 & -9 & 5 \\ -5 & -10 & 7 \\ -9 & -21 & 14 \end{pmatrix}$

15.97. Determine sets of unit eigenvectors corresponding to the matrices of Problem 15.96.

15.98. (a) Prove that if the eigenvalues of a matrix A are $\lambda_1, \lambda_2, \ldots$, then the eigenvalues of A^2 are $\lambda_1^2, \lambda_2^2, \ldots$. (b) Generalize the result in (a).

15.99. Prove that the eigenvalues of a skew-Hermitian matrix [or skew-symmetric real matrix] are either zero or pure imaginary.

15.100. Illustrate the result of Problem 15.99 by means of an example.

15.101. Prove that the eigenvalues of a unitary [or real orthogonal matrix] all have absolute value equal to one.

15.102. Illustrate the result of Problem 15.101 by means of an example.

15.103. Find matrices which transform those of Problem 15.96 to diagonal form.

15.104. Prove that the eigenvalues of A and $B^{-1}AB$ are the same and illustrate by an example.

15.105. If the eigenvector corresponding to a given eigenvalue of a matrix A is X, prove that the eigenvector corresponding to the same eigenvalue of $B^{-1}AB$ [see Problem 15.104] is $B^{-1}X$. Illustrate by an example.

15.106. (a) Write the quadratic form $5x_1^2 - 2x_2^2 - 3x_3^2 + 12x_1x_2 - 8x_1x_3 + 20x_2x_3$ in the matrix form X^TAX. (b) Find the eigenvalues and eigenvectors of A. (c) Find the matrix B and the transformation equations $X = BU$ when $U = (u_1 \ u_2 \ u_3)^T$ so that the quadratic form in (a) is reduced to canonical form. (d) Write the new quadratic form.

15.107. (a) Find a transformation which removes the xy term in $x^2 + xy + y^2 = 16$ and (b) give a geometric interpretation to the result.

15.108. Discuss the relationship between Problem 15.107 and the problem of finding the maximum or minimum of $x^2 + y^2$ subject to the condition $x^2 + xy + y^2 = 16$. [Hint. Use the method of Lagrange multipliers.]

15.109. (a) Discuss the relationship between the problem of removing the cross product terms in $5x_1^2 - 2x_2^2 - 3x_3^2 + 12x_1x_2 - 8x_1x_3 + 20x_2x_3 = 100$ [Problem 15.106] and finding the maximum or minimum of $x_1^2 + x_2^2 + x_3^2$ subject to this constraint. Give geometric interpretations. (b) What is a corresponding problem for reduction of any quadratic form to canonical form?

15.110. (a) Verify that the eigenvalues of $\begin{pmatrix} 2 & 1 & -1 \\ -1 & 0 & 1 \\ -1 & -1 & 2 \end{pmatrix}$ are $\lambda = 1, 1, 2$.

(b) Show that an eigenvector corresponding to $\lambda = 2$ is $\begin{pmatrix} -1 \\ 1 \\ 1 \end{pmatrix}$ and that there are *two* linearly independent eigenvectors, namely $\begin{pmatrix} 1 \\ 0 \\ 1 \end{pmatrix}, \begin{pmatrix} 0 \\ 1 \\ 1 \end{pmatrix}$, corresponding to the single eigenvalue $\lambda = 1$.

(c) Can Theorem 15-15 be used to transform the matrix in (a) to diagonal form? Justify your answer.

15.111. Reduce $4x_1^2 + 5x_2^2 + 5x_3^2 + 2x_2x_3$ to canonical form.

15.112. Explain how to reduce a Hermitian matrix to diagonal form and illustrate by an example.

15.113. Explain how to reduce a Hermitian quadratic form [see Problem 15.60] to canonical form and illustrate by an example.

15.114. (a) Prove that the sum of the diagonal elements of any square matrix, i.e. the *trace*, is equal to the sum of the eigenvalues of the matrix. (b) Illustrate by using the matrices of Problems 15.33, 15.35, 15.96. (c) Is there a relationship between the trace of a matrix A and the matrix $B^{-1}AB$? Explain your answer.

15.115. (a) Verify the Cayley-Hamilton theorem for a third order matrix. (b) Prove the Cayley-Hamilton theorem for any nth order matrix.

Answers to Supplementary Problems

15.49. $x = -33/5$, $y = -26/5$

15.76. (a) $\begin{pmatrix} 1 & 1 \\ 1 & 2 \end{pmatrix}$ (b) $\begin{pmatrix} \cos\theta & -\sin\theta \\ \sin\theta & \cos\theta \end{pmatrix}$ (c) $\begin{pmatrix} \frac{2}{3} & \frac{1}{3} & \frac{1}{3} \\ \frac{1}{3} & \frac{2}{3} & -\frac{1}{3} \\ -\frac{1}{3} & \frac{1}{3} & \frac{1}{3} \end{pmatrix}$ (d) $\begin{pmatrix} 1 & -1 & 0 \\ -1 & 2 & 0 \\ 0 & 0 & 1 \end{pmatrix}$ (e) $\begin{pmatrix} -2 & -1 & 2 \\ -1 & 1 & 0 \\ 2 & 0 & -1 \end{pmatrix}$

15.77. (b) $\begin{pmatrix} -1 & -3 & 2 \\ 2 & 5 & -3 \\ -1 & -1 & 1 \end{pmatrix}$

15.89. (a) $x_1 = 3$, $x_2 = -4$ (b) $x_1 = 4$, $x_2 = -1$, $x_3 = 2$ (c) $x_1 = -2$, $x_2 = 4$, $x_3 = -3$

15.90. 40

15.91. (a) inconsistent. (b) has unique solution $x_1 = -1$, $x_2 = 2$, $x_3 = -4$. (c) has only trivial solution $x_1 = 0$, $x_2 = 0$, $x_3 = 0$. (d) dependent. (e) dependent and has solutions other than trivial one.

15.96. (a) $3, 4; \begin{pmatrix} 2 \\ 1 \end{pmatrix}, \begin{pmatrix} 1 \\ 1 \end{pmatrix}$ (b) $0, 5; \begin{pmatrix} 2 \\ 1 \end{pmatrix}, \begin{pmatrix} 1 \\ -2 \end{pmatrix}$ (c) $5, 3 \pm 4i; \begin{pmatrix} 0 \\ 0 \\ 1 \end{pmatrix}, \begin{pmatrix} 1 \\ -i \\ 0 \end{pmatrix}, \begin{pmatrix} 1 \\ i \\ 0 \end{pmatrix}$

(d) $1, -1, 2; \begin{pmatrix} 2 \\ 1 \\ 3 \end{pmatrix} \begin{pmatrix} 1 \\ 1 \\ 2 \end{pmatrix} \begin{pmatrix} 1 \\ -1 \\ -1 \end{pmatrix}$

15.106. (a) $(x_1 \ x_2 \ x_3) \begin{pmatrix} 5 & 6 & -4 \\ 6 & -2 & 10 \\ -4 & 10 & -3 \end{pmatrix} \begin{pmatrix} x_1 \\ x_2 \\ x_3 \end{pmatrix}$ (b) $5, 9, -15 \begin{pmatrix} -2 \\ 1 \\ 2 \end{pmatrix}, \begin{pmatrix} 2 \\ 2 \\ 1 \end{pmatrix}, \begin{pmatrix} 1 \\ -2 \\ 2 \end{pmatrix}$

(c) $B = \begin{pmatrix} -\frac{2}{3} & \frac{2}{3} & \frac{1}{3} \\ \frac{1}{3} & \frac{2}{3} & -\frac{2}{3} \\ \frac{2}{3} & \frac{1}{3} & \frac{2}{3} \end{pmatrix}$, $x_1 = \frac{1}{3}(-2u_1 + 2u_2 + u_3)$, $x_2 = \frac{1}{3}(u_1 + 2u_2 - 2u_3)$,

$x_3 = \frac{1}{3}(2u_1 + u_2 + 2u_3)$ (d) $6u_1^2 + 9u_2^2 - 15u_3^2$

Chapter 16

Calculus of Variations

MAXIMUM OR MINIMUM OF AN INTEGRAL

One of the main problems of the calculus of variations is to determine that curve connecting two given points which either minimizes or maximizes some given integral. For example, the problem of determining that curve connecting two points (x_1, y_1) and (x_2, y_2) whose length is a minimum is the same as that of finding the curve $Y = y(x)$ where $y(x_1) = y_1$, $y(x_2) = y_2$ such that

$$\int_{x_1}^{x_2} \sqrt{1 + y'^2}\, dx \tag{1}$$

is a minimum.

In the general case we want to find the curve $Y = y(x)$ where $y(x_1) = y_1$, $y(x_2) = y_2$ such that for some given function $F(x, y, y')$,

$$\int_{x_1}^{x_2} F(x, y, y')\, dx \tag{2}$$

is either a maximum or minimum, also called an *extremum* or *stationary value*. A curve which satisfies this property is called an *extremal*. An integral such as (2) which assumes a numerical value for some class of functions $y(x)$ is often called a *functional*.

EULER'S EQUATION

In order to find the required curve $Y = y(x)$, we consider the effect on the integral (2) of neighboring curves [see Fig. 16-1]

$$Y = y(x) + \epsilon\eta(x) \tag{3}$$

where $\eta(x)$ is an arbitrary function and ϵ is an arbitrary parameter. In order for the curve (3) to pass through (x_1, y_1) and (x_2, y_2), we require that

$$\eta(x_1) = 0, \quad \eta(x_2) = 0 \tag{4}$$

We can show [see Problems 16.1 and 16.2] that the required curve $Y = y(x)$ satisfies the equation

$$\frac{d}{dx}\left(\frac{\partial F}{\partial y'}\right) - \frac{\partial F}{\partial y} = 0 \tag{5}$$

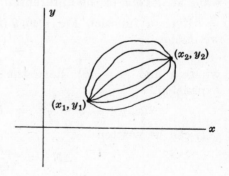

Fig. 16-1

which is called *Euler's equation*. The condition (5) is a necessary condition for $y = y(x)$ to be an extremal but it is not a sufficient condition.

In case $F(x, y, y')$ does not contain x explicitly, a first integral of (5) is found to be [see Problem 16.4]

$$F - y'\frac{\partial F}{\partial y'} = c \tag{6}$$

375

CONSTRAINTS

In certain problems we want to find that curve which makes a given integral

$$\int_{x_1}^{x_2} F(x, y, y')\, dx \tag{7}$$

a maximum or minimum but at the same time keeps the integral

$$\int_{x_1}^{x_2} G(x, y, y')\, dx \tag{8}$$

equal to some constant. This type of problem in the calculus of variations is one which involves a *constraint condition*, namely that the integral (8) is a constant. A special case of this is the problem of determining that curve having a given perimeter which encloses the largest area [Problem 16.8]. Because of this special case, we often refer to this class of problems as *isoperimetric problems*.

Such problems can generally be solved by using the method of Lagrange multipliers. To do this we consider the integral formed by adding (7) to λ multiplied by (8), whose λ is the Lagrange multiplier. The resulting integral given by

$$\int_{x_1}^{x_2} (F + \lambda G)\, dx \tag{9}$$

must be an extremum and leads to the Euler equation

$$\frac{d}{dx}\left(\frac{\partial H}{\partial y'}\right) - \frac{\partial H}{\partial y} = 0 \qquad \text{where} \quad H = F + \lambda G \tag{10}$$

Using this we can find the required extremal. Extensions can be made to cases where there are more constraint conditions.

THE VARIATIONAL NOTATION

It is often convenient to use a variational symbol δ having properties which are in many ways analogous to the differential d of the calculus.

Given a function $F(x, y(x), y'(x))$, or briefly $F(x, y, y')$ where we consider x as fixed, we define

$$\Delta F = F(x, y + \epsilon\eta, y' + \epsilon\eta') - F(x, y, y') \tag{11}$$

where ϵ and $\eta = \eta(x)$ have the same meaning given on page 375. Using the Taylor expansion

$$F(x, y + \epsilon\eta, y' + \epsilon\eta') = F(x, y, y') + \frac{\partial F}{\partial y}\epsilon\eta + \frac{\partial F}{\partial y'}\epsilon\eta' + \text{terms in } \epsilon^2, \epsilon^3, \ldots \tag{12}$$

(11) can be written

$$\Delta F = \frac{\partial F}{\partial y}\epsilon\eta + \frac{\partial F}{\partial y'}\epsilon\eta' + \text{terms in } \epsilon^2, \epsilon^3, \ldots \tag{13}$$

The sum of the first two terms on the right of (13) is denoted by δF called the *variation* of F, i.e.

$$\delta F = \frac{\partial F}{\partial y}\epsilon\eta + \frac{\partial F}{\partial y'}\epsilon\eta' \tag{14}$$

If in particular $F = y$ or $F = y'$ in (14), we have

$$\delta y = \epsilon\eta, \qquad \delta y' = \epsilon\eta' \tag{15}$$

so that (14) can be written

$$\delta F = \frac{\partial F}{\partial y}\delta y + \frac{\partial F}{\partial y'}\delta y' \tag{16}$$

From (15) we see that

$$\delta\left(\frac{dy}{dx}\right) \;=\; \epsilon\eta' \;=\; \frac{d}{dx}(\epsilon\eta) \;=\; \frac{d}{dx}(\delta y) \tag{17}$$

i.e.

$$\delta\left(\frac{dy}{dx}\right) \;=\; \frac{d}{dx}(\delta y) \quad\text{or}\quad \delta y' = (\delta y)' \tag{18}$$

showing that the operators δ and d/dx are commutative.

The variational symbol and its properties provide approaches alternative to those involving ϵ and $\eta(x)$ for dealing with problems of finding extrema of integrals. Thus we can show for example that a necessary condition for the integral (2) to be an extremum is

$$\delta\int_{x_1}^{x_2} F(x, y, y')\,dx \;=\; 0 \tag{19}$$

which in turn leads to the Euler equation. See Problem 16.11.

GENERALIZATIONS

The ideas above can be extended. An example is the problem of finding curves $X_1 = x_1(t)$, $X_2 = x_2(t)$, ..., $X_n = x_n(t)$ such that using $\dot{x}_1 = dx_1/dt, \ldots, \dot{x}_n = dx_n/dt$,

$$\int_{t_1}^{t_2} F(t, x_1, x_2, \ldots, x_n, \dot{x}_1, \dot{x}_2, \ldots, \dot{x}_n)\,dt \tag{20}$$

is a maximum or minimum. A necessary condition for this is that the Euler equations

$$\frac{d}{dt}\left(\frac{\partial F}{\partial \dot{x}_k}\right) - \frac{\partial F}{\partial x_k} \;=\; 0 \qquad k = 1, 2, \ldots, n \tag{21}$$

are satisfied. A solution of these equations leads to the required curves. See Problem 16.12.

Generalizations to cases where there are constraints are also possible. See Problem 16.13.

It is also possible to generalize to cases where multiple integrals are used rather than single integrals, and also where endpoints may not be fixed. See Problems 16.14-16.16.

HAMILTON'S PRINCIPLE

According to Newton's laws, a particle of mass m moves in a path according to the equation

$$\mathbf{F} \;=\; m\frac{d^2\mathbf{r}}{dt^2} \tag{22}$$

where $\mathbf{F}$ is the external force acting on the particle, $\mathbf{r}$ is the position vector with respect to the origin of some fixed coordinate system, and t is the time.

Now if the force field is conservative, then there exists a *potential function* V such that

$$\mathbf{F} \;=\; -\nabla V \tag{23}$$

The *kinetic energy* of the particle is defined as

$$T \;=\; \frac{1}{2}m\left(\frac{d\mathbf{r}}{dt}\right)^2 \;=\; \frac{1}{2}m\left[\left(\frac{dx}{dt}\right)^2 + \left(\frac{dy}{dt}\right)^2 + \left(\frac{dz}{dt}\right)^2\right] \tag{24}$$

if $\mathbf{r} = x\mathbf{i} + y\mathbf{j} + z\mathbf{k}$.

We can then show that equation (22) follows as a consequence of the problem of finding the path of the particle such that the integral

$$\int_{t_1}^{t_2} (T - V)\, dt \tag{25}$$

is an extremum, actually a minimum. In this integral t_1 and t_2 are two specified times and the path is required to join the positions of the particles at these times.

The principle that a particle moves in such a way that (25) is a minimum is often called *Hamilton's principle*. It can be generalized to systems of two or more particles.

LAGRANGE'S EQUATIONS

Very often in practice the position of a particle at any time can be described by a certain minimum number of variables called *generalized coordinates*. For example, if we consider the pendulum bob of a pendulum to be a point mass [Fig. 16-2] then its position is defined by using the generalized coordinate θ given by the angle made between the pendulum rod and the vertical.

Fig. 16-2   Fig. 16-3

Similarly in the case of a double pendulum with two masses m_1, m_2, as in Fig. 16-3, the positions are specified by using two angles θ_1, θ_2 which are the generalized coordinates.

The potential and kinetic energy can be expressed in terms of these generalized coordinates which are often denoted by $q_1, q_2, \ldots, q_N$. The number N of required coordinates is often called the number of *degrees of freedom* of the system.

According to Hamilton's principle the system moves so that

$$\int_{t_1}^{t_2} L\, dt \tag{26}$$

is an extremum where $$L = T - V \tag{27}$$

is called the *Lagrangian* of the system. Euler's equations then become

$$\frac{d}{dt}\left(\frac{\partial L}{\partial \dot{q}_k}\right) - \frac{\partial L}{\partial q_k} = 0 \qquad k = 1, 2, \ldots, N \tag{28}$$

which are then called *Lagrange's equations*. From these equations the motion of the system can be obtained.

STURM-LIOUVILLE SYSTEMS AND RAYLEIGH-RITZ METHODS

The calculus of variations often provides important methods for solving boundary-value problems. For example, the eigenvalues and eigenfunctions of the Sturm-Liouville system

$$\frac{d}{dx}\left[p(x)\frac{dy}{dx}\right] + [q(x) + \lambda r(x)]y = 0 \qquad (29)$$

$$a_1 y(a) + a_2 y'(a) = 0, \qquad b_1 y(b) + b_2 y'(b) = 0 \qquad (30)$$

considered in Chapter 11 can be formulated as a problem in finding extrema of suitable functionals.

Methods for finding approximate solutions of boundary-value problems by use of variational principles are called *Rayleigh-Ritz methods*. See Problems 16.22-16.25.

Solved Problems

EULER'S EQUATION AND APPLICATIONS

16.1. Let $Y = y(x)$ be the curve joining points (x_1, y_1) , (x_2, y_2) which makes $\displaystyle\int_{x_1}^{x_2} F(x, y, y')\, dx$ an extremum and $Y = y(x) + \epsilon\eta(x)$, $\eta(x_1) = 0$, $\eta(x_2) = 0$ be a neighboring curve joining these points. Prove that a necessary condition for this extremum is

$$\int_{x_1}^{x_2}\left[\frac{d}{dx}\left(\frac{\partial F}{\partial y'}\right) - \frac{\partial F}{\partial y}\right]\eta(x)\, dx = 0$$

The value of the integral along the neighboring curve is

$$I(\epsilon) = \int_{x_1}^{x_2} F(x,\, y(x) + \epsilon\eta(x),\, y'(x) + \epsilon\eta'(x))\, dx \qquad (1)$$

Now this function of ϵ is a maximum or minimum for the curve $Y = y(x)$ when

$$\frac{dI}{d\epsilon} = 0 \qquad \text{at} \quad \epsilon = 0 \qquad (2)$$

If we denote the integrand in (1) by F_ϵ, we have by differentiating under the integral sign [Leibnitz' rule]

$$\frac{dI}{d\epsilon} = \int_{x_1}^{x_2}\left[\frac{\partial F_\epsilon}{\partial y}\eta(x) + \frac{\partial F_\epsilon}{\partial y'}\eta'(x)\right]dx$$

At $\epsilon = 0$ we thus have

$$\begin{aligned}
\frac{dI}{d\epsilon}\bigg|_{\epsilon=0} &= \int_{x_1}^{x_2}\left[\frac{\partial F}{\partial y}\eta(x) + \frac{\partial F}{\partial y'}\eta'(x)\right]dx = 0 \\[2mm]
&= \int_{x_1}^{x_2}\frac{\partial F}{\partial y}\eta(x)\, dx + \frac{\partial F}{\partial y'}\eta(x)\bigg|_{x_1}^{x_2} - \int_{x_1}^{x_2}\frac{d}{dx}\left(\frac{\partial F}{\partial y'}\right)\eta(x)\, dx \qquad (3) \\[2mm]
&= \int_{x_1}^{x_2}\left[\frac{\partial F}{\partial y} - \frac{d}{dx}\frac{\partial F}{\partial y'}\right]\eta(x)\, dx
\end{aligned}$$

using integration by parts and the fact that $\eta(x_1) = 0$, $\eta(x_2) = 0$. Since this is equal to zero from (2), the required result follows.

16.2. Prove that a necessary condition for the extremum in Problem 16.1 is

$$\frac{d}{dx}\left(\frac{\partial F}{\partial y'}\right) - \frac{\partial F}{\partial y} = 0$$

From Problem 16.1 we must show that the condition

$$\int_{x_1}^{x_2}\left[\frac{d}{dx}\left(\frac{\partial F}{\partial y'}\right) - \frac{\partial F}{\partial y}\right]\eta(x)\,dx = 0 \qquad (1)$$

where $\eta(x)$ is arbitrary implies that the integrand is also zero, i.e.

$$\frac{d}{dx}\left(\frac{\partial F}{\partial y'}\right) - \frac{\partial F}{\partial y} = 0 \qquad (2)$$

To show this we suppose the contrary, i.e. the coefficient of $\eta(x)$ is not identically zero. Then since $\eta(x)$ is arbitrary, we can always choose it to be positive where $\frac{d}{dx}\left(\frac{\partial F}{\partial y'}\right) - \frac{\partial F}{\partial y} > 0$ and negative where $\frac{d}{dx}\left(\frac{\partial F}{\partial y'}\right) - \frac{\partial F}{\partial y} < 0$. In such case the left side of (1) will be positive and so will give us a contradiction. Thus (2) follows.

16.3. Show that Euler's equation can be written in the form

$$\frac{d}{dx}\left[F - y'\frac{\partial F}{\partial y'}\right] - \frac{\partial F}{\partial x} = 0$$

We have
$$\frac{dF}{dx} = \frac{\partial F}{\partial x} + \frac{\partial F}{\partial y}\frac{dy}{dx} + \frac{\partial F}{\partial y'}\frac{dy'}{dx} = \frac{\partial F}{\partial x} + \frac{\partial F}{\partial y}y' + \frac{\partial F}{\partial y'}y''$$

Also
$$\frac{d}{dx}\left[y'\frac{\partial F}{\partial y'}\right] = y'\frac{d}{dx}\left(\frac{\partial F}{\partial y'}\right) + \frac{\partial F}{\partial y'}y''$$

Then by subtraction
$$\frac{d}{dx}\left[F - y'\frac{\partial F}{\partial y'}\right] = \frac{\partial F}{\partial x} + y'\left[\frac{\partial F}{\partial y} - \frac{d}{dx}\left(\frac{\partial F}{\partial y'}\right)\right]$$

and using Euler's equation gives the required result.

16.4. If F does not involve x explicitly, show that the Euler equation can be integrated to yield

$$F - y'\frac{\partial F}{\partial y'} = c$$

If F does not depend explicitly on x, then $\partial F/\partial x = 0$ so that from Problem 16.3 we have

$$\frac{d}{dx}\left[F - y'\frac{\partial F}{\partial y'}\right] = 0 \qquad \text{or} \qquad F - y'\frac{\partial F}{\partial y'} = c$$

16.5. A curve C joining points (x_1, y_1) and (x_2, y_2) [see Fig. 16-4] is revolved about the x axis. Find the shape of the curve so that the surface thus generated is a minimum.

The surface area is given by

$$2\pi\int_{x_1}^{x_2} y\,ds = 2\pi\int_{x_1}^{x_2} y\sqrt{1+y'^2}\,dx$$

Since the integrand is independent of x, we can use Problem 16.4 with $F = y\sqrt{1+y'^2}$ to show that the required curve is a solution of

$$y\sqrt{1+y'^2} - y'\left[\frac{y}{2}(1+y'^2)^{-1/2}2y'\right] = c$$

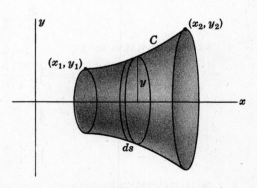

Fig. 16-4

which becomes on simplification

$$\frac{y}{\sqrt{1+y'^2}} = c \quad \text{or} \quad y' = \frac{dy}{dx} = \frac{\sqrt{y^2-c^2}}{c} \tag{1}$$

Separating the variables and integrating,

$$\int \frac{dy}{\sqrt{y^2-c^2}} = \int \frac{dx}{c} \quad \text{or} \quad \cosh^{-1}\frac{y}{c} = \frac{x+k}{c}$$

i.e.

$$y = c \cosh\left(\frac{x+k}{c}\right) \tag{2}$$

where the constants c and k are determined from points (x_1, y_1), (x_2, y_2). The curve (2) is often called a *catenary* from the Latin meaning *chain*, since this is the shape in which a chain would hang if suspended from the points (x_1, y_1) (x_2, y_2). The problem is also of importance in connection with *soap films* which are known to take shapes having minimum surfaces.

It should be mentioned that we have not proved that the surface is actually a minimum, which requires further analysis.

16.6. Work Problem 16.5 without using the results of Problems 16.3 and 16.4.

In this case we use $F = y\sqrt{1+y'^2}$ in Euler's equation to obtain

$$\frac{d}{dx}\left(\frac{yy'}{\sqrt{1+y'^2}}\right) - \sqrt{1+y'^2} = 0$$

which simplifies to

$$1 + y'^2 - yy'' = 0 \tag{1}$$

Letting $y' = p$ so that $y'' = \frac{dp}{dx} = \frac{dp}{dy}\frac{dy}{dx} = p\frac{dp}{dy}$, (1) becomes on separating variables and integrating,

$$\int \frac{p\,dp}{1+p^2} = \int \frac{dy}{y} \quad \text{or} \quad \tfrac{1}{2}\ln(1+p^2) = \ln y + c_1$$

Solving for p,

$$p = \frac{dy}{dx} = \frac{\sqrt{y^2-c^2}}{c}$$

and the result can be found as in Problem 16.5.

16.7. A wire in a vertical plane connects origin O and point $P_2(x_2, y_2)$ as indicated in Fig. 16-5. A bead of mass m placed at O slides without friction down the wire to P_2 under the influence of gravity. Find the shape of the wire so that the bead goes from O to P_2 in the least time.

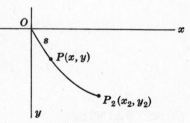

Fig. 16-5

Assume that at time t the bead is at $P(x, y)$ and that arc $OP = s$. Then from mechanics we have

Kinetic energy at O + Potential energy at O = Kinetic energy at P + Potential energy at P

or

$$0 + mgy_1 = \frac{1}{2}m\left(\frac{ds}{dt}\right)^2 + mg(y_1 - y)$$

i.e.

$$\left(\frac{ds}{dt}\right)^2 = 2gy \quad \text{or} \quad \frac{ds}{dt} = \sqrt{2gy}$$

Thus the time for the bead to go from O to P_1 is

$$\int_0^T dt = T = \int_{x=0}^{x_2} \frac{ds}{\sqrt{2gy}} = \frac{1}{\sqrt{2g}}\int_0^{x_2} \frac{\sqrt{1+y'^2}}{\sqrt{y}}\,dx$$

The time will be a minimum when the integral is a minimum. Letting $F = \sqrt{1 + y'^2}/\sqrt{y}$ and using Problem 16.4,

$$\frac{\sqrt{1 + y'^2}}{\sqrt{y}} - y'\left[\frac{y'}{\sqrt{1 + y'^2}\sqrt{y}}\right] = c$$

which simplifies to

$$\sqrt{y}\sqrt{1 + y'^2} = \frac{1}{c}$$

Letting $1/c = \sqrt{a}$ and solving for y',

$$y' = \frac{dy}{dx} = \sqrt{\frac{a - y}{y}}$$

Separating variables and integrating,

$$\int dx = \int \sqrt{\frac{y}{a - y}}\, dy \tag{1}$$

Letting

$$y = a \sin^2\theta = \frac{a}{2}(1 - \cos 2\theta) \tag{2}$$

in (1), we have

$$x = 2a \int \sin^2\theta\, d\theta = a \int (1 - \cos 2\theta)\, d\theta = \frac{a}{2}(2\theta - \sin 2\theta) + k$$

Thus the parametric equations of the curve are given by

$$x = b(\phi - \sin\phi) + k, \qquad y = b(1 - \cos\phi) \tag{3}$$

where $b = a/2$, $\phi = 2\theta$. Since the curve passes through the origin we must have $k = 0$ so that the required equations are

$$x = b(\phi - \sin\phi), \qquad y = b(1 - \cos\phi) \tag{4}$$

The constant b is determined from the fact that the curve must pass through (x_2, y_2).

The curve represented by (4) is a *cycloid* and is the path of a fixed point A on a circle of radius b as it rolls along the x axis [see Fig. 16-6].

The problem is often called the *brachistochrone problem* from the Greek words *brachistos* meaning *shortest* and *chronos* meaning *time*.

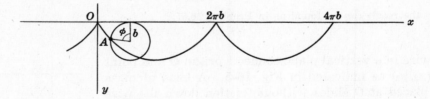

Fig. 16-6

CONSTRAINTS

16.8. Find that curve C having given length l which encloses a maximum area.

By Problem 6.17, page 164, the area bounded by C is

$$A = \frac{1}{2}\int_C (x\, dy - y\, dx) = \frac{1}{2}\int_C (xy' - y)\, dx \tag{1}$$

while the arc length is

$$s = \int_C \sqrt{1 + y'^2}\, dx = l \tag{2}$$

Using the method of Lagrange multipliers, we consider

$$H = \int_C \left[\tfrac{1}{2}(xy' - y) + \lambda\sqrt{1 + y'^2}\right] dx \tag{3}$$

From the Euler equation

$$\frac{d}{dx}\left(\frac{\partial F}{\partial y'}\right) - \frac{\partial F}{\partial y} = 0 \tag{4}$$

where

$$F = \tfrac{1}{2}(xy' - y) + \lambda\sqrt{1+y'^2} \tag{5}$$

we find

$$\frac{d}{dx}\left(\frac{1}{2}x + \frac{\lambda y'}{\sqrt{1+y'^2}}\right) + \frac{1}{2} = 0$$

or

$$\frac{d}{dx}\left(\frac{\lambda y'}{\sqrt{1+y'^2}}\right) = -1 \tag{6}$$

i.e.

$$\frac{\lambda y'}{\sqrt{1+y'^2}} = -X + c_1$$

Solving for y',

$$y' = \frac{dy}{dx} = \pm\frac{x - c_1}{\sqrt{\lambda^2 - (x-c_1)^2}}$$

Then on integrating,

$$y - c_2 = \pm\sqrt{\lambda^2 - (x-c_1)^2}$$

i.e.

$$(x-c_1)^2 + (y-c_2)^2 = \lambda^2 \tag{7}$$

which is a circle. A problem of this type is often referred to as an *isoperimetric problem*.

Another method. By carrying out the integration in (6), we have

$$\frac{y''}{(1+y'^2)^{3/2}} = -\frac{1}{\lambda}$$

which shows that the curvature of C must be constant, i.e. C is a circle.

THE VARIATIONAL NOTATION

16.9. Let F_1 and F_2 be functions of x, y, y'. Prove that (a) $\delta(F_1 + F_2) = \delta F_1 + \delta F_2$, (b) $\delta(F_1 F_2) = F_1 \delta F_2 + F_2 \delta F_1$.

(a) By definition,

$$\delta(F_1+F_2) = \frac{\partial(F_1+F_2)}{\partial y}\delta y + \frac{\partial(F_1+F_2)}{\partial y'}\delta y' = \frac{\partial F_1}{\partial y}\delta y + \frac{\partial F_2}{\partial y}\delta y + \frac{\partial F_1}{\partial y'}\delta y' + \frac{\partial F_2}{\partial y'}\delta y'$$

$$= \left(\frac{\partial F_1}{\partial y}\delta y + \frac{\partial F_1}{\partial y'}\delta y'\right) + \left(\frac{\partial F_2}{\partial y}\delta y + \frac{\partial F_2}{\partial y'}\delta y'\right) = \delta F_1 + \delta F_2$$

(b) By definition,

$$\delta(F_1 F_2) = \frac{\partial(F_1 F_2)}{\partial y}\delta y + \frac{\partial(F_1 F_2)}{\partial y'}\delta y' = \left(F_1\frac{\partial F_2}{\partial y} + F_2\frac{\partial F_1}{\partial y}\right)\delta y + \left(F_1\frac{\partial F_2}{\partial y'} + F_2\frac{\partial F_1}{\partial y'}\right)\delta y'$$

$$= F_1\left(\frac{\partial F_2}{\partial y}\delta y + \frac{\partial F_2}{\partial y'}\delta y'\right) + F_2\left(\frac{\partial F_1}{\partial y}\delta y + \frac{\partial F_1}{\partial y'}\delta y'\right) = F_1\delta F_2 + F_2\delta F_1$$

16.10. Prove that $\delta\int_{x_1}^{x_2} F(x,y,y')\,dx = \int_{x_1}^{x_2}\delta F(x,y,y')\,dx.$

Method 1.

$$\Delta\int_{x_1}^{x_2} F(x,y,y')\,dx = \int_{x_1}^{x_2} F(x, y+\epsilon\eta, y'+\epsilon\eta')\,dx - \int_{x_1}^{x_2} F(x,y,y')\,dx$$

$$= \int_{x_1}^{x_2}[F(x, y+\epsilon\eta, y'+\epsilon\eta') - F(x,y,y')]\,dx$$

$$= \int_{x_1}^{x_2}\left[\frac{\partial F}{\partial y}\epsilon\eta + \frac{\partial F}{\partial y'}\epsilon\eta' + \text{terms in } \epsilon^2, \epsilon^3, \ldots\right]dx$$

$$= \int_{x_1}^{x_2}\delta F\,dx + \text{terms in } \epsilon^2, \epsilon^3, \ldots$$

Then by definition we have as required,

$$\delta \int_{x_1}^{x_2} F(x, y, y')\, dx \;=\; \int_{x_1}^{x_2} \delta F\, dx$$

Method 2.

$$\delta \int_{x_1}^{x_2} F(x, y, y')\, dx \;=\; \frac{\partial}{\partial y}\left[\int_{x_1}^{x_2} F(x, y, y')\, dx\right]\delta y \;+\; \frac{\partial}{\partial y'}\left[\int_{x_1}^{x_2} F(x, y, y')\, dx\right]\delta y'$$

$$=\; \left[\int_{x_1}^{x_2} \frac{\partial F}{\partial y}\, dx\right]\delta y \;+\; \left[\int_{x_1}^{x_2} \frac{\partial F}{\partial y'}\, dx\right]\delta y'$$

$$=\; \int_{x_1}^{x_2}\left[\frac{\partial F}{\partial y}\,\delta y + \frac{\partial F}{\partial y'}\,\delta y'\right] dx \;=\; \int_{x_1}^{x_2} \delta F\, dx$$

where we have used Leibnitz's rule for differentiating under the integral sign.

16.11. Show that a necessary condition for $\displaystyle \int_{x_1}^{x_2} F(x, y, y')\, dx$ to be an extremum is

$$\delta \int_{x_1}^{x_2} F(x, y, y')\, dx \;=\; 0$$

By Problem 16.1, equation (3), we see on multiplying by ϵ that a necessary condition for an extremum is

$$\int_{x_1}^{x_2}\left[\frac{\partial F}{\partial y}\,\epsilon\eta + \frac{\partial F}{\partial y'}\,\epsilon\eta'\right] dx \;=\; 0$$

But this can be written

$$\int_{x_1}^{x_2}\left[\frac{\partial F}{\partial y}\,\delta y + \frac{\partial F}{\partial y'}\,\delta y'\right] dx \;=\; \int_{x_1}^{x_2} \delta F\, dx \;=\; \delta \int_{x_1}^{x_2} F\, dx \;=\; 0$$

as required.

Note that by retracing these steps, i.e. starting with $\displaystyle \delta \int_{x_1}^{x_2} F\, dx = 0$, we can arrive at the Euler equation as in Problem 16.2.

GENERALIZATIONS

16.12. Show that a necessary condition for

$$\int_{t_1}^{t_2} F(t, x_1, x_2, \ldots, x_n, \dot{x}_1, \dot{x}_2, \ldots, \dot{x}_n)\, dt$$

to be an extremum [maximum or minimum] is that

$$\frac{d}{dt}\left(\frac{\partial F}{\partial \dot{x}_k}\right) - \frac{\partial F}{\partial x_k} \;=\; 0 \qquad k = 1, 2, \ldots, n$$

As in the one dimensional case, a necessary condition for an extremum is

$$\delta \int_{t_1}^{t_2} F\, dt \;=\; 0 \qquad \text{or} \qquad \int_{t_1}^{t_2} \delta F\, dt \;=\; 0$$

i.e.
$$\int_{t_1}^{t_2}\left[\left(\frac{\partial F}{\partial x_1}\,\delta x_1 + \frac{\partial F}{\partial \dot{x}_1}\,\delta \dot{x}_1\right) + \cdots + \left(\frac{\partial F}{\partial x_n}\,\delta x_n + \frac{\partial F}{\partial \dot{x}_n}\,\delta \dot{x}_n\right)\right] dt \;=\; 0$$

Using the fact that $\delta \dot{x}_1 = d(\delta x_1)/dt, \ldots, \delta \dot{x}_n = d(\delta x_n)/dt$ and integrating by parts, we have

$$\int_{t_1}^{t_2}\left\{\left[\frac{d}{dt}\left(\frac{\partial F}{\partial \dot{x}_1}\right) - \frac{\partial F}{\partial x_1}\right]\delta x_1 + \cdots + \left[\frac{d}{dt}\left(\frac{\partial F}{\partial \dot{x}_n}\right) - \frac{\partial F}{\partial x_n}\right]\delta x_n\right\} dt \;=\; 0$$

and since $\delta x_1, \ldots, \delta x_n$ are arbitrary we have

$$\frac{d}{dt}\left(\frac{\partial F}{\partial \dot{x}_k}\right) - \frac{\partial F}{\partial x_k} \;=\; 0 \qquad k = 1, 2, \ldots, n$$

The result can also be found without the use of δ.

16.13. Show that a necessary condition for

$$\int_{t_1}^{t_2} F(t, x_1, x_2, \dot{x}_1, \dot{x}_2)\, dt$$

to be an extremum subject to the constraint condition $G(x_1, x_2) = 0$ is

$$\frac{d}{dt}\left(\frac{\partial F}{\partial \dot{x}_1}\right) - \frac{\partial F}{\partial x_1} + \lambda \frac{\partial G}{\partial x_1} = 0, \qquad \frac{d}{dt}\left(\frac{\partial F}{\partial \dot{x}_2}\right) - \frac{\partial F}{\partial x_2} + \lambda \frac{\partial G}{\partial x_2} = 0$$

where λ is a Lagrange multiplier which may be a function of t.

We must have as in Problem 16.12,

$$\delta \int_{t_1}^{t_2} F\, dt = \int_{t_1}^{t_2} \delta F\, dt = \int_{t_1}^{t_2} \left[\left(\frac{\partial F}{\partial x_1}\delta x_1 + \frac{\partial F}{\partial \dot{x}_1}\delta \dot{x}_1\right) + \left(\frac{\partial F}{\partial x_2}\delta x_2 + \frac{\partial F}{\partial \dot{x}_2}\delta \dot{x}_2\right)\right] dt = 0$$

or

$$\int_{t_1}^{t_2} \left\{\left[\frac{d}{dt}\left(\frac{\partial F}{\partial \dot{x}_1}\right) - \frac{\partial F}{\partial x_1}\right]\delta x_1 + \left[\frac{d}{dt}\left(\frac{\partial F}{\partial \dot{x}_2}\right) - \frac{\partial F}{\partial x_2}\right]\delta x_2\right\} dt = 0 \qquad (1)$$

Also from $G(x_1, x_2) = 0$ we have

$$\delta G = \frac{\partial G}{\partial x_1}\delta x_1 + \frac{\partial G}{\partial x_2}\delta x_2 = 0 \qquad (2)$$

Multiplying (2) by λ, which may be a function of t, and integrating we have

$$\int_{t_1}^{t_2} \lambda \left[\frac{\partial G}{\partial x_1}\delta x_1 + \frac{\partial G}{\partial x_2}\delta x_2\right] dt = 0 \qquad (3)$$

Adding (1) and (3) gives

$$\int_{t_1}^{t_2} \left\{\left[\frac{d}{dt}\left(\frac{\partial F}{\partial \dot{x}_1}\right) - \frac{\partial F}{\partial x_1} + \lambda\frac{\partial G}{\partial x_1}\right]\delta x_1 + \left[\frac{d}{dt}\left(\frac{\partial F}{\partial \dot{x}_2}\right) - \frac{\partial F}{\partial x_2} + \lambda\frac{\partial G}{\partial x_2}\right]\delta x_2\right\} dt = 0$$

Then since δx_1 and δx_2 are arbitrary the required equations follow. Generalizations to any number of variables $x_1, \ldots, x_n$ are immediate.

16.14. If $G = G(x, y)$ and $\mathcal{R}$ is a region bounded by a simple closed curve C in the xy plane, show that a necessary condition for

$$\iint_{\mathcal{R}} (G_x^2 + G_y^2)\, dx\, dy$$

to be an extremum is that G satisfy Laplace's equation, i.e. $\nabla^2 G = 0$, in $\mathcal{R}$.

A necessary condition is

$$\delta \iint_{\mathcal{R}} (G_x^2 + G_y^2)\, dx\, dy = \iint_{\mathcal{R}} \delta(G_x^2 + G_y^2)\, dx\, dy = 0 \qquad (1)$$

Now

$$\delta(G_x^2 + G_y^2) = 2G_x\, \delta G_x + 2G_y\, \delta G_y$$
$$= 2G_x\frac{\partial}{\partial x}(\delta G) + 2G_y\frac{\partial}{\partial y}(\delta G)$$

Let $\mathcal{R}$ be a region such as indicated in Fig. 16-7 where lines parallel to the x and y axes meet C in no more than two points.

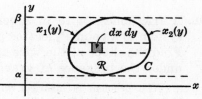

Fig. 16-7

We have

$$\iint_{\mathcal{R}} \delta G_x^2\, dx\, dy = \int_{y=\alpha}^{\beta} dy \int_{x=x_1(y)}^{x_2(y)} 2G_x\frac{\partial}{\partial x}(\delta G)\, dx$$

$$= \int_{y=\alpha}^{\beta} dy \left[2G_x\, \delta G \,\Big|_{x_1(y)}^{x_2(y)} - \int_{x=x_1(y)}^{x_2(y)} 2\frac{\partial G_x}{\partial x}\delta G\, dx\right]$$

$$= -2\int_{y=\alpha}^{\beta} \int_{x=x_1(y)}^{x_2(y)} \frac{\partial G_x}{\partial x}\delta G\, dx\, dy = -2\iint_{\mathcal{R}} \frac{\partial^2 G}{\partial x^2}\delta G\, dx\, dy$$

using the fact that $\delta G = 0$ on C.   Similarly we find

$$\iint\limits_{\mathcal{R}} \delta G_y^2 \; dx \, dy \;\; = \;\; -2 \iint\limits_{\mathcal{R}} \frac{\partial^2 G}{\partial y^2} \, \delta G \, dx \, dy$$

It follows from (1) that

$$\delta \iint\limits_{\mathcal{R}} (G_x^2 + G_y^2) \, dx \, dy \;\; = \;\; -2 \iint\limits_{\mathcal{R}} \left(\frac{\partial^2 G}{\partial x^2} + \frac{\partial^2 G}{\partial y^2} \right) \delta G \, dx \, dy \;\; = \;\; 0$$

so that since δG is arbitrary,

$$\frac{\partial^2 G}{\partial x^2} + \frac{\partial^2 G}{\partial y^2} \;=\; 0 \qquad \text{or} \qquad \nabla^2 G \;=\; 0 \quad \text{in } \mathcal{R}$$

16.15. If $y(x)$ makes $\displaystyle\int_{x_1}^{x_2} F(x, y, y') \, dx$ an extremum where $y(x_1)$ is fixed but $y(x_2)$ may vary, prove that we must have

$$\frac{d}{dx}\left(\frac{\partial F}{\partial y'}\right) - \frac{\partial F}{\partial y} \;=\; 0, \qquad \frac{\partial F}{\partial y'}\bigg|_{x=x_2} \;=\; 0$$

As in Problem 16.1 we find the necessary condition for an extremum to be

$$\int_{x_1}^{x_2} \frac{\partial F}{\partial y} \, \eta(x) \, dx + \frac{\partial F}{\partial y'} \, \eta(x) \bigg|_{x_1}^{x_2} - \int_{x_1}^{x_2} \frac{d}{dx}\left(\frac{\partial F}{\partial y'}\right) \eta(x) \, dx \;=\; 0 \qquad (1)$$

Then since x_1 is fixed but x_2 is not, we have $\eta(x_1) = 0$ while $\eta(x_2)$ is not necessarily zero. Thus (1) becomes

$$\int_{x_1}^{x_2} \left[\frac{\partial F}{\partial y} - \frac{d}{dx}\left(\frac{\partial F}{\partial y'}\right) \right] \eta(x) \, dx + \frac{\partial F}{\partial y'}\bigg|_{x=x_2} \eta(x_2) \;=\; 0 \qquad (2)$$

Now since η is arbitrary subject only to $\eta(x_1) = 0$, it must in particular hold for $\eta(x_2) = 0$, i.e.

$$\int_{x_1}^{x_2} \left[\frac{\partial F}{\partial y} - \frac{d}{dx}\left(\frac{\partial F}{\partial y'}\right) \right] \eta(x) \, dx \;=\; 0 \qquad (3)$$

from which

$$\frac{d}{dx}\left(\frac{\partial F}{\partial y'}\right) - \frac{\partial F}{\partial y} \;=\; 0 \qquad (4)$$

Using (4) in (2), we must then have

$$\frac{\partial F}{\partial y'}\bigg|_{x=x_2} \eta(x_2) \;=\; 0 \qquad (5)$$

Thus if we are to satisfy (5) for $\eta(x_2)$ not necessarily zero, we must have

$$\frac{\partial F}{\partial y'}\bigg|_{x=x_2} \;=\; 0$$

16.16. A frictionless wire in a vertical plane connects the origin O of an xy coordinate system to a point P_2 located somewhere on a given vertical line $x = x_2$ [see Fig. 16-8]. Find the shape of the wire so that a bead of mass m placed on it at O will slide down under the influence of gravity to P_2 in the least time.

The problem is identical with that of Problem 16.7 except that instead of being fixed at (x_2, y_2) the endpoint P_2 can vary along the line $x = x_2$ where x_2 is prescribed.

As in Problem 16.7 the total time taken to go from 0 to P_2 is

$$T \;=\; \frac{1}{\sqrt{2g}} \int_0^{x_2} \frac{\sqrt{1 + y'^2}}{\sqrt{y}} \, dx \qquad (1)$$

Then by Problem 16.15, taking $F = \sqrt{1 + y'^2}/\sqrt{y}$, we have

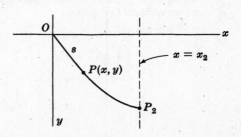

Fig. 16-8

$$\frac{d}{dx}\left(\frac{\partial F}{\partial y'}\right) - \frac{\partial F}{\partial y} = 0, \qquad \frac{\partial F}{\partial y'}\bigg|_{x=x_2} = 0 \tag{2}$$

The first condition in (2) leads as in Problem 16.7 to the fact that the shape of the wire must be a cycloid. The second condition in (2), i.e.

$$\frac{y'}{\sqrt{1+y'^2}\,\sqrt{y}}\bigg|_{x=x_2} = 0$$

shows that $y' = 0$ at $x = x_2$ or that the tangent to the cycloid at P_2 must be parallel to the x axis or, in other words, that the cycloid must be perpendicular to the line $x = x_2$ at P_2. These conditions are enough to enable us to obtain the equations of the cycloid.

HAMILTON'S PRINCIPLE AND LAGRANGE'S EQUATIONS

16.17. Derive Hamilton's principle for a system of n particles from Newton's laws.

Assume the n particles have masses m_k, $k = 1, 2, \ldots, n$, that they have position vectors $\mathbf{r}_k$, $k = 1, 2, \ldots, n$, relative to an xyz coordinate system and that the forces acting on them are $\mathbf{F}_k$, $k = 1, 2, \ldots, n$. Then the path C_k of the kth particle is determined from the equation

$$m_k \frac{d^2\mathbf{r}_k}{dt^2} = \mathbf{F}_k \qquad k = 1, 2, \ldots, n \tag{1}$$

Assume that we vary the path of the kth particle without changing the endpoints and let this variation, often called *virtual displacement*, be $\delta\mathbf{r}_k$; then from (1),

$$\left(m_k \frac{d^2\mathbf{r}_k}{dt^2}\right) \cdot \delta\mathbf{r}_k = \mathbf{F}_k \cdot \delta\mathbf{r}_k \tag{2}$$

Summing over all particles, we obtain

$$\sum_{k=1}^{n} m_k \frac{d^2\mathbf{r}_k}{dt^2} \cdot \delta\mathbf{r}_k = \sum_{k=1}^{n} \mathbf{F}_k \cdot \delta\mathbf{r}_k \tag{3}$$

where the right hand side of (3) is the total work δW done under the displacement of the path, i.e.

$$\delta W = \sum_{k=1}^{n} \mathbf{F}_k \cdot \delta\mathbf{r}_k = \sum_{k=1}^{n} m_k \frac{d^2\mathbf{r}_k}{dt^2} \cdot \delta\mathbf{r}_k \tag{4}$$

on using (1).

Now the total kinetic energy of the system is

$$T = \tfrac{1}{2} \sum_{k=1}^{n} m_k \left(\frac{d\mathbf{r}_k}{dt}\right)^2 \tag{5}$$

from which we have

$$\delta T = \sum_{k=1}^{n} m_k \frac{d\mathbf{r}_k}{dt} \cdot \delta\left(\frac{d\mathbf{r}_k}{dt}\right) = \sum_{k=1}^{n} m_k \frac{d\mathbf{r}_k}{dt} \cdot \frac{d}{dt}(\delta\mathbf{r}_k) \tag{6}$$

But

$$\frac{d}{dt}\left[\frac{d\mathbf{r}_k}{dt} \cdot \delta\mathbf{r}_k\right] = \frac{d^2\mathbf{r}_k}{dt^2} \cdot \delta\mathbf{r}_k + \frac{d\mathbf{r}_k}{dt} \cdot \frac{d}{dt}(\delta\mathbf{r}_k) \tag{7}$$

so that on multiplying by m_k and summing from $k = 1$ to n,

$$\sum_{k=1}^{n} m_k \frac{d}{dt}\left[\frac{d\mathbf{r}_k}{dt} \cdot \delta\mathbf{r}_k\right] = \sum_{k=1}^{n} m_k \frac{d^2\mathbf{r}_k}{dt^2} \cdot \delta\mathbf{r}_k + \sum_{k=1}^{n} m_k \frac{d\mathbf{r}_k}{dt} \cdot \frac{d}{dt}(\delta\mathbf{r}_k)$$

$$= \delta T + \delta W$$

where we have used (4) and (6). Integrating this with respect to t from t_1 to t_2 which represent the times at the endpoints of the path C, we have

$$\int_{t_1}^{t_2} (\delta T + \delta W)\,dt = \sum_{k=1}^{n} m_k \frac{d\mathbf{r}_k}{dt} \cdot \delta\mathbf{r}_k\bigg|_{t_1}^{t_2} = 0 \tag{8}$$

since $\delta\mathbf{r}_k = 0$ at t_1 and t_2. If the force field is conservative there is a potential V so that $W = -V$. Thus (8) becomes

$$\delta \int_{t_1}^{t_2} (T - V)\, dt \;=\; 0$$

which is *Hamilton's principle*, i.e. a system moves from time t_1 to time t_2 in such a way that

$$\int_{t_1}^{t_2} (T - V)\, dt$$

is an extremum, actually a minimum.

16.18. Derive Lagrange's equations from Hamilton's principle.

If the generalized coordinates specifying the position of a body are given by $q_1, q_2, \ldots, q_N$, then the position vector $\mathbf{r}_k$ of each particle of the body is a function of $q_1, q_2, \ldots, q_N$ so that its velocity $\dot{\mathbf{r}}_k = d\mathbf{r}_k/dt$ is a function of $q_1, q_2, \ldots, q_N$ and $\dot{q}_1, \dot{q}_2, \ldots, \dot{q}_N$. Thus the kinetic energy T is a function of $q_1, q_2, \ldots, q_N$, $\dot{q}_1, \dot{q}_2, \ldots, \dot{q}_N$. Also the potential energy, which we shall assume depends only on position, is a function of $q_1, q_2, \ldots, q_N$.

Now by Hamilton's principle the body moves so that

$$\int_{t_1}^{t_2} (T - V)\, dt$$

is an extremum. Thus by Euler's equations we have on writing $L = T - V$, called the *Lagrangian*,

$$\frac{d}{dt}\left(\frac{\partial L}{\partial \dot{q}_k}\right) - \frac{\partial L}{\partial q_k} \;=\; 0 \qquad k = 1, \ldots, N$$

which are often also referred to as *Lagrange's equations*.

16.19. A mass m, suspended at the end of a vertical spring which has spring constant κ and negligible mass, is set into vertical vibration [see Fig. 16-9]. Find the equation of motion of the mass.

If we let x be the displacement of m from the equilibrium position, then by Hooke's law the force is given by

$$\mathbf{F} \;=\; -\kappa x \mathbf{i} \qquad\qquad (1)$$

where $\mathbf{i}$ is a unit vector in the downward direction. Since

$$\mathbf{F} \;=\; -\nabla V \;=\; -\frac{\partial V}{\partial x}\mathbf{i} \qquad\qquad (2)$$

where V is the potential energy, we have from (1) and (2)

$$\frac{\partial V}{\partial x} \;=\; \kappa x \quad\text{or}\quad V \;=\; \tfrac{1}{2}\kappa x^2$$

taking the arbitrary constant as zero.

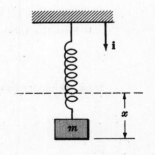

Fig. 16-9

The kinetic energy of the mass is

$$T \;=\; \tfrac{1}{2}m\left(\frac{dx}{dt}\right)^2 \;=\; \tfrac{1}{2}m\dot{x}^2$$

Thus the Lagrangian is $\qquad L \;=\; T - V \;=\; \tfrac{1}{2}m\dot{x}^2 - \tfrac{1}{2}\kappa x^2$

Then Lagrange's equation describing the motion of the mass is

$$\frac{d}{dt}\left(\frac{\partial L}{\partial \dot{x}}\right) - \frac{\partial L}{\partial x} \;=\; 0 \quad\text{or}\quad m\ddot{x} + \kappa x \;=\; 0$$

which agrees with the result obtained from Newton's laws.

16.20. A particle of mass m moves in the xy plane under the influence of a force of attraction to the origin O of magnitude $F(\rho) > 0$ where ρ is the distance of the mass from O. Set up the equations describing the motion.

Use polar coordinates (ρ, ϕ) to locate the position of m [Fig. 16-10]. Since the rectangular coordinates (x, y) of m are related to the polar coordinates by

$$x = \rho \cos \phi, \quad y = \rho \sin \phi$$

the position vector is given by

$$\boldsymbol{\rho} = \rho \cos \phi \, \mathbf{i} + \rho \sin \phi \, \mathbf{j}$$

where $\mathbf{i}$ and $\mathbf{j}$ are unit vectors in the x and y directions respectively. Thus

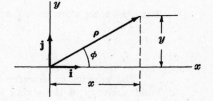

Fig. 16-10

$$\frac{d\boldsymbol{\rho}}{dt} = \dot{\boldsymbol{\rho}} = (\dot{\rho} \cos \phi - \rho \sin \phi \, \dot{\phi})\mathbf{i} + (\dot{\rho} \sin \phi + \rho \cos \phi \, \dot{\phi})\mathbf{j}$$

so that the kinetic energy is

$$T = \tfrac{1}{2}m\left(\frac{d\boldsymbol{\rho}}{dt}\right)^2 = \tfrac{1}{2}m\dot{\boldsymbol{\rho}}^2 = \tfrac{1}{2}m[(\dot{\rho} \cos \phi - \rho \sin \phi \, \dot{\phi})^2 + (\dot{\rho} \sin \phi + \rho \cos \phi \, \dot{\phi})^2]$$

$$= \tfrac{1}{2}m(\dot{\rho}^2 + \rho^2 \dot{\phi}^2)$$

since the force is given by $\qquad\qquad \mathbf{F} = -F(\rho)\boldsymbol{\rho}_1$

where $\boldsymbol{\rho}_1$ is a unit vector in the direction of $\boldsymbol{\rho}$ and since

$$\mathbf{F} = -\nabla V = -\frac{\partial V}{\partial \rho}\boldsymbol{\rho}_1$$

we have $\qquad\qquad \dfrac{\partial V}{\partial \rho} = F(\rho) \quad \text{or} \quad V = \displaystyle\int F(\rho)\,d\rho$

Then the Lagrangian is

$$L = T - V = \tfrac{1}{2}m(\dot{\rho}^2 + \rho^2 \dot{\phi}^2) - \int F(\rho)\,d\rho$$

Thus the Lagrange equations

$$\frac{d}{dt}\left(\frac{\partial L}{\partial \dot{\rho}}\right) - \frac{\partial L}{\partial \rho} = 0, \quad \frac{d}{dt}\left(\frac{\partial L}{\partial \dot{\phi}}\right) - \frac{\partial L}{\partial \phi} = 0$$

become $\qquad m\ddot{\rho} - m\rho\dot{\phi}^2 + F(\rho) = 0, \quad \dfrac{d}{dt}(m\rho^2\dot{\phi}) = 0 \qquad\qquad (1)$

From the second equation of (1) we have

$$\rho^2 \dot{\phi} = \kappa \qquad\qquad\qquad (2)$$

where κ is a constant. Then using $\dot{\phi} = \kappa/\rho^2$ in the first of equations (1), we obtain

$$m\left(\ddot{\rho} - \frac{\kappa^2}{\rho^3}\right) = -F(\rho) \qquad\qquad\qquad (3)$$

These equations can be used to describe the motion if $F(\rho)$ is known.

This problem is useful in discussing the motion of the planets around the sun.

16.21. Use Hamilton's principle to find the equation for the small vibrations of a flexible stretched string of length l and tension τ fixed at its endpoints.

In the vibration [see Fig. 16-11] an element of length dx of the string is stretched into an element of length ds where

$$ds = \sqrt{1 + \left(\frac{\partial y}{\partial x}\right)^2}\,dx = \left[1 + \frac{1}{2}\left(\frac{\partial y}{\partial x}\right)^2\right]dx$$

approximately, using the binomial theorem and the fact that $(\partial y/\partial x)^2$ is small compared with 1.

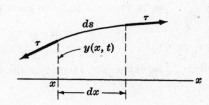

Fig. 16-11

The work done per unit length against the force of tension τ is then given by

$$\frac{\tau(ds - dx)}{dx} = \frac{1}{2}\tau\left(\frac{\partial y}{\partial x}\right)^2$$

Thus the total work done for the whole string [which by definition is the potential energy] is given by

$$V = \frac{1}{2}\int_0^l \tau\left(\frac{\partial y}{\partial x}\right)^2 dx \qquad (1)$$

The velocity of the string is $\partial y/\partial t$ and if the density [mass per unit length] is μ then the total kinetic energy of the string is

$$T = \frac{1}{2}\int_0^l \mu\left(\frac{\partial y}{\partial t}\right)^2 dx \qquad (2)$$

Then Hamilton's principle states that

$$\delta \int_{t_1}^{t_2} (T - V)\, dt \qquad (3)$$

or equivalently $\qquad \delta\left\{\frac{1}{2}\int_{t_1}^{t_2}\int_0^l \left[\tau\left(\frac{\partial y}{\partial x}\right)^2 - \mu\left(\frac{\partial y}{\partial t}\right)^2\right] dx\, dt\right\} = 0 \qquad (4)$

Thus by Problem 16.52 with $y = t$, $z = Y$ we have

$$\frac{\partial F}{\partial Y} - \frac{\partial}{\partial x}\left(\frac{\partial F}{\partial y_x}\right) - \frac{\partial}{\partial y}\left(\frac{\partial F}{\partial y_t}\right) = 0 \qquad (5)$$

where F is the integrand in (4) given by

$$F = \tau y_x^2 - \mu y_t^2 \qquad (6)$$

Using (6) in (5) we obtain $\qquad \dfrac{\partial}{\partial x}\left(\tau\dfrac{\partial y}{\partial x}\right) = \dfrac{\partial}{\partial t}\left(\mu\dfrac{\partial y}{\partial t}\right) \qquad (7)$

or if τ and μ are constants and $a^2 = \tau/\mu$ we have the required equation

$$\frac{\partial^2 y}{\partial t^2} = a^2\frac{\partial^2 y}{\partial x^2} \qquad (8)$$

STURM-LIOUVILLE SYSTEMS AND RAYLEIGH-RITZ METHODS

16.22. Show that the extremals of $I = \displaystyle\int_{x_1}^{x_2} [p(x)y'^2 - q(x)y^2]\, dx$ subject to the constraint $J = \displaystyle\int_{x_1}^{x_2} r(x)y^2\, dx = 1$ are solutions of the Sturm-Liouville equation

$$\frac{d}{dx}\left[p(x)\frac{dy}{dx}\right] + [q(x) + \lambda r(x)]y = 0$$

We shall find it convenient to use the Lagrange multiplier $-\lambda$ rather than λ. Then proceeding as in Problem 16.8, we must make

$$H = I - \lambda J = \int_{x_1}^{x_2} [p(x)y'^2 - q(x)y^2 - \lambda r(x)y^2]\, dx$$

an extremum. Denoting the integrand by F and using the Euler equation

$$\frac{d}{dx}\left(\frac{\partial F}{\partial y'}\right) - \frac{\partial F}{\partial Y} = 0$$

we obtain the required Sturm-Liouville equation.

16.23. Show that the extremals of

$$\lambda \;=\; \frac{\displaystyle\int_{x_1}^{x_2} [p(x)y'^2 - q(x)y^2]\, dx}{\displaystyle\int_{x_1}^{x_2} r(x)y^2\, dx}$$

where the denominator need not be equal to 1 as in Problem 16.22, are also solutions of the same Sturm-Liouville system of Problem 16.22.

The ratio of the integrals can be denoted by

$$\lambda \;=\; I/J$$

Then we have if λ is an extremal,

$$\delta\lambda \;=\; \frac{J\,\delta I - I\,\delta J}{J^2} \;=\; \frac{\delta I - \lambda\,\delta J}{J} \;=\; \frac{1}{J}\,\delta\,(I - \lambda J)$$

$$=\; \frac{-2\displaystyle\int_{x_1}^{x_2} [(py')' + qy + \lambda ry]\,\delta y\, dx}{\displaystyle\int_{x_1}^{x_2} ry^2\, dx} \;=\; 0$$

using the fact that $\delta y = 0$ at x_1 and x_2. Since δy is arbitrary we are again led to Sturm-Liouville equation

$$(py')' + qy + \lambda ry \;=\; 0$$

16.24. Show how to use Problem 16.23 to find eigenvalues and eigenfunctions of the Sturm-Liouville system

$$\frac{d}{dx}\left[p(x)\frac{dy}{dx}\right] + [q(x) + \lambda r(x)]y \;=\; 0$$

$$a_1 y(a) + a_2 y'(a) \;=\; 0, \qquad b_1 y(b) + b_2 y'(b) \;=\; 0$$

Problem 16.23 with $x_1 = a$, $x_2 = b$ shows that we can find the eigenfunctions of the Sturm-Liouville equation by determining the non-trivial functions $y(x)$ which satisfy the conditions $a_1 y(a) + a_2 y'(a) = 0$, $b_1 y(b) + b_2 y'(b) = 0$ and which make the ratio $\lambda = I/J$ an extremum. The corresponding values of λ are then the eigenvalues. That this is true can be seen by noting that if an eigenfunction and eigenvalue are given by $y_1(x)$ and λ_1 respectively, then for this eigenfunction y_1

$$\lambda \;=\; \frac{\displaystyle\int_a^b (py_1'^2 - qy_1^2)\, dx}{\displaystyle\int_a^b ry_1^2\, dx} \tag{1}$$

On integrating the numerator by parts, we have

$$\int_a^b (py_1'^2 - qy_1^2)\, dx \;=\; \int_a^b (py_1')y_1'\, dx - \int_a^b qy_1^2\, dx$$

$$=\; py_1'y_1 \Big|_a^b - \int_a^b y_1\frac{d}{dx}(py_1') - \int_a^b qy_1^2\, dx$$

$$=\; -\int_a^b \left[\frac{d}{dx}(py_1') + qy_1\right] y_1\, dx \;=\; \lambda_1 \int_a^b ry_1^2\, dx$$

using the boundary conditions to show that $\; py_1'y_1 \Big|_a^b = 0\;$ and the fact that

$$\frac{d}{dx}(py_1') + qy_1 \;=\; -\lambda r_1 y_1$$

Thus it follows from (1) that $\lambda = \lambda_1$.

It can be proved that the smallest eigenvalue is the minimum value of the ratio in (1) for all possible functions satisfying the boundary conditions in the Sturm-Liouville problem. This is often called *Rayleigh's principle*.

16.25. (a) Find approximately the smallest eigenvalue of the Sturm-Liouville system

$$y'' + \lambda y = 0, \qquad y(0) = 0, \qquad y(1) = 0$$

and (b) obtain the corresponding approximate eigenfunction.

(a) This is a special case of Problem 16.24 where $p = 1$, $q = 0$, $r = 1$, $a = 0$, $b = 1$, $a_1 = 1$, $a_2 = 0$, $b_1 = 1$, $b_2 = 0$. To find the smallest eigenvalue, we consider

$$\lambda = \frac{\int_0^1 y'^2 \, dx}{\int_0^1 y^2 \, dx} \tag{1}$$

Assume that $$y = A_0 + A_1 x + A_2 x^2 + A_3 x^3 \tag{2}$$

To satisfy $y(0) = 0$, $y(1) = 0$ we require $A_0 = 0$, $A_3 = -(A_1 + A_2)$ so that (2) becomes

$$y = A_1(x - x^3) + A_2(x^2 - x^3) \tag{3}$$

Substituting into (1) we find

$$\lambda = \frac{168A_1^2 + 126A_1A_2 - 35A_2^2}{16A_1^2 + 11A_1A_2 + 2A_2^2} = \frac{168B^2 + 126B - 35}{16B^2 + 11B + 2} \tag{4}$$

on dividing numerator and denominator by A_2^2 and writing $B = A_1/A_2$. Then λ is a minimum when $d\lambda/dB = 0$, which leads to

$$168B^2 - 1792B - 637 = 0 \tag{5}$$

with solutions $$B = 11.01102, \ -.3443521 \tag{6}$$

The second value leads to a negative value for λ which is clearly impossible from the form of (1). The first value leads to $\lambda = 10.5289$. As is easily found, the true eigenvalue is $\pi^2 = 9.8696$ approximately so that the error is less than 7%. Note that instead of (3) we could have assumed directly that

$$y = x(1 - x)(a_0 + a_1 x)$$

which automatically satisfied the boundary conditions and then proceeded as above. Better approximations could then be arrived at by choosing a convenient number of terms in the series

$$y = x(1 - x)(a_0 + a_1 x + a_2 x^2 + \cdots)$$

(b) Since $B = A_1/A_2 = 11.01102$, we find from (3)

$$y = 11.01102A_2(x - x^3) + A_2(x^2 - x^3) = A_2(11.01102x + x^2 - 12.01102x^3)$$

Supplementary Problems

EULER'S EQUATION AND APPLICATIONS

16.26. If $F(x, y, y')$ does not involve y explicitly, show that the extremals of

$$\int_{x_1}^{x_2} F(x, y, y')\, dx$$

are solutions of $\quad\quad\quad\quad \dfrac{\partial F}{\partial y'} = c$

16.27. Find the extremals of $\displaystyle\int_{x_1}^{x_2} \sqrt{1 + y'^2}\, dx$ and thus show that the shortest distance between two points in a plane is a straight line.

16.28. (a) Show that the shortest distance between two points expressed in polar coordinates (ρ, ϕ) by (ρ_1, ϕ_1) and (ρ_2, ϕ_2) is

$$\int_{\phi_1}^{\phi_2} \sqrt{\rho^2 + \left(\frac{d\rho}{d\phi}\right)^2}\, d\phi$$

(b) By minimizing the integral in (a) obtain the equation of a straight line in polar coordinates.

16.29. Work Problem 16.7, page 381, without using the result of Problem 16.4, showing that the differential equation to be solved is $1 + y'^2 - 2yy'' = 0$.

16.30. Show that if F is a function of y' alone, then the extremals of $\displaystyle\int_{x_1}^{x_2} F(x, y, y')\, dx$ are straight lines.

16.31. The shortest distance between two points on any surface is called a *geodesic* of the surface. Show that the geodesics on the surface of a sphere of radius a are the arcs of great circles. [*Hint*: First show that the element of arc length on the sphere is given by $ds^2 = a^2(d\theta^2 + \sin^2\theta\, d\phi^2)$ where θ and ϕ are spherical coordinates.]

16.32. Find the geodesics for (a) a right circular cylinder, (b) a right circular cone.

16.33. Find the extremals of $\displaystyle\int_0^\pi (y'^2 - y^2)\, dx$ such that $y(0) = 0,\ y(\pi) = 0$.

16.34. According to *Fermat's principle*, a light ray travels in a medium from one point to another so that the time of travel given by

$$\int \frac{ds}{v}$$

where s is arc length and v is velocity, is a minimum. Show that the path of travel is given by

$$vy'' + (1 + y'^2)\frac{\partial v}{\partial y} - (1 + y'^2)\frac{\partial v}{\partial x} = 0$$

16.35. Work Problem 16.34 if (a) $v = y$, (b) $v = 1/y$, (c) $v = 1/\sqrt{y}$ and show that the same types of curves are obtained if we use x in place of y.

THE VARIATIONAL NOTATION

16.36. Prove that (a) $\delta\left(\dfrac{F_1}{F_2}\right) = \dfrac{F_2\,\delta F_1 - F_1\,\delta F_2}{F_2^2}$, (b) $\delta F^n = nF^{n-1}\delta F$, (c) $\delta y^{(n)} = (\delta y)^{(n)}$.

16.37. If F is a function of x, y, y', we have $dF = \dfrac{\partial F}{\partial x}\,dx + \dfrac{\partial F}{\partial y}\,dy + \dfrac{\partial F}{\partial y'}\,dy'$ so that by analogy we would expect to have $\delta F = \dfrac{\partial F}{\partial x}\,\delta x + \dfrac{\partial F}{\partial y}\,\delta y + \dfrac{\partial F}{\partial y'}\,\delta y'$. Determine whether this analogy is correct or not by comparing with the result in equation (16), page 376.

16.38. Use the variational notation to justify the method of Lagrange multipliers given on page 376.

16.39. Work (a) Problem 16.5, (b) Problem 16.7, (c) Problem 16.33 by direct use of the variational symbol.

CONSTRAINTS

16.40. If the curve c of Problem 16.5 is required to have a given length l, what is the shape of c so that when it is revolved about the x axis the surface generated will be a minimum? Compare the result with that of Problem 16.5 and explain.

16.41. A curve c of given length l connects points (x_1, y_1) and (x_2, y_2). Find the shape of c so that the area bounded by c, the lines $x = x_1$, $x = x_2$ and the x axis will be a maximum.

16.42. A rope of length l suspended vertically from two fixed points hangs so that its center of gravity is as low as possible. Prove that the curve in which the rope hangs is a catenary. [The principle involved here is equivalent to the *principle of minimum potential energy* in mechanics.]

16.43. Find a function $y(x)$ for which $\int_0^\pi (y'^2 - y^2)\, dx$ is an extremum if $\int_0^\pi y\, dx = 1$ and $y(0) = 0$, $y(\pi) = 1$.

16.44. Of all curves enclosing a region of given area A which one will have the minimum length? Discuss the relationship of this problem with Problem 16.8.

GENERALIZATIONS

16.45. Use the results of equations (20) and (21), page 377, to find extremals of $\int_{t_1}^{t_2} \sqrt{\dot{x}^2 + \dot{y}^2}\, dt$ and interpret the results.

16.46. Find extremals of $\int_{t_1}^{t_2} \sqrt{\dot{x}^2 + \dot{y}^2 + \dot{z}^2}\, dt$ and interpret the results.

16.47. Show that a necessary condition for
$$\int_{t_1}^{t_2} F(t, x_1, x_2, \dot{x}_1, \dot{x}_2)\, dt$$
subject to the two constraint conditions
$$\int_{t_1}^{t_2} F_1(t, x_1, x_2, \dot{x}_1, \dot{x}_2)\, dt = K_1, \qquad \int_{t_1}^{t_2} F_2(t, x_1, x_2, \dot{x}_1, \dot{x}_2)\, dt = K_2$$
to be a maximum or minimum is
$$\frac{d}{dt}\left(\frac{\partial G}{\partial \dot{x}_1}\right) - \frac{\partial G}{\partial x_1} = 0, \qquad \frac{d}{dt}\left(\frac{\partial G}{\partial \dot{x}_2}\right) - \frac{\partial G}{\partial x_2} = 0$$
where λ_1, λ_2 are Lagrange multipliers and $G = F_1 + \lambda_1 F_1 + \lambda_2 F_2$.

16.48. Show that a necessary condition for
$$\int_{x_1}^{x_2} F(x, y, y', y'')\, dx$$
to be an extremum is
$$\frac{\partial F}{\partial y} - \frac{d}{dx}\left(\frac{\partial F}{\partial y'}\right) + \frac{d^2}{dx^2}\left(\frac{\partial F}{\partial y''}\right) = 0$$
and explain how the result can be generalized.

16.49. Find a function $y(x)$ such that $\int_0^\pi y^2\, dx = 1$ which makes $\int_0^\pi y''^2\, dx$ a minimum if $y(0) = 0$, $y(\pi) = 0$, $y''(0) = 0$, $y''(\pi) = 0$.

16.50. If $G = G(x, y, z)$ in a closed region $\mathcal{R}$, show that a necessary condition for
$$\iiint_{\mathcal{R}} (G_x^2 + G_y^2 + G_z^2)\, dx\, dy\, dz$$
to be an extremum [in this case a minimum] is that $\nabla^2 G = 0$ in $\mathcal{R}$.

16.51. (a) Show that in orthogonal curvilinear coordinates u_1, u_2, u_3

$$|\nabla G|^2 = G_x^2 + G_y^2 + G_z^2 = \frac{1}{h_1^2}\left(\frac{\partial G}{\partial u_1}\right)^2 + \frac{1}{h_2^2}\left(\frac{\partial G}{\partial u_2}\right)^2 + \frac{1}{h_3^2}\left(\frac{\partial G}{\partial u_3}\right)^2$$

where the element of arc length is $ds^2 = h_1^2\,du_1^2 + h_2^2\,du_2^2 + h_3^2\,du_3^2$.

(b) Use Problem 16.51 and the fact that $dx\,dy\,dz = h_1h_2h_3\,du_1\,du_2\,du_3$ to show that Laplace's equation in these curvilinear coordinates is

$$\frac{\partial}{\partial u_1}\left(\frac{h_2h_3}{h_1}\frac{\partial G}{\partial u_1}\right) + \frac{\partial}{\partial u_2}\left(\frac{h_1h_3}{h_2}\frac{\partial G}{\partial u_2}\right) + \frac{\partial}{\partial u_3}\left(\frac{h_1h_2}{h_3}\frac{\partial G}{\partial u_3}\right) = 0$$

16.52. A surface S having equation $z = z(x,y)$ is bounded by a closed curve C whose projection on the xy plane is C' which forms the boundary of a region $\mathcal{R}$ [see Fig. 16-12]. Show that a necessary condition for $z = z(x,y)$ to be an extremal of

$$\iint_{\mathcal{R}} F(x,y,z,p,q)\,dx\,dy$$

where $p = \partial z/\partial x$, $q = \partial z/\partial y$, is

$$\frac{\partial F}{\partial z} - \frac{\partial}{\partial x}\left(\frac{\partial F}{\partial p}\right) - \frac{\partial}{\partial y}\left(\frac{\partial F}{\partial q}\right) = 0$$

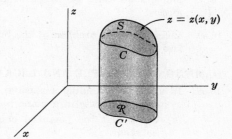

Fig. 16-12

16.53. (a) Show that the surface area of S in Problem 16.52 is

$$\iint_{\mathcal{R}} \sqrt{1 + p^2 + q^2}\,dx\,dy$$

(b) Show that the surface with minimum surface area must be a solution of the equation

$$\frac{\partial}{\partial x}\left(\frac{p}{\sqrt{1 + p^2 + q^2}}\right) + \frac{\partial}{\partial y}\left(\frac{q}{\sqrt{1 + p^2 + q^2}}\right) = 0$$

which can be written as

$$(1 + q^2)r - 2pqs + (1 + p^2)t \qquad \text{where} \qquad r = \frac{\partial^2 z}{\partial x^2},\ \ s = \frac{\partial^2 z}{\partial x\,\partial y},\ \ t = \frac{\partial^2 z}{\partial y^2}$$

(c) Show that $\dfrac{p\,dy - q\,dx}{\sqrt{1 + p^2 + q^2}}$ is an exact differential.

16.54. Generalize Problem 16.52 by showing that a necessary condition for $z = z(x,y)$ to be an extremal of

$$\iint_{\mathcal{R}} F(x,y,z,p,q,r,s,t)\,dx\,dy$$

where r, s, t are defined in Problem 16.53, is

$$\frac{\partial F}{\partial z} - \frac{\partial}{\partial x}\left(\frac{\partial F}{\partial p}\right) - \frac{\partial}{\partial y}\left(\frac{\partial F}{\partial q}\right) + \frac{\partial^2}{\partial x^2}\left(\frac{\partial F}{\partial r}\right) + \frac{\partial^2}{\partial x\,\partial y}\left(\frac{\partial F}{\partial s}\right) + \frac{\partial^2}{\partial y^2}\left(\frac{\partial F}{\partial t}\right) = 0$$

16.55. Generalize Problem 16.52 to the case where $F = F(x,y,z,u,z_x,z_y,u_x,u_y)$.

16.56. (a) Find a necessary condition for $z = z(x,y)$ to be an extremal of

$$\iint_{\mathcal{R}} \sqrt{1 + p^2 + q^2}\,dx\,dy$$

if

$$\iint_{\mathcal{R}} z\,dx\,dy = V \qquad \text{(a constant)}$$

(b) Give a geometric interpretation to part (a).

16.57. Show that the problem of finding geodesics on a surface $\phi(x, y, z) = 0$ joining points (x_1, y_1, z_1) and (x_2, y_2, z_2) can be found by obtaining the minimum of

$$\int_{x_1}^{x_2} \left[\sqrt{1 + \left(\frac{dy}{dx}\right)^2 + \left(\frac{dz}{dx}\right)^2} + \lambda(x)\, \phi(x, y, z) \right] dx$$

and thus show that these geodesics can be found by solving simultaneously the equations

$$\frac{d}{dx}\left(\frac{y'}{\sqrt{1 + y'^2 + z'^2}}\right) = \lambda(x)\frac{\partial \phi}{\partial y}, \qquad \frac{d}{dx}\left(\frac{z'}{\sqrt{1 + y'^2 + z'^2}}\right) = \lambda(x)\frac{\partial \phi}{\partial z}, \qquad \phi(x, y, z) = 0$$

16.58. Assuming that the line $x = x_2$ in Problem 16.16 is replaced by a curve $g(x, y) = 0$ in the vertical plane, show that the shape of the wire should be a cycloid which intersects the curve at right angles.

16.59. Investigate the problem of the brachistochrone (a) from a curve to a fixed point and (b) between two curves.

HAMILTON'S PRINCIPLE AND LAGRANGE'S EQUATIONS

16.60. In a simple pendulum a mass m is suspended from a rod of negligible mass and length l and the system can vibrate in a plane [see Fig. 16-13].

(a) Show that the potential energy of the mass is $V = mgl(1 - \cos\theta)$ apart from an additive constant.

(b) Show that the kinetic energy is $T = \tfrac{1}{2}ml^2\dot{\theta}^2$.

(c) Use Lagrange's equations to show that $\ddot{\theta} + (g/l)\sin\theta = 0$.

(d) Show that if θ is small so that $\sin\theta = \theta$ very nearly, then the period of vibrations is $2\pi\sqrt{l/g}$ approximately.

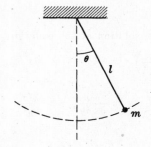

Fig. 16-13

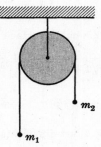

Fig. 16-14

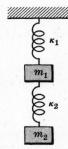

Fig. 16-15

16.61. Two masses m_1 and m_2 are suspended vertically from an inextensible string which in turn passes around a fixed pulley as shown in Fig. 16-14. Use Lagrange's equations to show that the acceleration of either mass is given numerically by

$$\frac{(m_1 - m_2)g}{m_1 + m_2}$$

16.62. A system consisting of masses m_1, m_2 connected to massless springs having stiffness factors κ_1, κ_2 [Fig. 16-15] is free to vibrate vertically under the influence of gravity.

(a) If x_1 and x_2 are the displacements of m_1 and m_2 respectively from their equilibrium positions, show that the potential energy of the system is

$$V = \tfrac{1}{2}\kappa_1 x_1^2 + \tfrac{1}{2}\kappa_2(x_2 - x_1)^2$$

(b) Write the equations of motion of the masses and solve.

16.63. (a) Show that for small angles θ_1, θ_2 the equations of motion of the double pendulum of Fig. 16-3, page 378, are given by

$$(m_1 + m_2)l_1\ddot{\theta}_1 + m_2 l_2\ddot{\theta}_2 = -(m_1 + m_2)g\theta_1$$

$$l_1\ddot{\theta}_1 + l_2\ddot{\theta}_2 = -g\theta_2$$

and (b) find the natural frequencies of the motion.

16.64. Use Lagrange's equations to determine the motion of a mass m sliding without friction down an inclined plane of angle α.

16.65. Show that the equations of motion of a particle of mass m in spherical coordinates (r, θ, ϕ) if the potential is $V(r, \theta, \phi)$ are given by

$$m[\ddot{r} - r\dot{\theta}^2 - r\sin^2\theta\,\dot{\phi}^2] = -V_r$$

$$m\left[\frac{d}{dt}(r^2\dot{\theta}) - r^2\sin\theta\cos\theta\,\dot{\phi}^2\right] = -V_\theta$$

$$m\frac{d}{dt}[r^2\sin^2\theta\,\dot{\phi}] = -V_\phi$$

16.66. Use Hamilton's principle to prove that a system which is in equilibrium has minimum potential energy.

16.67. A particle of mass m resting on top of a fixed sphere of radius a is given a slight displacement so that it slides down the sphere. Assuming no friction, determine the place where it leaves the sphere.

16.68. (a) A membrane situated in the xy plane executes small transverse vibrations denoted by $z(x, y, t)$. Assuming that the density and tension are given by μ and τ respectively, show that the membrane vibrates so that

$$\delta \int_{t_1}^{t_2} \iint_{\mathcal{R}} \left\{\tfrac{1}{2}\mu\left(\frac{\partial z}{\partial t}\right)^2 - \tfrac{1}{2}\tau\left[\left(\frac{\partial z}{\partial x}\right)^2 + \left(\frac{\partial z}{\partial y}\right)^2\right]\right\} dx\,dy\,dt = 0$$

where $\mathcal{R}$ is the region in the xy plane occupied by the membrane.

(b) Use part (a) to show that

$$\frac{\partial^2 z}{\partial t^2} = a^2\left(\frac{\partial^2 z}{\partial x^2} + \frac{\partial^2 z}{\partial y^2}\right)$$

where $a^2 = \tau/\mu$.

STURM-LIOUVILLE SYSTEMS AND RAYLEIGH-RITZ METHODS

16.69. Compare the approximate and exact eigenfunctions of Problem 16.25(b) by first normalizing them so that $\int_0^1 y^2\,dx = 1$ and then plotting their graphs on the same set of axes.

16.70. Express in variational form the problem of solving $y'' + \lambda y = 0$, $y(0) = 0$, $y'(1) = 0$ and find the lowest eigenvalue and the corresponding eigenfunction. Compare with the exact solution.

16.71. Obtain an approximate solution for the extremals of

$$\int_0^\pi (y'^2 - y^2 + 4xy)\,dx \qquad \text{if} \qquad y'(0) = y(\pi) = 0$$

and compare with the exact solution.

16.72. Show that $y = b_1 \sin \pi x + b_2 \sin 2\pi x$ satisfies the boundary conditions in Problem 16.25 and use this to find the approximate eigenvalue and eigenfunction. Discuss the significance of your findings.

16.73. Use an appropriate trigonometric expansion to obtain an approximate solution to Problem 16.71.

16.74. The equation for a vibrating spring [see Problem 16.19] is $m\ddot{x} + \kappa x = 0$. Assuming the boundary conditions $x(0) = 0$, $\dot{x}(0) = v_0$, express the problem in variational form and show how to find the period approximately by a Rayleigh-Ritz method.

16.75. Express the equation $(1 - x^2)y'' - 2xy' + 12y = 0$, $y(0) = 0$, $y(1) = 1$ in variational form and obtain approximate solutions. Compare with the exact solution.

16.76. Is it possible to solve Problem 16.25 by finding extremals of $\int_0^1 (y'^2 - \lambda y^2)\,dx$ of the form $y = x(1 - x)(a_0 + a_1 x)$? Justify your answer.

16.77. Show that eigenvalues and eigenfunctions for the fourth order Sturm-Liouville type differential equation

$$\frac{d^2}{dx^2}\left[s(x)\frac{d^2y}{dx^2}\right] + \frac{d}{dx}\left[p(x)\frac{dy}{dx}\right] + [q(x) + \lambda r(x)]y = 0$$

can be obtained by considering the extremals of

$$\lambda = \frac{\displaystyle\int_{x_1}^{x_2}(py'^2 - qy^2 - sy''^2)\,dx}{\displaystyle\int_{x_1}^{x_2}ry^2\,dx}$$

16.78. In the periodic transverse vibrations $y(x,t) = v(x)\sin\omega t$ of a beam which is simply supported at its endpoints $x = 0$, $x = l$, the boundary-value problem which arises is given by

$$\frac{d^4v}{dx^4} - \frac{\omega^2}{b^2}v = 0, \qquad v(0) = v(l) = v''(0) = v''(l) = 0$$

(a) Express this in variational form using Problem 16.77 and (b) obtain the lowest "critical" frequency for the beam.

Answers to Supplementary Problems

16.28. (b) $\rho\cos(\phi + c_1) = c_2$

16.33. $y = B_n\sin nx$, $n = 1, 2, 3, \ldots$

16.35. (a) Circles $(x - c_1)^2 + (y - c_2)^2 = c_3^2$ (b) Catenaries $y = c_1\cosh\left(\dfrac{x + c_2}{c_1}\right)$

(c) Parabolas $y = \dfrac{(x - c_1)^2}{c_2^2} + \tfrac{1}{4}c_2^2$

16.43. $y = \tfrac{1}{2}(1 - \cos x) + \tfrac{1}{4}(2 - \pi)\sin x$

16.49. $y = B_n\sin nx$, $n = 1, 2, 3, \ldots$

16.70. Exact solution $y = B\sin\dfrac{\pi x}{2}$, exact eigenvalue $\dfrac{\pi^2}{4}$

16.71. $y = 2\pi\cos x - 2\sin x + 2x$

16.75. $y = \tfrac{1}{2}(5x^3 - 3x)$

INDEX

Schaum's Outlines and the Power of Computers... The Ultimate Solution!

Now Available! An electronic, interactive version of *Theory and Problems of Electric Circuits* from the **Schaum's Outline Series.**

MathSoft, Inc. has joined with McGraw-Hill to offer you an electronic version of the *Theory and Problems of Electric Circuits* from the **Schaum's Outline Series.** Designed for students, educators, and professionals, this resource provides comprehensive interactive on-screen access to the entire Table of Contents including over 390 solved problems using Mathcad technical calculation software for PC Windows and Macintosh.

When used with Mathcad, this "live" electronic book makes your problem solving easier with quick power to do a wide range of technical calculations. Enter your calculations, add graphs, math and explanatory text anywhere on the page and you're done – Mathcad does the calculating work for you. Print your results in presentation-quality output for truly informative documents, complete with equations in real math notation. As with all of Mathcad's Electronic Books, *Electric Circuits* will save you even more time by giving you hundreds of interactive formulas and explanations you can immediately use in your own work.

Topics in *Electric Circuits* cover all the material in the **Schaum's Outline** including circuit diagramming and analysis, current voltage and power relations with related solution techniques, and DC and AC circuit analysis, including transient analysis and Fourier Transforms. All topics are treated with "live" math, so you can experiment with all parameters and equations in the book or in your documents.

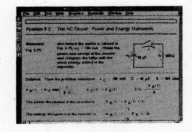

To obtain the latest prices and terms and to order Mathcad and the electronic version of *Theory and Problems of Electric Circuits* from the **Schaum's Outline Series,** call 1-800-628-4223 or 617-577-1017.